Books are to be returned on or before
the last date below.

YOUR REF
4WS0183
NE 17060
OUR REF:
061173
Due
30/8/07.

# Polymer Processing

# Polymer Processing Principles and Design

**Donald G. Baird**
**Dimitris I. Collias**

A Wiley-Interscience Publication

**JOHN WILEY & SONS, INC.**

New York • Chichester • Weinheim • Brisbane • Singapore • Toronto

*Library of Congress Cataloging-in-Publication Data:*

Baird, Donald G.
    Polymer processing: principles and design / Donald G. Baird,
Dimitris I. Collias.
        p.   cm.
    "A Wiley-Interscience publication."
    Includes bibliographical references and index.
    ISBN 0-471-25453-3 (acid-free paper)
    1. Thermoplastics.   I. Collias, Dimitris I.   II. Title.
TP1180.T5B26   1998
668.4′23—dc21                                        98-2551

Printed in the United States of America.

10  9  8  7  6  5  4  3  2  1

# Contents

# Preface

This book is intended to serve as an introduction to the design of processes for thermoplastics. It is intended to meet the needs of senior chemical, mechanical, and materials engineers who have been exposed to fluid mechanics, heat transfer, and mass transfer. With some supplements, the book can also be used by graduate students. For example, by providing more sophisticated coverage of nonlinear constitutive equations and by adding a discussion on finite element methods in Chapters 2 and 3 the book can be used in more advanced courses.

A large number of chemical and mechanical engineers are employed in the polymer industry. They are asked to improve existing processes or to design new ones to provide polymeric materials with a certain level of properties (e.g., mechanical, optical, electrical, or barrier). Although one used to believe that when a given polymer system did not meet the desired requirements a new polymer had to be used, it is becoming more apparent that the properties of the given polymer can be altered by processing or by adding other materials such as other polymers, fillers, glass fibers, or plasticizers. Certainly, a large number of these activities are carried out by trial and error (Edisonian research) approaches. The time to carry out the experiments can be reduced considerably by quantitative design work aimed at estimating the processing conditions that will provide the desired properties. Yet, engineers receive little or no training in the design of polymer processes during their education, partly because they have an inappropriate background in transport phenomena and in the mathematical knowledge required to solve the equations that arise in the design of polymer processes. One aim of this book is to strengthen the background of engineering students in transport phenomena as applied to polymer processing, and the other is to introduce them to numerical simulation.

There are several books available on the processing of polymers with an emphasis on thermoplastics, and one may ask how this book meets the described needs any differently or better than existing books? First of all we cannot revolutionize the area of teaching polymer processing as the principles do not change. What we have done, however, is to make the material more accessible for solving polymer processing design problems. Many times there may be several theories available to use in the modeling of a process. Rather than discuss all the different approaches, we choose what we think is the best theory (but pointing out its limitations and shortcomings) and show how to use it in solving design problems. Another important feature is that we provide the mathematical tools for solving the equations. Other books leave the student with the equations and a description of how they were solved. This does not help someone who has a slightly different set of equations and needs an answer. In this book, as much as possible, we offer the student several methods for getting a solution. Included is also a selection of the subroutines from the International Mathematics and Statistical Libraries (IMSL) (Visual Numerics Inc., Houston, TX) to solve various types of equations that arise in the design of polymer processes. The subroutines have been made relatively "user-friendly," and by following the examples and the descriptions of each subroutine given in Appendix D, solutions are readily available to a number of complex problems. The book does not totally depend on the use of a computer, but certain problems just cannot be solved without resorting to numerical techniques. Rather than dwell on the numerical techniques, we choose to use them in somewhat of a "black box" form. However, sufficient documentation is available in the References if the reader needs to understand the numerical technique. Many will criticize this approach. However, in the time it takes them to voice their objection, the equations will be solved and an answer will be available. With

practice the student will learn when the "black box" has spit out senseless results.

The first five chapters of the book are concerned with the background needed to design polymer processes, and the last five chapters cover the specifics of various types of processes. Chapter 1 contains an overview of polymer processing techniques to help the reader understand the examples and problems used in the next four chapters. Furthermore, a case study presented at the end of Chapter 1 shows how the properties of blown film strongly depend on the processing conditions. Each of the remaining chapters starts with a design problem that provides insights into the material presented in the chapter. In Chapters 2 and 3 we present the basics of non-Newtonian fluid mechanics, which are crucial to the design of polymer processes. In Chapter 4 we introduce the topic of mass transfer as applied to polymeric systems. Finally, in Chapter 5 we discuss the nonisothermal aspects of polymer processing, and the interrelation between processing, structure, and properties is emphasized. These first five chapters contain all the background information the reader needs, including examples illustrating the use of the IMSL subroutines. Mixing is so important to the processing of polymers that we have devoted a full chapter, Chapter 6, to this topic. In the remaining chapters we present the factors associated with the design of various processing methods. We have tried to arrange the subject matter by process similarities. In each chapter we are careful to make it known what aspects of design the readers should be able to execute based on their educational level. In many books on polymer processing it is not clear to the readers just what part of the design they should be able to carry out.

All but the first chapter contain problem sets. The problems are grouped into four classes: *Class A*: These problems can be solved by using equations or graphs given in the chapter and usually involve arithmetic manipulations. *Class B*: These problems require the development of equations and serve to reinforce knowledge acquired about the major subject matter in the chapter. *Class C*: These problems require the use of the computer and are aimed at making direct use of the IMSL subroutines. *Class D*: These problems are design problems and, as such, have a number of solutions. They require the use of all the previous subject matter covered but with an emphasis on the material presented in the given chapter. We have attempted to integrate the problems with the subject matter in an effort to reinforce the material in the given chapter. Furthermore, most of the problems have been motivated by situations that can be encountered in industry.

The coverage of the material in this book requires from 45 to 60 lectures. The number of lectures depends on the background of the students and the depth in which one covers the last five chapters of the book. In most cases, it is recommended to teach the material in Chapter 5 first before teaching Chapter 4, as understanding the heat transfer topics facilitates the teaching of mass transfer. If only 30 lectures are available for teaching the material, then we recommend eliminating Chapters 4 and 6. However, this depends on the specific preference of the instructor.

The book has evolved out of teaching a senior level course in polymer processing at Virginia Polytechnic Institute and State University, Blacksburg, teaching numerical methods to undergraduate chemical engineers, and consulting experiences. First, it was evident that we needed to reinforce the students's knowledge of transport phenomena before we could begin to teach polymer processing. Second, we recognized that B.S. engineers are required to deliver answers and don't have time to weigh out all the variations and perturbations in the various theories. Third, undergraduate engineers are becoming computer literate and have less fear of using computers than many professors. With these ideas in mind we tried to write a book on polymer processing that provides the necessary tools to do design calculations and informs the students exactly what they can be expected to do with the level of the material at hand.

## ACKNOWLEDGEMENTS

Without the contributions of a number of people our efforts in writing this book would have been fruitless. First, one of us (D. G. B.) would specifically like to thank the Department of Chemical Engineering and the College of Engineering at Virginia Polytechnic Institute and State University for a study leave during the spring semester of 1992 so that a full effort could be devoted to writing the book. Diane Cannaday deserves our most sincere appreciation for typing the manuscript and enduring the continuous changes and modifications. Thanks go to Ms. Sylvan Chardon and Ms. Jennifer Brooks who produced the numerous figures and graphs. A number of graduate students in the polymer processing group have contributed to the text in various ways. In particular, we would like to thank Gerhard Guenther, Agnita Handlos, Will Hartt, Chris Robertson, Ed Sabol, David Shelby, Paulo de Souza, and Robert Young.

Finally, we would like to thank our families, especially our wives, Patricia and Eugenia, for their patience and consideration during times when it seemed that all that mattered was writing the book.

Blacksburg, VA

*Donald G. Baird*
*Dimitris I. Collias*

# 1

# IMPORTANCE OF PROCESS DESIGN

In this chapter we intend not merely to present the technology of polymer processing but also to initiate the concepts required in the design of polymer processes. A knowledge of the types of polymers available today and of the methods by which they are processed is certainly needed, but this is available in several sources such as *Modern Plastics Encyclopedia* (Green, 1992) and the *Plastics Engineering Handbook* (Frados, 1976). In this chapter we present primarily an overview of the major processes used in the processing of thermoplastics. In Section 1.1 we begin by classifying the various processes and point out where design is important. In Section 1.2 we present a case study on film blowing to illustrate how the final physical properties are related all the way back to the melt flow of a polymer through the die. Finally, in Section 1.3 we summarize the principles on which polymer process design and analysis are based.

## 1.1. CLASSIFICATION OF POLYMER PROCESSES

The major processes for thermoplastics can be categorized as follows: extrusion, postdie processing, forming, and injection molding. We describe specific examples of some of the more common of these processes here.

The largest volume of thermoplastics is probably processed by means of *extrusion*. The *extruder* is the main device used to melt and pump thermoplastics through the shaping device called a *die*. There are basically two types of extruders: single- and twin-screw. The *single-screw extruder* is shown in Fig. 1.1. The single-screw extruder basically consists of a screw (Fig. 1.2) that rotates within a metallic barrel. The length-to-diameter ratio (L:D) usually falls in the range of 20 to 24 with diameters between 1.25 cm and 50 cm. The primary design factors are the screw pitch (or helix angle, $\theta$) and the channel depth profile. The main function of the plasticating extruder is to melt solid polymer and to deliver a homogeneous melt to the die at the end

## Extruder screw geometry

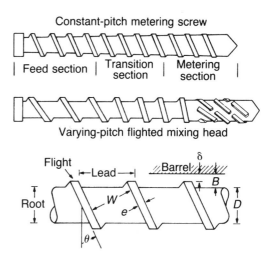

**Figure 1.2** Two different extruder screw geometries along with the various geometric factors describing the characteristics of the screw. (Reprinted by permission of the publisher from S. Middleman, *Fundamentals of Polymer Processing*, McGraw-Hill, Company, New York, 1977.)

of the extruder. The extruder can also be used as a mixing device, a reactor, and a devolatilization tool (see Chapter 8).

There are as many *twin-screw extruders* as single-screw extruders in use today. Many different configurations are available, including corotating and counterrotating screws (see Fig. 1.3) and intermeshing and nonintermeshing screws. These extruders are primarily adapted to handling difficult-to-process materials and are used for compounding and mixing operations. The analysis and design of these devices are quite complicated and somewhat out of the range of the material level in this text. However, some of the basic design elements are discussed in Chapter 8. The extruder feeds a shaping device called a die. The performance of the single- and corotating twin-screw extruders is affected by resistance to flow offered by the die. Hence, we cannot separate extruder design from die design. Problems in die design include distributing the melt flow uniformly over the width of a die, obtaining a uniform thermal history, predicting the die dimensions, which lead to the desired final shape, and producing a smooth extrudate free of surface irregularities. Some of these design problems are accessible at this level of material, but others are still research issues (see Chapter 7).

There are many types of extrusion die geometries, including those for producing sheet and film, pipe and tubing, rods and fiber, irregular cross sections (profiles), and coating wire. As an example, a wire-coating die is shown in Fig. 1.4. Here metal wire is pulled through the center of the die with melt being pumped through the opening to encapsulate

## Single-screw extruder

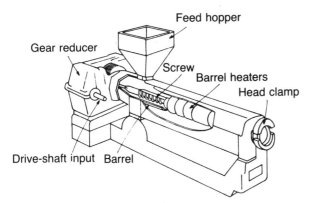

**Figure 1.1** Typical single-screw extruder. (Reprinted by permission of the publisher from S. Middleman., *Fundamentals of Polymer Processing*, McGraw-Hill, Company, New York, 1977.)

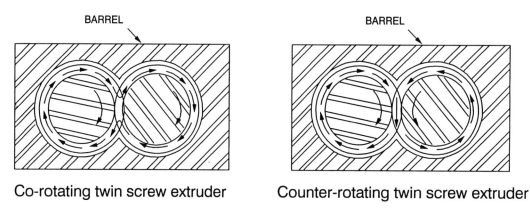

**Figure 1.3** Cross-sectional view of corotating and counterrotating twin-screw extruders.

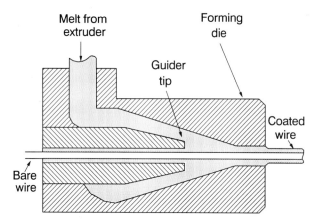

**Figure 1.4** Cross-head wire-coating die. (Reprinted by permission of the publisher from Tadmor & Gogos, 1979.)

the wire. The design problems encountered here are concerned with providing melt flowing under laminar flow conditions at the highest extrusion rate possible and to give a coating of polymer of specified thickness and uniformity. At some critical condition, polymers undergo a low Reynolds number flow instability, called *melt fracture*, which leads to a nonuniform coating. Furthermore, the melt expands on leaving the die, leading to a coating that can be several times thicker than the die gap itself (this is associated with the phenomenon of die swell). The problems are quite similar for other types of extrusion

processes even though the die geometry is different. The details associated with die design are presented in Chapter 7.

We next turn to *postdie* processing operations. Examples of these processes include *fiber spinning* (Fig. 1.5), film blowing (Fig. 1.6), and *sheet forming* (Fig. 1.7). These processes share a number of similarities. In particular they are free-surface processes in which the shape and thickness or diameter of the extrudate are determined by the rheological (flow) properties of the melt, the die dimensions, cooling conditions, and take-up speed relative to the extrusion rate. The physical and, in the case of film blowing and sheet forming, the optical properties are determined by both the conditions of flow in the die and the cooling rates and stretching conditions of the melt during the cooling process. Furthermore, slight changes in the rheological properties of the melt can have a significant effect on the final film or fiber properties. Design considerations must include predictions of conditions that provide not only the desired dimensions but also the optical and physical properties of the film, fiber, or sheet.

The third category of processing of thermoplastics is *forming*. Three examples of this type of process are blow molding, thermoforming, and compression molding. *Blow molding* (Fig. 1.8) is primarily employed for making containers used to package a wide variety of fluids. Although polyolefins, such as high-density polyethylene (HDPE) or polyethylene-terephthalate (PET), both of which can be considered as commodity resins, are commonly used, there is growing interest in using this technique for the processing of higher-performance engineering thermoplastics. Essentially, a parison that has been extruded or injection-molded is inflated with air until it fills the mold cavity. The inflated parison is held in contact with the cold mold walls until it is solidified.

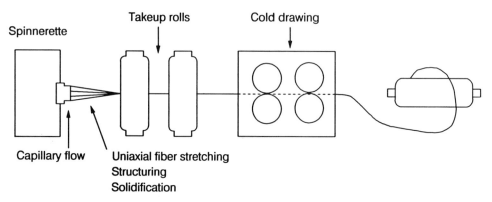

**Figure 1.5** Fiber melt-spinning process. (Reprinted by permission of the publisher from Tadmor & Gogos, 1979.)

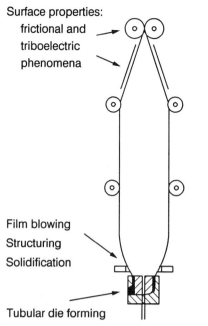

Surface properties:
frictional and
triboelectric
phenomena

Film blowing
Structuring
Solidification

Tubular die forming

**Figure 1.6** Film-blowing process. (Reprinted by permission of the publisher from P. N. Richardson, *Introduction to Extrusion*, Society of Plastics Engineers Inc., Greenwich, CT, 1974.)

In *thermoforming* (Fig. 1.9) a sheet of polymer is heated by radiation (and sometimes cooled intermittently by forced convection) to a temperature above its glass transition temperature or in some cases above the crystalline melting temperature and then pressed into the bottom part of the mold (female part) by mechanical force or pressure or by pulling a vacuum. The key flow property is the extensional flow behavior of the melt which controls the uniformity of the part's thickness. Sometimes the deformation is applied at a temperature just below the onset of melting, in which case the process is referred to as

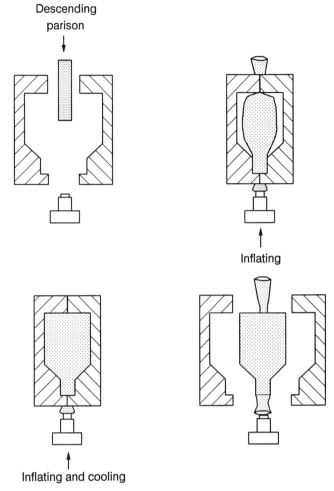

Descending
parison

Inflating

Inflating and cooling

**Figure 1.8** Blow-molding process. (Reprinted by permission of the publisher from W. A. Holmes-Walker, *Polymer Conversion*, Elsevier, London, 1975.)

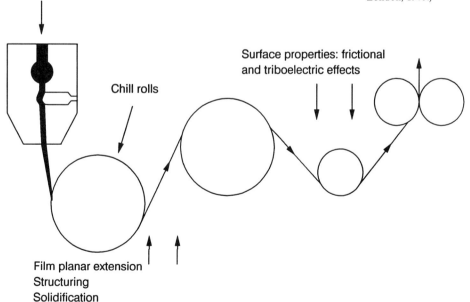

Chill rolls

Surface properties: frictional
and triboelectric effects

Film planar extension
Structuring
Solidification

**Figure 1.7** Flat film and sheet process. (Reprinted by permission of the publisher from Tadmor & Gogos, 1979.)

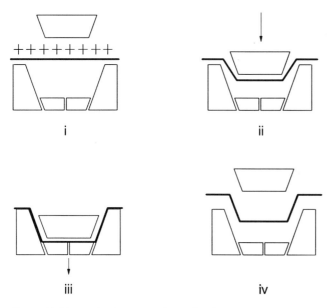

**Figure 1.9** Plug-assisted vacuum thermoforming. (Reprinted by permission of the publisher from R. Greene, ed., *Modern Plastics Encyclopedia*, Vol. 53, McGraw-Hill, New York, 1977.)

*solid-phase forming*. At other times the sheet is extruded directly to the forming unit and is formed before it cools down (this is called scrapless, or continuous, thermoforming). Some of the key design considerations are the time required to heat the sheet, the final thickness of the part, especially around sharp corners, and the cooling rate that controls the amount and type of crystallinity.

In *compression molding* (Fig. 1.10) a slug of polymer is heated, and then pressure is applied to squeeze the material into the remaining part of the mold. Some aspects of forming are discussed in Chapter 10.

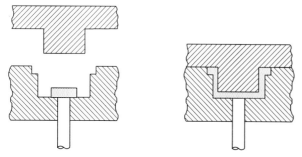

**Figure 1.10** Compression-molding process. (Reprinted by permission of the publisher from Tadmor & Gogos, 1979.)

The last general category is *injection molding* (see Fig. 1.11). Polymer is melted and pumped forward just as in a screw extruder. The screw is then advanced forward by a hydraulic system that pushes the melt into the mold. Because of the high deformation and cooling rates, a considerable degree of structuring and molecular orientation occurs during mold filling. The physical properties of injection-molded parts can be significantly affected by processing conditions. Design considerations include the required injection pressure to fill the mold cavity, the location of weld lines (places where two melt fronts come together), cooling rates, length of hold time in the mold, and distribution of molecular orientation. Some of these factors are out of the realm of the material that can be covered by this book, but accessible aspects are presented in Chapter 10.

Although the majority of the material in this book is concerned with the processing of thermoplastics, the processing of *thermosetting* systems should also be mentioned for the sake of completeness. We describe three types of processes involving reactive processing: reaction injection molding (RIM), compression molding, and pultrusion.

Reaction injection molding is a process in which two liquid intermediates are metered separately to a mixing head where they are combined by high-pressure impingement mixing and subsequently flow into a mold where they are polymerized to form a molded part (see Fig. 1.12). A typical process consists of the reaction of diisocyanate and a polyol to form polyurethane. The important design factors are the degree of mixing and the appropriate heat transfer conditions to ensure uniform curing conditions in the mold. This process is discussed in more detail in Becker (1979).

Thermosetting composites can be processed by means of *compression molding* of uncured resin. Usually, fiber reinforcement is used to provide additional strength and stiffness. The application of pressure pushes the resin into the fiber reinforcement, and heat cross-links the resin to form a solid material. The critical factors are the flow of the uncured resin into and around the reinforcement and the uniform and complete cure of the resin throughout the part. This technique is used primarily in the aerospace and automobile industries.

*Pultrusion* is a process used for making continuous filament-reinforced composite extruded profiles (see Fig. 1.13). Reinforcing filaments, such as glass fiber roving, are saturated with catalyzed resin and then pulled through an orifice similar to an extrusion die. As the two materials pass through the die, polymerization of the resin occurs to continuously form a rigid cured profile corresponding to the die orifice shape. The materials are pulled through the die rather than being pumped. Although the primary resins used are of the thermosetting type such as polyester, vinyl ester, and epoxy, thermoplastic resins can be utilized in the same process. The major design considerations for thermosetting systems consist of dispersion of the resin in the reinforcement and the conditions for complete cure of the resin. The processing of thermosetting systems is discussed in Macosko (1990).

The intention in Section 1.1 was not merely to review the technology of polymer processing, but also to point out factors that

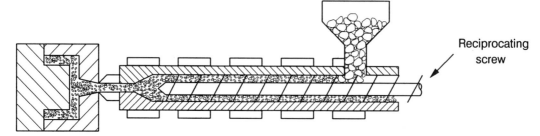

**Figure 1.11** Typical injection-molding unit. (Reprinted by permission of the publisher from Tadmor & Gogos, 1979.)

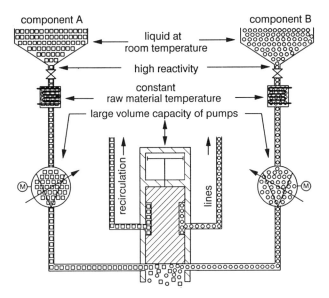

**Figure 1.12** Reaction injection-molding (RIM) process. (Reprinted by permission of the publisher from Becker, 1979.)

must be considered in the design of polymer processes. However, because most students have little knowledge of the technology of polymer processing, some general introduction is of value. Furthermore, a physical picture of the various processes is required to facilitate the discussion of the material presented in the next four chapters.

## 1.2. FILM BLOWING: A CASE STUDY

In the preceding section we described the technology of polymer processing. In this section we illustrate the role of processing in affecting the properties of polymeric systems. In particular, the properties of

films of polybutylene (PB1) generated by film blowing are shown to be highly sensitive to processing conditions.

Most blown film is made from some form of polyethylene (PE), but polybutylene, PB1, has been considered because it is slightly cheaper to use in the production of film. However, one does not obtain the same physical properties without changing the processing conditions. Identifying the appropriate processing conditions is usually done by either a trial-and-error approach or through statistically designed experiments. If a model of film blowing was available, or if one could apply dimensional analysis concepts, then it might be possible to find the appropriate processing conditions without carrying out a lengthy set of costly experiments. The following example illustrates the many factors that affect the properties of blown PB1 film.

The film-blowing process is shown in Fig. 1.6. Polymer pellets are fed to the extruder in which melting, homogenization, and pumping occur. The melt then passes through the die, which is designed to subject the melt to both a uniform deformation and thermal history. Air is blown through the center of the die to expand the molten bubble to impart orientation of the molecules in the hoop direction. At the same time, the bubble is being stretched as a result of the take up velocity being greater than the average velocity of the melt leaving the die. The stretching imparted in the two directions controls the degree to which the molecules orient and, hence, affects the physical properties. Cooling air is blown along the bubble by an air ring that is placed around the outside of the die. This causes the film to solidify or crystallize and lock in the orientation imparted by the biaxial stretching process. The film is then taken up on a roller and either slit to make flat film or sealed and cut to make bags.

We now look at some of the factors that affect the physical properties of the blown film. The recommendations for a desired die gap opening for a desired film thickness are given in Table 1.1. It is probably clear why the die gap is larger than the desired film thickness as the film is to be drawn down to create molecular orientation. What is not clear is why it is recommended that the die land (the annular portion of the die) be shortened as the die gap increases. The physical properties based on the tear strength of the film are found to be

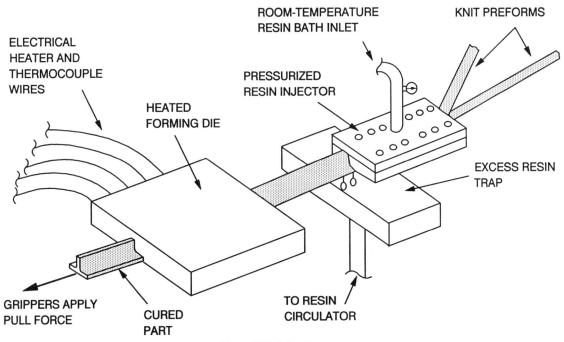

**Figure 1.13** Pultrusion process.

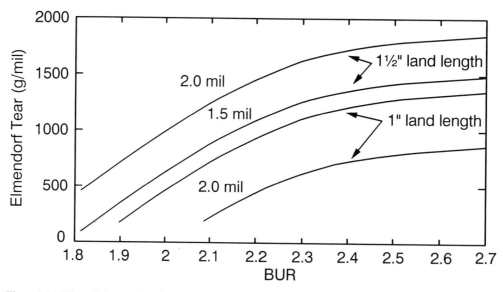

**Figure 1.14** Effect of the annular die land length on the film tear strength of polybutylene extruded at the rate of 10 m/min for two different film thicknesses.

**TABLE 1.1 Die Recommendations for the Blowing of PBI with a Blowup Ratio of 2.4 to 2.8 at a Melt Temperature of 370°F to 390°F**

| Film Thickness ($\times 10^3$ in.) | Die Gap (in.) | Land Length (in.) |
|---|---|---|
| 0.5–1 | 0.015 | $1\frac{1}{2}$–2 |
| 1–2 | 0.018–0.023 | $1\frac{1}{2}$ |
| 2–7 | 0.028–0.032 | 1 |
| 8–15 | 0.040–0.050 | 1 |
| 15–40 | 0.050–0.060 | 1 |

significantly affected by the length of the die land as shown in Fig. 1.14. Here the tear strength is plotted versus the blowup ratio (BUR), which is the ratio of the final film diameter to the die diameter (outer diameter). It is observed that there is of the order of a threefold difference in the tear strength for a $2.0 \times 10^{-3}$ in. thick film when the die land is decreased from 1.5 in. to 1.0 in. One reason for this result is that more "die swell" (the expansion of a polymer melt on leaving a die) occurs for the die with the 1.0 in. land length, and hence a higher stretch ratio is required to draw the film down to $2.0 \times 10^{-3}$ in. This leads to higher orientation of the molecules along the draw direction than in the case of the die with a 1.5 in. land length. Finally, the effect of the die gap on the tear strength measured along both the film length (the machine direction [MD]) and the circumference of the film (the transverse direction [TD]) is shown in Fig. 1.15. Here we see that the tear strength in the TD direction decreases significantly with an increase in die gap, whereas in the MD the effect is significant but nowhere near as large. Again, it is not clear what would cause the loss of properties in both directions as the die gap increases other than the longer time available for molecular relaxation due to the increase in time required for cooling of the film. Factors other than orientation must be involved in controlling the properties. For example, the amount of crystallinity and the size of the spherulitic regions may play a significant role.

The melt extrusion temperature is also observed (Fig. 1.16) to have a significant effect on the physical properties as the tear strength in both directions increases with increasing melt temperature. This is probably due to lower levels of orientation as the result of lower stress levels in the melt and shorter relaxation times allowing a rapid relaxation of molecular orientation.

The line speed, given in feet per minute (fpm), as shown in Table 1.2, has a very significant effect on the properties. For example, as the line speed increases from 14 fpm, the breaking strength in the MD increases from 4800 g/mil to 6600 g/mil but decreases in the TD from 5000 g/mil to 2300 g/mil (1 mil = 0.001 in.). This is mostly associated with the degree of molecular orientation. The more the molecules are oriented along the MD, the stronger the films are, but also the poorer the tear strength in this direction is. The other properties given here can be explained by similar arguments.

The BUR, can be used to obtain a better balance of properties, as shown in Fig. 1.17. As the BUR increases, the tear properties become more uniform in both directions. Biaxial orientation (i.e., orientation of molecules in two directions) is generated in the blowing process and leads to more uniform properties.

As one can imagine, the film-blowing process is very difficult to model, and hence very little quantitative design work has been done. Although the complete modeling of this process is beyond the level of the material in this book (or even an advanced book for that matter), the example serves to illustrate that the properties of a polymeric material highly depend on the processing conditions and to highlight some of the problems faced by the engineer. In designing a polymer process, one must be concerned not only with how much material per unit time can be produced but also with the quality of the properties of the material. In the next section we look at the fundamental principles on which the design and analysis of polymer processes are based.

## 1.3. BASICS OF POLYMER PROCESS DESIGN

In order to design and analyze polymer processes there are common steps associated with nearly every process. Following Tadmor and Gogos (1979), these basic steps are:

1. Handling of particulate solids
2. Melting, cooling, and crystallization
3. Pumping and pressurization
4. Mixing
5. Devolatilization and stripping
6. Flow and molecular orientation

**TABLE 1.2 Line Speed versus Properties for PBI**

| (FPM) Line Speed | Break Strength (g/mil) | | Yield Strength (g/mil) | | Ultimate Elongation (%) | | Dart Drop (g/m²/s²) | Tear Strength (g/mil) | |
|---|---|---|---|---|---|---|---|---|---|
| | MD | TD | MD | TD | MD | TD | | MD | TD |
| 14 | 4800 | 5000 | 2000 | 2000 | 220 | 260 | 350 | 1700 | 550 |
| 20 | 4600 | 3700 | 2100 | 2000 | 160 | 230 | 280 | 1500 | 550 |
| 30 | 5500 | 2800 | 2500 | 1900 | 110 | 170 | 190 | 680 | 390 |
| 40 | 6600 | 2300 | 3000 | 1900 | 80 | 150 | 90 | 80 | 270 |

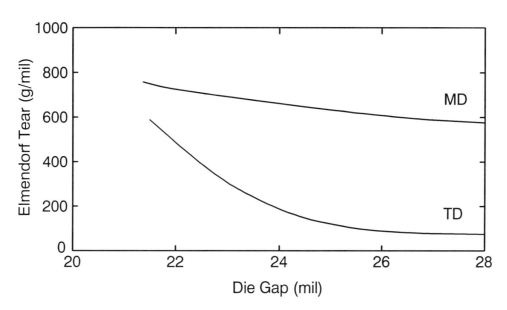

**Figure 1.15** Effect of die gap on the film tear strength of polybutylene extruded at the rate of 10 m/min. The film thickness is 0.002 in., and the blowup ratio is 2.8.

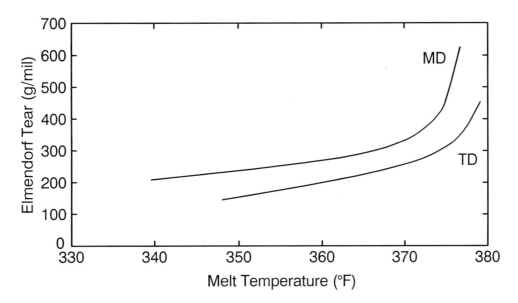

**Figure 1.16** Effect of melt temperature on the film tear strength of polybutylene film extruded at the rate of 10 m/min. The film thickness is 0.002 in., and the blowup ratio is 2.8.

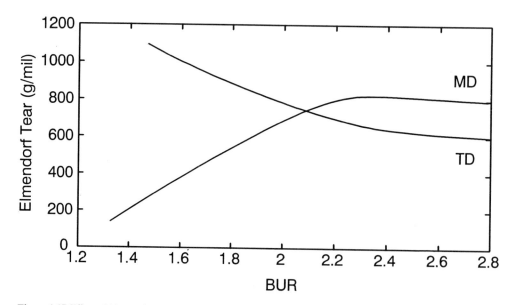

**Figure 1.17** Effect of blowup ratio on the film tear strength measured along the machine (MD) and transverse (TD) directions.

These basic steps are based on the following concepts:

1. Transport phenomena: fluid mechanics, heat transfer, and mass transfer
2. Polymer rheology
3. Solid mechanics and flow
4. Principles of mixing
5. Chemical reactions

In the first five chapters of this book we present the fundamental principles required in the design of polymer processes. In the last five chapters we analyze specific types of processes in detail.

## REFERENCES

Becker, W. E., Ed. 1979. *Reaction Injection Molding* (Van Nostrand Reinhold, New York).

Frados, J., Ed. 1976. *Plastics Engineering Handbook* (Van Nostrand Reinhold, New York).

E. Green, Ed. 1992. *Modern Plastics Encyclopedia* (McGraw-Hill, New York).

Macosko, C. W. 1989. *RIM: Fundamentals of Reaction Injection Molding* (Hanser, New York).

Tadmor, Z., and C. G. Gogos 1979. *Principles of Polymer Processing* (Wiley, New York).

# 2

# ISOTHERMAL FLOW OF PURELY VISCOUS NON-NEWTONIAN FLUIDS

## DESIGN PROBLEM 1
### Design of a Blow-Molding Die

A typical blow-molding die is shown in Fig. 2.1. The region of particular interest is shown in Fig. 2.2. The die exit is the region that controls the final dimensions of the parison, which is a cylindrically shaped tube of polymer. The parison consisting of high-density polyethylene (HDPE) is to have a weight of 90 g with an outside diameter of 0.127 m

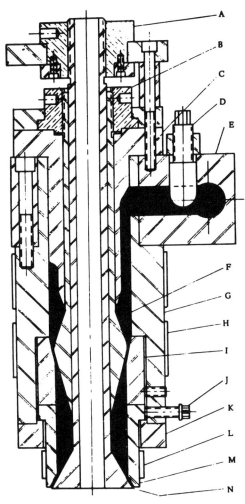

**Figure 2.1** Typical blow-molding die: A, choke-adjusting nut; B, mandrel adjustment; C, feed throat; D, choke screw; E, die head; F, plastic melt; G, die barrel; H, heater band; I, choke ring; J, centering screw; K, clamp ring; L, die heater; M, die; N, mandrel. (Reprinted by permission of the publisher from J. D. Frankland, "A High Speed Blow Molding Process", *Trans. Soc. Rheol. (J. Rheol.)*, *19*, 371, 1975.)

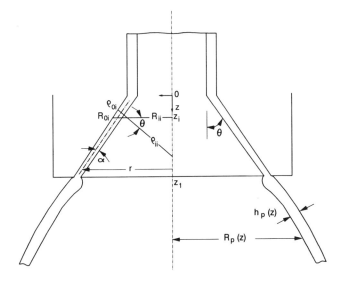

**Figure 2.2** Details of the conical region of the blow-molding die.

**TABLE 2.1 Power-Law and Ellis Model Parameters**

| | Graphically Obtained | From Nonlinear Regression Analysis |
|---|---|---|
| $m(\text{Pa}\cdot\text{s}^n)$ | 2.374 E + 04 | 1.616 E + 04 |
| $n$ | 0.424 | 0.520 |
| $\eta_0(\text{Pa}\cdot\text{s})$ | 1.33 E + 04 | |
| $\tau_{1/2}(\text{Pa})$ | 3.325 E + 04 | |
| $\alpha - 1$ | 1.54 | |

and a wall thickness of $3.81 \times 10^{-4}$ m. Consider only the conical region in your calculations. The angles $\alpha$ and $\theta$ are taken as $0°$ and $30°$, respectively. The distance $z_1$ should be 20 times the gap thickness. Determine the remaining dimensions of the die required to produce the desired extrudate. At this point neglect die swell (i.e., the increase in diameter and thickness due to elastic recovery) in your calculations. Determine the maximum extrusion rate ($m^3$/s) and pressure drop assuming that the limiting factor is melt fracture. (This occurs when the wall shear stress, $\tau_w$, reaches $1.4 \times 10^5$ Pa.) Use the rheological parameters given in Table 2.3. Determine the length of time required to extrude the parison.

**(a)** Use the lubrication approximation to determine a design equation (i.e., $Q$ versus $\Delta P$), and then provide the required information.

**(b)** Carry out the design calculations by breaking up the flow region into a series of cones of length $\Delta z$. Use the annular flow equations presented in Section 2.3.1 and the computer to get a solution. At $180°$C ($453°$K), $\rho = 965 \text{ kg/m}^3$.

The transport properties of polymeric materials, which distinguish them most from other materials, are their flow properties or rheological behavior. There are many differences between the flow properties of a polymeric fluid and typical low-molecular-weight fluids such as water, benzene, sulfuric acid, and other fluids that we classify as Newtonian. Newtonian fluids can be characterized by a single flow property called viscosity ($\mu$) and by density ($\rho$). Polymeric fluids, on the other hand, exhibit a viscosity function that depends on shear rate or shear stress, time-dependent rheological properties, viscoelastic behavior such as elastic recoil (memory), additional normal stresses in shear flow, and an extensional viscosity not simply related to the shear viscosity, to name a few differences.

Because of these vastly different rheological properties, polymeric fluids are known to exhibit flow behavior that cannot be accounted for merely through a single rheological parameter such as viscosity. Some of the differences in flow behavior include a nonlinear relation between pressure drop and volumetric flow rate for flow through a tube, swelling of the extrudate when emerging from a tube, the onset of a low Reynolds number flow instability, called melt fracture, gradual relaxation of stresses on cessation of flow, and the ability of the molecules to orient during flow. These phenomena are discussed in more detail in Bird et al. (1987).

The emphasis in this chapter is on the viscous behavior of polymeric fluids and in particular their pseudoplastic behavior. The chapter is arranged in the following manner. First, in Section 2.1 we review the definition of a Newtonian fluid, and then we present empiricisms for describing the viscosity of polymeric fluids. In Section 2.2 we use shell force or momentum balances to solve two one-dimensional flow problems commonly found in polymer processing. In Section 2.3 we generalize the force or momentum balances to give the equations of motion, and we generalize the constitutive equation presented in Section 2.1. In Section 2.4 we present two useful approximations for solving polymeric flow problems. Finally, in Section 2.5 the topics discussed in the preceding sections are used to solve Design Problem 1.

## 2.1. VISCOUS BEHAVIOR OF POLYMER MELTS

When a Newtonian fluid is placed between the two plates as shown in Fig. 2.3, in which the top plate is moved to the right with constant velocity, $V$, the relation between force, $F$, divided by the area of the plates, $A$, and the velocity divided by the separation distance, $H$, is given as follows:

$$F/A = \mu V/H. \tag{2.1}$$

The constant of proportionality, $\mu$, is called the *viscosity* of the fluid. The *force*, $F$, is the force required to keep the top plate moving at a constant velocity. The force per unit area acting in the $x$ direction on a fluid surface at a constant $y$ by the fluid in the region of lesser $y$ is the *shear stress*, $\tau_{yx}$. Because the velocity of the fluid particles varies in a linear manner with respect to the $y$ coordinate, it is clear that $V/H = dv_x/dy$ as shown:

$$\lim_{\Delta y \to 0} \frac{\Delta v_x}{\Delta y} = \frac{dv_x}{dy} = \frac{V-0}{H-0} = \frac{V}{H}. \tag{2.2}$$

Eq. 2.1 can be rewritten as:

$$\tau_{yx} = -\mu(dv_x/dy). \tag{2.3}$$

This states that the shear force per unit area is proportional to the negative of the local velocity gradient. This is known as *Newton's law of viscosity*. The sign convention used here follows that of Bird et al. (1960).

The definition of $\tau_{yx}$ can also be interpreted in another fashion. $\tau_{yx}$ may be considered as the viscous flux of $x$ momentum in the $y$ direction. The idea here is that the plate located at $y=H$ transmits its $x$ momentum to the layer below which in turn transmits momentum to the next layer. The momentum flux, $\tau_{yx}$, is negative in this case as the momentum is transferred in the negative $y$ direction. The sign convention follows the ideas used for heat flux in that heat flows from hot to cold or in the direction of a negative temperature gradient. This also makes the law of viscosity fit with the ideas of diffusion, in which matter flows in the direction of decreasing concentration.

Probably the most frequently used notation, however, is that found in mechanics in which material at a greater $y$ exerts force in the $x$ direction on a layer of fluid at a lesser $y$. The shear stress, $\tau_{xy}*$, is then related to that used above as follows:

$$\tau_{yx} = -\tau_{xy}*. \tag{2.4}$$

$\tau_{xy}*$ is then defined as the force per unit area acting in the $x$ direction by fluid at $y$ on a surface of lesser $y$ with a unit outward normal in the $+y$ direction.

The flow behavior of most thermoplastics does not follow Newton's law of viscosity. To quantitatively describe the viscous behavior of polymeric fluids, Newton's law of viscosity is generalized as follows:

$$\tau_{yx} = -\eta \, dv_x/dy, \tag{2.5}$$

### Non-Newtonian Viscosity

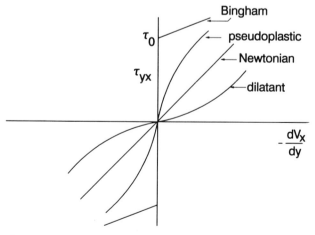

**Figure 2.4** Viscous response of non-Newtonian fluids.

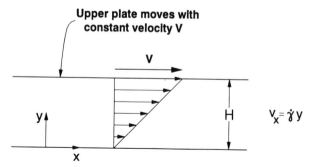

**Figure 2.3** Steady simple shear flow with shear rate $= V/b$.

where $\eta$ can be expressed as a function of either $dv_x/dy$ or $\tau_{yx}$. Some typical responses of polymeric fluids are shown in Fig. 2.4, where $\tau_{yx}$ is plotted versus the velocity gradient. For a pseudoplastic fluid, the slope of the line decreases with increasing magnitude of $dv_x/dy$, or in essence the viscosity decreases. Some polymeric fluids (in some cases polymer blends and filled polymers) exhibit a yield stress, which is the stress that must be overcome before flow can occur. When flow occurs, if the slope of the line is constant, then the fluid is referred to as a *Bingham fluid*. In many cases the fluid is still pseudoplastic once flow begins. Finally, in some cases the viscosity of the material increases with an increasing velocity gradient. The fluid is then referred to as *dilatant*.

Many empiricisms have been proposed to describe the steady-state relation between $\tau_{yx}$ and $dv_x/dy$, but we mention only a few that are most useful for polymeric fluids. The first is the *power-law* of Ostwald-de Waele:

$$\eta = m\left|\frac{dv_x}{dy}\right|^{n-1} \tag{2.6}$$

This is a two-parameter model in which $n$ describes the degree of deviation from Newtonian behavior. $m$, which has the units of Pa·s$^n$, is called the *consistency*. For $n=1$ and $m=\mu$, this model predicts Newtonian fluid behavior. For $n<1$, the fluid is pseudoplastic and for $n>1$ the fluid is dilatant. The *Ellis model* is a three-parameter model and is defined as:

$$\frac{\eta_0}{\eta} = 1 + \left(\frac{\tau_{yx}}{\tau_{1/2}}\right)^{\alpha-1} \tag{2.7}$$

Here $\eta_0$ is the zero shear viscosity, and $\tau_{1/2}$ is the value of $\tau_{yx}$ when $\eta = \frac{1}{2}\eta_0$. Actually, most polymeric fluids exhibit a constant viscosity at low shear rates and then shear thin at higher shear rates (see Fig. 2.5). A model that is often used in numerical calculations, because it fits the full flow curve, is the Carreau model:

$$\frac{\eta - \eta_\infty}{\eta_0 - \eta_\infty} = [1 + (\lambda\dot{\gamma})^2]^{\frac{n-1}{2}}. \tag{2.8}$$

This model contains four parameters: $\eta_0$, $\eta_\infty$, $\lambda$ and $n$. $\eta_0$ is the zero shear viscosity just as before. $\eta_\infty$ is the viscosity as the shear rate ($\dot{\gamma}$) or $dv_x/dy \to \infty$, and for polymer melts this can be taken as zero. $\lambda$ has units of seconds and approximately represents the reciprocal of the shear rate for the onset of shear thinning behavior. $n$ represents the degree of shear thinning and is nearly the same as the value in the power-law model. As a number of polymeric fluids exhibit yield stresses, models that include these are the Bingham and Hershel-Bulkley models. The Bingham model is given as:

$$\eta = \mu_0 + |\tau_0|\left/\left|\frac{dv_x}{dy}\right|\right. \qquad \text{if} \quad |\tau_{yx}| \geq |\tau_0| \tag{2.9}$$

$$\eta = \infty \qquad \text{if} \quad |\tau_{yx}| < |\tau_0|. \tag{2.10}$$

Here $\tau_0$ is the yield stress, and $\mu_0$ is the slope of the line of $\tau_{yx} - \tau_0$ versus $dv_x/dy$. The Hershel-Bulkley model is given as:

$$\eta = m'\left|\frac{dv_x}{dy}\right|^{n'-1} + |\tau_0|\left/\left|\frac{dv_x}{dy}\right|\right. \qquad \text{if} \quad |\tau_{yx}| \geq \tau_0. \tag{2.11}$$

Here $m'$ and $n'$ are power-law parameters determined from $\tau_{yx} - \tau_0$ versus $dv_x/dy$. This model describes fluids that are pseudoplastic once flow starts.

## 2.2. ONE-DIMENSIONAL ISOTHERMAL FLOWS

In this section we make use of a shell momentum or force balance plus the Newtonian fluid model and the generalized Newtonian fluid (GNF) model to solve some basic flow problems. Although some students have been exposed to transport phenomena, there are a number who have not. For this reason we start by considering the use of shell momentum balances rather than the use of the equations of change, which are presented in Section 2.3. The material is arranged in such a way that a number of useful processing flows can be analyzed without the use of the three-dimensional equations of change.

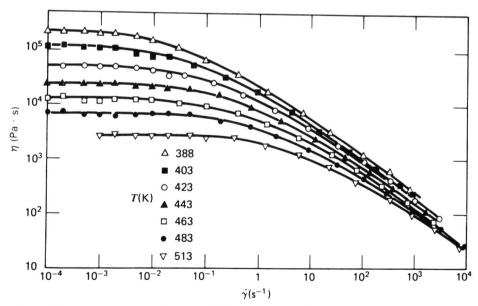

**Figure 2.5** Non-Newtonian viscosity of a LDPE melt at several different temperatures as shown in the figure (Reprinted by permission of the publisher from J. Meissner, *Kunststoffe*, *61*, 576, 1971.)

EXAMPLE 2.1
**Power Law and Ellis Model Parameters for Linear Low Density Polyethylene (LLDPE)**
Determine the power-law (Eq. 2.6) and Ellis model (Eq. 2.7) parameters for LLDPE (NTA 101) at 170°C using the rheological data given in Appendix Tables A.7–A.9.

**Solution**

To find the parameters $m$ and $n$ in Eq. 2.6 we first plot $\ln \eta$ versus $\ln \dot{\gamma}$ as shown in Fig. 2.6. The slope of the line in the linear region is $n-1$ and is estimated to be $-0.576$. Hence $n$ is 0.424. $m$ is found by taking the natural logarithm of both sides of Eq. 2.6:

$$\ln \eta = \ln m + (n-1)\ln \dot{\gamma} \qquad (2.12)$$

and then arbitrarily selecting values of $\eta$ and $\dot{\gamma}$ in the linear region. For example, substituting $\dot{\gamma} = 140\,\mathrm{s}^{-1}$ and $\eta = 1.45 \times 10^3$ Pa·s into Eq. 2.12 we find $m$ to be $2.374 \times 10^4$ Pa·s$^n$.

To find the Ellis model parameters we plot $\ln(\eta_0/\eta - 1.0)$ versus $\ln \tau_{yx}$ as shown in the upper right-hand corner of Fig. 2.6. $\tau_{1/2}$ is the value of $\tau_{yx}$ when $\eta = \frac{1}{2}\eta_0$ and is estimated from the graph in Fig. 2.6 to be $3.325 \times 10^4$ Pa. $\alpha - 1$ is estimated from the slope of the line to be 1.54. $\eta_0$ is read directly from the data and is $1.33 \times 10^4$ Pa·s. All the parameters for both models are summarized in Table 2.1.

EXAMPLE 2.2
**Power Law Parameters for LLDPE Using Nonlinear Regression Analysis**
Determine $m$ and $n$ in Eq. 2.6 for LLDPE using nonlinear regression analysis. In particular use the IMSL (International Mathematics and Statistics Libraries) subroutine RNLIN described in Appendix D.3 to find $m$ and $n$ for LLDPE at 170°C.

**Solution**

According to the description of the subroutine RNLIN given in Appendix D.3, we must differentiate Eq. 2.6 with respect to the parameters in the model. For the power-law model, these derivatives are:

$$\frac{\partial \eta}{\partial m} = \dot{\gamma}^{n-1} \qquad (2.13)$$

$$\frac{\partial \eta}{\partial n} = m\dot{\gamma}^{n-1}\ln \dot{\gamma}. \qquad (2.14)$$

The Fortran calling program is given in Table 2.2. The values obtained from nonlinear regression analysis for $m$ and $n$ are given in Table 2.1.

A few comments should be made regarding the selection of data points used in the regression analysis. In the case of the power law, only the data in the linear region are used. Certainly, the regression analysis could have been carried out on all the data, but the coefficients obtained would not lead to an accurate prediction of $\eta$ at intermediate shear rates.

It is also observed that the power-law parameters obtained graphically differ from those obtained by nonlinear regression analysis. The function with the two different sets of parameters is plotted in Fig. 2.6. Regression analysis basically removes the arbitrariness of finding $n$. It should also be noted that the Ellis model with the parameters even obtained graphically fits the data well over the full range.

Shear thinning or pseudoplastic behavior is an important property which must be taken into account in the design of polymer processes. However, it is not the only property, and in Chapter 3 we will discuss models that describe the viscoelastic response of polymeric fluids. However, first we would like to solve some basic one-dimensional isothermal flow problems using the shell momentum balance and the empiricisms for viscosity described in this section.

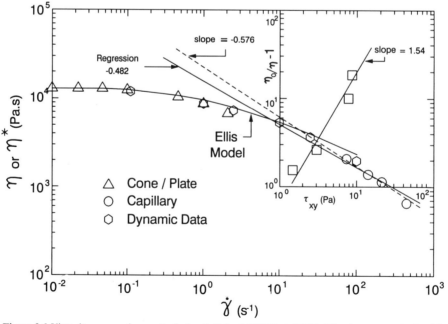

**Figure 2.6** Viscosity versus shear rate (ln-ln plot) for LLDPE at 170°C. The data were obtained by various types of rheometers as indicated in the figure. (---) Graphical fit of power law; (——) fit of power law using regression analysis. Inset shows plot of $(\eta/\eta_0)-1$ versus shear stress used to obtain Ellis model parameters.

**TABLE 2.2 Program Listing for Obtaining Power-Law Parameters**

```
C******************************ABSTRACT********************************
C  THIS PROGRAM SOLVES EXAMPLE 2.1 USING THE IMSL SUBROUTINE RNLIN
C  THE DESCRIPTION OF RNLIN IS GIVEN IN APPENDIX D.3. NONLINEAR REGRESSION
C  IS USED TO FIND THE PARAMETERS IN THE POWER LAW MODEL FOR
C  VISCOSITY. THE VISCOSITY FUNCTION IS FIT TO THE STEADY AND DYNAMIC
C  VISCOSITY DATA GIVEN IN APPENDIX A.3 FOR LLDPE.
C
      INTEGER LDR, NOBS, NPARM
      PARAMETER (NOBS=7, NPARM=2, LDR=NPARM)
      INTEGER IDERIV, IRANK, NOUT
      REAL DFE, R(LDR,NPARM), SSE, THETA(NPARM)
      EXTERNAL EXAMPL, RNLIN, UMACH, WRRRN
C
      DATA THETA/5000.0, 0.5/
C
      CALL UMACH (2,NOUT)
C
      IDERIV = 1
      CALL RNLIN (EXAMPL, NPARM, IDERIV, THETA, R, LDR, IRANK, DFE,
     &     SSE)
      WRITE (NOUT,*) 'THETA = ', THETA
      WRITE (NOUT,*) 'IRANK = ', IRANK, ' DFE,= ', DFE, ' SSE = ',
     &     SSE
      CALL WRRRN ('R', NPARM, NPARM, R, LDR, 0)
      END
C
      SUBROUTINE EXAMPL (NPARM,THETA, IOPT, IOBS, FRQ, WT, E, DE,
     &     IEND)
      INTEGER NPARM, IOPT, IOBS, IEND
      REAL     THETA(NPARM), FRQ, WT, E, DE(NPARM)
C
      INTEGER NOBS
      PARAMETER (NOBS=7)
      REAL      XDATA(NOBS), YDATA(NOBS)
C
C**********INPUT VISCOSITY VERSUS SHEAR RATE DATA*******
C
      DATA YDATA/5.270E+3, 3.790E+3, 2.090E+3, 2.03E+3, 1.450E+3,
     &     1.170E+3, 6.450E+2/
      DATA XDATA/10.0,25.1,70.0,100.0,140.0,209.0,419.0/
C
      IF (IOBS .LE. NOBS) THEN
        WT = 1.0E0
        FRQ=1.0E0
        IEND = 0
        IF (IOPT .EQ. 0) THEN
          E = YDATA(IOBS) - THETA(1)*XDATA(IOBS)**(THETA(2)-1)
        ELSE
          DE(1) = -XDATA(IOBS)**(THETA(2)-1)
          DE(2) = -THETA(1)*XDATA(IOBS)**(THETA(2)-1)*LOG(XDATA
     &       (IOBS))
        END IF
      ELSE
        IEND = 1
      END IF
      RETURN
      END
```

### 2.2.1. FLOW THROUGH AN ANNULAR DIE

The manufacture of pipe and the generation of parisons used in blow molding involve flow through an annular die. Our first goal is to design an annular die for extruding a pipe at 180°C with an outer diameter (O.D.) of 0.0762 m (3.0 in.) and an inner diameter (I.D.) of 0.0635 m (2.5 in.) for HDPE at the highest extrusion rate possible (m/min).

Assume that the limiting factor is the onset of melt fracture, which occurs at a wall shear stress, $\tau_w$, of $1.0 \times 10^5$ Pa (melt fracture is discussed in Section 7.1). The rheological data for this polymer (i.e., $\eta$ versus $dv_z/dr$) have been analyzed, and the parameters for the various models discussed in Section 2.1 are given in Table 2.3.

To solve this problem we must obtain relations between the wall shear stress, $\tau_w$, and the pressure drop, $\Delta P$, as well as the volumetric

**TABLE 2.3 Parameters for HDPE, Alathon, DuPont**

| Temperature (°K) | Power Law | | | | Carreau Model | | | Ellis Model | |
|---|---|---|---|---|---|---|---|---|---|
| | $\dot{\gamma}\,(s^{-1})$ | $m\,(Pa \cdot s^n)$ | $n$ | $\eta_0\,(Pa \cdot s)$ | $\dot{\gamma}\,(s^{-1})$ | $n$ | $\lambda\,(s)$ | $\alpha$ | $\tau_{1/2}\,(Pa)$ |
| 453 | 100–1000 | $6.19 \times 10^3$ | 0.56 | $2.1 \times 10^3$ | 100–1200 | 0.54 | 0.07 | 2.57 | $7.50 \times 10^4$ |
| 473 | 100–1000 | $4.68 \times 10^3$ | 0.59 | $1.52 \times 10^3$ | 100–1200 | 0.50 | 0.08 | 2.51 | $7.49 \times 10^4$ |
| 493 | 100–1000 | $3.73 \times 10^3$ | 0.61 | $1.17 \times 10^3$ | 186–1400 | 0.58 | 0.05 | 2.49 | $7.67 \times 10^4$ |

(Data from Z. Tadmor and C. G. Gogos, *Principles of Polymer Processing*, Wiley, New York, 1979, Table A.1.)

flow rate, $Q$. This is done by carrying out a momentum or force balance on a differential element of fluid to obtain a differential equation for the stress distribution. A constitutive equation is then substituted into the stress equation to obtain a differential equation for the velocity field. This is then integrated, and the velocity field is found when the appropriate boundary conditions are specified. The following assumptions are made:

1. The flow is steady, laminar, and isothermal,
2. There are no entry or exit effects,
3. Inertia is insignificant (i.e., the Reynolds number is negligible),
4. The fluid is inelastic, and hence die swell is not considered,
5. The fluid does not slip on the die surfaces.

A cross section of the annular die required to produce the pipe is shown in Fig. 2.7. The approximate velocity and stress profiles for this flow

are also sketched there. A thin cylindrical shell of length $L$ and thickness $\Delta r$ is now chosen as shown in Fig. 2.8. The shell is selected so that the surface is parallel to the flow direction. A force (or momentum) balance is now performed on the shell. Because the flow is under steady-state conditions, the forces in the $z$ direction must sum to zero, as shown:

$$\tau_{rz}|_r 2\pi r L - \tau_{rz}|_{r+\Delta r} 2\pi(r+\Delta r)L - \rho g 2\pi r \Delta r L + (p_0 - p_L)2\pi r \Delta r = 0.$$

(2.15)

$\tau_{rz}$ is the force per unit area acting in the $z$ direction on a surface at $r$ by a layer of fluid at a lesser $r$. It is customary to take the force per unit area acting at $r$ to be positive and that at $r+\Delta r$ to be negative.

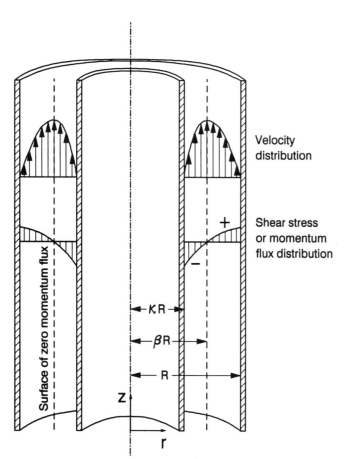

Figure 2.7 Flow through a cylindrical annulus.

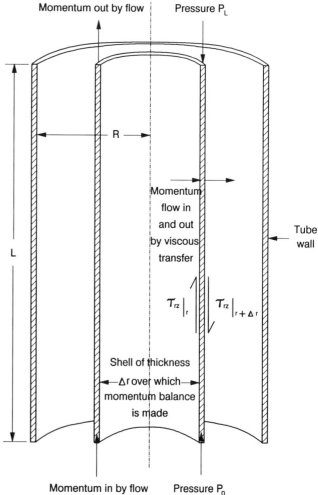

**Figure 2.8** Cylindrical shell of fluid over which the momentum or force balance is performed.

We now divide Eq. 2.15 by $2\pi\Delta rL$ and take the limit as $\Delta r\to 0$:

$$\lim_{\Delta r\to 0}\left(\frac{(r\tau_{rz})|_{r+\Delta r}-(r\tau_{rz})|_r}{\Delta r}\right)=\left(\frac{p_0-p_L}{L}-\rho g\right)r. \qquad (2.16)$$

But this first term is just the derivative, and this gives the following differential equation:

$$\frac{d}{dr}(r\tau_{rz})=+\left(\frac{P'_0-P'_L}{L}\right)r, \qquad (2.17)$$

where $P'_0=p_0$ and $P'_L=p_L+\rho gL$.* This gives a differential equation which we can solve for $\tau_{rz}$ by integration to give:

$$\tau_{rz}=\left(\frac{P'_0-P'_L}{L}\right)\frac{r}{2}+\frac{C_1}{r}. \qquad (2.18)$$

At some distance, $\beta R$, the velocity field must pass through a maximum, and $\tau_{rz}$ (which is proportional to $dv_z/dr$) must be zero. Utilizing this information $C_1$ is replaced by $-(P'_0-P'_L)(\beta R)^2/2L$, which leads to the following equation in place of Eq. 2.18:

$$\tau_{rz}=\frac{(P'_0-P'_L)R}{2L}\left[\frac{r}{R}-\beta^2\left(\frac{R}{r}\right)\right]. \qquad (2.19)$$

Our next goal is to find the velocity field. To do this, the type of fluid and an appropriate constitutive equation must be specified. In the first case the fluid is considered to be Newtonian. $\tau_{rz}$ is replaced with Newton's law of viscosity to obtain a differential equation for $v_z$:

$$\frac{dv_z}{dr}=-\frac{(P'_0-P'_L)}{2\mu L}\left[\left(\frac{r}{R}\right)-\beta^2\left(\frac{R}{r}\right)\right]. \qquad (2.20)$$

The flow is considered to consist of two parts:

$$\kappa R\leqslant r<\beta R \qquad v_z=v_z^< \qquad (2.21)$$

$$\beta R<r\leqslant R \qquad v_z=v_z^>. \qquad (2.22)$$

Furthermore, the no-slip boundary conditions (B.C.) are assumed at the walls:

B.C.1:   at $r=\kappa R$   $v_z=0$ $\qquad (2.23)$

B.C.2:   at $r=R$   $v_z=0.$ $\qquad (2.24)$

Equation 2.20 is integrated using the boundary conditions to obtain:

$$v_z^<=R\left[\frac{(P'_0-P'_L)R}{2\mu L}\right]\int_\kappa^\xi\left(\frac{\beta^2}{\xi'}-\xi'\right)d\xi', \qquad \kappa\leqslant\xi\leqslant\beta \qquad (2.25)$$

$$v_z^>=R\left[\frac{(P'_0-P'_L)R}{2\mu L}\right]\int_\xi^1\left(\xi'-\frac{\beta^2}{\xi'}\right)d\xi', \qquad \beta\leqslant\xi\leqslant 1, \qquad (2.26)$$

where $\xi'=r/R$ is a dummy variable of integration. At $r=\beta R$, $v_z^<=v_z^>$,

---

*The pressure, $P'_L$, represents the combined effect of dynamic pressure and the pressure due to gravity. In general $P'$ may be defined as $P'=p+\rho gh$, where $h$ is the distance upward from any chosen plane.

and one can equate Eqs. (2.25) and (2.26) to find $\beta$ as follows:

$$\int_\kappa^\beta\left(\frac{\beta^2}{\xi}-\xi\right)d\xi=\int_\beta^1\left(\xi-\frac{\beta^2}{\xi}\right)d\xi. \qquad (2.27)$$

For a Newtonian fluid these expressions can be integrated to give $\beta$:

$$2\beta^2=(1-\kappa^2)/\ln(1/\kappa). \qquad (2.28)$$

For a Newtonian fluid it is observed that $\beta$ is determined merely by geometric factors, and one can then write down the solutions for $\tau_{rz}$ and $v_z$:

$$\tau_{rz}=\frac{(P'_0-P'_L)R}{2L}\left[\frac{r}{R}-\frac{1}{2\ln(1/\kappa)}\left(\frac{R}{r}\right)\right] \qquad (2.29)$$

$$v_z=\frac{(P'_0-P'_L)R^2}{4\mu L}\left[1-\left(\frac{r}{R}\right)^2+\frac{1-\kappa^2}{\ln(1/\kappa)}\ln\left(\frac{r}{R}\right)\right]. \qquad (2.30)$$

We can now integrate Eq. (2.30) over the cross section of the annulus to find the volumetric flow rate, $Q$:

$$Q=\pi R^2(1-\kappa^2)\langle v_z\rangle=\frac{\pi(P'_0-P'_L)R^4}{8\mu L}\left((1-\kappa^4)-\frac{(1-\kappa^2)^2}{\ln(1/\kappa)}\right). \qquad (2.31)$$

One can now determine $Q$ and the extrusion rate which is the average velocity, $\langle v_z\rangle=Q/A$. With the assumption that the fluid leaves the die with the same dimensions as the die, then one can calculate that $\kappa R=0.03175$ m and $R=0.0381$ m. However, these values will not be quite correct as the melt will swell on leaving the die but shrink as it cools because of the change in density. These factors will be considered in Section 7.2.1. Furthermore, we have not specified the die length, which must be large enough to eliminate any effects from the entry but not so large as to create excessive pressures. For the time being we will take $L$ to be 10 times the die gap $[R(1-\kappa)]$, which makes $L=0.0635$ m. One now calculates $P'_0-P'_L$, $Q$, and $\langle v_z\rangle$ assuming $\tau_w=1\times 10^5$ Pa:

$$P'_0-P'_L=2.024\times 10^6\text{ Pa}$$

$$Q=7.24\times 10^{-5}\text{ m}^3/\text{s}$$

$$\langle v_z\rangle=3.12\text{ m/min}.$$

Next we consider the case in which the fluid is shear thinning, and the viscosity is described by the power-law model. The derivation leading to Eq. (2.19) is unchanged. Starting with this equation one now solves for the velocity field using the power-law model:

$$\tau_{rz}=-m|(dv_z/dr)|^{n-1}(dv_z/dr). \qquad (2.32)$$

For $\kappa R<r<\beta R$, $dv_z/dr>0$, and we express Eq. 2.32 as

$$\tau_{rz}=-m(dv_z^</dr)^n. \qquad (2.33)$$

For $\beta R<r<R$, $dv_z/dr<0$, and Eq. 2.32 becomes:

$$\tau_{rz}=m(-dv_z^>/dr)^n. \qquad (2.34)$$

Two different expressions for $\tau_{rz}$ are required to ensure that a negative number is not raised to a fractional exponent, which will lead to an imaginary number. Analogous to Eqs. 2.25 and 2.26 we solve for $v_z^>$

and $v_z^<$:

$$-m\left(\frac{dv_z^<}{dr}\right)^n = C\left[\frac{r}{R} - \beta^2\frac{R}{r}\right] \tag{2.35}$$

$$m\left(\frac{-dv_z^>}{dr}\right)^n = C\left[\frac{r}{R} - \beta^2\frac{R}{r}\right], \tag{2.36}$$

where $C = (P_0' - P_L')R^2/2L$. We now integrate these equations after taking the $1/n$ power of both sides:

$$v_z^< = R\left[\frac{(P_0' - P_L')R}{2mL}\right]^s \int_\kappa^\xi \left(\frac{\beta^2}{\xi'} - \xi'\right)^s d\xi', \quad \kappa \leqslant \xi \leqslant \beta \tag{2.37}$$

$$v_z^> = R\left[\frac{(P_0' - P_L')R}{2mL}\right]^s \int_\xi^1 \left(\xi' - \frac{\beta^2}{\xi'}\right)^s d\xi', \quad \beta \leqslant \xi \leqslant 1. \tag{2.38}$$

When $1/n$ is a whole number (e.g., 1, 2, 3, etc.), then these integrals can be integrated to directly obtain expressions for $v_z^<$ and $v_z^>$. However $n$ is usually some decimal number such as 0.52, in which case we must evaluate the integrals numerically. First we employ the expression for finding $\kappa$ by equating $v_z^>$ and $v_z^<$:

$$\int_\kappa^\beta \left(\frac{\beta^2}{\xi} - \xi\right)^s d\xi = \int_\beta^1 \left(\xi - \frac{\beta^2}{\xi}\right)^s d\xi. \tag{2.39}$$

Equation 2.39 gives $\beta$ as a function of $\kappa$ and $s$ but must be evaluated numerically unless $1/n$ is an integer. $Q$ is found by integrating Eqs. 2.37 and 2.38 over the cross section of the die:

$$Q = 2\pi \int_{\kappa R}^R v_z r \, dr = \pi R^3\left[\frac{(P_0' - P_L')R}{2mL}\right]^s \int_\kappa^1 |\beta^2 - \xi^2|^{s+1} \xi^{-s} d\xi. \tag{2.40}$$

If we treat each term within the integral as a function of $\xi'$, that is, $F(\xi')$, then Eq. 2.40 can be integrated by parts (*note*: let $u = F(\xi')$ and $dv = \xi' d\xi'$) to obtain:

$$Q = \frac{\pi R^3}{(1/n) + 3}\left(\frac{P_0' - P_L'}{2mL}\right)^{1/n}[(1 - \beta^2)^{1+(1/n)} - \kappa^{1-(1/n)}(\beta^2 - \kappa^2)^{1+(1/n)}]. \tag{2.41}$$

Once $\beta$ is known from Eq. 2.29, then Eq. 2.41 can be evaluated. We express Eq. 2.41 as follows:

$$Q = \pi R^3\left(\frac{(P_0' - P_L')R}{2mL}\right)^{1/n}\frac{(1 - \kappa)^{1/n+2}}{s+2}F(s, \kappa), \tag{2.42}$$

where $F(1/n, \kappa)$ is given for several values of $1/n$ in Fig. 2.9.

Now referring to Eq. 2.19 we determine that melt fracture will occur at the outer wall first, because the shear stress is higher there than at the inner wall:

$$\tau_w(\kappa R) = \left(\frac{(P_0' - P_L')\kappa R}{2L}\right)\left[1 - \frac{\beta^2}{\kappa^2}\right] \tag{2.43a}$$

$$\tau_w(R) = \left(\frac{(P_0' - P_L')R}{2L}\right)(1 - \beta^2). \tag{2.43}$$

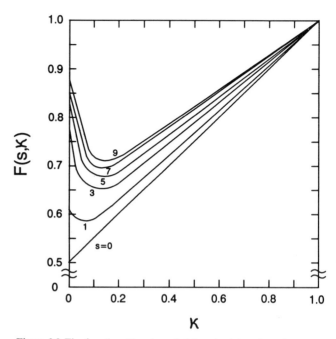

**Figure 2.9** The function $F(s, \kappa)$ needed for obtaining the volume rate of flow through an annulus for a power-law fluid.

This follows from the fact that $\kappa < 1$. From Eq. 2.39 and using $n = 0.56$, we find that $\beta$ is about 0.91. We can now calculate $P_0' - P_L'$, $Q$, and $\langle v_z \rangle$:

$$P_0' - P_L' = 1.91 \times 10^6 \, \text{Pa}$$

$$Q = 1.56 \times 10^{-4} \, \text{m}^3/\text{s}$$

$$\langle v_z \rangle = 6.72 \, \text{m/min}$$

We observe that although the pressure drop is only slightly lower than for the Newtonian case, the extrusion rate is more than twice as large.

### 2.2.2. FLOW IN A WIRE-COATING DIE

To further illustrate the use of the force balance on a differential element of fluid to obtain an equation for the shear stress distribution, we consider the design of a wire-coating die next. The problem is to design a wire-coating die to coat a $0.655 \times 10^{-2}$ m diameter wire with a $0.330 \times 10^{-2}$ m thick layer of HDPE at 200°C at the highest extrusion rate possible (assume again that the extrusion rate limit is the onset of melt fracture at $\tau_w = 1.4 \times 10^5$ Pa).

A typical wire-coating die was shown in Fig. 1.4. The design that is shown there is somewhat beyond our capabilities at this point. For this reason we consider only the annular flow region, as shown in Fig. 2.10. We make the following assumptions:

1. The flow is steady, incompressible, and isothermal,
2. The rheological properties of the fluid are described by the power-law model, and elastic effects (i.e., die swell) can be neglected,
3. The converging part of the die will be neglected,
4. There is no significant pressure drop ($\Delta P$) across the die. However, in practice a small $\Delta P$ is used to control the thickness as shown in Fig. 2.11.

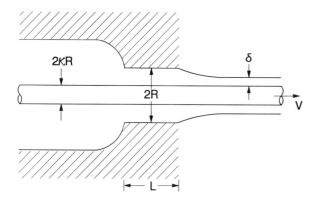

**Figure 2.10** Simplified view of a wire-coating die with no imposed pressure gradient.

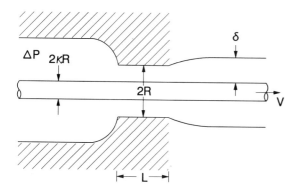

**Figure 2.11** Wire-coating die with an imposed pressure gradient.

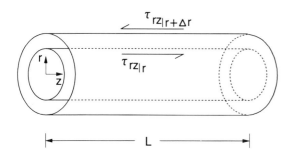

**Figure 2.12** Cylindrical element of differential thickness, $\Delta r$, on which momentum balance is made.

The shell force or momentum balance is performed on an element of fluid as shown in Fig. 2.12, which is similar to that shown in Fig. 2.8. Here the only terms that have to be considered are due to the stresses exerted by one layer of fluid on another as there are no effects due to gravity or pressure. The force or momentum balance is:

$$\tau_{rz}|_r 2\pi r - \tau_{rz}|_{r+\Delta r} 2\pi(r+\Delta r) = 0. \tag{2.44}$$

In the limit as $\Delta r \to 0$, Eq. 2.44 becomes:

$$-\frac{d}{dr}(r\tau_{rz}) = 0. \tag{2.45}$$

This can now be integrated to find $\tau_{rz}$:

$$\tau_{rz} = C_1/r. \tag{2.46}$$

We use the power-law model to find $v_z(r)$:

$$m(-dv_z/dr)^n = C_1/r. \tag{2.47}$$

Here we have used the fact that $dv_z/dr < 0$. This equation can be integrated to find $v_z$:

$$v_z \equiv (C_1/m)^{1/n} \frac{r^{-1/n+1}}{\left(\dfrac{1-n}{n}\right)} - C_2. \tag{2.48}$$

The following boundary conditions are used to find $C_1$ and $C_2$:

$$\text{B.C.1} \quad \text{at } r = R \qquad v_z = 0. \tag{2.49}$$

$$\text{B.C.2} \quad \text{at } r = \kappa R \qquad v_z = V. \tag{2.50}$$

Using these conditions, the velocity profile becomes:

$$\frac{v_z(r)}{V} = \frac{\xi^{1-s}-1}{\kappa^{1-s}-1}, \tag{2.51}$$

where $\xi = r/R$ and $s = 1/n$. $\tau_{rz}$ becomes:

$$\tau_{rz} = \frac{mV^n(s-1)^n}{R^n(\kappa^{1-s}-1)^n}\left(\frac{R}{r}\right). \tag{2.52}$$

At this point we cannot specify $V$ or $R$, until we relate the thickness of the coating to these variables. To do this we perform a mass balance on the region starting at the exit of the die and ending where the fluid takes on the same velocity as the wire (the density, $\rho$, is assumed not to change, although the melt and solid would in reality have different densities):

$$2\pi \int_{\kappa R}^{R} v_z r\, dr = V\pi[(\kappa R + \delta)^2 - (\kappa R)^2], \tag{2.53}$$

where $\delta$ is the coating thickness. On substituting Eq. 2.51 into the integral in Eq. 2.53 we obtain the following equation for the coating thickness, $\delta$:

$$\frac{2\pi VR^2}{\kappa^{1-s}-1}\left[\frac{1-\kappa^{3-s}}{3-s} - \frac{1-\kappa^2}{2}\right] = V\pi[(\kappa R + \delta)^2 - (\kappa R)^2]. \tag{2.54}$$

Eqs. 2.52 and 2.54 represent two nonlinear equations for $R$ and $V$. Actually $V$ drops out of Eq. 2.54, allowing us to solve for $R$. Once $R$ is found, it can be substituted back into Eq. 2.52 to find $V$. This is done in Ex. 2.3, which follows.

Finally, there are a number of flow geometries for which the shell balance approach can be used and that are found frequently in polymer processing. We summarize these results in Tables 2.5 through 2.7 for future reference. Here solutions are given for flow between parallel plates, tube flow, and flow through an annulus. Solutions are given for Newtonian, power-law, and Ellis models.

The use of the shell balance cannot accommodate all the flows we find in polymer processing. In the next section we summarize the isothermal equations of change and generalize the Newtonian and non-Newtonian constitutive equations to three dimensions.

**EXAMPLE 2.3**
**Die Radius and Wire Speed**

Equations 2.52 and 2.54 represent two nonlinear algebraic equations for finding the wire velocity and radius. Find $R$ and $V$ using the IMSL subroutine NEQNF described in Appendix D.4.

**Solution**

Because Eq. 2.54 does not contain $V$, we can first find $R$ and then substitute $R$ into Eq. 2.52 and find $V$. According to the program description for subroutine NEQNF, we must write Eq. 2.54 in the following form:

$$\frac{2R^2(1-\kappa^{3-s})}{(\kappa^{1-s}-1)(3-s)} - \frac{R^2(1-\kappa^2)}{(\kappa^{1-s}-1)} - 2\kappa R\delta - \delta^2 = 0. \tag{2.55}$$

Using the following information given in the problem,

$$s = 1.6949 \qquad \kappa = 3.275 \times 10^{-3}R^{-1} \qquad \delta = 3.30 \times 10^{-3} \, \text{m},$$

Eq. 2.55 becomes:

$$1.032R^2 - 2.608 \times 10^{-3}R^{s-1} + 4.323 \times 10^{-5} = 0. \tag{2.56}$$

Subroutine NEQNF solves a single nonlinear algebraic equation or a system of nonlinear algebraic equations for the roots. The calling program listing is given in Table 2.4. To use the subroutine one must provide an initial guess for $R$. As there is more than one root, obtaining a realistic value of $R$ depends on the initial guess. If $n$ is a value such as $\frac{1}{2}$, $\frac{1}{3}$, $\frac{1}{4}$, and so on, then $s$ is an integer. Taking $n = 0.5$, and hence $s = 2$ in Eq. 2.56, a quadratic equation is formed for which the physically realistic solution is $R = 4.33 \times 10^{-3}$ m. Another approach is to start with the solution for the Newtonian case ($s = 1$), but this requires additional algebraic manipulations. Using the initial guess as $4.33 \times 10^{-3}$ m, $R$ is calculated to be $4.549 \times 10^{-3}$ m. We can now substitute this value back into Eq. 2.52 to find $V$ subject to the condition that melt fracture will be visible at the outer surface at $\tau_{rz}(R) = 1.4 \times 10^5$ Pa:

$$V = \left[\frac{\tau_{rz}(R)R^n(1-\kappa^{s-1})}{m(s-1)^n}\right]^{1/n} = 0.141 \, \text{m s}^{-1}. \tag{2.57}$$

Line speeds in actual processes can run in the range of $10 \, \text{m s}^{-1}$ to $40 \, \text{m s}^{-1}$, which actually leads to conditions well beyond where melt fracture is initiated.

Several points should be made concerning this solution. The program was written so that the initial guess, the accuracy of the solution, and the maximum number of iterations could be input through the terminal. This made it easy to change these variables without having to compile the program each time. Furthermore, the program could have easily been written in a more general form by starting with Eq. 2.55 and then reading in quantities such as $s$, $\delta$ and $\kappa R$. This, however, has been saved for the homework problems.

Although we have made a number of simplifications to get a solution to this problem, the situation is still complicated, and computational techniques must be used. The addition of a pressure drop (see Pr. 2B.5) and the use of the converging geometry present even more difficulties. The addition of fluid elasticity (die swell) will be dealt with in Section 7.2.

**TABLE 2.4 Program Listing for Calling Subroutine NEQNF (WIRECT1.FOR)**

```
C••••••••••••••••••••••••••ABSTRACT••••••••••••••••••••••••••••••
C
C  THIS PROGRAM CALCULATES THE DIE RADIUS AS REQUESTED IN E. 2.3
C  USING THE IMSL SUBROUTINE NEQNF FOUND IN APPENDIX D.4
C
C              DECLARE VARIABLES
      INTEGER ITMAX, N
      REAL ERRREL
      PARAMETER (N=1)
      INTEGER K, NOUT
      REAL FCN, FNORM, X(N), XGUESS(N)
      EXTERNAL FCN, NEQNF, UMACH
C
      DATA XGUESS/.1/
C
      ERRREL = 0.001
      ITMAX = 20
C
      CALL UMACH(2,NOUT)
C
      CALL NEQNF(FCN, ERRREL, N, ITMAX, XGUESS, X, FNORM)
      WRITE (NOUT,100) (X(K),K=1,N), FNORM
  100 FORMAT(' THE SOLUTION TO THE SYSTEM IS',/,' X = (',E10.4,
     & ')',/,' WITH FNORM =',E10.4,//)
          END
C
      SUBROUTINE FCN(X,F,N)
      INTEGER N
      REAL X(N), F(N)
      F(1) = 1.032X(1)**2-2.608E-03*X(1)**.6949+4.325E-05
      RETURN
      END
C
```

**TABLE 2.4** *continued*

```
C   INPUT XGUESS, ERREL, AND ITMAX:
C   .005, .0005, 10000
C   OUTPUT
C   THE SOLUTION TO THE SYSTEM IS
C   X = (.4549E-02)
C   WITH FNORM = .9592E-11
```

---

**TABLE 2.5 Parallel Plate Pressure Flow**

$$\tau_{yz} = -\mu \frac{dv_z}{dy} \qquad\qquad \tau_{yz} = -m\left|\frac{dv_z}{dy}\right|^{n-1}\frac{dv_z}{dy}$$

$$\tau_{yz}(y) = \left(\frac{\Delta P}{L}\right)y \qquad\qquad \tau_{yz}(y) = \left(\frac{\Delta P}{L}\right)y$$

$$\tau_w = \tau_{yz}\left(\frac{H}{2}\right) = \frac{H\Delta P}{2L} \qquad\qquad \tau_w = \tau_{yz}\left(\frac{H}{2}\right) = \frac{H\Delta P}{2L}$$

$$-\dot\gamma_{yz}(y) = \left(\frac{\Delta P}{\mu L}\right)y \qquad\qquad -\dot\gamma_{yz}(y) = \left(\frac{\Delta P}{mL}y\right)^s \quad y \geqslant 0$$

$$\dot\gamma_w = -\dot\gamma_{yz}\left(\frac{H}{2}\right) = \frac{H\Delta P}{2\mu L} \qquad\qquad \dot\gamma_w = -\dot\gamma_{yz}\left(\frac{H}{2}\right) = \left(\frac{H\Delta P}{2mL}\right)^s$$

$$v_z(y) = \left(\frac{H^2\Delta P}{8\mu L}\right)\left[1 - \left(\frac{2y}{H}\right)^2\right] \qquad\qquad v_z(y) = \frac{H}{2(s+1)}\left(\frac{H\Delta P}{2mL}\right)^s\left[1 - \left(\frac{2y}{H}\right)^{s+1}\right] \quad y \geqslant 0$$

$$v_z(0) = v_{max} = \frac{H^2\Delta P}{8\mu L} \qquad\qquad v_z(0) = v_{max} = \frac{H}{2(s+1)}\left(\frac{H\Delta P}{2mL}\right)^s$$

$$\langle v_z\rangle = \frac{2}{3}v_{max} \qquad\qquad \langle v_z\rangle = \left(\frac{s+1}{s+2}\right)v_{max}$$

$$Q = \frac{WH^3\Delta P}{12\mu L} \qquad\qquad Q = \frac{WH^2}{2(s+2)}\left(\frac{H\Delta P}{2mL}\right)^s$$

---

$$\tau_{yz} = -\eta(\tau)\frac{dv_z}{dy} \qquad \eta(\tau) = \frac{\eta_0}{1 + (\tau/\tau_{1/2})^{\alpha-1}} \qquad \tau = |\tau_{yz}|$$

$$\tau_{yz}(y) = \left(\frac{\Delta P}{L}\right)y$$

$$\tau_w = \tau_{yz}\left(\frac{H}{2}\right) = \frac{H\Delta P}{2L}$$

$$-\dot\gamma_{yz} = \left(\frac{\Delta P}{\eta_0 L}\right)y\left[1 + \left(\frac{\Delta Py}{\tau_{1/2}L}\right)^{\alpha-1}\right]$$

$$\dot\gamma_w = -\dot\gamma_{yz}\left(\frac{H}{2}\right) = \frac{H\Delta P}{2\eta_0 L}\left[1 + \left(\frac{H\Delta P}{2\tau_{1/2}L}\right)^{\alpha-1}\right]$$

$$v_z(y) = \frac{H^2\Delta P}{8\eta_0 L}\left\{\left[1 - \left(\frac{2y}{H}\right)^2\right] + \left(\frac{2}{1+\alpha}\right)\left(\frac{H\Delta P}{2L\tau_{1/2}}\right)^{\alpha-1}\left[1 - \left(\frac{2y}{H}\right)^{\alpha+1}\right]\right\}$$

$$v_z(0) = v_{max} = \frac{H^2\Delta P}{8\eta_0 L}\left[1 + \left(\frac{2}{1+\alpha}\right)\left(\frac{H\Delta P}{2L\tau_{1/2}}\right)^{\alpha-1}\right]$$

$$\langle v_z\rangle = \frac{2}{3}v_{max}\left[1 + \left(\frac{3}{2+\alpha}\right)\left(\frac{H\Delta P}{2L\tau_{1/2}}\right)^{\alpha-1}\right]\bigg/\left[1 + \left(\frac{2}{1+\alpha}\right)\left(\frac{H\Delta P}{2L\tau_{1/2}}\right)^{\alpha-1}\right]$$

$$Q = \frac{WH^3\Delta P}{12\eta_0 L}\left[1 + \left(\frac{3}{2+\alpha}\right)\left(\frac{H\Delta P}{2L\tau_{1/2}}\right)^{\alpha-1}\right]$$

Data from Z. Tadmor and C. G. Gogos, *Principles of Polymer Processing*, Wiley, New York, 1979.)

---

**TABLE 2.6 Circular Tube Pressure Flow**

$$\tau_{rz} = -\mu \frac{dv_z}{dr} \qquad\qquad \tau_{rz} = -m\left|\frac{dv_z}{dr}\right|^{n-1}\frac{dv_z}{dr}$$

$$\tau_{rz}(r) = \left(\frac{\Delta P}{2L}\right)r \qquad\qquad \tau_{rz}(r) = \left(\frac{\Delta P}{2L}\right)r$$

$$\tau_w = \tau_{rz}(R) = \frac{R\Delta P}{2L} \qquad\qquad \tau_w = \tau_{rz}(R) = \frac{R\Delta P}{2L}$$

$$-\dot\gamma_{rz}(r) = \left(\frac{\Delta P}{2\mu L}\right)r \qquad\qquad -\dot\gamma_{rz}(r) = \left(\frac{\Delta Pr}{2mL}\right)^s$$

$$\dot\gamma_w = -\dot\gamma_{rz}(R) = \frac{R\Delta P}{2\mu L} \qquad\qquad \dot\gamma_w = -\dot\gamma_{rz}(R) = \left(\frac{R\Delta P}{2mL}\right)^s$$

$$v_z(r) = \frac{R^2\Delta P}{4\mu L}\left[1 - \left(\frac{r}{R}\right)^2\right] \qquad\qquad v_z(r) = \frac{R}{1+s}\left(\frac{R\Delta P}{2mL}\right)^s\left[1 - \left(\frac{r}{R}\right)^{s+1}\right]$$

$$v_z(0) = v_{max} = \frac{R^2\Delta P}{4\mu L} \qquad\qquad v_z(0) = v_{max} = \frac{R}{1+s}\left(\frac{R\Delta P}{2mL}\right)^s$$

$$\langle v_z\rangle = \frac{1}{2}v_{max} \qquad\qquad \langle v_z\rangle = \left(\frac{s+1}{s+3}\right)v_{max}$$

$$Q = \frac{\pi R^4\Delta P}{8\mu L} \qquad\qquad Q = \left(\frac{\pi R^3}{s+3}\right)\left(\frac{R\Delta P}{2mL}\right)^s$$

---

$$\tau_{rz} = -\eta(\tau)\frac{dv_z}{dr}, \quad \text{where } \eta(\tau) = \frac{\eta_0}{1 + (\tau/\tau_{1/2})^{\alpha-1}} \qquad \tau = |\tau_{rz}|$$

$$\tau_{rz}(r) = \left(\frac{\Delta P}{2L}\right)r$$

$$\tau_w = \tau_{rz}(R) = \frac{R\Delta P}{2L}$$

$$-\dot\gamma_{rz}(r) = \left(\frac{\Delta P}{2\eta_0 L}\right)r\left[1 + \left(\frac{\Delta Pr}{2L\tau_{1/2}}\right)^{\alpha-1}\right]$$

$$\dot\gamma_w = -\dot\gamma_{rz}(R) = \left(\frac{R\Delta P}{2\eta_0 L}\right)\left[1 + \left(\frac{R\Delta P}{2L\tau_{1/2}}\right)^{\alpha-1}\right]$$

$$v_z(r) = \frac{R^2\Delta P}{4L\eta_0}\left\{\left[1 - \left(\frac{r}{R}\right)^2\right] + \left(\frac{2}{1+\alpha}\right)\left(\frac{R\Delta P}{2L\tau_{1/2}}\right)^{\alpha-1}\left[1 - \left(\frac{r}{R}\right)^{\alpha+1}\right]\right\}$$

$$v_z(0) = v_{max} = \frac{R^2\Delta P}{4L\eta_0}\left[1 + \left(\frac{2}{1+\alpha}\right)\left(\frac{R\Delta P}{2L\tau_{1/2}}\right)^{\alpha-1}\right]$$

$$\langle v_z\rangle = \frac{1}{2}v_{max}\left[1 + \left(\frac{4}{3+\alpha}\right)\left(\frac{R\Delta P}{2L\tau_{1/2}}\right)^{\alpha-1}\right]\bigg/\left[1 + \left(\frac{2}{1+\alpha}\right)\left(\frac{R\Delta P}{2L\tau_{1/2}}\right)^{\alpha-1}\right]$$

$$Q = \frac{\pi R^4\Delta P}{8\eta_0 L}\left[1 + \left(\frac{4}{3+\alpha}\right)\left(\frac{R\Delta P}{2L\tau_{1/2}}\right)^{\alpha-1}\right]$$

Data from Z. Tadmor and C. G. Gogos, *Principles of Polymer Processing*, Wiley, New York, 1979.)

**TABLE 2.7 Concentric Annular Pressure Flows**

$$\tau_{rz} = -\mu\frac{dv_z}{dr}$$

$$\tau_{rz}(r) = \frac{\Delta PR}{2L}\left[\left(\frac{r}{R}\right) - \left(\frac{1-\kappa^2}{2\ln(1/\kappa)}\right)\left(\frac{R}{r}\right)\right]$$

$$\tau_{w1} = \tau_{rz}(R) = \frac{\Delta PR}{2L}\left[1 - \left(\frac{1-\kappa^2}{2\ln(1/\kappa)}\right)\right]$$

$$\tau_{w2} = \tau_{rz}(\kappa R) = \frac{\Delta PR}{2L}\left[\kappa - \left(\frac{1-\kappa^2}{2\ln(1/\kappa)}\right)\left(\frac{1}{\kappa}\right)\right]$$

$$\dot{\gamma}_{w1} = -\dot{\gamma}_{rz}(R) = \frac{\Delta PR}{2\mu L}\left[1 - \left(\frac{1-\kappa^2}{2\ln(1/\kappa)}\right)\right]$$

$$\dot{\gamma}_{w2} = \dot{\gamma}_{rz}(\kappa R) = \frac{\Delta PR}{2\mu L}\left[\kappa - \left(\frac{1-\kappa^2}{2\ln(1/\kappa)}\right)\left(\frac{1}{\kappa}\right)\right]$$

$$v_z(r) = \frac{\Delta PR^2}{4\mu L}\left[1 - \left(\frac{r}{R}\right)^2 + \left(\frac{1-\kappa^2}{\ln(1/\kappa)}\right)\ln\left(\frac{r}{R}\right)\right]$$

$$v_z(\lambda R) = v_{max} = \frac{\Delta PR^2}{4\mu L}\left\{1 - \left(\frac{1-\kappa^2}{2\ln(1/\kappa)}\right)\left[1 - \ln\left(\frac{1-\kappa^2}{2\ln(1/\kappa)}\right)\right]\right\} \qquad \lambda^2 = \frac{1-\kappa^2}{2\ln(1/\kappa)}$$

$$\langle v_z\rangle = \frac{\Delta PR^2}{8\mu L}\left[(1+\kappa^2) - \left(\frac{1-\kappa^2}{\ln(1/\kappa)}\right)\right]$$

$$Q = \frac{\pi\Delta PR^4}{8\mu L}\left[(1-\kappa^4) - \frac{(1-\kappa^2)^2}{\ln(1/\kappa)}\right]$$

---

$$\tau_{rz} = -\left|\frac{dv_z}{dr}\right|^{n-1}\left(\frac{dv_z}{dr}\right) \qquad \xi = \frac{r}{R} \qquad s = \frac{1}{n} \qquad \tau_{rz}(\beta R) = 0$$

---

$$v_z^{1}(r) = R\left(\frac{\Delta PR}{2mL}\right)^s\int_{\kappa}^{\xi}\left(\frac{\beta^2}{\xi'} - \xi'\right)^s d\xi' \qquad \kappa \leqslant \xi \leqslant \lambda$$

$$v_z^{11}(r) = R\left(\frac{\Delta PR}{2mL}\right)^s\int_{\xi}^{1}\left(\xi' - \frac{\beta^2}{\xi'}\right)^s d\xi' \qquad \lambda \leqslant \xi \leqslant 1$$

$\lambda$ is evaluated numerically for the equations above using the boundary condition

$$v_z^{1}(\beta R) = v_z^{11}(\beta R)$$

$$Q = \frac{\pi R^3}{s+2}\left(\frac{R\Delta P}{2mL}\right)^s(1-\kappa)^{s+2}F(s,\kappa)$$

(Data from Z. Tadmor and C. G. Gogos, *Principles of Polymer Processing*, Wiley, New York, 1979.)

## 2.3. EQUATIONS OF CHANGE FOR ISOTHERMAL SYSTEMS

As the process becomes more complex in terms of flow patterns, the shell balance approach is inadequate. Therefore, we introduce the equations of conservation of mass and momentum that are suited for handling multidimensional flows. We do not intend to rederive these equations, as they are discussed in detail in Bird et al. (1960, Chap. 3). Suffice it to say that the principles of conservation of mass and momentum applied to a differential cubic element of fluid lead to the equations of continuity and motion, respectively. We give these without derivation with emphasis on their use.

The equation of continuity is presented in Table 2.8 for three coordinate systems. Basically, this equation represents a constraint on

**TABLE 2.8 Equation of Continuity in Several Coordinate Systems**

Rectangular coordinates $(x, y, z)$

$$\frac{\partial\rho}{\partial t} + \frac{\partial}{\partial x}(\rho v_x) + \frac{\partial}{\partial y}(\rho v_y) + \frac{\partial}{\partial z}(\rho v_z) = 0 \tag{A}$$

Cylindrical coordinates $(r, \theta, z)$

$$\frac{\partial\rho}{\partial t} + \frac{1}{r}\frac{\partial}{\partial r}(\rho r v_r) + \frac{1}{r}\frac{\partial}{\partial\theta}(\rho v_\theta) + \frac{\partial}{\partial z}(\rho v_z) = 0 \tag{B}$$

Spherical coordinates $(r, \theta, \phi)$

$$\frac{\partial\rho}{\partial t} + \frac{1}{r^2}\frac{\partial}{\partial r}(\rho r^2 v_r) + \frac{1}{r\sin\theta}\frac{\partial}{\partial\theta}(\rho v_\theta\sin\theta) + \frac{1}{r\sin\theta}\frac{\partial}{\partial\phi}(\rho v_\phi) = 0 \tag{C}$$

(Data from Bird et al., 1960.)

the velocity field as a result of the fact that one cannot generate voids in the material during deformation. When $\rho$ is constant, then we consider the flow to be incompressible and for rectangular coordinates the continuity equation becomes:

$$\partial v_x/\partial x + \partial v_y/\partial y + \partial v_z/\partial z = 0. \tag{2.58}$$

The use of the continuity equation is illustrated in Ex. 2.4.

The equation of motion is obtained by generalizing the momentum or force balance for a three-dimensional element of fluid. The components of the equation of motion are given in Table 2.9. The terms on the left side of the equations are associated with the transport of momentum by bulk flow. For nearly all polymer processes these terms are negligible compared to the terms on the right side. This is equivalent to saying that the Reynolds number, Re, is negligible where $Re = \rho D\langle v\rangle/\eta$. In this expression $D$ is a characteristic length and $\langle v\rangle$ is the average velocity.

To solve these equations one needs to relate the stresses to the velocity gradients through a relation called the *constitutive equation*. For a Newtonian fluid the components of the constitutive equation are given by multiplying the components of the rate of deformation tensor by the viscosity, $\mu$. The components of the rate of deformation tensor are given in Table 2.10. In taking the constitutive equation as the product of the components of the rate of deformation tensor and $\mu$, expressions are generated for fluids that are incompressible and in which the bulk viscosity has been neglected (Bird et al., 1960, p. 79). In general tensor notation we write Newton's law of viscosity for an incompressible isotropic fluid as:

$$\tau = -\mu\dot{\gamma}. \tag{2.59}$$

This notation will not be explained in this text, but suffice it to say that it represents all the components given in Table 2.10 multiplied by $\mu$. If we substitute the components of Eq. 2.59 into the equation of motion, the Navier-Stokes equations are obtained.

In order to deal with the viscous response of non-Newtonian fluids, it is necessary to generalize Eq. 2.59 to include a viscosity function, $\eta$, that must depend on the magnitude of the rate of deformation or stress tensors. In actuality $\eta$ is a scalar quantity that must be a scalar function of the rate of deformation tensor, $\dot{\gamma}$, or of the stress tensor, $\tau$. The scalar quantities associated with any tensorial quantity (i.e., with any second-ranked tensor) are the invariants of the tensor, which for $\dot{\gamma}$ are given as (Bird et al., 1987, Chap. 4):

$$I_1 = \sum\dot{\gamma}_{ii} = 2[(\partial v_x/\partial x) + (\partial v_y/\partial y) + (\partial v_z/\partial z)] \tag{2.60}$$

**TABLE 2.9 The Equation of Motion in Terms of τ**

Rectangular coordinates $(x, y, z)$

$$\rho\left(\frac{\partial v_x}{\partial t} + v_x \frac{\partial}{\partial x} v_x + v_y \frac{\partial}{\partial y} v_x + v_z \frac{\partial}{\partial z} v_x\right) = -\left[\frac{\partial}{\partial x}\tau_{xx} + \frac{\partial}{\partial y}\tau_{yx} + \frac{\partial}{\partial z}\tau_{zx}\right] - \frac{\partial p}{\partial x} + \rho g_x \quad \text{(A)}$$

$$\rho\left(\frac{\partial v_y}{\partial t} + v_x \frac{\partial}{\partial x} v_y + v_y \frac{\partial}{\partial y} v_y + v_z \frac{\partial}{\partial z} v_y\right) = -\left[\frac{\partial}{\partial x}\tau_{xy} + \frac{\partial}{\partial y}\tau_{yy} + \frac{\partial}{\partial z}\tau_{zy}\right] - \frac{\partial p}{\partial y} + \rho g_y \quad \text{(B)}$$

$$\rho\left(\frac{\partial v_x}{\partial t} + v_x \frac{\partial}{\partial x} v_z + v_y \frac{\partial}{\partial y} v_z + v_z \frac{\partial}{\partial z} v_z\right) = -\left[\frac{\partial}{\partial x}\tau_{xz} + \frac{\partial}{\partial y}\tau_{yz} + \frac{\partial}{\partial z}\tau_{zz}\right] - \frac{\partial p}{\partial z} + \rho g_z \quad \text{(C)}$$

Cylindrical coordinates $(r, \theta, z)$

$$\rho\left(\frac{\partial v_r}{\partial t} + v_r \frac{\partial v_r}{\partial r} + \frac{v_\theta}{r}\frac{\partial v_r}{\partial \theta} - \frac{v_\theta^2}{r} + v_z \frac{\partial v_r}{\partial z}\right) = -\left[\frac{1}{r}\frac{\partial}{\partial r}(r\tau_{rr}) + \frac{1}{r}\frac{\partial}{\partial \theta}\tau_{\theta r} + \frac{\partial}{\partial z}\tau_{zr} - \frac{\tau_{\theta\theta}}{r}\right] - \frac{\partial p}{\partial r} + \rho g_r \quad \text{(D)}$$

$$\rho\left(\frac{\partial v_\theta}{\partial t} + v_r \frac{\partial v_\theta}{\partial r} + \frac{v_\theta}{r}\frac{\partial v_\theta}{\partial \theta} - \frac{v_r v_\theta}{r} + v_z \frac{\partial v_\theta}{\partial z}\right) = -\left[\frac{1}{r^2}\frac{\partial}{\partial r}(r^2\tau_{r\theta}) + \frac{1}{r}\frac{\partial}{\partial \theta}\tau_{\theta\theta} + \frac{\partial}{\partial z}\tau_{z\theta} + \frac{\tau_{\theta r} - \tau_{r\theta}}{r}\right] - \frac{1}{r}\frac{\partial p}{\partial \theta} + \rho g_\theta \quad \text{(E)}$$

$$\rho\left(\frac{\partial v_z}{\partial t} + v_r \frac{\partial v_z}{\partial r} + \frac{v_\theta}{r}\frac{\partial v_z}{\partial \theta} + v_z \frac{\partial v_z}{\partial z}\right) = -\left[\frac{1}{r}\frac{\partial}{\partial r}(r\tau_{rz}) + \frac{1}{r}\frac{\partial}{\partial \theta}\tau_{\theta z} + \frac{\partial}{\partial z}\tau_{zz}\right] - \frac{\partial p}{\partial z} + \rho g_z \quad \text{(F)}$$

Spherical coordinates $(r, \theta, \phi)$

$$\rho\left(\frac{\partial v_r}{\partial t} + v_r \frac{\partial v_r}{\partial r} + \frac{v_\theta}{r}\frac{\partial v_r}{\partial \theta} + \frac{v_\phi}{r\sin\theta}\frac{\partial v_r}{\partial \phi} - \frac{v_\theta^2 + v_\phi^2}{r}\right)$$
$$= -\left[\frac{1}{r^2}\frac{\partial}{\partial r}(r^2\tau_{rr}) + \frac{1}{r\sin\theta}\frac{\partial}{\partial \theta}(\tau_{\theta r}\sin\theta) + \frac{1}{r\sin\theta}\frac{\partial}{\partial \phi}\tau_{\phi r} - \frac{\tau_{\theta\theta} + \tau_{\phi\phi}}{r}\right] - \frac{\partial p}{\partial r} + \rho g_r \quad \text{(G)}$$

$$\rho\left(\frac{\partial v_\theta}{\partial t} + v_r \frac{\partial v_\theta}{\partial r} + \frac{v_\theta}{r}\frac{\partial v_\theta}{\partial \theta} + \frac{v_\phi}{r\sin\theta}\frac{\partial v_\theta}{\partial \phi} - \frac{v_r v_\theta}{r} - \frac{v_\phi^2 \cot\theta}{r}\right)$$
$$= -\left[\frac{1}{r^3}\frac{\partial}{\partial r}(r^3\tau_{r\theta}) + \frac{1}{r\sin\theta}\frac{\partial}{\partial \theta}(\tau_{\theta\theta}\sin\theta) + \frac{1}{r\sin\theta}\frac{\partial}{\partial \phi}\tau_{\phi\theta} + \frac{\tau_{\theta r} - \tau_{r\theta} - \tau_{\phi\phi}\cot\theta}{r}\right] - \frac{1}{r}\frac{\partial p}{\partial \theta} + \rho g_\theta \quad \text{(H)}$$

$$\rho\left(\frac{\partial v_\phi}{\partial t} + v_r \frac{\partial v_\phi}{\partial r} + \frac{v_\theta}{r}\frac{\partial v_\phi}{\partial \theta} + \frac{v_\phi}{r\sin\theta}\frac{\partial v_\phi}{\partial \phi} - \frac{v_\phi v_r}{r} + \frac{v_\theta v_\phi}{r}\cot\theta\right)$$
$$= -\left[\frac{1}{r^3}\frac{\partial}{\partial r}(r^3\tau_{r\phi}) + \frac{1}{r\sin\theta}\frac{\partial}{\partial \theta}(\tau_{\theta\phi}\sin\theta) + \frac{1}{r\sin\theta}\frac{\partial}{\partial \phi}\tau_{\phi\phi} + \frac{\tau_{\phi r} - \tau_{r\phi} + \tau_{\phi\theta}\cot\theta}{r}\right] - \frac{1}{r\sin\theta}\frac{\partial p}{\partial \phi} + \rho g_\phi \quad \text{(I)}$$

(Data from Bird et al., 1987.)

**TABLE 2.10 The Rate-of-Strain Tensor $\gamma = \nabla v + (\nabla v)^t$**

Rectangular coordinates $(x, y, z)$

$$\dot\gamma_{xx} = 2\frac{\partial v_x}{\partial x} \qquad\qquad \dot\gamma_{xy} = \dot\gamma_{yx} = \frac{\partial v_y}{\partial x} + \frac{\partial v_x}{\partial y}$$

$$\dot\gamma_{yy} = 2\frac{\partial v_y}{\partial y} \qquad\qquad \dot\gamma_{yz} = \dot\gamma_{zy} = \frac{\partial v_z}{\partial y} + \frac{\partial v_y}{\partial z}$$

$$\dot\gamma_{zz} = 2\frac{\partial v_z}{\partial z} \qquad\qquad \dot\gamma_{zx} = \dot\gamma_{xz} = \frac{\partial v_x}{\partial z} + \frac{\partial v_z}{\partial x}$$

Cylindrical coordinates $(r, \theta, z)$

$$\dot\gamma_{rr} = 2\frac{\partial v_r}{\partial r} \qquad\qquad \dot\gamma_{r\theta} = \dot\gamma_{\theta r} = r\frac{\partial}{\partial r}\left(\frac{v_\theta}{r}\right) + \frac{1}{r}\frac{\partial v_r}{\partial \theta}$$

$$\dot\gamma_{\theta\theta} = 2\left(\frac{1}{r}\frac{\partial v_\theta}{\partial \theta} + \frac{v_r}{r}\right) \qquad \dot\gamma_{\theta z} = \dot\gamma_{z\theta} = \frac{1}{r}\frac{\partial v_z}{\partial \theta} + \frac{\partial v_\theta}{\partial z}$$

$$\dot\gamma_{zz} = 2\frac{\partial v_z}{\partial z} \qquad\qquad \dot\gamma_{zr} = \dot\gamma_{rz} = \frac{\partial v_r}{\partial z} + \frac{\partial v_z}{\partial r}$$

Spherical coordinates $(r, \theta, \phi)$

$$\dot\gamma_{rr} = 2\frac{\partial v_r}{\partial r} \qquad\qquad \dot\gamma_{r\theta} = \dot\gamma_{\theta r} = r\frac{\partial}{\partial r}\left(\frac{v_\theta}{r}\right) + \frac{1}{r}\frac{\partial v_r}{\partial \theta}$$

$$\dot\gamma_{\theta\theta} = 2\left(\frac{1}{r}\frac{\partial v_\theta}{\partial \theta} + \frac{v_r}{r}\right) \qquad \dot\gamma_{\theta\phi} = \dot\gamma_{\phi\theta} = \frac{\sin\theta}{r}\frac{\partial}{\partial \theta}\left(\frac{v_\phi}{\sin\theta}\right) + \frac{1}{r\sin\theta}\frac{\partial v_\theta}{\partial \phi}$$

$$\dot\gamma_{\phi\phi} = 2\left(\frac{1}{r\sin\theta}\frac{\partial v_\phi}{\partial \phi} + \frac{v_r}{r} + \frac{v_\theta\cot\theta}{r}\right) \qquad \dot\gamma_{\phi r} = \dot\gamma_{r\phi} = \frac{1}{r\sin\theta}\frac{\partial v_r}{\partial \phi} + r\frac{\partial}{\partial r}\left(\frac{v_\phi}{r}\right)$$

$$I_2 = \sum_i \sum_j \dot\gamma_{ij}\dot\gamma_{ji}$$
$$= \dot\gamma_{xx}^2 + \dot\gamma_{xy}^2 + \dot\gamma_{xz}^2 + \dot\gamma_{yz}^2 + \dot\gamma_{yy}^2 + \dot\gamma_{yz}^2 + \dot\gamma_{zx}^2 + \dot\gamma_{zy}^2 + \dot\gamma_{zz}^2, \quad (2.61)$$

and

$$I_3 = \det\dot\gamma, \quad (2.62)$$

where $\det\dot\gamma$ is the derminant of the matrix consisting of the components of $\dot\gamma$. For incompressible flow $I_1 = 0$,

$$\eta = \eta(I_2, I_3). \quad (2.63)$$

For flows dominated by shear flow rather than extensional flow (see Section 3.1), $I_3$ is not very significant and $\eta$ is taken to be a function of $I_2$ only (Bird et al., 1987, p. 171). Actually we use $\sqrt{\frac{1}{2}I_2}$, which is called the shear rate, $\dot\gamma$. Likewise for models for which $\eta$ depends on the invariants of $\tau$, $\eta$ is taken to be a function of $\sum_i\sum_j \tau_{ij}\tau_{ji}$, which can be found by replacing terms such as $\dot\gamma_{yx}$ by $\tau_{yx}$ in Eq. 2.61.

One can generalize the expression for $\tau$ as follows:

$$\tau = -\eta(\dot\gamma)\dot\gamma. \quad (2.64)$$

Equation 2.64 is called the *Generalized Newtonian Fluid (GNF) model.* The empiricisms for $\eta(\dot\gamma)$ analogous to those given in Eqs. 2.6 to 2.9 become, respectively:

$$\eta = m\dot\gamma^{n-1} \tag{2.65}$$

$$\frac{\eta_0}{\eta} = 1 + \left(\frac{\tau}{\tau_{1/2}}\right)^{\alpha-1} \tag{2.66}$$

$$\frac{\eta-\eta_\infty}{\eta_0-\eta_\infty} = [1+(\lambda\dot\gamma)^2]^{\frac{n-1}{2}} \tag{2.67a}$$

$$\eta = \mu_0 + (\tau_0/\dot\gamma) \qquad \tau \geqslant \tau_0 \tag{2.67b}$$

$$\eta = \infty \qquad \tau < \tau_0, \tag{2.67c}$$

where $\tau = \sqrt{\frac{1}{2}\tau:\tau}$ and $\dot\gamma = \sqrt{\frac{1}{2}I_2}$. In one-dimensional flows such as those considered in Section 2.2 these equations reduce to the same form with $\dot\gamma$ becoming $dv_x/dy$ or $\tau$ becoming $\tau_{yx}$.

Before considering two examples that illustrate the use of the equations of motion, it is necessary to discuss the equation of mechanical energy. For single particles the work done on a particle is given by taking the dot product of Newton's second law of motion with the velocity, that is,

$$\mathbf{F}\cdot\mathbf{v} = m\mathbf{a}\cdot\mathbf{v} = \tfrac{1}{2}m(dv^2/dt) = d/dt(\tfrac{1}{2}mv^2) \tag{2.68}$$

This equation tells us that the work done on the particle is just equal to the change in kinetic energy. For a continuum most mechanical and chemical engineers have solved the macroscopic mechanical energy balance, which is shown below for a fluid of constant $\rho$:

$$\frac{d}{dt}(K_{\text{tot}}+\Phi_{\text{tot}}) = -\Delta\left[\left(\frac{1\langle\bar{v}^3\rangle}{2\langle\bar{v}\rangle}+\hat\Phi+\frac{p}{\rho}\right)w\right]-W-E_v, \tag{2.69}$$

where $K_{\text{tot}}$ and $\Phi_{\text{tot}}$ are the total kinetic energy and potential energy, respectively, $\bar{v}$ is the time-averaged velocity, $w$ is the mass flow rate, $W$

is the work input to the system, $E_v$ is the friction loss, $\hat\Phi$ is the potential energy per unit mass, and $\Delta$ represents the change in the quantities (Bird et al., 1960, p. 212). By taking the dot or scalar product of the equation of motion with the velocity field, one obtains the mechanical energy equation for a continuum:

$$\frac{\partial}{\partial t}(\tfrac{1}{2}\rho v^2) \qquad = \qquad -(\nabla\cdot\tfrac{1}{2}\rho v^2\mathbf{v})$$

| Rate of increase in kinetic energy per unit volume | Net rate of input of kinetic energy by virtue of bulk flow |
|---|---|

$$-(\nabla\cdot p\mathbf{v}) \qquad\qquad -p(-\nabla\cdot\mathbf{v})$$

| Rate of work done by pressure of surroundings on volume element | Rate of *reversible* conversion to internal energy |
|---|---|

$$-(\nabla\cdot[\tau\cdot\mathbf{v}]) \qquad\qquad -(-\tau:\nabla\mathbf{v})$$

| Rate of work done by viscous forces on volume element | Rate of *irreversible* conversion to internal energy |
|---|---|

$$+\rho(\mathbf{v}\cdot\mathbf{g})$$

Rate of work done by gravity force on volume element
$\qquad\qquad\qquad\qquad\qquad\qquad(2.70)$

The various quantities that make up this equation have been written using tensor notation. As we do not intend to evaluate these terms, we have given the physical significance of each term. By Integrating Eq. 2.70 over the volume of a region, one can obtain Eq. 2.69. Equation 2.70 is used to determine the power requirements for a piece of processing equipment.

---

## EXAMPLE 2.4
### Radial Flow of a Newtonian Fluid between Two Parallel Disks

Determine the velocity field and the pressure distribution for the flow of an incompressible Newtonian fluid between two disks as shown in Fig. 2.13. Flows similar to this occur in several polymer processes, including injection moulding in a center-gated disk mold or compression molding.

**Solution**

We start with the postulates for the velocity and pressure fields:

$$v_r = v_r(z,r) \qquad p = p(r). \tag{2.71}$$

Next one decides which components of stress arise for a given constitutive equation. Here the Newtonian fluid model is selected, and one uses Table 2.10 to determine the components of $\dot\gamma$. As $\tau_{ij} = -\mu\dot\gamma_{ij}$, then based on Eq. 2.71 one finds the following components for $\tau$:

$$\tau_{rr} = -2\mu(\partial v_r/\partial r) \tag{2.72}$$

$$\tau_{\theta\theta} = -2\mu v_r/r \tag{2.73}$$

$$\tau_{rz} = \tau_{zr} = -\mu(\partial v_r/\partial z). \tag{2.74}$$

The equation of continuity reduces to:

$$\frac{1}{r}\frac{\partial}{\partial r}(rv_r) = 0. \tag{2.75}$$

It is indicated by Eq. 2.75 that $rv_r$ must be a function of $z$ only, $\Phi(z)$. Hence from Eq. 2.75 one finds:

$$v_r = \Phi(z)/r. \tag{2.76}$$

This reduces the expressions for the stresses to the following equations:

$$\tau_{\theta\theta} = -2\mu\frac{\Phi(z)}{r^2} \qquad \tau_{rr} = +2\mu\frac{\Phi(z)}{r^2} \qquad \tau_{zr} = -\frac{\mu}{r}\frac{d\Phi(z)}{dz}. \tag{2.77}$$

The equation of motion, using Table 2.9, becomes (note that the inertial terms are neglected):

$$-\frac{1}{r}\frac{\partial}{\partial r}(r\tau_{rr})-\frac{\partial}{\partial z}\tau_{zr}+\frac{\tau_{\theta\theta}}{r}-\frac{\partial p}{\partial r}=0. \tag{2.78}$$

Substituting Eq. 2.77 into Eq. 2.78 and making use of Eq. 2.75, one obtains the following differential equation for $\Phi(z)$:

$$0=-\frac{dp}{dr}+\frac{\mu}{r}\frac{d^2\Phi}{dz^2}. \tag{2.79}$$

We can integrate this equation to find $\Phi$, because $dp/dr$ is a function of $r$, and the constants of integration can be solved for using the following boundary conditions:

B.C.1:   at $z=+b$     $\Phi(z)=0$   for all $r$ (2.80)

B.C.2:   at $z=-b$     $\Phi(z)=0$   for all $r$. (2.81)

Furthermore, one can find the pressure difference using:

B.C.3:   at $r=r_1$     $p=p_1$ (2.87)

B.C.4:   at $r=r_2$     $p=p_2$. (2.83)

The velocity field is given, then, as:

$$v_r(r,z)=\frac{b^2\Delta p}{2\mu r\ln(r_2/r_1)}[1-(z/b)^2]. \tag{2.84}$$

The volumetric flow rate, $Q$, is found by integrating $v_r$ over the cross sectional area:

$$Q=\int_0^{2\pi}\int_{-b}^{b}v_r(r,z)r\,d\theta\,dz=\frac{4\pi b^2\Delta p}{3\mu\ln(r_2/r_1)}. \tag{2.85}$$

The key points of this example are that the continuity equation helps us find the form of $v_r(r,z)$ and that there are normal stresses for this flow. The values of $\tau_{rr}$ and $\tau_{\theta\theta}$ come from the extensional deformations generated in this flow. Because of the sign convention, one notes that tensile stresses are negative.

## EXAMPLE 2.5
### Radial Flow of a Power-law Fluid.
Do Example 2.4 for a power-law fluid.

## Solution

For a power-law fluid the same postulates and assumptions that are made for the Newtonian case are used. In particular Eq. 2.76 holds as well as those for the stress components. The difference is that $\mu$ must be replaced by $\eta$, which depends on $\dot\gamma$. $\dot\gamma$ for this flow is:

$$\dot\gamma=\sqrt{2\left(\frac{\partial v_r}{\partial r}\right)^2+2\left(\frac{v_r}{r}\right)^2+\left(\frac{\partial v_r}{\partial z}\right)^2} \tag{2.86a}$$

$$=\sqrt{2\frac{\Phi^2}{r^4}+\frac{2\Phi^2}{r^4}+\frac{1}{r^2}\left(\frac{d\Phi}{dz}\right)^2}. \tag{2.86b}$$

The stresses are then:

$$\tau_{rr}=+2m|\dot\gamma|^{n-1}\frac{\Phi(z)}{r^2} \tag{2.87}$$

$$\tau_{\theta\theta}=-2m|\dot\gamma|^{n-1}\frac{\Phi(z)}{r^2} \tag{2.88}$$

$$\tau_{zr}=-m|\dot\gamma|^{n-1}\frac{d\Phi(z)}{dz}\frac{1}{r}. \tag{2.89}$$

When these are substituted into Eq. 2.78, one obtains a nonlinear differential equation for $\Phi(z)$. Hence, the use of the GNF model for the stress components leads to a complex differential equation.

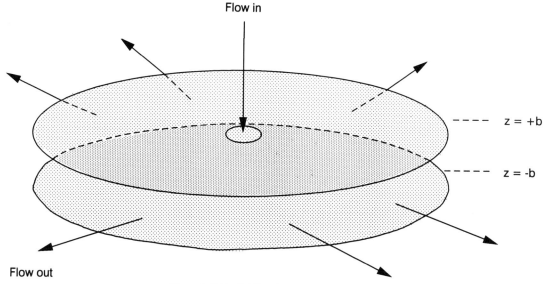

Flow in

z = +b

z = -b

Flow out

**Figure 2.13** Radial flow between two parallel disks.

## 2.4. USEFUL APPROXIMATIONS

The radial flow problem illustrates how rapidly problems involving the flow of non-Newtonian fluids become mathematically complicated. There are a number of times when the analysis can be simplified. Two useful approximations simplify the differential equations that arise out of the equations of motion: neglecting the effect of curvature and the lubrication approximation. The solution of several problems that have already been dealt with are used to illustrate these approximations. First, we illustrate how neglecting curvature can lead to a simplification of the differential equations. This is followed by the handling of geometries in which a variation in the dimension transverse to the flow direction occurs, such as the case of a tapered tube.

The extrusion of a polymer melt for film blowing is usually carried out by using an annular die with a thin gap such that $\kappa > 0.90$. For example, a film blowing die may have an outer diameter of 0.045 m and an inner diameter of 0.0449 m. Hence, $\kappa$ is about 0.998. The equation of motion for this flow is:

$$\frac{d}{dr}(r\tau_{rz}) = \frac{(P_0 - P_L)r}{L}. \tag{2.90}$$

For a power-law fluid, $\tau_{rz}$ is given by Eq. 2.32. One now expands the derivative in Eq. 2.90 and compares the order of each term:

$$\frac{d\tau_{rz}}{dr} + \frac{\tau_{rz}}{r} = \frac{P_0' - P_L'}{L}, \tag{2.91}$$

where $\tau_{rz}$ is given by the following expression

$$\tau_{rz} = -m(dv_z/dr)^n \tag{2.92}$$

in the region $\kappa R < r < \beta R$. We now approximate the derivatives as follows:

$$\frac{dv_z}{dr} \approx \frac{v_{max} - 0}{\beta R - \kappa R} = \frac{v_{max}}{(\beta - \kappa)R} \tag{2.93}$$

$$\tau_{rz} \approx -m\left(\frac{v_{max}}{(\beta - \kappa)R}\right)^n \tag{2.94}$$

$$\frac{d\tau_{rz}}{dr} \approx -mn\left(\frac{v_{max}}{(\beta - \kappa)R}\right)^{n-1} \frac{v_{max}}{R^2(\beta - \kappa)^2}. \tag{2.95}$$

One can compare the order of each term on the left side of Eq. 2.91 to determine if one or the other is dominant. Given the conditions in the example, $\beta R$ will be about $0.5R(1 - \kappa)$ or $\beta = 0.999$. Hence,

$$\frac{\tau_{rz}}{r} \approx -m\left(\frac{v_{max}}{(\beta - \kappa)R}\right)^n \frac{1}{R} \tag{2.96}$$

$$\frac{d\tau_{rz}}{dr} \approx -mn\left(\frac{v_{max}}{(\beta - \kappa)R}\right)^n \frac{1}{R(0.001)}. \tag{2.97}$$

Therefore, $d\tau_{rz}/dr$ is about 1000 times greater than $\tau_{rz}/r$ and one neglects the second term in Eq. 2.91. Now, we integrate Eq. 2.91 to find the stress field:

$$\tau_{rz} = ((P_0' - P_L')/L)r + C_1. \tag{2.98}$$

At $r = \beta R$, $\tau_{rz} = 0$ and $C_1 = -(P_0' - P_L')\beta R/L$. Before continuing it is noted that Eq. 2.98 is exactly that which is solved for flow between flat plates. One now substitutes the expression for $\tau_{rz}$ based on the power-law fluid into Eq. 2.98 to find $v_z(r)$:

$$-m\left(\frac{dv_z^<}{dr}\right)^n = \left(\frac{P_0' - P_L'}{L}\right)[r - \beta R] \qquad \text{for } \kappa R < r < \beta R \tag{2.99}$$

$$+m\left(\frac{-dv_z^>}{dr}\right)^n = \left(\frac{P_0' - P_L'}{L}\right)[r - \beta R] \qquad \text{for } \beta R < r < R \tag{2.100}$$

These equations are integrated using the no-slip boundary conditions as before to give $v_z(r)$:

$$v_z^<(\xi) = \left(\frac{(P_0' - P_L')R}{mL}\right)^{1/n} \frac{Rn}{n+1}[(\beta - \kappa)^{1/n+1} - (\beta - \xi)^{1/n+1}] \tag{2.101}$$

$$v_z^>(\xi) = \left(\frac{(P_0' - P_L')R}{mL}\right)^{1/n} \frac{Rn}{n+1}[(1 - \beta)^{1/n+1} - (\xi - \beta)^{1/n+1}]. \tag{2.102}$$

With the condition that $v_z^< = v_z^>$ at $r = \beta R$, one finds that $\beta = (1 + \kappa)/2$ or that for small gaps $\beta$ is at the center of the gap. Hence, for small gaps one finds an analytical solution for $v_z(r)$. This solution is expected to apply for values of $\kappa$ down to about 0.8. Obviously the expressions in Eqs. 2.101 and 2.102 can be integrated to find $Q$:

$$Q = \left(\frac{\pi R^3}{s+2}\right)\left(\frac{R\Delta P}{2mL}\right)^s (1 - \kappa)^{s+2}\frac{(1 - \kappa)}{2}, \tag{2.103}$$

where $s = 1/n$ and $\Delta P = P_0' - P_L'$.

This same expression could be found by using the expression for slit flow in Table 2.5 and replacing the width, $W$, by $\pi R(1 + \kappa)$ and the height, $H$, by $R(1 - \kappa)$ (See Pr. 2B.4). In essence, the annular region is opened up and treated as plane slit flow. One will find this a useful approach throughout the design of polymer processes.

There are many geometries in polymer processing where the dimensions change along the flow direction: for example, in extrusion through a tapered die, in calendering, and the channel of an extrusion screw. Solving the differential equations associated with these geometries would lead to nonlinear partial differential equations. However, by introducing an approximation known as the *lubrication approximation*, one can simplify the differential equation. To illustrate this method, flow through a tapered tube, as shown in Fig. 2.14, is considered. Based on this geometry one would postulate velocity and pressure fields as follows:

$$v_z = v_z(r, z) \qquad p = p(z). \tag{2.104}$$

This would lead to the following form of the equation of motion:

$$-\frac{1}{r}\frac{\partial}{\partial r}(r\tau_{rz}) - \frac{\partial}{\partial z}\tau_{zz} - \frac{\partial p}{\partial z} = 0. \tag{2.105}$$

If one evaluated $(\partial/\partial z)\tau_{zz}$ relative to the other term, it would be found that for small amounts of taper this term would be small compared to the other term. One would follow the same analysis as in the preceding example to show that the derivative is small. Hence, one neglects $(\partial/\partial z)\tau_{zz}$ and solves the following equation:

$$\frac{1}{r}\frac{\partial}{\partial r}(r\tau_{rz}) - \frac{\partial p}{\partial z} = 0. \tag{2.106}$$

This is just the equation that would be solved for a straight tube. Hence

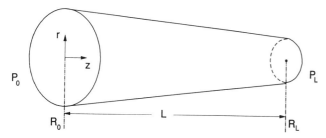

**Figure 2.14** Flow through a tapered tube.

the solution to this equation is:

$$v_z(r) = \frac{R}{1+s}\left(\frac{R\Delta P}{2mL}\right)^{1/n}\left[1 - \left(\frac{r}{R}\right)^{\frac{1}{n}+1}\right].$$
(2.107)

However, the solution is applied at each distance $z$ down the tube. One replaces $\hat{R}$ by $R(z)$ which is obtained from the geometry of the tube as follows:

$$R(z) = -[(R_0 - R_L)/L]z + R_0.$$
(2.108)

At any $z$ position, one can integrate Eq. 2.107 over the cross section of the tube to find $Q$:

$$Q = \left(\frac{\pi R^3}{s+3}\right)\left(\frac{R\Delta P}{2mL}\right)^s.$$
(2.109)

$\Delta P/L$ is replaced by $-dp/dz$ and $R$ by Eq. 2.108. This gives a first-order differential equation for finding $p$:

$$\left(\frac{Q(s+3)}{\pi}\right)^n \frac{R^{-3n-1}}{2m} = -\frac{dp}{dz}$$
(2.110)

Equation 2.110 can be integrated using the conditions that at $z=0$, $p=\mathrm{P}_0$ and at $z=L$, $p=\mathrm{P}_L$ to give:

$$P_0 - P_L = \frac{2mL}{3n}\left[\frac{Q}{\pi}\left(\frac{1}{n+3}\right)\right]^n\left(\frac{R_L^{-3n} - R_0^{-3n}}{R_0 - R_L}\right).$$
(2.111)

This approximation is probably adequate for tapers of less than 30°, but for more abrupt contractions the viscoelastic nature of polymeric fluids may make the pressure drop higher (see Section 7.2). However, for the most part the lubrication approximation is extremely useful in the design of extrusion processes (i.e., the design of extrusion dies and screw design).

## 2.5. SOLUTION TO DESIGN PROBLEM 1

The solution to Design Problem 1 is presented in this section. The lubrication approximation is used first to obtain a solution. This is followed by a numerical approach in which the die is broken into a series of annuli.

### 2.5.1. LUBRICATION APPROXIMATION SOLUTION

The solution will proceed as follows: (1) the dimensions of the die will be determined; (2) an expression for $\Delta P$ versus $Q$ will be determined using the lubrication approximation; (3) from this relation one can calculate $Q$ for the given $\Delta P$ and the average velocity $\langle v_z \rangle$; and (4) the time required to extrude the parison is calculated.

The following information is given in the problem statement:

Parison weight: 90 g
Parison thickness ($t_p$): $3.81 \times 10^{-4}$ m
Parison diameter ($D_p$): 0.127 m
Parison density ($\rho_p$): 965 kg/m$^3$

From this information and the equation for the mass of the parison ($m_p$)

$$m_p = \rho_p L_p \pi (R_{0,p}^2 - R_{i,p}^2),$$
(2.112)

one can calculate the desired length of the parison ($L_p$), which is 0.615 m. Because the phenomenon of die swell (see Section 3.2) is neglected, the final thickness and diameter of the die are assumed to be the same as those of the parison. The dimensions of the die geometry are then as shown in Fig. 2.15.

The next goal is to determine an expression that will allow one to calculate $Q$ and the average velocity $\langle v_z \rangle$, for the maximum allowable pressure drop. To do this the lubrication approximation is used. The flow region of interest is shown in Fig. 2.16. With the assumption that the gap is small relative to the radius (which is acceptable over most of the length of the die), the annular region can be opened up to the planar region shown in Fig. 2.16. For a Newtonian fluid the expression for $Q$ for flat plates (see Table 2.5) is

$$Q = \frac{WH^3 \Delta P}{12\mu L}.$$
(2.113)

$\Delta P/L$ is replaced by $-dp/dz$. $H$ is the gap, which is constant and is $R_0(1-\kappa_0)$ or $R_L(1-\kappa_L)$ where $R_0$ and $R_L$ are the outer radii of the tapered annulus at $z=0$ and $z=L$, respectively. It is also noted that although the gap is constant, the ratio of the inner to outer radii, $\kappa$, varies slightly with the distance $z$ (e.g., the value of $\kappa$ at $z=L$ is defined as $\kappa_L$). $W$ varies linearly with $z$ as given:

$$W = W_0 + \frac{W_L - W_0}{L}z,$$
(2.114)

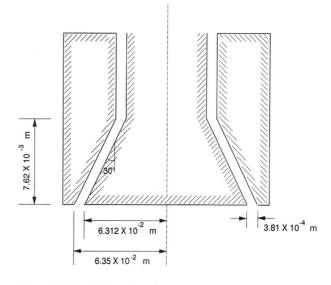

**Figure 2.15** Conical region of the blow-molding die showing the dimensions required to produce the parison as requested in Design Problem 1.

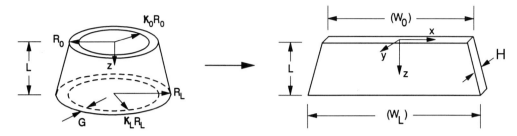

**Figure 2.16** Three-dimensional view of a conical die unfolded into a planar die.

where $W_0 = \pi R_0(1 + \kappa_0)$ and $W_L = \pi R_L(1 + \kappa_L)$. On substituting the expression for $W$ into Eq. 2.113 and replacing $\Delta P/L$ by $-dp/dz$, Eq. 2.113 becomes:

$$Q = \left(W_0 + \frac{W_L - W_0}{L}z\right)\frac{H^3}{12\mu}(-dp/dz). \tag{2.115}$$

Separating variables and integrating, Eq. 2.115 becomes

$$\int_{P_0}^{P_L} -dp = \int_0^L \frac{12\mu Q}{H^3[W_0 + (W_L - W_0)z/L]}\,dz. \tag{2.116}$$

Carrying out the integration one obtains the following expression

$$P_L - P_0 = \frac{12\mu QL}{H^3(W_L - W_0)}\ln\left(\frac{W_L}{W_0}\right). \tag{2.117}$$

Equation 2.117 is now rearranged to give the following expression for $Q$:

$$Q = \frac{(P_L - P_0)H^3(W_L - W_0)}{12\mu L \ln(W_L/W_0)}. \tag{2.118}$$

Before using Eq. 2.118 to answer the questions associated with Design Problem 1, a similar expression is obtained for a fluid whose viscosity function is described by the power law. Again, using Table 2.5, the expression for pressure-driven flow of a power-law fluid through parallel plates is:

$$Q = \frac{WH^2}{2(s+2)}\left(\frac{H\Delta P}{2mL}\right)^s, \tag{2.119}$$

where $s = 1/n$. Following the same approach as for the Newtonian fluid, one obtains

$$(-dp/dz)^s = \frac{2Q(s+2)}{(W_0 + (W_L - W_0)z/L)H^2}\left(\frac{2m}{H}\right)^s. \tag{2.120}$$

The $n$th power is taken off both sides of Eq. 2.120 to obtain:

$$-\frac{dp}{dz} = \left(\frac{2Q(s+2)}{(W_0 + (W_L - W_0)z/L)H^2}\right)^{1/s}\left(\frac{2m}{H}\right). \tag{2.121}$$

Equation 2.121 is integrated to find $P_0 - P_L$:

$$P_0 - P_L = \left(\frac{2Q(s+2)}{H^2}\right)^n\left(\frac{2mL}{H}\right)\left(\frac{1}{1-n}\right)\left(\frac{W_L^{1-n} - W_0^{1-n}}{W_L - W_0}\right). \tag{2.122}$$

Equation 2.122 is rearranged to give the following expression for $Q$:

$$Q = (P_0 - P_L)^s H^2 \left/ \left[2(s+2)(H/2mL)^s\left(\frac{1}{1-n}\right)^s\left(\frac{W_L^{1-n} - W_0^{1-n}}{W_L - W_0}\right)^s\right]\right.. \tag{2.123}$$

Equations 2.118 and 2.123 are now used to make the decisions required to solve Design Problem 1. The pressure drop at the onset of melt fracture is obtained from the expression for the wall shear stress in Table 2.5.

$$P_0 - P_L = \frac{\tau_w L 2}{\cos\theta H}$$

$$= \frac{(1.4 \times 10^5\,\text{Pa})(3.81 \times 10^{-4}\,\text{m})40}{0.866(3.81 \times 10^{-4}\,\text{m})} = 6.467 \times 10^6\,\text{Pa}. \tag{2.124}$$

The value of $P_0 - P_L$ is the same whether the fluid is assumed to be described by a Newtonian or power-law model. Equation 2.118 is used to find $Q$ and $\langle v_z \rangle = Q/A$ for the Newtonian case:

$$Q = \frac{(6.467 \times 10^6\,\text{Pa})(3.81 \times 10^{-4})^3(3.978 \times 10^{-1}\,\text{m} - 3.632 \times 10^{-1}\,\text{m})}{12(2.108 \times 10^3\,\text{Pa·s})20(3.81 \times 10^{-4})(0.091)}$$

$$= 7.08 \times 10^{-7}\,(\text{m}^3/\text{s})$$

$$Q/A = \frac{7.08 \times 10^{-3}\,\text{m}^3\,\text{s}^{-1}}{(3.805 \times 10^{-1}\,\text{m})(3.81 \times 10^{-4}\,\text{m})} = 4.8 \times 10^{-3}\,\text{m/s}. \tag{2.125}$$

For the power law Eq. 2.123 is used to find $Q$ and $\langle v_z \rangle$:

$$Q = 2.47 \times 10^{-6}\,\text{m}^3\,\text{s}^{-1}$$

$$Q/A = \frac{2.47 \times 10^{-6}}{1.45 \times 10^{-4}}\,\text{ms}^{-1} = 1.704 \times 10^{-2}\,\text{ms}^{-1}. \tag{2.126}$$

Hence one can see that there is a significant difference in the values for the Newtonian and power-law cases.

Finally, the length of time to "hang the parison" is determined. Based on the parison weight, the required parison length is calculated to be 0.615 m. If it is assumed that the parison does not sag under its own weight, then the hang time, $t_H$, is

Newtonian: $t_H = \dfrac{L}{\langle v_z \rangle} = \dfrac{0.615\,\text{m}}{4.88 \times 10^{-3}\,\text{ms}^{-1}} = 126.0\,\text{s}$

Power law: $t_H = \dfrac{0.615\,\text{m}}{1.704 \times 10^{-2}\,\text{ms}^{-1}} = 36.1\,\text{s}. \tag{2.127}$

Again, there is a significant difference between the hang times for the Newtonian and power-law cases.

In practice, the polymer swells as it leaves the die, giving a wall thickness and parison diameter greater than would be expected. Furthermore, the weight of the parison causes the extruded material to sag under its own weight. This reduces the wall thickness. If the sag is too great, the parison may fail when blown or the wall thickness will vary considerably over the length of the parison. These factors will be dealt with in the next chapter.

### 2.5.2. COMPUTER SOLUTION

In the second part of Design Problem 1, a computer solution is requested. This solution is outlined here, and the program listing is given in Table 2.11. The computer solution allows one to deal with dies having a variable gap (In Fig. 2.2, this corresponds to angle $\alpha \neq 0$). Although one could extend the parallel plate solution to annular dies of variable gap, the computer solution outlined here uses the solution for annular die flow (Table 2.7) applied to segments of length $\Delta z$.

The solution proceeds as described. $\Delta P/\Delta z$ is calculated using Eq. 2.41 for a segment of length $\Delta z$:

$$\Delta P/\Delta z' = \left(\frac{Q(s+3)}{\pi R^3(z')}\right)^n \frac{2m}{(1-\beta^2)^{1+s} - \kappa^{1-s}(\beta^2 - \kappa^2)^{1+s})^n}, \qquad (2.128)$$

where $\Delta z' = \Delta z/\cos\theta$ and $z' = z/\cos\theta$. $R^o$ is the outer radius of the annular segment and is given by the following:

$$R^o = R_0^o + \tan\theta z, \qquad (2.129)$$

and $\kappa$ is ratio of the inner radius to outer radius at each segment given by:

$$\kappa = \frac{R^i(z)}{R^o(z)} = \frac{R_0^i + \tan\theta z}{R_0^o + \tan\theta z}. \qquad (2.130)$$

Here $R_0^o$ is $R^o$ at $z=0$. $\beta$ must also be determined for each step, and this is done for each segment using Eq. 2.39. Because neither $P_0 - P_L$ nor $Q$ are known, one must use the equation for $\tau_{rz}(R)$ in Table 2.7 to determine $\Delta P/\Delta z'$ over the last segment of the annulus. One can then guess at $Q$ and calculate $\Delta P$ over each segment using Eq. 2.118. If $\Delta P$ does not match the value at the end segment based on the limit due to melt fracture, then $Q$ is changed, and the iteration procedure is started all over. When $\alpha = 0$ and the gap is small, one would expect that this improved method of computation would not be necessary.

The computer solution is compared against the plate solution for the case when $\alpha = 10°$ (note that in Section 2.5.1 $\alpha$ was taken as zero) in Table 2.12. First we observe that $\langle v_z \rangle$ or $Q$ is about 10 times greater for the power-law case than for the Newtonian case. The computer solution based on breaking the die up into annular segments (100 sections) leads to values of $\langle v_z \rangle$ that are only about one-third of those calculated using the flat plate. Hence, at the beginning of the conical

**TABLE 2.11 Program Listing for Solution to Design Problem 1**

```
        PROGRAM DADA
C
C   Die design analysis : Discrete annular die approximation
C
C
C   This program approximates a conical extrusion die (for extruding
C   cylindrical parisons) as a series of annular die sections. The user
C   inputs the die geometry, fluid parameters, the die inlet pressure
C   (exit pressure is atmospheric), and the number of annular die
C   sections desired for the approximation. Convergence of the solution
C   is determined by comparison of the calculated pressure drop to the
C   actual pressure drop.
C
C   Program determines the flow rate, the die exit pressure gradients
C   and wall shear stresses, the average exit velocity, and the time
C   required to extrude a parison of cylindrical shape.
C
        INTEGER IFLAG,ICOUNT,NDIV,NUMIT,ANS1,ANS2,IGUESS
        REAL PARDIAM,PARTH,PARL,THETA,ALPHA,VMU,M,N,PIN,PI
        REAL ROZ1,RIZ1,Z1,GZ1,K,RO,RI,ROAVE,RIAVE
        REAL A,B,ERRABS,ERRREL,RESULT,ERREST
        REAL Q,OMEGA,DELP,TAUWI,TAUWO,PSUM,PGRAD,VAVE,TAUMAX
        REAL TIME,DELZ,UERR,PERR,LHS,RHS,LAMBDA
        REAL DIFF,PDIFF,OLDDIFF,OLDLAM,SLOPE
        REAL PE,OLDPE,OLDQ,QSLOPE
        DIMENSION DELP(250),TAUWI(250),TAUWO(250)
C
        COMMON/FP/ LAMBDA,N,IFLAG
        INTRINSIC TAN, COS
        EXTERNAL F, QDAGS
        PI = 3.1415927
        ANS1 = 0
C   Set error tolerances for integration subroutine.
        ERRABS = 0.0
        ERRREL = 0.0001
C
C   Input parison dimensions, material parameters, and die parameters.
```

**TABLE 2.11** *continued*

```
      C
        5 WRITE(*,*) 'Input the diameter,thickness,and length of parison (in
          1 m):'
          READ(*,*) PARDIAM, PARTH, PARL
          WRITE(*,*)
          WRITE(*,*) 'Input m and n for power-law fluid:'
          READ(*,*) M,N
          WRITE(*,*)
          WRITE(*,*) 'Input the angles THETA and ALPHA (degrees):'
          READ(*,*) THETA, ALPHA
          WRITE(*,*)
       10 WRITE(*,*) 'Input the die inlet pressure (Pa):'
          READ(*,*) PIN
          WRITE(*,*)
          WRITE(*,*) 'Input the number of divisions:'
          READ(*,*) NDIV
          WRITE(*,*)
          WRITE(*,*) 'Input the desired accuracy (%) (for accuracy within 1%
          1, input a 1):'
          READ(*,*) UERR
          UERR = UERR / 100.0
          WRITE(*,*)
          IF (ANS1 .EQ. 1) GOTO 30
      C
      C   Compute the basic dimensions of the die at the exit.
      C
          ROZ1 = PARDIAM / 2.0
          RIZ1 = ROZ1 - PARTH
          GZ1 = ROZ1 - RIZ1
          Z1 = 20.0 * PARTH
      C
      C   Set up ROAVE and RIAVE for initial guess for Q. Set up flag IGUESS
      C   such that integration subroutines can be used to return appropriate
      C   values to determine an "intelligent" guess for Q based on the known
      C   inlet pressure and averages of the die dimensions.
      C
          ROAVE = ROZ1 - 0.5*TAN((THETA-ALPHA)*PI/180.0)*Z1
          RIAVE = RIZ1 - 0.5*TAN(THETA*PI/180.0)*Z1
          K = RIAVE / ROAVE
      C   Determine LAMBDA and OMEGA
          IGUESS = 0
          GOTO 50
      C   Determine initial guess for Q using returned OMEGA value.
       20 Q = PI*(ROAVE)**3.0*((PIN*ROAVE)/(2.0*M*Z1))**(1.0/N)*OMEGA
          IGUESS = 1
      C   Initialize before start of loop.
       30 NUMIT = 0
      C
      C   Calculate DELZ and begin loop structure.
      C
          DELZ = Z1 / REAL(NDIV)
       40 PSUM = 0.0
          DO 80 I = 1,NDIV
      C   Calculate mean radii for annulus(i).
          IF (I .EQ. 1) THEN
             RO = ROZ1 - 0.5*TAN((THETA - ALPHA)*PI/180.0) * DELZ
             RI = RIZ1 - 0.5*TAN((THETA)*PI/180.0) * DELZ
          ELSE
             RO = RO - TAN((THETA - ALPHA)*PI/180.0) * DELZ
             RI = RI - TAN((THETA)*PI/180.0) * DELZ
          END IF
          K = RI / RO
      C   Use loop structure together with IMSL subroutines to determine the
      C   appropriate value for LAMBDA.
       50 LAMBDA = (1.0 + K)/2.0
          ICOUNT = 0
          IFLAG =1
      C   Set up limits of integration—start loop structure.
       60 ICOUNT = ICOUNT + 1
```

**TABLE 2.11** *continued*

```
C   Integrate left-hand side
        A = K
        B = LAMBDA
        CALL QDAGS(F,A,B,ERRABS,ERRREL,RESULT,ERREST)
        LHS = RESULT
C   Integrate right-hand side
        A = LAMBDA
        B = 1.0
        CALL QDAGS(F,A,B,ERRABS,ERRREL,RESULT,ERREST)
        RHS = RESULT
C   Compare sides, determine error, exit loop or update LAMBDA
        DIFF = LHS - RHS
        PDIFF = (ABS(DIFF))/(0.5 * (LHS + RHS))
        IF (PDIFF .LE. 0.0001) GOTO 70
C   Update LAMBDA value and go back through loop to check equation
        IF (ICOUNT .EQ. 1) THEN
          OLDLAM = LAMBDA
          OLDDIFF = DIFF
          LAMBDA = LAMBDA - (DIFF/ABS(DIFF))*0.001*LAMBDA
          GOTO 60
        ELSE
          SLOPE = (DIFF - OLDDIFF)/(LAMBDA - OLDLAM)
          OLDDIFF = DIFF
          OLDLAM = LAMBDA
          LAMBDA = LAMBDA - (DIFF/SLOPE)
        END IF
        GOTO 60
C   Use value of LAMBDA to integrate term needed for calculations
C
C   Set limits of integration.
     70 IFLAG = 2
        A = K
        B = 1.0
        CALL QDAGS(F,A,B,ERRABS,ERRREL,RESULT,ERREST)
        OMEGA = RESULT
        IF (IGUESS .EQ. 0) GOTO 20
C   Determine the incremental pressure drop
        DELP(I) = ((Q/(PI*OMEGA))**N*(2*M)/(RO**(3*N+1)))*DELZ
C   Calculate inner and outer wall shear stresses
        TAUWI(I) = 0.5*DELP(I)/DELZ*K*RO*(1-(LAMBDA/K)**2.0)
        TAUWO(I) = 0.5*DELP(I)/DELZ*RO*(1-LAMBDA*LAMBDA)
C   Sum pressure drops
        PSUM = PSUM + DELP(I)
     80 CONTINUE
C   Compare computed pressure drop through die to the known pressure.
C   drop—correct Q accordingly and recalculate pressure drop.
        NUMIT = NUMIT + 1
        PE = PSUM - PIN
        PERR = PE/PIN
        IF (ABS(PERR).LE.UERR) GOTO 90
        WRITE(*,*) ' Iteration ',NUMIT,' Q = ',Q
        WRITE(*,*) ' Calculated Pressure Drop is',PSUM
C   Use linear interpolation to converge on Q such that computed
C   pressure drop is nearly equal to the true pressure drop.
        IF (NUMIT .EQ. 1) THEN
          OLDQ = Q
          OLDPE = PE
          Q = Q - (PE/ABS(PE))*0.01*Q
          GOTO 40
        ELSE
          QSLOPE = (PE - OLDPE)/(Q - OLDQ)
          OLDQ = Q
          OLDPE = PE
          Q = Q - (PE/QSLOPE)
        END IF
        GOTO 40
C   Determine other quantities of interest.
C   Determine the maximum shear stress at the exit.
     90 IF (TAUWI(1).GE.TAUWO(1)) TAUMAX = TAUWI(1)
```

**TABLE 2.11** *continued*

```
              IF (TAUWI(1).LT.TAUWO(1)) TAUMAX = TAUWO(1)
C   Determine pressure gradient at die exit.
          PGRAD = DELP(1)/DELZ
C   Determine the average exit velocity of the melt.
          VAVE = Q / (PI*(RO*RO - RI*RI))
C   Determine the time required to extrude the parison.
          TIME = PARL / VAVE
C   Print out the parison and die dimensions and the input parameters
          WRITE(*,*) 'PARISON DIMENSIONS (m):'
          WRITE(*,100) PARDIAM, PARTH, PARL
          WRITE(*,*)
          WRITE(*,*) 'DIE DIMENSIONS (m):'
          WRITE(*,*)
          WRITE(*,200) THETA, ALPHA
          WRITE(*,300) ROZ1, RIZ1, GZ1
          WRITE(*,400) PIN,PSUM
          WRITE(*,*) 'RHEOLOGICAL PROPERTIES:'
          WRITE(*,*)
          WRITE(*,500) M, N
          WRITE(*,*) 'RESULTS - DISCRETE ANNULAR DIE APPROXIMATION:'
          WRITE(*,*) '     (Tabulated values for die exit)'
          WRITE(*,*)
          WRITE(*,600) NDIV
C   Print out the results
          WRITE(*,*) 'Fluid Model    Q (m3/s)   dP/dz (Pa/m)   TAUw (Pa)
         1 Vave (m/s) Time (s)'
          WRITE(*,*) '---------------------------------------------------
         1-----------------------'
          WRITE(*,700) Q, PGRAD, TAUMAX, VAVE, TIME
          WRITE(*,*) '---------------------------------------------------
         1-----------------------'
          WRITE(*,*)
C   Format statements.
     100 FORMAT(1X,'Diameter = ',E11.5,4X,' Thickness = ',E11.5,4X,' Length
         1 = ',E10.4,/)
     200 FORMAT(1X,'Angles of inclination: Theta = ',F5.1,5X,' Alpha = ',
         1 F5.1,)
     300 FORMAT(1X,'Exit plane : Ro = ',E11.5,4X,' Ri = ',E11.5,4X,
         1 ' Gap = ',E11.5,/)
     400 FORMAT(1X,'Extrusion pressure = ',E11.5,' Calculated drop = ',
         1 E11.5,/)
     500 FORMAT(1X,'Power law : m = ',F11.5,' Pa/s ^n',4X,' n = ',F4.2,//)
     600 FORMAT(6X,'Number of annular sections: ',I3,/)
     700 FORMAT(1X,'Power Law',5X,5(E11.5,2X))
C   Run program again?
          WRITE(*,*) 'Run program again (1=Y or 0=N)?'
          READ(*,*) ANS1
          IF (ANS1 .EQ. 1) THEN
            WRITE(*,*) 'Use same die dimensions and fluid parameters?'
            READ(*,*) ANS2
            IF (ANS2 .EQ. 1) GOTO 10
            GOTO 5
          END IF
          STOP
          END
C
          REAL FUNCTION F(X)
          REAL X, LAMBDA, N
          INTEGER IFLAG
          COMMON/FP/ LAMBDA,N,IFLAG
C   F(X) for determination of LAMBDA
          IF (IFLAG.EQ.1) THEN
            F = (ABS(LAMBDA*LAMBDA/X - X)**(1.0/N))
C   F(X) for determination of OMEGA
          ELSE
            F = (ABS(LAMBDA*LAMBDA - X*X))**(1.0/N+1.0)*(X**(-1.0/N))
          END IF
          RETURN
          END
```

**TABLE 2.12 Comparison of Predictions for Different Methods of Computation**

| Approximation | Flat-Plate*(variable width and separation) | | Series of Annuli† | |
|---|---|---|---|---|
| Fluid | Newtonian | Power law | Power law | Power law |
| | | | 20 sections: | 100 sections: |
| $Q\,(\mathrm{m^3/s})$ | $5.608 \times 10^{-6}$ | $6.638 \times 10^{-5}$ | $1.174 \times 10^{-4}$ | $1.1652 \times 10^{-4}$ |
| $\frac{dP}{dz}\big\vert_{\mathrm{EXIT}}[\mathrm{Pa/m}]$ | $-8.630 \times 10^{9}$ | $-5.397 \times 10^{9}$ | $-4.8504 \times 10^{9}$ | $-5.7050 \times 10^{9}$ |
| $\tau_{w\mathrm{EXIT}}\,(\mathrm{Pa})$ | $1.490 \times 10^{5}$ | $9.318 \times 10^{5}$ | $10.217 \times 10^{5}$ | $11.091 \times 10^{5}$ |
| Melt fracture? | Yes | Yes | Yes | Yes |
| $\langle v \rangle (\mathrm{m/s})$ | $4.083 \times 10^{-2}$ | $4.832 \times 10^{-1}$ | $1.589 \times 10^{-1}$ | $1.533 \times 10^{-1}$ |
| $t$, time required to extrude parison (s) | 15.11 | 1.277 | 3.884 | 3.974 |

* Results obtained from Computer Code DIEFPA (flat plate approximation).
† Results obtained from Computer Code SERADA (series annular dies approximation).

region there must be a large error in neglecting the curvature in the geometry. We also observe that the taper ($\alpha = 10°$) has a significant effect on the flow rate. The parison hang times vary significantly from the flat plate to the annular segment approach. Hence, more accurate solutions may only be obtained by use of the computer.

## PROBLEMS

### A. Applications

**2A.1. Power-Law Parameters from the Ellis Model**
Ellis model parameters for a polypropylene sample at 200°C are: $\eta_0 = 1.24\mathrm{E}+04$ Pa·s, $\tau_{1/2} = 6.90\mathrm{E}+03$ Pa, and $\alpha = 2.82$. Estimate $m$ and $n$ in the power-law model using these values of the parameters in the Ellis model.

**2A.2 Flow of HDPE through Parallel Plates**
(a) Compare the predictions at $\tau_w = 1.0\mathrm{E}+05$ Pa for the volumetric flow rate for HDPE at 180°C for flow through parallel plates (Table 2.5) for Newtonian, power-law, and Ellis models. Use the parameters given in Table 2.3. The plate dimensions are: $H = 2.54\mathrm{E}-03$ m, $W = 2.54\mathrm{E}-02$ m, $L = 5.08\mathrm{E}-02$ m.
(b) Determine the pressure drop in each case.
(c) Determine the wall shear rate, $\dot{\gamma}_w$, for each model.

**2A.3. Pressure Transducer Selection**
It is desired to select pressure transducers to be mounted on the upper wall of the plates described in Pr. 2A.2. The accuracy of the transducers depends on the range of pressure that must be measured. For the conditions described in Pr. 2A.2, what is the maximum pressure that would have to be measured for pressure transducers mounted at the entrance, halfway down the channel, and at a distance of $H$ from the exit?

**2A.4. Change in Wire-Coating Conditions**
For the wire-coating problem described in Section 2.2.2 it is found that one can actually pass through melt fracture by increasing the wire speed to the point where the wall shear stress $(\tau_w) = 2.2\mathrm{E}+05$ Pa (on increasing to higher values of $\tau_w$ another form of melt fracture occurs). Using the same die radius as calculated in Ex. 2.3 find the wire-coating thickness, $\delta$, and the wire speed, $V$, for $\tau_w = 2.2\mathrm{E}+05$ Pa. Use the rheological parameters given for HDPE at 200° in Table 2.3.

**2A.5. Flow through a Tubing Die**
A polypropylene is extruded through an annular die at 210°C to make tubing for various biomedical applications. The die dimensions are: $D = 3.175\mathrm{E}-03$ m, $\kappa = 0.7$, $L = 3.175\mathrm{E}-02$ m. The extruder can feed a

maximum of 22 kg/hr to the die. With the following rheological parameters given calculate $\Delta P$, and determine the maximum wall shear stress: $m = 3.21\mathrm{E}+04$ Pa·s$^n$, $n = 0.25$, $\eta_0 = 3.5\mathrm{E}+04$ Pa·s, $\tau_{1/2} = 3.35\mathrm{E}+04$ Pa, and $\alpha = 4.19$. Density values can be determined using Table 5.6

**2A.6 Pressure Drop Across a Pelletizing Die**
Polypropylene is mixed with pigments (1.0 wt%) in a single screw-extruder and then pumped through a pelletizing die consisting of 10 capillaries ($D = 3.175\mathrm{E}-03$ m and $L = 3.175\mathrm{E}-02$ m). The 10 strands are quenched in a water bath and then cut into pellets $6.35\mathrm{E}-03$ m in length. Given that the polymer has the same properties as described in Pr. 2A.5 and that the extruder is capable of delivering 50 kg/hr determine the pressure drop, $\Delta P$, across the die (neglect the pressure drop across the distribution system) and the wall shear stress at the maximum mass flow rate.

### B. Principles

**2B.1. Flow between Parallel Plates of a Newtonian Fluid**
The solution to the flow between parallel plates as shown in Fig. 2.17 is frequently used in the analysis of polymer processes.
(a) Make a differential force (or momentum) balance, and obtain expressions for the distributions of shear stress (or momentum flux) and velocity for a Newtonian fluid (note: these expressions are given in Table 2.5).
(b) Start with the equations of motion (See Tables 2.9 and 2.10), and obtain the velocity distribution.
(c) Obtain an expression for the volumetric flow rate, $Q$.

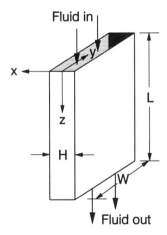

**Figure 2.17** Flow through a set of parallel plates.

**2B.2**  Flow between Parallel Plates of a Non-Newtonian Fluid

(a) Use the equations of motion (Table 2.9) and the power-law empiricism for viscosity to find the shear stress and velocity distributions for the flow of a non-Newtonian fluid through parallel plates (See Fig. 2.17). Also determine expressions for $\dot{\gamma}$ and $Q$.

(b) Do the same as in Part **a**, but use the Ellis model (*note*: both solutions are given in Table 2.5).

**2B.3.**  Flow of a Non-Newtonian Fluid through a Tube

(a) Obtain an expression for $Q$ for the flow of a power-law fluid through a tube of radius, $R$ (confirm your expression with that in Table 2.6).

(b) Derive an express for $Q$ for the Ellis model (see Table 2.6).

(c) Show that in the limit as $s$ goes to 1 and $m = \mu$ one obtains the Newtonian solution.

**2B.4.**  Adapting the Parallel Plate Solution to Annular Flow

For small annular gaps (e.g., $\kappa = 0.9$) the expression for $Q$ for flow through a slit of a power-law fluid (Pr. 2B.3) can be used to obtain an expression for $Q$ for annular flow. Adapt the parallel plate flow solution for the power-law model to that for flow through an annulus with a small gap (you should obtain the expression in Eq. 2.103).

**2B.5.**  Wire Coating with an Imposed Pressure

In practice there is always a pressure drop across a wire-coating die as the result of polymer melt being pumped by an extruder. By imposing a pressure drop, the coating thickness can be controlled independently of the wire speed.

(a) For the situation shown in Fig. 2.11 obtain the following expression for the volumetric flow rate for a Newtonian fluid:

$$Q = \frac{\pi R^2 V}{2}\left(\frac{1-\kappa^2}{\ln(1/\kappa)} - 2\kappa^2\right) + \frac{\pi \Delta P R^4}{8\mu L}\left[1 - \kappa^4 - \frac{(1-\kappa^2)^2}{\ln(1/\kappa)}\right].\text{(2.131)}$$

(b) Show that the coating thickness for a Newtonian fluid is given by

$$\delta' = \frac{\delta}{R_i} = (1 + f_d + f_p)^{1/2} - 1, \tag{2.132}$$

where

$$f_d = \frac{\rho'}{\rho}\left(\frac{1-\kappa^2}{2\kappa^2 \ln(1/\kappa)} - 1\right) \qquad f_p = \frac{\rho'}{\rho}\Phi\frac{1}{8\kappa^2}\left[1 - \kappa^4 - \frac{(1-\kappa^2)^2}{\ln(1/\kappa)}\right], \tag{2.133}$$

and where

$$\Phi = \frac{\Delta P R^2}{\mu U L},$$

and $\rho'$ is the density of the polymer at 25°C.

(c) Solve the problem for the power-law model. (Can one obtain an analytical expression for $Q$ in this case?)

(d) Obtain a solution for the power-law model for the case of small gaps.

**2B.6.**  Tangential Annular Flow

(a) Determine the stress and velocity field for tangential annular flow (see Fig. 2.18) of a power-law fluid.

(b) Determine the torque required to turn the inner cylinder and the power required.

(c) Obtain the velocity field for the case of small gaps.

Inner cylinder rotating

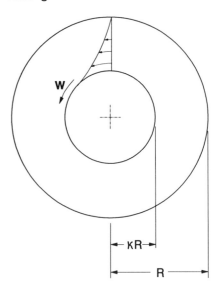

**Figure 2.18** Flow between two coaxial cylinders with the inner cylinder rotating.

**Answer**

(a)  $\dfrac{v_\theta}{Wr} = \dfrac{(R/r)^{2/n} - 1}{(1/\kappa)^{2/n} - 1}.$

(b)  $T = 2\pi(\kappa R)^2 mL\left(\dfrac{2W/n}{1-\kappa^{2/n}}\right)^n.$

**2B.7.**  Flow through an Annulus with a Rotating Mandrel

In some processes involving pressure-driven flow of polymer melts through an annulus, the outer or inner cylinder (mandrel) is rotated as shown in Fig. 2.19.

(a) Show that for a Newtonian fluid the velocity field consists of two independent components $v_z(r)$ and $v_\theta(r)$.

(b) Show that the expression for $Q$ for a Newtonian fluid is identical to that for flow through an annulus given in Eq. 2.31.

(c) Show that for a power-law fluid $v_z(r)$ and $v_\theta(r)$ are coupled and cannot be obtained analytically.

(d) Find $v_z(r)$ and $v_\theta(r)$ for a power-law fluid for the case of a small gap (i.e., when $\kappa$ approaches 1.0), when the inner cylinder is rotated.

**2B.8.**  Force for Pulling a Wire through a Coating Die

In Section 2.2.2 we discussed an approximation to flow in a wire-coating die. Use the velocity field given in Eq. 2.51 to determine the force required to pull a wire through a die such as that shown in Fig. 2.10.

(a) Obtain the expression for $\tau_{rz}$ given in Eq. 2.52 from the velocity field.

(b) Obtain an expression for the force, $F_z$, required to pull the wire through the die by integrating the shear stress over the area of the wire.

(c) For HDPE at 180°C (use power-law parameters in Table 2.3) calculate $F_z$ for the following conditions: $V = 0.5\,\text{m/s}$, $R = 5.0\text{E} - 03\,\text{m}$, $L = 0.1\,\text{m}$, and $\kappa = 0.8$.

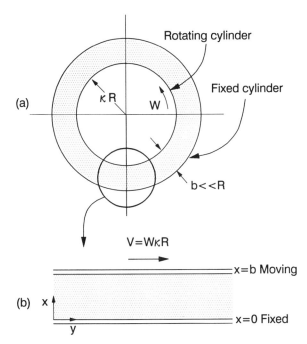

**Figure 2.19** (a) Helical flow in a thin annulus where the fluid flows axially because of a pressure gradient and tangentially because of the rotation of the inner cylinder. (b) Planar approximation of helical flow.

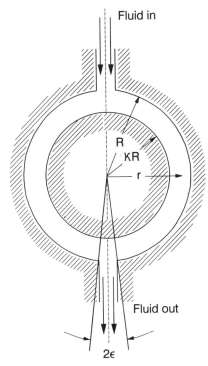

**Figure 2.20** Creeping flow between two stationary concentric spheres.

**2B.9.** Stress Components for Flow between Two Concentric Spheres (See Bird et al. [1960, p. 117] for this problem.) An idealized flow of a Newtonian fluid between two concentric spheres is shown in Fig. 2.20. The pressure drop and the velocity field for this flow are given as:

$$\Delta p = B \ln\left(\frac{1-\cos\varepsilon}{1+\cos\varepsilon}\right) = -BE(\varepsilon)$$

$$v_\theta \sin\theta = u = \frac{R\Delta P}{2\mu E(\varepsilon)}\left[\left(1-\frac{r}{R}\right)+\kappa\left(1-\frac{R}{r}\right)\right].$$

(a) From the velocity field determine expressions for the stress components, and specify those acting in tension and those in compression.

(b) Obtain an expression for the net $z$ force acting on the inner sphere.

**2B.10.** Flow between Tapered Plates
Obtain an expression for the volumetric flow rate for tapered plates similar to that for a tapered tube (Eq. 2.111) using the lubrication approximation. The plates have an initial height $H_0$ and a final height of $H_L$ as shown in Fig. 2.21.

**2B.11** Power Input for Annular Flow with a Rotating Mandrel
The integration of the mechanical energy equation (Eq. 2.70) over a fixed volume with an inlet and an outlet stream gives Eq. 2.69.

(a) Show that for flow through an annulus of length L with a rotating inner mandrel that Eq. 2.70 reduces to:

$$Q\Delta P - W - E_v = 0,$$

where $\Delta P = P_0 - P_L$, $W$ is the energy input into the system through the mandrel, and $E_v$ is the viscous dissipation. (*Hint*: Integrate Eq. 2.70 over the volume and use the divergence theorem).

(b) Determine an expression for $W$ for a Newtonian fluid (*Hint*:

$$W = \iint (\tau\cdot\mathbf{n})\cdot\mathbf{v}\,dA = \int_0^{2\pi}\int_0^L \tau_{r\theta}(\kappa R)v_\theta(\kappa R)\kappa R\,d\theta\,dz.$$

(c) Determine an expression for $W$ for a power-law fluid for a small gap. (*Note*: Evaluation of the expression will require numerical integration.)

## C. Numerical Problems

**2C.1.** Ellis Model Parameters from Regression Analysis
Use the IMSL subroutine RNLIN (Ex. 2.1) to find the Ellis model parameters for LLDPE at 170°C (viscosity data are given in Appendix. A.3), and compare the results with those given in Table 2.3.

**2C.2.** Best Fit of Viscosity Data for a Glass-Filled Nylon 6,6
Use the IMSL subroutine RNLIN (Ex. 2.1) to determine whether the Bingham or power-law model gives the best fit to the viscosity data given in Appendix. A.11 for a mineral-filled nylon 6,6 melt at 285°C.

**2C.3.** Velocity Maximum in Annular Flow for Integer Values of the Power-Law Index.
The velocity passes through a maximum in annular flow at $\beta R$. $\beta$ is found by using Eq. 2.39 for specified values of $\kappa$. When $s$ is an integer, Eq. 2.39 can be integrated analytically, giving a nonlinear algebraic equation for determining $\beta$. For $s=2$ (i.e., $n=0.5$) and $\kappa=0.5$ find a polynomial expression for determining $\beta$, and then use the IMSL subroutine NEQNF to find $\beta$.

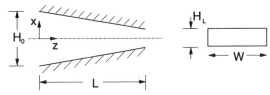

**Figure 2.21** Pressure-driven flow through tapered plates.

**2C.4** Velocity Maximum in Annular Flow for Non-integer Values of the Power-law Index

Equation 2.39 is used to find the position of the maximum in the velocity field for flow through an annulus. For $n=0.59$ and $\kappa=0.5$ and 0.8 determine $\beta$ using the numerical integration subroutine QDAGS described in Appendix D.5.

**2C.5.** Velocity Profile for Flow in a Capillary

Although the Carreau model describes the viscosity behavior of polymer melts accurately, it is not possible to obtain analytical expressions for the velocity field for one-dimensional flows such as occur in a capillary. Obtain the velocity field for HDPE at 180°C (use the Carreau model parameters in Table 2.3) by using the subroutine BVPFD described in Appendix D.9.

(a) Show that the following non-linear ordinary differential equation is obtained by substituting the GNF model with the Carreau empiricism for viscosity into the momentum balance:

$$\left(\frac{-dv_z}{dr}\right)^{\frac{2}{n-1}}+\lambda^2,\left(\frac{-dv_z}{dr}\right)^{\frac{2n}{n-1}}=\left(\frac{\Delta P}{2\eta_0 L}\right)\frac{2}{n-1}^{\frac{2}{n-1}}$$

(b) At $\tau_w=1.0E+05$ Pa and for a capillary of $D=3.175E-03$ m and $L=3.175E-01$ m use the subroutine BVPFD to solve the differential equation shown.

(c) Compare the velocity profile to that obtained for the power-law case.

**2C.6.** Flow through a Rectangular Channel

(a) Show that for the flow of a Newtonian fluid through a rectangular channel (See Fig. 2.22) that the equations of motion along with the use of the constitutive equation for a Newtonian fluid lead to the following differential equation:

$$\frac{\partial^2 v_z}{x^2}+\frac{\partial^2 v_z}{\partial y^2}=-\frac{1}{\mu}\frac{\Delta P}{L}.$$

(b) For the rectangular channel shown in Fig. 2.22 use the IMSL subroutine FPS2H (Appendix. D.10) to find the velocity profile for $\Delta P/L=1.0E+08$ Pa/m. $W/B=5$, and $B=0.2$ cm.

(c) Calculate the volumetric flow rate by using the subroutine

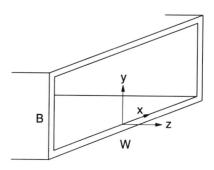

**Figure 2.22** Pressure-driven flow through a rectangular channel having an aspect ratio $W/B<10$.

TWODQ (Appendix D.6) to numerically integrate the velocity profile obtained in Part **b**.

(d) Compare the results to the series solution for flow in a rectangular channel in Chapter 7.

**2C.7.** Numerical Solution of Problem 2B.7

For the case of a thin gap the components of the velocity field are given by the following two integrals:

$$\bar{v}_y=\int_0^{\bar{x}} C_1[C_1^2+(C_2-a\bar{x})^2]^{\frac{1-n}{2n}}d\bar{x} \tag{2.134}$$

$$\bar{v}_z=\int_0^{\bar{x}}(C_2-a\bar{x})[C_1^2+(C_2-a\bar{x})^2]^{\frac{1-n}{2n}}d\bar{x} \tag{2.135}$$

where $\bar{x}=x/b$, $\bar{v}_i=v_i/V$ and $a=(\Delta P/mL)(b^{n+1}/V^n)$. From the boundary condition that $d\bar{v}_z/dx=0$ at $\bar{x}=1/2$ it is found that $C_2=a/2$. $V=\kappa R W$

(a) Use the IMSL subroutine QDAGS to numerically integrate Eqs. 2.134 and 2.135 to find $\bar{v}_y$ and $\bar{v}_z$. Assume that $n=0.59$, and take arbitrary values of $C_1$ from 0.01 to 10 and of $a$ from 0.5 to 30.

(b) Numerically integrate the results to find $Q$.

$$\left(\textit{note:}\quad Q=2\pi WbR^2\int_0^1 \bar{v}_z d\bar{x}.\right)$$

## REFERENCES

Bird, R. B., R. C. Armstrong, and O. Hassager. 1987. *Dynamics of Polymeric Liquids. Vol. I: Fluid Mechanics*, 2nd Ed. (Wiley, New York).

Bird, R. B., W. E. Stewart, and E. N. Lightfoot, 1960. *Transport Phenomena* (Wiley, New York).

# VISCOELASTIC RESPONSE OF POLYMERIC FLUIDS

## DESIGN PROBLEM 2
### Design of a Parison Die for a Viscoelastic Fluid

In extrusion blow molding a cylindrical tube of polymer is formed by extrusion of polymer melt through an annular die as shown in Fig. 3.1. As discussed in Chapter 1, the tube of polymer is expanded by gas pressure into a mold to form a shaped object such as a bottle. Design a die that will allow one to extrude at the highest rate possible a parison of low-density polyethylene (LDPE) (NPE 953) at 170°C with a diameter, $D_p$, of 6.13 cm and a thickness, $H_p$, of 0.565 cm. In particular, specify the diameter, $D_0$, and the gap, $H_0$. Take the length to be 10 $H_0$. The extruder feeding melt to the die is capable of delivering a maximum of 300 lb/hr. As a result of a flow instability, called *melt fracture* (discussed in Chapter 7), the maximum wall shear stress that can be reached is $1.13 \times 10^5$ Pa. In your calculations consider the swell of the extrudate as shown in Fig. 3.1 (i.e. consider the increase in the diameter and thickness of the parison relative to the die dimensions as a result of the viscoelastic nature of the melt). Rheological data for the polymer are given in the Tables in Appendix A.1. $\rho$ is 772 kg m$^{-3}$ at 170°C.

In the last chapter the pseudoplastic behavior of polymeric fluids was emphasized. In this chapter the viscoelastic behavior of polymer melts is discussed. It is this property that not only allows one to process these materials by a number of different ways such as blow molding and film blowing but also causes many problems in the design of polymer processes. By viscoelastic behavior it is meant that polymeric fluids can exhibit a response which resembles that of an elastic solid under some circumstances, while under others they can act as viscous liquids. The macromolecular nature of polymeric molecules along with physical interactions called entanglements leads to the elastic behavior. Deformed molecules are driven by thermal motions to return to their undeformed states giving the bulk fluid elastic recovery. Phenomena associated with the viscoelastic nature of polymers such as die swell

or extrudate swell, rod-climbing, and elastic recoil will not be discussed here. A qualitative description of the flow of polymeric fluids can be found in Bird et al. (1987).*

In Section 3.1 we define two basic flows used in the characterization of polymeric fluids along with the appropriate material functions. These basic flows are also found in polymer processes. In Section 3.2 we present several constitutive equations capable of describing the viscoelastic behavior of polymer melts. The emphasis in this section is on manipulating these equations for flows in which the deformation history is known. In Section 3.3 we introduce the methods for measuring rheological properties. In Section 3.4 several useful relationships between material functions are explained. These relationships (or correlations) are important as they allow one to obtain estimates, for example, of steady shear material functions from linear viscoelastic data. Because quantitative design work requiring the use of nonlinear viscoelastic constitutive equations is mathematically very difficult, we discuss in Section 3.5 the value of making qualitative decisions about the processability of polymers through the measurement of the nonlinear rheological properties. Finally in Section 3.6 we offer a solution to Design Problem 2.

## 3.1 MATERIAL FUNCTIONS FOR VISCOELASTIC FLUIDS
### 3.1.1 KINEMATICS

There are two basic flows used to characterize polymers: shear and shear-free flows (it so happens that processes are usually a combination of these flows or sometimes are dominated by one type or the other). The velocity field for rectilinear shear flow is given as:

$$v_x = \dot{\gamma}(t)y \qquad v_y = v_z = 0, \qquad (3.1)$$

where $\dot{\gamma}(t)$ may be constant or a function of time. The velocity field for shear-free flows can be given in a general form as:

$$v_x = -\tfrac{1}{2}\dot{\varepsilon}(1+b)x \qquad v_y = -\tfrac{1}{2}\dot{\varepsilon}(1-b)y \qquad v_z = +\dot{\varepsilon}z, \qquad (3.2)$$

*where $\dot{\varepsilon}$ is the extension rate and $b$ is a constant that is either 0 or 1. When $b=0$ and $\dot{\varepsilon}>0$, the flow is a *uniaxial* extensional flow. When $b=0$ but $\dot{\varepsilon}<0$, the flow is *equibiaxial* extensional flow. When $b=1$ and $\dot{\varepsilon}>0$, the flow is called a *planar* extensional flow.*

The deformational types are shown for a unit cube of incompressible material in Fig. 3.2. In shear flow, the unit cube is merely skewed with the degree of strain given by the angle, $\dot{\gamma}(t_2 - t_1)$, that the edge makes with the $y$ axis. $\dot{\gamma}(t_2 - t_1)$ is the shear strain. Three types of shear-free flow are described in Fig. 3.2. In uniaxial extensional flow the unit cube is stretched along the $z$ axis while it contracts uniformly along the $x$ and $y$ axes in such a manner that mass is conserved. The elongational strain is given by $\dot{\varepsilon}(t_2 - t_1)$. In biaxial elongational flow, the unit cube is stretched equally along the $x$ and $y$ directions but must contract in

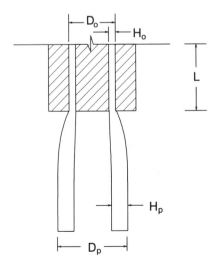

**Figure 3.1** Quantities used to define annular extrudate swell.

---

*Two movies describing the viscoelastic nature of polymers are available: H. Markovich, *Rheological Behavior of Fluids*, Educational Services, Inc., Watertown, MA (1965); and K. Walters and J. M. Broadbent, *Non-Newtonian Fluids*, Department of Applied Mathematics, University College of Wales, Aberystwyth, UK (1980).

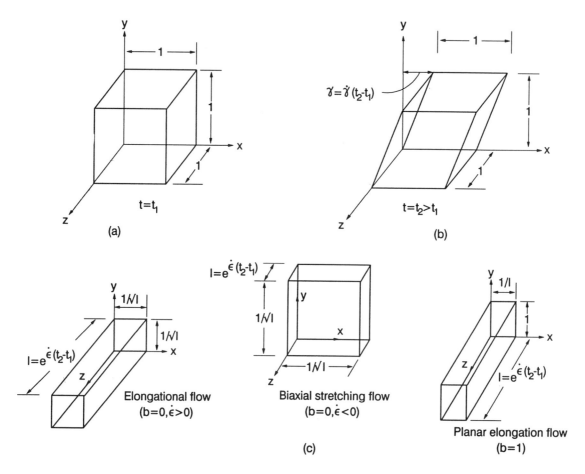

**Figure 3.2** The deformation of (a) a unit cube of material from time $t_1$ to $t_2$ ($t_2 > t_1$) in (b), steady simple shear flow and (c), three kinds of shear-free flow. The volume of material is preserved in all of these flows. (Reprinted by permission of the publisher from Bird, Armstrong & Hassager, 1987, © Wiley and Sons.)

the $z$ direction in such a way that mass is conserved. In planar extensional flow, the unit cube is stretched along the $z$ axis but is constrained so that it contracts only in the $x$ direction.

There are significant differences in the behavior of polymeric fluids in these two types of deformation, and each type of deformation has a different effect on the orientation of macromolecules. For example, uniaxial and planar extensional flows impart significant molecular orientation in polymers during flow compared to shear flows. On the other hand, biaxial extensional flow is a weak flow and does not lead to a strong degree of molecular orientation. Furthermore, the rheological response can be significantly different for a polymer in extensional flow versus shear flow. We demonstrate these differences later in this chapter.

For these two types of flows, the components of the rate of deformation tensor (Table 2.10) take on a distinct form. For shear flow the components of the rate of deformation tensor are:

$$\dot{\gamma}_{ij} = \dot{\gamma}(t) \begin{pmatrix} 0 & 1 & 0 \\ 1 & 0 & 0 \\ 0 & 0 & 0 \end{pmatrix}. \tag{3.3}$$

It is noted that only the off-diagonal components of this tensor exist. For shearfree flow the rate of deformation tensor also takes on a distinct

form. In particular the components of $\dot{\gamma}$ are:

$$\dot{\gamma}_{ij} = \begin{pmatrix} -\dot{\varepsilon}(1+b) & 0 & 0 \\ 0 & -\dot{\varepsilon}(1-b) & 0 \\ 0 & 0 & 2\dot{\varepsilon} \end{pmatrix}. \tag{3.4}$$

Here it is seen that only diagonal components exist. The physical significance of these matrices is that in shear flow the velocity gradient is transverse to the flow direction but in shear-free flow it goes in the same direction as the flow.

### 3.1.2 STRESS TENSOR COMPONENTS

In general the state of stress in a flowing material in rectangular Cartesian coordinates is:

$$\begin{pmatrix} \tau_{xx} + p & \tau_{xy} & \tau_{xz} \\ \tau_{yx} & \tau_{yy} + p & \tau_{yz} \\ \tau_{zx} & \tau_{zy} & \tau_{zz} + p \end{pmatrix}, \tag{3.5}$$

where $p$ is the isotropic pressure. The components such as $\tau_{xx} + p$, $\tau_{yy} + p$, and $\tau_{zz} + p$ are the normal stresses, and when the fluid is incompressible,

$p$ is unknown. Furthermore, $\tau_{xx}$, $\tau_{xy}$, and so on are referred to as the *extra*, or *molecular*, stresses, and

$$\pi_{xx} = \tau_{xx} + p \qquad \pi_{xy} = \tau_{xy} \qquad \pi_{xz} = \tau_{xz}, \quad \text{etc.} \tag{3.6}$$

are called the *total stress components*. If one lets $x$, $y$, and $z$ correspond to 1, 2, and 3, respectively, then we can also write out the stress components in terms of the following notation:

$$\pi_{ij} = \tau_{ij} + p\delta_{ij} \qquad \text{for} \quad i, j = 1, 2, \text{ or } 3, \tag{3.7}$$

and $\pi_{ij}$ is interpreted as the $ij$th component of the total stress tensor, and $\tau_{ij}$ is the $ij$th component of the extra stress tensor. $\delta_{ij}$ is the Kronecker Delta and is defined as:

$$\begin{aligned} \delta_{ij} &= 1 \qquad \text{if} \quad i = j \\ &= 0 \qquad \text{if} \quad i \neq j. \end{aligned} \tag{3.8}$$

The number of components in the matrix in Eq. 3.5 is reduced for an incompressible Newtonian fluid in shear flow. Referring to Table 2.10, the definition of a Newtonian fluid, and Eq. 3.1 (the kinematics for shear flow) one can show that the stress components are:

$$\begin{pmatrix} p & \tau_{xy} & 0 \\ \tau_{yx} & p & 0 \\ 0 & 0 & p \end{pmatrix}, \tag{3.9}$$

where $\tau_{xy} = \tau_{yx} = -\mu\dot{\gamma}(t)$.

On the other hand, we have not yet discussed a constitutive equation for viscoelastic fluids, and we must resort to another method to find the stress components. Without proof it can be shown by using symmetry arguments that in general for a viscoelastic fluid in shear flow the stress tensor must be of the form:

$$\begin{pmatrix} \tau_{xx} + p & \tau_{xy} & 0 \\ \tau_{yx} & \tau_{yy} + p & 0 \\ 0 & 0 & \tau_{zz} + p \end{pmatrix}. \tag{3.10}$$

We note that in shear flow additional normal stresses are generated do not appear for a Newtonian fluid. Because polymeric fluids are considered to be incompressible, the components $\tau_{ii} + p$ have no direct rheological significance Therefore, we define three independent quantities of stress of rheological significance:

$$\tau_{yx} = \tau_{xy}, \qquad \pi_{xx} - \pi_{yy} = \tau_{xx} - \tau_{yy} = N_1, \tag{3.11}$$

$$\pi_{yy} - \pi_{zz} = \tau_{yy} - \tau_{zz} = N_2, \tag{3.12}$$

where $N_1$ and $N_2$ are called the primary and secondary normal stress differences, respectively. These additional stresses are thought to be related to phenomena such as die swell, elastic recoil, and rod-climbing (the climbing of polymer fluid up a mixing blade) and hence are associated with the ideas of elasticity.

For shearfree flows it can be shown by using symmetry arguments again that the extra stress tensor is of the form:

$$\begin{pmatrix} \tau_{xx} + p & 0 & 0 \\ 0 & \tau_{yy} + p & 0 \\ 0 & 0 & \tau_{zz} + p \end{pmatrix}. \tag{3.12}$$

For incompressible polymeric fluids there are two normal stress differences of rheological interest:

$$\tau_{zz} - \tau_{xx} \qquad \text{and} \qquad \tau_{yy} - \tau_{xx}. \tag{3.13}$$

When $b = 0$ in Eq. 3.2, then there is only one quantity of rheological significance:

$$\tau_{zz} - \tau_{xx}. \tag{3.14}$$

It is important to realize that these normal stress differences are not $N_1$ or $N_2$ as they are generated under different flow conditions.

### 3.1.3. MATERIAL FUNCTIONS FOR SHEAR FLOW

Various types of shear flow experiments are used in the characterization of polymeric fluids, and some of the more commonly used ones are shown in Fig. 3.3. When $\dot{\gamma}(t)$ is a constant, that is, $\dot{\gamma}_{yx} = \dot{\gamma}_0$, then we can define three material functions in steady shear flow:

$$\tau_{xy} = -\eta(\dot{\gamma})\dot{\gamma}_0, \tag{3.15}$$

$$\tau_{xx} - \tau_{yy} = -\Psi_1(\dot{\gamma})\dot{\gamma}_0^2, \tag{3.16}$$

$$\tau_{yy} - \tau_{zz} = -\Psi_2(\dot{\gamma})\dot{\gamma}_0^2, \tag{3.17}$$

where $\eta$ is the viscosity, $\Psi_1$ is the primary normal stress difference coefficient, and $\Psi_2$ is the secondary normal stress difference coefficient. Some representative values for $\eta$ and $\Psi_1$ are presented in Fig. 3.4. We observe that $\Psi_1$ is more shear rate–sensitive than is $\eta$. Furthermore, it is reported that $-\Psi_2/\Psi_1$ is in the range of 0.1 to 0.2. Because there is no concrete evidence that $\Psi_2$ plays a direct role in processing, we do not include $\Psi_2$ in any further discussions in this book.

There are numerous transient shear flows in which $\dot{\gamma}(t)$ varies in a specific way with time. One of the most frequently used experiments is when $\dot{\gamma}(t)$ varies sinusoidally with time:

$$\dot{\gamma}_{yx} = \dot{\gamma}_0 \cos \omega t, \tag{3.18}$$

where $\dot{\gamma}_0$ is the amplitude and $\omega$ is the angular frequency. Because polymeric fluids are viscoelastic, the stress lags behind the input frequency. One component of the stress is in phase with the rate of deformation given by Eq. 3.18, and one is out of phase. Mathematically we represent the shear stress as.

$$\tau_{yx} = -B(\omega)\dot{\gamma}_0 \cos(\omega t - \Phi) \qquad (0 \leqslant \Phi \leqslant \pi/2). \tag{3.19}$$

By expanding Eq. 3.19 using a trigonometric identity (i.e., $\cos(A - B) = \cos A \cos B + \sin A \sin B$), we find that

$$\tau_{yx} = -B(\omega)\dot{\gamma}_0 \cos\Phi \cos\omega t - B(\omega)\dot{\gamma}_0 \sin\Phi \sin\omega t. \tag{3.20}$$

This allows us to define a complex viscosity, $\eta^*$, as follows:

$$\eta^* = \eta' - i\eta'', \tag{3.21}$$

where $\eta' = B(\omega)\dot{\gamma}_0 \cos\Phi$ is the dynamic viscosity (viscous contribution) and $\eta'' = B(\omega)\dot{\gamma}_0 \sin\Phi$ is the elastic contribution associated with energy storage per cycle of deformation. Hence, we see that $\eta'$ is in-phase with $\dot{\gamma}_{yx}(t)$ and that $\eta''$ is out of phase. Representative values for $(\eta^*)$ are given in Fig. 3.4. We note here the close correlation between $\eta^*$ and $\eta$, which will be discussed in Section 3.4.

Some prefer to treat polymeric fluids as viscoelastic solids and

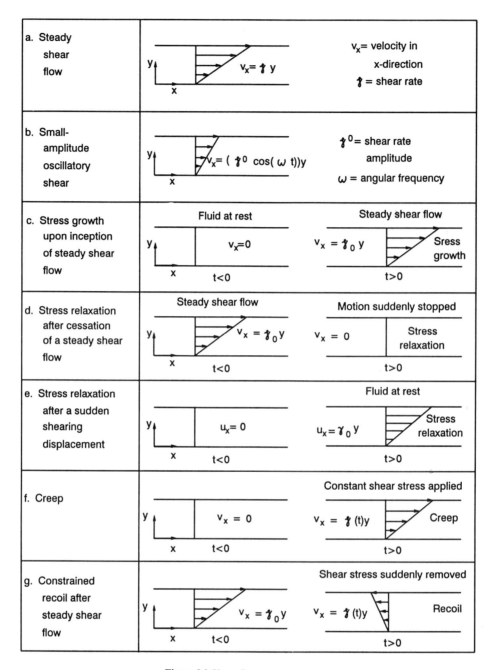

**Figure 3.3** Shear flow experiments.

thereby represent $\tau_{yx}$ as:

$$\tau_{yx} = -A(\omega)\gamma_0 \sin(\omega t + \delta), \tag{3.22}$$

where $\gamma_0$ is the strain amplitude given as:

$$\gamma_0 = \dot{\gamma}_0/\omega. \tag{3.23}$$

Again, using trigonometric identities and defining the complex shear modulus as:

$$G^* = i\omega\eta^* = G' + iG'', \tag{3.24}$$

we find $G' = A(\omega)\gamma_0 \sin \delta$ and $G'' = A(\omega)\gamma_0 \cos \delta$, where $G'$ is the storage modulus and $G''$ is the loss modulus. Representative values of $G'$ and $G''$ for polyphenylene sulfide (PPS) are given in Fig. 3.5. Part of the value of the $G'$ measurements rests on the fact there is a good correlation between $2G'$ and $N_1$ as shown in Fig. 3.6 and discussed in Section 3.4. It should also be pointed out that $\eta'$ and $G''$ and $\eta''$ and $G'$ are interrelated: that is, $\eta'\omega = G''$ and $\eta''\omega = G'$.

The stress growth experiment (Fig. 3.3c) is also used to characterize polymeric fluids. In this experiment the fluid that is at rest is suddenly set in motion and the stresses are measured as a function of time. $\dot{\gamma}_{yx}(t)$ is given mathematically as:

$$\dot{\gamma}_{yx} = \dot{\gamma}_0 H(t), \tag{3.25}$$

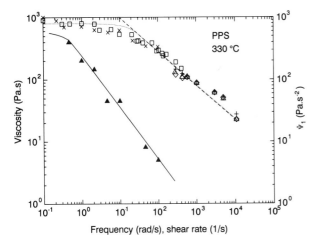

**Figure 3.4** Viscosity (steady and complex) versus shear rate ($\dot{\gamma}$) or angular frequency ($\omega$) and the primary normal stress difference coefficient versus shear rate for polyphenylenesulfide (PPS) at 330°C. Values of $\Psi_1$ (▲), $\eta^*$ (□), $\Psi_1$ and $\eta$ (X) were obtained by means of a cone and plate device. All other values of $\eta$ were obtained by means of a capillary rheometer.

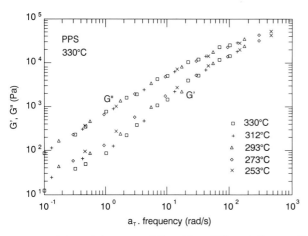

**Figure 3.5** Values of the storage ($G'$) and loss ($G''$) moduli versus the product of the shift factor and angular frequency ($\omega$) for polyphenylene-sulfide (PPS) shifted to a reference temperature of 330°C.

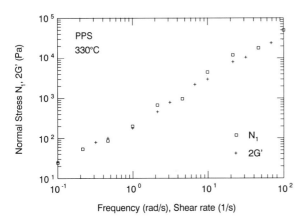

**Figure 3.6** Values of $N_1$ and $2G'$ versus shear rate and angular frequency, respectively, for PPS at 330°C.

where $H(t)$ is the unit step function which is defined as:

$$H(t) = 0 \qquad t < 0$$
$$H(t) = 1 \qquad t \geq 0. \tag{3.26}$$

The following material functions are defined for this flow:

$$\tau_{xy} = -\eta^+(t, \dot{\gamma}_0)\dot{\gamma}_0, \tag{3.27}$$

$$\tau_{xx} - \tau_{yy} = -\Psi_1^+(t, \dot{\gamma}_0)\dot{\gamma}_0^2, \tag{3.28}$$

$$\tau_{yy} - \tau_{zz} = -\Psi_2^+(t, \dot{\gamma}_0)\dot{\gamma}_0^2. \tag{3.29}$$

Representative data are shown in Fig. 3.7 for a PPS melt. We note that $\tau_{yx}$ and $N_1$ overshoot their equilibrium values and that the maximum in $N_1$ usually occurs later than that in $\tau_{yx}$ at the same value of $\dot{\gamma}_0$. Because many processes take place in short intervals, the transient properties may be more important than the steady shear ones. More will be said about the importance of transient behavior in Section 3.2, when we define a dimensionless group called the *Deborah number*.

Another important experiment is that of stress relaxation following steady shear flow. This experiment is shown schematically in Fig. 3.3.d, and the deformation history is given mathematically as follows:

$$\dot{\gamma}_{yx}(t) = \dot{\gamma}_0[1 - H(t)]. \tag{3.30}$$

In this experiment on cessation of steady shear flow, the stresses are monitored with time. The material functions are defined as:

$$\tau_{yx} = -\eta^-(\dot{\gamma}_0, t)\dot{\gamma}_0, \tag{3.31}$$

$$\tau_{xx} - \tau_{yy} = -\Psi_1^-(\dot{\gamma}_0, t)\dot{\gamma}_0^2, \tag{3.32}$$

$$\tau_{yy} - \tau_{zz} = -\Psi_2^-(\dot{\gamma}_0, t)\dot{\gamma}_0^2. \tag{3.33}$$

Representative behavior is also presented in Fig. 3.7. It is observed that $\tau_{yx}$ relaxes faster than $\tau_{xx} - \tau_{yy}$. Furthermore, it is known that as $\dot{\gamma}_0$ increases, the time for $\tau_{yx}$ and $N_1$ to relax to zero is shorter. After relaxation is complete, the stress growth and relaxation experiment are repeated as shown in Fig. 3.7, and it is observed that the behavior is repeatable. Although the stresses relax within 2.0 s, higher-molecular-

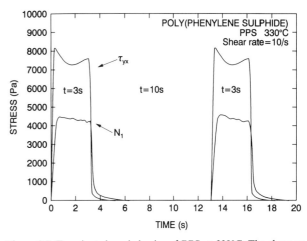

**Figure 3.7.** Transient shear behavior of PPS at 330°C. The shear and primary normal stress difference are recorded at the startup of shear flow and on cessation of flow. After 10 s the stress growth experiment is repeated.

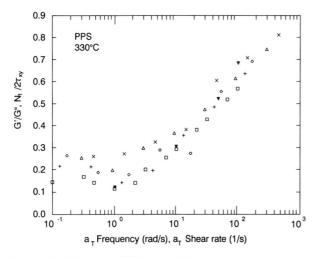

**Figure 3.8** Values of $G'/G''$ and $N_1/2\tau_{xy}$ versus $a_T\omega$ and $a_T\dot{\gamma}$, respectively, for PPS at 330°C. Dynamic data were obtained at various temperatures and shifted to a reference temperature of 330°C: (□) 330; (+) 312; (◇) 393; (△) 273; (X) 253°C.

weight polymers may take 10 to 20 s before the stresses relax completely. Residual stresses in an injection-molded part can lead to the warpage of parts.

The fact that the normal stresses are relatively large compared to $\tau_{yx}$ is shown in Fig. 3.8, where the ratio of $(\tau_{xx} - \tau_{yy})/2\tau_{yx}$ and $G'/G''$ are plotted versus shear rate and $\omega$, respectively, for PPS. Here we see that $\tau_{xx} - \tau_{yy}$ increases monotonically with $\dot{\gamma}$, and that it reaches values where $\tau_{xx} - \tau_{yy}$ is about twice as large as $\tau_{yx}$. Hence, $\tau_{xx} - \tau_{yy}$ can exceed the magnitude of $\tau_{yx}$. This dimensionless ratio divided by a factor of 2 is equivalent to a quantity called the *Weissenberg number*, We, and is a measure of fluid elasticity. For most polymers it has been observed that We reaches a plateau at higher values of $\dot{\gamma}$ leveling off from 1.0 to 2.0 (Bird et al., 1987).

### 3.1.4. SHEAR-FREE FLOW MATERIAL FUNCTIONS

Similar flow histories for shear-free flows as described for shear flows in Fig. 3.3 can also be used. Here we discuss only steady and stress growth shear-free flows. For steady simple (i.e., homogeneous deformation) shear-free flows two viscosity functions, $\eta_1$ and $\eta_2$, are defined based on the two normal stress differences given in Eq. 3.13:

$$\tau_{zz} - \tau_{xx} = -\bar{\eta}_1(\dot{\varepsilon}, b)\dot{\varepsilon}, \tag{3.34}$$

$$\tau_{yy} - \tau_{xx} = -\bar{\eta}_2(\dot{\varepsilon}, b)\dot{\varepsilon}. \tag{3.35}$$

For a uniaxial extensional flow where $b=0$ and $\dot{\varepsilon}>0$, $\bar{\eta}_2=0$, and $\bar{\eta}_1$ is called the elongational viscosity, $\bar{\eta}$:

$$\bar{\eta}(\dot{\varepsilon}) = \bar{\eta}_1(\dot{\varepsilon}, 0). \tag{3.36}$$

For stress growth the two viscosity functions are defined as:

$$\tau_{zz} - \tau_{xx} = -\bar{\eta}_1^+ \dot{\varepsilon}, \tag{3.37}$$

$$\tau_{yy} - \tau_{xx} = -\bar{\eta}_2^+ \dot{\varepsilon}. \tag{3.38}$$

Likewise for a uniaxial extensional flow $\bar{\eta}_2^+ = 0$ and $\bar{\eta}_1^+ = \bar{\eta}^+$.

Representative data for $\bar{\eta}$ and $\bar{\eta}^+$ are shown in Figures 3.9 and 3.10, respectively. In Fig. 3.9 $\bar{\eta}$-versus-tensile stress and $\eta$-versus-shear-

stress values are compared for a polystyrene melt (see Münstedt, 1980, for detailed molecular weight [MW] features of the polymers). At low stress values, $\bar{\eta} = 3\eta_0$. However, when $\eta$ shear thins, $\bar{\eta}$ tends to increase slightly with stress and then decrease. At higher values of stress, $\eta$ is several decades lower than $\bar{\eta}$. Linear polymers tend to reach an equilibrium stress in the stress growth experiment as shown in Fig. 3.10 for PS III. However, if there is branching or small amounts of high-molecular-weight tail in the polymer, then the stresses tend not to reach a steady state (at least for the strains experimentally accessible). As shown in Fig. 3.10 for PS IV and Fig. 3.11 for PS II, $\bar{\eta}^+$ tends to increase without bound (especially for PS II). In this case the cause of the "strain hardening" (i.e., the increase in $\bar{\eta}^+$ with time or strain, $\dot{\varepsilon}t$) is due to the presence of a small amount of high-molecular-weight PS in the MW distribution.

## 3.2. NONLINEAR CONSTITUTIVE EQUATIONS

In Section 3.1 we learned that viscosity is not adequate to characterize polymeric fluids but that many different material functions must be used. Furthermore, the generalized Newtonian fluid (GNF) model is not adequate to describe the rheological properties of polymer melts as it can only describe their shear-thinning viscosity behavior. In this section we describe several constitutive equations capable of describing some of the nonlinear behavior of polymer melts. We do not intend to cover completely the topic of constitutive equations as this is done elsewhere (Bird et al., 1987; Larson, 1988). Our only intention is to present several realistic possibilities, show how to manipulate them algebraically, and then illustrate their use.

### 3.2.1. DESCRIPTION OF SEVERAL MODELS

The deformational behavior of polymeric fluids is qualitatively described by the spring and dashpot model shown in Fig. 3.12. The ratio of the dashpot resistance, $\mu$, to the spring modulus, $G$, has the units of time. Hence $\mu/G$ is equivalent to a relaxation time, $\lambda$. If the deformation rate, $\dot{\gamma}$, is a lot less than $1/\lambda$, then the dashpot dominates and the material flows like a Newtonian fluid. On the other hand, if $\dot{\gamma} \gg 1/\lambda$, then the spring dominates, and the material behaves like an elastic solid. In one dimension the relation between force (per unit area) and rate of deformation is $(F/A) + \lambda[d(F/A)/dt] = -\mu\dot{\gamma}$. This one-dimensional model can be generalized to give the linear viscoelastic model for the spring and dashpot model shown in Fig. 3.12:

$$\tau_{ij} + \lambda \partial(\tau_{ij})/\partial t = -\mu\dot{\gamma}_{ij} \tag{3.39}$$

Note that this is again a shorthand notation for representing the six components of the stress tensor. For steady shear flow this model predicts no normal stresses and a constant viscosity. The material functions predicted by this model are summarized in Table 3.1. Obviously this model as it stands is wholly inadequate to describe the behavior of polymeric melts under processing conditions.

By replacing the partial time derivative with a nonlinear time derivative, which is based on a co-deforming and translating coordinate system, the constitutive equation given in Eq. 3.39 now becomes (Bird et al., 1987):

$$\tau_{ij} + \lambda \overset{\triangledown}{\tau}_{ij} = -\mu\dot{\gamma}_{ij}. \tag{3.40}$$

$\overset{\triangledown}{\tau}_{ij}$ is given as follows:

$$\overset{\triangledown}{\tau}_{ij} = \frac{\partial\tau_{ij}}{\partial t} + (\mathbf{v}\cdot\nabla\tau)_{ij} - [(\nabla\mathbf{v})_{jk}\tau_{ki} + \tau_{ik}(\nabla\mathbf{v})_{kj}]. \tag{3.41}$$

This constitutive equation is referred to as the upper convected Maxwell (UCM) model. In Ex. 3.1 this time derivative will be written out for

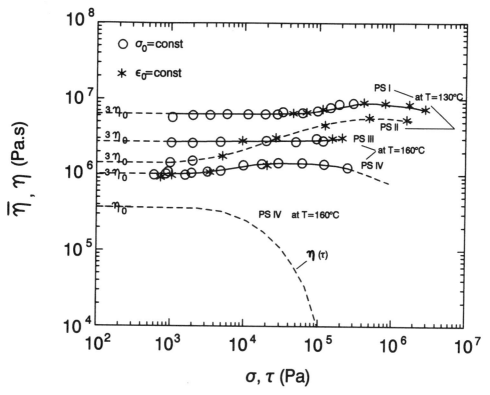

**Figure 3.9** Steady shear ($\eta$) and elongational ($\bar{\eta}$) viscosity versus shear stress ($\tau$) and tensile stress ($\sigma$) for four polystyrene melts:

|        | $\bar{M}_w$          | $\bar{M}_w/\bar{M}_n$ |
|--------|----------------------|-----------------------|
| PS I   | $7.4 \times 10^4$    | 1.2                   |
| PS II  | $3.9 \times 10^4$    | 1.1                   |
| PS III | $2.53 \times 10^5$   | 1.9                   |
| PS IV  | $2.19 \times 10^5$   | 2.3                   |

(Reprinted by permission of the publisher from Münstedt, 1980.)

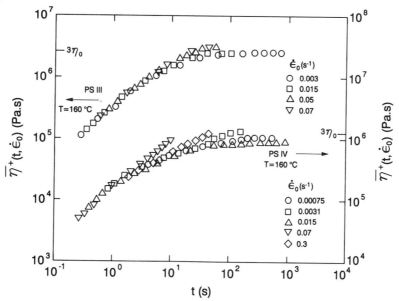

**Figure 3.10** Time dependence of the elongational viscosity at the startup of simple elongational flow for two polystyrene samples (PS III and PS IV). (Reprinted by permission of the publisher from Mündstedt, 1980.)

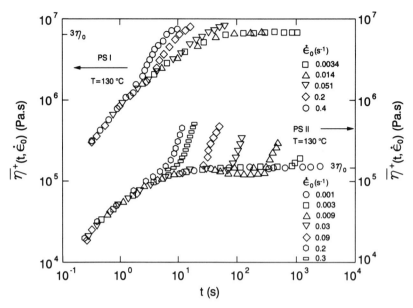

**Figure 3.11** Time dependence of the elongational viscosity at the startup of simple elongational flow for two polystyrene samples (PS I and PS II). (Reprinted by permission of the publisher from Mündstedt, 1980.)

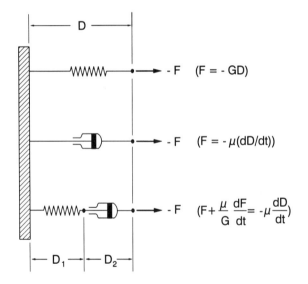

**Figure 3.12** Spring and dashpot analogs for rheological equations: top, spring element; *middle*, dashpot element; *bottom*, spring and dashpot element in series.

**TABLE 3.1 Predictions of Viscoelastic Models for Steady Shear and Elongational Flow**

| Model | Steady Shear Flow | | Steady Extensional Flow |
|---|---|---|---|
| Maxwell | $\eta = \mu$ | $\Psi_1 = \Psi_2 = 0$ | $\bar{\eta} = 3\mu$ |
| Upper Convected Maxwell (UCM) | $\eta = \mu$    $\Psi_1 = 2\mu\lambda$ $\Psi_2 = 0$ | | $\bar{\eta} = \dfrac{3\mu}{(1 + \lambda\dot{\varepsilon})(1 - 2\lambda\dot{\varepsilon})}$ |
| White-Metzner | $\eta = \eta(\dot{\gamma})$ $\Psi_1 = 2\eta(\dot{\gamma})\lambda(\dot{\gamma})$ $\psi_2 = 0 \quad \lambda = \eta(\dot{\gamma})/G$ | | $\bar{\eta} = \dfrac{3\eta}{(1 + \lambda\dot{\varepsilon})(1 - 2\lambda\dot{\varepsilon})}$ |
| Phan Thien-Tanner (single relaxation time) | $\eta = \dfrac{\eta_0}{1 + \xi(2 - \xi)(\lambda\dot{\gamma})^2}$ $\Psi_1 = \dfrac{2\eta\lambda}{1 + \xi(2 - \xi)(\lambda\dot{\gamma})^2}$ $\Psi_2 = -\dfrac{\xi}{2}\Psi_1$ | | No analytical solution is available* |

*The following nonlinear algebraic equatioms must be solved to find $\dot{\eta}$:

$$\exp\left[\frac{-\varepsilon\lambda}{\eta_0}(\tau_{11} + 2\tau_{22})\right]\tau_{11} - 2\lambda\dot{\varepsilon}(1 - \xi)\tau_{11} = -2\eta_0\dot{\varepsilon}$$

$$\exp\left[\frac{-\varepsilon\lambda}{\eta_0}(\tau_{11} + 2\tau_{22})\right]\tau_{22} + \lambda\dot{\varepsilon}(1 - \xi)\tau_{22} = +\eta_0\dot{\varepsilon}$$

simple shear and shear-free flows. The predictions of this model for steady shear and shear-free flows are summarized in Table 3.1, and the predicted material functions are fit to shear flow data in Figures 3.13 and 3.14. The most important points to be made are that the small amplitude dynamic mechanical properties are the same as those predicted by the linear viscoelastic model (Eq. 3.39). However, for shear flow one now sees that the model predicts normal stresses, but that both $\eta$ and $\Psi_1$ are constant rather than functions of $\dot{\gamma}$. Furthermore, $\bar{\eta}$ is observed to be equal to $3\eta_0$ at low $\dot{\varepsilon}$, but when $\lambda\dot{\varepsilon} = 1/2$, then $\bar{\eta}$ rises in an unbounded manner (see Fig. 3.15). Yet, we see that this

model is not adequate either, as it fails to predict even the most basic property of polymeric fluids, shear-thinning viscosity.

One way to at least make the UCM model predict realistic steady shear flow properties is to replace the material properties $\mu$ and $\lambda = \mu/G$ by functions that depend on $\dot{\gamma} = \sqrt{\frac{1}{2}\Pi_{\dot{\gamma}}}$. If one replaces $\mu$ by any of the empiricisms for $\eta$ discussed in Section 2.1, then the constitutive equation, which is known as the White-Metzner (WM) model, can be written as:

$$\tau_{ij} + \lambda\overset{\triangle}{\tau}_{ij} = -\eta\dot{\gamma}_{ij}. \tag{3.42}$$

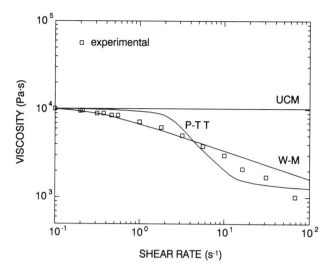

**Figure 3.13** Comparison of the predictions for viscosity for the White-Metzner (WM), upper convected Maxwell (UCM), and Phan Thien-Tanner (PTT) models with experimental data for polystyrene (Styron 678, Dow Chemical Company) at 190°C. (Data from A. D. Gotsis, "Study of the Numerical Simulation of Viscoelastic Flow: Effect of the Rheological Model and the Mesh," Ph.D. Thesis, Department of Chemical Engineering, Virginia Polytechnic Institute and State University, Blacksburg, VA, 1987.)

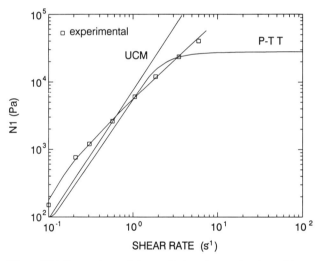

**Figure 3.14** Comparison of the predictions for $N_1$ for three different constitutive equations (WM, PTT, and UCM) with experimental data for polystyrene (Styron 678, Dow Chemical Company) at 190°C. (Data from A. D. Gotsis, "Study of the Numerical Simulation of Viscoelastic Flow: Effect of the Rheological Model and the Mesh," Ph.D. Thesis, Department of Chemical Engineering, Virginia Polytechnic Institute and State University, Blacksburg, VA, 1987.)

The material properties predicted by this model are summarized in Table 3.1. In particular, in shear flow it is observed that the following stresses are predicted:

$$\tau_{yx} = -\eta\dot{\gamma}_{yx} \tag{3.43}$$

$$\tau_{xx} - \tau_{yy} = -2\lambda\eta\dot{\gamma}_{yx}^2. \tag{3.44}$$

For this model, $\lambda = \eta/G$, and $G$ is a constant shear modulus. This model

predicts a value of $\Psi_1$ that is proportional to $\eta^2$. However, $\Psi_1$ usually shear-thins more rapidly than this. For numerical calculations it is best to fit the $\Psi_1$ data exactly and then obtain $\lambda$ as a function of $\dot{\gamma}$. In the case of extensional flow, $\bar{\eta}$ is predicted to rise in an unbounded manner, but the critical extension rate is postponed to higher values of $\dot{\varepsilon}$ relative to that for the UCM model (See Fig. 3.15).

The choice of a constitutive equation cannot be determined solely by whether it predicts the appropriate steady shear flow behavior. The only way that one can really assess the value of a constitutive relation is to use it in conjunction with the equations of motion to predict stress and velocity fields. Then these results must be confirmed by experiments using flow visualization and birefringence techniques to obtain the velocity fields or streamlines and the stress fields, respectively. This has been done by White and Baird (1988) for flow of two polymer melts through a planar contraction. They found that the observed behavior could be predicted if the constitutive equation used in the computations predicted accurately both the shear and extensional flow properties of the melt.

For this reason we now discuss the Phan Thien-Tanner (PTT) model. This model allows one to describe the various types of behavior observed for $\bar{\eta}$ while still describing the shear-thinning viscosity. The PTT model (Phan Thien-Tanner, 1977) is as follows:

$$Z(tr\tau)\tau + \lambda\overset{\triangle}{\tau} + \frac{\xi}{2}\lambda(\dot{\gamma}\cdot\tau + \tau\cdot\dot{\gamma}) = -\eta_0\dot{\gamma} \tag{3.45}$$

The function $Z(tr\tau)$, where $tr\tau$ is the trace of the stress tensor, is given by one of the following functions:

$$Z = \begin{matrix} 1 - \varepsilon\lambda tr\tau/\eta_0 \\ \text{or} \\ \exp(-\varepsilon\lambda tr\tau/\eta_0) \end{matrix} \tag{3.46}$$

Here $\lambda$, $\eta_0$, and $\xi$ are found from steady shear and dynamic data. The parameter $\varepsilon$ is obtained from $\bar{\eta}$ data and has little effect on the prediction of the shear flow properties. Actually, $\eta_0$ can be replaced

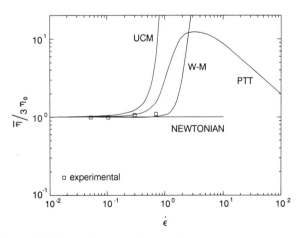

**Figure 3.15** Comparison of predictions for extensional viscosity ($\bar{\eta}$) for three constitutive equations (WM, UCM, and PTT) with experimental data for polystyrene (Styron 678, Dow Chemical Company) at 190°C (Data from Gotsis, "Study of the Numerical Simulation of Viscoelastic Flow: Effect of the Rheological Model and the Mesh," Ph.D. Thesis, Department of Chemical Engineering, Virginia Polytechnic Institute and State University, Blacksburg, VA, 1987.)

by $\eta(\dot{\gamma})$ or a set of values for $\eta_i$ and $\lambda_i$. Predictions of the model for steady shear and uniaxial extensional flows are summarized in Table 3.1. The choice of the exponential function allows one to predict values of $\bar{\eta}$ that increase with $\dot{\varepsilon}$ and then pass through a maximum.

A few additional comments should be made pertaining to this model. The model as it stands has a problem in that the shear stress passes through a maximum at critical value of $\dot{\gamma}$ (this leads to instabilities in any numerical calculation). This problem is overcome by several different methods. One is to use a spectrum of relaxation times and viscosities, $\lambda_i$ and $\eta_i$, respectively (referred to as a multimode approach). Another is to add a Newtonian viscosity term (i.e., $\tau = \tau_s + \tau_p$ where $\tau_p$ is given by Eq. 3.45 and $\tau_s = -\mu_s\dot{\gamma}$). Furthermore, one can replace $\eta$ by one of the empiricisms for viscosity discussed in Section 2.4. Finally, no problems can be solved analytically using this model (or with the others for that matter), and numerical methods are usually required.

The predictions of these models and their fit to rheological data for a polystyrene melt are summarized in Figures 3.13, 3.14, and 3.15. Here it can be seen how well each model fits experimental data. Certainly if a spectrum of relaxation times were used, the PTT model would fit the steady shear data even better. In Fig. 3.15 the predictions of $\bar{\eta}$ are presented. In fact, for the PTT model the predictions for $\bar{\eta}$ versus $\dot{\varepsilon}$ can be changed by the values chosen for $\varepsilon$. With $\varepsilon = 0.01$, $\bar{\eta}$ strain rate hardens, passes through a maximum, and then decreases with increasing $\dot{\varepsilon}$. If $\varepsilon > 0.1$, $\bar{\eta}$ starts at $3\eta_0$ and then decreases with increasing $\dot{\varepsilon}$. It is believed that these two different types of behavior

for $\bar{\eta}$ are related to differences in the processability of polymers, especially in cases where extensional deformations are important. Over the range for which $N_1$ data are available all the models predict reasonable values of $N_1$ (see Fig. 3.14). However, it is observed that the PTT model predicts values of $N_1$ that become independent of shear rate (when a Newtonian limiting viscosity is included) and may offer more realistic predictions at high shear rates. (*Note*: values of $N_1$ at values of $\dot{\gamma} > 100$ are not readily available.)

It should be noted that the constitutive equations that are discussed here represent only a few of the plethora of equations to choose from. We chose the PTT model because of its ability to fit the extensional flow behavior of polymeric fluids in a manner nearly independent of shear flow behavior. There are several other possibilities for constitutive equations that might behave in a manner similar to the PTT model. In particular we note the Giesekus model (see Pr. 3B.5) and a couple of integral models (Bird et al., 1987). Because we do not intend to include the additional mathematical complexities associated with integral models, these will not be discussed in this book.

We do not intend to use nonlinear constitutive equations in conjunction with equations of motion to solve polymer processing problems either. Therefore, we at least show in the next several examples how one determines the predictions of a nonlinear model for flows in which the kinematics are known. In particular, we consider shear and shear-free flows. Furthermore, we show how one goes about finding the material parameters in a constitutive equation from rheological data for a polymer melt.

---

**EXAMPLE 3.1**
**Shear Flow Predictions for the White-Metzner Model**
Calculate the material functions for the WM model in simple shear flow (i.e., $v_x = \dot{\gamma}_{yx}(t)y$, $v_y = v_z = 0$), including the startup of simple shear flow, steady simple shear flow, and stress relaxation following steady shear flow.

**Solution**

For an unsteady simple shear flow the terms in Eq. 3.42 are:

$$\tau_{ij} = \begin{pmatrix} \tau_{xx} & \tau_{yx} & 0 \\ \tau_{yx} & \tau_{yy} & 0 \\ 0 & 0 & \tau_{zz} \end{pmatrix} \tag{3.47}$$

$$\overset{\triangle}{\tau} = \frac{\partial}{\partial t}\begin{pmatrix} \tau_{xx} & \tau_{yx} & 0 \\ \tau_{yx} & \tau_{yy} & 0 \\ 0 & 0 & \tau_{zz} \end{pmatrix} + \underline{v_x\frac{\partial}{\partial x}\begin{pmatrix} \tau_{xx} & \tau_{yx} & 0 \\ \tau_{yx} & \tau_{yy} & 0 \\ 0 & 0 & \tau_{zz} \end{pmatrix}}$$

$$-\begin{pmatrix} 0 & \dot{\gamma}_{yx} & 0 \\ 0 & 0 & 0 \\ 0 & 0 & 0 \end{pmatrix}\begin{pmatrix} \tau_{xx} & \tau_{yx} & 0 \\ \tau_{yx} & \tau_{yy} & 0 \\ 0 & 0 & \tau_{zz} \end{pmatrix}$$

$$-\begin{pmatrix} \tau_{xx} & \tau_{yx} & 0 \\ \tau_{yx} & \tau_{yy} & 0 \\ 0 & 0 & \tau_{zz} \end{pmatrix}\begin{pmatrix} 0 & 0 & 0 \\ \dot{\gamma}_{yx} & 0 & 0 \\ 0 & 0 & 0 \end{pmatrix}$$

$$= \frac{\partial}{\partial t}\begin{pmatrix} \tau_{xx} & \tau_{yx} & 0 \\ \tau_{yx} & \tau_{yy} & 0 \\ 0 & 0 & \tau_{zz} \end{pmatrix} - \dot{\gamma}_{yx}\begin{pmatrix} 2\tau_{yx} & \tau_{yy} & 0 \\ \tau_{yy} & 0 & 0 \\ 0 & 0 & 0 \end{pmatrix} \tag{3.48}$$

$$\dot{\gamma}_{ij} = \begin{pmatrix} 0 & \dot{\gamma}_{yx} & 0 \\ \dot{\gamma}_{yx} & 0 & 0 \\ 0 & 0 & 0 \end{pmatrix} \tag{3.49}$$

Note that the underlined term in Eq. 3.48 is zero for a homogeneous flow (i.e., a flow in which $\dot{\gamma}_{yx}$ is not a function of $x$, $y$, or $z$). This leads to the following matrix equation for determining $\tau_{ij}$:

$$\begin{pmatrix} \tau_{xx} & \tau_{yx} & 0 \\ \tau_{yx} & \tau_{yy} & 0 \\ 0 & 0 & \tau_{zz} \end{pmatrix} + \frac{\eta(\dot{\gamma})}{G}\left[\frac{d}{dt}\begin{pmatrix} \tau_{xx} & \tau_{yx} & 0 \\ \tau_{yx} & \tau_{yy} & 0 \\ 0 & 0 & \tau_{zz} \end{pmatrix}\right.$$
$$\left. -\dot{\gamma}_{yx}\begin{pmatrix} 2\tau_{yx} & \tau_{yy} & 0 \\ \tau_{yy} & 0 & 0 \\ 0 & 0 & 0 \end{pmatrix}\right] = -\eta(\dot{\gamma})\begin{pmatrix} 0 & \dot{\gamma}_{yx} & 0 \\ \dot{\gamma}_{yx} & 0 & 0 \\ 0 & 0 & 0 \end{pmatrix}. \tag{3.50}$$

From this matrix equation one obtains a set of coupled ordinary differential equations for the stress components:

$$\left(1 + \frac{\eta(\dot{\gamma})}{G}\frac{d}{dt}\right)\tau_{xx} - \frac{2\eta(\dot{\gamma})}{G}\tau_{yx}\dot{\gamma}_{yx} = 0, \tag{3.51}$$

$$\left(1 + \frac{\eta(\dot{\gamma})}{G}\frac{d}{dt}\right)\tau_{yy} = 0, \tag{3.52}$$

$$\left(1 + \frac{\eta(\dot{\gamma})}{G}\frac{d}{dt}\right)\tau_{yx} - \frac{\eta(\dot{\gamma})}{G}\dot{\gamma}_{yx}\tau_{yy} = -\eta(\dot{\gamma})\dot{\gamma}_{yx}, \tag{3.53}$$

$$\left(1 + \frac{\eta(\dot{\gamma})}{G}\frac{d}{dt}\right)\tau_{zz} = 0. \tag{3.54}$$

From Eqs. 3.52 and 3.54 one sees that $\tau_{yy}$ and $\tau_{zz}$ are zero for all time-dependent simple shear flows (this follows from $\tau_{zz} = \tau_{yy} = 0$ at $t = -\infty$). One can now solve for $\tau_{xx}$ and $\tau_{xy}$. First using Eq. 3.53 one solves for $\tau_{yx}$ using an integrating factor:

$$d(e^{t/\lambda}\tau_{yx})/dt = -G\exp(t/\lambda)\dot{\gamma}_{yx}(t). \tag{3.55}$$

This gives the following expression for $\tau_{yx}$ on integration:

$$\tau_{yx}(t) = -G\int_{-\infty}^{t} \exp[-(t-t')/\lambda]\dot{\gamma}_{yx}(t')\,dt'. \tag{3.56}$$

Note that the integration starts at $-\infty$, where the stresses are assumed to be zero. Likewise, one can find an expression for $\tau_{xx}(t)$:

$$\tau_{xx}(t) = -2G\int_{-\infty}^{t} e^{-(t-t')/\lambda}\tau_{yx}(t')\,dt'. \tag{3.57}$$

One now considers the three special cases requested in this example.

**Stress Growth.** For start up of shear flow (the stress growth experiment), $\dot{\gamma}_{yx}(t)$ is given by:

$$\dot{\gamma}_{yx}(t) = \dot{\gamma}_0 H(t), \tag{3.58}$$

where $H(t)$ is the unit step function. When Eq. 3.58 is substituted into Eqs. 3.56 and 3.57, one obtains the following expressions for the stress components:*

$$\tau_{yx} = -\eta(\dot{\gamma})\dot{\gamma}_0[1 - e^{-t/\lambda}], \tag{3.59}$$

$$\tau_{xx}(t) = -2\eta(\dot{\gamma})\dot{\gamma}_0^2\lambda[1 - e^{-t/\lambda}]. \tag{3.60}$$

It is observed from these equations that the stresses rise monotonically to their steady-state values.

**Steady Shear Flow.** For steady shear flow one starts with Eqs. 3.56 and 3.57 and with $\dot{\gamma}_{yx}(t) = \dot{\gamma}_0$. One finds:

$$\tau_{yx}(t) = -G\int_{-\infty}^{t} e^{-(t-t')/\lambda}\dot{\gamma}_0\,dt'. \tag{3.61}$$

Next one replaces the integration variable $t'$ by $s = t - t'$ to give:

$$\tau_{yx}(t) = -G\int_{-\infty}^{t} e^{-s/\lambda}\dot{\gamma}_0\,ds, \tag{3.62}$$

---

*Eq. 3.59 is obtained on integrating by parts in the following manner:

$$\tau_{yx} = -G\dot{\gamma}_0 e^{-t/\lambda}\lambda\left[ e^{t'/\lambda}H(t')|_{-\infty}^{t} - \int_{-\infty}^{t} e^{t'/\lambda}\delta(t')\,dt' \right]$$

$$= -\eta(\dot{\gamma})\dot{\gamma}[1 - e^{-t/\lambda}].$$

which can be further integrated to give:

$$\tau_{yx} = -\eta(\dot{\gamma})\dot{\gamma}_0. \tag{3.63}$$

One can carry out the same procedure to find $\tau_{xx}$:

$$\tau_{xx} = -2\eta(\dot{\gamma})\lambda\dot{\gamma}_0^2. \tag{3.64}$$

Hence, it is seen by replacing $\eta(\dot{\gamma})$ by an appropriate function that one can fit the viscosity data exactly. The normal stress is predicted to be proportional to $\eta^2$ if $\lambda$ is replaced by $\eta/G$. However, this is not quite what is observed experimentally.

**Stress Relaxation Following Steady Shear Flow.** For this flow one uses $\dot{\gamma}_{yx} = \dot{\gamma}_0(1 - H(t))$. Returning to Eqs. 3.56 and 3.57 one obtains the following expressions:

$$\tau_{yx}(t) = -G\int_{-\infty}^{t} e^{-(t-t')/\lambda}\dot{\gamma}_0(1 - H(t'))\,dt', \tag{3.65}$$

$$\tau_{xx}(t) = -2\int_{-\infty}^{t} e^{-(t-t')/\lambda}\tau_{yx}(t')\dot{\gamma}_0(1 - H(t'))\,dt'. \tag{3.66}$$

Equation 3.65 is integrated by parts to obtain $\tau_{yx}(t)$:

$$\tau_{yx}(t) = -\eta\dot{\gamma}_0 e^{-t/\lambda}. \tag{3.67}$$

Now one substitutes Eq. 3.67 into Eq. 3.66 to obtain:

$$\tau_{xx}(t) = -2\eta(\dot{\gamma})\lambda\dot{\gamma}_0^2 e^{-t/\lambda}. \tag{3.68}$$

It is indicated by these equations that on cessation of flow the stresses decay exponentially with time. At higher $\dot{\gamma}_{yx}$, the stresses relax faster as a result of a decrease in $\lambda$. This is in agreement with experimental observations. However, the model predicts that $\tau_{yx}$ and $\tau_{xx}$ relax at the same rate, which is not what is observed in general.

**EXAMPLE 3.2**
**Predictions of the PTT Model in Steady Simple Shear and Steady Shear-free Flow**

(a) Determine the predictions for the material functions in steady simple shear flow for the PTT model.

(b) Do the same for steady shear-free flow, and then show the results for simple elongational flow.

**Solution**

(a) The PTT model requires additional manipulations relative to the WM model. Quantities such as $tr\tau$ and $\dot{\gamma}\cdot\tau$ must be determined. $\tau$ is given in Eq. 3.48. The trace of $\tau$ (i.e., $tr\tau$) is the sum of the diagonal components of $\tau$, which is:

$$tr\tau = \tau_{xx} + \tau_{yy} + \tau_{zz}. \tag{3.69}$$

Hence the first term in Eq. 3.45 is;

$$Z(tr\tau)\tau = \exp[-\varepsilon\lambda(\tau_{xx} + \tau_{yy} + \tau_{zz})/\eta_0]\tau. \tag{3.70}$$

The third term in Eq. 3.45 is:

$$(\xi\lambda/2)[\dot{\gamma}\cdot\tau+\tau\cdot\dot{\gamma}] = \left(\frac{\xi\lambda}{2}\right)\begin{pmatrix} 0 & \dot{\gamma}_0 & 0 \\ \dot{\gamma}_0 & 0 & 0 \\ 0 & 0 & 0 \end{pmatrix}\begin{pmatrix} \tau_{xx} & \tau_{xy} & 0 \\ \tau_{yx} & \tau_{yy} & 0 \\ 0 & 0 & \tau_{zz} \end{pmatrix}$$

$$+\left(\frac{\xi\lambda}{2}\right)\begin{pmatrix} \tau_{xx} & \tau_{xy} & 0 \\ \tau_{yx} & \tau_{yy} & 0 \\ 0 & 0 & \tau_{zz} \end{pmatrix}\begin{pmatrix} 0 & \dot{\gamma}_0 & 0 \\ \dot{\gamma}_0 & 0 & 0 \\ 0 & 0 & 0 \end{pmatrix}$$

$$=\frac{\xi\lambda}{2}\dot{\gamma}_0\begin{pmatrix} 2\tau_{yx} & \tau_{yy}+\tau_{xx} & 0 \\ \tau_{xx}+\tau_{yy} & 2\tau_{xy} & 0 \\ 0 & 0 & 0 \end{pmatrix}. \tag{3.71}$$

Finally we put all the terms together to obtain the following matrix equation:

$$Z(tr\tau)\begin{pmatrix} \tau_{xx} & \tau_{xy} & 0 \\ \tau_{yx} & \tau_{yy} & 0 \\ 0 & 0 & \tau_{zz} \end{pmatrix}+\frac{\xi\lambda\dot{\gamma}_0}{2}\begin{pmatrix} 2\tau_{yx} & \tau_{yy}+\tau_{xx} & 0 \\ \tau_{xx}+\tau_{yy} & 2\tau_{yx} & 0 \\ 0 & 0 & 0 \end{pmatrix}$$

$$-\lambda\dot{\gamma}_0\begin{pmatrix} 2\tau_{yx} & \tau_{yy} & 0 \\ \tau_{yy} & 0 & 0 \\ 0 & 0 & 0 \end{pmatrix}=-\eta_0\begin{pmatrix} 0 & \dot{\gamma}_0 & 0 \\ \dot{\gamma}_0 & 0 & 0 \\ 0 & 0 & 0 \end{pmatrix}. \tag{3.72}$$

From this matrix equation we obtain a set of coupled algebraic equations for the stress tensor components:

$$\tau_{xx}:\quad Z(tr\tau)\tau_{xx}+\xi\lambda\dot{\gamma}_0\tau_{yx}-2\lambda\dot{\gamma}_0\tau_{yx}=0 \tag{3.73}$$

$$\tau_{xy}:\quad Z(tr\tau)\tau_{xy}+\frac{\xi\lambda}{2}\dot{\gamma}_0(\tau_{yy}+\tau_{xx})-\lambda\dot{\gamma}_0\tau_{yy}=-\eta_0\dot{\gamma}_0 \tag{3.74}$$

$$\tau_{yy}:\quad Z(tr\tau)\tau_{yy}+\xi\lambda\dot{\gamma}_0\tau_{xy}=0 \tag{3.75}$$

$$\tau_{zz}:\quad Z(tr\tau)\tau_{zz}=0. \tag{3.76}$$

In that form above the equations are coupled nonlinear algebraic equations, and one would have to use Newton's method to solve them. However, they can be simplified by the fact that for shear flow (this is called a weak flow in the continuum mechanics literature) $tr\tau$ is small and the term $\varepsilon\lambda tr\tau/\eta_0$ approaches zero. Hence $Z(tr\tau)\approx 1.0$ (where $\approx$ means approximately). We also note that $\tau_{xy}=\tau_{yx}$, because the stress tensor is symmetric. Solving the foregoing equations we find that:

$$\tau_{xy}=\frac{-\eta_0\dot{\gamma}_0}{[1+\xi(2-\xi)(\lambda\dot{\gamma}_0)^2]} \tag{3.77}$$

$$\tau_{xx}=\frac{-\lambda\dot{\gamma}_0^2(2-\xi)\eta_0}{[1+\xi(2-\xi)(\lambda\dot{\gamma}_0)^2]} \tag{3.78}$$

$$\tau_{yy}=\frac{+\xi\lambda\eta_0\dot{\gamma}_0^2}{[1+\xi(2-\xi)(\lambda\dot{\gamma}_0)^2]} \tag{3.79}$$

$$\tau_{zz}=0. \tag{3.80}$$

Using the definitions in Section 3.1 for $\eta$, $\Psi_1$, and $\Psi_2$, we obtain the entries that are given in Table 3.1.

**(b)** For simple shear-free flow the approach is the same. The matrix equation for the stresses is:

$$Z(tr\tau)\begin{pmatrix} \tau_{xx} & 0 & 0 \\ 0 & \tau_{yy} & 0 \\ 0 & 0 & \tau_{zz} \end{pmatrix}-\lambda\dot{\varepsilon}_0\begin{pmatrix} -(1-b)\tau_{xx} & 0 & 0 \\ 0 & -(1-b)\tau_{yy} & 0 \\ 0 & 0 & 2\tau_{zz} \end{pmatrix}$$

$$+\xi\lambda\dot{\varepsilon}_0\begin{pmatrix} -(1+b)\tau_{xx} & 0 & 0 \\ 0 & -(1-b)\tau_{yy} & 0 \\ 0 & 0 & 2\tau_{zz} \end{pmatrix}$$

$$=-\eta_0\dot{\varepsilon}_0\begin{pmatrix} -(1+b) & 0 & 0 \\ 0 & -(1-b) & 0 \\ 0 & 0 & 2 \end{pmatrix}. \tag{3.81}$$

In this case the factor $Z(tr\tau)$ does not reduce to one and hence must be maintained. The stress components are:

$$\exp\left[\frac{-\varepsilon\lambda}{\eta_0}(\tau_{xx}+\tau_{yy}+\tau_{zz})\right]\tau_{xx}+\lambda\dot{\varepsilon}_0(1+b)\tau_{xx}$$
$$-\xi\lambda\dot{\varepsilon}_0(1+b)\tau_{xx}=+\eta_0\dot{\varepsilon}_0(1+b) \tag{3.82}$$

$$\exp\left[\frac{-\varepsilon\lambda}{\eta_0}(\tau_{xx}+\tau_{yy}+\tau_{zz})\right]\tau_{yy}+\lambda\dot{\varepsilon}_0(1+b)\tau_{yy}$$
$$-\xi\lambda\dot{\varepsilon}_0(1-b)\tau_{yy}=+\eta_0\dot{\varepsilon}_0(1-b) \tag{3.83}$$

$$\exp\left[\frac{-\varepsilon\lambda}{\eta_0}(\tau_{xx}+\tau_{yy}+\tau_{zz})\right]\tau_{zz}-2\lambda\dot{\varepsilon}_0\tau_{zz}$$
$$+2\xi\lambda\dot{\varepsilon}_0\tau_{zz}=-2\eta_0\dot{\varepsilon}_0 \tag{(3.84)}$$

For steady elongational flow, $b=0$, $\dot{\varepsilon}_0>0$, and we see that $\tau_{xx}=\tau_{yy}$. Hence, the equations become:

$$\exp\left[\frac{-\varepsilon\lambda}{\eta_0}(2\tau_{xx}+\tau_{zz})\right]\tau_{xx}+\lambda\dot{\varepsilon}_0(1-\xi)\tau_{xx}=-\eta_0\dot{\varepsilon}_0 \tag{3.85}$$

$$\exp\left[\frac{-\varepsilon\lambda}{\eta_0}(2\tau_{xx}+\tau_{zz})\right]\tau_{zz}+2\lambda\dot{\varepsilon}_0(1-\xi)\tau_{zz}=-2\lambda_0\dot{\varepsilon}_0. \tag{3.86}$$

In this case we cannot obtain an analytical expression for $\bar{\eta}$, but Eqs. 3.85 and 3.86 must be solved numerically for specified values of $\dot{\varepsilon}_0$, $\varepsilon$, $\lambda$, and $\eta_0$.

### EXAMPLE 3.3
**Material Parameters for the Phan Thien–Tanner Model**

Obtain the material parameters (i.e., $\eta_0$, $\lambda$, $\xi$, and $\varepsilon$) in the PTT model for LDPE (NPE 953) at 170°C for which rheological data are given in Appendix A.1.

(a) Obtain the parameters by any method desired.

(b) Use the IMSL nonlinear least squares subroutine RNLIN found on the disk containing the IMSL subroutines (see Appendix D.3).

(c) Compare how well the methods fit the data, and comment on the ability of the PTT model to fit the data.

**Solution**

**(a)** From the stress quantities given in Eqs. 3.77 to 79 the viscosity function is:

$$\eta=\frac{\eta_0}{1+\xi(2-\xi)(\lambda\dot{\gamma}_0)^2}, \tag{3.87}$$

and $N_1$ is given by:

$$N_1 = \frac{-2\eta_0 \lambda \dot{\gamma}_0^2}{1 + \xi(2-\xi)(\lambda \dot{\gamma}_0)^2}. \qquad (3.88)$$

Working with Eq. 3.87, we can obtain $\eta_0$ directly from the viscosity data in Appendix A.1. In this case we take $\eta_0$ to be 23,100 Pa·s (i.e., we used $\eta$ at the lowest $\dot{\gamma}_0$). To obtain $\xi$ and $\lambda$, we plot $\eta/\eta_0$ versus $\lambda \dot{\gamma}_0$, which is a dimensionless shear rate, for different values of $\xi$. The dimensionless viscosity curves (or "master curves") for each value of $\xi$ are then superimposed on plots of $\eta/\eta_0$ versus $\dot{\gamma}_0$ or $\eta^*/\eta_0$ versus $\omega$ based on data taken from Appendices Tables A.1 and A.2, respectively. The curves are then shifted horizontally until the "best fit" is obtained between the experimental results and the theoretical curves. The parameter that shifts the experimental data on to the master curve is $\lambda$. Curves for three different values of $\lambda$ are shown in Fig. 3.16. The best fit seems to be obtained by selecting a value of $\lambda = 0.5$.

In Fig. 3.17 we plot predicted values of $N_1$ (Eq. 3.88) versus $\dot{\gamma}_0$ and compare them against experimental values of $N_1$ and $2G'$. At low $\dot{\gamma}_0$ the predicted values of $N_1$ agree well with the experimental values. The agreement between $N_1$ and $2G'$ is not good, and hence one would not expect good agreement between the predicted values of $N_1$ and $2G'$ at high $\dot{\gamma}_0$ (for an alternate relation between $N_1$ and $G'$ see Eq. 3.118).

It is difficult to fit Eq. 3.87 to the viscosity data because the shape of the curve is not quite what is observed experimentally. In particular, the slope of $\ln \eta$ versus $\ln \dot{\gamma}_0$ is less than $-1.0$, which is physically unrealistic (this means that $\tau_{yx}$ passes through a maximum, and for flow through a capillary it would mean there could be more than one flow rate possible for a given pressure drop). However, in an effort to improve the fit of Eq. 3.87 to the viscosity data, we use nonlinear regression analysis to obtain a set of coefficients that fit the data in a least-squares sense.

**(b)** The method used in Part **a** to obtain the coefficients is in essence a least-squares approach, but it is somewhat subjective. In this solution, we use the IMSL subroutine RNLIN to find $\eta_0$, $\xi$, and $\lambda$. A description of the subroutine is given in Appendix D.3, and the calling program is shown in Table 3.2. The values for $\eta_0$, $\xi$, and $\lambda$ obtained in a least-squares sense, called theta (1), theta (2), and theta (3) in the computer code, respectively, but listed as theta in the solution, are 23,100 Pa·s, 0.038, and 5.475 s, respectively. It is also observed that the subroutine found it difficult to fit the PTT viscosity function to the experimental data. The function does not fit the data well as the sum of the squares of the errors (SSE) is very large. Using the values for $\eta_0$, $\xi$, and $\lambda$, Eq. 3.87 is evaluated, and $\eta/\eta_0$ is plotted versus $\lambda \dot{\gamma}_0$ in Fig. 3.16. Here we see that the fit is very good at low $\dot{\gamma}_0$, but at high $\dot{\gamma}_0$ there is considerable disagreement. Furthermore, this curve is very close to the "eyeballed" curve for $\xi = 0.05$ and $\lambda = 2.7$.

Values of $N_1$ calculated from Eq. 3.88 using the coefficients determined by means of the subroutine RNLIN are plotted in Fig. 3.17 and compared against the experimental values. Values based on Eq. 3.88 and the eyeballed values of $\eta_0$, $\xi$, and $\lambda$ are also plotted in Fig. 3.17. The statistically determined values lead to a better prediction of $N_1$ at low $\dot{\gamma}_0$, but the values seem to converge at higher $\dot{\gamma}_0$.

The parameter $\varepsilon$ is determined from extensional data and requires the solution of ordinary differential equations for various values of $\varepsilon$. This is done in the solution to Pr. 3C.3.

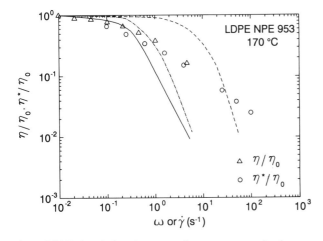

**Figure 3.16** Reduced viscosity versus shear rate or angular frequency ($\omega$) for LDPE (NPE 953) at 170°C. The curves represent the PTT model viscosity function for three sets of parameters: (—) $\xi = 0.05$, $\lambda = 2.7$; (---) $\xi = 0.05$, $\lambda = 0.5$; (·—·) $\xi = 0.038$, $\lambda = 5.47$.

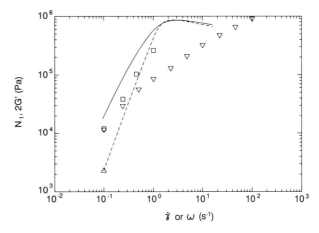

**Figure 3.17** Comparison of predicted values of $N_1$ (PTT model) with experimental values of $N_1$ and $2G'$ for LDPE (NPE 953) at 170°C: (—) $\xi = 0.038$, $\lambda = 5.47$ s; (---) $\xi = 5.00E\text{-}02$, $\lambda = 2.70$ s; ($\nabla$) $2$ $G'$, ($\square$) $N_1$.

**TABLE 3.2 Program Listing for Calling the IMSL Subroutine RNLIN (PTTCOEF.FOR)**

```
C********************ABSTRACT************************
C  THIS PROGRAM SOLVES EXAMPLE 3.3(B) USING THE IMSL SUBROUTINE
C  RNLIN. THE DESCRIPTION OF RNLIN IS GIVEN IN APP. D.3. NONLINEAR
C  REGRESSION IS USED TO FIND THE PARAMETERS IN THE PTT MODEL FOR
C  VISCOSITY GIVEN IN TABLE 3.1. THE VISCOSITY FUNCTION IS FIT TO THE
C  STEADY AND DYNAMIC VISCOSITY DATA GIVEN IN APP. A.1 FOR LDPE.
C
      INTEGER  LDR, NOBS, NPARM
      PARAMETER (NOBS=17, NPARM=3, LDR=NPARM)
```

**TABLE 3.2** *continued*

```
      INTEGER   IDERV, IRANK, NOUT
      REAL DFE, R(LDR,NPARM), SSE, THETA(NPARM)
      EXTERNAL   EXAMPL, RNLIN, UMACH, WRRRN
C
      DATA THETA/23100.0, 0.05, 0.5/
C
      CALL UMACH (2,NOUT)
C
      IDERIV = 1
      CALL RNLIN (EXAMPL, NPARM, IDERIV, THETA, R, LDR, IRANK, DFE,
     &    SSE)
      WRITE (NOUT,*) 'THETA = ', THETA
      WRITE (NOUT,*) 'IRANK = ', IRANK, ' DFE,= ', DFE, ' SSE = ',
     &    SSE
      CALL WRRRN ('R', NPARM, NPARM, R, LDR, 0)
      END
C
      SUBROUTINE EXAMPL (NPARM,THETA, IOPT, IOBS, FRQ, WT, E, DE,
     &    IEND)
      INTEGER   NPARM, IOPT, IOBS, IEND
      REAL   THETA(NPARM), FRQ, WT, E, DE(NPARM)
C
      INTEGER NOBS
      PARAMETER (NOBS=17)
      REAL   XDATA(NOBS), YDATA(NOBS)
C
C**********INPUT VISCOSITY VERSUS SHEAR RATE DATA*******
C
      DATA YDATA/2.310E+4, 2.215E+4, 2.013E+4, 1.693E+4, 1.437E+4,
     &    1.122E+4, 8.192E+3, 1.461E+4, 1.131E+4, 8.597E+3, 6.349E+3,
     &    4.542E+3, 3.123E+3, 2.108E+3, 1.394E+4, 9.008E+2, 5.651E+2/
      DATA XDATA/.01, .02154, .0464, .1, .215, .464, 1.0, .1, .215,
     &    .464, 1.0, 2.154, 4.641, 10.0, 21.54, 46.41, 100.0/
C
      IF (IOBS .LE. NOBS) THEN
      WT = 1.0E0
      FRQ=1.0E0
      IEND = 0
      IF (IOPT .EQ. 0) THEN
        E = YDATA(IOBS) - THETA(1)/(1.0+THETA(2)*(2.0-THETA(2))*
     &    THETA(3)**2*XDATA(IOBS)**2)
      ELSE
        DE(1) = -1.0/(1.0+THETA(2)*(2.0-THETA(2))*(THETA(3)**2)*
     &    (XDATA(IOBS)**2)
        DE(2) = THETA(1)*2.0*(THETA(3)**2)*(XDATA(IOBS)**2)*(1.0-
     &    THETA(2))/(1.0 + THETA(2)*(2.0-THETA(2))*(THETA(3)**2)*
     &    (XDATA(IOBS)**2))
        DE(3) = 2.0*THETA(1)*THETA(3)*THETA(2)*(2.0-THETA(2))*
     &    (XDATA(IOBS)**2)/((1.0+THETA(2)*(2.0-THETA(2))*THETA(3)**2
     &    *XDATA(IOBS)**2)**2)
      END IF
      ELSE
      IEND = 1
      END IF
      RETURN
      END
```

Finally, we recognize that the PTT model in its present form does not fit the viscosity data well because it predicts a function for $\eta$ that is too highly dependent on shear rate. By adding a Newtonian viscosity to the function in Eq. 3.87, the fit can be made better and $\eta$ does not shear-thin in an unrealistic manner.

It is clear that viscoelastic fluids require a constitutive equation that is capable of describing time-dependent rheological properties, normal stresses, elastic recovery, and an extensional viscosity that is independent of the shear viscosity. It is not clear at this point exactly how a constitutive equation for a viscoelastic fluid, when coupled with the equations of motion, leads to the prediction of behavior (i.e., velocity and stress fields), that is any different from that calculated for a Newtonian fluid. As the constitutive relations for polymeric fluids lead to nonlinear differential equations that cannot be solved easily, it is difficult to show how their use affects calculations. Furthermore, it is not clear how using a constitutive equation that predicts normal stress differences leads to predictions of velocity and stress fields that are significantly different from those predicted by using a Newtonian fluid model. Finally, there are numerous possibilities of constitutive relations to choose from. The question is, then, when and how one uses a viscoelastic constitutive relation in design calculations, especially when sophisticated numerical methods such as finite element methods are not available to the student at this point. For the time being, it appears that the most important material functions that the model must describe accurately are shear and extensional viscosity. The design calculations for which one must use a viscoelastic constitutive relation are those involving shear-free flows, such as film blowing, expansion of a parison in blow molding, squeezing flow, film casting, and fiber spinning. Furthermore, a viscoelastic model can be useful for estimating extrudate swell, which is the expansion of a polymeric fluid on leaving a die. It has been suggested that die swell can be correlated to $N_1$. One commonly used relation proposed by Tanner (1970) is:

$$\frac{D_p}{D_0} = 0.1 + \left[ 1.0 + \frac{1}{2}\left(\frac{N_{1,w}}{2\tau_w}\right)^2 \right]^{\frac{1}{6}}, \tag{3.89}$$

where $D_p$ is the diameter of the extrudate and $D_0$ is the capillary diameter. (Further discussion of extrudate or "die" swell is presented in Chapter 7.) Hence, we will primarily use nonlinear viscoelastic models for estimating values of $N_1$ at high shear rates and when the kinematics are known. It is beyond the level of this book to solve problems in which nonlinear constitutive equations are used in conjunction with equations of motion.

Let us add a few comments about when and under what conditions one must use a nonlinear viscoelastic constitutive equation. At this time it seems that whenever the flow is unsteady in either a Lagrangian ($Dv/Dt \neq 0$) or an Eulerian ($\partial v/\partial t \neq 0$) sense, then viscoelastic effects become important. In the former case one finds flows of this nature whenever nonhomogeneous shear-free flows arise (e.g., flow through a contraction) and in the latter case in the startup of flows. However, even in simple flows such as in capillaries or slit dies, viscoelastic effects can be important, especially if the residence time of the fluid in the die is less than the longest relaxation time of the fluid. Then, factors such as stress overshoot could lead to an apparent viscosity that is higher than the steady-state viscosity. In line with these ideas one defines a dimensionless group referred to as the *Deborah number*:

$$\mathrm{De} = \lambda/t_{av}, \tag{3.90}$$

where $t_{av}$ is the process time and $\lambda$ is the relaxation time. If $\mathrm{De} \geq 1$, then transient effects are important. Another place where viscoelastic effects are important would be in injection molding where the stresses in the melt may relax slowly relative to the heat transfer rate, and in which case residual stresses are frozen into the part. Hence, even when the flows are steady, the viscoelastic nature of the fluid can be important. Certainly for the most part to consider viscoelastic effects in design calculations, it will be necessary to use numerical techniques.

## 3.3. RHEOMETRY

Of all the transport properties of polymeric materials, the rheological properties are probably the most important to the design of polymer processes. Whereas the other transport properties such as thermal conductivity, heat capacity, and density remain nearly constant with changes in molecular structure such as molecular weight and branching, slight changes in molecular structure can alter the rheological properties and, hence, processing behavior of a polymer significantly. As one can imagine, variations in molecular structure from batch to batch of a polymer are quite common. Hence, one cannot expect to have available a set of rheological data that could be used for design calculations such

as would be found in a handbook. It is necessary that the rheological behavior of polymeric fluids be determined on a regular basis. For this reason it is necessary to know how rheological properties are measured.

It is not our intention here to give an in-depth description of the techniques used in measuring rheological properties of polymer melts as these details can be found elsewhere (Dealy 1982; Walters 1975). The goal is to make sure that the student be aware of at least the most common methods, of how data are manipulated to obtain material functions, and of the limitations of various techniques. We discuss methods for measuring shear flow properties first, followed by methods for measuring shear-free flow properties.

### 3.3.1.  SHEAR FLOW MEASUREMENTS

Measurements of rheological properties at low shear rates are usually carried out in rotary rheometers such as the cone-and-plate (C-P) or plate–plate (P–P) systems shown in Figures 3.18 and 3.19, respectively. In rotary rheometers one of the members of the system is driven and transmits forces through the fluid to the bottom plate. The torque, $T$, and the normal force, $F$, are recorded at the bottom member by means of transducers. The cone-and-plate has the advantage that the shear rate, $\dot{\gamma}$, is nearly uniform through the gap, and hence the material properties of the polymeric fluid can be measured at each $\dot{\gamma}$ directly. The equations for determining the material functions, $\eta$ and $\Psi_1$, are given in the caption of Fig. 3.18. Because $\dot{\gamma}$ is uniform throughout the gap, it is possible to use the C-P to measure the transient response of polymeric fluids. For the case of the P–P device (Fig. 3.19), $\dot{\gamma}$ is found to vary with the distance $r$ from the center of the plates. Hence, one must make a series of measurements at various shear rates before obtaining values of $\eta$ and $\Psi_1 - \Psi_2$ at specific values of $\dot{\gamma}$. For the C-P device the maximum $\dot{\gamma}$ for which measurements are possible (the melt usually fractures and comes out of the gap) is about $10\,\mathrm{s}^{-1}$ and slightly higher values of $\dot{\gamma}$ are possible with the P–P device.

The capillary rheometer (Fig. 3.20) is commonly used to obtain $\eta$

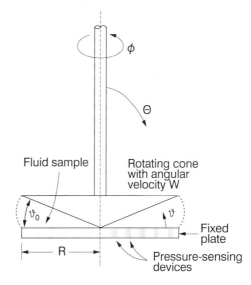

**Figure 3.18** Cone-and-plate rheometer. The cone turns with an angular velocity, $W$. The torque, $T$, and normal force, $F$, measurements are used to determine $\eta$ and $\Psi_1$, respectively, as given:

$$\eta = \frac{3T}{2\pi R^3 \dot{\gamma}}; \qquad \Psi_1 = \frac{2F}{\pi R^2 \dot{\gamma}^2}; \qquad \dot{\gamma} = \frac{W}{\phi_0}.$$

In some instances pressure transducers are mounted along the bottom plate to measure the pressure distribution from which $\Psi_2$ is obtained.

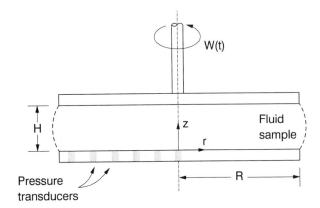

**Figure 3.19** Plate-plate rheometer. The torque, $T$, and normal force, $F$, measurements are used to determine $\eta$ and $\Psi_2 - \Psi_2$:

$$\eta = (T/2\pi R^3 \dot{\gamma}_R)[3 + d\ln(T/2\pi R^3)/d\ln\dot{\gamma}_R].$$
$$\Psi_1 - \Psi_2 = (F/\pi R^2 \dot{\gamma}_R^2)[2 + d\ln(F/\pi R^2)/d\ln\dot{\gamma}_R]$$
$$\dot{\gamma}_R = \frac{WR}{H}.$$

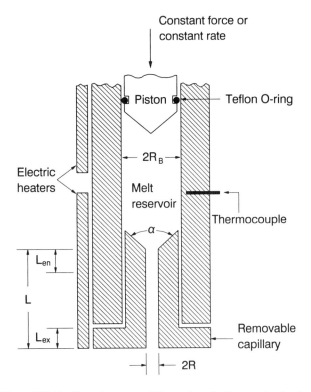

**Figure 3.20** Capillary rheometer. Polymer is melted by conduction in the reservoir and then pushed through the capillary by the plunger. Viscosity data are obtained from $\Delta P$ and $Q$ measurements.

at high shear rates. Basically, the device consists of a barrel for melting the polymer and a plunger that pushes the melt through the capillary. The data obtained from this device consist of the pressure required to push the melt through the capillary and the volumetric flow rate (plunger speed and cross-sectional area). Two corrections are applied to these data. First, the pressure drop must be corrected for the additional pressure required for the melt to pass through the contraction

between the barrel and the capillary. For any fluid, the wall shear stress is given by:

$$\tau_R = \left(\frac{-dp}{dz}\right)\frac{R}{2}, \tag{3.91}$$

where $dp/dz$ is the pressure gradient in the capillary. Usually $dp/dz$ is approximated by $-\Delta P/L$, where $\Delta P$ is the pressure drop across the whole capillary including the entrance, and $L$ is the capillary length. For a Newtonian fluid the pressure gradient is nearly constant over the length of the capillary. For polymeric fluids the pressure drop is shown schematically in Fig. 3.21. The pressure gradient is nonlinear for polymeric materials, and approximating it as $-\Delta P/L$ would lead to large errors in the determination of $\tau_R$. The difference between the pressure extrapolated from the linear region and the true pressure is called the entrance pressure, $\Delta P_{ent}$. There may be residual pressure at the die exit, called the exit pressure, $\Delta P_{ex}$, but it is quite small relative to $\Delta P_{ent}$ and hence is neglected. If there is additional pressure at the die exit, then the method used to obtain $\Delta P_{ent}$ actually includes $\Delta P_{ex}$. The total pressure correction for exit and entrance regions is called the end pressure, $\Delta P_{end}$, that is,

$$\Delta P_{end} = \Delta P_{ex} + \Delta P_{ent}. \tag{3.92}$$

The true wall shear stress, $\tau_R$, is then obtained by plotting the total pressure, $\Delta P_{tot}$, versus $L/D$ at each value of $\dot{\gamma}$ for several $L/D$'s (see Fig. 3.22). The extrapolation of $\Delta P_{tot}$ to $L/D = 0$ is $\Delta P_{end}$. One now obtains $\tau_R$ as follows:

$$\tau_R = \left(\frac{\Delta P_{tot} - \Delta P_{end}}{L}\right)\frac{R}{2}. \tag{3.93}$$

One may also correct for $\Delta P_{end}$ by calculating the equivalent die length required to produce $\Delta P_{tot}$. Referring again to Fig. 3.22, one finds the additional length by extrapolating $\Delta P_{tot}$ versus $L/D$ to $\Delta P_{tot} = 0$. The additional length required is given as a factor, $N_{ent}$, times the radius:

$$(L/D)_c = (L + N_{ent}D/2)/D \tag{3.94}$$

where $(L/D)_c$ is the corrected value of $L/D$. $\tau_R$ is now given as:

$$\tau_R = \Delta P_{tot}/(L/D)_c/4. \tag{3.95}$$

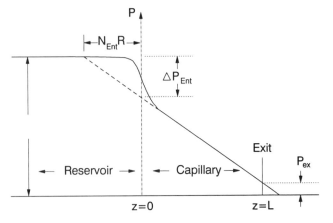

**Figure 3.21** Pressure profile in a capillary rheometer. The various pressures are defined in the figure including the exit pressure, $P_{ex}$, and the entrance pressure, $\Delta P_{ent}$.

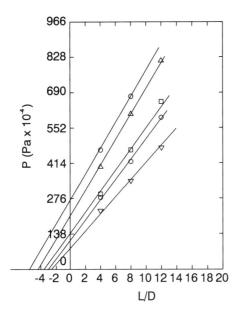

**Figure 3.22** Total pressure drop, $\Delta P_{tot}$, versus $L/D$ for various shear rates for PP at 200°C: ($\triangledown$) 131; ($\bigcirc$) 235; ($\square$) 289; ($\triangle$) 569; (o) 724 s$^{-1}$. These plots are sometimes referred to as *Bagley plots*.

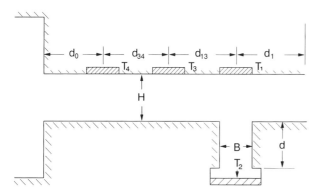

**Figure 3.23** Slit-die rheometer showing the position of the pressure transducers ($T_i$).

Because the velocity profile is nonparabolic, one must correct the apparent wall shear rate, $\dot{\gamma}_a$, defined as $4Q/\pi R^3$. The correction is obtained by integrating the volumetric flow rate, $Q$, by parts:

$$Q = 2\pi \int_0^R v_z(r) r \, dr \qquad (3.96)$$

and then differentiating with respect to $\tau_R$ using Leibnitz's rule to give (see Pr. 3B.1):

$$\dot{\gamma}_w = \frac{\dot{\gamma}_a}{4}\left(3 + \frac{d \ln \dot{\gamma}_a}{d \ln \tau_R}\right). \qquad (3.97)$$

Hence, by plotting $\tau_R$ versus $\dot{\gamma}_a$ on a ln–ln plot, one obtains the reciprocal of the required correction factor. It turns out that this value is just $1/n$, where $n$ is the power-law index.

Slit-die rheometers (See Fig. 3.23) are useful devices for measuring the viscosity of polymer melts because it is possible to measure the

pressure gradient directly. The geometry is that of two flat plates with a rectangular cross section. If the aspect ratio, $W/H$, is greater or equal to 10, then there is no side wall effect. The wall shear stress, $\tau_w$, is then:

$$\tau_w = \left(\frac{-dp}{dz}\right)\frac{H}{2} = \frac{P_3 - P_1}{d_{31}}\frac{H}{2}, \qquad (3.98)$$

where $P_3$ and $P_1$ are the pressures recorded by transducers $T_3$ and $T_1$, respectively, and $d_{13}$ is the distance between the center of the transducers. The wall shear rate, $\dot{\gamma}_w$, is obtained from the following relation:

$$\dot{\gamma}_w = \frac{\dot{\gamma}_a}{3}\left[2 + \frac{d \ln \dot{\gamma}_a}{d \ln \tau_w}\right], \qquad (3.99)$$

where $\dot{\gamma}_a = 6Q/WH^2$ (This is just the wall shear rate for a Newtonian fluid) for flow through flat plates.

The slit-die rheometer also offers the possibility of obtaining values of $N_1$ at high shear rates. The method is based on the measurement of a quantity called the *hole pressure*, $P_H$, (Kaye et al., 1968), which is the difference of pressures $P_1$ and $P_2$, where $P_2$ is the pressure measured by transducer $T_2$ mounted at the bottom of a rectangular slot placed perpendicular to the flow direction, that is:

$$P_H = P_1 - P_2. \qquad (3.100)$$

$N_1$ is obtained from the following equation (Baird, 1975):

$$N_1 = 2\tau_w(dP_H/d\tau_w). \qquad (3.101)$$

Hence from data of $P_H$ versus $\tau_w$ one can obtain $N_1$ as a function of $\tau_w$. (Accurate measurements of $P_H$ are quite difficult, and great care must be taken to make these measurements.)

### 3.3.2. SHEAR-FREE FLOW MEASUREMENTS

Two techniques for measuring uniaxial extensional viscosity of polymer melts are shown in Fig. 3.24. In the first technique (Ballman method) polymer melt is either glued or clamped at both ends, and then one end is moved in such a manner as to either generate a uniform extension rate, $\dot{\varepsilon}$, or a constant tensile stress. In the Meissner method, both ends of the melt are pulled at a constant velocity to achieve either a uniform extension rate or a constant stress.

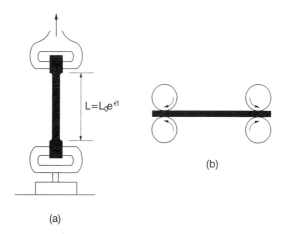

**Figure 3.24** Two methods for generating uniaxial extensional flow.

In the Ballman method, to generate a uniform $\dot{\varepsilon}$ throughout the sample, one end of the sample must be deformed such that the length of the sample is increased exponentially with time: that is, $L = L_0 e^{\dot{\varepsilon}t}$. An apparatus based on the Ballman method is shown in Fig. 3.25. The sample must be supported in an oil of similar density as shown in Fig. 3.25. The limitations of the technique include (1) the availability of a suitable adhesive for gluing the sample to metal clips; (2) the ability of the sample to deform uniformly without necking; (3) the availability of an oil of similar density as the polymer melt (unless the sample is deformed in a horizontal plane); and (4) low strains ($\dot{\varepsilon}t$) of only about 3.9. In spite of these limitations, valuable data, especially for polyolefins, can be obtained.

The Meissner method has several advantages and disadvantages relative to the Ballman method. First, it is possible to reach very high strains (of the order of 7.0). Second, the sample is usually deformed horizontally, so that the matching of the oil density with that of the polymer melt is not as critical. Finally, finding a suitable glue is not necessary. On the other hand, the construction of the apparatus is more complicated and expensive. Larger samples are required, and they must be nearly free of inhomogeneities.

With both methods the technique for obtaining the extensional viscosity, $\eta^+(\dot{\varepsilon}, t)$, is similar. With the assumption that the surroundings of the sample are at atmospheric pressure, $p_a$, the total force per unit area exerted by the load cell and atmospheric pressure on the sample must be balanced by $\Pi_{zz}$, so that:

$$\Pi_{zz} = -(F(t)/A(t)) + p_a, \tag{3.102}$$

where $A$ is the instantaneous cross sectional area of the sample. A force balance in the radial direction gives the normal stress difference as a function of time:

$$\tau_{zz} - \tau_{rr} = -F(t)/A(t). \tag{3.103}$$

## Extensional Rheometer

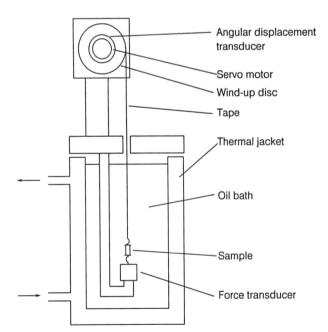

- Angular displacement transducer
- Servo motor
- Wind-up disc
- Tape
- Thermal jacket
- Oil bath
- Sample
- Force transducer

**Figure 3.25** Extensional rheometer for polymer melts based on the Ballman method. The length of the sample is increased exponentially with time to generate a constant extension rate.

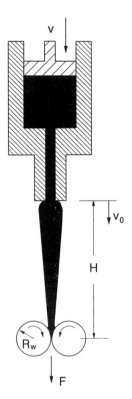

**Figure 3.26** Rheotens apparatus for estimating the uniaxial extensional viscosity.

$\bar{\eta}^+$ is then obtained from

$$\bar{\eta}^+ = -\frac{\tau_{zz} - \tau_{rr}}{\dot{\varepsilon}} = \frac{F(t)e^{\dot{\varepsilon}t}}{A_0\dot{\varepsilon}}, \tag{3.104}$$

where $A_0$ is the initial cross-sectional area.

In the event the devices for measuring $\bar{\eta}$ are not available, there are two methods for obtaining approximate values of $\bar{\eta}$. The first method is based on the fiber-spinning technique as shown in Fig. 3.26 (the device is known as a *rheotens*). Without any discussion of the theory the extension rate is given by:

$$\dot{\varepsilon} = \frac{2\pi v R_w}{H} \ln\left[\frac{8\pi v R_w}{\dot{\gamma}_a R}\right], \tag{3.105}$$

where $v$ is the angular velocity (radians per s.) of the wheel of radius, $R_w$, $H$ is the distance of the capillary to the wheel, $\dot{\gamma}_a$ is the apparent shear rate in the capillary, and $R$ is the radius of the capillary. The normal stress difference is given by:

$$\tau_{zz} - \tau_{rr} = -8v R_w F/R^3 \dot{\gamma}_a, \tag{3.106}$$

where $F$ is the tensile force determined by the tension transmitted through the melt strand to the takeup wheels. Finally, the stretch ratio, $\Lambda$, is given as

$$\Lambda = 8\Pi v R_w/R\dot{\gamma}_a. \tag{3.107}$$

Obviously $\bar{\eta}$ is obtained from the values calculated using Eqs. 3.105 and 3.106.

The second method for estimating $\bar{\eta}$ is based on entrance pressure data (Cogswell, 1978). The extension rate is given as:

$$\dot{\varepsilon} = \frac{4\dot{\gamma}_a^2 \eta_a}{3(n+1)\Delta P_{ent}}, \qquad (3.108)$$

where $n$ is the power-law index and $\eta_a$ is the apparent viscosity $(\eta_a = \tau_w/\dot{\gamma}_a)$. The normal stress difference is

$$\tau_{zz} - \tau_{rr} = -3(n+1)\Delta P_{ent}/8 \qquad (3.109)$$

and hence

$$\bar{\eta} = 9(n+1)^2 \Delta P_{ent}^2 / 32 \dot{\gamma}_a^2 \eta_a.$$

It must be emphasized that these two methods will give only approximate values for $\bar{\eta}$.

## 3.4. USEFUL RELATIONS FOR MATERIAL FUNCTIONS

In this section three topics are discussed: (1) the molecular weight dependence of the rheological properties; (2) the interrelation between steady shear and dynamic oscillatory shear measurements; and (3) the effect of branching. The importance of the second topic rests on the fact that dynamic oscillatory properties are easier to measure and can be obtained at higher equivalent shear rates than are possible for the steady shear flow properties obtained by means of rotary rheometers.

### 3.4.1. EFFECT OF MOLECULAR WEIGHT

Molecular weight, $M$, has a significant effect on the magnitude of the rheological properties. At low molecular weight, that is, below some critical molecular weight ($M_c$), for flexible chain polymers $\eta_0$ depends on $M$, and on $M$ to the 3.4 power above $M_c$:

$$\eta_0 \propto M \qquad (M < M_c)$$

$$\eta_0 \propto M^{3.4} \qquad (M > M_c). \qquad (3.110)$$

The 3.4 power dependence has been observed experimentally and predicted theoretically. Furthermore, the primary normal stress difference coefficient in the limit as $\dot{\gamma}$ goes to zero, $\Psi_{1,0}$, is observed to be proportional to $M$ raised to the 7.0 power, that is,

$$\Psi_{1,0} \propto \eta_0^2 \propto M^{7.0}. \qquad (3.111)$$

Usually $M$ is replaced by $\bar{M}_w$, which is the weight-average molecular weight (which is the second moment of the molecular weight distribution). In the case of $\Psi_{1,0}$ there is not as much experimental confirmation as there is for $\eta_0$. For rodlike molecules there is some evidence that the following relations hold (Baird & Ballman, 1979):

$$\eta_0 \propto \bar{M}_w^{6.8} \qquad (M > M_c)$$

$$\Psi_{1,0} \propto \bar{M}_w^{13.0}. \qquad (3.112)$$

The change in the linear dependence of $\eta_0$ on $M$ to the 3.4 power dependence for flexible chain polymers is believed to be due to the formation of an "entanglement" network or temporary physical junctions between the polymer chains. In the case of rod-like molecules, the hindrance of free rotation of the rodlike molecule by neighboring molecules serves as the entanglements.

In addition to the dependence of the magnitude of $\eta_0$ and $\Psi_{1,0}$ on $M$, the onset of shear-thinning behavior is affected by $M$. In particular, as $M$ increases, the shear rate at which shear-thinning

behavior starts, $\dot{\gamma}_0$, decreases. The relation between $\dot{\gamma}_0$ at $M_1$ and $M_2$ where $M_2 > M_1$ is:

$$\dot{\gamma}_0(M_1) = a_M \dot{\gamma}_0(M_2), \qquad (3.113)$$

where

$$a_M = \frac{\eta_0(M_2)}{\eta_0(M_1)}. \qquad (3.114)$$

Temperature leads to a similar effect, and in Section 5.1 we discuss the temperature dependence of $a_T$. Hence, within a given series of the same polymer, it is possible to generate the flow curves at all other molecular weights given the relation between $\eta_0$ and $M$ and the flow curve of a sample at a given $M$. (*Note*: the breadth of the molecular weight distribution must be the same.)

### 3.4.2. RELATIONS BETWEEN LINEAR VISCOELASTIC PROPERTIES AND VISCOMETRIC FUNCTIONS

It has been observed experimentally for many polymers that the magnitude of the complex viscosity, $|\eta^*(\omega)|$, and the shear viscosity, $\eta(\dot{\gamma})$, evaluated at the same values of $\omega$ and $\dot{\gamma}$, respectively, are identical:

$$\eta(\dot{\gamma}) = |\eta^*(\omega)||_{\omega = \dot{\gamma}}. \qquad (3.115)$$

This relation is known as the Cox-Merz rule (Cox & Merz, 1958). When $\eta(\dot{\gamma})$ is not available, the Cox-Merz rule serves as a useful way to obtain $\eta(\dot{\gamma})$, especially for linear polymers (i.e., those without branching). When dealing with filled polymers, polymer blends, or highly branched polymers, the Cox-Merz rule may not hold.

An alternative to the Cox-Merz rule is Gleissele's mirror relation (Gleissele, 1980):

$$\eta(\dot{\gamma}) = \eta^+(t)||_{t = 1/\dot{\gamma}}, \qquad (3.116)$$

where $\eta^+(t)$ is the limiting curve of $\eta^+(\dot{\gamma}, t)$ as $\dot{\gamma} \to 0$. This relation has been tested for a wide variety of polymers, including polyethylene (PE), polyisobutylene (PIB), and silicone oils.

It is desirable to be able to estimate $N_1$, as measurements of this quantity at high $\dot{\gamma}$ are also difficult. At low values of $\dot{\gamma}$ and $\omega$ it is observed for a number of polymers that:

$$(N_1/2)||_{\dot{\gamma} = \omega} = G'. \qquad (3.117)$$

At higher shear rates it is observed that this relation fails for some polymers. Laun (1986) suggested another empiricism for $N_1$ that seems to fit data over a wider range of shear rates:

$$N_1 = 2\omega\eta''(\omega)\left[1 + \left(\frac{\eta''}{\eta'}\right)^2\right]^{0.7}\Bigg|_{\omega = \dot{\gamma}}, \qquad (3.118)$$

where $\eta'$ is the dynamic viscosity and $\eta''$ is associated with the elastic energy stored per cycle of deformation (it is noted that $G'$, $G''$, $\eta'$, and $\eta''$ are interrelated with $G' = \omega\eta''$). The empiricism in Eq. 3.118 has been tested for PS, LDPE, HDPE, and PP, and the agreement between $N_1$ and the value estimated from linear viscoelastic data was found to be excellent (Laun, 1986).

### 3.4.3. BRANCHING

*Branching* is known to have a significant effect on the rheology of polymeric fluids, especially on the extensional behavior. Polyethylene is known to have various degrees of branching depending on the method

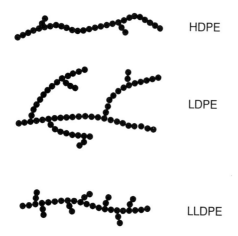

**Figure 3.27** Branching characteristics exhibited by various types of polyolefins.

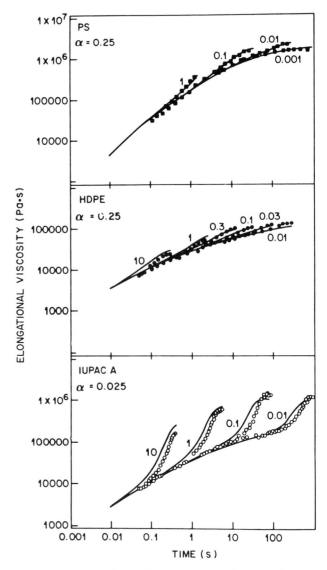

**Figure 3.28** Extensional growth viscosity versus time for polystyrene (*top*), LDPE, and HDPE. (Reprinted by permission of the publisher from S. A. Khan, R. K. Prud'homme, and R. G. Larson, "Comparison of the Rheology of Polymer Melts in Shear, and Biaxial and Uniaxial Extensions," *Rheol. Acta, 26,* 44, 1987.)

of polymerization as shown in Fig. 3.27. LDPE is believed to have very long branches, and LLDPE is believed to have numerous short branches. On the other hand, HDPE has only a few short branches. Values of $\bar{\eta}^+$ are shown for LDPE and HDPE in Fig. 3.28. Here we see that $\bar{\eta}^+$ for LDPE deviates drastically from the linear viscoelastic limit, whereas for HDPE it shows only a slight deviation. LLDPE tends to approach equilibrium with very little deviation from the linear viscoelastic behavior, much in the same manner reported for PS in Fig. 3.28. It is not certain how the shear viscosity is affected by branching, but it is believed that $\eta_0$ for the same molecular weight polymer is enhanced by branching.

In summary, with the relations given in this section it is possible to obtain steady shear rheological data from linear viscoelastic data over a wide range of shear rates by using Eqs. 3.116 and 3.118. Furthermore, within a given series of polymers of different molecular weights it is possible to obtain $\eta(\dot{\gamma})$ and $N_1(\dot{\gamma})$ data at any $M$ using Eq. 3.114 and data at two values of $M$. Branching has a significant effect on the extensional rheology of polymer melts but there are no specific correlations available between the rheology and the degree and length of the branches.

## 3.5. RHEOLOGICAL MEASUREMENTS AND POLYMER PROCESSABILITY

The emphasis so far in this chapter has been on quantitative relations between stress and deformation rate or constitutive equations and methods for measuring rheological properties. It is clear that any attempts to carry out quantitative design work using a nonlinear constitutive equation quite often meets with mathematical difficulties. However, certain rheological measurements of a polymeric fluid may provide a tool for assessing differences in processability when most standard methods fail. To illustrate this idea let us consider a case study (Meissner, 1979) concerned with the effect of branching on the processing performance of LDPE.

The following case study is about processing LDPE using the technique of film blowing (Meissner, 1979). As shown in Table 3.3 two of the resins (identified as *B* and *C*) could be drawn down at the same critical rate of 23 m/min, leading to a film thickness at break of 10 μm. Resin A, on the other hand, could not be drawn down as much (critical draw down rate of 13 to 18 m/min depending on which group made the measurements). Once a processing difference was detected the next goal was to characterize the three resins using standard techniques to

determine if any differences could be detected. As far as the molecular weight distribution was concerned (Fig. 3.29), the samples were identical. The standard rheological measurements usually made in industry as shown in Table 3.4 provided no clue as to the differences in the samples. For example, it was indicated by the melt flow index* that *A* and *B* are similar. It was suggested by the melt memory index, which is a measure of die swell measured under specified conditions, that *A* and *C* were similar polymers. It was indicated by the zero shear viscosity that *A* and *B* were identical but that *C* was slightly different. Hence,

---

*The melt flow index (MFI) is a characterization parameter provided by companies and is the amount of polymer in grams that passes through a capillary of specified radius in 10 minutes when subjected to a fixed pressure drop. More details are found regarding this topic in Pr. 3A.1.

**TABLE 3.3 Critical Film Drawdown Speed and Thickness for LDPE**

| Participant | Melt Temperature | | A | B | C |
|---|---|---|---|---|---|
| IV | 180°C | Critical film drawdown (m/min) | 18 | 23 | 23 |
| | | film thickness at break (µm) | 15 | 10 | 10 |
| IV | 150°C | Critical film drawdown (m/min) | 13 | 23 | 23 |
| | | Film thickness at break (µm) | 20 | 10 | 10 |

(Data from Meissner, 1979.)

as far as the more standard measurements were concerned, there was no way to differentiate between the three samples.

The measurement of the full flow curve is less commonly made in industry, but it might provide more information as to differences in polymer systems. However, as shown in Fig. 3.30, the flow curves for the three samples are, for all practical purposes, identical. One difference in the samples is noted, the onset of melt fracture. (Melt fracture is associated with flow instabilities and is discussed in Chapter 7.) It appears that $B$ and $C$ undergo melt fracture at a higher shear rate than does sample $A$. The significance of this observation is not clear at this point, but the difference may be due to differences in the flow behavior of these resins in the die entry (see Section 7.1).

Whereas the standard measurements provided no clue as to the differences in the samples, the non-linear measurements provided some

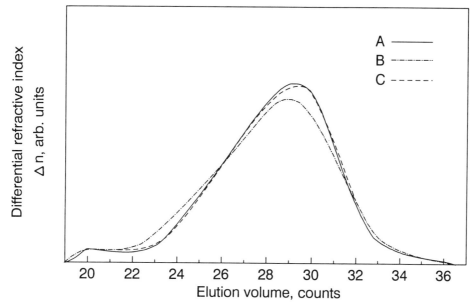

**Figure 3.29** Molecular weight distribution of three LDPE samples as determined by means of gel permeation chromatography. (Reprinted by permission of the publisher from Meissner, 1979.)

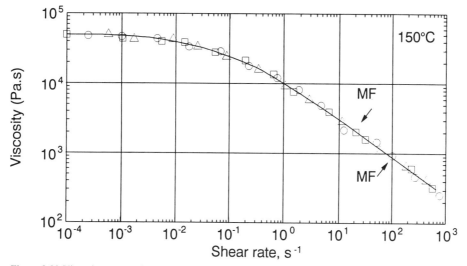

**Figure 3.30** Viscosity versus shear rate data for three LDPE samples obtained by various investigators. The letters MF indicate the onset of melt fracture: symbols the same as in Fig. 3.32. (Reprinted by permission of the publisher from Meissner, 1979.)

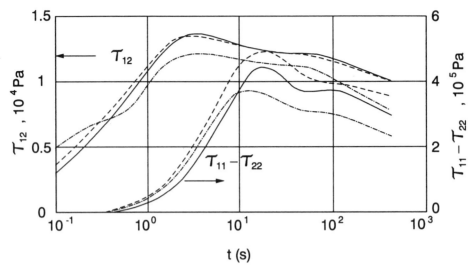

**Figure 3.31** Stress growth behavior of three LDPE samples at 150°C: (---) A; (—) B; (·–·–) C. (Reprinted by permission of the publisher from Meissner, 1979.)

**TABLE 3.4 Melt Flow Characteristics for Three LDPE Samples**

|  | A | B | C |
|---|---|---|---|
| Melt flow index (g/10 min) | 1.37 | 1.41 | 1.59 |
| Melt memory index (% die swell) | 54 | 59 | 53 |
| Zero shear viscosity (150°C) ($\times 10^5$ Poise) | 4.7 | 4.7 | 4.9 |

(Data from Meissner, 1979.)

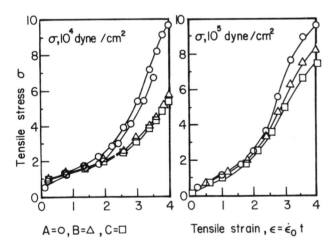

**Figure 3.32** Extensional stress growth versus strain at two extension rates ($\dot{\varepsilon}_0 = 0.001\,\text{s}^{-1}$ and $0.01\,\text{s}^{-1}$) for three LDPE samples: (○) A; (△) B; (□) C. (Reprinted by permission of the publisher from Meissner, 1979.)

insight into the differences in the samples. In Fig. 3.31 is shown the stress growth behavior of the three polymers. The shear stress growth curves of the three samples at the same shear rate are essentially the same. However, $N_1^+(\dot{\gamma}, t)$ for sample A apparently rises to a higher value then it does for either sample B or C. This is the first material property that indicated that there was a difference between the three samples.

The most enlightening difference occurred in the extensional viscosity and $\Delta P_{ent}$ data. In Fig. 3.32 values of the tensile stress (i.e., $\tau_{zz} - \tau_{rr}$), are plotted versus strain for two values of $\dot{\varepsilon}$. Sample A was observed to exhibit higher values of stress than did either sample B or C, especially at lower extension rates. Furthermore, entrance pressure behavior in the form of end correction ($N_{ent}$) data (Fig. 3.33) reflected the same tendency as the extensional data; i.e. $N_{ent}$ of sample A was higher than $N_{ent}$ of samples B and C. This is reasonable in light of the discussion in Section 3.4 in which $\bar{\eta}$ can be estimated from $\Delta P_{ent}$ data. Hence, the quantities related to extensional viscosity seem to be the most sensitive in distinguishing differences in these polymers.

At a later date it was revealed what the difference was in the three samples. The difference was that the degree of long chain branching was varied with sample A having longer branches than polymers B and C. Hence, extensional viscosity is apparently sensitive to changes in the degree of chain branching.

### 3.6. SOLUTION TO DESIGN PROBLEM 2

Design Problem 2 at the beginning of this chapter requires you to design an annular die for LDPE at 170°C considering the fact that the extrudate increases in both thickness and diameter on leaving the die. The die dimensions must therefore be less than those of the extrudate. The constraints are the output from the extruder or the onset of melt fracture.

The basic idea is to determine the dimensions of the annular die that will yield the extrudate with the desired dimensions at the highest extrusion rate possible. The extrudate leaves the die with dimensions (i.e., diameter and thickness) greater than those of the die. This increase in dimensions is due to a phenomenon called extrudate swell, which is associated with elastic recovery. For flow through a capillary, Eq. 3.89 has been proposed for predicting extrudate swell. The questions we are faced with in carrying out this design are: (1) can one extend Eq. 3.89 to other geometries; (2) what is the relation between extrudate swell and diameter and thickness swell for an annular geometry; and (3) how can one obtain the appropriate rheological data at high shear rates?

The starting point is Eq. 3.89 which gives the increase in diameter of an extrudate emerging from a capillary as a function of $N_1(\tau_w)$ and $\tau_w$. It is assumed that Eq. 3.89 applies to both the increase in diameter and in thickness of the annular extrudate. Based on the dimensions of

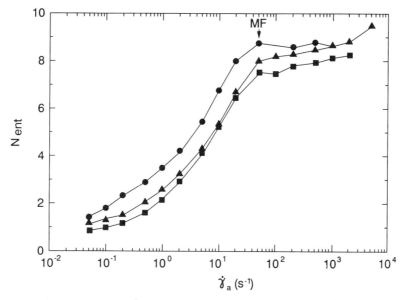

**Figure 3.33** End-correction versus apparent shear rate for three LDPE samples at 150°C. The arrow indicates the onset of melt fracture: symbols the same as in Fig. 3.32. (Reprinted by permission of the publisher from Meissner, 1979.)

the desired extrudate, we can expect the ratio of the inner radius, $\kappa R_o$, to outer radius, $R_o$, to be in the range of 0.8 (2.5 cm/3.065 cm), and hence we can treat the annulus to a first approximation as a slit of width, $W_o = \pi(R_o + \kappa R)$, and height, $H_o = R_o - \kappa R_o$. If an appropriate constitutive equation was available, one could in principle substitute expressions for $N_1$ and $\tau_w$ (e.g. see entries in Table 3.1) into Eq. 3.89 and thereby obtain values for the thickness swell, $H_p/H_0$, as a function of the wall shear rate, $\dot{\gamma}_w$, or the wall shear stress, $\tau_w$. The PTT model in its present form (see Table 3.1 for the predictions in shear and shear-free flows) cannot be used because $\tau_{xy}$ passes through a maximum and then decreases, which leads to abnormally high values of extrudate swell. The viscosity can be described by the following expression (Crochet & Bezy, 1979).

$$\eta = \frac{\eta_1}{1 + \xi(2 - \xi)(\lambda \dot{\gamma})^2} + \eta_2, \qquad (3.119)$$

where $\eta_2$ is a high shear rate limit viscosity. The addition of $\eta_2$ makes $\tau_{xy}$ increase monotonically with $\dot{\gamma}$. Using the IMSL subroutine RNLIN (Ex. 3.3) the following parameters were obtained:

$$\eta_1 = 17,753 \text{ Pa·s}, \quad \eta_2 = 1975 \text{ Pa·s}, \quad \xi = 0.0379, \quad \lambda = 5.47.$$

Values of $N_1/2\tau_{yx}$ and $H_p/H_0$ are calculated and listed in Table 3.5, but they are unreasonably low at high shear rates. This is due to the predictions of the PTT model at high shear rates when only a single relaxation time is used. Hence, the PTT model does not allow us to properly estimate thickness swell.

One would like to go directly to experimental data, but values of $N_1$ are usually not available at values of $\dot{\gamma}$ greater than $10 \text{ s}^{-1}$. Hence, we use the approximation given in Eq. 3.118 along with dynamic mechanical data given in Appendix A.1, Table A.2. Estimated values of $N_1$ presented in Table 3.6 are used to calculate values of $N_1/2\tau_{yx}$ and $H_p/H_0$. These values seem to be more realistic and in line with present knowledge. However, values of $H_p/H_0$ at shear rates higher than $100 \text{ s}^{-1}$ are not available. To obtain these values, one can only extrapolate values of $N_1/2\tau_{yx}$ versus $\dot{\gamma}$ to higher shear rates. In fact,

**TABLE 3.5** Predicted Values of $N_1$, $N_1/2\tau_{yx}$, and Extrudate Swell ($H_p/H_o$) for LDPE (NPE 953) at 170°C Using the Phan Thien-Tanner Model*

| $\dot{\gamma} (\text{s}^{-1})$ | $\tau_{yx} (\text{Pa})$ | $N_1 (\text{Pa})$ | $N_1/2\tau_{yx}$ | $H_p/H_o^\dagger$ |
|---|---|---|---|---|
| 0.1 | 1.74E+03 | 1.90E+03 | 0.55 | 1.12 |
| 1.0 | 7.50E+03 | 2.09E+03 | 1.40 | 1.22 |
| 10.0 | 2.77E+04 | 8.69E+04 | 1.57 | 1.24 |
| 100.0 | 1.90E+05 | 8.73E+04 | 0.23 | 1.10 |
| 553.0 | 1.98E+05 | 8.73E+04 | 0.22 | 1.10 |

*$\eta = 17{,}753$ Pa·s, $\eta_2 = 1975$ Pa·s, $\lambda = 5.475$, and $\xi = 0.0379$
†Calculated using Eq. 3.89.

**TABLE 3.6** Experimental Values of $N_1$ and $N_1/2\tau_{yx}$ for LDPE and Calculated Values of $H_p/H_0$ (Eq. 3.89)

| $\dot{\gamma} (\text{s}^{-1})$ | $\tau_{yx} (\text{Pa})$ | $N_1 (\text{Pa})$ | $N_1 (\text{Pa})^*$ | $N_1/2\tau_{yx}$ | $H_p/H_o$ |
|---|---|---|---|---|---|
| 0.1 | 6.9E+02 | 1.48E+03 | 1.32E+03 | 1.07 | 1.18 |
| 1.0 | 6.35E+03 | 1.32E+04 | 1.084E+04 | 0.85 | 1.15 |
| 10.0 | 2.11E+04 | 8.66E+04 | 5.40E+04 | 1.28 | 1.20 |
| 100.0 | 5.65E+04 | | 1.875E+05 | 1.66 | 1.26 |

*$N_1$ calculated from Eq. 3.118 and dynamic oscillatory shear data.

values of $N_1$ versus $\tau_{xy}$ are plotted in Fig. 3.34, and it is observed that on a plot of $\ln N_1$ versus $\ln \tau_{xy}$ the relation is linear. Hence, we fit a function of the form $N_1 = A\tau_{xy}^b$ to the data.

Finally, one other approach is to use the WM model (see Table 3.1), as at least the viscosity function can be fit to the viscosity data (parameters for the Carreau viscosity model are found using RNLIN and are $\eta_0 = 23{,}000$ Pa·s, $n = 0.587$, and $\lambda = 19.7$ s). Values of $N_1$ cannot be predicted directly from $\eta$ but the modulus, $G$, is required. $G$ is determined usually at each $\dot{\gamma}$ by fitting the predictions for $N_1$ ($= 2\eta^2\dot{\gamma}^2/G$) to data. For $\dot{\gamma}$ greater than $100 \text{ s}^{-1}$ this data is not available, and hence we use $G$ determined at $100 \text{ s}^{-1}$ for predicting values of $N_1$ at higher shear rates. As shown in Table 3.7, the values

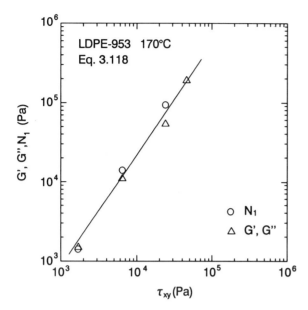

**Figure 3.34** $N_1$ versus $\tau_{xy}$ for LDPE (NPE 953) at 170°C: (○) measured values of $N_1$; (△) estimated values of $N_1$ from Equation 3.118 and values of $G'$ and $G''$.

**TABLE 3.7 Predicted Values of $N_1$, $N_1/2\tau_{yx}$, and Thickness Swell ($H_p/H_0$) for LDPE (NPE 953) Using the White-Metzner Model**

| $\dot{\gamma}(s^{-1})$ | $\tau_{yx}(Pa)$ | $N_1(Pa)$ | $N_1/2\tau_{rr}$ | $H_p/H_o$ |
|---|---|---|---|---|
| 0.1 | 1.66E+03 | 1.45E+03 | 0.44 | 1.12 |
| 1.0 | 6.71E+03 | 1.08E+04 | 0.80 | 1.15 |
| 10.0 | 2.59E+04 | 5.40E+04 | 1.04 | 1.18 |
| 100.0 | 1.00E+05 | 1.88E+05 | 0.94 | 1.16 |
| 553.0 | 2.74E+05 | 1.40E+06 | 2.6 | 1.37 |

of $N_1/2\tau_{yx}$ at least increase monotonically with increasing $\tau_{xy}$ but they are lower than values based on experimental results.

We are now in a position to estimate the diameter and gap of the die. First we observe that $H_0$ and $D_0$ are functions of shear rate and hence the die dimensions will change with changes in the mass throughput. Based on the information given in the problem, our limits are the extruder output or the onset of melt fracture that occurs at a wall shear stress of $1.13 \times 10^5$ Pa for LDPE. Based on the results in Table 3.6 (*Note*: We arbitrarily choose the experimental data over the estimated values given in Table 3.7), then we predict the following dimensions:

$$H_0 = H_p/1.317 = 0.429 \text{ cm}$$

$$D_0 = D_p/1.317 = 4.655 \text{ cm}.$$

Using these dimensions and the wall shear stress of $1.13 \times 10^5$ Pa, the mass flow rate is found using the appropriate expression for the power-law case in Table 2.5:

$$\rho Q = \frac{\rho W_0 H_0^2}{2(s+2)} \left(\frac{\tau_w}{m}\right)^s = \frac{772(4.29 \times 10^{-3})^2(0.1327)}{2(3.704)} \times$$

$$\left(\frac{1.13 \times 10^5}{5.17 \times 10^3}\right)^{1.704} = 4.88 \times 10^{-2} \text{ kg s}^{-1}$$

$$= 3.87 \times 10^2 \text{ lb hr}^{-1}. \qquad (3.120)$$

Here we have used $m = 5.17 \times 10^3$ Pa·s$^n$ and $n = 0.587$ which were obtained by fitting the power-law expression to viscosity data using the subroutine RNLIN. The value of $\rho Q$ exceeds that which is possible with the given extruder. Hence, it is necessary to recalculate the die dimensions using the maximum flow rate of 300 lb/hr. From the expression for $\rho Q$ (Eq. 3.120 or Table 2.4) $\tau_w$ at $\rho Q = 300$ lb/hr$^{-1}$ is calculated:

$$\tau_w = m\left[\frac{2\rho Q(s+2)}{\rho W_0 H_0^2}\right]^n$$

$$= (5.17 \times 10^3)\left[\frac{(2)(3.78 \times 10^{-2})(3.704)}{(772)(0.1327)(4.29 \times 10^{-3})^2}\right]^{0.587}$$

$$= 9.735 \times 10^4 \text{ Pa} \qquad (3.121)$$

Using the relation $N_1 = A\tau_{xy}^b$ with $A = 0.119$ and $b = 1.304$ (These values were obtained by regression analysis of data plotted in Fig. 3.34) we calculate $N_1$ to be $3.8 \times 10^5$ Pa. Using Eq. 3.89 we calculate the thickness and diameter swell to be:

$$H_p/H_0 = 1.294 \qquad D_p/D_0 = 1.294.$$

Using the required extrudate dimensions, $H_0 = 0.436$ cm and $D_0 = 4.737$ cm and the expression for $\rho Q$ (Eq. 3.120) we calculate $\rho Q$ to be 315.9 lb/hr$^{-1}$. This still exceeds the maximum flow rate possible, and hence we repeat the process. $\tau_w$ is recalculated and found to be $9.45 \times 10^4$ Pa. $N_1$ is estimated to be $3.66 \times 10^5$ Pa. The new dimensions based on an extrudate swell of 1.292 are $H_0 = 0.437$ cm and $D_0 = 4.743$ cm. The mass flow rate for these conditions is 302 lb/hr$^{-1}$. Another iteration of this process would be required to determine the exact dimensions. However, one is now close enough, for all practical considerations, to the final die dimensions.

Before leaving Design Problem 2, a few additional comments are required. The die design typically used is more similar to that shown in Design Problem 1. For this type of die the flow is more complex than just shear flow, and hence extrudate swell will be different from what is expected. Extrudate swell, which is described in Chapter 7, is a complex function of several variables, not just $N_1/2\tau_{yx}$. Furthermore, there is not always a simple relation between capillary swell, thickness swell, and diameter swell. The approach we have used in solving Design Problem 2 is at best a crude approximation.

**PROBLEMS**

**A. Applications**

**3A.1.** Viscosity from the Melt Flow Index for LDPE

The melt flow index (MFI) is in essence a single-point viscosity used by industry to characterize a polymer within a family of resins. Most resins are bought and sold based on the MFI. The device used to measure the MFI, which is shown in Fig. 3.35, is basically a capillary rheometer in which a known force is applied to the polymer melt and the mass of polymer leaving the capillary of known dimensions is measured over a 10 min. period. Determine the viscosity and wall shear rate corresponding to a 2.0 MFI LDPE (NPE 953) at 190°C. The following information is given: 2.0 MFI = 2.0 g/10 min; $D = 2.095$ mm; $L = 8.00$ mm; weight = 2.16 kg; reservoir diameter = 9.53 mm, and $\rho = 820$ kg/m$^3$. (*Hint*: The data in Appendix A.1, Table A.3 could be useful.)

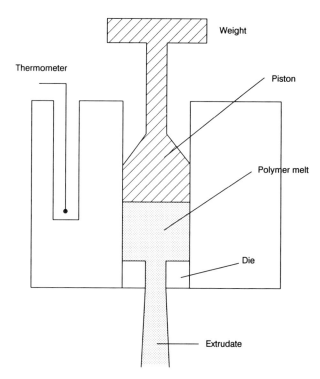

**Figure 3.35** Melt flow index (MFI) device.

**3A.2.** Increase in Wire Coating Thickness Due to Extrudate Swell
A LDPE (NPE 953) melt at 170°C is used to coat a wire as shown in Fig. 2.10 Given that $R=5.0E-03$ m, $\kappa=0.85$, and the wire speed, $V=15$ m/min, calculate the thickness of the coating at the die exit taking into account extrudate swell. List any assumptions you make.

**3A.3.** Viscosity from Capillary Rheometer Data for LLDPE
From values of the apparent wall shear stress, $\tau_a$, and the apparent wall shear rate, $\dot\gamma_a$, given in Appendix A.3, Table A.9, calculate the viscosity as a function of shear rate, and compare your results to those that can be obtained directly from the data in Table A.9 (i.e., $\tau_c$ and $\dot\gamma_c$ where $\tau_c$ and $\dot\gamma_c$ are the corrected wall shear stress and rates, respectively). In particular, use the values of $\tau_a$ at each $L/D$ to obtain $\Delta P_{ent}$ at each shear rate. Correct the values of $\Delta P_{tot}$ to obtain $\tau_w$. Determine $\dot\gamma_w$ by correcting $\dot\gamma_a$ for the nonparabolic velocity profile using Eq. 3.97.

**3A.4.** Estimate the Primary Normal Stress Difference from Dynamic Data
Use Eq. 3.118 and the dynamic oscillatory data given in Appendix A.3, Table A.8, to estimate $N_1$ for LLDPE at 170°C. Compare your values with those given in Appendix A.3, Table A.7.

**3A.5.** Estimate Extensional Viscosity from Capillary Rheometer Data
(a) Use Eq. 3.109 and capillary rheometer data given in Appendix A.1, Table A.3, for LDPE and Appendix A.3, Table A.8, for LLDPE to calculate the extensional viscosity, $\bar\eta$, as a function of extension rate, $\dot\varepsilon$.
(b) Normalize the values of $\bar\eta$ by dividing them by $\eta_0$, and compare the normalized values for the two polymers. Is there any significant difference between the values for the two polymers?

**B. Principles**

**3B.1.** Corrected Wall Shear Rate in a Capillary (No-Slip Case)

The wall shear rate for a Newtonian fluid is given by $4Q/\pi R^3$, which is called the apparent shear rate, $\dot\gamma_a$, for a polymeric fluid. Because of the shear-thinning viscosity exhibited by a polymeric fluid, a$\dot\gamma_a$ does not represent the wall shear rate, $\dot\gamma_w$, for a polymeric fluid. A correction procedure is needed that allows one to determine $\dot\gamma_w$ without any knowledge of the viscosity of the fluid.

(a) Show that if there is no slip of the fluid at the capillary walls the integral for the volume rate of flow may be integrated by parts to give:

$$Q=-\pi\int_0^R \frac{dv_z}{dr}r^2\,dr. \tag{3.122}$$

(b) Introduce the change of variable $r/R=\tau_{rz}/\tau_R$, where $\tau_R=(P_0-P_L)R/2L$ is the wall shear stress, and rewrite the integral in Part **a** in terms of the integration variable $\tau_{rz}$.
(c) Differentiate the integral obtained in (b) with respect to $\tau_R$ to obtain the following equation:

$$\left(-\frac{dv_z}{dr}\right)_{r=R}=\frac{1}{\pi R^3\tau_R^2}\frac{d}{d\tau_R}(\tau_R^3 Q). \tag{3.123}$$

(d) Show how to obtain Eq. 3.97 in the text from Eq. 3.123.

**3B.2.** Wall Shear Rate in a Capillary with Slip
Very viscous polymeric materials such as rubber, highly filled polymers, and polymers containing processing aids do not readily adhere to the walls of a capillary, leading to slip.
(a) Assuming the slip velocity is $v_s$ (and is independent of $\tau_R$) show that the integral for the volume rate of flow may be integrated by parts to give:

$$Q=\pi R^2 v_s-\pi\int_0^R \frac{dv_z}{dr}r^2\,dr$$

(b) Introduce a change of variable as in Part **b** of Pr. 3B.1, and rewrite the integral in Part **a** in terms of the variable $\tau_{rz}$.
(c) Differentiate the integral in Part **b** with respect to $\tau_R$ to obtain the following equation:

$$-\dot\gamma_w=(\dot\gamma_a/4)[3+d\ln\dot\gamma_a/d\ln\tau_R]-3v_s/R. \tag{3.124}$$

(d) Explain how you would use Eq. 3.124 to obtain $v_s$.

**3B.3.** Squeezing Flow between Lubricated Disks
Biaxial stretching flow can be generated in lubricated squeezing flow (see Fig. 3.36). A thin layer of lubricant applied to the upper and lower disks prevents the fluid from sticking to the plates.
(a) Assuming that the squeezing rate, $\dot\varepsilon$, is constant throughout the gap, use the equation of continuity to show that the velocity field is:

$$v_z=-\dot\varepsilon z \qquad v_r=\frac{1}{2}\dot\varepsilon r.$$

(b) How must the gap, H(t), change with time to make $\dot\varepsilon$ constant?
(c) Determine the components of the rate of deformation tensor for this flow.
(d) Which components of stress exist for any fluid?
(e) Find $\bar\eta_1$ and $\bar\eta_2$ for a Newtonian fluid. In particular, show that $\bar\eta_1=6\eta_0$ ($\bar\eta_1$ is called the biaxial extensional viscosity in this case).
(f) Show that the normal stress difference is related to the force

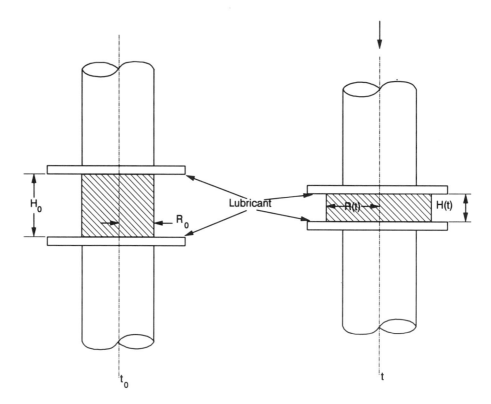

**Figure 3.36** Lubricated squeezing flow between two parallel disks. *Left*, initial configuration, *right*, material at some other time $t > t_0$. (Reprinted by permission of the publisher from P. R. Soskey and H. H. Winter, "Equibiaxial Extension of Two Polymer Melts: Polystyrene and Low-Density Polyethylene," *J. Rheol.*, *29*, p. 493, 1985.)

$F$ by the following expression:

$$\tau_{zz} - \tau_{rr} = \frac{F(t)}{\pi[R(t)]^2}.$$

**3B.4.** Tangential Annular Flow for a Polymeric Liquid

Tangential annular flow was analyzed in Pr. 2B.6 for a power-law fluid. Carry out a similar analysis for a viscoelastic fluid.

**(a)** Show that the components of the equation of motion are:

$$\frac{\partial}{\partial r}(r^2 \tau_{r\theta}) = 0 \tag{3.125}$$

$$-\rho v_\theta^2 / r = -\frac{\partial p}{\partial r} - \frac{1}{r}\frac{\partial}{\partial r}(r\tau_{rr}) + \frac{\tau_{\theta\theta}}{r}. \tag{3.126}$$

**(b)** What is the relation between torque, $T$, and $\tau_{r\theta}$?

**(c)** Use Eq. 3.126 to show how one can obtain the primary normal stress difference.

**3B.5.** Steady Shear Material Functions for the Giesekus Model

The Giesekus (1982) model is a non-linear constitutive equation with a quasi-molecular basis as follows:

$$\tau + \lambda_1 \overset{\triangledown}{\tau} - a\frac{\lambda_1}{\eta_0}\{\tau \cdot \tau\} - a\lambda_2\{\dot{\gamma} \cdot \tau + \tau \cdot \dot{\gamma}\}$$

$$= -\eta_0\left[\dot{\gamma} + \lambda\overset{\triangledown}{\dot{\gamma}} - a\frac{\lambda_2^2}{\lambda_1}\{\dot{\gamma} \cdot \dot{\gamma}\}\right]. \tag{3.127}$$

The model contains four parameters: $\alpha$, $\lambda_1$, $\lambda_2$, and $\eta_0$ and $a = \alpha/(1 - \lambda_2/\lambda_1)$. The model is capable of describing many of the observed rheological properties of polymeric fluids. Show that the steady shear material functions are:

$$\frac{\eta}{\eta_0} = \frac{\lambda_2}{\lambda_1} + \left(1 - \frac{\lambda_2}{\lambda_1}\right)\frac{(1-f)^2}{1 + (1 - 2\alpha)f} \tag{3.128}$$

$$\frac{\Psi_1}{2\eta_0(\lambda_1 - \lambda_2)} = \frac{f(1 - \alpha f)}{(\lambda_1\dot{\gamma})^2\alpha(1 - f)} \tag{3.129}$$

$$\frac{\Psi_2}{\eta_0(\lambda_1 - \lambda_2)} = \frac{-f}{(\lambda_1\dot{\gamma})^2}, \tag{3.130}$$

where

$$f = \frac{1 - \chi}{1 + (1 - 2\chi)\chi^2}; \qquad \chi^2 = \frac{[1 + 16\alpha(1 - \alpha)(\lambda_1\dot{\gamma})^2]^{1/2} - 1}{8\alpha(1 - \alpha)(\lambda_1\dot{\gamma})^2}.$$

**3B.6.** Steady Elongational Flow for the Giesekus Model

Show that in steady elongational flow the extensional viscosity for the Giesekus model is given by:

$$\frac{\bar{\eta}}{3\eta_0} = \frac{\lambda_2}{\lambda_1} + \left(1 - \frac{\lambda_2}{\lambda_1}\right)\frac{1}{6\alpha}\left[3 + \frac{1}{\lambda_1\dot{\epsilon}}\{[1 - 4(1 - 2\alpha)\lambda_1\dot{\epsilon} + 4\lambda_1^2\dot{\epsilon}^2]^{1/2}\right.$$

$$\left. - [1 + 2(1 - 2\alpha)\lambda_1\dot{\epsilon} + \lambda_1^2\dot{\epsilon}^2]^{1/2}\}\right].$$

## C. Numerical Problems

**3C.1.**  Steady Extensional Viscosity for the Phan Thien–Tanner Model
Two nonlinear algebraic equations must be solved as shown in Table 3.1 to determine $\bar{\eta}$ for the PTT model. Using the values of $\eta_0 = 17{,}753$ Pa·s and $\lambda = 5.47$ s for LDPE at 170°C determine $\bar{\eta}$ at $\dot{\varepsilon} = 0.01, 0.1, 1, 10,$ and $100\,\text{s}^{-1}$ for values of $\varepsilon = 0.001, 0.01,$ and $1.0$. Use the IMSL subroutine NEQNF described in Appendix D.4.

**3C.2.**  Transient Extensional Stress Growth for the PTT Model
For the startup of uniaxial extensional flow two coupled ordinary differential equations are obtained for the PTT model as shown

$$\lambda \frac{d\tau_{11}}{dt} + \exp\left[-\frac{\varepsilon\lambda}{\eta_0}(\tau_{11} + 2\tau_{22})\right]\tau_{11} - 2\lambda\dot{\varepsilon}(1-\xi)\tau_{11} = -2\eta_0\dot{\varepsilon}$$
(3.131)

$$\lambda \frac{d\tau_{22}}{dt} + \exp\left[-\frac{\varepsilon\lambda}{\eta_0}(\tau_{11} + 2\tau_{22})\right]\tau_{22} + \lambda\dot{\varepsilon}(1-\xi)\tau_{22} = \eta_0\dot{\varepsilon}.$$
(3.132)

Using the IMSL subroutine IVPAG described in Appendix D.7, solve these equations for $\tau_{11}$ and $\tau_{22}$ as a function of time, and calculate $\bar{\eta}$ as a function of time. Take $\dot{\varepsilon} = 0.2$, $\varepsilon = 0.01$, and the rest of the parameters as given in Pr. 3C.1. Compare the predicted values with the experimental ones for LDPE given in Appendix A.1, Table A.4.

**3C.3**  Fit of Giesekus Model to Rheological Data
The steady shear material functions for the Giesekus model are given in Pr. 3B.5. Find the parameters in this model that give the best fit of the steady shear and dynamic oscillatory data at 170°C given for LDPE in Appendixes A.1, Table A.1 and A.2.

**3C.4.**  Regression Analysis of Capillary Rheometer Measurements
Capillary rheometer data are given for LDPE in Appendix A.1,

Table A.3. Using the apparent values of wall shear stress, $\tau_a$, and shear rate, $\dot{\gamma}_a$, for three different $L/D$ capillaries, determine $\Delta P_{ent}$. Use these values to find the corrected values of the wall shear stress. Correct the apparent wall shear rate values. Compare your values with those given in Appendix A.1, Table A.3. Use the linear regression analysis to determine values of $\Delta P_{ent}$. Correction of the apparent shear rate data may also be best carried out using RNLIN.

## D. Design Problems

**3D.1.**  Slit-die Rheometer Design for a Viscoelastic Fluid
Design a slit-die rheometer for measuring the viscosity and $N_1$ for HDPE (rheological data are given in Appendix A.2, Tables A.5 and A.6) for shear rates from 1.0 to $100\,\text{s}^{-1}$ at 170°C. Pressure transducers with full pressure ranges of 500, 1000, and 3000 psi are available. If the pressure transducers are calibrated in the die, then pressures can be measured as accurately as 1.0% of the reading. The extruder and gear pump system used to feed the die can provide up to 25 kg/hr. In your design specify the placement of the transducers and the dimensions of the die.

**3D.2.**  Design of a Parison Stretching Process
A parison of LDPE (NPE 953 at 170°C) with an outside diameter of 0.127 m, thickness of $3.81 \times 10^{-4}$ m, and length of 0.615 m is stretched to twice its original length before being blown. The end of the parison is stretched to its final length in 1.0 s by pulling on the end of the parison at a constant velocity. Determine the maximum force required to stretch the parison under the conditions given and the minimum clamping force required to hold the parison if the friction coefficient between the grips and the polymer is 0.3.

## REFERENCES

Baird, D. G. 1975. "A Possible Method for Determining Normal Stress Differences from Hole Pressure Error Data." *J. Rheol.*, **19**(1), 147–51.

Baird, D. G., and R. L. Ballman. 1979. "Comparison of the Rheological Properties of Concentrated Solutions of a Rodlike and a Flexible Chain Polyamide." *J. Rheol.*, **23**(4), 505–24.

Bird, R. B., R. C. Armstrong, and O. Hassager. 1987. *Dynamics of Polymeric Liquids: Vol. 1: Fluid Mechanics* (Wiley, New York.)

Cogswell, F. N. 1978. "Converging Flow and Stretching Flow: A Compilation." *J. Non-Newt. Fluid Mech.*, **4**, 23.

Cox, W. P., and E. H. Merz. 1958. "Correlation of the Complex Viscosity with Steady Shear Viscosity." *J. Polym. Sci.*, **28**, 619.

Crochet, M. J., and M. Bezy 1979. "Numerical Solution for the Flow of Viscoelastic Fluids." *J. Non-Newtonian Fluid Mech.*, **5**, 201.

Dealy, J. M. 1982. *Rheometers for Molten Plastics* (Van Nostrand Reinhold, New York.)

Giesekus, H. 1982. "A Simple Constitutive Equation for Polymer Fluids Based on the Concept of Deformation-Dependent Tensorial Mobility." *J. Non-Newtonian Fluid Mech.*, **11**, 69–109.

Gleissele, W. 1980. *Rheology*, Vol. 2 (Plenum, New York), p. 457.

Gotsis, A. D. 1987. "Study of the Numerical Simulation of Viscoelastic Flow: Effect of the Rheological Model and the Mesh." Ph.D. Thesis, Department of Chemical Engineering, Virginia Polytechnic Institute and State University, Blacksburg, VA.

Han, C. D. 1976. *Rheology in Polymer Processing* (Academic Press, New York,).

Kaye, A., A. S. Lodge, and D. G. Vale. 1968. "Determination of Normal Stress Differences in Steady Shear Flow." *Rheol. Acta.*, **7**, 368–79.

Larson, R. G. 1988. *Constitutive Equations for Polymer Melts and Solutions.* (Butterworth–Heinemann, Newton, MA).

Laun, H. M. 1986. "Prediction of Elastic Strains of Polymer Melts in Shear and Elongation." *J. Rheol.*, **30**, 459.

Meissner, J. 1979. "Basic Parameters, Melt Rheology, Processing, and End-Use Properties of Three Low-Density Polyethylene Samples." *J. Pure Appl. Chemistry*, **42**, 553–612.

Münstedt, H. 1980. "Dependence of the Elongational Behavior of Polystyrene Melts on Molecular Weight and Molecular Weight Distribution." *J. Rheol.*, **24**, 847–67.

Phan Thien, N., and R. I. Tanner 1977. "A New Constitutive Equation Derived from Network Theory." *J. Non-Newtonian Fluid Mech.*, **2**, 255–70.

Tanner, R. I. 1970. "Theory of Die Swell." *J. Polym. Sci.*, **A8**, 2067.

Walters, K. 1975. *Rheometry* (Chapman and Hall, London).

White, S. A., and D. G. Baird 1988. "Numerical Simulation Studies of the Planar Entry Flow of Polymer Melts." *J. Non-Newtonian Fluid Mech.* **30**, 47–71.

# 4

# DIFFUSION AND MASS TRANSFER

**DESIGN PROBLEM 3**
**Design of a Dry-Spinning System**

One of the production methods for fiber formation includes the evaporation of a solvent from the spinning line and the resulting solidification of the fiber. This method is called dry-spinning (Ohzawa et al., 1969; Ziabicki, 1976), and it finds application to polymers that do not form thermally stable and viscous melts. These polymers are dissolved in low-molecular-weight volatile solvents (ether, acetone, dimethylformamide, alcohols, etc.). The method consists of extruding the polymer solution into a vertical cell of jets. These jets after leaving the spinneret come into contact with hot air, Fig. 4.21 (p. 86), in which the solvent evaporates; and thus the concentration of the polymer increases, and the spinning line solidifies.

Consider the system of polyacrylonitrile (PAN) and dimethylformamide (DMF). This solution is fed into the dry-spinning apparatus (Fig. 4.21), and the solvent DMF evaporates into the hot air. Three mechanisms account for the mass transfer of the solvent to the air: flash vaporization, diffusion within the spinning line, and convective mass transfer from the spinning line surface to the air. It is expected that the first and the third mechanisms are important in the region close to the spinneret. Analyze the region where the diffusion within the spinning line is the controlling mass transfer mechanism, and calculate the axial distance the fiber travels before all the solvent is removed. For the system of PAN-DMF the following data are given (Ohzawa & Nagano, 1970): dope output $\dot{m}_P = 2.0 \times 10^{-2}$ g/s; dope solvent mass fraction $\bar{\omega}_{Ad} = 0.74$; dope temperature $T_d = 100°C$; air temperature $T_\infty = 200°C$; velocity of cross air flow $V_{C\infty} = 2$ m/s; velocity of parallel air flow $V_{P\infty} = 50$ cm/s; solvent mole fraction in the air $x_{A\infty} = 0$; diffusivity of DMF in PAN in cm²/s, $Ð_{AP} = 9.03 \times 10^{-4} \exp[-2,360/T]$; dope cross-sectional area = 0.001 cm²; and final cross-sectional area = $1.56 \times 10^{-5}$ cm².

The mass transfer coefficients for parallel and cross air flow, respectively, are given below

$$k_{c,P} = 0.26 \left[ Ð_{Aair} \left(\frac{\mu}{\rho}\right)^{-1/3} \right] \left(\frac{Sc}{Pr}\right)^{1/2} R(z)^{-2/3} V_{P\infty}^{1/3}$$

$$k_{c,C} = 0.52 \left[ Ð_{Aair} \left(\frac{\mu}{\rho}\right)^{-1/3} \right] \left(\frac{Sc}{Pr}\right)^{1/2} R(z)^{-2/3} V_{C\infty}^{1/3},$$

where Sc and Pr are the Schmidt and Prandtl numbers, respectively, and $R(z)$ is the radius of the spinning line at every $z$ distance. For DMF in air for the conditions previously described: Sc = 1.81 and Pr = 0.69.

---

Chapters 2 and 3 dealt with momentum transfer and rheological equations of state and their applications to polymer processing. In this chapter and the next one we are concerned with the other two transfers: mass and heat. A substantial number of polymer processes involve changes of composition of the component materials through mass diffusion and convection methods. In many cases these changes of composition do not necessarily involve chemical reactions. We describe some of these polymer processes below.

Mass transfer operations can be found in a number of polymer processing operations. In the dry-spinning of polymer solutions to form fibers we find operations such as diffusion of the solvent within the filaments and convective mass transfer from the filament surfaces to a flowing gas. In the wet-spinning of polymer solutions, the viscous polymer solution is extruded through a spinneret submerged in a bath of a nonsolvent. As the filament passes through the bath of liquid, the nonsolvent diffuses into the filament, while the solvent diffuses out. As a result, the fiber is solidified. Foaming of polymer materials in an extruder involves diffusion of the physical blowing agent to the nucleation sites and subsequent growth of gas bubbles. The formation of a microfoam, that is, a foam with a final bubble size of about 10 μm or less, is controlled by the diffusional aspects of the blowing agent (usually nitrogen or carbon dioxide gas). Drying of polymer pellets before processing is an example of the use of mass transfer in preprocessing stages. The presence of extensive moisture in the pellets can render the final product unacceptable. The size of the dispersed phase in a polymer blend and consequently its mechanical properties are controlled by the mutual diffusivity and time of the mixing step. Finally, mass transfer occurs in the removal of volatiles from polymers inside an extruder (this process is called devolatilization), where residual monomers or volatile solvents diffuse out of the polymer.

In addition to polymer processing operations mass transfer is involved in many applications of polymers. For example, successful packaging of food and beverages is made possible through the use of barrier polymers, which impede the diffusion of gases, such as oxygen and carbon dioxide, as well as flavors, aroma, and odors. Plastic fuel tanks are made possible by exploiting the barrier properties of some polymers to hydrocarbons. Similarly, drug packages contain layers of various polymers that can provide controlled drug release. Welding and crack healing of polymers can be modeled as a diffusional process.

We do not intend in this chapter to present an extensive analysis of mass transfer concepts but, rather, to summarize the basics of mass transfer as required in the design and analysis of polymer processing operations. In this regard, we will give only an extensive overview of the estimation techniques for the diffusivity, solubility, and permeability of solvents in polymers. The laws of diffusional mass transfer as well as the relationships for convective mass transfer remain the same as applied to any material. The books by Brandrup and Immergut (1989), Perry and Chilton (1973), and Reid et al. (1977) provide an extensive overview of experimental data and formulas for the calculation of diffusivity, solubility, and permeability of various polymer systems.

This chapter is organized as follows. In Section 4.1 we describe the fundamentals of mass transfer, such as the various definitions for concentrations and velocities, Fick's first law of diffusion, and the microscopic mass balance principle. In this section the analogy between heat and mass transfer is introduced and used to solve problems. The specific estimation relationships for permeants in polymers are discussed in Section 4.2 with the emphasis placed on gas–polymer systems. This section provides the necessary formulas for a first approximation of the diffusivity, solubility, and permeability and their dependence on temperature. Non-Fickian transport, which is frequently present in high-activity permeants in glassy polymers, is introduced in Section 4.3. Convective mass transfer coefficients are discussed in Section 4.4, and the analogies between mass and heat transfer are used to solve problems involving convective mass transfer. Finally, in Section 4.5 the solution to Design Problem 3 is presented.

## 4.1. MASS TRANSFER FUNDAMENTALS

This section includes the terminology for concentrations, velocities, and fluxes and their relationships. Although the discussion of new physical situations is limited, knowledge of the definitions is necessary for the next sections. Fick's first and second laws and the microscopic mass balance principle are introduced. Finally, simple cases based on the analogy between heat and mass transfer are analyzed.

### 4.1.1. DEFINITIONS OF CONCENTRATIONS AND VELOCITIES

The concentrations of the species in a multicomponent system can be expressed in various forms. Four of these forms, which are the most frequently used, are the following. *Mass concentration*, $\rho_i$, is the mass of species $i$ per unit volume of solution. Similarly, *molar concentration*, $C_i$, is the number of moles of species $i$ per unit volume of solution. *Mass fraction*, $\omega_i$, is the mass of species $i$ divided by the total mass of the system. Finally, *mole fraction*, $x_1$, is the number of moles of species $i$ divided by the total number of moles of the solution. The total mass and molar concentrations are $\rho$ and $C$, respectively, so that $\omega_i$ and $x_i$ are given as:

$$\omega_i = \frac{\rho_i}{\rho}; \qquad x_i = \frac{C_i}{C}. \qquad (4.1)$$

Table 4.1 presents these definitions and some of their relations for a binary system.

**TABLE 4.1 Definitions for Concentrations in Binary Systems**

| | |
|---|---|
| $\rho = \rho_A + \rho_B$ = mass density of solution (g/cm³) | (A) |
| $\rho_A = C_A M_A$ = mass concentration of $A$ (g of $A$/cm³ of solution) | (B) |
| $\omega_A = \dfrac{\rho_A}{\rho}$ = mass fraction of $A$ | (C) |
| $C = C_A + C_B$ = molar density of solution (g-moles/cm³) | (D) |
| $C_A = \dfrac{\rho_A}{M_A}$ = molar concentration of $A$ (g-moles of $A$/cm³ of solution) | (E) |
| $x_A = \dfrac{C_A}{C}$ = mole fraction of $A$ | (F) |
| $M = \dfrac{\rho}{C}$ = number-mean molecular weight of mixture | (G) |

| | | | |
|---|---|---|---|
| $x_A + x_B = 1$ | (H) | $\omega_A + \omega_B = 1$ | (I) |
| $x_A M_A + x_B M_B = M$ | (J) | $\dfrac{\omega_A}{M_A} + \dfrac{\omega_B}{M_B} = \dfrac{1}{M}$ | (K) |
| $x_A = \dfrac{\dfrac{\omega_A}{M_A}}{\dfrac{\omega_A}{M_A} + \dfrac{\omega_B}{M_B}}$ | (L) | $\omega_A = \dfrac{x_A M_A}{x_A M_A + x_B M_B}$ | (M) |

(Data from Bird et al., 1960.)

The species in a diffusing mixture move with different velocities. If $\mathbf{v}_i$ is the velocity of the $i$th species with respect to a fixed coordinate system, then the *mass average bulk velocity*, $\mathbf{v}$, is defined as:

$$\mathbf{v} \equiv \frac{\sum_{i=1}^{n} \rho_i \mathbf{v}_i}{\sum_{i=1}^{n} \rho_i} = \frac{\sum_{i=1}^{n} \rho_i \mathbf{v}_i}{\rho}, \qquad (4.2)$$

where $n$ is the total number of species in the system. Similarly, the *molar average bulk velocity*, $\mathbf{v}^\star$, is defined as:

$$\mathbf{v}^\star \equiv \frac{\sum_{i=1}^{n} C_i \mathbf{v}_i}{\sum_{i=1}^{n} C_i} = \frac{\sum_{i=1}^{n} C_i \mathbf{v}_i}{C}. \qquad (4.3)$$

Finally, the *volume average bulk velocity*, $\mathbf{v}^\blacksquare$, is defined as:

$$\mathbf{v}^\blacksquare \equiv \sum_{i=1}^{n} \rho_i \mathbf{v}_i \frac{\bar{V}_i}{M_i}, \qquad (4.4)$$

where $\bar{V}_i$ and $M_i$ are the molar volume and molecular weight of component $i$, respectively. Note that $\rho\mathbf{v}$ and $C\mathbf{v}^\star$ represent the local rates of mass and molar transport through planes perpendicular to $\mathbf{v}$ and $\mathbf{v}^\star$, respectively.

The velocity of the $i$th species, $\mathbf{v}_i$, can be also written as:

$$\mathbf{v}_i = (\mathbf{v}_i - \mathbf{v}) + \mathbf{v}. \qquad (4.5)$$

This implies that the total mass flux of the $i$th species relative to a fixed coordinate system consists of two fluxes: one is due to the molecular diffusion (diffusion velocity: $\mathbf{v}_i - \mathbf{v}$), and the other is due to bulk movement (bulk velocity: $\mathbf{v}$). Similar arguments hold for the total molar and volume fluxes. This partitioning is necessary, because interdiffusion of unequal-size molecules of two components causes bulk flow, even in the absence of an external bulk flow. Table 4.2 summarizes the definitions of the average velocities and their relations. The following example (basic features of which are drawn from Bird et al., 1960) illustrates the meaning of the various velocities for a binary mixture.

**TABLE 4.2 Definitions for Velocities in Binary Systems**

| Basic Definitions | |
|---|---|
| $\mathbf{v}_A$ = velocity of species $A$ relative to fixed coordinates | (A) |
| $\mathbf{v}_A - \mathbf{v}$ = diffusion velocity of species $A$ relative to $\mathbf{v}$ | (B) |
| $\mathbf{v}_A - \mathbf{v}^\star$ = diffusion velocity of species $A$ relative to $\mathbf{v}^\star$ | (C) |
| $\mathbf{v}_A - \mathbf{v}^\blacksquare$ = diffusion velocity of species $A$ relative to $\mathbf{v}^\blacksquare$ | (D) |
| $\mathbf{v}$ = mass average velocity = $(1/\rho)(\rho_A \mathbf{v}_A + \rho_B \mathbf{v}_B) = \omega_A \mathbf{v}_A + \omega_B \mathbf{v}_B$ | (E) |
| $\mathbf{v}^\star$ = molar average velocity = $(1/C)(C_A \mathbf{v}_A + C_B \mathbf{v}_B) = x_A \mathbf{v}_A + x_B \mathbf{v}_B$ | (F) |
| $\mathbf{v}^\blacksquare$ = volume average velocity = $\rho_A \mathbf{v}_A \bar{V}_A / M_A + \rho_B \mathbf{v}_B \bar{V}_B / M_B$ | (G) |
| Additional relations | |
| $\mathbf{v} - \mathbf{v}^\star = \omega_A(\mathbf{v}_A - \mathbf{v}^\star) + \omega_B(\mathbf{v}_B - \mathbf{v}^\star)$ | (H) |
| $\mathbf{v}^\star - \mathbf{v} = x_A(\mathbf{v}_A - \mathbf{v}) + x_B(\mathbf{v}_B - \mathbf{v})$ | (I) |

(Data from Bird et al., 1960.)

## EXAMPLE 4.1
### Velocities and Their Meaning

Consider a long tube that contains liquid $A$ and vapor $B$ (Fig. 4.1). The liquid starts evaporating, and it moves in the region initially filled with $B$. Calculate the velocity vectors $\mathbf{v}$ and $\mathbf{v}_B$, for $x_A = 1/6$, $\mathbf{v}^\star = 12$, $\mathbf{v}_A = 15$, and $M_A = 5M_B$. How do $\mathbf{v}$ and $\mathbf{v}_B$ change if $M_A = M_B$?

### Solution

Figure 4.1a is shows the schematic of the system. $A$ and $B$ diffuse along the $z$ axis in the positive and negative directions, respectively. At the $z$ position, where $x_A = 1/6$, the velocities are found as follows:

$$\mathbf{v}_B = \frac{\mathbf{v}^\star - x_A \mathbf{v}_A}{x_B} = \frac{57}{5}. \tag{4.6}$$

The ratios of the component molecular weight to the number-mean molecular weight of the mixture are:

$$M_A/M = 3; \qquad M_B/M = 3/5, \tag{4.7}$$

and thus, the mass average velocity $\mathbf{v}$ is calculated as:

$$\mathbf{v} = 3 x_A \mathbf{v}_A + \frac{3}{5} x_B \mathbf{v}_B = \frac{66}{5}. \tag{4.8}$$

In Fig. 4.1.b the velocity vectors for both components are shown. For component $A$, the diffusion velocity accounts for only 12% of the

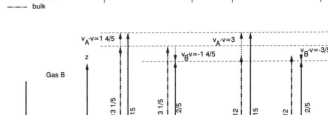

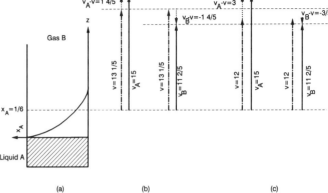

**Figure 4.1** (a) Long tube for diffusion experiments. (b) Velocity vectors for the case of $M_A/M_B = 5$. (c) Velocity vectors for the case of $M_A/M_B = 1$.

velocity of that component with respect to a fixed coordinate system with the remaining 88% being attributed to the bulk flow. The diffusion velocity for component $B$ is negative (about 14% of $\mathbf{v}$), because it flows in the negative $z$ direction. $\mathbf{v} = \mathbf{v}^\star$ when the molecular weights of the components are equal. In that case, the diffusion velocity for component $A$ accounts for 20% of $\mathbf{v}_A$ (Fig. 4.1.c), and the diffusion velocity for component $B$ is negative and close to zero (about 5% of $\mathbf{v}$).

### 4.1.2. FLUXES AND THEIR RELATIONSHIPS

Once the velocities are known, the fluxes, both mass and molar, can be evaluated. The flux is a vector quantity, and its magnitude denotes the mass (or moles) passing through a unit area per unit time. Depending on the velocity we choose, we can define mass and molar fluxes relative to stationary coordinates, relative to the mass average velocity, $\mathbf{v}$, and relative to the molar average velocity, $\mathbf{v}^\star$. Thus, the mass and molar fluxes relative to fixed coordinates are defined as:

$$\mathbf{n}_i = \rho_i \mathbf{v}_i \quad \text{mass}$$

$$\mathbf{N}_i = C_i \mathbf{v}_i \quad \text{molar.} \tag{4.9}$$

Similarly, the fluxes with respect to the mass average velocity, $\mathbf{v}$, are defined as:

$$\mathbf{j}_i = \rho_i (\mathbf{v}_i - \mathbf{v}) \quad \text{mass}$$

$$\mathbf{J}_i = C_i (\mathbf{v}_i - \mathbf{v}) \quad \text{molar,} \tag{4.10}$$

and the fluxes with respect to the molar average velocity are defined as:

$$\mathbf{j}_i^\star = \rho_i (\mathbf{v}_i - \mathbf{v}^\star) \quad \text{mass}$$

$$\mathbf{J}_i^\star = C_i (\mathbf{v}_i - \mathbf{v}^\star) \quad \text{molar.} \tag{4.11}$$

The mass and molar fluxes with respect to the volume average velocity $\mathbf{v}^\blacksquare$ can be similarly formulated, but their uses are limited, and

thus we omit them. Table 4.3 summarizes the various definitions and relations for the mass and molar fluxes of a binary system. The most frequently used definitions are those of molar fluxes $\mathbf{N}_i$ and $\mathbf{J}_i^\star$ and mass flux $\mathbf{j}_i$. Actually, $\mathbf{N}_i$ is used in engineering applications, since it offers the advantage of a fixed coordinate system, whereas the fluxes $\mathbf{j}_i$ and $\mathbf{J}_i^\star$ are the usual measures of diffusion rates. Both definitions will be used in the subsequent analysis of mass transfer.

### 4.1.3. FICK'S FIRST LAW OF DIFFUSION

In analogy with momentum and heat transfer, mass transfer is governed by a simple law of diffusion, Fick's first law. This law states that component $A$ moves relative to the bulk motion (diffuses) in the direction of decreasing mole fraction of $A$:

$$\mathbf{J}_A^\star = -\mathcal{D}_{AB} \nabla C_A, \tag{4.12}$$

where $\mathcal{D}_{AB}$ is the *mutual diffusion coefficient* or *mass diffusivity* or *interdiffusion coefficient* or simply *diffusion coefficient* in the binary system of $A$ and $B$. Note that in a binary system $\mathcal{D}_{AB} = \mathcal{D}_{BA}$.

Table 4.4 summarizes some of the different expressions for Fick's first law. In principle, any of these expressions can be used to solve a diffusion problem, but the right choice of the expression helps in reducing the mathematical complexities involved. The right choice depends on the specific problem. One of the most frequently used expressions relates the molar flux to the mole fraction of component $A$ as follows (see also Pr. 4A.1):

$$\mathbf{N}_A = x_A (\mathbf{N}_A + \mathbf{N}_B) - C\mathcal{D}_{AB} \nabla x_A. \tag{4.13}$$

**TABLE 4.3 Mass and Molar Fluxes in Binary Systems**

| Quantity | With Respect to Stationary Axes | | With Respect to $\mathbf{v}$ | | With Respect to $\mathbf{v}^\star$ | |
|---|---|---|---|---|---|---|
| Basic definitions | | | | | | |
| Velocity of species $A$ (cm/s) | $\mathbf{v}_A$ | (A) | $\mathbf{v}_A - \mathbf{v}$ | (B) | $\mathbf{v}_A - \mathbf{v}^\star$ | (C) |
| Mass flux of species $A$ (g/cm²·s) | $\mathbf{n}_A = \rho_A \mathbf{v}_A$ | (D) | $\mathbf{j}_A = \rho_A(\mathbf{v}_A - \mathbf{v})$ | (E) | $\mathbf{j}_A^\star = \rho_A(\mathbf{v}_A - \mathbf{v}^\star)$ | (F) |
| Molar flux of species $A$ (g-moles/cm²·s) | $\mathbf{N}_A = C_A \mathbf{v}_A$ | (G) | $\mathbf{J}_A = C_A(\mathbf{v}_A - \mathbf{v})$ | (H) | $\mathbf{J}_A^\star = C_A(\mathbf{v}_A - \mathbf{v}^\star)$ | (I) |
| Relations among the fluxes | | | | | | |
| Sum of mass fluxes (g/cm²·s) | $\mathbf{n}_A + \mathbf{n}_B = \rho\mathbf{v}$ | (J) | $\mathbf{j}_A + \mathbf{j}_B = 0$ | (K) | $\mathbf{j}_A^\star + \mathbf{j}_B^\star = \rho(\mathbf{v} - \mathbf{v}^\star)$ | (L) |
| Sum of molar fluxes (g-moles/cm²·s) | $\mathbf{N}_A + \mathbf{N}_B = C\mathbf{v}^\star$ | (M) | $\mathbf{J}_A + \mathbf{J}_B = C(\mathbf{v}^\star - \mathbf{v})$ | (N) | $\mathbf{J}_A^\star + \mathbf{J}_B^\star = 0$ | (O) |
| Fluxes in terms of $\mathbf{n}_A$ and $\mathbf{n}_B$ | $\mathbf{N}_A = \dfrac{\mathbf{n}_A}{M_A}$ | (P) | $\mathbf{j}_A = \mathbf{n}_A - \omega_A(\mathbf{n}_A + \mathbf{n}_B)$ | (Q) | $\mathbf{j}_A^\star = \mathbf{n}_A - x_A\left(\mathbf{n}_A + \dfrac{M_A}{M_B}\mathbf{n}_B\right)$ | (R) |
| Fluxes in terms of $\mathbf{N}_A$ and $\mathbf{N}_B$ | $\mathbf{n}_A = \mathbf{N}_A M_A$ | (S) | $\mathbf{J}_A = \mathbf{N}_A - \omega_A\left(\mathbf{N}_A + \dfrac{M_B}{M_A}\mathbf{N}_B\right)$ | (T) | $\mathbf{J}_A^\star = \mathbf{N}_A - x_A(\mathbf{N}_A + \mathbf{N}_B)$ | (U) |
| Fluxes in terms of $\mathbf{j}_A$ and $\mathbf{v}$ | $\mathbf{n}_A = \mathbf{j}_A + \rho_A \mathbf{v}$ | (V) | $\mathbf{J}_A = \dfrac{\mathbf{j}_A}{M_A}$ | (W) | $\mathbf{j}_A^\star = \dfrac{M}{M_B}\mathbf{j}_A$ | (X) |
| Fluxes in terms of $\mathbf{J}_A^\star$ and $\mathbf{v}^\star$ | $\mathbf{N}_A = \mathbf{J}_A^\star + C_A \mathbf{v}^\star$ | (Y) | $\mathbf{J}_A = \dfrac{M_B}{M}\mathbf{J}_A^\star$ | (Z) | $\mathbf{j}_A^\star = \mathbf{J}_A^\star M_A$ | (AA) |

(Reprinted by permission of the publisher from Bird et al., 1960.)

**TABLE 4.4 Forms of Fick's First Law of Diffusion in Binary Systems**

| Flux | Gradient | Form of Fick's First Law | |
|---|---|---|---|
| $\mathbf{n}_A$ | $\nabla\omega_A$ | $\mathbf{n}_A - \omega_A(\mathbf{n}_A + \mathbf{n}_B) = -\rho \mathcal{D}_{AB}\nabla\omega_A$ | (A) |
| $\mathbf{N}_A$ | $\nabla x_A$ | $\mathbf{N}_A - x_A(\mathbf{N}_A + \mathbf{N}_B) = -C\mathcal{D}_{AB}\nabla x_A$ | (B) |
| $\mathbf{j}_A$ | $\nabla\omega_A$ | $\mathbf{j}_A = -\rho \mathcal{D}_{AB}\nabla\omega_A$ | (C) |
| $\mathbf{J}_A^\star$ | $\nabla x_A$ | $\mathbf{J}_A^\star = -C\mathcal{D}_{AB}\nabla x_A$ | (D) |
| $\mathbf{j}_A$ | $\nabla x_A$ | $\mathbf{j}_A = -\left(\dfrac{C^2}{\rho}\right)M_A M_B \mathcal{D}_{AB}\nabla x_A$ | (E) |
| $\mathbf{J}_A^\star$ | $\nabla\omega_A$ | $\mathbf{J}_A^\star = -\left(\dfrac{\rho^2}{CM_A M_B}\right)\mathcal{D}_{AB}\nabla\omega_A$ | (F) |

(Data from Bird et al., 1960.)

This equation shows that the molar flux of one component is the sum of two contributions: one from the bulk motion of the fluid, and the other from the molar flux of the component due to diffusion.

The equations for mass, heat, and momentum transfer, i.e., Eqs. 4.12, 5.18 and 2.3, are analogous. All state that mass, energy, and momentum are transferred because of a gradient in concentration, temperature, and velocity, respectively. Also, the proportionality constants in all these equations (mass and thermal diffusivity and kinematic viscosity) have the same dimensions of length²/time. These analogies break down in two- and three-dimensional problems, because stress is a tensorial quantity, whereas heat and mass (or molar) fluxes are vectors.

**EXAMPLE 4.2**
**Mass Flux in a Polymer Membrane-Penetrant System**

Calculate the mass flux of a penetrant diffusing through a polymer membrane as a function of the penetrant mass fraction. Start with the definition of the mass flux with respect to the mass average velocity.

**Solution**

For the binary system of the polymer membrane ($P$) and the diffusing species ($A$) Eq. 4.10 yields:

$$\mathbf{j}_A = \rho_A(\mathbf{v}_A - \mathbf{v}) = \rho_A\mathbf{v}_A - \frac{\rho_A}{\rho}(\rho_A\mathbf{v}_A + \rho_P\mathbf{v}_P). \tag{4.14}$$

The velocity of the membrane with respect to a fixed reference system, $\mathbf{v}_P$, is zero. Hence,

$$\mathbf{j}_A = \mathbf{n}_A - \omega_A\mathbf{n}_A = -\rho\mathcal{D}_{AP}\nabla\omega_A, \tag{4.15}$$

and thus

$$\mathbf{n}_A = \frac{-\rho\mathcal{D}_{AP}}{1-\omega_A}\nabla\omega_A. \tag{4.16}$$

In typical polymer membrane–gas systems, the term $1/(1-\omega_A)$ is approximately equal to 1, so that Eq. 4.16 is further simplified. However, in polymer membrane-solvent systems, $\omega_A$ can be significant, and it cannot be neglected in Eq. 4.16.

## 4.1.4. MICROSCOPIC MATERIAL BALANCE

*Microscopic material balance* is based on the conservation of mass. The law of conservation of mass for material $A$ flowing in and out of a differential volume element $dxdydz$ (Fig. 4.2) in its rate form states that

$$\begin{pmatrix} \text{Rate of accumulation} \\ \text{in the element} \end{pmatrix} = \begin{pmatrix} \text{Rate of transport} \\ \text{into the element} \end{pmatrix}$$
$$- \begin{pmatrix} \text{Rate of transport} \\ \text{out of the element} \end{pmatrix} + \begin{pmatrix} \text{Rate of generation} \\ \text{in the element} \end{pmatrix}$$
$$- \begin{pmatrix} \text{Rate of consumption} \\ \text{in the element} \end{pmatrix}. \qquad (4.17)$$

This mass balance is similar to Eqs. 2.15 and 5.15, which contain the momentum and energy balances, respectively.

The rate of generation and consumption (net rate of mass production) of material $A$ in Eq. 4.17, refers to chemical reactions, and it will be denoted as $\dot{r}_A$ (dimensions: mass/volume-time). If we consider that the volume element of Fig. 4.2 is fixed in space (*Eulerian approach*) Eq. 4.17 can be written as:

$$dxdydz \frac{\partial \rho_A}{\partial t} = (n_{Ax|x} - n_{Ax|x+dx})dydz + (n_{Ay|y} - n_{Ay|y+dy})dxdz$$
$$+ (n_{Az|z} - n_{Az|z+dz})dxdy + \dot{r}_A dxdydz. \qquad (4.18)$$

Dividing this equation by $dxdydz$, and shrinking the differential volume to zero, we get the microscopic mass balance (or *continuity*) equation as follows:

$$\frac{\partial \rho_A}{\partial t} + \frac{\partial n_{Ax}}{\partial x} + \frac{\partial n_{Ay}}{\partial y} + \frac{\partial n_{Az}}{\partial z} - \dot{r}_A = 0, \qquad (4.19)$$

or in vector notation:

$$\frac{\partial \rho_A}{\partial t} + \nabla \cdot \mathbf{n}_A - \dot{r}_A = 0. \qquad (4.20)$$

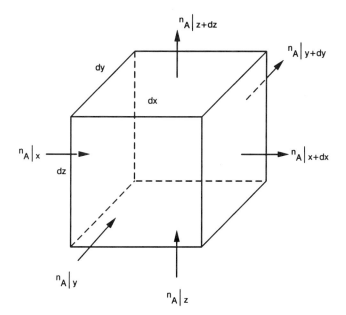

**Figure 4.2** Differential volume element for the microscopic material balance.

Obviously, a similar expression holds for every component $i$ of a multicomponent system. Table 4.5 summarizes the continuity equations in terms of mass fluxes for component $A$ in rectangular, cylindrical, and spherical coordinates. In terms of molar fluxes Eq. 4.20 becomes:

$$\frac{\partial C_A}{\partial t} + \nabla \cdot \mathbf{N}_A - \dot{R}_A = 0, \qquad (4.21)$$

where $\dot{R}_A$ is the net molar production of component $A$ in mol/volume-time. Table 4.6 summarizes the continuity equations in terms of molar fluxes for component $A$ in rectangular, cylindrical, and spherical coordinates.

By summing all the continuity equations for each component $i$ and noting that the sum of all net mass productions due to reaction, $\sum \dot{r}_i$, is equal to zero, we get:

$$\frac{\partial \rho}{\partial t} + \nabla \cdot (\rho \mathbf{v}) = 0, \qquad (4.22)$$

where use of Table 4.3 was made. This is the *continuity equation for the mixture*. For a fluid mixture with constant density we note that Eq. 4.22 becomes:

$$\nabla \cdot \mathbf{v} = 0, \qquad (4.23)$$

which is the familiar continuity equation for incompressible flow (e.g.

**TABLE 4.5 Forms of Continuity Equation of Species $A$ with Mass Fluxes (No Chemical Reactions)**

A. Continuity equation of species $A$ in various coordinate systems
   Rectangular coordinates

$$\frac{\partial \rho_A}{\partial t} + \left( \frac{\partial n_{Ax}}{\partial x} + \frac{\partial n_{Ay}}{\partial y} + \frac{\partial n_{Az}}{\partial z} \right) = 0 \qquad (A)$$

   Cylindrical coordinates

$$\frac{\partial \rho_A}{\partial t} + \left( \frac{1}{r} \frac{\partial (rn_{Ar})}{\partial r} + \frac{1}{r} \frac{\partial n_{A\theta}}{\partial \theta} + \frac{\partial n_{Az}}{\partial z} \right) = 0 \qquad (B)$$

   Spherical coordinates

$$\frac{\partial \rho_A}{\partial t} + \left( \frac{1}{r^2} \frac{\partial (r^2 n_{Ar})}{\partial r} + \frac{1}{r \sin\theta} \frac{\partial (n_{A\theta} \sin\theta)}{\partial \theta} + \frac{1}{r \sin\theta} \frac{\partial n_{A\phi}}{\partial \phi} \right) = 0 \qquad (C)$$

B. Continuity equation of species $A$ for constant $\rho$ and $Ð_{AB}$
   Rectangular coordinates

$$\frac{\partial \rho_A}{\partial t} + \left( v_x \frac{\partial \rho_A}{\partial x} + v_y \frac{\partial \rho_A}{\partial y} + v_z \frac{\partial \rho_A}{\partial z} \right) = Ð_{AB} \left( \frac{\partial^2 \rho_A}{\partial x^2} + \frac{\partial^2 \rho_A}{\partial y^2} + \frac{\partial^2 \rho_A}{\partial z^2} \right) \qquad (D)$$

   Cylindrical coordinates

$$\frac{\partial \rho_A}{\partial t} + \left( v_r \frac{\partial \rho_A}{\partial r} + v_\theta \frac{1}{r} \frac{\partial \rho_A}{\partial \theta} + v_z \frac{\partial \rho_A}{\partial z} \right)$$
$$= Ð_{AB} \left( \frac{1}{r} \frac{\partial}{\partial r} \left( r \frac{\partial \rho_A}{\partial r} \right) + \frac{1}{r^2} \frac{\partial^2 \rho_A}{\partial \theta^2} + \frac{\partial^2 \rho_A}{\partial z^2} \right) \qquad (E)$$

   Spherical coordinates

$$\frac{\partial \rho_A}{\partial t} + \left( v_r \frac{\partial \rho_A}{\partial r} + v_\theta \frac{1}{r} \frac{\partial \rho_A}{\partial \theta} + v_\phi \frac{1}{r \sin\theta} \frac{\partial \rho_A}{\partial \phi} \right)$$
$$= Ð_{AB} \left( \frac{1}{r^2} \frac{\partial}{\partial r} \left( r^2 \frac{\partial \rho_A}{\partial r} \right) + \frac{1}{r^2 \sin\theta} \frac{\partial}{\partial \theta} \left( \sin\theta \frac{\partial \rho_A}{\partial \theta} \right) + \frac{1}{r^2 \sin^2\theta} \frac{\partial^2 \rho_A}{\partial \phi^2} \right) \qquad (F)$$

**TABLE 4.6 Forms of Continuity Equation of Species *A* with Molar Fluxes (No Chemical Reactions)**

A. Molar flux of species *A* in various coordinate systems
   Rectangular coordinates

$$\frac{\partial C_A}{\partial t} + \left(\frac{\partial N_{Ax}}{\partial x} + \frac{\partial N_{Ay}}{\partial y} + \frac{\partial N_{Az}}{\partial z}\right) = 0 \qquad \text{(A)}$$

   Cylindrical coordinates

$$\frac{\partial C_A}{\partial t} + \left(\frac{1}{r}\frac{\partial(rN_{Ar})}{\partial r} + \frac{1}{r}\frac{\partial N_{A\theta}}{\partial \theta} + \frac{\partial N_{Az}}{\partial z}\right) = 0 \qquad \text{(B)}$$

   Spherical coordinates

$$\frac{\partial C_A}{\partial t} + \left(\frac{1}{r^2}\frac{\partial(r^2 N_{Ar})}{\partial r} + \frac{1}{r\sin\theta}\frac{\partial(N_{A\theta}\sin\theta)}{\partial \theta} + \frac{1}{r\sin\theta}\frac{\partial N_{A\phi}}{\partial \phi}\right) = 0 \qquad \text{(C)}$$

B. Continuity equation of species *A* for constant $\rho$ and $\text{\DH}_{AB}$
   Rectangular coordinates

$$\frac{\partial C_A}{\partial t} + \left(v_x\frac{\partial C_A}{\partial x} + v_y\frac{\partial C_A}{\partial y} + v_z\frac{\partial C_A}{\partial z}\right) = \text{\DH}_{AB}\left(\frac{\partial^2 C_A}{\partial x^2} + \frac{\partial^2 C_A}{\partial y^2} + \frac{\partial^2 C_A}{\partial z^2}\right) \qquad \text{(D)}$$

   Cylindrical coordinates

$$\frac{\partial C_A}{\partial t} + \left(v_r\frac{\partial C_A}{\partial r} + v_\theta\frac{1}{r}\frac{\partial C_A}{\partial \theta} + v_z\frac{\partial C_A}{\partial z}\right)$$
$$= \text{\DH}_{AB}\left(\frac{1}{r}\frac{\partial}{\partial r}\left(r\frac{\partial C_A}{\partial r}\right) + \frac{1}{r^2}\frac{\partial^2 C_A}{\partial \theta^2} + \frac{\partial^2 C_A}{\partial z^2}\right) \qquad \text{(E)}$$

   Spherical coordinates

$$\frac{\partial C_A}{\partial t} + \left(v_r\frac{\partial C_A}{\partial r} + v_\theta\frac{1}{r}\frac{\partial C_A}{\partial \theta} + v_\phi\frac{1}{r\sin\theta}\frac{\partial C_A}{\partial \phi}\right)$$
$$= \text{\DH}_{AB}\left(\frac{1}{r^2}\frac{\partial}{\partial r}\left(r^2\frac{\partial C_A}{\partial r}\right) + \frac{1}{r^2\sin\theta}\frac{\partial}{\partial \theta}\left(\sin\theta\frac{\partial C_A}{\partial \theta}\right) + \frac{1}{r^2\sin^2\theta}\frac{\partial^2 C_A}{\partial \phi^2}\right) \qquad \text{(F)}$$

Eq. 2.58 or Table 2.8). If we use molar fluxes in place of mass fluxes, the equivalent continuity equation for the mixture becomes:

$$\frac{\partial C}{\partial t} + \nabla\cdot(C\mathbf{v}^\star) - \sum_{i=1}^{n}\dot{R}_i = 0. \qquad \text{(4.24)}$$

Note that, in general, the sum of the rates of molar productions by chemical reactions cannot be set equal to zero unless the reaction is equimolar (i.e., for every mole of component *i* consumed one mole of component *j* is produced, etc.). For a fluid of constant molar density *C*, Eq. 4.24 yields:

$$C\nabla\cdot\mathbf{v}^\star = \sum_{i=1}^{n}\dot{R}_i. \qquad \text{(4.25)}$$

Neither form of the continuity equation for species *A*, Eqs. 4.20 or 4.21, are very useful as they stand. They can be written in terms of the mass and molar average velocities by combining Eqs. 4.10, 4.11, 4.20, and 4.21 as follows:

$$\frac{\partial \rho_A}{\partial t} + \nabla\cdot\mathbf{j}_A + \nabla\cdot(\rho_A\mathbf{v}) - \dot{r}_A = 0, \qquad \text{(4.26)}$$

and

$$\frac{\partial C_A}{\partial t} + \nabla\cdot\mathbf{J}_A^\star + \nabla\cdot(C_A\mathbf{v}^\star) - \dot{R}_A = 0. \qquad \text{(4.27)}$$

Combining these equations and the equations for $\mathbf{n}_A$ and $\mathbf{N}_A$ from Table 4.4, we get:

$$\frac{\partial \rho_A}{\partial t} + \nabla\cdot(\rho_A\mathbf{v}) - \nabla\cdot(\text{\DH}_{AB}\nabla\rho_A) - \dot{r}_A = 0, \qquad \text{(4.28)}$$

and

$$\frac{\partial C_A}{\partial t} + \nabla\cdot(C_A\mathbf{v}^\star) - \nabla\cdot(\text{\DH}_{AB}\nabla C_A) - \dot{R}_A = 0. \qquad \text{(4.29)}$$

These equations are valid for variable mass or molar density, $\rho$ or *C*, and variable diffusion coefficient $\text{\DH}_{AB}$. Their generality can be reduced in certain cases as is shown in the following.

***Constant $\rho$ and $\text{\DH}_{AB}$ and No Chemical Reaction.*** For this case, Eq. 4.28 becomes (using also the continuity Eq. 4.23):

$$\frac{\partial \rho_A}{\partial t} + \mathbf{v}\cdot\nabla\rho_A = \text{\DH}_{AB}\nabla^2\rho_A. \qquad \text{(4.30)}$$

This equation is usually used for diffusion in *dilute solutions*, and it is similar to equations used in momentum (Table 2.9 or Eq. 5.58) and heat transfer (Table 5.3 or Eq. 5.59). Table 4.5 summarizes the expressions for Eq. 4.30 for the three coordinate systems. Division of all terms of Eq. 4.30 by the molecular weight of *A* gives the forms of the continuity equation shown in Table 4.6. Finally, note that the left side of Eq. 4.30 can be written as $D(\rho_A)/Dt$, where $D/Dt$ notes the material derivative.

***Zero Mass or Molar Average Velocity and No Chemical Reaction.*** For this case, Eq. 4.28 (or Eq. 4.29) becomes:

$$\frac{\partial \rho_A}{\partial t} = \nabla\cdot(\text{\DH}_{AB}\nabla\rho_A), \qquad \text{(4.31)}$$

which for constant $\text{\DH}_{AB}$ yields:

$$\frac{\partial \rho_A}{\rho t} = \text{\DH}_{AB}\nabla^2\rho_A, \qquad \text{(4.32)}$$

which is called *Fick's second law of diffusion*, or simply the *diffusion equation*. Note again that this equation is used in cases where **v** is zero (i.e., diffusion in solids or stationary liquids) or $\mathbf{v}^\star$ is zero (i.e., equimolar counterdiffusion in gases). The similarity between Eq. 4.32 and the heat conduction equation is the basis for the similarity of solutions to these equations.

### 4.1.5. SIMILARITY WITH HEAT TRANSFER: SIMPLE APPLICATIONS

The majority of diffusion problems can be solved by recognizing their similarity to heat transfer problems. For example, Eqs. 4.30 and 4.32 are directly analogous to the heat transfer equations, as is illustrated in Table 4.7. Note for this analogy to hold one should make the following substitutions: $\rho_A$ or $C_A$ in place of *T* and $\text{\DH}_{AB}$ in place of $\alpha$ (thermal diffusivity). More specifically, Figs. 5.10, 5.11, 5.12 and 5.13 (pp. 108 and 109) which present the graphical solutions of the unsteady heat conduction equation (see Table 4.7) for infinite slabs and cylinders, are equally well applicable to the unsteady diffusion equation (Eq. 4.32) for the same geometries and boundary conditions. For example, Figs.

**TABLE 4.7 Analogy between Heat Conduction and Mass Diffusion**

| Unsteady-State Nonflow | Steady-State Flow | Steady-State Nonflow |
|---|---|---|
| $\dfrac{\partial T}{\partial t} = \alpha \nabla^2 T$ | $(\mathbf{v} \cdot \nabla T) = \alpha \nabla^2 T$ | $\nabla^2 T = 0$ |
| Heat conduction in solids | Heat conduction in laminar incompressible flow | Steady heat conduction in solids |
| 1. $k$ = constant<br>2. $\mathbf{v} = 0$ | 1. $k$, $\rho$ = constants<br>2. No viscous dissipation<br>3. Steady state | 1. $k$ = constant<br>2. $\mathbf{v} = 0$<br>3. Steady state |
| $\dfrac{\partial C_A}{\partial t} = Ð_{AB} \nabla^2 C_A$ | $(\mathbf{v} \cdot \nabla C_A) = Ð_{AB} \nabla^2 C_A$ | $\nabla^2 C_A = 0$ |
| Diffusion of traces of $A$ through $B$ | Diffusion in laminar flow (dilute solutions of $A$ in $B$) | Steady diffusion in solids |
| 1. $Ð_{AB}$, $\rho$ = constants<br>2. $\mathbf{v} = 0$<br>3. No chemical reactions | 1. $Ð_{AB}$, $\rho$ = constants<br>2. Steady state<br>3. No chemical reactions | 1. $Ð_{AB}$, $\rho$ = constants<br>2. Steady state<br>3. No chemical reactions<br>4. $\mathbf{v} = 0$ |
| OR Equimolar counter-diffusion in low-density gases<br>1. $Ð_{AB}$, $C$ = constants<br>2. $\mathbf{v}^\star = 0$<br>3. No chemical reactions | | |

(Reprinted by permission of the publisher from Bird et al., 1960.)

5.10 and 5.11 can also be considered to show the dimensionless concentration as a function of time and relative position in a slab and a cylinder, respectively, when their surfaces are kept at a constant concentration.

Furthermore, Figs. 5.12 and 5.13 (p. 109) can also be used to show the dimensionless concentration as a function of dimensionless time and position for the case in which there is resistance to mass transfer at the interface between a solid and fluid: $-Ð_{AB}(\partial C_A/\partial z) = k_c(C_{Ai} - C_{A\infty})$, where $k_c$ is the convective mass transfer coefficient (Section 4.4), $C_{Ai}$ is the concentration of species $A$ at the interface in the fluid side, and $C_{A\infty}$ is the concentration of species $A$ in the fluid far away from the interface. Note that the constant concentration boundary condition referred to in the previous paragraph could be considered as a special case of the convective-type boundary condition for a Sherwood number, Sh, (or Nusselt number for diffusion) equal to $\infty$. Also, in Figs. 5.12 and 5.13 (p. 109), the Biot (or Nusselt) number for heat transfer should be replaced by $(k_c b / Ð_{AB})(1/K) = (\text{Sh}/K) = \text{Sh}'$, where $K$ is the ratio of the equilibrium concentration in the solid to the concentration in the surrounding fluid (also called partition coefficient or equilibrium distribution coefficient; it is a form of the Henry's law constant of Subsection 4.2.1), and $b$ is the half thickness of the slab.

Figures 4.3 and 4.4 show the graphical solutions of the unsteady diffusion problem for a sphere, which is not covered in the heat transfer chapter 5. $C_{A0}$ and $C_{A\infty}$ in Fig. 4.3 correspond to $T_0$ and $T_1$, respectively, in Figs. 5.10 and 5.11. Similarly, $T_0$ in Figs. 5.12 and 5.13 should be replaced by $C_{A0}$. Finally, in Fig. 4.4, $C_{A\infty}^*$, which is the concentration of $A$ in the solid in equilibrium to the concentration of $A$ in the fluid $C_{A\infty}$, i.e., $C_{A\infty}^* = K C_{A\infty}$, is used in place of $C_{A\infty}$.

One additional feature of mass transfer is the calculation of the amount sorbed or desorbed at time $t$, $M_t$, and its comparison to the total possible amount sorbed or desorbed, $M_\infty$ (see also Pr. 4B.1).

Such comparisons as a function of the diffusivity, time, and length scale of the geometry are shown in Fig. 4.5, for a slab, cylinder, and sphere. The parameter in this figure is the Sherwood number (or Sh'). The following examples illustrate the similarity between heat and mass transfer for the case of unsteady transfer in simple geometries.

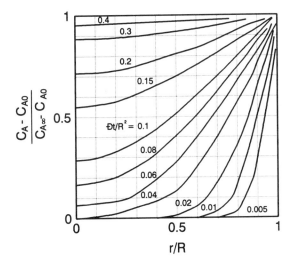

**Figure 4.3** Concentration profiles for unsteady mass diffusion in a sphere of radius $R$. $C_A(R, t) = C_{A\infty}$; $C_A(r, 0) = C_{A0}$. (Reprinted by permission of the publisher from J. Crank, *The Mathematics of Diffusion*, 1st Ed., Oxford University Press, Oxford, 1956.)

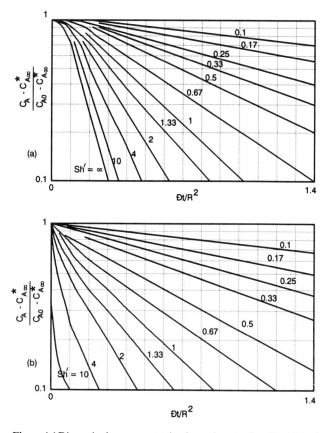

**Figure 4.4** Dimensionless concentration in a sphere (radius $R$) subjected to step change in surface concentration. $Sh' = k_c R/\mathcal{D}K$; $C_{A\infty}^* = KC_{A\infty}$ (a) Center of sphere. (b) Surface of sphere. (Reprinted by permission of the publisher from L. M. K. Boelter, V. H. Cherry, H. A. Johnson, and R. C. Martinelli, *Heat Transfer Notes*, McGraw-Hill, New York, 1965.)

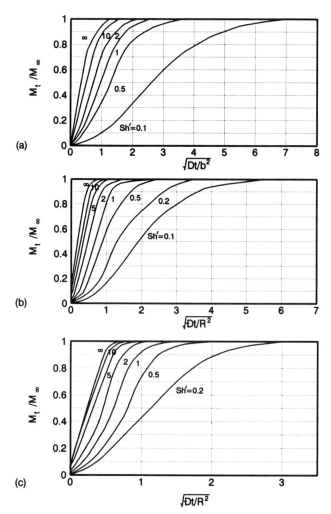

**Figure 4.5** Ratio of the amount absorbed (or desorbed) at time $t$, $M_t$, to the total amount absorbed, $M_\infty$, as a function of time, for various values of $Sh' = k_c b/\mathcal{D}K$. (a) Slab of thickness $2b$. (b) Cylinder of radius $R = b$. (c) Sphere of radius $R = b$. (Reprinted by permission of the publisher from Crank, 1956.)

---

**EXAMPLE 4.3**
**Diffusion in a Slab**

An infinite slab of polypropylene (PP) is exposed to high-pressure nitrogen at time equal to zero. Calculate the exposure time required for the nitrogen concentration at the slab's axis to reach 90% of its equilibrium value. The slab thickness is 0.318 cm, and the diffusivity of nitrogen in PP is $3.87 \times 10^{-8}$ cm²/s at room temperature.

**Solution**

Figure 5.8 (p. 107) shows the PP slab exposed to constant concentration nitrogen gas. We assume that the slab edges are kept at this constant nitrogen concentration, which is also the equilibrium concentration of the whole slab. From Fig. 5.10 the 90% concentration at the centerline corresponds to $\mathcal{D}t/b^2 = 1.0$. Thus,

$$t = \frac{b^2}{\mathcal{D}} = \frac{(0.318/2)^2}{3.87 \times 10^{-8}} \, \text{s} = 7.56 \, \text{days}. \tag{4.33}$$

One way to reduce this long time is to increase the temperature of the experiment so that the diffusivity increases.

---

The graphical solutions in Figs. 5.10, 5.11, and 4.3 (pp. 108, 109, and 69) can be combined to give solutions to multidimensional problems. For example, unsteady diffusion into a short cylindrical pellet is considered to consist of mass transfer in both the radial and axial directions. Thus, the solution for a short cylinder is

$$\left[\frac{C_A - C_{A\infty}}{C_{A0} - C_{A\infty}}\right]_{\text{short cylinder}} = \left[\frac{C_A - C_{A\infty}}{C_{A0} - C_{A\infty}}\right]_{\text{infinite cylinder}}$$
$$\times \left[\frac{C_A - C_{A\infty}}{C_{A0} - C_{A\infty}}\right]_{\text{flat plate}} \tag{4.34}$$

where $C_{A0}$ is the concentration in the short cylinder at time equal to zero and $C_{A\infty}$ is the concentration at equilibrium (or infinite time) with the surrounding medium (Hines and Maddox, 1985). Note that the ratio $(C_A - C_{A\infty})/(C_{A0} - C_{A\infty})$ determines the difference of the concentration of species $A$ at a certain point inside the cylinder from the equilibrium value in a percentage form. Similarly, diffusion in a parallelepiped consists of mass transfer in all three directions $x$, $y$, and $z$, and, as such, the solution is represented by the product of the flat plate solutions in each direction.

**EXAMPLE 4.4**
**Unsteady Diffusion in Multidimensional Objects**
PP is extruded and pelletized into small cylindrical pellets. Pellets can be produced of three different sizes or surface areas. Pellets I have a diameter of $2R = 3$ mm and a length of $2b = 3$ mm, pellets II have respective dimensions 2 and 6.75 mm, and pellets III have respective dimensions 2 and 5.75 mm. Thus, pellets I have the same surface area and larger volume than pellets III, and pellets I have the same volume and larger surface area than pellets II. These PP pellets should be saturated with nitrogen gas and then processed. Which type pellet will have a concentration closer to equilibrium saturation after 86,400 s (24 hrs) exposure to nitrogen? $Ð$ for nitrogen in PP at room temperature is $3.87 \times 10^{-8}$ cm$^2$/s.

**Solution**

For type I pellets at 86,400 s

$$\frac{Ðt}{R^2} = 0.15; \quad \frac{Ðt}{b^2} = 0.15 \tag{4.35}$$

and therefore the corresponding concentrations at the centerline are:

$$\left[\frac{C_A - C_{A0}}{C_{A\infty} - C_{A0}}\right]_{\text{infinite cylinder}} = 0.34$$

$$\left[\frac{C_A - C_{A0}}{C_{A\infty} - C_{A0}}\right]_{\text{plate}} = 0.15. \tag{4.36}$$

The infinite cylinder value was obtained from Fig. 5.11 (p. 109) and the plate value was obtained from Fig. 5.10 (p. 108). Thus,

$$\left[\frac{C_A - C_{A\infty}}{C_{A0} - C_{A\infty}}\right]_{\text{pellets I}} = (1 - 0.34) \times (1 - 0.15) = 0.56. \tag{4.37}$$

Similarly,

$$\left[\frac{C_A - C_{A\infty}}{C_{A0} - C_{A\infty}}\right]_{\text{pellets II}} = (1 - 0.7) \times (1 - 0) = 0.3 \tag{4.38}$$

and

$$\left[\frac{C_A - C_{A\infty}}{C_{A0} - C_{A\infty}}\right]_{\text{pellets III}} = (1 - 0.7) \times (1 - 0) = 0.3. \tag{4.39}$$

Type I pellets are the closer to equilibrium saturation, and then come pellets II and III.

**EXAMPLE 4.5**
**Diffusion into a Falling Polymer Film**
Consider the sorption of a dye, $A$, into a falling thin polymer film as shown in Fig. 4.6. Assume that the flow of the film is fully developed and laminar; the process is at steady-state conditions; the solubility of the dye in the polymer is low, and no chemical reaction is present. Calculate the dye concentration profile in the polymer film as a function of time.

**Solution**

In this example we illustrate the use of the microscopic balances in

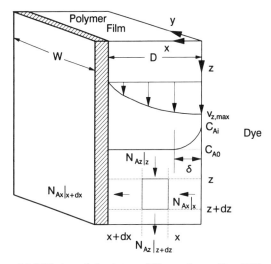

**Figure 4.6** Diffusion of dye into a falling polymer film. Differential element and fluxes.

solving mass transfer problems. Fig. 4.6 shows the falling polymer film, its velocity profile, the concentration of the dye, and a microscopic element, $dxdz$. It is evident from this figure that dye is transferred both in the $x$ and $z$ directions, so that a molar balance on the differential element yields:

$$(N_{Ax|x} - N_{Ax|x+dx})W\,dz + (N_{Az|z} - N_{Az|z+dz})W\,dx = 0, \tag{4.40}$$

where $W$ is the width of the film. Equation 4.40 becomes on dividing by the volume of the element and taking the limit as $dx$ and $dz$ go to zero:

$$\frac{\partial N_{Ax}}{\partial x} + \frac{\partial N_{Az}}{\partial z} = 0. \tag{4.41}$$

Note that this equation could have been taken directly from Table 4.6.

The concentration of the dye at the film–dye solution interface is $C_{Ai}$, and in the main bulk of the film it is equal to $C_{A0}$ (in this example $C_{A0} = 0$). If the dye is considered to be slightly soluble in the polymer (low concentrations are adequate for dyeing), it is logical to consider the thickness, $\delta$, over which the concentration changes from $C_{Ai}$ to $C_{A0}$, to be small relative to the film thickness, $D$. Consequently, the boundary conditions are:

B.C.1: $\quad C_A = C_{A0} \quad$ at $\quad z = 0 \quad$ for all $x$ $\qquad$ (4.42)

B.C.2: $\quad C_A = C_{A0} \quad$ at $\quad x = \infty \quad$ for all $z$ $\qquad$ (4.43)

B.C.3: $\quad C_A = C_{Ai} \quad$ at $\quad x = 0 \quad$ for all $z$. $\qquad$ (4.44)

Equation 4.13 can be written for the $x$ coordinate as:

$$N_{Ax} = -Ð_{AP}\frac{\partial C_A}{\partial x} + x_A(N_{Ax} + N_{Px}). \tag{4.45}$$

Because the dye is slightly soluble, $x_A \approx 0$, and hence we can write:

$$N_{Ax} \simeq -Ð_{AP}\frac{\partial C_A}{\partial x}. \tag{4.46}$$

Similarly, in the $z$ direction:

$$N_{Az} = -Ð_{AP}\frac{\partial C_A}{\partial z} + C_A v_z^{\star}. \tag{4.47}$$

In this direction we postulate that the convective flow is the predominant one, so that:

$$N_{Az} \simeq C_A v_z^{\star}. \tag{4.48}$$

Also, because the penetration, $\delta$, of the dye into the film is small, we will be concerned with only the velocity at the surface, so that:

$$N_{Az} \simeq C_A v_{z,max}. \tag{4.49}$$

Equations 4.41, 4.46, and 4.49 can be combined as

$$Ð_{AP}\frac{\partial^2 C_A}{\partial x^2} = \frac{\partial C_A}{\partial t}, \tag{4.50}$$

where $t = z/v_{z,max}$. Note that this equation represents unsteady-state diffusion and that we could have obtained this equation if we had allowed the differential element $dxdz$ to move with velocity $v_{z,max}$. Also, note that Eq. 4.42 should be considered now as an initial condition, because $z = 0$ is equivalent to $t = 0$. The solution to Eq. 4.50 can be obtained by transforming it to an ordinary differential equation (see Bird et al., 1960, p. 125). After some algebraic manipulations this technique gives:

$$\frac{C_A - C_{A0}}{C_{Ai} - C_{A0}} = 1 - \frac{2}{\sqrt{\pi}}\int_0^{\eta} e^{-\eta^2}\, d\eta = 1 - \text{erf}(\eta), \tag{4.51}$$

where $\eta = x/\sqrt{4Ð_{AP}t}$ and $\text{erf}(\cdot)$ is the *error function*, which is tabulated in mathematical handbooks. Another type of problem treated with the same technique is that of diffusion in a semi-infinite slab (see Pr. 4C.3a). Finally, note that this approach can be applied to the problem of devolatilization of a falling polymer film.

## 4.2. DIFFUSIVITY, SOLUBILITY, AND PERMEABILITY IN POLYMER SYSTEMS

In this section we analyze diffusive mass transport, as applied to polymer processing applications. The focus is on the estimation of the diffusion coefficient and solubility in the following three systems: gas–polymer, liquid–polymer, and polymer–polymer. The goal of this section is for the reader to be able to estimate the parameters of diffusional mass transfer for any polymer system using formulas and tables. Note that the diffusion coefficient is one of the two primary parameters of mass transfer, the other being the convective mass transfer coefficient discussed in Section 4.4. Typical values of diffusivity, solubility, and permeability can be found in the extensive collection by Brandrup and Immergut (1989).

### 4.2.1. DIFFUSIVITY AND SOLUBILITY OF SIMPLE GASES

Simple gases in the following context are the noneasily condensable gases (low boiling and critical points) with weak molecular interactions. Typical examples are nitrogen, oxygen, helium, and so on at relatively low pressures or equivalently low activities. These gases can also be called *permanent gases*. Carbon dioxide ($CO_2$), sulfur dioxide ($SO_2$), methane ($CH_4$), and other similar gases can also be considered simple gases at relatively low pressures, although they are condensable. The model law, which finds direct application to the diffusivity and solubility of simple gases, is the *Lennard-Jones* equation:

$$\phi(r) = 4\varepsilon\left[\left(\frac{\sigma}{r}\right)^{12} - \left(\frac{\sigma}{r}\right)^{6}\right], \tag{4.52}$$

where $\phi(r)$ is the intermolecular energy of two molecules $r$ distance apart, $\varepsilon$ is the potential energy constant, and $\sigma$ is the potential length constant. Note that $\varepsilon$ and $\sigma$ are also called the *Lennard-Jones scaling factors*. Division of $\varepsilon$ by the Boltzmann constant $k$ gives the *Lennard-Jones temperature* $\varepsilon/k$ in units of Kelvin, and $\sigma$ is also called the *collision diameter* of the molecule. Properties of simple gases can be found in Table 4.8 and an extensive collection appears in Reid et al. (1977).

Simple gases interact weakly with polymers with a consequence that their diffusion behavior is *Fickian*. This means that the diffusion coefficient $Ð$ does not depend on concentration and time but only on temperature. The dependence on temperature is expressed by an

Arrhenius-type equation as

$$Ð = Ð_0\exp(-E_Ð/R_gT), \tag{4.53}$$

because diffusion of gases is considered to be a thermally activated process. $Ð_0$ and $E_Ð$ are constants for the particular gas–polymer system. $E_Ð$ is the activation energy of diffusion, $R_g$ is the universal gas constant, and $Ð_0$ is the preexponential factor in the same units as $Ð$. The estimation of these two parameters follows.

The activation energy of diffusion, $E_Ð$, is the most important parameter in the diffusion process. It is the energy necessary for the

**TABLE 4.8 Boiling ($T_b$, °K), Critical ($T_{cr}$, °K) and Lennard-Jones ($\varepsilon/k$, °K) Temperatures and Collision Diameter ($\sigma$, nm) of Various Gases**

| Gas | $T_b$ (°K) | $T_{cr}$ (°K) | $\varepsilon/k$ (°K) | $\sigma$ (nm) |
|---|---|---|---|---|
| He | 4.2 | 5.2 | 10.22 | 25.51 |
| Air | | | 78.6 | 37.11 |
| $N_2$ | 77.4 | 126.2 | 71.4 | 37.98 |
| $O_2$ | 90.2 | 154.6 | 106.7 | 34.67 |
| $H_2$ | 20.4 | 33.2 | 59.7 | 28.27 |
| $CO_2$ | 194.7 | 304.2 | 195.2 | 39.41 |
| CO | 81.7 | 132.9 | 91.7 | 36.90 |
| Ar | 87.3 | 150.8 | 93.3 | 35.42 |
| $CCl_4$ | 349.7 | 556.4 | 322.7 | 59.47 |
| $CH_4$ | 111.7 | 190.6 | 148.6 | 37.58 |
| HCl | 188.1 | 324.6 | 344.7 | 33.39 |
| $Cl_2$ | 238.7 | 417.0 | 316.0 | 42.17 |
| $H_2O$ | 373.2 | 647.3 | 809.1 | 26.41 |
| $NH_3$ | 239.7 | 405.6 | 558.3 | 29.00 |
| o-xylene | 417.6 | 630.2 | 532.3 | 60.00 |
| $C_6H_6$ | 353.3 | 562.1 | 412.3 | 53.49 |
| $n$-$C_6H_{14}$ | 341.9 | 507.4 | 399.3 | 59.49 |
| $n$-$C_4H_{10}$ | 272.7 | 425.2 | 531.4 | 46.87 |
| $C_3H_8$ | 231.1 | 369.8 | 237.1 | 51.18 |
| $n$-$C_5H_{12}$ | 309.2 | 469.6 | 341.1 | 57.84 |
| $CClF_3$ (CFC-13) | 191.7 | 302.0 | 248.24 | 46.40 |
| $CCl_2F_2$ (CFC-12) | 243.4 | 385.0 | 316.2 | 49.48 |
| $C_2Cl_2F_4$ (CFC-114) | 276.9 | 418.9 | 349.64 | 54.87 |
| $CHClF_2$ (CFC-22) | 232.4 | 369.2 | 305.66 | 45.77 |

(Data from Reid et al., 1977.)

and $Ð_0$ is the preexponential factor in the same units as $Ð$. The estimation of these two parameters follows.

The activation energy of diffusion, $E_Ð$, is the most important parameter in the diffusion process. It is the energy necessary for the gas molecule to jump into a new position ("hole"). It is thus obvious that the larger the size of the diffusant molecule, the higher the activation energy and the lower the diffusivity are. The size of the gas molecule $x$ is determined by its collision diameter, $\sigma_x$. If nitrogen gas is taken as the standard diffusing gas, then the activation energy is given by the following relations (Van Krevelen, 1990) for elastomers (and for polymers in the rubbery state):

$$10^{-3}\frac{E_Ð}{R_g}=\left(\frac{\sigma_x}{\sigma_{N_2}}\right)^2[7.5-2.5\times10^{-4}(298-T_g)^2]\pm0.6, \qquad (4.54)$$

and for glassy amorphous polymers:

$$10^{-3}\frac{E_Ð}{R_g}=\left(\frac{\sigma_x}{\sigma_{N_2}}\right)^2[7.5-2.5\times10^{-4}(T_g-298)^{3/2}]\pm1.0. \qquad (4.55)$$

In these equations $R_g=8.314$ J/mol $^\circ$K, and $E_Ð$ is in kJ, so that the ratio $E_Ð/R_g$ is in 1,000 $^\circ$K. Also, $T_g$ is in $^\circ$K.

The other important parameter is the constant $Ð_0$. This was found to correlate with $E_Ð$ rather well (Van Krevelen, 1990). For elastomers

and glassy amorphous materials the correlations are:

$$\log Ð_0=\frac{E_Ð\times10^{-3}}{R_g}-4.0\pm0.4 \text{ for elastomers} \qquad (4.56)$$

and

$$\log Ð_0\simeq\frac{E_Ð\times10^{-3}}{R_g}-5.0\pm0.8$$

for glassy amorphous polymers. (4.57)

In these equations $Ð_0$ is in cm$^2$/s.

Finally, the diffusivity of a semi-crystalline material, $Ð_{sc}$, can be approximated as follows:

$$Ð_{sc}\simeq Ð_a(1-\phi_c), \qquad (4.58)$$

where $\phi_c$ is the crystallinity of the material, and $Ð_a$ is the diffusivity of the corresponding completely amorphous material. In other words, the crystalline regions are considered impermeable to the gas. A number of diffusivity data for common gas–polymer systems at room temperature (25°C) are compiled in Table 4.9. Brandrup and Immergut (1989), Duda et al. (1973), and Durrill and Griskey (1966) present some other data of gas diffusivities in polymers at elevated temperatures.

**TABLE 4.9 Diffusivity ($Ð$, $10^{-6}$ cm$^2$/s) and Activation Energy of Diffusion ($E_Ð/R_g$, $10^3$ $^\circ$K) for Various Gas–Polymer Systems at 298°K**

| Gas / Polymer | N$_2$ | | O$_2$ | | CO$_2$ | |
|---|---|---|---|---|---|---|
| | $Ð$ | $E_Ð/R_g$ | $Ð$ | $E_Ð/R_g$ | $Ð$ | $E_Ð/R_g$ |
| Polybutadiene | 1.1 | 3.60 | 1.5 | 3.40 | 1.05 | 3.65 |
| Polychloroprene (Neoprene rubber) | 0.24 | 5.18 | 0.38 | 4.74 | 0.23 | 5.40 |
| cis-1,4-polyisoprene (natural rubber) | 1.1 | 4.35 | 1.73 | 4.03 | 1.25 | 4.13 |
| Silicone rubber | 15 | 1.35 | 25 | 1.10 | 15 | 1.35 |
| HDPE | 0.10 | 4.5 | 0.17 | 4.40 | 0.12 | 4.25 |
| LDPE | 0.35 | 4.95 | 0.46 | 4.80 | 0.37 | 4.60 |
| PET | 0.0014 | 5.25 | 0.0036 | 5.50 | 0.0015 | 5.95 |
| PS | 0.06 | 4.25 | 0.11 | 4.15 | 0.06 | 4.35 |
| PVC, unplasticized | 0.0038 | 7.45 | 0.012 | 6.55 | 0.0025 | 7.75 |
| PVAc | 0.03 | 6.15 | 0.056 | 7.30 | – | – |
| PC | 0.015 | 4.35 | 0.021 | 3.85 | 0.005 | 4.50 |
| Poly(ethyl methacrylate) | 0.17 | 5.14 | 0.89 | 3.82 | 3.79 | 3.98 |
| Poly(tetrafluoro ethylene) | 0.088 | 3.60 | 0.152 | 3.16 | 0.095 | 3.44 |
| Teflon (FEP) | 0.0948 | 4.63 | 0.184 | 4.17 | 0.105 | 4.40 |

(Data from Brandrup & Immergut, 1989.)

---

**EXAMPLE 4.6**
**Estimation of $D$**

Estimate the diffusivity of oxygen in polycarbonate (PC) at 298°K.

**Solution**

The collision diameter of oxygen is 34.67 nm (from Table 4.8) and the $T_g$ of PC is 150°C=423°K (glassy amorphous polymer). From Eq. 4.55 we estimate the activation energy as:

$$10^{-3}\frac{E_Ð}{R_g}=\left(\frac{34.67}{37.98}\right)^2[7.5-2.5\times10^{-4}(423-298)^{3/2}]\pm1.0$$

$$=5.96\pm1.0, \qquad (4.59)$$

which gives $E_Ð=49.6\pm8.3$ kJ/mol. This value is higher than the experimentally observed value of 32.2 kJ/mol (Brandrup & Immergut 1989). $Ð_0$ (in cm$^2$/s) is calculated from Eq. 4.57 as: $\log Ð_0=0.96\pm0.8$. Finally, Eq. 4.53 gives the diffusivity at 298°K as:

$$Ð(298)=10^{0.96}\exp\left(\frac{-5960}{298}\right)=1.8\times10^{-8} \text{ cm}^2/\text{s}, \qquad (4.60)$$

which compares very well with the value given in Table 4.9.

Solubility, $S$, is the amount of gas dissolved in a polymer matrix at equilibrium with a partial pressure, $P$, and it is defined as:

$$\frac{V_A}{V_P} = S(P)P, \tag{4.61}$$

where $V_A$ is the volume of gas (at standard temperature, 298°K and pressure, 0.1013 MPa = 1 atm, conditions; STP) dissolved into the polymer per unit volume of the solution, and $V_P$ is the volume of polymer per unit volume of the solution. $V_A$ is related to $\rho_A$ via the expression: $\rho_A = d_A V_A$, where $d_A$ is the density of species $A$ in g/cm$^3$. The unit of $S$ is: cm$^3$(STP)/cm$^3 \cdot$ Pa. Eq. 4.61 for $S(P)$ = const. is also called *Henry's law*, and it is generally followed by substances at low concentrations. For organic vapors, the solubility is usually expressed in weight of vapor per weight of polymer per unit pressure, and the definition is similar to Eq. 4.61 with the substitution of $\rho_A$ and $\rho_P$ for $V_A$ and $V_1$, respectively. The solubility, $S$, follows an Arrhenius-type expression with temperature:

$$S = S_0 \exp(-\Delta \bar{H}_S/R_g T), \tag{4.62}$$

where $S_0$ is the preexponential factor and $\Delta \bar{H}_S$ is the molar heat of sorption. Note that the ratio $\Delta \bar{H}_S/R_g$ has the unit of temperature (°K).

The diffusivity was described as a function of the glass transition temperature, $T_g$, and degree of crystallinity, $\phi_c$, of the polymer and the Lennard-Jones temperature, $\varepsilon/k$, of the gas. Similarly, the solubility can be estimated with good accuracy from the same variables. The pertinent expressions are (Van Krevelen, 1990): for elastomers (and for polymers in the rubbery state):

$$10^{-3}\frac{\Delta \bar{H}_S}{R_g} = 1.0 - 0.010\frac{\varepsilon}{k} \pm 0.5 \tag{4.63}$$

$$\log S_0 = -5.5 - 0.005\frac{\varepsilon}{k} \pm 0.8, \tag{4.64}$$

and for glassy amorphous polymers:

$$10^{-3}\frac{\Delta \bar{H}_S}{R_g} = 0.5 - 0.010\frac{\varepsilon}{k} \pm 1.2 \tag{4.65}$$

$$\log S_0 = -6.65 - 0.005\frac{\varepsilon}{k} \pm 1.8. \tag{4.66}$$

Finally, the solubility for semi-crystalline polymers, $S_{sc}$, depends on the degree of crystallinity, $\phi_c$, and the solubility for the completely amorphous polymers, and it can be approximated as follows:

$$S_{sc} = S_a(1 - \phi_c). \tag{4.67}$$

In other words, the gas is soluble in the amorphous regions only.

At 298°K, the solubility of gases in elastomers and glassy amorphous polymers can be expressed as ($S$ in cm$^3$/cm$^3 \cdot$ Pa):

$$\log S(298) = -7.0 + 0.010\frac{\varepsilon}{k} \pm 0.25 \tag{4.68}$$

for elastomers and

$$\log S(298) = -7.4 + 0.010\frac{\varepsilon}{k} \pm 0.6 \tag{4.69}$$

for glassy amorphous polymers. The nature of the polymer affects the solubility slightly, and the size of the gas molecules affects the sign of the heat of sorption: dissolution of small gas molecules is endothermic (positive $\Delta \bar{H}_S$), whereas larger gas molecules cause exothermic dissolution (negative $\Delta \bar{H}_S$). Table 4.10 summarizes some data on solubility for common gas–polymer systems. More data can be found in Brandrup and Immergut (1989), Gorski et al. (1985), Cheng and Bonner (1978), Stiel and Harnish (1976), Duda et al. (1973), Durrill and Griskey (1966) and (1969). As a rule of thumb with the solubility of nitrogen taken as 1, that of oxygen is about 2 and that of carbon dioxide is 25.

**TABLE 4.10 Solubility ($S$, $10^{-6}$ cm$^3$ (STP)/cm$^3 \cdot$ Pa) and Heat of Sorption ($\Delta \bar{H}_S/R_g$, $10^3$ °K) for Various Gas–Polymer Systems at 298°K**

| Gas / Polymer | N$_2$ | | O$_2$ | | CO$_2$ | |
|---|---|---|---|---|---|---|
| | $S$ | $\Delta \bar{H}_S/R_g$ | $S$ | $\Delta \bar{H}_S/R_g$ | $S$ | $\Delta \bar{H}_S/R_g$ |
| Polybutadiene | 0.44 | 0.51 | 0.96 | 0.14 | 9.87 | −1.06 |
| Polychloroprene (Neoprene rubber) | 0.36 | 0.16 | 0.74 | 0.25 | 8.19 | −1.15 |
| cis-1,4-polyisoprene (natural rubber) | 0.55 | 0.25 | 1.02 | −0.51 | 9.20 | −1.50 |
| Silicone rubber | 0.81 | — | 1.26 | — | 4.30 | — |
| HDPE | 0.15 | 0.24 | 0.18 | −0.20 | 0.22 | −0.66 |
| LDPE | 0.23 | 0.95 | 0.47 | 0.30 | 2.54 | 0.05 |
| PET | 0.39 | −1.37 | 0.69 | −1.56 | 13.0 | −3.78 |
| PS | — | — | 0.55 | — | 6.5 | — |
| PVC, unplasticized | 0.23 | 0.85 | 0.29 | 0.14 | 4.7 | −0.94 |
| PVAc | 0.2 | — | 0.64 | −0.55 | — | — |
| PC | 0.28 | — | 5.03 | −1.55 | 1.24 | −2.62 |
| Poly(ethyl methacrylate) | 0.57 | −0.25 | 0.84 | 0.55 | 11.3 | −0.51 |
| Poly(tetrafluoro ethylene) | 1.20 | −0.66 | 2.1 | −0.87 | 9.2 | −1.76 |
| Teflon (FEP) | 1.25 | −0.95 | 2.02 | −1.11 | 9.08 | −1.89 |

(Data from Brandrup & Immergut, 1989.)

**EXAMPLE 4.7**
**Estimation of $S$**

Estimate the solubility of oxygen in PC and poly(vinyl acetate) (PVAc) at 298°K.

**Solution**

From Table 4.8 we get $\varepsilon/k = 106.7°K$ for oxygen. Thus, Eq. 4.69 gives for both polymers (they are in the glassy amorphous state) the mean

value for solubility as:

$$S(298) = 4.64 \times 10^{-7} \, cm^3(STP)/cm^3 \cdot Pa \qquad (4.70)$$

with the values at the extremes being $1.9 \times 10^{-6}$ and $1.2 \times 10^{-7} \, cm^3/cm^3 \cdot Pa$. Table 4.10 shows that the theoretical and experimental values for PVAc are very close, and that the experimental value for PC is a little higher than the upper limit of the theoretical value.

---

### 4.2.2. PERMEABILITY OF SIMPLE GASES AND PERMACHOR

The product of the diffusivity and the solubility is called permeability, $\bar{P}$:

$$\bar{P} \equiv Ð S. \qquad (4.71)$$

It determines the amount of the diffusing species passing through a polymer film of unit thickness per unit cross-sectional area, per unit time and at a unit pressure difference. Its primary applications are in polymer membrane separations and polymer film packaging. More specifically, the ability of certain polymer films to allow only minimal amounts of gases (e.g., oxygen, carbon dioxide, moisture, flavors, and odors) to be transported across them makes these films useful for food and beverage packaging. In this case the polymers are called *barrier* polymers.

The physical meaning of permeability is better understood from the schematic in Fig. 4.7. Consider a polymer film (or membrane) of thickness $\ell = 2b$ and a diffusing gas $A$. This film separates two regions of different pressures (or concentrations) of species $A$. The left region contains high pressure and the right low pressure, so that the diffusion takes place from left to right. At steady state and in cases of low mass fractions of $A$ inside the film, Eq. 4.16 simplifies to:

$$n_{Az} = -\rho Ð_{AP} \frac{d\omega_A}{dz} = \frac{\rho Ð_{AP}}{2b}(\omega_{A0} - \omega_{A\ell}). \qquad (4.72)$$

By transforming $\omega$ to $V$ and using Eq. 4.61, Eq. 4.72 becomes:

$$n_{Az} = Ð_{AP} \frac{d_A}{2b}(V_{A0} - V_{A\ell}) = Ð_{AP} S \frac{d_A V_P}{2b}(P_{A0} - P_{A\ell}). \qquad (4.73)$$

Finally, combining Eqs. 4.71 and 4.72 yields:

$$n_{Az} = \bar{P} \frac{d_A V_P}{2b} \Delta P \simeq \bar{P} \frac{d_A}{2b} \Delta P, \qquad (4.74)$$

where $\Delta P$ is the difference in pressures of species $A$ of the left and the right regions, and the approximate relation holds for $V_P \simeq 1$.

Equation 4.74 shows clearly that the mass transfer through the membrane or film depends on the permeability $\bar{P}$ of the gas $A$ through the polymer. The units of $\bar{P}$ are: $cm^3(STP)/s \cdot cm^3 \cdot Pa$ or $cm^2/s \cdot Pa$. Also, the unit of *Barrer* is frequently used in the literature: $1 \, Barrer = 10^{-10} \, cm^3(STP) \cdot cm/cm^2 \cdot s \cdot cm \, Hg = 7.5 \times 10^{-14} \, cm^2/s \cdot Pa$. Combining Eqs. 4.53, 4.62 and 4.71 we get:

$$\bar{P} = \bar{P}_0 \exp\left(\frac{-E_P}{R_g T}\right) = Ð_0 S_0 \exp\left(-\frac{E_Ð + \Delta\bar{H}_s}{R_g T}\right), \qquad (4.75)$$

so that the activation energy for permeability is the sum of the activation energies for diffusion and solubility, that is, $E_P = E_Ð + \Delta\bar{H}_S$, and $\bar{P}_0 = Ð_0 S_0$. From available experimental data the following relations can provide an approximation for the permeability values (Van Krevelen, 1990): for elastomers:

$$\log \bar{P}_0 = -10.1 + 10^{-3} \frac{E_P}{R_g} \pm 0.25, \qquad (4.76)$$

and for glassy amorphous polymers:

$$\log \bar{P}_0 = -11.25 + 10^{-3} \frac{E_{\bar{P}}}{R_g} \pm 0.75. \qquad (4.77)$$

Experimental values of permeability of nitrogen at room temperature through various elastomers, semi-crystalline, and amorphous polymers showed great variation depending on the polymer. Thus, silicone rubber showed the highest permeability of $10^{-11} \, cm^2/s \cdot Pa$, whereas poly(vinylidene chloride) showed the lowest of $4 \times 10^{-17} \, cm^2/s \cdot Pa$. Table 4.11 provides rules of thumb on relative permeability and activation energy.

A more accurate way of predicting the barrier properties of polymers is to correlate polymer structure and morphology to permeability. Salame (1986) introduced the idea of the *permachor*, $\pi$, which assigns a value for the polymer chain based on the specific values for the various structural units of that chain. Although three gases were studied extensively ($N_2$, $O_2$, and $CO_2$), the approach can be extended to other gases as well. The permeability is defined as

$$\bar{P}(298) = \bar{P}^\star(298) \exp(-s\pi), \qquad (4.78)$$

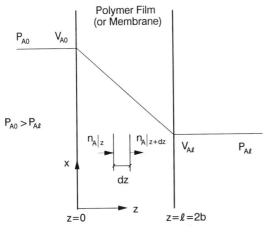

**Figure 4.7** Permeation across a polymer film (or membrane).

**TABLE 4.11 Relative Permeability Parameters for Various Gases**

| Gas / Parameter | $\bar{P}$ | $E_P$ |
|---|---|---|
| $N_2$ | 1 | 1 |
| CO | 1.2 | 1 |
| $CH_4$ | 3.4 | 1 |
| $O_2$ | 3.8 | 0.86 |
| He | 15 | 0.62 |
| $H_2$ | 22.5 | 0.70 |
| $CO_2$ | 24 | 0.75 |
| $H_2O$ | (550) | 0.75 |

(Data from Van Krevelen, 1990.)

**TABLE 4.12 Preexponential Permeability Factor, $\bar{P}\star$, and Scaling Factor, $s$**

| Gas / Factor | $\bar{P}\star$ (298) ($cm^3$[STP]·$cm/cm^2$·s·Pa) | $s$(298) ($cm^3$/cal) |
|---|---|---|
| $N_2$ | $1.35 \times 10^{-12}$ | 0.121 |
| $O_2$ | $3.98 \times 10^{-12}$ | 0.112 |
| $CO_2$ | $2.48 \times 10^{-11}$ | 0.122 |

(Data from Salame, 1986.)

**TABLE 4.13 Group Contributions to the Molar Permachor**

| Group | $\pi_i$(cal/$cm^3$) | Group | $\pi_i$(cal/$cm^3$) |
|---|---|---|---|
| $-CH_2-$ | 15 | $-CH(OH)-$ | 255 wet 100 |
| $\underset{|}{-\overset{|}{C}-}$ | −50 | $-CH(CN)-$ | 205 |
| $-CH(CH_3)-$ | 15 | $-CHF-$ | 85 |
| $-CH(C_6H_5)-$ | 39 | $-CF_2-$ | 120 |
| $-CH(i \text{ butyl})-$ | −1 | $-CHCl-$ | 108 |
| $-C(CH_3)_2-$ | −20 | $-CCl_2-$ | 155 |
| $-CH=CH-$ | −12 | $-CH(CH_2Cl)-$ | 50 |
| $-CH=C(CH_3)-$ | −30 | $-Si(CH_3)_2-$ | −116 |
| $-CH=C(Cl)-$ | 33 | $-O-$ | 70 |
| | | $\overset{O}{\underset{\|}{-C-O-}}$ | 102 |
| ⬡(H) | −54 | $\overset{O}{\underset{\|}{-O-C-O-}}$ | 24 |
| ⬡ | 60 | $\overset{O}{\underset{\|}{-C-NH-}}$ | 309 wet 210 |
| (dimethyl benzene ring with CH₃, CH₃) | −44 | $\underset{\|}{-Si-}$ | −146 |
| $\underset{\|}{-CH-}$ | 0 | | |

(Data from Salame, 1986.)

where $s$ is a scaling factor and $\bar{P}\star$ is a preexponential permeability factor. These factors are constant for a specific gas (Table 4.12). The values reported in Table 4.12 refer to both elastomers and glassy amorphous materials. The method of calculating $\pi_a$, that is, the permachor for completely amorphous materials, is simply:

$$\pi_a = \frac{1}{n}\sum_{i=1}^{n}\pi_i, \tag{4.79}$$

**TABLE 4.14 Values of the Permachor, $\pi$, for Various Polymers**

| Polymer | $\pi$, cal/$cm^3$ |
|---|---|
| Elastomers | |
| Silicone rubber | −23 |
| Butyl rubber (polyisobutylene) | −2 |
| *Natural rubber* (cis-1,4-polyisoprene) | *0* |
| Butadiene rubber (polybutadiene) | 6 |
| Poly-4-methyl-1-pentene | 7 |
| EPR elastomer (unmodified) | 15 |
| Neoprene rubber (polychloroprene) | 21 |
| Glassy amorphous polymers | |
| Hydropol (hydrogenated polybutadiene) | 15 |
| Polystyrene | 27 |
| Bisphenol-A polycarbonate | 31 |
| Poly(vinyl fluoride)(quenched) | 50 |
| Poly(ethylene terephthalate) | 59 |
| Poly(vinyl acetate) | 61 |
| Poly(vinyl chloride) | 61 |
| Poly(vinylidene chloride) | 86 |
| Polyacrylonitrile | 110 |
| Semi-crystalline polymers | |
| Low-density polyethylene ($\phi_c = 0.43$) | 25 |
| Polypropylene ($\phi_c = 0.60$) | 31 |
| High-density polyethylene ($\phi_c = 0.74$) | 39 |
| Poly(vinyl fluoride) ($\phi_c = 0.40$) | 59 |
| Poly(ethylene terephthalate) ($\phi_c = 0.30$) | 65 |
| Nylon 6,6 ($\phi_c = 0.40$) | 73 (dry) 60 (wet) |
| Nylon 6 ($\phi_c = 0.60$) | 80 (dry) 67 (wet) |
| Poly(*vinylidene chloride*) (oriented and crystalline) | *100* |
| Poly(vinyl alcohol) (dry, $\phi_c = 0.70$) | 157 |

(Data from Salame, 1986.)

where $n$ is the total number of individual groups per structural unit of the macromolecule, and $\pi_i$ is the individual group value of the permachor. Typical $\pi_i$ values are shown in Table 4.13.

For semi-crystalline materials, with crystallinity $\phi_c$, the permachor $\pi_{sc}$ should be calculated based on the amorphous $\pi_a$ as follows:

$$\pi_{sc} \simeq \pi_a - 18\ln(1-\phi_c), \tag{4.80}$$

and for oriented crystallites the following correction should be made:

$$\bar{P}_{oriented,sc} = \frac{\bar{P}_{sc}}{\tau_0} \simeq \bar{P}_{sc}\frac{(1-\phi_c)^{1/2}}{1.13}, \tag{4.81}$$

where $\tau_0$ is the *tortuosity* of crystallites. Based on these expressions one can calculate the total permachor values, $\pi$, for various polymers (Table 4.14). Note that the permachor value scale has a zero value assigned to natural rubber (cis-1,4-polyisoprene) and a value of 100 assigned to oriented and crystalline poly(vinylidene chloride). The following example illustrates the application of the permachor algorithm to estimate the permeability of gas–polymer systems.

Finally, there exists another one technique that predicts the transport properties of polymers by performing summations over atom and bond contributions defined in terms of connectivity indices. For more about this technique the reader is referred to Bicerano (1993).

**EXAMPLE 4.8**
**Permeability of Nylon 6 and Nylon 6,6 Films to $CO_2$**
Calculate the $CO_2$ permeability of nylon 6 and nylon 6,6 films with 40% crystallinity at room temperature and compare it with the literature value (Brandrup & Immergut, 1989) of $5.2 \times 10^{-15}$ cm$^2$/s·Pa. Use the permachor estimation algorithm.

**Solution**

The structural unit of nylon 6 is:

$$[-(CH_2)_5-CONH-],$$

and that of nylon 6,6 is:

$$[-NH(CO)-(CH_2)_4-(CO)NH-(CH_2)_6-].$$

Their permachor is easily proved to be the same, and so we continue

our calculations with nylon 6,6 only. The permachor is calculated as follows (Table 4.13):

| | | |
|---|---|---|
| 10 | $-CH_2-$ | $= 10 \times 15 = 150$ cal/cm$^3$ |
| 2 | $-NH-CO-$ | $= 2 \times 309 = 618$ cal/cm$^3$ |
| $n = 12$ | | $\sum \pi_i = 768$ cal/cm$^3$. |

Thus $\pi_a = 768/12 = 64$ cal/cm$^3$. The value for the 40% crystalline nylon 6,6 is calculated from Eq. 4.80 as: $\pi_{sc} = 64 - 18 \ln(1 - 0.4) = 73$ cal/cm$^3$ (see also Table 4.14). The permeability of this film to $CO_2$ is calculated with the aid of Eq. 4.78 and Table 4.12 as:

$$\bar{P} = 2.48 \times 10^{-11} \exp(-0.122 \times 73) = 3.4 \times 10^{-15} \text{ cm}^2/\text{s·Pa}. \quad (4.82)$$

That value compares well with the literature value of $5.2 \times 10^{-15}$ cm$^2$/s·Pa taking into consideration that the literature value does not refer to any specific crystallinity.

## 4.2.3. MOISTURE SORPTION AND DIFFUSION

The diffusivity and solubility of water in polymers differs from that of simple gases to the extent that the water molecule interacts with the polymer. Based on this interaction, the polymer molecules are classified as *hydrophilic* or *hydrophobic*. The solubility of water in hydrophobic polymers, such as polyolefins and some polyesters, is very low and the diffusivity follows the rules of the other simple gases mentioned in Subsection 4.2.1. On the other hand, hydrophilic polymers, such as cellulose, poly(vinyl alcohol) and proteins, interact strongly with water, and water diffusivity increases with its content. A good approximation for this situation is (Van Krevelen, 1990):

$$\log Ð_w = \log Ð_{w=0} + 0.08w, \quad (4.83)$$

where $w$ is the water content in weight percent. The preexponential factor, $Ð_{0w}$, is taken as:

$$\log Ð_{0w} \simeq \frac{E_Ð \times 10^{-3}}{R_g} - 7, \quad (4.84)$$

where $E_Ð$ is in J/mol. In between the extremes of hydrophobic and hydrophilic polymers are less hydrophilic polymers, such as polyethers and polymethacrylates. Although the preexponential factor $Ð_{0w}$ depends on $E_Ð$ as in simple gases, the diffusivity itself decreases with water content as:

$$\log Ð_w = \log Ð_{w=0} - 0.08w. \quad (4.85)$$

The solubility of water in polymers can be estimated from contributions from individual groups of the structural unit. Table 4.15 presents these molar water contributions of various groups (Van Krevelen, 1990) in a manner similar to the permachor. The solubility in g/g or cm$^3$(STP)/g can be easily calculated from the contributions of Table 4.15. The molar volume of the structural unit is needed in case the solubility is needed in cm$^3$(STP)/cm$^3$. Finally, the heat of sorption is about 25 kJ/mol for nonpolar polymers and about 40 kJ/mol for polar polymers.

**EXAMPLE 4.9**
**Estimation of the Saturation Moisture Content of Polymers**
Estimate the saturation moisture content of poly(methyl acrylate) (PMA) and polystyrene (PS) at 25°C and relative humidity of 50%. Compare these values with experimental evalues from Crank and Park (1968, p. 263).

**Solution**

The structural unit of PMA is:

$$[-CH(COOCH_3)-CH_2-]$$

and from Table 4.15 the molar moisture content at 50% relative humidity is $(0.05 + 7.5 \times 10^{-5}) \approx 0.05$, or equivalently $0.05 \times 18/100 = 0.9/100$ g of water/g of polymer. This value compares really well with the experimental value of 0.99/100. Similarly the weight percent of water in PS is calculated to be 0.032%, which is identical to the experimental value of 0.032%.

## 4.2.4. PERMEATION OF HIGHER-ACTIVITY PERMEANTS

We considered the ideal cases in terms of diffusion and solubility, that is, Fickian diffusion and constant solubility (Henry's law), in Subsections 4.2.1, 4.2.2, and 4.2.3. However, complexities might arise at higher penetrant activities, because the high solubility in these cases can cause swelling of the polymer and nonlinearities of the coefficients of diffusion and solubility. In this subsection, we restrict ourselves to outlining the nonlinear behavior of some systems without going into details.

A typical example for amorphous rubbery polymers is given in Fig. 4.8 with the sorption isotherms of ethylene, carbon dioxide, methane and nitrogen in silicone rubber at 35°C (Koros & Hellums, 1989). Methane and nitrogen obey Henry's law at all pressures, whereas

**TABLE 4.15 Molar Water Content per Structural Group at Various Relative Humidities at 25°C**

| Group | Relative Humidity | | | | |
|---|---|---|---|---|---|
| | 0.3 | 0.5 | 0.7 | 0.9 | 1.0 |
| $-CH_3$ / $-CH_2-$ / $-CH\!\!<$ | $(1.5 \times 10^{-5})$ | $(2.5 \times 10^{-5})$ | $(3.3 \times 10^{-5})$ | $(4.5 \times 10^{-5})$ | $(5 \times 10^{-5})$ |
| ⬡ | 0.001 | 0.002 | 0.003 | 0.004 | 0.005 |
| $>C=O$ | 0.025 | 0.055 | (0.11) | (0.20) | (0.3) |
| $-C(\!\!\stackrel{O}{\scriptstyle O-}\!\!)$ | 0.025 | 0.05 | 0.075 | 0.14 | 0.2 |
| $>O$ | 0.006 | 0.01 | 0.02 | 0.06 | 0.1 |
| $-OH$ | 0.35 | 0.5 | 0.75 | 1.5 | 2 |
| $-NH_2$ | 0.35 | 0.5 | 0.75 | (1.5) | (2) |
| $-NH_3^+$ | | | 2.8 | 5.3 | |
| $-COOH$ | 0.2 | 0.3 | 0.6 | 1.0 | 1.3 |
| $-COO^-$ | 1.1 | 2.1 | 4.2 | | |
| $-C(\!\!\stackrel{O}{\scriptstyle NH-}\!\!)$ | 0.35 | 0.5 | 0.75 | 1.5 | 2 |
| $-Cl$ | 0.003 | 0.006 | 0.015 | 0.06 | (0.1) |
| $-CN$ | 0.015 | 0.02 | 0.065 | 0.22 | (0.3) |

(Reprinted by permission of the publisher from Van Krevelen, 1990.)

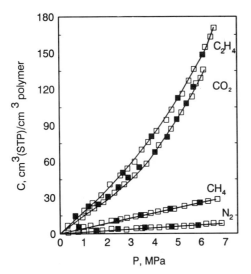

**Figure 4.8** Sorption isotherms of various gases in silicone rubber at 35°C. □ sorption; ■ desorption. (Reprinted by permission of the publisher from Koros & Hellums, 1989.)

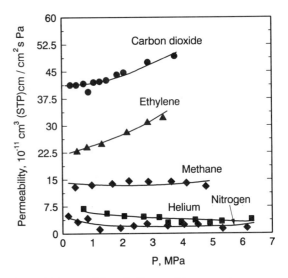

**Figure 4.9** Permeability of silicon rubber to various gases at 35°C. (Reprinted by permission of the publisher from Koros & Hellums, 1989.)

ethylene and carbon dioxide obey it at low pressures and deviate positively (from the initial slope) at high pressures. This positive deviation is called the *Flory-Huggins swelling behavior*. This behavior is mathematically expressed as:

$$\ln \frac{P}{P_0} = \ln \phi_A + (1 - \phi_A) + \chi(1 - \phi_A)^2, \qquad (4.86)$$

where $P$ is the solvent partial pressure, $P_0$ is the pure solvent vapor pressure, $\phi_A$ is the solvent volume fraction and $\chi$ is the interaction parameter of the solvent–polymer system. Equation 4.86 is valid in the absence of cross-linking and for a high degree of polymerization of the polymer. Furthermore, the permeability of silicon rubber to carbon dioxide and ethylene (Fig. 4.9) increases with pressure, while for helium and nitrogen it decreases with increasing pressure, whereas pressure has no effect on the permeability to methane. Note that the critical temperature of the first two gases is high, of the next two is low, and that of methane is intermediate.

The strong dependence of the diffusivity on the mass fraction of the diffusant, for intermediate size molecules such as organic solvents in rubbery materials, is shown in Fig. 4.10. The sorption isotherms of various gases in PC are shown in Fig. 4.11. The solubility of helium and nitrogen in PC varies linearly with pressure even at relatively high pressures, whereas the solubility of carbon dioxide and methane varies linearly with pressure only at relatively low and high pressures. Argon

shows intermediate behavior. Note that the shape of the $CO_2$ isotherm in PC is different from the shape of the $CO_2$ isotherm in silicone rubber (Fig. 4.8), that is, it shows negative deviation from the initial slope in PC and positive deviation in silicone rubber. The sorption isotherm of $CO_2$ in PC is called *Langmuir isotherm* (see also Pr. 4C.6b, or Vieth, 1991). Finally, the permeability of PC to carbon dioxide is shown in Fig. 4.12.

Diagrammatically, the effect of the penetrant activity on the transport properties is shown in Fig. 4.13 (Hopfenberg & Frisch, 1969), which is valid for amorphous polymers. Concentration-independent diffusion takes place at any temperature and low penetrant activities or at low temperatures and high penetrant activities. At high penetrant activities and temperatures lower than $T_g$, the transport of the penetrant causes high stresses and thus crazing and failure of the polymer. The so-called case-II (or relaxation-controlled; see also Section 4.3) transport takes place at medium to high activities and at temperatures below $T_g$. Finally, at lower temperatures, both relaxation and diffusion mechanisms are present, causing anomalous behavior.

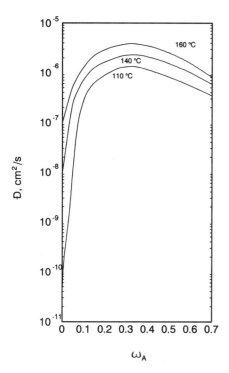

**Figure 4.10** Diffusion coefficient of toluene in the toluene-polystyrene system as a function of the toluene mass fraction. (Reprinted by permission of the publisher from J. S. Vrentas, J. L. Duda, and M. K. Lau, "Solvent Diffusion in Molten Polyethylene." *J. Appl. Polym. Sci.*, **27**, 3987-97, 1982.)

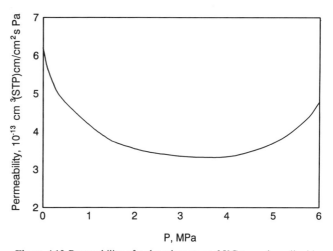

**Figure 4.12** Permeability of polycarbonate at 35°C to carbon dioxide. (Reprinted by permission of the publisher from Koros, & Hellums, 1989.)

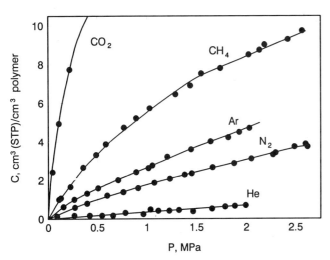

**Figure 4.11** Sorption isotherms of various gases in polycarbonate at 35°C. (Reprinted by permission of the publisher from Koros & Hellums, 1989.)

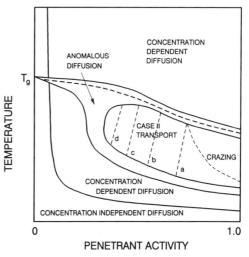

**Figure 4.13** Transport regions in the temperature-penetrant activity plane. Lines a, b, c, and d are lines of constant activation energy and decreasing in that order. (Reprinted by permission of the publisher from Hopfenberg & Frisch, 1969.)

In conclusion, permeation at relatively high pressures of higher-activity permeants, such as condensable gases, intermediate-size vapors, and liquids, presents deviations from the simple Henry's law. The Flory-Huggins and the Langmuir isotherms are two such deviations observed in rubbery and glassy materials, respectively. In terms of permeability as a function of pressure, a single gas can exhibit all possible behaviors, such as increasing, decreasing, and being constant, in a single graph. Finally, as a rule of thumb, the diffusivity of liquids in polymers is inversely proportional to the viscosity of the liquid.

### 4.2.5. POLYMER–POLYMER DIFFUSION

Diffusion in polymer–polymer systems is important in many practical applications, such as crack healing, polymer blending, welding, adhesion, elastomer tack, and polymer fusion. However, diffusion coefficient data are lacking because of the complexity of the polymer diffusion process. Typical techniques employed are infrared spectroscopy, spectroscopic ellipsometry and optical schlieren. Only a few studies of interdiffusion in polymer pairs are reported. Depending on the temperature and other physical parameters of the system, Fickian and Case-II diffusions (see Section 4.3) can be observed with varying relative importance (Jabbari & Peppas, 1993).

In polymer melts it is customary to measure the *self-diffusion*, $Đ_s$, and the *tracer diffusion*, $Đ^*$, coefficients (Kausch & Tirell, 1989). The self-diffusion refers to the diffusion of a macromolecule in a background of identical macromolecules. On the other hand, tracer diffusion refers to the diffusion of a small number of identical macromolecules (i.e., at the tracer level; usually in a tagged state) in a background which consists of macromolecules all of which are not identical to the diffusing macromolecules. Note that in a binary system of species $A$ and $B$, which approaches infinite dilution of species $A$, the diffusion coefficient of $A$ and $B$ becomes equal to the tracer diffusion coefficient of $A$, i.e., $Đ_{AB} = Đ_A^*$. Finally, the self-diffusion coefficients can be combined to yield the mutual diffusion coefficient of species $A$ and $B$.

Typical diffusivity values are in the range from $10^{-16}$ to $10^{-6}$ cm$^2$/s. Figures 4.14 and 4.15 present the self-diffusion data for the two polymers studied extensively in the literature, PE and PS, respectively. The correlations in these two figures show that the self-diffusion coefficient in an entangled linear polymer depends on the

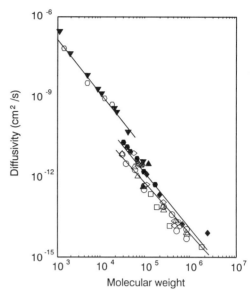

**Figure 4.15** Self-diffusion coefficient of molten polystyrene (PS) as a function of molecular weight at or near 175°C. Lines drawn have slope −2.

molecular weight of the diffusing species as follows (see also Pr. 4B.6)

$$Đ_s \propto (\text{MW})^{-2}. \tag{4.87}$$

The activation energy for self-diffusion of PE is about 23 kJ/mol, whereas that of PS varies widely, from 62.2 to 167 kJ/mol (Tirrell, 1984). There are no more extensive data in the literature for any other polymer system.

Equation 4.87 agrees with the predictions of the reptation theory applied to entangled polymer systems. According to this theory, each polymer chain is confined to a tube surrounding its own contour, whose walls are made up by the neighboring chains. The diffusion of the polymer chain is assumed to proceed primarily by reptation, which is similar to the motion of snakes as they move through a fixed set of obstacles. This diffusion path is the one that offers the least resistance, since the tube walls impede the lateral motions of the polymer chain. Finally, the molecular-weight dependence of the self-diffusion coefficient is calculated from the molecular-weight dependences of the radius of gyration, $R_G$, and the longest relaxation time, $\tau$, and by recognizing that the center of mass of the polymer chain will move a distance approximately equal to $R_G$ in time equal to $\tau$ (Tirrell, 1984, and Pr. 4B.6).

### 4.2.6. MEASUREMENT TECHNIQUES AND THEIR MATHEMATICS

This section is concerned with measurement techniques of the diffusivity and solubility from which the permeability can be easily calculated. In the following analysis we restrict ourselves to the measurement of constant values of $Đ$. Concentration- and position-dependent diffusivities are analyzed in Crank and Park (1968) and Crank (1975). Generally speaking, the techniques are, for permeability, steady state and time lag, and for diffusivity, sorption and desorption kinetics and concentration–distance curves. For self-diffusivity in polymer melts the techniques are (Tirrell, 1984): nuclear magnetic resonance, neutron scattering, radioactive tracer, and infrared spectroscopy.

In the steady-state permeation method, the surfaces of a polymer film of thickness $\ell = 2b$ are kept at constant gas pressures $P_{A0}$ and $P_{A\ell}$.

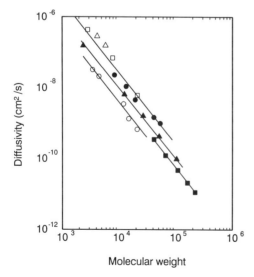

**Figure 4.14** Self-diffusion coefficient of molten polyethylene (PE) as a function of molecular weight at 176°C. Lines drawn have slope −2. (Reprinted by permission of the publisher from Tirrell, 1984.)

At steady state Eq. 4.75 applies, and the permeability $\bar{P}$ is found from this equation. If the solubility is known, then the diffusivity can be found from Eq. 4.71. For hollow cylinders and spheres, expressions similar to Eq. 4.74 hold. On the other hand, in the time lag method we deal with the unsteady state of the permeation of a diffusant through a slab of thickness $2b$. The surfaces of the slab are kept at concentrations $C_{A0}$ and at zero. The accumulated amount of gas that has passed through the slab in time $t$, $Q_t = -\mathcal{D}\int_0^t (\partial C_A/\partial z)_{z=0}\, dt$, is given by:

$$\frac{Q_t}{2bC_{A0}} = \frac{\mathcal{D}t}{(2b)^2} - \frac{1}{6} - \frac{2}{\pi^2}\sum_{n=1}^{\infty}\frac{(-1)^n}{n^2}\exp\left(\frac{-\mathcal{D}n^2\pi^2 t}{4b^2}\right). \tag{4.88}$$

At long times (i.e., when steady state is achieved) the exponential term becomes negligible, so that:

$$Q_t = \frac{\mathcal{D}C_{A0}}{2b}\left(t - 4\frac{b^2}{6\mathcal{D}}\right). \tag{4.89}$$

The intercept of the above equation with the time axis gives the so-called time lag as:

$$t_{\text{lag}} = \frac{4b^2}{6\mathcal{D}}. \tag{4.90}$$

Similar relations hold for cylinders and spheres.

In the sorption kinetics techniques, the mass uptake of a slab of thickness $2b$ at time $t$, $M_t = 2\int_0^b (C_A - C_{A0})\, dz$, relative to the maximum mass uptake, $M_\infty = 2b(C_{A\infty} - C_{A0})$, is given by:

$$\frac{M_1}{M_\infty} = 4\left(\frac{\mathcal{D}t}{(2b)^2}\right)^{1/2}\left(\frac{1}{\sqrt{\pi}} + 2\sum_{n=0}^{\infty}(-1)^n \mathrm{ierfc}\left(\frac{2nb}{\sqrt{4\mathcal{D}t}}\right)\right) \tag{4.91}$$

where $C_{A0}$ is the initial concentration in the slab, $C_{A\infty}$ is the final concentration in the slab, and the function $\mathrm{ierfc}(\cdot)$ is given by:

$$\mathrm{ierfc}(x) = \int_x^\infty \mathrm{erfc}(x')\, dx', \tag{4.92}$$

where $\mathrm{erfc}(\cdot)$ is the *complementary error function*, which is equal to $1\text{-erf}(\cdot)$. Eq. 4.91 is more accurate at short times (i.e., for $M_t/M_\infty \leq 0.6$, even if the infinite summation term is neglected), whereas the expression given in Pr. 4B.1 is more accurate for $M_t/M_\infty > 0.6$, even if only the first term in the summation is used. Fig. 4.16 shows a typical graph of mass uptake in a slab. At $t\to 0$, Eq. 4.91 yields:

$$\mathcal{D} = \frac{\pi(2b)^2}{16t}\left(\frac{M_t}{M_\infty}\right)^2. \tag{4.93}$$

Thus, $\mathcal{D}$ can be estimated from the slope of Fig. 4.16 at $t=0$. Another way to determine $\mathcal{D}$ using the data from the same figure is to evaluate it at $M_t/M_\infty = 1/2$ as follows:

$$\mathcal{D} = 0.04919\frac{(2b)^2}{t_{1/2}}. \tag{4.94}$$

A desorption experiment is similar to the sorption experiment just described. Experimentally, the mass gain or loss are plotted as a function of time, and the diffusivity is calculated from that plot.

Finally, the diffusion coefficient can also be deduced from the concentration–distance curves in the case of two semi-infinite media brought together and interdiffusing. For example, a polymer sheet and a solvent or even two polymer sheets are brought in contact, and the concentration–distance curves are calculated by measuring the refractive index or by analyzing the radioactivity of a tracer (Crank & Park, 1968). Problem 4C.3 presents an example of the concentration–distance

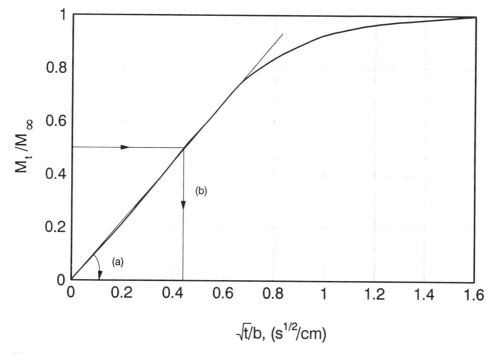

**Figure 4.16** Relative mass uptake, $M_t/M_\infty$, of a slab with constant surface concentration. The diffusivity can be calculated by either (a) the slope at $t=0$ or (b) at $M_t/M_\infty = 1/2$

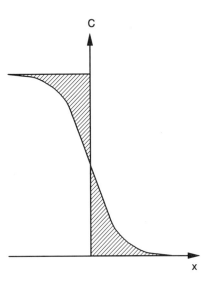

**Figure 4.17** Schematic of the concentration–distance curve used in the calculation of the diffusivity.

curve, which is also shown in Fig. 4.17 and mathematically given by:

$$C_A = \frac{1}{2} C_{A0} \operatorname{erfc}\left(\frac{x}{2\sqrt{\mathcal{D}t}}\right), \tag{4.95}$$

where $C_{A0}$ is the initial concentration of the radioactive tracer in one of the semi-infinite media.

## 4.3. NON-FICKIAN TRANSPORT

In some cases, solvents cause swelling of the polymers, and consequently the penetrant transport cannot be described by Fick's law even with a concentration-dependent diffusivity. Generally, this case is most frequently encountered during sorption in glassy polymers, whereas desorption from glassy polymers and either desorption or sorption in rubbery polymers exhibit mostly Fickian diffusion. To better understand the effect of the type of polymer on the diffusion process, we need to visualize the diffusion process as a sequence of creation of bulges between two macromolecules, diffusional jumps of the penetrant molecules to these bulges, and relaxation of the original bulges containing the penetrant molecules before the jumps. In the rubbery state, both creation of a new bulge (by activation) and the relaxation of the original bulge are completed in short time scales (short relaxation times), so that they do not cause any diffusion anomalies. However, in the glassy state the relevant time scales are long and the behavior is time-dependent. As temperature and/or penetrant concentration increase, the relaxation times of the glassy material shorten, and the diffusional behavior tends to resemble that of a rubbery material.

A convenient method of predicting whether the transport of a solvent in an amorphous polymer is Fickian or non-Fickian is to examine the diffusional Deborah number, $\mathrm{De}_{\mathcal{D}}$. This number is defined as the ratio of a characteristic relaxation time for the polymer–solvent system to the characteristic time of the diffusion process (Vrentas et al., 1975). Fickian transport is observed when either $\mathrm{De}_{\mathcal{D}} < 0.1$ or $\mathrm{De}_{\mathcal{D}} > 10$, whereas non-Fickian transport is observed when $\mathrm{De}_{\mathcal{D}} \approx 1$.

Alfrey et al. (1966) proposed the following classification of the diffusional processes:

1. Case-I, or Fickian, transport: The diffusion time scale is much longer than the relaxation time scale (Rubbery polymers usually exhibit such behavior).

2. Case-II diffusion: This is the other extreme of Case I, and it refers to the diffusion time scale being much shorter than the relaxation time scale.

3. Non-Fickian, or anomalous, transport: This process occurs when the diffusion and relaxation time scales are comparable.

Case-II diffusion has been associated with advancing fronts of the penetrant material, where these fronts also mark the regions of swollen and glassy polymers. An exponential dependence of the diffusivity on the concentration for large values of the exponential constant can produce advancing fronts. Both Cases I and II are considered simple diffusion cases, because they are described by a constant, which is the diffusivity in Case I and the constant velocity of the advancing front of the penetrant in Case II (see also Pr. 4C.8). In terms of the sorption kinetics, the mass uptake of a semi-infinite slab in Case I was shown (Fig. 4.16) to be proportional to $t^{1/2}$, in the early stages of the sorption process, whereas in Case II it is primarily proportional to $t$ (although at the very beginning it may include a term proportional to the square root of time). Non-Fickian systems lie between those two cases with the mass uptake being proportional to $t^n$, where $1/2 < n \leqslant 1$, or changing from one case to another. Non-Fickian systems also require two or more parameters for their description.

A typical example of Case-II diffusion is observed in the system PMMA–methanol (Thomas & Windle, 1978, 1982; Sarti & Doghieri, 1994). At 30°C and below, the system shows linear mass uptake with time (Fig. 4.18.a) and sharp concentration profiles (Fig. 4.18.b), whereas

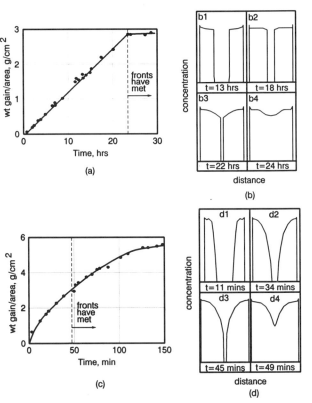

**Figure 4.18** Case II diffusion in the system PMMA-methanol. PMMA sheets are submerged into liquid methanol. (a) Mass uptake per unit area against time at 30°C. (b) Concentration of methanol in the PMMA sheets as a function of time at 30°C. (c) Mass uptake per unit area against time at 60°C. (d) Concentration of methanol in the PMMA sheets as a function of time at 60°C. (Reprinted by permission of the publisher from Thomas & Windle, 1978.)

at 60°C (boiling point of methanol is 65°C) the mass uptake is not linear but curved, and PMMA continues to sorb methanol after the two fronts meet (Fig. 4.18.c). Also, at 60°C the concentration profiles are not as sharp as before (Fig. 4.18.d). Finally, an example of combination of Fickian and Case-II diffusions, with relative importance depending on the temperature, is presented by Jabbari and Peppas (1993) for the interdiffusion of polystyrene and poly(vinyl methyl ether) at either 85°C or 105°C.

In conclusion, Case II and non-Fickian transport behaviors are frequently present in glassy polymer systems. Case-II transport, in particular was found to be associated with sharp penetrant fronts and linear mass uptake with time, whereas in the non-Fickian transport the mass uptake is proportional to $t^n$, where $1/2 < n \leqslant 1$. Methanol absorbed in PMMA exhibits Case-II diffusion characteristics at relatively low temperatures, whereas at higher temperatures a more peculiar behavior is noticed.

## 4.4. MASS TRANSFER COEFFICIENTS

In the preceding sections of this chapter we considered mass transfer inside a medium, and we presented correlations for the solubility and diffusivity of various penetrants in polymers. This approach is the most general and fundamental. However, mass transfer can also be described in an empirical and a simpler way using a mass transfer coefficient. The two approaches (i.e., the first one based on the diffusivity and the second one on the mass transfer coefficient) are practically equivalent and the choice between them depends on the difficulty of obtaining the respective data. A more thorough discussion of this topic is given by Bird et al. (1960) and Hines and Maddox (1985).

### 4.4.1. DEFINITIONS

Consider the steady-state diffusion of solute $A$ through a polymer membrane and into a stream with concentration $C_{A\infty}$ as shown in Fig. 4.19.a. The interface concentration is $C_{Ai}$. Similar to the definition of

the heat transfer coefficient, Eq. 5.122, the mass transfer coefficient is defined as:

$$k_c^{\bullet} \equiv \frac{J_{Az}^{\bigstar}}{C_{Ai} - C_{A\infty}} = -\frac{\mathcal{D}(\partial C_A/\partial z)_{z=0}}{C_{Ai} - C_{A\infty}}, \qquad (4.96)$$

where $k_c^{\bullet}$ is the convective local mass transfer coefficient (or simply the mass transfer coefficient). This coefficient also depends on the molar transfer rate of species $A$ and $B$ as (Table 4.3, Eq. U):

$$k_c^{\bullet} = \frac{N_A - x_{Ai}(N_A + N_B)}{C_{Ai} - C_{A\infty}}, \qquad (4.97)$$

where we have dropped the $z$ dependence of the molar rate, for simplicity. A similar expression holds for the convective local mass transfer coefficient $k_x^{\bullet}$ (based on $x_{Ai}$ and $x_{A\infty}$). The units of $k_c^{\bullet}$ and $k_x^{\bullet}$ are cm/s and mol/cm²·s, respectively.

The dependence of the mass transfer coefficients on the mass transfer rate distinguishes them from the heat transfer coefficients. However, in the majority of applications, this dependence is negligible and the mass transfer coefficients $k_c^{\bullet}$ and $k_x^{\bullet}$ can be replaced for $k_c$ and $k_x$, respectively. In such cases,

$$N_A = k_c(C_{Ai} - C_{A\infty}) = k_x(x_{Ai} - x_{A\infty}). \qquad (4.98)$$

The units of $k_c$ and $k_x$ are the same as the respective units of $k_c^{\bullet}$ and $k_x^{\bullet}$. Finally, the average mass transfer coefficients, $\bar{k}_c$ and $\bar{k}_x$, are defined as:

$$\mathbb{N}_A - x_{Ai}(\mathbb{N}_A + \mathbb{N}_B) = \bar{k}_c A_s(C_{Ai} - C_{A\infty}) \qquad (4.99)$$

$$\mathbb{N}_A - x_{Ai}(\mathbb{N}_A + \mathbb{N}_B) = \bar{k}_x A_s(x_{Ai} - x_{A\infty}), \qquad (4.100)$$

where $A_s$ is the area across which the mass transfer takes place, and

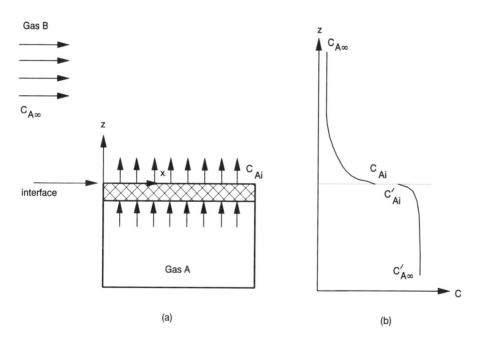

(a)

(b)

**Figure 4.19** (a) Mass transfer across a polymer film–fluid interface. (b) Mass transfer across an interface with two resistances present: the first in one side of the interface ($C'_{A\infty}$ to $C'_{Ai}$), and the second in the other side of the interface ($C_{Ai}$ to $C_{A\infty}$). Henry's law is applicable at the interface, between $C_{Ai}$ and $C'_{Ai}$.

$N_A$ and $N_B$ are the molar rates over the entire interface, in mol/s. The units of $\bar{k}_c$ and $\bar{k}_x$ are cm/s and mol/cm²·s, respectively.

The definitions for the mass transfer coefficients can be used to theoretically predict them using the diffusivity, concentrations, length scales, and fluid flow characteristics, thus rendering the two mass transfer approaches equivalent. This can be easily done in the cases of equimolar counter diffusion ($N_{Az} + N_{Bz} = 0$) and diffusion of $A$ through a stagnant film ($N_{Bz} = 0$) (Hines & Maddox, 1985, p. 140). Also, theoretical models of film, penetration, surface renewal and film penetration have been proposed for the estimation of the mass transfer coefficients at a fluid–fluid interface (Hines & Maddox, 1985, pp. 146–151).

Finally, Fig. 4.19.b shows the concentration profile across an interface where a concentration difference ($C_{A\infty}$ to $C_{Ai}$) is followed by a change in interfacial concentrations (Henry's law) and another concentration difference ($C_{Ai}$ to $C_{A\infty}$). Thus, the interface offers two resistances, one in every side of the interface. A typical example that fits into this description is the drying of polymer pellets by air flowing over them. The concentration difference inside the pellet is due to diffusional resistance, and the same difference in the air is due to convective mass transfer. The discussion of Design Problem 3 (Section 4.5) elaborates on the two resistances and their relative importance.

### 4.4.2. ANALOGIES BETWEEN HEAT AND MASS TRANSFER

Similarly to the heat transfer problems, the dimensionless groups in the mass transfer processes are:

Sherwood number $\quad \text{Sh} = \dfrac{k_c \ell}{\text{Đ}}; \quad \dfrac{k_x \ell}{C\text{Đ}}$ $\qquad$ (4.101)

Schmidt number $\quad \text{Sc} = \dfrac{\mu}{\rho \text{Đ}}$ $\qquad$ (4.102)

Grashof number $\quad \text{Gr}_{\text{Đ}} = \dfrac{\ell^3 \rho^2 g \zeta (x_{Ai} - x_{A\infty})}{\mu^2}$ $\qquad$ (4.103)

Stanton number $\quad \text{St}_{\text{Đ}} = \dfrac{\text{Sh}}{\text{ReSc}} = \dfrac{k_c}{v}; \quad \dfrac{k_x}{Cv}$ $\qquad$ (4.104)

Peclet number $\quad \text{Pe}_{\text{Đ}} = \text{ReSc} = \dfrac{\ell v}{\text{Đ}}$ $\qquad$ (4.105)

and

$$j_{\text{Đ}} = \text{ShRe}^{-1} \text{Sc}^{-1/3} = \frac{k_c}{v}\left(\frac{\mu}{\rho\text{Đ}}\right)^{2/3}; \quad \frac{k_x}{Cv}\left(\frac{\mu}{\rho\text{Đ}}\right)^{2/3}, \qquad (4.106)$$

where Sh in mass transfer corresponds to Nu in heat transfer, $j_{\text{Đ}}$ is the mass transfer Chilton-Colburn $j$-factor, $\ell$ is the length scale of the system and $\zeta$ is the concentration coefficient of volumetric expansion (see Table 4.16 for the definition). Table 4.16 summarizes all the

**TABLE 4.16 Analogies between Heat and Mass Transfer at Low Mass Transfer Rates** ($\ell$ = length scale)

| | Heat Transfer Quantities | Binary Mass Transfer Quantities |
|---|---|---|
| Profiles | $T$ | $x_A$ |
| Diffusivity | $\alpha = \dfrac{k}{\rho \bar{C}_p}$ | $\text{Đ}_{AB}$ |
| Effect of profiles on density | $\beta = -\dfrac{1}{\rho}\left(\dfrac{\partial \rho}{\partial T}\right)_{p, x_A}$ | $\zeta = -\dfrac{1}{\rho}\left(\dfrac{\partial \rho}{\partial x_A}\right)_{p, T}$ |
| Flux | $q$ | $\mathbf{J}_A^{\bigstar} = N_A - x_A(N_A + N_B)$ |
| Transfer rate | $Q$ | $N_A - x_{Ai}(N_A + N_B)$ |
| Transfer coefficient | $h = \dfrac{Q}{A_s \Delta T}$ | $\bar{k}_x = \dfrac{N_A - x_{Ai}(N_A + N_B)}{A_s(x_{Ai} - x_{A\infty})}$ |
| Dimensionless groups that are the same in both correlations | $\text{Re} = \dfrac{\ell v \rho}{\mu}$ | $\text{Re} = \dfrac{\ell v \rho}{\mu}$ |
| | $\text{Fr} = \dfrac{v^2}{g\ell}$ | $\text{Fr} = \dfrac{v^2}{g\ell}$ |
| Basic dimensionless groups that are different | $\text{Nu} = \dfrac{h\ell}{k}$ | $\text{Sh} = \dfrac{k_x \ell}{C\text{Đ}_{AB}}; \quad \dfrac{k_c \ell}{\text{Đ}_{AB}}$ |
| | $\text{Pr} = \dfrac{\bar{C}_p \mu}{k} = \dfrac{\mu}{\rho \alpha}$ | $\text{Sc} = \dfrac{\mu}{\rho \text{Đ}_{AB}}$ |
| | $\text{Gr}_H = \dfrac{\ell^3 \rho^2 g \beta \Delta T}{\mu^2}$ | $\text{Gr}_{\text{Đ}} = \dfrac{\ell^3 \rho^2 g \zeta \Delta x_A}{\mu^2}$ |
| | $\text{St}_H = \dfrac{\text{Nu}}{\text{RePr}} = \dfrac{h}{\rho \bar{C}_p v}$ | $\text{St}_{\text{Đ}} = \dfrac{\text{Sh}}{\text{ReSc}} = \dfrac{k_x}{Cv}; \quad \dfrac{k_c}{v}$ |
| Special combinations of dimensionless groups | $\text{Pe}_H = \text{RePr} = \dfrac{\ell v \rho \bar{C}_p}{k}$ | $\text{Pe}_{\text{Đ}} = \text{ReSc} = \dfrac{\ell v}{\text{Đ}_{AB}}$ |
| | $j_H = \text{NuRe}^{-1}\text{Pr}^{-1/3}$ | $j_{\text{Đ}} = \text{ShRe}^{-1}\text{Sc}^{-1/3}$ |
| | $= \dfrac{h}{\rho \bar{C}_p v}\left(\dfrac{\bar{C}_p \mu}{k}\right)^{2/3}$ | $= \dfrac{k_x}{Cv}\left(\dfrac{\mu}{\rho\text{Đ}_{AB}}\right)^{2/3}; \quad \dfrac{k_c}{v}\left(\dfrac{\mu}{\rho\text{Đ}_{AB}}\right)^{2/3}$ |

(Reprinted by permission of the publisher from Bird et al., 1960.)

definitions of variables and dimensionless numbers and their correspondence to their respective entities in the heat transfer area. Note that the analogy of Table 4.16 holds for low mass transfer rates.

The analogies between heat and mass transfer in either form, forced or free convection, can be visualized as follows (Bird et al., 1960):

Forced convection:

Nu = a function of (Re, Pr, geometry)

Sh = the same function of (Re, Sc, geometry);     (4.107)

Free convection around submerged objects:

Nu = a function of ($Gr_H$, Pr)

Sh = the same function of ($Gr_D$, Sc).     (4.108)

As examples of the application of these analogies we illustrate their use in flow over a flat plate, around a submerged cylinder, and in free convection problems. Thus, Eq. 5.123 can be directly used for mass transfer problems for flow over a smooth flat plate as:

$$Sh = 0.332\,Re^{1/2}\,Sc^{1/3}.$$     (4.109)

For transverse flow over a long cylinder (e.g., fiber spinning), the relevant equations (analogous to Eqs. 5.125 and 5.126) are:

$$Sh = (0.43 + 0.50\,Re^{1/2})Sc^{0.38} \qquad \text{for} \quad 1 < Re < 10^3$$

$$Sh = 0.25\,Re^{0.6}\,Sc^{0.38} \qquad \text{for} \quad 10^3 < Re < 2 \times 10^5. \quad (4.110)$$

For free convection problems, Eqs. 5.130, 5.131 and 5.132 can be easily translated to mass transfer applications using Table 4.16.

---

**EXAMPLE 4.10**
**Solution Casting**

Fig. 4.20 shows the process of production of solution-cast films. A polymer solution is cast onto a rotating roll (Middleman, 1977), and the solvent is removed by air flowing in the opposite direction. Then, the dry film is removed from the roll and goes into the next processing step. Estimate the convective mass transfer coefficient of the solvent into the air and the time needed for 80% solvent removal. The air speed, $V$, is 1 m/s, the radius of the roll, $R$, is 30 cm, and the thicknesses of the film, $L$, and of the air duct, $L'$, are 2 mm. The polymer solution is in contact with the roll for one-fifth of its periphery. The average air temperature is 50°C, a typical value for the diffusivity of the solvent into the air stream is $D_{Aair} = 1 \times 10^{-5}$ cm²/s, and the diffusivity of the solvent in the solution is $D_{Aliq} = 1 \times 10^{-6}$ cm²/s.

**Solution**

In this problem, two mass transfer operations take place: diffusion of the solvent in the polymer solution (I) and into the air stream (II). The importance of these two operations is evaluated from the value of the Sherwood number, Sh:

$$Sh = \frac{k_c \ell}{D_{Aair}}$$     (4.111)

We first calculate the mass transfer coefficient, and then we will solve the transient diffusion equation for the solvent in the solution.

Because the thickness of the solution on the roll is very small compared to the radius of the roll, we assume that the solution forms a plate over which an air stream flows and removes the solvent. Then, Eq. 4.109 is directly applicable to our case. The concentration of the solvent in the air is assumed small, so that the properties of the air remain unaltered. The density and viscosity of the air at 50°C are: $\rho = 0.0011$ g/cm³ and $\mu = 0.019$ mPa·s (Perry & Chilton, 1973). The contact length of the solution onto the roll is $l = 2\pi R/5 = 37.7$ cm. The Reynolds number is calculated as:

$$Re = \frac{100 \times 37.7 \times 0.00019}{0.00019} = 21{,}826.$$     (4.112)

Similarly, the Schmidt number is:

$$Sc = \frac{0.00019}{0.0011 \times 0.00001} = 17{,}273.$$     (4.113)

Thus, the Sherwood number for the solvent transfer into the air, Sh, is calculated from Eq. 4.109 as:

$$Sh = 0.332 \times 21{,}826^{1/2} \times 17{,}273^{1/3} = 1{,}268,$$     (4.114)

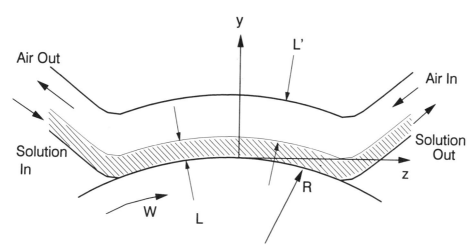

**Figure 4.20** Solution-cast film geometry.

and from this value the mass transfer coefficient $k_c$ is calculated as

$$k_c = \frac{Sh\mathcal{D}}{\ell} = 3.36 \times 10^{-4} \text{ cm/s}. \tag{4.115}$$

Assuming that the solution is moving in the $z$ direction with a flat velocity profile of magnitude $V$, the term for the convective mass transfer in the $z$ direction in Fick's second law, $V(\partial/\partial z)$, is equivalent to the

unsteady term, $(\partial/\partial t)$. Furthermore, the boundary condition at $y = L$ is of the convective type, with Sh equal to 1,268 (Eq. 4.114), which is equivalent to $\infty$ in Fig. 4.5. From the same figure, for 80% removal

$$\frac{\mathcal{D}_{Aliq}t}{L^2} = 0.6^2 \Rightarrow t = 14,400 \text{ s}. \tag{4.116}$$

Thus, the time is approximately 4 hours and some changes should be made to shorten this time.

## 4.5. SOLUTION TO DESIGN PROBLEM 3

Fig. 4.21 shows the schematic of one jet in dry-spinning conditions of fiber production. In the following analysis we solve the problem of one jet, and assume that the solution is valid for·all the jets together. We assume that the fiber at any axial distance is round, and thus we will solve the problem using cylindrical coordinates. As the polymer solution exits the spinneret, it expands, and at the point of maximum cross sectional area (dope) we put the origin of the coordinate system. As $z$ increases, the cross sectional area decreases because of the loss of solvent and the tension from the wind-up roll. At $z_w$ (i.e., at the wind-up roll) the solvent should have been evaporated and the fiber solidified. The complete solution of the dry-spinning problem should include the momentum, energy, and mass balance equations along with the appropriate boundary conditions. The resulting system of equations is very complex, and in the present analysis we will use some simplifying assumptions. We will consider isothermal conditions and that the flash vaporization takes place at the spinneret exit only, and thus the vaporization will not affect our discussion of the later stages.

The solvent diffuses from the spinning line (or filament) to the

spinning line surface with diffusivity $\mathcal{D}_{AP}$, and it is carried away by the flowing air in a convective mode. Air is flowing at a cross direction to the fiber movement, but because of the high speed of the fiber air is also entrained in the parallel direction. Consequently, the convective mass transfer is taking place in the $r$, as well as in the $z$, directions. Because the cross-flow velocity is four times the parallel-flow velocity, the mass transfer coefficient of the cross flow is 3.17 ($=0.52 \times 4^{1/3}/0.26$) times that of the parallel flow. Consequently, the controlling mass transfer coefficient is that of parallel flow.

To calculate the mass transfer coefficient we need to estimate the properties of the air—DMF mixture at 200°C. We further assume that the properties of the mixture are very close to those of plain air. From Perry and Chilton (1973) we get that $\mu = 2.5 \times 10^{-5}$ Pa·s, and from simple calculations from the law of ideal gases we get that $\rho = 7.47 \times 10^{-4}$ g/cm³. Based on these values and the value of Sc we get that $\mathcal{D}_{Aair} = 0.18$ cm²/s. Thus,

$$k_{c,P} = 0.402R(z)^{-2/3} \text{ cm/s}. \tag{4.117}$$

The extreme values of $R(z)$ are 0.0178 cm at the dope and 0.0022 cm at the windup roll, so that the convective mass transfer coefficient varies from 5.9 to 23.8 cm/s. Therefore, the resistance to mass transfer in the air flow scales from 0.17 ($=1/5.9$) to 0.04 ($=1/23.8$).

The resistance to diffusion inside the spinning line is scaled as $R/\mathcal{D}_{AP}$. For an average spinning-line temperature of 90°C, the diffusivity of DMF in PAN is calculated as:

$$\mathcal{D}_{AP} = 9.03 \times 10^{-4} \exp\left(\frac{-2360}{273+90}\right) = 1.36 \times 10^{-6} \text{ cm}^3. \tag{4.118}$$

Note that because of the lack of more data we assume that the polymer fraction of the spinning line has no effect on the diffusivity of the DMF within the spinning line. Therefore, the diffusive resistance varies from 1,620 to 13,130, which represents from 4 to 6 orders of magnitude higher resistance within the spinning line than in the air flow. It is thus permissible to neglect the mass transfer in the air and to solve the diffusion equation within the spinning line with a boundary condition at the spinning line surface of zero DMF concentration.

The region where the solution that follows is applicable is somewhat away from the spinneret, and its beginning is denoted by $z_0$ in Fig. 4.21. A fluid element is located at $z_0$ at time $t = 0$. Finally, we arbitrarily set the average mass fraction of solvent at that point equal to 0.25.

As a fluid element in the spinning line moves in the $z$ direction, it changes its radius because of two reasons: tension from the windup roll, and mass change due to solvent removal. Assuming that the mass change due to solvent removal is negligible, we can consider the DMF diffusion within the spinning line as taking place in a stationary liquid ($v_r = 0$ or very small compared to $v_z$). Consequently, the diffusion

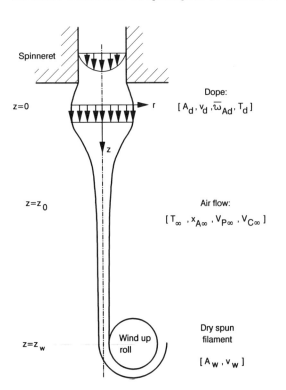

**Figure 4.21** Dry-spinning system.

equation in cylindrical coordinates for constant $\rho$ becomes (Table 4.5):

$$\frac{\partial \omega_A}{\partial t} = \frac{1}{r}\frac{\partial}{\partial r}\left(r \mathcal{D}_{AP}\frac{\partial \omega_A}{\partial r}\right), \qquad (4.119)$$

with initial and boundary conditions

I.C.:    $\omega_A(r,0)=\omega_{A0}$

B.C.1:   $\omega_A(R,t)=0$

B.C.2:   $\left(\dfrac{\partial \omega_A}{\partial r}\right)_{r=0}=0.$         (4.120)

The solution to Eq. 4.119 subjected to the boundary and initial conditions given in Eq. 4.120 is given many places (e.g. Crank, 1975, p. 66). If we average the solvent mass fraction over the cross-sectional area, we get:

$$\frac{\bar{\omega}_A}{\bar{\omega}_{A0}} = 4\sum_{k=1}^{\infty} \mu_k^{-2}\exp\left(\frac{-\mu_k^2 \mathcal{D}_{AP}t}{R^2}\right), \qquad (4.121)$$

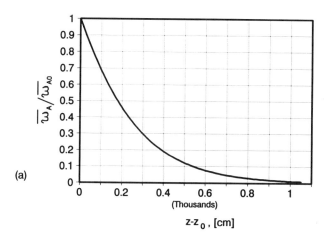

(a)

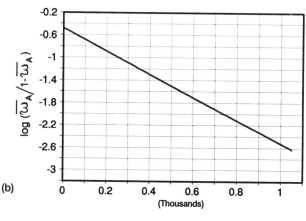

(b)

**Figure 4.22** (a) Average mass fraction of the solvent DMF divided by the initial average mass fraction (at $z_0$) as a function of the axial distance from the reference point $z_0$, for the system PAN–DMF. (b) The logarithm of the relative solvent concentration $\bar{\omega}_A/(1-\bar{\omega}_A)$ as a function of the axial distance from the reference point $z_0$, for the system PAN–DMF.

where $\mu_k$ are the roots of the Bessel function $J_0(\mu_k)=0$. If the series in Eq. 4.121 converges rapidly, we can keep only the first term, so that we get:

$$\frac{\bar{\omega}_A}{\bar{\omega}_{A0}} = \frac{4}{\mu_1^2}\exp\left(\frac{-\mu_1^2 \mathcal{D}_{AP}t}{R^2}\right), \qquad (4.122)$$

where $\mu_1=2.4048$.

Time, $t$, in Eq. 4.122 can be replaced by:

$$t=\frac{z-z_0}{\bar{V}}, \qquad (4.123)$$

where $\bar{V}$ is the velocity of the filament, which is considered to move as in plug flow. The mass flow rate of the polymer, $\dot{m}_P$, follows the relation:

$$\dot{m}_P=\pi\rho R^2 \bar{V}(1-\bar{\omega}_A). \qquad (4.124)$$

Combining Eqs. 4.122–4.124, we get the following equation:

$$\frac{d\bar{\omega}_A}{dz} = -\left(\frac{\pi\rho\mu_1^2 \mathcal{D}_{AP}}{\dot{m}_P}\right)(1-\bar{\omega}_A)\bar{\omega}_A, \qquad (4.125)$$

which is subject to the following boundary condition:

B.C.:   $\bar{\omega}_A(z_0)=\bar{\omega}_{A0}.$         (4.126)

Equation 4.125 subject to the boundary condition in Eq. 126 was solved using the IMSL subroutine IVPAG and the results are shown in Fig. 4.22. In Fig. 4.22.a the ratio of the average solvent mass fraction at the axial position $z$ relative to that at $z_0$ is plotted against the axial distance, and in Fig. 4.22.b the same data are formed in terms of the logarithm of the ratio of the average solvent mass fraction to the polymer mass fraction. The straight line in Fig. 4.22.b has also been experimentally observed (Ziabicki, 1976) for the region away from the spinneret. Fig. 4.22.a shows that the complete removal of the solvent is achieved at about 10 m away from the point where $\bar{\omega}_{A0}=0.25$.

## PROBLEMS
### A. Applications

**4A.1.** Relationships between Fluxes
Prove that

    **(a)** the sum of the molar fluxes with respect to the molar average velocity is equal to zero in a multicomponent system,

    **(b)** the sum of the mass fluxes with respect to the mass average velocity is equal to zero, and

    **(c)** $\mathbf{J}_i^{\star}=\mathbf{N}_i-x_i\sum_{j=1}^{n}\mathbf{N}_j.$

**4A.2.** Saturation of PP with Nitrogen Gas
A very thin disk (thickness: 0.159 cm) of PP is used for microcellular foaming experiments. The disk sits at the bottom of a cylindrical pressure vessel, and it is exposed to high-pressure nitrogen gas. Calculate the time necessary for nitrogen gas to diffuse in PP so that its concentration at the PP surface facing the bottom of the vessel reaches 90% of its equilibrium value. The diffusion coefficient of nitrogen in PP is $\mathcal{D}=3.87\times10^{-8}\,cm^2/s$, and assume that the high pressure does not affect the type of the diffusion process (i.e., it is still Fickian) or the value of the diffusivity.

**4A.3.** Diffusion of Gas into Polymer Droplets and Fibers
Consider a molten polymer blend with the minor component in the form of small spherical droplets of radius $5\mu m$. Calculate the time it takes for the diffusion of gas into these spheres from the major component to reach 90% equilibrium. Repeat the same calculations when the minor component is in the form of long fibers with the same radius as the spherical droplets. $Đ = 10^{-8}\,cm^2/s$.

**4A.4.** Economics of Barrier Polymers
Four polymers are being considered for an oxygen barrier application. The polymers are: (1) Barex (trademark of Vistron Co.), which is an amorphous copolymer of acrylonitrile 70% mol/mol [–CH(CN)–CH$_2$–] and methyl methacrylate [–C(CH$_3$)(COOCH$_3$)–CH$_2$–]; (2) EVAL-F (trademark of Kuraray Co.), which is a copolymer of ethylene (23% weight) and vinyl alcohol [–CH(OH)–CH$_2$–] with 70% crystallinity; (3) amorphous poly(vinylidene chloride) (PVDC) [–C(Cl$_2$)–CH$_2$–] and (4) nylon 6,6 with 40% crystallinity. Their density and cost per weight are:

| | | |
|---|---|---|
| Barex | 1.15 g/cm$^3$ | 2.76 $/kg |
| EVAL-F | 1.16 ,, | 4.85 ,, |
| PVDC | 1.70 ,, | 2.76 ,, |
| Nylon 6,6 | 1.19 ,, | 3.75 ,, |

Which is the most advantageous material for this application?

**4A.5.** Plastic Automotive Fuel Tanks
The North American auto industry is converting to plastic automotive fuel tanks from blow-molded HDPE. The HDPE tanks are lighter than steel tanks, easily shaped to fit in the car, more puncture resistant, and they do not corrode. However, the HDPE tanks suffer from high permeability to liquid gasoline.

(a) Calculate the average mass loss of gasoline per day at steady state from an HDPE 60-liter tank with wall thickness of 1 mm, and compare it with the EPA standard of 2 g/day. Assume Fickian diffusion of gasoline in the HDPE wall and that the gasoline uptake by the wall via Langmuir sorption is zero. Simulate the tank with an equal-volume cubic structure and assume that the tank is emptied with equal gasoline consumption each day. At 29°C the following data is given (Kathios & Ziff, 1991): constant diffusivity and solubility of gasoline in HDPE equal to $7 \times 10^{-8}\,cm^2/s$ and 0.066 g/cm$^3 \cdot$ atm, respectively. Assume that the pressure in the tank is 1 atm and that the density of gasoline is 1 g/cm$^3$.

(b) The most promising barrier technology to correct this high permeability is the coextrusion of HDPE and polyamide to form a laminated structure. Thus, consider two adjacent layers of HDPE and polyamide. If the permeability of gasoline in polyamide is one-tenth of that in HDPE, what should the thickness of the polyamide layer be for the tank to conform with the EPA regulation?

**4A.6.** Unsteady Diffusion in Cubes and Spheres
Consider the unsteady diffusion of a gas into two polymer pellets: The first is a cube, and the other is a sphere. Calculate the time needed for 90% saturation of the pellets at their centers, for the following cases: (a) same linear dimension, (b) same surface area and (c) same volume of both pellets.

**4A.7.** Poly(ethylene terephthalate) (PET) Pellet Drying
PET is a relatively hydrophobic polymer (Myers et al., 1961) that still needs to be dried before being processed. Calculate the temperature of the drying oven, so that the drying of spherical pellets, 3 mm in diameter, does not take more than 11 hours.

## B. Principles

**4B.1.** Mass Uptake by Slabs and Spheres
Consider a polymer slab at initial solute concentration $C_{A0}$ and thickness $2b$. A solution with solute concentration $C_{A\infty}$ is brought in contact with the slab, so that the slab surface solute concentration is considered to be equal to $C_{A\infty}$. Prove that the mass uptake of solute, $M_t$, at time $t$ is equal to:

$$\left[\frac{M_t}{M_\infty}\right]_{slab} = 1 - \frac{8}{\pi^2}\sum_{n=0}^{\infty}\frac{1}{(2n+1)^2}\exp\left[-\frac{(2n+1)^2\pi^2 Đt}{4b^2}\right],$$

where $M_\infty$ is the maximum uptake which is equal to $2b(C_{A\infty} - C_{A0})$. Note that the left-hand side of the equation is also equal to $(\bar{C}_A - C_{A0})/(C_{A\infty} - C_{A0})$, where $\bar{C}_A$ is the average concentration of $A$ in the slab. Also prove that a similar expression holds for a polymer sphere of radius $R$, that is,

$$\left[\frac{M_t}{M_\infty}\right]_{sphere} = 1 - \frac{6}{\pi^2}\sum_{n=1}^{\infty}\frac{1}{n^2}\exp\left[-\frac{n^2\pi^2 Đt}{R^2}\right].$$

**4B.2.** Saturation of Polyetheretherketone (PEEK) with Carbon Dioxide Gas
PEEK (see Table 5.8, p. 106) is used in foaming experiments with carbon dioxide as the foaming agent. The description of the experiment is the same as in Pr. 4A.2. Calculate the diffusion coefficient of $CO_2$ in PEEK at room temperature and then calculate the time needed for 90% saturation of the bottom surface. At which temperature is the time span of the experiment one day? Assume $\phi_c = 50\%$.

**4B.3.** PET Bottles for Carbonated Beverages—Steady State Simulation
Amorphous PET is extensively used in the production of bottles for carbonated beverages. Consider such a bottle with radius of 6 cm and with a headspace equal to 6% of the volume of the beverage. Assume that the bottle loses carbon dioxide through the sides only and that the wall thickness is much smaller than the bottle radius. The beverage inside the bottle is considered non-marketable if the partial pressure differential between inside and outside of the bottle drops to about 90% of the initial pressure differential in three months. Calculate the minimum wall thickness to achieve such partial pressure drop at steady state if the bottle is left on the shelf for three months at 25°C. The structural unit of PET is [–O–CH$_2$–CH$_2$–O–CO–C$_6$H$_4$–CO–]. Assume that Henry's law is applicable, and neglect any carbon dioxide uptake by the bottle walls via Langmuir sorption. The solubility of carbon dioxide into the beverage obeys Henry's law with a coefficient of $3.28 \times 10^{-5}\,mol/cm^3 \cdot$ atm.

**4B.4.** Permeability of Composite Laminates
Consider a laminar composite structure, in which the laminae, $L_i$, are normal to the direction of permeation. These laminae can be either slabs or hollow cylinders or spherical shells (Fig. 4.23). Prove that the composite permeability, $\bar{P}$, is given by the relationship:

$$\frac{I_v(R_0, R_n)}{\bar{P}} = \sum_{i=1}^{n}\frac{I_v(R_{i-1}, R_i)}{\bar{P}_i},$$

where $n$ is the total number of laminates and $R$ represents the linear dimension of the composite structure. $I_v$ is given by:

$$I_v(R_{i-1}, R_i) = \int_{R_{i-1}}^{R_i}\frac{1}{r^{v-1}}\,dr,$$

where $v$ is the *order* of the structure, that is, 1 for slabs, 2 for cylinders, and 3 for spheres.

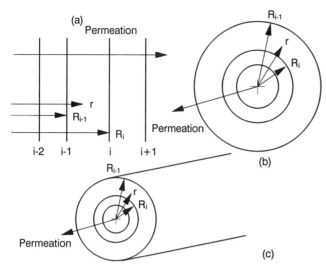

**Figure 4.23** Structures of laminar composites in permeation experiments. (a) Slabs. (b) Hollow cylinders. (c) Spherical Shells.

**4B.5.** Diffusivity of a Polymer Blend
Model a polymer blend by a lattice of rectangular parallelepipeds (Barrer & Petropoulos, 1961) suspended in a polymer matrix, Fig. 4.24. Using geometrical arguments and Fick's first law, calculate the blend diffusivity, $D_{blend}$, as a function of the diffusivities of both phases and geometrical constants. Simplify the expression for the cases of: **(a)**

impermeable dispersed phase, **(b)** dispersion of platelets and **(c)** extremely permeable dispersed phase in the form of platelets.

**4B.6.** Crack Healing of PMMA
The hot-melt adhesive bonding of polyethylene (see Pr. 4C.4) is similar to the problem of welding (healing) a crack at elevated temperatures with the application of a slight pressure (Jud et al., 1981). The success of the healing process is judged by comparing the fracture toughness of the healed sample, $K_{Ii}$, to the original fracture toughness, Fig. 4.25. Calculate the penetration depth of the PMMA macromolecules for successful healing, and compare it to the radius of gyration of PMMA. According to Graessley (1980) and based on the reptation theory for the diffusion of macromolecules ("snakelike movements"), the self-diffusion coefficient is given by

$$D_s = \frac{G_N^0}{135}\left(\frac{\rho R_g T}{G_N^0}\right)^2\left(\frac{R^2}{MW}\right)\frac{M_{cr}}{(MW)^2\eta_0(M_{cr})},$$

where the symbols, their names, and values for PMMA at $390°$K are (Jud et al., 1981):

| | | |
|---|---|---|
| $G_N^0$ | Plateau modulus | $6.36 \times 10^4\,N/m^2$ |
| $\rho$ | Density | $1.14\,g/cm^3$ |
| MW | Molecular weight | 120,000 |
| $R^2/MW$ | Mean square | $4.56 \times 10^{-19}\,m^2 \cdot mol/g$ |
| | end to end distance/MW | |
| $M_{cr}$ | Critical MW for | |
| | entanglements | 30,000 |
| $\eta_0(M_{cr})$ | Zero shear viscosity for $M_{cr}$ | $2.14 \times 10^7\,Pa \cdot s$ |

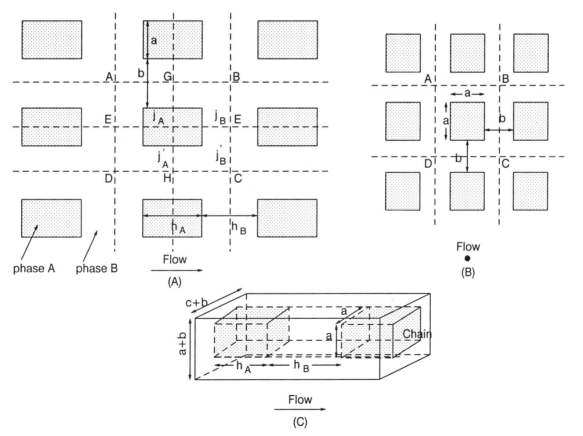

**Figure 4.24** Lattice of rectangular parallelepipeds. (a) Side view. (b) End-on view. (c) Three-dimensional view. (Reprinted by permission of the publisher from Barrer & Petropoulos, 1961.)

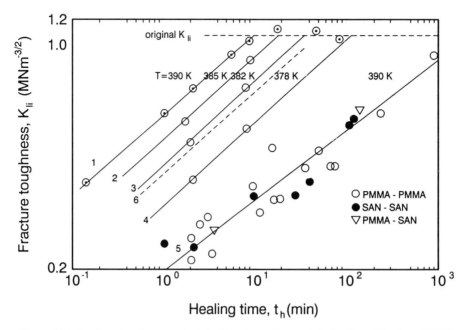

**Figure 4.25** Fracture toughness against healing time. Curves 1–4: healing of broken PMMA immediately after fracture. Curve 5: Surfaces welded after vacuum drying and polishing. Curve 6: healing at 390°K immediately after fracture of dried PMMA samples. (Reprinted by permission of the publisher from Jud et al., 1981.)

## C. Numerical Problems

**4C.1.** Dissolution of a Bubble into a Molten Polymer Matrix
Consider a molten polymer matrix with dispersed bubbles of radius $R_0 = 10\,\mu\text{m}$. Calculate the dissolution time of the bubble, $t_{\text{dis}}$, using Fick's second law of diffusion in spherical coordinates and moving (shrinking) boundaries. As the gas (species $A$) diffuses out the radius of the bubble decreases. Also, consider that (Epstein & Plesset, 1950) the initial mass fraction of the gas into the polymer matrix is equal to zero, and neglect any surface tension effects as well as the hydrodynamic response of the polymer melt to the bubble shrinkage. Then

1. Write Fick's second law of diffusion.
2. Introduce a new dependent variable $u = r(\rho_A - \rho_{As})$ where $\rho_{As}$ is the gas concentration at the interface.
3. Solve the resulting equation and calculate the flux of gas at the surface of the bubble.
4. Relate the change in radius to the flux.
5. Solve the resulting differential equation, using either a simplifying assumption (constant term in the equation vanishes) or the IMSL subroutine IVPAG. Show that the approximate complete dissolution time is given by:

$$t_{\text{dis}} = \frac{R_0^2}{2 Ð K},$$

where $K$ is the partition coefficient of the gas at the interface and the gas inside the bubble. Also show that $t_{\text{dis}} = 250\,\text{s}$ for $Ð = 10^{-7}\,\text{cm}^2/\text{s}$ and $K = SR_g T/\text{MW} = 0.02$, where $S$ is the Henry's law constant for the gas-polymer system in $\text{g/cm}^3\cdot\text{atm}$.

**4C.2.** Nylon 6 Pellet Drying
Nylon 6 is a relatively hydrophilic polymer. Asada and Onogi (1963) measured the diffusion coefficient of water vapor in nylon 6, and their data are shown in Fig. 4.26. The polymer pellets are to be dried to 0.15% g/g of polymer with dehumidified air at temperatures under than

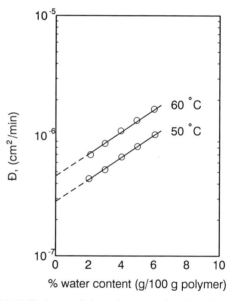

**Figure 4.26** Diffusion coefficient of moisture in nylon 6 as a function of moisture content. (Reprinted by permission of the publisher from Asada & Onogi, 1963.)

80°C. Calculate the dehumidification time for spherical pellets 3 mm in diameter. Nylon 6 has 70% crystallinity and in this particular case it was in equilibrium with air at 70% relative humidity. $\rho = 1.13\,\text{g/cm}^3$ for nylon 6.

**4C.3.** Mutual Diffusion of Polymers in Contact (Adhesion)
Diffusion theory is one of the theories of adhesion. According to this theory, the molecules of two polymers in mutual contact diffuse across

the interface (Fig. 4.27), so that after some time the distinct interface ceases to exist. The strength of the joint will then depend on the distance the macromolecules have interpenetrated each other. Crack healing is also based on the theory of diffusion.

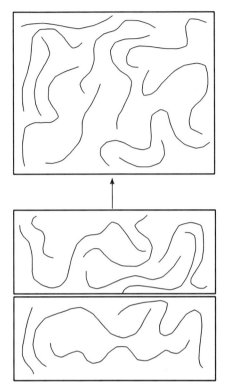

**Figure 4.27** Diffusion across the interface between two polymers in contact. The interface becomes diffuse as the mutual diffusion takes place.

**(a)** Consider two semi-infinite polymer slabs. One of the slabs is doped with a diffusant at concentration $C_{A0}$, whereas the other one is diffusant free. At time $t = 0$, the two slabs are brought into contact at $x = 0$. Calculate the diffusant concentration profile as a function of time as shown in Fig. 4.28.

**(b)** An infrared spectrometer is used to measure the interdiffusion of poly(vinyl methyl ether) (PVME), and polystyrene (PS) (Jabbari & Peppas, 1993). The assembly consists of a solution-cast PS thin film on top of a germanium crystal, and a solution-cast PVME film on top of the PS film. The thickness of the PS film is 1 μm and that of the PVME film is 3 μm. Calculate the time it takes for the PVME molar concentration at the crystal-PS interface to reach 80% of its equilibrium value. Assume Fickian diffusion and diffusivity equal to $1.1 \times 10^{-12}$ cm²/s.

**4C.4.** Adhesive Bond Strength and Diffusion
Yamakawa (1976) studied the effect of bonding temperature on the peel strength of two polyethylene sheets bonded with an ethylene copolymer as the bonding agent (Fig. 4.29). Assume that (1) the thicknesses of the polyethylene sheets and the adhesive bond are considered as infinite compared to the penetration depth of the adhesive; (2) $Ð_{AP}(140°C) = 5 \times 10^{-11}$ cm²/s and $E_D = 50$ kJ/mol; and (3) the adhesive bonding is satisfactory (leveled portion of curves in Fig. 4.29) when the penetration depth of the adhesive (at 10% concentration change) exceeds a threshold value. Calculate that threshold value and the bonding time for successful bonding of the two polyethylene sheets at 110°C.

**4C.5.** Mass Uptake by Slabs, Spheres and Cylinders
Solve Pr. 4B.1 for slabs, spheres, and cylinders using the IMSL subroutine MOLCH.

**4C.6.** Transient Diffusion and Dual-Mode Sorption
**(a)** Rework Pr. 4B.3 in the transient mode, assuming again that the carbon dioxide uptake by the PET wall via Langmuir sorption is zero. Assume $Ð = 1.10 \times 10^{-9}$ cm²/s, and wall thickness, $b$, equal to 0.57 mm.

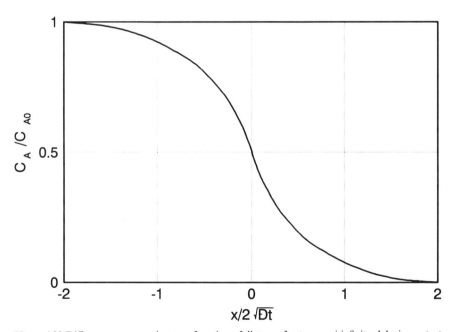

**Figure 4.28** Diffusant concentration as a function of distance for two semi-infinite slabs in contact. Initially only one of them had nonzero diffusant concentration.

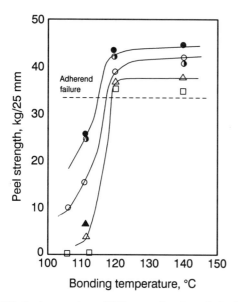

**Figure 4.29** Peel strength at 23°C as a function of the bonding temperature. The symbols represent the various types of thermoplastic adhesives (ethylene copolymers) used to adhere two sheets of polyethylene for 300 s. (Reprinted by permission of the publisher from Yamakawa, 1976.)

**(b)** Sorption and diffusion of gases in glassy polymers might be described by the dual-mode sorption model (Vieth, 1991). The basic assumptions of the model are:

1. Two modes of sorption, Henry's law sorption, $C_D = S_D P$, and Langmuir sorption (or "microvoid filling" sorption; Section 4.2.4), $C_H = C'_H b'P/(1 + b'P)$, occur simultaneously. $S_D$ is the Henry's law constant in mol/cm$^3 \cdot$ atm, $C'_H$ is the microvoid saturation constant in mol/cm$^3$, $b'$ is the microvoid affinity constant in atm$^{-1}$, and $P$ is the pressure in atm.
2. The two modes are in local equilibrium throughout the glassy polymer.
3. The gas sorbed in the Langmuir mode is completely immobilized.
4. Henry's mode is the only diffusion mode.
5. The diffusion coefficient is constant.

Based on these assumptions show that Fick's second law is now expressed as:

$$ D \frac{\partial^2 C_D}{\partial x^2} = \frac{\partial C_D}{\partial t} \left[ 1 + \frac{C'_H b'/S_D}{(1 + C_D b'/S_D)^2} \right] $$

and then rework part **(a)** of this problem, using the following data (Masi & Paul, 1982): the initial pressure in the headspace is equal to 4.0 atm, $S_D = 1.48 \times 10^{-5}$ mol/cm$^3 \cdot$ atm, $C'_H = 3.235 \times 10^{-4}$ mol/cm$^3$, and $b' = 0.351$ atm$^{-1}$.

### 4C.7. Interdiffusion in a Polymer Blend
A dilute polymer blend consists of a minor component $A$, which is dispersed in the form of droplets 2 μm in diameter, and a major component $B$. The volume fraction of $A$ is $\phi_A = 1.56\%$. Calculate the time it takes for the molar concentration of $A$ at the center of a droplet to decrease by 95% of its original concentration, assuming Fickian interdiffusion and constant $D = 1 \times 10^{-12}$ cm$^2$/s.

### 4C.8. Combination of Fickian and Case-II Transports
During the sorption of acetone in a poly(vinyl chloride) (PVC) slab of thickness $2b$, submerged into a large acetone bath at room temperature, the slab is divided into two regions. The first one is the gel-like region, which is in direct contact with the acetone bath and where the transport of acetone can be described by a combination of a Fickian diffusion, with diffusion coefficient $D$, and a Case-II diffusion, with velocity of the advancing front $V$. The second region is the glass region, where the transport is described by a Fickian diffusion, with diffusion coefficient $D_g$. For $D \gg D_g$ only the transport of acetone in the gel-like region is considered. According to Kwei et al. (1972) the one-dimensional transport equation for the gel-like region is:

$$ D \frac{\partial^2 C}{\partial x^2} - V \frac{\partial C}{\partial x} = \frac{\partial C}{\partial t} $$

where $C$ is the molar concentration of acetone. The advancing acetone front is located at $\theta_x = C_x/C_0 = 0.068$, where $C_0 = 8.62 \times 10^{-3}$ mol/cm$^3$ is the equilibrium molar concentration of acetone in the gel-like region. Calculate the acetone molar concentration profile and the mass uptake as a function of distance and time for swelling number Sw $= Vb/D$ equal to 3.87, and compare it with the simple Fickian case (i.e., where Sw $= 0$).

### D.  Design Problems

**4D.1.** Microcellular Foaming
Microcellular foam is a kind of plastic foam with bubble size on the order of 10 μm, which is much less than the conventional foam. Indications of improvement of the properties of the polymer matrix by microcellular foaming have given importance to this relatively new process. One technique to produce microcellular foam is the gas supersaturation technique, which for amorphous materials consists of three steps: gas saturation, supersaturation, and foaming. In the first step the polymer matrix is saturated with the foaming gas (nitrogen, carbon dioxide, etc.). In the second step the gas pressure is released, creating supersaturation conditions, but the polymer matrix does not foam because the temperature is low (above room temperature but

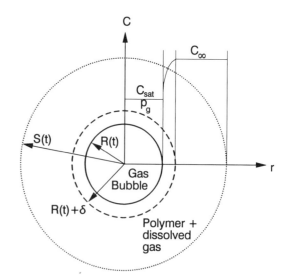

**Figure 4.30** Gas bubble growing in an infinite sea of polymer matrix. $S(t)$ presents the radius of the cell assigned to each growing bubble ($\infty$ in our problem). $\delta$ presents the thickness of the concentration layer, where the gas concentration changes from its undisturbed value $C_\infty$ to $C_{sat}$.

below the glass transition temperature of the polymer matrix). Foaming takes place in the final step, during which the sample is submerged into an oil bath kept at a temperature above the glass transition temperature of the polymer matrix. During this step the bubble can be initiated either at a heterogeneous nucleation site or at a microcrack of the matrix created during the forming process.

**(a)** Follow the steps outlined in Pr. 4C.1, and calculate the radius of the bubble as a function of foaming time for a diffusion-controlled process (Fig. 4.30), that is, the hydrodynamic force that resists bubble growth is considered negligible compared to the diffusion force. Neglect the effects of surface tension, and assume that the bubble is suspended in an infinite sea of polymer matrix ($S(t) \rightarrow \infty$).

**(b)** As a numerical example consider PS as the polymer matrix at 188°C, and the following gases with their corresponding Henry's law constants (Durill & Griskey, 1966) in $cm^3(STP)/g \cdot atm$:

| | |
|---|---|
| Nitrogen | 0.049 |
| Carbon dioxide | 0.220 |
| CFC-22 | 0.388 |
| Argon | 0.093 |
| Helium | 0.029 |

The saturation pressure is 1,000 psi (6.9 MPa), and the pressure inside the growing bubble is the ambient pressure. Neglect the surface tension forces between the bubble and the polymer. Which gas is the most promising for microcellular foaming?

**(c)** Propose changes in the experimental procedure that might bring more controllability to the foaming process.

**4D.2.** Precipitation Foams

The insulation of a wire is achieved by a solution-coating process, which also involves a precipitation foaming process. The whole process consists of following four steps.

1. *Coating (coating of the conductor).* A copper conductor is passed through a crosshead die where it is coated with a solution of about 30% HDPE in xylene at 130°C.
2. *Whitening.* The coated conductor passes through water at about 80°C, and thus the HDPE crystallizes, precipitates out, and whitens the solution.
3. *Drying.* The whitened insulation on the conductor is dried by passing through an air duct where the air temperature and the air speed are controlled at about 55°C and 4 m/s, respectively.
4. *Inversion.* The final foamed structure is created by sintering and fusion of the polyethylene particles.

If the radius of the copper wire is 150 μm and the radius of the insulation is 350 μm calculate the time of the drying step, neglecting any shrinkage of the insulation during this step.

# REFERENCES

Alfrey, T., E. F. Gurnee, and W. G. Lloyd. 1966. "Diffusion in Glassy Polymers." *J. Polym. Sci., Part C*, **12**, 249–61.

Asada, T., and S. Onogi. 1963. "The Diffusion Coefficient for the Nylon 6 and Water System." *J. Colloid Sci.*, **18**, 784–92.

Barrer, R. M., and J. H. Petropoulos. 1961. "Diffusion in Heterogeneous Media: Lattices of Parallelepipeds in a Continuous Phase." *Brit. J. Appl. Phys.*, **12**(12), 691–7.

Biceramo, J. 1993. *Prediction of Polymer Properties* (Marcel Dekker, New York).

Bird, R. B., W. E. Stewart, and E. N. Lightfoot. 1960. *Transport Phenomena* (Wiley, New York).

Brandrup J., and E. H. Immergut, Eds. 1989. *Polymer Handbook*, 3rd Ed. (Wiley, New York).

Cheng, Y. L., and D. C. Bonner. 1978. "Solubility of Nitrogen and Ethylene in Molten Low-Density Polyethylene to 69 Atmospheres." *J. Polym. Sci.*, **16**, 319–33.

Comyn, J., Ed. 1985. *Polymer Permeability* (Elsevier, New York).

Crank, J. 1975. *The Mathematics of Diffusion*, 2nd Ed. (Oxford University Press, Oxford).

Crank, J., and G. S. Park, Eds. 1968. *Diffusion in Polymers* (Academic Press, New York).

Duda, J. L., G. K. Kimmerly, W. L. Sigelco, and J. S. Vrentas. 1973. "Sorption Apparatus for Diffusion Studies with Molten Polymers." *Ind. Eng. Chem. Fundam.*, **12**(1), 133–6.

Durill, P. L., and R. G. Griskey. 1966. "Diffusion and Solution of Gases in Thermally Softened or Molten Polymers: Part I: Development of Technique and Determination of Data." *AIChE J.*, **12**(6), 1147–51.

Epstein, P. S., and M. S. Plesset. 1950. "On the Stability of Gas Bubbles in Liquid-Gas Solutions." *J. Chem. Phys.*, **18**(11), 1505–9.

Gorski, R. A., R. B. Ramsey, and K. T. Dishart. 1985. "Physical Properties of Blowing Agent Polymer Systems-I. Solubility of Fluorocarbon Blowing Agents in Thermoplastic Resins." 29th Annual Technical/Marketing. Conference, SPI Polyurethane Division, pp. 286–99.

Graessley, W. W. 1980. "Some Phenomenological Consequences of the Doi-Edwards Theory of Viscoelasticity." *J. Polym. Sci.: Polym. Phys. Ed.*, **18**, 27–34.

Hines, A. L., and R. N. Maddox. 1985. *Mass Transfer* (Prentice Hall, Englewood Cliffs, NJ).

Hopfenberg, H. B., and H. L. Frisch. 1969. "Transport of Organic Macromolecules in Amorphous Polymers." *Polym. Letters*, **7**, 405–9.

Jabbari, E., and N. A. Peppas. 1993. "Use of ATR-FTIR to Study Interdiffusion in Polystyrene and Poly(vinyl methyl ether)." *Macromolecules*, **26**, 2175–86.

Jud, K., H. H. Kausch, and J. G. Williams. 1981. "Fracture Mechanics Studies of Crack Healing and Welding of Polymers." *J. Mater. Sci.*, **16**, 204–10.

Kathios, D. J., and R. M. Ziff. 1991. "Permeation of Gasoline-Alcohol Fuel Blends through High Density Polyethylene Fuel Tanks." 49th SPE Annual Technical Conference, Montreal, **37**, 1509–11.

Kausch, H. H., and M. Tirrell. 1989. "Polymer Interdiffusion." *Ann. Rev. Mater. Sci.* **19**, 341–77.

Koros, W. J. and M. W. Hellums. 1989. "Transport Properties." In H. F. Mark et al., Eds., *Encyclopedia of Polymer Science and Technology*, Supplement, pp. 724–802 (Wiley, New York).

Kwei, T. K., T. T. Wang, and H. M. Zupko. 1972. "Diffusion in Glassy Polymers. V. Combination of Fickian and Case II Mechanisms". *Macromolecules*, **5**(5), 645–6.

Masi, P., and D. R. Paul. 1982. "Modeling Gas Transport in Packaging Applications." *J. Membr. Sci.*, **12**, 137–51.

Middleman, S. 1977. *Fundamentals of Polymer Processing* (McGraw-Hill, New York).

Myers, A. W., J. A. Meyer, C. E. Rogers, V. Stannett, and M. Szwarc. 1961. "Studies on the Gas and Vapor Permeability of Plastic Films and Coated Papers." *Tappi*, **44**(1), 58–64.

Ohzawa, Y., and Y. Nagano. 1970. "Studies on Dry Spinning. II. Numerical Solutions for Some Polymer-Solvent Systems Based on the Assumption that Drying is Controlled by Boundary-Layer Mass Transfer." *J. Appl. Polym. Sci.*, **14**, 1879–99.

Ohzawa, Y., Y. Nagano, and T. Matsuo. 1969. "Studies on Dry Spinning. I. Fundamental Equations." *J. Appl. Polym. Sci.*, **13**, 257–83.

Perry, R. H., and C. H. Chilton, Eds. 1973. *Chemical Engineers' Handbook*, 5th Edn. (McGraw-Hill, New York).

Reid, R. C., J. M. Prausnitz, and T. K. Sherwood. 1977. *The Properties of Gases and Liquids*, 3rd Edn. (McGraw-Hill, New York).

Salame, M. 1986. "Prediction of Gas Barrier Properties of High Polymers." *Polym. Eng. Sci.*, **26**(22), 1543–6.

Sarti, G. C., and F. Doghieri. 1994. "Non-Fickian Transport of Alkyl Alcohols Through Prestretched PMMA." *Chem. Eng. Sci.*, **49**(5), 733–48.

Stiel, L. I., and D. F. Harnish. 1976. "Solubility of Gases and Liquids in Molten Polystyrene." *AIChE J.*, **22**(1), 117–22.

Thomas, N., and A. H. Windle. 1978. "Case II Swelling of PMMA Sheet in Methanol," *J. Membr. Sci.*, **3**, 337–42.

Thomas, N., and A. H. Windle. 1982. "A Theory of Case II Diffusion." *Polymer*, **23**, 529–42.

Tirrel, M. 1984. "Polymer Self-Diffusion in Entangled Systems." *Rub. Chem. Techn.*, **57**, 523–56.

Van Krevelen, D. W. 1990. *Properties of Polymers*, 3rd Edn. (Elsevier, New York).

Vieth, W. R. 1991. *Diffusion In and Through Polymers* (Hanser Publ., Munich).

Vrentas, J. S., C. M. Jarzebski, and J. L. Duda. 1975. "A Deborah Number for Diffusion in Polymer-Solvent Systems." *AIChE J.*, **21**, 894–901.

Yamakawa, S. 1976. "Hot-Melt Adhesive Bonding of Polyethylene with Ethylene Copolymers." *Polym. Eng. Sci.*, **16**(6), 411–18.

Ziabicki, A. 1976. *Fundamentals of Fibre Formation* (Wiley, New York).

# 5

# NONISOTHERMAL ASPECTS OF POLYMER PROCESSING

## DESIGN PROBLEM 4
### Casting of Polypropylene Film

Polypropylene is extruded at 200°C from a film die having lips 76.2 cm wide and 0.1016 cm thick (See Fig. 5.1). The extruded film is drawn down to a width of 60.96 cm and a thickness of 0.005 cm. The distance from the die face to the casting drum is 2.54 cm. The film is in contact with the drum over a length of 70% of the circumference of the drum. The air temperature is taken as 25°C, and the line speed is 60 m/min. The radius of the drum is 0.45 m. Determine the heat transfer coefficient required at the drum surface to produce a clear film. The requirement for a clear film is based on keeping the crystallinity at the center to be less than 3% and the spherulite size less than 5000 μm. Tap water at 12°C is available for cooling.

Most polymer processes involve heat transfer. Polymers must usually be heated above their melting points before shaping and then cooled to maintain the desired shape. It is during the cooling phase of the

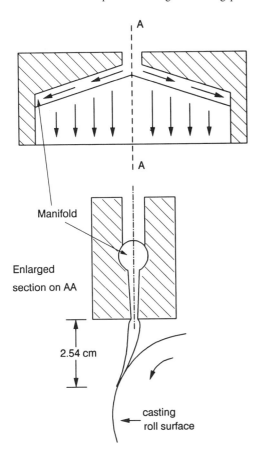

Manifold

Enlarged

section on AA

2.54 cm

casting
roll surface

**Figure 5.1** Film-casting process showing the polymer film being cast onto a rotating drum.

process that the physical properties of the polymer can be drastically altered. Because the thermodynamic and thermal properties of most polymers are rather similar to other materials, it is not necessary to develop any new laws as it was for the flow of polymers. Hence, this chapter serves mostly as a review of heat transfer with emphasis on those topics pertinent to polymer processing. The main aspects that require additional discussion and set polymers apart from other materials are their crystallization behavior and the ability to control molecular orientation during processing.

We begin by summarizing how one handles the temperature dependence of rheological properties of polymer melts in Section 5.1. In Section 5.2 the shell balance approach is used to set up one-dimensional nonisothermal problems encountered in polymer processing. In this same section we summarize the nonisothermal equations of change and their use in polymer processing. The thermal transport properties that occur in these equations include heat capacity, thermal conductivity, density, and for semi-crystalline polymers, heat of fusion. These topics are addressed in Section 5.3. In Section 5.4 the solutions to well-known problems in heat transfer are presented as well as a summary of charts for heat transfer coefficients for commonly encountered geometries. As radiation heat transfer is encountered in a number of processes, this form of heat transfer is also reviewed in Section 5.4. The mechanical properties of polymers depend on the morphology and orientation generated during the cooling process, and in Section 5.5 these topics are discussed. Finally, in Section 5.6 the solution to Design Problem 4 is presented.

## 5.1. TEMPERATURE EFFECTS ON RHEOLOGICAL PROPERTIES

The rheological properties of polymer melts and solutions highly depend on temperature. This is clearly illustrated by the data presented in Fig. 2.5 (p. 11), where the zero shear rate viscosity, $\eta_0$, drops by two orders of magnitude as the temperature is raised from 115 to 240°C. In general, as illustrated in this same figure, the shape of the curves remains nearly unchanged at each temperature. Because of this similarity in the shape of the flow curves, it is possible to represent the viscosity versus $\dot{\gamma}$ data by a single curve as shown in Fig. 3.4 (p. 39), by plotting the reduced viscosity, $\eta_r$, versus the reduced shear rate, $\dot{\gamma}_r$, where:

$$\eta_r = \eta(\dot{\gamma}, T) \frac{\eta_0(T_0)}{\eta_0(T)} \qquad (5.1)$$

and

$$\dot{\gamma}_r = a_T \dot{\gamma}. \qquad (5.2)$$

$\eta_0(T_0)$ and $\eta_0(T)$ are the zero shear viscosities measured at temperatures $T_0$ and $T$, respectively. The "shifting factor," $a_T$, is given as:

$$a_T = \frac{\eta_0(T) T_0 \rho_0}{\eta_0(T_0) T \rho}, \qquad (5.3)$$

where $\rho_0$ and $\rho$ are the densities of the melt at $T_0$ and $T$, respectively.

Actually the ratio $T_0\rho_0/T\rho$ is about unity so that:

$$a_T = \frac{\eta_0(T)}{\eta_0(T_0)}. \tag{5.4}$$

The value of the reduced variables approach is that, given the flow curve at one temperature, we can find the complete flow curve at any other temperature if we know the ratio of the zero shear viscosities at the two temperatures.

If $\eta_0$ at $T$ or $T_0$ is not known, then one can use the insensitivity of shear stress to temperature to find $a_T$. From Eqs. 5.2 and 5.3 we define the reduced shear stress, $\tau_r$, as:

$$\tau_r(\dot{\gamma}, T_0) = \tau_{yx}(\dot{\gamma}, T)\frac{T_0\rho_0}{T\rho}. \tag{5.5}$$

Hence, because $T_0\rho_0/T\rho$ is about unity, this implies that $\tau_{yx}$ is insensitive to temperature. The horizontal shifting of different $\tau_r(\dot{\gamma}, T_0)$ curves gives a master curve of $\tau_r(\dot{\gamma}_r, T_0)$ with the amount of shifting along the shear rate axis at each temperature being $a_T$.

The temperature dependence of $a_T$ is shown in Figure 5.2 for two polymer melts. In this figure $\ln a_T$ is plotted versus $1/T$ (note $T$ is in °K), which suggests that $a_T$ has the following form:

$$a_T = \exp\left[\frac{\Delta E}{R}\left(\frac{1}{T} - \frac{1}{T_0}\right)\right], \tag{5.6}$$

where $\Delta E$ is the activation energy for flow. Most polymer melts seem to follow this behavior. Values of $\Delta E/R$ are reported to be $4.5 \times 10^3$°K, $2.83 \times 10^3$°K, $5.14 \times 10^3$°K, $6.34 \times 10^3$°K, and $4.98 \times 10^3$°K for LDPE, HDPE, polypropylene (PP), polyphenylenesulfide and polyetheretherketone, respectively.

Another way to estimate $a_T$ is through the Williams-Landel-Ferry (WLF) equation, which has been found to hold for a wide variety of polymers for temperatures between the glass-transition temperature, $T_g$, and $T_g + 100$ (Williams et al., 1955). This equation is given as:

$$\log a_T = \frac{-C_1^\circ(T - T_0)}{C_2^\circ + (T - T_0)}. \tag{5.7}$$

If $T_0$ is taken as $T_g$, then $C_1^\circ = 17.44$ and $C_2^\circ = 51.6$°K for a wide variety of polymers. When some data are available, then it is recommended to take $C_1^\circ = 8.86$ and $C_2^\circ = 101.6$°K and then choose $T_0$ to give the best fit of the data.

The primary normal stress difference and other quantities are handled in a manner similar to that used for $\tau_r$:

$$N_{1,r}(\dot{\gamma}, T_0) = N_1(\dot{\gamma}, T)\frac{T_0\rho_0}{T\rho}. \tag{5.8}$$

Thus, $\Psi_1$ is reduced as follows:

$$\Psi_{1,r}(\dot{\gamma}, T_0) = \Psi_1 T_0/a_T^2 T, \tag{5.9}$$

where $T_0\rho_0/T\rho = 1$. The dynamic oscillatory functions can be reduced in the same manner.

$$G_r'(\omega, T_0) = G'(\omega, T)\frac{T_0\rho_0}{T\rho} \tag{5.10}$$

$$G_r''(\omega, T_0) = G''(\omega, T)\frac{T_0\rho_0}{T\rho}, \tag{5.11}$$

with $\omega_r = a_T\omega$.

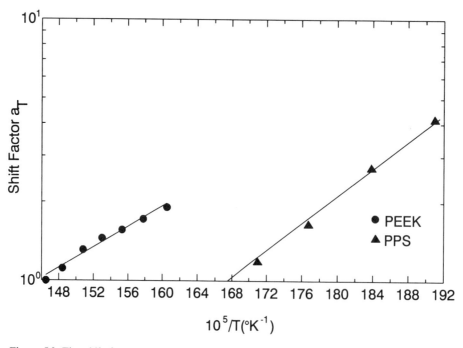

**Figure 5.2** The shift factor versus reciprocal temperature for polyphenylene sulfide (PPS) and polyetheretherketone (PEEK).

The temperature dependence of viscosity can be incorporated into the empiricisms for viscosity. For example, the temperature dependence of the power law coefficients is given as:

$$m = m^0 \exp[-B(T - T_0)] \tag{5.12}$$

$$n = n^0 + C(T - T_0), \tag{5.13}$$

where $m^0$ and $n^0$ are the values of the parameters at the reference temperature, $T_0$, and $B$ and $C$ are constants. Although $m$ is a strong function of $T$, $n$ is not. It is customary to assume that $n$ is constant for most computations. For the Carreau model we can replace $\eta_0(T)$ by $a_T\eta_0(T_0)$ and $\lambda$ by $\lambda(T_0) a_T$. Hence, the Carreau model becomes

$$\eta(T, \dot{\gamma}) = \eta_0(T_0)a_T[1 + (\lambda(T_0)a_T\dot{\gamma})^2]^{\frac{n-1}{2}} \tag{5.14}$$

From Eq. 5.14 one can calculate the flow curve at any other temperature provided $\eta_0$ and $\lambda$ are known at the reference temperature, $T_0$, and $a_T$ is known. Computations must be done numerically with this model.

## 5.2. THE ENERGY EQUATION
### 5.2.1. SHELL ENERGY BALANCES

In this section we set up shell energy balances for flowing polymeric fluids. This material should be helpful in conceptualizing the non-isothermal equations of change. The basic principle that is used is the conservation of thermal energy as applied to a thin shell of fluid stated as:

$$\begin{pmatrix} \text{Net rate of gain of} \\ \text{thermal energy} \end{pmatrix} = \begin{pmatrix} \text{Rate of thermal} \\ \text{energy in} \end{pmatrix}$$
$$- \begin{pmatrix} \text{Rate of thermal} \\ \text{energy out} \end{pmatrix} + \begin{pmatrix} \text{Rate of thermal} \\ \text{energy production} \end{pmatrix}. \tag{5.15}$$

Applying this principle to a differential volume element and taking the limit as the volume element goes to zero lead to a differential equation for the temperature distribution. The procedure is described in more detail in Bird et al. (1960, Chapter 9). We illustrate the use of Eq. 5.15 through the following example.

---

**EXAMPLE 5.1**
**Cooling of Polypropylene Film**
As shown in the figure associated with Design Problem 4 (Fig. 5.1), a film of polypropylene (PP) 0.1016 cm in thickness is extruded from a 0.162-m-wide film die on to a casting drum 2.54 cm below the die. The temperature of the film drops as a result of forced convection at the film surface. Determine the temperature of the film surface when it contacts the casting drum. The melt temperature as it leaves the die is 200°C. The heat transfer coefficient is 100 W m$^{-2}$K$^{-1}$. Neglect die swell and use an average film thickness of $2.54 \times 10^{-2}$ cm (The film thickness is not uniform, and this aspect is discussed in Chapter 9). In Fig. 5.3 is shown a model of the region of interest.

**Solution**

The model of the region which is to be analyzed is shown in Fig. 5.3. The heat is removed primarily by forced convection at the surfaces because of air at 25°C moving over the sheet (but in principle one should consider radiation effects as is discussed in Section 5.4). An energy balance is performed on the element shown in Fig. 5.3 to give:

$$V_0\rho\bar{C}_p(T - T_R)|_x W\Delta y - V_0\rho\bar{C}_p(T - T_R)|_{x+\Delta x}$$
$$W\Delta y + q_x|_x W\Delta y - q_x|_{x+\Delta x}(W\Delta y)$$
$$+(q_y|_y - q_y|_{y+\Delta y})W\Delta x = 0, \tag{5.16}$$

where $T_R$ is a reference temperature, $\bar{C}_p$ is the constant pressure heat capacity per unit mass, $V_0$ is the film velocity, $W$ is the film width, and $\rho$ is the density. We divide Eq. 5.16 by the volume of the element, $W\Delta x\Delta y$ and take the limit as $\Delta x$ and $\Delta y$ go to zero to give:

$$-V_0\rho\bar{C}_p\frac{\partial T}{\partial x} - \frac{\partial q_x}{\partial x} - \frac{\partial q_y}{\partial y} = 0. \tag{5.17}$$

To obtain a differential equation in terms of temperature rather than the heat fluxes, $q_x$ and $q_y$, we use Fourier's law of heat conduction (Bird et al., 1960, Chapter 9):

$$q_x = -k\frac{\partial T}{\partial x} \qquad q_y = -k\frac{\partial T}{\partial y}, \tag{5.18}$$

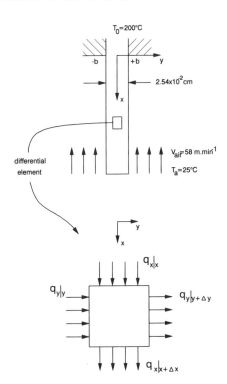

**Figure 5.3** Model of region between the film die and casting drum (*top*) and expanded element on which energy balance is performed (*bottom*).

where $k$ is the thermal conductivity. Substitution of the expressions in Eq. 5.18 into Eq. 5.17 gives:

$$V_0\rho\bar{C}_p\frac{\partial T}{\partial x} - k\frac{\partial^2 T}{\partial x^2} - k\frac{\partial^2 T}{\partial y^2} = 0, \tag{5.19}$$

where it has been assumed that $k$ is constant. Eq. 5.19 must be solved along with the following boundary and initial conditions:

B.C.1 at $y = +b$ $\qquad q_y = -k\dfrac{\partial T}{\partial y} = h(T(b) - T_a)$

B.C.2 at $y = -b$ $\qquad q_y = -k\dfrac{\partial T}{\partial y} = h(T_a - T(-b))$

I.C. at $x = 0$ $\qquad T = T_0 = 200°C.$ $\qquad\qquad$ (5.20)

Here $T_a$ is the air temperature.

We next try to determine whether Eq. 5.19 can be reduced or simplified. By writing Eq. 5.19 and the boundary and initial conditions in dimensionless form it is much easier to determine which terms in the differential equation are most important. The dependent and independent variables are written in dimensionless form by dividing them by an appropriate characteristic quantity. In particular we introduce the following dimensionless quantities:

$$\zeta = \frac{x}{b} \qquad \xi = \frac{y}{b} \qquad \theta = \frac{T - T_a}{T_0 - T_a}. \qquad (5.21)$$

Equation 5.19 in dimensionless form becomes

$$\frac{bV_0\rho\bar{C}_p}{k}\frac{\partial\theta}{\partial\zeta} = \frac{\partial^2\theta}{\partial\zeta^2} + \frac{\partial^2\theta}{\partial\xi^2}. \qquad (5.22)$$

The term multiplying $\partial\theta/\partial\zeta$ is also dimensionless and is called the Peclet number, Pe, and represents the ratio of the heat transfer by forced convection to that by conduction. The boundary and initial conditions given in Eq. 5.20 become:

B.C.1 at $\xi = 1$ $\qquad \dfrac{\partial\theta}{\partial\xi} = \dfrac{-hb}{k}\theta|_{\xi=1}$

B.C.2 at $\xi = -1$ $\qquad \dfrac{\partial\theta}{\partial\xi} = \dfrac{hb}{k}\theta|_{\xi=-1}$

I.C. at $\zeta = 0.$ $\qquad \theta = 1.$ $\qquad\qquad$ (5.23)

The combination of terms, $hb/k$, is called the Nusselt number, Nu, and it is basically a dimensionless temperature gradient averaged over the heat transfer surface.

We are now in a position to evaluate which terms could be eliminated from Eq. 5.22. First we compare the order of $\partial\theta/\partial\zeta$ and $\partial^2\theta/\partial\zeta^2$. Evaluating Pe using data in Table 5.4 we find Pe = 2100. Hence, Pe $\partial\theta/\partial\zeta \gg \partial^2\theta/\partial\zeta^2$, and we can eliminate $\partial^2\theta/\partial\zeta^2$ from Eq. 5.22. The equation to be solved is:

$$\text{Pe}\frac{\partial\theta}{\partial\zeta} = \frac{\partial^2\theta}{\partial\xi^2}. \qquad (5.24)$$

Except when dealing with molten metals it is common to neglect conduction in the flow direction as heat transfer by this manner is small relative to convection. The solution to this equation requires the use of the separation of variables, and the solution is presented in graphical form in Fig. 5.12.

We can try to reduce the equation further in order to obtain an analytical solution. Since Pe $\gg$ 1, then heat transfer is dominated by forced convection rather than conduction, and it is possible to assume that the temperature distribution through the thickness of the sheet is nearly uniform, except near the outer edges. In order to reduce

Eq. 5.24 to an ordinary differential equation (ODE), we first define the mean temperature, $\bar{\theta}$:

$$\bar{\theta} = \int_{-1}^{1}\theta\,d\xi \bigg/ \int_{-1}^{1}d\xi. \qquad (5.25)$$

Next, we integrate Eq. 5.24 over the thickness of the sheet and use Eq. 5.25 to obtain:

$$\text{Pe}\frac{\partial\bar{\theta}}{\partial\zeta} = \frac{1}{2}\int_{-1}^{1}\frac{\partial}{\partial\xi}\frac{\partial\theta}{\partial\xi}d\xi. \qquad (5.26)$$

As

$$\frac{\partial\theta}{\partial\xi} = \frac{-q_y b}{k(T_0 - T_a)}, \qquad (5.27)$$

Eq. 5.26 becomes:

$$\text{Pe}\frac{\partial\bar{\theta}}{\partial\zeta} = \frac{(-b)}{k(T_0 - T_a)}\int_{q_y(-1)}^{q_y(+1)}dq_y. \qquad (5.28)$$

Using the boundary conditions in Eq. 5.23, we now obtain

$$\text{Pe}\frac{\partial\bar{\theta}}{\partial\zeta} = -\frac{hb}{k}\bar{\theta}. \qquad (5.29)$$

The solution to Eq. 5.29 can be found by either using an integrating factor or separating variables and is:

$$\bar{\theta} = C_1 e^{-C\zeta}, \qquad (5.30)$$

where $C_1$ is an integration constant and $C = h/\rho\bar{C}_p V_0$. $C_1$ is obtained from the I.C. (Eq. 5.23) and is 1.0. Using the material properties for PP given in Table 5.4 and the conditions given in the problem, $\theta = 0.999$, or the change in the film temperature over a distance of 2.54 cm is insignificant.

### EXAMPLE 5.2
### Temperature Rise Due to Viscous Dissipation for HDPE in a Cone-and-Plate Rheometer

Determine the maximum temperature rise for HDPE in a cone-and-plate rheometer at a shear rate of $10\,\text{s}^{-1}$. The diameter of the plate is 2.54 cm, and the cone angle is 0.1 radian. The properties are assumed to be independent of temperature and are given below:

$\rho = 782\,\text{kg m}^{-3}$ $\qquad\qquad k = 0.255\,\text{W m}^{-1}\,°\text{K}^{-1}$

$m = 4.68 \times 10^3\,\text{Pa·s}^n$ $\qquad n = 0.54$

$\bar{C}_p = 2650\,\text{J kg}^{-1}\,°\text{K}^{-1}$

The plate and melt temperatures are taken to be 200°C, and no heat transfer is assumed to occur at the free surface.

### Solution

Based on the description of the cone-and-plate rheometer in Section 3.3 we can consider the flow to be as shown in Fig. 5.4. Based on the dimensions given, the height, $H$, at the edge is 0.125 cm. An energy

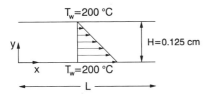

**Figure 5.4** Approximation of flow in a cone-and-plate rheometer.

balance is performed on the slab of thickness $\Delta y$ and unit width. It is assumed that heat is generated by viscous dissipation, which is the conversion of mechanical energy into heat. The viscous dissipation per unit volume is:

$$\dot{S} = -\tau_{yx}\frac{dv_x}{dy} \tag{5.31}$$

and is analogous to the rate of work in moving a single particle (i.e., $W = F_x v_x$). The heat that is generated is conducted out of the melt through the metal plates by means of conduction in the $y$-direction. The energy balance per unit width is:

$$q_y|_y L - q_y|_{y+\Delta y} L + \dot{S}L\Delta y = 0. \tag{5.32}$$

Next, we divide through by the volume of the element, that is $1 \cdot \Delta y \cdot L$, and take the limit as $\Delta y \to 0$. This gives the following differential equation:

$$-\frac{dq_y}{dy} + \dot{S} = 0. \tag{5.33}$$

We now replace $q_y$ with $-kdT/dy$, which is Fourier's law, and $\dot{S}$ with Eq. 5.31:

$$k\frac{d^2T}{dy^2} - \tau_{yx}\frac{dv_x}{dy} = 0. \tag{5.34}$$

Substituting in for $\tau_{yx}$, which is

$$\tau_{yx} = +m\left(-\frac{dv_x}{dy}\right)^n = +m\dot{\gamma}^n, \tag{5.35}$$

we obtain the following differential equation:

$$k\frac{d^2T}{dy^2} + m\dot{\gamma}^{n+1} = 0. \tag{5.36}$$

Equation 5.36 can be integrated to give:

$$T = -\frac{m}{k}(\dot{\gamma})^{n+1}\left(\frac{y^2}{2}\right) + C_1 y + C_2. \tag{5.37}$$

The boundary conditions required for finding $C_1$ and $C_2$ are:

B.C.1: at $y = 0$    $T = T_w = 200°C$ $\tag{5.38}$

B.C.2: at $y = H$    $T = T_w = 200°C.$ $\tag{5.39}$

Using Eqs. 5.39 and 5.38, Eq.5.37 becomes:

$$T = -\frac{H^2 m\dot{\gamma}^{n+1}}{2k}\left(\frac{y}{H}\right)^2 + \frac{H^2 m\dot{\gamma}^{n+1}}{2k}\frac{y}{H} + T_w. \tag{5.40}$$

From this equation it is found that the temperature is a maximum at $H/2$ and for $\dot{\gamma} = 10\,s^{-1}$ $T - T_w$ is 6.50°C. Hence, at this shear rate there is a significant rise in the melt temperature and the viscosity measurements would be affected. Furthermore, this would be enough of a temperature rise to significantly affect the measurement of $N_1$ because of the increase in volume that causes the plates to be pushed apart. There is very little one can do to improve the design of the cone-and-plate system to eliminate this problem.

The next example is more complicated and involves the flow of HDPE through a pipe die. It is important to minimize the temperature increase in a melt due to viscous dissipation. In this case the velocity profile may not be known a priori as it may be affected by the temperature distribution.

**EXAMPLE 5.3**
**Nonisothermal Flow of HDPE through a Pipe Die**

Returning to Section 2.2.1 we now ask whether the melt temperature will remain at 453°K as HDPE passes through the pipe die at the upper extrusion limit. We assume that the melt enters the die at $T_0 = 453°K$ and that the wall temperature at $R$ is 453°K. The inner wall is also maintained at 453°K. The rheological properties for this melt were given in Table 2.3 (p. 14). The thermal properties are given in Table 5.6 (p. 105). For this problem we take these properties to be independent of temperature and use $\rho = 782\,kg/m^3$, $C_p = 2650\,J\,kg^{-1}\,°K^{-1}$, and $k = 0.255\,W\,m^{-1}\,°K^{-1}$.

**Solution**

For pedagogical purposes we will solve this problem using the forced convection assumption in which it is assumed that the velocity profile is unaffected by changes in the viscosity as a result of changes in temperature. With this assumption, the velocity profile remains unchanged from that of the isothermal case and the equation of motion can be used independently of the energy balance. The equation of motion for this problem was given in Eq. 2.10 and is repeated:

$$\frac{d}{dr}(r\tau_{rz}) = -\left(\frac{dp}{dz}\right)r = \frac{(P_0-P_L)}{L}r. \tag{5.41}$$

For a generalized Newtonian fluid (GNF) model using the power-law empiricism for viscosity the velocity field was given in Eqs. 2.28 and 2.29. Taking $n = 0.5$ (actually it is 0.56, but this allows us to obtain an analytical solution), we find $v_z^<$ and $v_z^>$ to be, respectively:

$$v_z^< = R\left[\frac{(P_0-P_L)R}{2mL}\right]^s\left[(\xi-\kappa)\left(\frac{\beta^4}{\xi} - 2\beta^2\right) + \frac{1}{3}(\xi^3 - \kappa^3)\right]$$

$$v_z^> = R\left[\frac{(P_0-P_L)R}{2mL}\right]^s\left[(1-\xi)\left(\frac{\beta^4}{\xi} - 2\beta^2\right) + \frac{1}{3}(1-\xi^3)\right]. \tag{5.42}$$

We next carry out an energy balance on the differential (doughnut) element shown in Fig. 5.5. The sources of energy transport consist of heat transfer into and out of the element by convection due to flow in the $z$ direction, conduction of heat into and out of the element in both the $z$ and $r$ directions and heat generated by viscous dissipation. When an energy balance is applied to the element shown in Fig. 5.5 the following equation is obtained:

$$\rho\bar{C}p(T - T_R)|_z v_z 2\pi r\Delta r - \rho\bar{C}_p(T - T_R)|_{z+\Delta z} v_z 2\pi r\Delta r$$
$$+ (q_z|_z - q_z|_{z+\Delta z})2\pi r\Delta r$$
$$+ (q_r r|_r - q_r r|_{r+\Delta r})2\pi r\Delta z + \dot{S}2\pi r\Delta r\Delta z = 0. \tag{5.43}$$

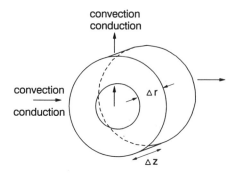

**Figure 5.5** Cylindrical element on which the energy balance is performed in Example 5.3.

Here $\bar{C}_p$ is the constant pressure heat capacity per unit mass, $T_R$ is a reference temperature, $q_z$ and $q_r$ are the heat fluxes due to conduction in the $z$ and $r$ directions, respectively, and $\dot{S}$ is a source term in units of energy per unit time per unit volume. In this example there is no chemical reaction, but because of the high viscosity of the fluid, there may be significant viscous dissipation. Viscous dissipation is given by:

$$\dot{S} = -\tau_{rz}\frac{dv_z}{dr}. \tag{5.44}$$

We now divide all the terms in Eq. 5.43 by the volume of the element, $2\pi r \Delta r \Delta z$, and then take the limit as $\Delta r$ and $\Delta z \rightarrow 0$. This leads to the following differential equation:

$$\rho\bar{C}_p v_z \frac{\partial T}{\partial z} = -\frac{\partial q_z}{\partial z} - \frac{1}{r}\frac{\partial}{\partial r}(rq_r) + \dot{S}. \tag{5.45}$$

In order to put Eq. 5.45 into a form which can be solved for $T$, two substitutions are required. We first replace $q_z$ and $q_r$ by the following expressions that come from Fourier's law of heat conduction (Bird et al., 1960):

$$q_z = -k\frac{\partial T}{\partial z} \qquad q_r = -k\frac{\partial T}{\partial r}. \tag{5.46}$$

Furthermore, we replace $\dot{S}$ by $-\tau_{rz}(dv_z/dr)$, which for a power-law fluid becomes:

$$-\tau_{rz}\frac{dv_z}{dr} = +m\left|\frac{dv_z}{dr}\right|^{n-1}\left(\frac{dv_z}{dr}\right)^2. \tag{5.47}$$

This leads to a partial differential equation for the temperature distribution

$$\rho\bar{C}_p v_z(r)\frac{\partial T}{\partial z} = k\frac{\partial^2 T}{\partial z^2} + \frac{k}{r}\frac{\partial}{\partial r}\left(r\frac{\partial T}{\partial r}\right) + m\left|\frac{dv_z}{dr}\right|^{n-1}\left(\frac{dv_z}{dr}\right)^2. \tag{5.48}$$

Because of the forced convection assumption, we can solve Eq. 5.48 independently of the equation of motion. However, these two equations would ordinarily be coupled because of the dependence of $v_z(r)$ on $\eta$,

which is highly temperature dependent, and hence numerical methods would be required.

In order to estimate whether a significant temperature rise will occur we make three further simplifications. Because $\kappa = 0.83$, we can neglect curvature and treat the annular die as if it is parallel plate geometry of height $H = R(1-\kappa)$ and width $W = \pi R(1+\kappa)$. Eq. 5.48 becomes:

$$\rho\bar{C}_p v_z(y)\frac{\partial T}{\partial z} = k\frac{\partial^2 T}{\partial y^2} + m\left|\frac{dv_z}{dy}\right|^{n-1}\left(\frac{dv_z}{dy}\right)^2, \tag{5.49}$$

where again it is assumed that the conduction term in the flow direction ($z$ direction) is insignificant relative to the convection term. The highest possible temperature rise will occur if there is no heat conducted through the walls of the die (this is referred to as *adiabatic conditions*). This is equivalent to saying that $q_y = 0$ at both walls. Using the velocity profile for flow through parallel plates given in Table 2.5 (p. 19), Eq. 5.49 becomes:

$$\rho\bar{C}_p v_{max}\left[1 - \left(\frac{y}{H/2}\right)^{s+1}\right]\frac{\partial T}{\partial z}$$
$$= k\frac{\partial^2 T}{\partial y^2} + m\left[\frac{v_{max}(s+1)}{H/2}\right]^{n+1}\left(\frac{y}{H/2}\right)^{1+s} \tag{5.50}$$

subject to the following boundary and initial conditions

$$\text{B.C.1 at } y=0 \qquad q_y = 0 \tag{5.51}$$

$$\text{B.C.2 at } y=H/2 \qquad q_y = 0 \tag{5.52}$$

$$\text{I.C. at } z=0 \qquad T = T_0. \tag{5.53}$$

We next average Eq. 5.50 over the cross sectional area using the boundary conditions above to yield:

$$\rho\bar{C}_p\langle v_z\rangle HW\frac{dT}{dz} = \frac{WHm}{s+2}\left[\frac{v_{max}(s+1)}{H/2}\right]^{n+1}. \tag{5.54}$$

This is a first-order ordinary differential equation that we can now integrate to find the temperature rise at the exit of the die:

$$T - T_0 = \frac{m}{\rho\bar{C}_p\langle v_z\rangle(s+2)}\left[\frac{v_{max}(s+1)}{H/2}\right]^{n+1}z. \tag{5.55}$$

Using the results from Section 2.2.2 and the values given in the problem statement, we find

$$T - T_0 = \frac{6.19E+03\ \text{Pa·s}^n(3.33E+03)}{(782\ \text{kg m}^{-3})(2650\ \text{J kg}^{-1}\ ^\circ\text{K}^{-1})(0.112\ \text{ms}^{-1})(3.786)}z.$$
$$= 14.6z. \tag{5.56}$$

With the length of the die given as $6.35 \times 10^{-3}$ m, the temperature rise is only $0.92^\circ$K. Hence, for the case at hand there is no significant rise in temperature as a result of viscous heating. However, if the die length was increased to $L = 0.635$ m, then there would be a $9.2^\circ$K increase in temperature that could be significant.

### 5.2.2. EQUATION OF THERMAL ENERGY

As in Chapter 2 for the equation of motion it is more convenient to use the general form of the thermal energy equation rather than use the shell energy balance approach. The thermal energy equation is given in terms of energy and momentum fluxes in Table 5.1. When the components of the heat flux, $\mathbf{q}$, which are given in Table 5.2 and the GNF model are substituted into the equations given in Table 5.1, we obtain the energy equation in terms of transport properties. This form of the equation with the assumption of constant properties is given in Table 5.3 (In this table we replace $\mu$ by $\eta$). These tables are much more convenient to use than the shell energy balance approach.

We solve the following set of equations written in vector notation:

Continuity:   $\mathbf{V} \cdot \rho \mathbf{v} = 0$ $\qquad\qquad$ (5.57)

Motion: $\qquad \rho \dfrac{D\mathbf{v}}{Dt} = -\mathbf{V}p + \mathbf{V} \cdot \eta \dot{\gamma} + \rho \mathbf{g}$ $\qquad$ (5.58)

Energy: $\qquad \rho \bar{C}_p \dfrac{DT}{Dt} = (\mathbf{V} \cdot k \mathbf{V} T) + \dfrac{1}{2}\eta(\dot{\gamma} : \dot{\gamma}) + \dot{S}.$ $\qquad$ (5.59)

Here $D/Dt$ is the material time derivative or the time derivative following the fluid motion. These equations have been written in a form that demonstrates the temperature dependence of the viscosity and thermal conductivity. Furthermore, the equations have been written with the assumption that the rheological properties are described by the GNF model. We now use the nonisothermal equations of change to resolve the examples in Section 5.2.1.

### TABLE 5.1 Equation of Thermal Energy in Terms of Energy and Momentum Fluxes

Rectangular coordinates

$$\rho\hat{C}_v\left(\frac{\partial T}{\partial t} + v_x\frac{\partial T}{\partial x} + v_y\frac{\partial T}{\partial y} + v_z\frac{\partial T}{\partial z}\right) = -\left[\frac{\partial q_x}{\partial x} + \frac{\partial q_y}{\partial y} + \frac{\partial q_z}{\partial z}\right]$$

$$-T\left(\frac{\partial p}{\partial T}\right)_\rho\left(\frac{\partial v_x}{\partial x} + \frac{\partial v_y}{\partial y} + \frac{\partial v_z}{\partial z}\right) - \left\{\tau_{xx}\frac{\partial v_x}{\partial x} + \tau_{yy}\frac{\partial v_y}{\partial y} + \tau_{zz}\frac{\partial v_z}{\partial z}\right\}$$

$$-\left\{\tau_{xy}\left(\frac{\partial v_x}{\partial y} + \frac{\partial v_y}{\partial x}\right) + \tau_{xz}\left(\frac{\partial v_x}{\partial z} + \frac{\partial v_z}{\partial x}\right) + \tau_{yz}\left(\frac{\partial v_y}{\partial z} + \frac{\partial v_z}{\partial y}\right)\right\}$$

(A)

Cylindrical coordinates

$$\rho\hat{C}_v\left(\frac{\partial T}{\partial t} + v_r\frac{\partial T}{\partial r} + \frac{v_\theta}{r}\frac{\partial T}{\partial \theta} + v_z\frac{\partial T}{\partial z}\right) = -\left[\frac{1}{r}\frac{\partial}{\partial r}(rq_r) + \frac{1}{r}\frac{\partial q_\theta}{\partial \theta} + \frac{\partial q_z}{\partial z}\right]$$

$$-T\left(\frac{\partial p}{\partial T}\right)_\rho\left(\frac{1}{r}\frac{\partial}{\partial r}(rv_r) + \frac{1}{r}\frac{\partial v_\theta}{\partial \theta} + \frac{\partial v_z}{\partial z}\right) - \left\{\tau_{rr}\frac{\partial v_r}{\partial r} + \tau_{\theta\theta}\frac{1}{r}\left(\frac{\partial v_\theta}{\partial \theta} + v_r\right) + \tau_{zz}\frac{\partial v_z}{\partial z}\right\}$$

$$-\left\{\tau_{r\theta}\left[r\frac{\partial}{\partial r}\left(\frac{v_\theta}{r}\right) + \frac{1}{r}\frac{\partial v_r}{\partial \theta}\right] + \tau_{rz}\left(\frac{\partial v_z}{\partial r} + \frac{\partial v_r}{\partial z}\right) + \tau_{\theta z}\left(\frac{1}{r}\frac{\partial v_z}{\partial \theta} + \frac{\partial v_\theta}{\partial z}\right)\right\}$$

(B)

Spherical coordinates

$$\rho\hat{C}_v\left(\frac{\partial T}{\partial t} + v_r\frac{\partial T}{\partial r} + \frac{v_\theta}{r}\frac{\partial T}{\partial \theta} + \frac{v_\phi}{r\sin\theta}\frac{\partial T}{\partial \phi}\right)$$

$$= -\left[\frac{1}{r^2}\frac{\partial}{\partial r}(r^2 q_r) + \frac{1}{r\sin\theta}\frac{\partial}{\partial \theta}(q_\theta\sin\theta) + \frac{1}{r\sin\theta}\frac{\partial q_\phi}{\partial \phi}\right]$$

$$-T\left(\frac{\partial p}{\partial T}\right)_\rho\left(\frac{1}{r^2}\frac{\partial}{\partial r}(r^2 v_r) + \frac{1}{r\sin\theta}\frac{\partial}{\partial \theta}(v_\theta\sin\theta) + \frac{1}{r\sin\theta}\frac{\partial v_\phi}{\partial \phi}\right)$$

$$-\left\{\tau_{rr}\frac{\partial v_r}{\partial r} + \tau_{\theta\theta}\left(\frac{1}{r}\frac{\partial v_\theta}{\partial \theta} + \frac{v_r}{r}\right) + \tau_{\phi\phi}\left(\frac{1}{r\sin\theta}\frac{\partial v_\phi}{\partial \phi} + \frac{v_r}{r} + \frac{v_\theta\cot\theta}{r}\right)\right\}$$

$$-\left\{\tau_{r\theta}\left(\frac{\partial v_\theta}{\partial r} + \frac{1}{r}\frac{\partial v_r}{\partial \theta} - \frac{v_\theta}{r}\right) + \tau_{r\phi}\left(\frac{\partial v_\phi}{\partial r} + \frac{1}{r\sin\theta}\frac{\partial v_r}{\partial \phi} - \frac{v_\phi}{r}\right) + \tau_{\theta\phi}\left(\frac{1}{r}\frac{\partial v_\phi}{\partial \theta} + \frac{1}{r\sin\theta}\frac{\partial v_\theta}{\partial \phi} - \frac{\cot\theta}{r}v_\phi\right)\right\}$$

(C)

(Reprinted by permission of the publisher from Bird et al., 1960.)

### TABLE 5.2 Components of the Energy Flux $q$

| Rectangular | Cylindrical | Spherical |
|---|---|---|
| $q_x = -k\dfrac{\partial T}{\partial x}$ | $q_r = -k\dfrac{\partial T}{\partial r}$ | $q_r = -k\dfrac{\partial T}{\partial r}$ |
| $q_y = -k\dfrac{\partial T}{\partial y}$ | $q_\theta = -k\dfrac{1}{r}\dfrac{\partial T}{\partial \theta}$ | $q_\theta = -k\dfrac{1}{r}\dfrac{\partial T}{\partial \theta}$ |
| $q_z = -k\dfrac{\partial T}{\partial z}$ | $q_z = -k\dfrac{\partial T}{\partial z}$ | $q_\phi = -k\dfrac{1}{r\sin\theta}\dfrac{\partial T}{\partial \phi}$ |

**TABLE 5.3 Equation of Thermal Energy in Terms of the Transport Properties (for Newtonian fluids of constant $\rho$ and $k$)**

Rectangular coordinates

$$
\rho\hat{C}_p\left(\frac{\partial T}{\partial t} + v_x\frac{\partial T}{\partial x} + v_y\frac{\partial T}{\partial y} + v_z\frac{\partial T}{\partial z}\right) = k\left[\frac{\partial^2 T}{\partial x^2} + \frac{\partial^2 T}{\partial y^2} + \frac{\partial^2 T}{\partial z^2}\right]
$$
$$
+ 2\mu\left\{\left(\frac{\partial v_x}{\partial x}\right)^2 + \left(\frac{\partial v_y}{\partial y}\right)^2 + \left(\frac{\partial v_z}{\partial z}\right)^2\right\}
$$
$$
+ \mu\left\{\left(\frac{\partial v_x}{\partial y} + \frac{\partial v_y}{\partial x}\right)^2 + \left(\frac{\partial v_x}{\partial z} + \frac{\partial v_z}{\partial x}\right)^2 + \left(\frac{\partial v_y}{\partial z} + \frac{\partial v_z}{\partial y}\right)^2\right\} \tag{A}
$$

Cylindrical coordinates

$$
\rho\hat{C}_p\left(\frac{\partial T}{\partial t} + v_r\frac{\partial T}{\partial r} + \frac{v_\theta}{r}\frac{\partial T}{\partial \theta} + v_z\frac{\partial T}{\partial z}\right) = k\left[\frac{1}{r}\frac{\partial}{\partial r}\left(r\frac{\partial T}{\partial r}\right) + \frac{1}{r^2}\frac{\partial^2 T}{\partial \theta^2} + \frac{\partial^2 T}{\partial z^2}\right]
$$
$$
+ 2\mu\left\{\left(\frac{\partial v_r}{\partial r}\right)^2 + \left[\frac{1}{r}\left(\frac{\partial v_\theta}{\partial \theta} + v_r\right)\right]^2 + \left(\frac{\partial v_z}{\partial z}\right)^2\right\}
$$
$$
+ \mu\left\{\left(\frac{\partial v_\theta}{\partial z} + \frac{1}{r}\frac{\partial v_z}{\partial \theta}\right)^2 + \left(\frac{\partial v_z}{\partial r} + \frac{\partial v_r}{\partial z}\right)^2 + \left[\frac{1}{r}\frac{\partial v_r}{\partial \theta} + r\frac{\partial}{\partial r}\left(\frac{v_\theta}{r}\right)\right]^2\right\} \tag{B}
$$

Spherical coordinates

$$
\rho\hat{C}_p\left(\frac{\partial T}{\partial t} + v_r\frac{\partial T}{\partial r} + \frac{v_\theta}{r}\frac{\partial T}{\partial \theta} + \frac{v_\phi}{r\sin\theta}\frac{\partial T}{\partial \phi}\right) = k\left[\frac{1}{r^2}\frac{\partial}{\partial r}\left(r^2\frac{\partial T}{\partial r}\right) + \frac{1}{r^2\sin\theta}\frac{\partial}{\partial \theta}\left(\sin\theta\frac{\partial T}{\partial \theta}\right) + \frac{1}{r^2\sin^2\theta}\frac{\partial^2 T}{\partial \phi^2}\right]
$$
$$
+ 2\mu\left\{\left(\frac{\partial v_r}{\partial r}\right)^2 + \left(\frac{1}{r}\frac{\partial v_\theta}{\partial \theta} + \frac{v_r}{r}\right)^2 + \left(\frac{1}{r\sin\theta}\frac{\partial v_\phi}{\partial \phi} + \frac{v_r}{r} + \frac{v_\theta\cot\theta}{r}\right)^2\right\}
$$
$$
+ \mu\left\{\left[r\frac{\partial}{\partial r}\left(\frac{v_\theta}{r}\right) + \frac{1}{r}\frac{\partial v_r}{\partial \theta}\right]^2 + \left[\frac{1}{r\sin\theta}\frac{\partial v_r}{\partial \phi} + r\frac{\partial}{\partial r}\left(\frac{v_\phi}{r}\right)\right]^2\right.
$$
$$
\left. + \left[\frac{\sin\theta}{r}\frac{\partial}{\partial \theta}\left(\frac{v_\phi}{\sin\theta}\right) + \frac{1}{r\sin\theta}\frac{\partial v_\theta}{\partial \phi}\right]^2\right\} \tag{C}
$$

(Reprinted by permission of the publisher from Bird et al., 1960.)

---

## EXAMPLE 5.4
### Use of the Nonisothermal Equations of Change

Reformulate Examples 5.1, 2, and 3 using the nonisothermal equations of change.

### Solution to Example 5.1

We start by making postulates pertaining to the velocity and temperature fields:

$$
v_x = \text{const.} = v_0 \qquad T = T(x, y). \tag{5.60}
$$

From these postulates and Fourier's law (Table 5.2) the following fluxes exist:

$$
q_x = -k\frac{\partial T}{\partial x} \qquad q_y = -k\frac{\partial T}{\partial y}. \tag{5.61}
$$

The equation of thermal energy becomes:

$$
\rho\bar{C}_p v_x\frac{\partial T}{\partial x} = -\frac{\partial q_x}{\partial x} - \frac{\partial q_y}{\partial y}. \tag{5.62}
$$

Similar arguments as presented in Ex. 5.1 can be used to reduce Eq. 5.62 to Eq. 5.29.

### Solution to Example 5.2

Again postulates pertaining to the velocity and temperature fields are made:

$$
v_x = v_x(y) \qquad T = T(y). \tag{5.63}
$$

The equation of motion is

$$
-\frac{\partial \tau_{yx}}{\partial x} = 0. \tag{5.64}
$$

For the GNF model with the power-law empiricism for viscosity

$$
\tau_{yx} = -m\left(\frac{dv_x}{dy}\right)^n. \tag{5.65}
$$

Hence, the velocity profile is

$$
v_x = \frac{V_0}{H}y, \tag{5.66}
$$

and for this flow $v_x$ is unaffected by the temperature profile. The energy equation for constant properties becomes:

$$k\frac{\partial^2 T}{\partial y^2}+\eta\left(\frac{\partial v_x}{\partial y}\right)^2=0. \tag{5.67}$$

This equation can be solved as in Ex. 5.2. We again assume that all the physical properties are constant.

**Solution to Example 5.3**

The following postulates are made for $v_z$ and $T$:

$$v_z=v_z(r) \qquad T=T(r,z). \tag{5.68}$$

The equation of motion becomes:

$$-\frac{dp}{dz}=+\frac{1}{r}\frac{d}{dr}(r\tau_{rz}), \tag{5.69}$$

or with $\tau_{rz}=-\eta\left(\dfrac{dv_z}{dr}\right)$, we obtain:

$$\frac{dp}{dz}=\frac{1}{r}\frac{d}{dr}\left(r\eta\frac{dv_z}{dr}\right). \tag{5.70}$$

The energy equation is:

$$\rho\bar{C}_p v_z\frac{\partial T}{\partial z}=-\frac{1}{r}\frac{\partial}{\partial r}(rq_r)-\frac{\partial q_z}{\partial z}-\tau_{rz}\frac{dv_z}{dr}. \tag{5.71}$$

With the substitution of Fourier's law of heat conduction and the GNF model, we obtain:

$$\rho\bar{C}_p v_z\frac{\partial T}{\partial z}=\frac{1}{r}\frac{\partial}{\partial r}\left(kr\frac{\partial T}{\partial r}\right)+\frac{\partial}{\partial z}\left(k\frac{\partial T}{\partial z}\right)+\eta\left(\frac{dv_z}{dr}\right)^2. \tag{5.72}$$

With the assumption of constant physical properties, Eq. 5.72 would be the same as that obtained by means of the shell energy balance (Eq. 5.48).

---

## 5.3. THERMAL TRANSPORT PROPERTIES

The material properties that appear in the thermal energy equation are the density, $\rho$, the constant pressure heat capacity, $\bar{C}_p$, (note: when $\rho$ is constant, $\bar{C}_p\approx\bar{C}_v$), and the thermal conductivity. In addition, because a number of polymers are semicrystalline, energy can be absorbed during melting or given up when crystallization occurs on cooling. The energy absorbed is referred to as the heat of fusion ($\Delta\bar{H}_f$), and the energy released is the heat of crystallization, $\Delta\bar{H}_c$. In this section we present representative values for both common commercial polymers as well as high-performance engineering thermoplastics. Procedures for handling composite systems (i.e., filled polymers and polymer blends) are also discussed.

### 5.3.1. HOMOGENEOUS POLYMER SYSTEMS

Representative thermal properties for an amorphous polymer, in this case polycarbonate, are shown in Fig. 5.6 as a function of temperature. Here it is observed that all the quantities except $\bar{C}_p$ change continuously with increasing temperature. At about 153°C, there is a discontinuity in $\bar{C}_p$ that is associated with the glass transition temperature, $T_g$. Above $T_g$, the polymer becomes more easily deformable and is usually processed above $T_g$. It is observed that above $T_g$ there is very little change in the properties. For example, $\bar{C}_p$ changes from 0.46 kcal/g°K at 435°K to 0.5 kcal/g°K at 480°K. The thermal conductivity even changes less.

Representative values of $\rho$, $k$, and $\bar{C}_p$ are shown in Fig. 5.7 for a

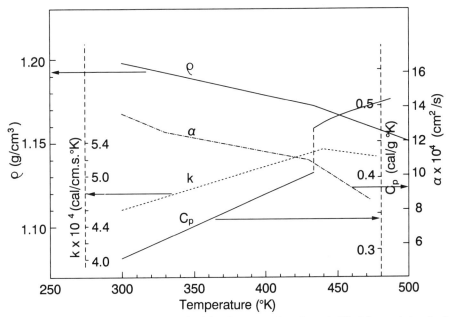

**Figure 5.6** Coefficient of thermal conductivity, heat capacity, thermal diffusivity, and density for polycarbonate, a glassy polymer. (Reprinted with permission of the publisher from Tadmor Z. and C. G. Gogos, *Principles of Polymer Processing*, Wiley, New York, p. 132, 1979.)

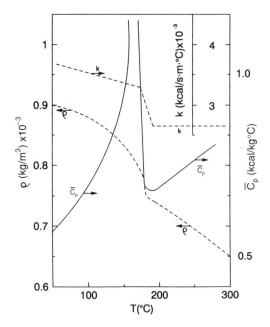

**Figure 5.7** Thermal property data for a semicrystalline polymer, polypropylene. (Data from W. H. Wanger, Jr., *Cooling of Polymer filaments during Melt-Spinning*, Ph.D. Dissertation, University of Denver, 1969.)

semi-crystalline polymer. Here it is observed that $\bar{C}_p$ increases rapidly with temperature passing through a maximum and then decreasing with temperature. The temperature at the peak value is taken as the melting point, $T_m$. The area under the curve is associated with the melting of the crystalline phase and is referred to as the heat of fusion, $\Delta \bar{H}_f$. Above $T_m$ the thermal properties are observed not to change significantly with temperature. For computational purposes, above $T_m$, we take $\rho$, $\bar{C}_p$, $k$, and $\alpha$ ($= k/\rho \bar{C}_p$) to be independent of temperature.

Values of $\rho$, $\bar{C}_p$ and $k$ are presented for a number of commercially available polymers at 25°C in Table 5.4 and at 150°C in Table 5.5. Here it is observed that the values all lie within a fairly narrow range. For more rigorous computations the temperature dependence of $\bar{C}_p$, $\rho$, and $k$ is presented for a number of commercial polymers in Table 5.6. Here all properties are given as polynomial functions of temperature. Thermal properties can be difficult to obtain, but many values can be found in the book by Van Krevelen (1990) and *The Polymer Handbook* (Brandrup & Immergut, 1989). A method for estimating $\bar{C}_p$ is given in Appendix C. Data are presented for three high-performance engineering thermoplastics in Table 5.7. These polymers typically have higher $T_g$'s, $T_m$'s and densities than those of the commodity resins but similar values of $\bar{C}_p$ and $k$. They are primarily used in applications where high strength and stiffness are required at elevated temperatures. Furthermore, they are frequently used in combination with carbon or other stiff fibers to form thermoplastic composites. A more complete list of engineering thermoplastics is given in Table 5.8. In this table their chemical structures along with their $T_g$'s and $T_m$'s are given. Additional information on these polymers as well as others can be found in the book by Van Krevelen (1990).

**TABLE 5.4 Density, Thermal Conductivity, and Heat Capacity of Some Polymers at Room Temperature**

| Polymer | Density (kg/m³) × 10⁻³ | Thermal Conductivity (J/m·s·°K) | $T_g$ (°C) | $T_m$ (°C) | Heat Capacity (kJ/kg·°K) |
|---|---|---|---|---|---|
| ABS | 1.16 | 0.188–0.335 | 80 | | 1.25–1.67 |
| Nylon 6,6 | 1.13–1.15 | 0.243 | 57 | 240 | 0.46 |
| Polycarbonate | 1.2 | 0.192 | 149 | | 1.25 |
| Polyester (PETP) | 1.37 | 0.289 | 80 | 249 | 1.25 |
| LDPE | 0.910–0.925 | 0.335 | | | 2.30 |
| HDPE | 0.940–0.965 | 0.460–0.519 | −78 | 141 | 2.30 |
| PMMA | 1.17–1.20 | 0.167–0.251 | | | 1.46 |
| Polyoxymethylene | 1.42 | 0.230 | −82 | 183 | 1.46 |
| PS | 1.04–1.09 | 0.100–0.138 | 107 | | 1.34 |
| PTFE | 2.0–2.14 | 0.250 | | | 1.05 |
| Polyurethane (thermoplastic) | 1.05–1.25 | 0.070–0.310 | −46 to −18 | | 1.67–1.88 |
| PVC (rigid) | 1.30–1.45 | 0.125–0.293 | | | 0.84–1.25 |
| PP | 0.91 | 0.172 | −10 | 165 | 2.14 |

(Data from Van Krevelen, 1990, and H. H. Winter, *Adv. Heat Transfer*, **13**, 225–67, 1977.)

**TABLE 5.5 Thermal Conductivity and Heat Capacity of Some Polymers at 150°C**

| Polymer | $k$ (W/m·°K) | $\bar{C}_p$ (kJ/kg·°K) | $\rho$ (10⁻³ kg/m³) |
|---|---|---|---|
| LDPE (low-density polyethylene) | 0.241 | 2.57 | 0.782 |
| HDPE (high-density polyethylene) | 0.255 | 2.65 | 0.782 |
| PP* (polypropylene) | 0.142 | 2.80 | 0.867 |
| PVC (polyvinyl chloride) | 0.166 | 1.53 | 1.31 |
| PS (polystyrene) | 0.167 | 2.04 | 0.997 |
| PMMA (polymethylmethacrylate) | 0.195 | | 1.11 |

*PP data are at 180°C ($T_m$=165°C).
(Data from H. H. Winter, *Adv. Heat Transfer*, **13**, 205–67, 1977.)

For semicrystalline polymers it is observed that melting occurs leading to the absorption of energy. The energy associated with the change from the crystalline phase to the completely amorphous state is obtained by integrating the area under the curve of $\bar{C}_p$ versus temperature data.* The energy associated with this phase transition is called the heat of fusion, $\Delta \bar{H}_f$. Values of $\Delta \bar{H}_f$ are presented in Table

*Values of $\bar{C}_p$ versus temperature are most often obtained by means of differential scanning calorimetry (DSC). This technique is based on monitoring the heat flow to or from the polymer sample from a standard, both of which are increased in temperature at the same rate. The properties of the standard are rigorously known.

**TABLE 5.6 Temperature Dependence of Physical Properties of Several Polymers***

| Property | Polymer | Temperature Range (°C) | Coefficients in Polynominal Representation | | | | | |
|---|---|---|---|---|---|---|---|---|
| | | | A | B | C | D | E | F |
| $k$ (W/m °K) | HDPE | 10–143 | 0.453 | $-8.59 \times 10^{-4}$ | $-5.29 \times 10^{-6}$ | $4.12 \times 10^{-8}$ | $-1.98 \times 10^{-8}$ | |
| | | 143–200 | 0.26 | | | | | |
| | LDPE | 10–126 | 0.365 | $-4.07 \times 10^{-4}$ | $-7.34 \times 10^{-6}$ | $8.28 \times 10^{-8}$ | $-5.53 \times 10^{-8}$ | |
| | | 126–200 | 0.223 | | | | | |
| | PVC | 0–200 | 0.168 | | | | | |
| $\bar{C}_p$ (kJ/kg °K) | HDPE | 10–88 | 1.597 | $3.61 \times 10^{-3}$ | $5.96 \times 10^{-5}$ | $-3.44 \times 10^{-8}$ | $9.77 \times 10^{-9}$ | |
| | | 88–121 | $-1.983 \times 10^2$ | 6.17 | $-6.34 \times 10^{-2}$ | $2.19 \times 10^{-4}$ | | |
| | | 121–130 | $-2.837 \times 10^2$ | 2.41 | | | | |
| | | 130–133 | $1.208 \times 10^3$ | $-9.07$ | | | | |
| | | 133–200 | 1.984 | $3.88 \times 10^{-3}$ | | | | |
| | LDPE | 10–90 | 1.943 | $5.39 \times 10^{-2}$ | $2.56 \times 10^{-2}$ | $-3.23 \times 10^{-6}$ | $3.53 \times 10^{-8}$ | |
| | | 90–105 | $8.497 \times 10^1$ | $-1.84$ | $1.04 \times 10^{-2}$ | | | |
| | | 105–110 | $-1.29 \times 10^2$ | 1.3 | | | | |
| | | 110–113.5 | $3.786 \times 10^2$ | $-3.31$ | | | | |
| | | 113.5–200 | 1.98 | $3.70 \times 10^{-3}$ | | | | |
| | PVC | 10–67 | 0.75 | $4.66 \times 10^{-3}$ | $1.21 \times 10^{-1}$ | $-9.71 \times 10^{-4}$ | $2.90 \times 10^{-6}$ | |
| | | 67–96 | $1.361 \times 10^2$ | $-6.64$ | | | | |
| | | 96–200 | 1.208 | $2.96 \times 10^{-3}$ | | | | |
| $\rho^{-1}$ (cm³/g) | HDPE | 10–133 | 1.033 | $17.87 \times 10^{-4}$ | $-7.19 \times 10^{-5}$ | $16.11 \times 10^{-7}$ | $-15.45 \times 10^{-9}$ | $5.58 \times 10^{-11}$ |
| | | 133–200 | 1.158 | $8.09 \times 10^{-4}$ | | | | |
| | LDPE | 10–113.5 | 1.078 | $1.24 \times 10^{-4}$ | $2.68 \times 10^{-5}$ | $-3.95 \times 10^{-7}$ | $2.35 \times 10^{-9}$ | |
| | | 113.5–200 | 1.158 | $8.09 \times 10^{-4}$ | | | | |
| | PVC | 10–110 | 0.7154 | $1.02 \times 10^{-4}$ | $0.0781 \times 10^{-5}$ | $-0.0167 \times 10^{-7}$ | $0.0524 \times 10^{-9}$ | |
| | | 110–200 | 0.6791 | $5.67 \times 10^{-4}$ | | | | |

*All properties are given as polynomials of the form $A + BT + CT^2 + DT^3 + \cdots$ in which $T$ is the temperature in degrees centigrade. The coefficients in this table were taken from U. Kleindienst, Ph.D. Dissertation, Stuttgart University, 1976.

**TABLE 5.7 Thermal Properties for Three Semi-crystalline Engineering Thermoplastics at Room Temperature**

| Polymer | $T_g$ (°C) | $T'_m$ (°C) | $T_m$ (°C) | $k$ (W m$^{-1}$ °K$^{-1}$) | $\bar{C}_p$ (kJ kg$^{-1}$ °K$^{-1}$) | $\rho$ (10$^{-3}$ kg m$^{-3}$) |
|---|---|---|---|---|---|---|
| PEEK (polyetheretherketone) | 144 | 390 | 334–343 | 0.251 | 1.34 | 1.401 (cry)* 1.263 (amor)** |
| PPS (polyphenylenesulfide) | 88 | 315 | 285–290 | 0.289 | 1.09 | 1.34 (amor) 1.44 (cry) |
| PEKK (polyaryletherketoneketone) | 156 | 354 | 330–339 | — | — | — |

*cry = crystalline
**amor = amorphous

5.9 for a number of commercially available polymers as well as two high-performance polymers used in the formation of thermoplastic composites. It should also be noted that the melting point is not really distinct but covers a broad temperature range. When carrying out calculations involving melting, it may be more appropriate to use $\bar{C}_p$ as a function of temperature rather than treat the polymer as having a distinct phase change.

The cooling of a semicrystalline polymer from a temperature above $T_m$ to some lower temperature leads to recrystallization. The energy associated with crystallization, called the heat of crystallization ($\Delta \bar{H}_c$), is affected by the temperature at which crystallization takes place. Representative data for PPS are presented in Table 5.10. Here it is observed that the heat of crystallization, $\Delta \bar{H}_c$, depends on $\bar{M}_w$ and temperature. Furthermore, the values of $\Delta \bar{H}_c$ are somewhat lower than those of $\Delta \bar{H}_f$. Unless data are available, it is customary to consider $\Delta \bar{H}_f = \Delta \bar{H}_c$. Values of $\Delta \bar{H}_c$ for some common polymers are given in Appendix C and can be found for other polymers in the books by Van Krevelen (1990) and Brandrup and Immergut (1989).

## 5.3.2. THERMAL PROPERTIES OF COMPOSITE SYSTEMS

By composite systems we mean any combination of a polymer system (called the matrix) with another polymer, fillers such as glass fibers or carbon black (particulates), or long continuous fibers used in the formation of thermoplastic prepregs (a prepreg is a sheet of polymer reinforced with fiber and can be later processed by forming techniques). Blending rules are usually used to weight the contribution of each component. The rules consider the thermal properties to be either in a series or a parallel arrangement of the matrix and second component (Richardson, 1977). The series arrangement is:

$$1/\Gamma_b = \phi_1/\Gamma_1 + \phi_2/\Gamma_2, \tag{5.73}$$

where

$\Gamma_b$ = bulk composite property ($\bar{C}_p$ or $k$)
$\Gamma_1$ = matrix property ($\bar{C}_p$ or $k$)
$\Gamma_2$ = second component property ($\bar{C}_p$ or $k$)
$\phi_1$ = volume fraction of the matrix
$\phi_2$ = volume fraction of the second component.

The parallel arrangement of the matrix and second component properties is:

$$\Gamma_b = \phi_1 \Gamma_1 + \phi_2 \Gamma_2. \tag{5.74}$$

**TABLE 5.8 Structures of a Number of Engineering Thermoplastics, Their Manufacturers, and Their Glass Transition and Melt Temperatures**

PEEK—Polyetheretherketone (ICI)

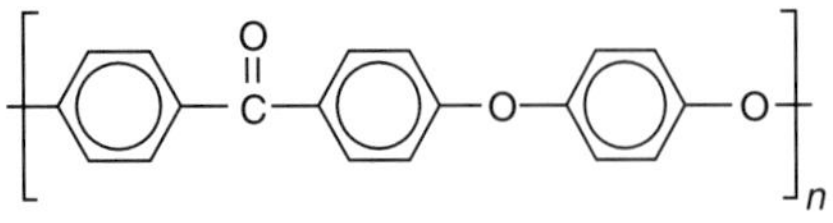

$T_g(°C) = 144$
$T_m(°C) = 334$

PEK—Polyetherketone (ICI)

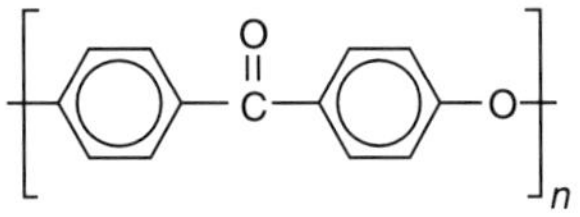

$T_g(°C) = 154$
$T_m(°C) = 367$

VITREX PES 200P—Polyethersulfone (ICI)

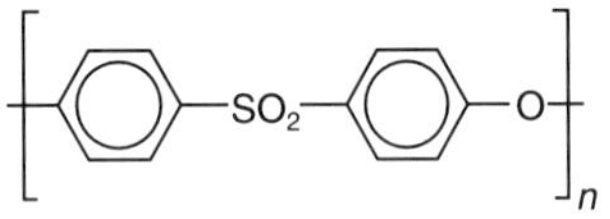

$T_g(°C) = 220$

ULTEM—Polyetherimide (General Electric)

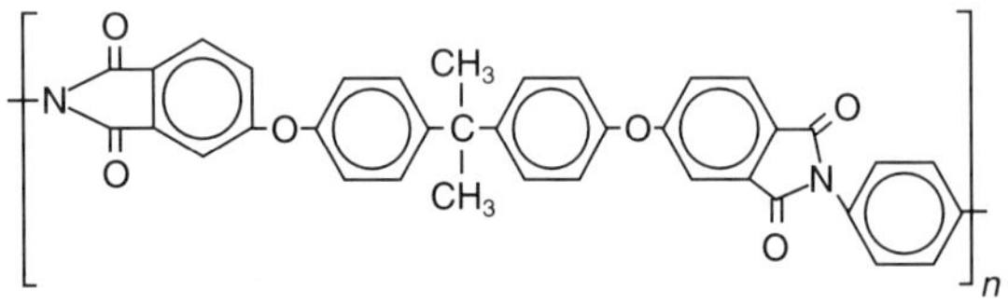

$T_g(°C) = 220$

UDEL P-1700—Polysulfone (Amoco)

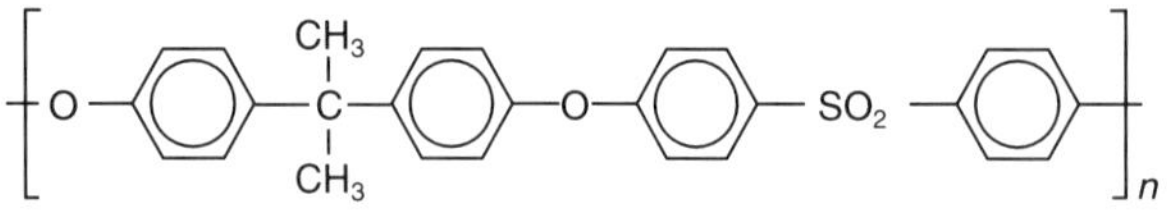

$T_g(°C) = 190$

RADEL—Polyarylsulfone (Amoco)

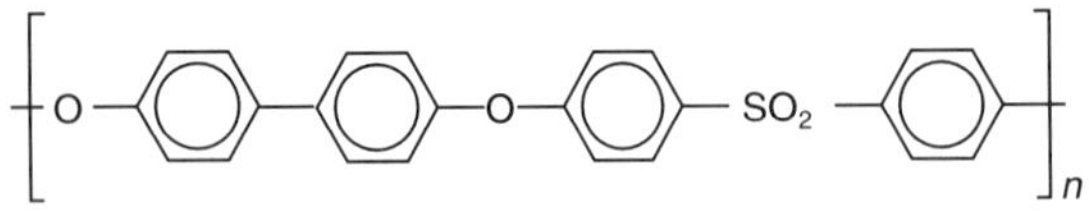

$T_g(°C) = 220$

RYTON—Polyphenylenesulfide (Phillips)

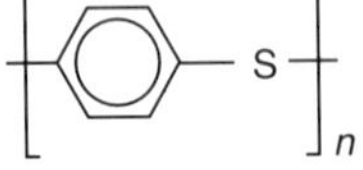

$T_g(°C) = 85$
$T_m(°C) = 285$

PEKK—Polyetherketoneketone

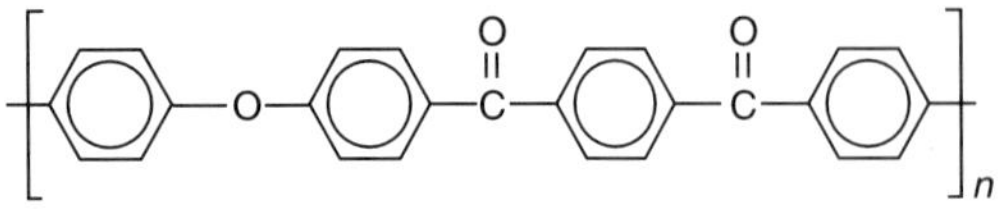

**TABLE 5.9 Heat of Fusion of Some Polymers**

| Polymer | $\Delta \bar{H}_f$ (J/kg) $\times 10^{-4}$ | Reference |
|---|---|---|
| Polyoxymethylene | 24.9 | Starkweather & Boyd, 1960 |
| Polybutene-1 | 24.7 | Nielsen, 1962 |
| HDPE ("Super Dylan") | 24.5 | Ke, 1960 |
| PP | 23.4 | Nielsen, 1962 |
| HDPE ("Marlex 50") | 21.8 | Ke, 1960 |
| Nylon 6,6 | 20.5 | Nielsen, 1962 |
| LDPE | 13.8 | Nielsen, 1962 |
| PET | 13.7 | Nielsen, 1962 |
| Natural rubber (cis-polyisoprene) | 6.4 | Nielsen, 1962 |
| PTFE | 5.7 | Starkweather & Boyd, 1960 |
| PEEK | 13.0 | Velisaris & Seferis, 1986 |
| PPS | 10.5 | Lopez & Wilkes, 1988 |

**TABLE 5.10 Heats of Crystallization for Polyphenylenesulfide as a Function of $\bar{M}_w$ and Crystallization Temperature, $T_c$**

| $T_c$ (°C) | $\Delta \bar{H}_c$ ($M_w = 24{,}000$) (J kg$^{-1}$ $\times 10^{-4}$) | $\Delta \bar{H}_c$ ($M_w = 63{,}000$) (J kg$^{-1}$ $\times 10^{-4}$) |
|---|---|---|
| 225 | 4.3 | 3.7 |
| 235 | 4.6 | 4.2 |
| 245 | 4.7 | 4.4 |
| 255 | 5.1 | 4.7 |

(Data from Lopez & Wilkes, 1988.)

Actually the series arrangement represents the highest limit for the bulk composite property, and the parallel arrangement represents the lowest limit.

For a PEEK and carbon fiber composite, the thermal conductivity was found to be best determined using Eq. 5.73 and the heat capacity was best determined using Eq. 5.74 but with mass fractions instead of volume fractions (Velisaris & Seferis, 1988). Values for ρ, $k$, and $\bar{C}_p$ are given in Table 5.11. It is interesting to note that the values of $\bar{C}_p$ for the matrix and carbon fiber are similar and that the values of $k$ are lower for the matrix. The bulk values of $k$ for the composite are then increased somewhat over those of the matrix. Although it is not certain that these rules apply to polymer blends, filled polymers, or other composite structures, they at least represent the starting point for estimating the thermal transport properties of composite systems.

## 5.4. HEATING AND COOLING OF NONDEFORMING POLYMERIC MATERIALS

Heat transfer is as important as polymer rheology in the processing of polymeric materials. Polymers usually start as solids and then are heated to temperatures above $T_g$ or $T_m$ before being shaped. During or immediately after the shaping process the cooling process starts, and it is here that the morphology and structure and associated physical properties are developed. In this section three methods of heat transfer are summarized: conduction, convection, and radiation. First, solutions to well-known problems are presented in graphical form in Section 5.4.1, as many processes can be modeled as one of these basic heat transfer processes. From the graphical solutions we proceed directly to numerical solutions of heat transfer problems. In Section 5.4.2 the important heat transfer coefficients commonly found in polymer processing are summarized. Radiation heat transfer is discussed in Section 5.4.3 as this technique is used frequently to heat polymers up rapidly before applying shaping operations such as thermoforming and blow molding.

### 5.4.1. TRANSIENT HEAT CONDUCTION IN NONDEFORMING SYSTEMS

When no deformation occurs, the thermal energy equation reduces to the following form:

$$\rho \bar{C}_p \frac{\partial T}{\partial t} = \nabla \cdot k \nabla T + \dot{S}. \tag{5.75}$$

Here the source term, $\dot{S}$, could represent the rate of energy generated per unit volume due to a phase change, absorbed radiation, or a chemical reaction. For a number of problems encountered in polymer processing this equation takes on two relative simple forms for planar and cylindrical geometries. For a planar geometry (see Fig. 5.8), Eq. 5.75 becomes

$$\rho \bar{C}_p \frac{\partial T}{\partial t} = k \frac{\partial^2 T}{\partial x^2} + \dot{S}, \tag{5.76}$$

which is the one-dimensional heat conduction equation for a slab.

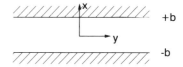

**Figure 5.8** One-dimensional heat transfer in a slab with the heat flux in the $x$-direction. Common boundary conditions are (1) constant wall temperatures at $x = +b$ and $-b$; (2) heat flux at $x = +b$ and $-b$ due to thermal resistance between a fluid and the slab surface.

**TABLE 5.11 Thermal Properties of PEEK and PEEK/Carbon Fiber composites**

| Material | Density (kg/m³) | Mass Fraction (kg matrix/kg) | Volume Fraction (m³ matrix/m³) | $\bar{C}_p$ (kJ kg$^{-1}$ °K$^{-1}$) | $k$ (W m$^{-1}$ °K$^{-1}$) | $\alpha \times 10^7$ (m² s$^{-1}$) |
|---|---|---|---|---|---|---|
| Matrix | 1263 (amor) 1401 (cryst) | | | 1.34 | 0.251 | 1.485 |
| Carbon fiber | 1790 | | | 1.26 | 0.427 | 1.899 |
| Composite APC-1 | 1534 | 0.40 | 0.48 | 1.30 | 0.318 | 1.598 |
| Composite APC-2 | 1579 | 0.32 | 0.32 | 1.30 | 0.339 | 1.655 |

**Figure 5.9** One dimensional heat transfer in a cylinder in the $r$-direction. Common boundary conditions are (1) constant wall temperature at $r = R$ and a zero heat flux in the $r$ direction at the center (i.e., $q_r = 0$); (2) heat flux at $r = R$ due to thermal resistance between a fluid and the cylinder surface.

Examples of processes described by this equation include the cooling of an expanded parison as it contacts a cold mold wall, the cooling of an injection-molded part, cooling of a laminated thermoplastic composite prepreg in a mold, and cooling of a cast film on a metal drum. For cylindrical geometries (see Fig. 5.9), Eq. 5.71 reduces to:

$$\rho \bar{C}_p \frac{\partial T}{\partial t} = k \frac{1}{r} \frac{\partial}{\partial r}\left[ r \frac{\partial T}{\partial r}\right] + \dot{S}, \tag{5.77}$$

which is the heat conduction equation for a cylinder in which heat transfer occurs only in the $r$ direction. Examples of processes described by Eq. 5.77 include cooling of a strand of polymer in a pelletizing process, of a polymer-coated metal wire, and of an injection-molded part of cylindrical cross section (e.g., a cylindrical preform that is later blow-molded).

The solutions of Eqs. 5.76 and 5.77 are subject to various boundary conditions. For a slab of finite thickness (thickness $2b$, with the axis at the center of the slab), the boundary conditions are usually given as constant surface temperatures or a step change in the surface temperature due to convection at the free surfaces. Mathematically, for the first case the boundary and initial conditions are given as:

B.C.1 at $x = +b$     $T(b,t) = T_w$      (5.78)

B.C.2 at $x = -b$     $T(-b,t) = T_w$      (5.79)

I.C. at $t = 0$     $T(x,0) = T_0$.      (5.80)

In the second case the boundary and initial conditions are given as:

B.C.1 at $x = -b$    $q_x(-b,t) = -k\left.\dfrac{\partial T}{\partial x}\right|_{-b} = h[T_a - T(-b)]$    (5.81)

B.C.2 at $x = +b$    $q_x(b,t) = -k\left.\dfrac{\partial T}{\partial x}\right|_{+b} = h[T(+b) - T_a]$    (5.82)

I.C. at $t = 0$     $T(x,0) = T_0$.      (5.83)

$T_a$ here is the temperature of the cooling fluid. In the case of the cylindrical geometry the corresponding boundary and initial conditions for the constant surface temperature (step change in temperature) or the step change in surface temperature due to convection are written, respectively, as:

B.C.1     $T(R,t) = T_1$      (5.84)

B.C.2     $\dfrac{\partial T}{\partial t}(0,t) = 0$      (5.85)

I.C.     $T(r,0) = T_0$      (5.86)

or

B.C.1     $q_r(R,t) = -k\dfrac{\partial T}{\partial r}(R,t) = h(T(R) - T_a)$      (5.87)

B.C.2     $q_r(0,t) = -k\dfrac{\partial T}{\partial r}(0,t) = 0$      (5.88)

I.C.     $T(r,0) = T_0$.

The solutions of Eqs. 5.76 and 5.77 depend on the boundary conditions. For constant coefficients (i.e., constant values of $\rho$, $\bar{C}_p$, and $k$) the solutions are well-known and can be found in many books on heat transfer. We note here, in particular, the book by Carslaw and Jaeger (1973). Rather than reproduce the solutions, which are based on the method of separation of variables, we provide graphical solutions of some of the more commonly encountered cases. For example, Fig. 5.10 shows the transient temperature profiles for a slab with constant surface temperatures. The parameter is the dimensionless time, $t^* = \alpha t/b^2$. In Fig. 5.11 the corresponding case is given for the infinite cylinder. For a slab in which there is resistance to heat transfer at the interface between a solid and liquid the solution is graphically presented in Fig. 5.12. Graphs are given for two positions, $x/b = 0$ and 1.0. The parameter is the reciprocal of the Biot number, $Bi^{-1} = k/hb$. Similar plots are given in Fig. 5.13 for an infinite cylinder. These figures can be used to make estimates for heating and cooling times for polymeric materials.

For constant coefficients, the temperature profiles for various geometries such as flat plates, infinite cylinders, and spheres are given in terms of infinite series. For example, for flat plates with surface resistance to heat transfer the equation to be solved is the one-dimensional heat transfer equation subject to the following boundary conditions:

B.C.1 at $x = b$     $-\dfrac{\partial T}{\partial x}(b,t) = (h/k)[T(b,t) - T_1]$      (5.89)

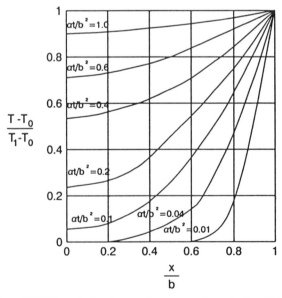

**Figure 5.10** Dimensionless temperature versus dimensionless distance for a slab subjected to a step change in surface temperature (no thermal resistance). The parameter is the dimensionless time, $t^*$. $T_0$ is the initial temperature, and $T_1$ is the wall temperature.

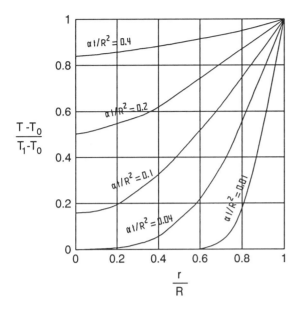

**Figure 5.11** Dimensionless temperature profiles versus dimensionless distance for a cylinder subjected to a step change in surface temperature (no thermal resistance). The parameter is the dimensionless time, $t^*$. $T_0$ is the initial temperature, and $T_1$ is the wall temperature.

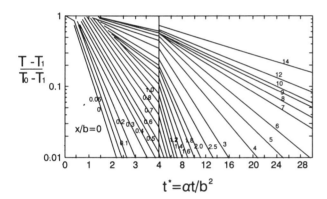

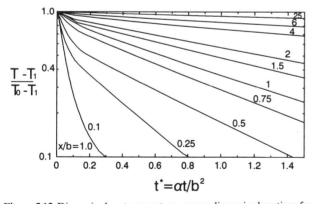

**Figure 5.12** Dimensionless temperature versus dimensionless time for a slab subjected to a step change in surface temperature with thermal resistance at the interface between a fluid and a solid surface. $x/b = 0$ is at the centerline and, $x/b = 1$ is at the surface. The parameter is the reciprocal of the Biot number, $\mathrm{Bi}^{-1} = k/hb$. $T_0$ is the initial temperature, and $T_1$ is the temperature of the cooling fluid.

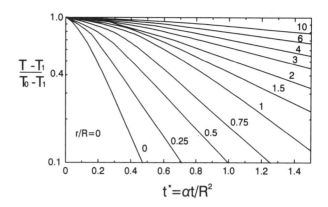

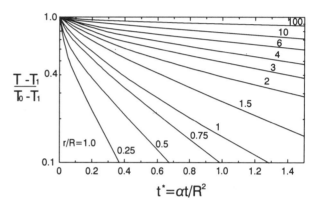

**Figure 5.13** Dimensionless temperature versus dimensionless time for a cylinder subjected to a step change in surface temperature with thermal resistance. The parameter is $\mathrm{Bi}^{-1} = k/hR$. $T_0$ is the initial temperature, and $T_1$ is the temperature of the cooling fluid.

B.C.2 at $x = -b$ $\qquad -\dfrac{\partial T}{\partial x}(-b, t) = (h/k)[T_1 - T(-b, t)]$ $\qquad$ (5.90)

I.C. at $t = 0$ $\qquad T(x, 0) = T_0.$ $\qquad$ (5.91)

The·solution is

$$\frac{T_1 - T}{T_1 - T_0} = \sum_{n=1}^{\infty} \frac{2\left(\dfrac{hb}{k}\right)\cos\left(\beta_n \dfrac{x}{b}\right)}{\left[\beta_n^2 + \dfrac{hb}{k} + \left(\dfrac{hb}{k}\right)^2\right]\cos\beta_n} \exp\left(-\frac{\beta_n^2 \alpha t}{b^2}\right),$$ (5.92)

where the eigenvalues, $\beta_n$, are given by

$$\beta_n \tan \beta_n = (hb/k).$$ (5.93)

To use the series solution, numerical techniques are needed to find $\beta_n$.

Rather than reproduce the series solutions for the various cases encountered in polymer processing, we use numerical techniques. Once we add factors such as temperature-dependent properties, chemical reactions, viscous dissipation, or enthalpy changes due to a phase change, then the series solutions become of little use. The IMSL subroutine MOLCH (see Appendix D.8) solves one-dimensional partial

**TABLE 5.12 Program Listing (SLABHT1.FOR) for Example 5.5 for the Case of a Step Change in Surface Temperature**

```
C********************ABSTRACT************************************
C  THIS PROGRAM CALLS SUBROUTINE MOLCH DESCRIBED IN APPENDIX D.8 TO
C  SOLVE THE ONE-DIMENSIONAL TRANSIENT HEAT CONDUCTION EQUATION
C  WITH A STEP CHANGE IN THE SURFACE TEMPERATURE. IN ORDER TO USE
C  MOLCH IT IS NECESSARY THAT THE INITIAL CONDITIONS SATISFY THE
C  BOUNDARY CONDITIONS. THE INITIAL CONDITIONS ARE GIVEN AS THETA
C  = 1.0 - X**20.THIS FUNCTION NEARLY SATISFIES THE REQUIREMENT THAT
C  THETA = 1.0 AT T = 0. OVER ALL X WHILE AT X = 0.0, DTHETA/DX = 0.0,
C  AND THETA = 0 AT X = 1.0.
C  THE RESULTS ARE OUTPUT TO A FILE CALLED "DATA541.R1."
C
C      SPECIFICATIONS FOR LOCAL VARIABLES
       INTEGER LDY, NPDES, NX
       PARAMETER (NPDES=1, NX=6, LDY=NPDES)
C
       INTEGER I, IDO, J, NOUT, NSTEP, MT
       REAL FCNBC, FCNUT, FLOAT, HINIT, T, TEND, TOL, XBREAK(NX),
     &    Y(LDY,NX)
       INTRINSIC FLOAT
       EXTERNAL FCNBC, FCNUT, MOLCH, UMACH, WRRRN
       OPEN(6,FILE = 'DATA541.R1', STATUS = 'NEW')
C
C
C      SET BREAKPOINTS AND INITIAL CONDITIONS
       DO 10 I=1,NX
         XBREAK(I) = FLOAT(I-1)/FLOAT(NX-1)
         Y(1,I) = 1.0-XBREAK(I)**20
   10 CONTINUE
C      SET PARAMETERS FOR MOLCH
       TOL = 1.0E-04
       HINIT = .01
       T = 0.0
       IDO = 1
       NSTEP = 10
       DO 20 J=1, NSTEP
         TEND =FLOAT(J)
C  SOLVE PROBLEM
       CALL MOLCH (IDO,FCNUT, FCNBC, NPDES,T,TEND, NX, XBREAK,TOL,
     &    HINIT, Y, LDY)
C  PRINT RESULTS
       WRITE (6,15) T,(N*1,N = 1,NX),((Y(I,M),M = 1,NX),I = 1,LDY)
   15 FORMAT(25X,'THE SOLUTION AT TIME T = ',F5.2,/,6I10,/,6F10.5,//)
   20 CONTINUE
C  FINAL CALL TO RELEASE WORKSPACE
       IDO = 3
       CALL MOLCH (IDO, FCNUT, FCNBC, NPDES, T, TEND, NX, XBREAK, TOL,
     &    HINIT, Y, LDY)
       END
       SUBROUTINE FCNUT (NPDES, X, T, U, UX, UXX, UT)
       INTEGER NPDES
       REAL    X, T, U(1), UX(1), UXX(1), UT(1)
C  DEFINE DIFFERENTIAL EQUATION
       UT(1) =4.421E-02*UXX(1)
       RETURN
       END
C
C  INPUT BOUNDARY CONDITIONS
       SUBROUTINE FCNBC (NPDES, X, T, ALPHA, BETA, GAMP)
       INTEGER NPDES
       REAL    X, T, ALPHA(1), BETA(1), GAMP(1)
       IF(X .LT. 1.0)THEN
         ALPHA(1) = 0.0
         BETA(1) = 1.0
         GAMP(1) = 0.0
       ELSE
         ALPHA(1) = 1.0
         BETA(1) = 0.0
         GAMP(1) = 0.0
       END IF
       RETURN
       END
```

differential equations of the form found in heat and mass transfer. This subroutine is capable of handling the case of a step change in surface temperature when there is no resistance to heat transfer. The only shortcoming is that the initial condition must satisfy the boundary conditions, which is difficult for situations in which there is a step change in temperature. The use of MOLCH is illustrated in Ex. 5.5. In particular, one should note how we handle the initial condition so that it satisfies the boundary conditions.

---

**EXAMPLE 5.5**
**Cooling of an Injection-Molded Slab of HDPE**
HPDE is injection-molded into a rectangular cavity having dimensions of 10 cm by 10 cm by 0.32 cm thick. HDPE enters the cavity at 200°C, and the filling process occurs so rapidly that the drop in temperature during filling can be considered to be negligible. Determine the time required to drop the centerline temperature of the melt to 130°C (neglect crystallization) if the mold temperature is 25°C and the heat transfer coefficient between the mold wall and the polymer is 25 W m$^{-2}$ °K$^{-1}$.

**Solution**

The temperature variation in the polymeric melt is described by the one-dimensional heat transfer equation:

$$\rho \bar{C}_p \frac{\partial T}{\partial t} = k \frac{\partial^2 T}{\partial x^2}. \tag{5.94}$$

If it is assumed that there is perfect thermal contact between the cooling medium and the melt, then the boundary and initial conditions are:

$$\text{B.C.1 at } x = b \qquad T(b,t) = 25°C \tag{5.95}$$

$$\text{B.C.2 at } x = 0 \qquad \frac{\partial T}{\partial x}\bigg|_0 = 0 \tag{5.96}$$

$$\text{I.C. at } t = 0 \qquad T(x,0) = 200°C. \tag{5.97}$$

When thermal contact is poor, then there is a flux of thermal energy at the surface given by:

$$q_x(b,t) = -k \frac{\partial T}{\partial x}\bigg|_b = h[T(b,t) - 25°C]. \tag{5.98}$$

We now introduce the following dimensionless variables:

$$\xi = x/b \qquad \theta = \frac{T - T_1}{T_0 - T_1}, \tag{5.99}$$

where $T_1 = 25°C$ and $T_0 = 200°C$. Eq. 5.94 becomes:

$$\frac{\partial \theta}{\partial t} = \frac{k}{\rho \bar{C}_p b^2} \frac{\partial^2 \theta}{\partial \xi^2}, \tag{5.100}$$

with boundary and initial conditions

$$\text{B.C.1 at } \xi = 1 \qquad \theta(1,t) = 0 \tag{5.101}$$

$$\text{B.C.2 at } \xi = 0 \qquad \frac{\partial \theta}{\partial \xi}\bigg|_0 = 0 \tag{5.102}$$

$$\text{I.C. at } t = 0 \qquad \theta(\xi,0) = 1 \tag{5.103}$$

or for a heat flux at the surface:

$$\text{B.C.1 at } \xi = 1 \qquad \frac{\partial \theta}{\partial \xi}\bigg|_1 = \frac{hb}{k} \theta(1,t) \tag{5.104}$$

$$\text{B.C.2 at } \xi = 0 \qquad \frac{\partial \theta}{\partial \xi}\bigg|_0 = 0 \tag{5.105}$$

$$\text{I.C. at } t = 0 \qquad \theta(\xi,0) = 1. \tag{5.106}$$

Time can be made dimensionless by introducing

$$t^* = tk/\rho \bar{C}_p b^2 \tag{5.107}$$

The solutions are given in Figures 5.10 and 5.12 for the two cases. For the temperatures given,

$$\theta = \frac{130 - 25}{200 - 25} = 0.6. \tag{5.108}$$

At $x/b = 0$ and $\theta = 0.6$,

$$\alpha t/b^2 = t^* \approx 0.44. \tag{5.109}$$

We note that it was necessary to estimate the value of $t^*$ because it falls between the values of 0.4 and 0.6 given on the graph. Using values of $\rho$, $\bar{C}_p$, and $k$ given in Table 5.5, which appear to be nearly independent of temperature over the range of interest, we find:

$$t = \frac{b^2(0.44)}{\alpha} = \frac{(2.56 \times 10^{-6})(0.44)}{1.1317 \times 10^{-7}} = 9.95 \text{ s}. \tag{5.110}$$

For the case when there is a heat flux at the surface, we use Fig. 5.12. In Fig. 5.12 the dimensionless temperature is plotted versus dimensionless time with the parameter being the reciprocal of the Biot number, $Bi^{-1} = k/hb$. With a dimensionless temperature of $\theta = 0.6$:

$$t = b^2 t^*/\alpha = \frac{(2.56 \times 10^{-6})(4.0)}{1.1317 \times 10^{-7}} = 90.5 \text{ s}. \tag{5.111}$$

Hence, it takes about 13 times as long to cool the polymer melt down to 130°C if there is significant resistance at the surface.

The problem is also solved numerically using the subroutine MOLCH which is described in Appendix D.8. The subroutine has been made as "user-friendly" as possible so that one does not have to really understand the numerical method. However, because of this, there is one limitation that requires that the initial condition satisfy the boundary condition. For the case at hand this presents a problem (i.e., the initial condition of $T(x,0) = T_0$ satisfies B.C.2 but does not satisfy B.C.1, which is $T(1,t) = T_1$). For the dimensionless temperature, $\theta$, we introduce the following function:

$$\theta(\xi,0) = 1 - \xi^{20}. \tag{5.112}$$

**TABLE 5.13 Numerical Output (File DATA541.R1) Corresponding to Program Listing in Table 5.12**

|          | 1       | 2       | 3       | 4       | 5       | 6       |
|----------|---------|---------|---------|---------|---------|---------|
| $t=1.00$  | 0.99618 | 0.97742 | 0.91180 | 0.74588 | 0.43256 | 0.00006 |
| $t=2.00$  | 0.94835 | 0.91236 | 0.80682 | 0.61546 | 0.33646 | 0.00006 |
| $t=3.00$  | 0.87437 | 0.83301 | 0.71885 | 0.53252 | 0.28449 | 0.00006 |
| $t=4.00$  | 0.79491 | 0.75395 | 0.64380 | 0.47096 | 0.24912 | 0.00006 |
| $t=5.00$  | 0.71863 | 0.68009 | 0.57804 | 0.42055 | 0.22150 | 0.00006 |
| $t=6.00$  | 0.64834 | 0.61275 | 0.51967 | 0.37716 | 0.19827 | 0.00006 |
| $t=7.00$  | 0.58462 | 0.55195 | 0.46754 | 0.33892 | 0.17801 | 0.00006 |
| $t=8.00$  | 0.52720 | 0.49728 | 0.42089 | 0.30489 | 0.16007 | 0.00006 |
| $t=9.00$  | 0.47562 | 0.44820 | 0.37909 | 0.27448 | 0.14405 | 0.00006 |
| $t=10.00$ | 0.42935 | 0.40418 | 0.34161 | 0.24723 | 0.12972 | 0.00006 |

This function satisfies $\partial\theta/\partial\xi=0$ at $\xi=0$ and $\theta=0$ at $\xi=1$ and is approximately 1.0 over the region $0\leqslant\xi\leqslant1$. Other exponents can be used without affecting the results significantly. The calling program is given in Table 5.12 and the results are shown in Table 5.13. The first column corresponds to the dimensionless temperature at the centerline. At $t=7.0$ s, $\theta=0.58462$, which corresponds to a temperature of 128°C. Actually $t$ is less than 7.0 s.

A better way to solve Eq. 5.100 subject to a step change in the surface temperature, especially when there is resistance at the interface, is described next. By introducing a finite difference operator for the spatial derivative (i.e., $\partial^2 T/\partial x^2$), the partial differential equation becomes a system of ordinary differential equations that can be solved using the IMSL subroutine IVPAG or DIVPAG (Appendix D.7). The procedure for converting Eq. 5.100 to a system of ordinary differential equations is as follows. The region $0\leqslant\xi\leqslant1$ is divided up into $N$ nodes or spatial points. In the case of the slab, the node at the center is taken as 0, and the node at the mold wall is numbered as $NEQ+1=N$. Using a central difference approximation (Riggs 1988), the differential equation (Eq. 5.100) at the interior nodes becomes:

$$\frac{d\theta_i}{dt}=\alpha'\left[\frac{\theta_{i-1}-2\theta_i+\theta_{i+1}}{\Delta\xi^2}\right], \tag{5.113}$$

where $\alpha'=k/\rho\bar{C}_p b^2$ and $\Delta\xi$ is the distance between the nodes and i runs from 2 to $NEQ-1$. At node 0, using Eq. 5.105 and a forward difference approximation for $\partial\theta/\partial\xi$, we find that:

$$\frac{\partial\theta}{\partial\xi}=0=\frac{-3\theta_0+4\theta_1-\theta_2}{2\Delta\xi} \tag{5.114}$$

or

$$\theta_0=\frac{1}{3}(4\theta_1-\theta_2). \tag{5.115}$$

At the mold wall Eq. 5.100 can be expressed using a backward difference approximation:

$$\frac{\partial\theta}{\partial\xi}=\frac{3\theta_i-4\theta_{i-1}+\theta_{i-2}}{2\Delta\xi} \tag{5.116}$$

to give:

$$\theta_{NEQ+1}=\frac{4\theta_{NEQ}-\theta_{NEQ-1}}{3+(2\Delta\xi hb)/k}. \tag{5.117}$$

The PDE at node 1 becomes:

$$\frac{d\theta_1}{dt}=\left(\frac{\alpha'}{\Delta\xi^2}\right)\left[-\frac{2}{3}\theta_1+\frac{2}{3}\theta_2\right], \tag{5.118}$$

and that at node NEQ is:

$$\frac{d\theta_{NEQ}}{dt}=(\alpha'/\Delta\xi^2)\left[\theta_{NEQ-1}-2\theta_{NEQ}+\frac{4\theta_{NEQ}-\theta_{NEQ-1}}{3+(2\Delta\xi hb)/k}\right]. \tag{5.119}$$

Hence a system of NEQ ordinary differential equations is solved. The temperature at each boundary node is calculated at each time step using the algebraic expressions in Eqs. 5.115 and 5.117.

The calling program is shown in Table 5.14. To solve Eq. 5.100 the region from the centerline of the slab to the surface of the slab is broken up into 10 segments (one should try other nodal spacings). The node at the center is numbered 0 and that at the surface is numbered 10. The differential equation at node 1 is given by Eq. 5.118 while that at node 9 is given by:

$$\frac{d\theta_9}{dt}=(\alpha'/\Delta\xi^2)\left[\theta_8-2\theta_9+\frac{4\theta_9-\theta_8}{3+(2\Delta\xi hb)/k}\right]. \tag{5.120}$$

The differential equations for the remaining nodes (2 through 8) are given by Eq. 5.113. These equations are written in subroutine FCN. The values of $\theta$ at the boundaries, $\theta_0$ and $\theta_{10}$, are calculated after each time step (or call to DIVPAG) using Eq. 5.115 for $\theta_0$ and the following equation for $\theta_{10}$ (see Eq. 5.117):

$$\theta_{10}=\frac{4\theta_9-\theta_8}{3+(2\Delta\xi hb)k}. \tag{5.121}$$

The output from this program is given in Table 5.15. The nodal values for dimensionless temperature are given at each time step in the output. A centerline temperature of 130°C corresponds to $\theta=0.6$. From Table 5.15 we see that it takes about 75 s to reach this temperature, which is somewhat different from the value obtained using Fig. 5.12. However, this difference is due to the inability to accurately extrapolate values from the graph in Fig. 5.12.

**TABLE 5.14 Program Listing (Ex 541) for Example 5.5 for the Case of Resistance at the Interface (i.e., Heat Transfer Coefficient)**

```
C·················ABSTRACT·······························
C
C   THIS PROGRAM CALLS THE SUBROUTINE DIVPAG DESCRIBED IN APPENDIX D7
C   TO SOLVE THE ONE-DIMENSIONAL TRANSIENT HEAT CONDUCTION EQUATION
C   WITH A FLUX DEFINED AT THE WALL BOUNDARY. THE METHOD OF LINES
C   IS USED TO TRANSFORM THE PARTIAL DIFFERENTIAL EQUATION INTO A
C   SYSTEM OF FIRST-ORDER ORDINARY DIFFERENTIAL EQUATIONS.
C   THIS IS A SOLUTION TO EXAMPLE 5.5.
```

**TABLE 5.14** *continued*

```
C••••••••••••••••••••••••••••••••••••••••••••••••••••••••••
C
C   SPECIFICATION OF LOCAL VARIABLES, PARAMETERS, AND FILES
C
     INTEGER NEQ, NPARAM
     PARAMETER (NEQ=9, NPARAM=50)
     INTEGER IDO, IEND, IMETH, INORM, NOUT
     DOUBLE PRECISION A(1,1),FCN,FCNJ,HINIT,PARAM(NPARAM),
    & TOL,X,XEND,Y(NEQ),T,DX,H,AL,B,T0,T10,K,DELTAT
     EXTERNAL FCN,DIVPAG,SSET,UMACH,FCNJ
     OPEN(6,FILE='DATA5C1.R1',STATUS=NEW)
C
C   SPECIFICATION OF DIVPAG PARAMETERS
C
     HINIT=1.0D-5
     INORM=2
     IMETH=2
     CALL SSET (NPARAM,0.0,PARAM,1)
     PARAM(1)=HINIT
     PARAM(10)=INORM
     PARAM(12)=IMETH
     PARAM(4)=5000
     IDO=1
     TOL=1.0D-4
C
C   SPECIFICATION OF INITIAL CONDITIONS
C
     DO 1 II=1,9
        Y(II)=1.0
   1 CONTINUE
     T0=1.0
     T10=1.0
C
C   MATERIAL PROPERTIES
C
     K=0.26
     AL=K/(749.0*2624.0)
     H=25.0
     DX=0.1
     B=0.0016
     DELTAT=10.0
C
C   T=TIME(SEC)
C
     T=0.0
     DO 10 IEND=1,11
          WRITE(6,20)T,(N,N=0,NEQ+1),T0,(Y(I),I=1,9),T10
  20      FORMAT(25X,'THE SOLUTION AT TIME T = ',F5.1,
    &        /,11I6,/,11F6.3,//)
     T=T+DELTAT
     XEND=T
     CALL DIVPAG (IDO,NEQ,FCN,FCNJ,A,X,XEND,TOL,
    &     PARAM,Y)
     T0=(1.0/3.0)*(4.0*Y(1)-Y(2))
     T10=-(Y(8)-4.0*Y(9))/((2.0*DX*B*H)/.26+3.0)
  10 CONTINUE
C
C   FINAL CALL TO RELEASE WORKSPACE
C
     IDO=3
     CALL DIVPAG (IDO,NEQ,FCN,FCNJ,A,X,XEND,TOL,PARAM,Y)
C
     STOP
     END
C
C
C
C   SUBROUTINE CALLED BY DIVPAG TO CALCULATE TIME DERIVATIVES
C
```

**TABLE 5.14** *continued*

```
      SUBROUTINE FCN (NEQ,X,Y,YPRIME)
C
      INTEGER NEQ,JJ
      DOUBLE PRECISION X,Y(NEQ),YPRIME(NEQ),K,DX,H,AL,B
      K=0.26
      AL=K/(749.0*2624.0)
      DX=0.1
      H=25.0
      B=0.0016
      DO 2 JJ=1,7
C
C YPRIME FOR INTERIOR NODES
C
      YPRIME(JJ+1)=(AL/DX**2/B**2)*(Y(JJ)-2.0*Y(JJ+1)+Y(JJ+2))
  2 CONTINUE
C
C YPRIME FOR NODAL POINT NEXT TO WALL WITH ZERO ENERGY FLUX
C
      YPRIME(1)=(AL/DX**2/B**2)*(-(2.0/3.0)*Y(1)+(2.0/3.0)*Y(2))
C
C YPRIME FOR NODAL POINT NEXT TO WALL WITH FINITE ENERGY FLUX
C
      YPRIME(9)=(AL/DX**2/B**2)*Y(8)-2.0*Y(9)+(4.0*Y(9)-Y(8))/
     &  (3.0+(2.0*DX*B*H)/0.26))
C
C
      RETURN
      END
C
C DUMMY SUBROUTINE FOR DIVPAG (JACOBIAN)
C
      SUBROUTINE FCNJ (NEQ,X,Y,DYPDY)
      INTEGER NEQ
      DOUBLE PRECISION X,Y(NEQ),DYPDY(*)
      RETURN
      END
```

**TABLE 5.15** Numerical Output (data 541.ex) Corresponding to Program Listing in Table 5.14

|  | 0 | 1 | 2 | 3 | 4 | 5 | 6 | 7 | 8 | 9 | 10 |
|---|---|---|---|---|---|---|---|---|---|---|---|
| $t=0.0$ | 1.000 | 1.000 | 1.000 | 1.000 | 1.000 | 1.000 | 1.000 | 1.000 | 1.000 | 1.000 | 1.000 |
| $t=10.0$ | 0.953 | 0.953 | 0.950 | 0.947 | 0.942 | 0.936 | 0.928 | 0.919 | 0.909 | 0.897 | 0.884 |
| $t=20.0$ | 0.888 | 0.887 | 0.885 | 0.882 | 0.877 | 0.871 | 0.864 | 0.855 | 0.846 | 0.835 | 0.822 |
| $t=30.0$ | 0.826 | 0.826 | 0.824 | 0.821 | 0.817 | 0.811 | 0.804 | 0.796 | 0.787 | 0.777 | 0.766 |
| $t=40.0$ | 0.769 | 0.769 | 0.767 | 0.764 | 0.760 | 0.755 | 0.749 | 0.741 | 0.733 | 0.723 | 0.713 |
| $t=50.0$ | 0.716 | 0.716 | 0.714 | 0.711 | 0.708 | 0.703 | 0.697 | 0.690 | 0.682 | 0.673 | 0.664 |
| $t=60.0$ | 0.667 | 0.666 | 0.665 | 0.662 | 0.659 | 0.654 | 0.649 | 0.643 | 0.635 | 0.627 | 0.618 |
| $t=70.0$ | 0.621 | 0.620 | 0.619 | 0.617 | 0.613 | 0.609 | 0.604 | 0.598 | 0.591 | 0.584 | 0.575 |
| $t=80.0$ | 0.578 | 0.577 | 0.576 | 0.574 | 0.571 | 0.567 | 0.562 | 0.557 | 0.551 | 0.543 | 0.535 |
| $t=90.0$ | 0.538 | 0.538 | 0.536 | 0.534 | 0.532 | 0.528 | 0.524 | 0.518 | 0.513 | 0.506 | 0.498 |

### 5.4.2. HEAT TRANSFER COEFFICIENTS

The equation of thermal energy presented in the preceding section is usually solved along with prescribed boundary conditions. We either specify (1) the surface (or boundary) temperature; or (2) the heat flux at the surface. This section is concerned with empiricisms for heat transfer coefficients that allow us to deal with the difference in temperature between a fluid and a solid interface as a result of thermal resistance.

The heat flux is usually given in terms of heat transfer coefficients, defined by Newton's law of cooling as:

$$q_n|_s = -k\frac{\partial T}{\partial n}\bigg|_s = h[T(s)-T_a] \tag{5.122}$$

where $q_n|_s$ is the heat flux in the direction normal to the surface and evaluated at the surface and $h$ is the heat transfer coefficient. $T_a$ is the

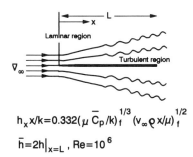

$$h_x x/k = 0.332 (\bar{\mu} \bar{C}_p/k)_f^{1/3} (v_\infty \varrho x/\mu)_f^{1/2}$$

$$\bar{h} = 2h|_{x=L}, \quad Re = 10^6$$

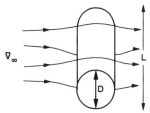

$$Nu = hD/k = (0.43 + 0.5 Re^{0.5}) Pr^{0.38} (Pr_f/Pr_w)^{0.25} \quad 1 < Re < 10^3$$

$$Nu = 0.25 Re^{0.6} Pr^{0.38} (Pr_f/Pr_w)^{0.25} \quad 10^3 < Re < 2 \times 10^5$$

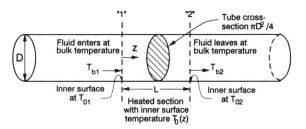

$$h_{\ln} D/k_b = 0.026 (DG/\mu_b)^{0.9} (\bar{C}_p\mu/k)^{0.5} (\mu_b/\mu_0)^{0.14} \quad Re > 20{,}000$$

$$h_{\ln} D/k_b = 0.026 (Re_b Pr_b D/L)^{0.5} (\mu_b/\mu_0)^{0.14} \quad Re < 2200$$

**Figure 5.14** Forced convection heat transfer coefficients for three commonly encountered geometries.

temperature of the ambient fluid (gas or liquid) and $T(s)$ is the local surface temperature of the solid or molten polymer. In the case of forced convection, correlations for $h$ with the Reynolds number, Re, are well-known for certain common geometries.

In particular we note several cases which occur frequently in polymer processing. For flow of a fluid over a smooth flat plate (see Fig. 5.14), the local heat transfer coefficient, $h_x$, is given in terms of the Prandtl number, Pr, and local Reynolds number, Rex, as:

$$\frac{h_x x}{k} = 0.332 \, Pr^{1/3} \, Rex^{1/2} \tag{5.123}$$

where $Pr = (\mu \bar{C}_p/k)_f$ and $Rex = (v_\infty \rho x/\mu)_f$. The subscript, $f$, implies to evaluate the properties of the fluid at the film temperature, $T_f$, where $T_f = \frac{1}{2}(T_0 + T_\infty)$ where $T_0$ is the surface temperature of the solid and $T_\infty$ is the temperature of the approaching fluid. The relation in Eq. 5.123 is valid for $Re < 10^6$. The average heat transfer coefficient, $\bar{h}$, over the length of the plate is found by averaging Eq. 5.123 over the length of

the plate, $L$, and is:

$$\bar{h} = 2h|_{x=L}. \tag{5.124}$$

In the processing of polymers the relations in Eq. 5.123 and 5.124 are usually applied for situations where the temperature along the plate is not constant but the fluid temperature is. Hence, with the availability of numerical methods we can use $h_x$. The correlation in Eq. 5.123 should be applicable to processes such as film blowing and the extrusion of flat sheet.

Another common geometry encountered in polymer processing is the cooling of long cylinders such as might occur in fiber spinning. The heat transfer coefficient for transverse flow over a long cylinder (see Fig. 5.14) is given in terms of Re and Pr as:

$$Nu = \frac{hD}{k_f} = (0.43 + 0.50 \, Re^{1/2}) Pr^{0.38} \left(\frac{Pr_f}{Pr_0}\right)^{0.25}$$

for $1 < Re < 10^3$ \hfill (5.125)

$$Nu = 0.25 \, Re^{0.6} \, Pr^{0.38} \left(\frac{Pr_f}{Pr_0}\right)^{0.25}$$

for $10^3 < Re < 2 \times 10^5$, \hfill (5.126)

where the subscript $f$ implies to evaluate the fluid properties at the film temperature, $T_f = (T_0 + T_\infty)/2$. $T_\infty$ is the temperature of the approaching fluid, and $T_0$ is the surface temperature of the cylinder. For gases the ratio $Pr_f/Pr_0$, where $Pr_0$ is Pr evaluated at $T_0$, is dropped, and the fluid properties are evaluated at $T_f$. For fluids the ratio is retained with the fluid properties evaluated at $T_\infty$. Correlations for tube banks are also available (Holman, 1981). This situation might more closely resemble fiber spinning processes in which many filaments are spun from a spinneret.

Cooling of metal surfaces, such as occurs in the cooling of injection-molding tooling, is usually done by circulating cooling fluid through channels located just beneath the metal surface. This requires the knowledge of heat transfer coefficients for fully developed flow in smooth pipes (see Fig. 5.14). For highly turbulent flow, for $L/D > 10$ and $Re > 20{,}000$, the logarithmic mean heat transfer coefficient is given by:

$$\frac{h_{\ln} D}{k_b} = 0.026 \left(\frac{DG}{\mu_b}\right)^{0.9} \left(\frac{\bar{C}_p\mu}{k}\right)^{1/2} \left(\frac{\mu_b}{\mu_0}\right)^{0.14} \tag{5.127}$$

For laminar flow $h_{\ln}$ is given by:

$$\frac{h_{\ln} D}{k_b} = 1.86 (Re_b \, Pr_b D/L)^{1/2} (\mu_b/\mu_0)^{0.14}, \tag{5.128}$$

where $D$ is the tube diameter and $G$ is $\langle \rho v_z \rangle$. The subscript $b$ means to evaluate $\mu$ at $\frac{1}{2}(T_{b1} + T_{b2})$ while the subscript 0 means to evaluate $\mu$ at $T = \frac{1}{2}(T_{01} + T_{02})$. For highly turbulent flow the results in Eq. 5.127 can be extended to noncircular cross sections by replacing the diameter with $4R_H$, where $R_H$ is the mean hydraulic radius.

There are situations in which heat transfer by free convection could be important such as the cooling of a slowly moving strand of polymer in a water bath and the cooling of a parison as it hangs from a die before being blown. For those who have studied heat transfer, it is known that the dimensionless heat transfer coefficient, Nu, is given as a function of the Grashof number, Gr, and of the Prandtl number:

$$Nu_m = Nu(Gr, Pr), \tag{5.129}$$

where $Nu_m$ is based on the heat transfer coefficient for the total surface, $h_m$, of the submerged object, $Gr = L^3 \rho^2 g \beta \Delta T / \mu^2$ and $Pr = \bar{C}_p \mu / k$. $\beta$ is the volume coefficient of expansion of the fluid and $\Delta T = T_0 - T_\infty$. For a long horizontal cylinder in an infinite fluid

$$Nu_m = 0.518 (Gr_f Pr_f)^{0.25}, \tag{5.130}$$

where $Gr_f = D^3 \rho_f^2 g \beta_f \Delta T / \mu_f^2$ and $Pr_f = \bar{C}_p \mu_f / k_f$. Eq. 5.130 is valid for $Gr\,Pr > 10^4$. An equation similar to Eq. 5.130 can be applied to horizontal flat plates:

$$Nu_m = 0.6 (Gr_f Pr_f)^{1/4}, \tag{5.131}$$

with $D$ replaced by $L = L_n L_v / (L_n + L_v)$, where $L_n$ is the horizontal length

and $L_v$ is the thickness of the plate. For vertical plates and cylinders suspended in air:

$$Nu_m = 0.59 (Gr_f Pr_f)^{1/4} \tag{5.132}$$

for $10^4 < Gr\,Pr < 10^9$. For a flat plate $D^3$ is replaced by $L^3$ (length of the plate).

We have attempted to list only a few of the more pertinent heat transfer coefficients. Certainly correlations for other geometries and situations will be used in polymer processing. Furthermore, the use of correlations between Nu and Gr and Pr have ranges of applicability. There are numerous references that should be consulted for more details concerning the use of heat transfer coefficients and for other correlations (Bird et al., 1960; Holman, 1981; McAdams, 1954; and Whitaker, 1977).

---

**EXAMPLE 5.6**
**Cooling of a Strand in a Pelletizing Bath**
Ten strands of PP with a diameter of $3.175 \times 10^{-3}$ m are extruded at 200°C from a pelletizing die into a water bath at the rate of $3\,m\,min^{-1}$. In order to determine the length of the bath required to drop the temperature of a strand to 75°C, it is necessary to determine a heat transfer coefficient. Determine the heat transfer coefficient if the water temperature is 12°C.

**Solution**

Heat transfer may occur by both free convection transverse to the filament surface and by forced convection as the strand moves through the water. We first estimate $h$ for free convection from a horizontal cylinder using Eq. 5.130. The properties of water are needed at $T_f$, but $T_f$ changes along the length of the strand. We calculate $Nu_m$ at the beginning and end of the bath. At the beginning,

$$T_f = \frac{200 + 12}{2} = 106°C.$$

From Table C.6 (Appendix C),

$$Pr\,Gr = 7.40 \times 10^5.$$

Using Eq. 5.130,

$$Nu_m = 0.518(7.40 \times 10^5)^{0.25} = 15.2$$

$$h = \frac{(15.2)(0.684)}{3.175 \times 10^{-3}} = 3273\,Wm^{-2}C^{-1}.$$

At the end of the bath,

$$T_f = \frac{75 + 12}{2} = 43.5°C.$$

Again referring to Table C.6, we find:

$$Pr\,Gr = 9.86 \times 10^4,$$

and hence $Nu_m = 9.18$ and $h_m = 1842\,W\,m^{-2}\,°C^{-1}$.

For forced convection there are no correlations for flow along a cylinder, and hence we use the correlation for flow over a flat plate (Eqs. 5.123 and 5.124). Using the data in Appendix C.6, we calculate Pr and Re at $T_f = 106°C$:

$$Pr_f (\mu \bar{C}_p / k)_f = (2.67 \times 10^{-4})(4.216 \times 10^3)/0.684 = 1.66$$

$$Re_f = (v_\infty \rho L / \mu)_f = (0.05)(955)(1)/2.67 \times 10^{-4} = 1.79 \times 10^5$$

We arbitrarily estimate $L$ to be 1 m in the above calculation. Using Eq. 5.123 we calculate:

$$h|_{x=L} = (k/1)(0.332)(1.66)^{1/3}(179.10^5)^{1/2}$$
$$= 113.4\,W\,m^{-2}\,°C^{-1}$$

$$\bar{h} = 2h|_{x=L} = 226.8\,Wm^{-2}\,°C^{-1}.$$

Hence, it is apparent that the dominant form of heat transfer for the conditions given is that of free convection.

If one were required to design a pelletizing bath it would be necessary to solve the equation of energy using a heat flux boundary condition at the strand surface. $h$ changes significantly along the length of the strand, as shown in the calculations. Certainly for the most rigorous calculations we would attempt to calculate $h$ as a function of temperature. As a first approximation we could use a simple average value based on the two extremes.

---

### 5.4.3. RADIATION HEAT TRANSFER

Thermal radiation is often used to heat up preforms used in blow molding or plastic sheets used in thermoforming. Furthermore, in processes such as fiber spinning considerable cooling of the outer filaments can occur through radiation heat transfer. We review here the basic ideas of radiation heat transfer.

The heat flux at a surface associated with radiation heat transfer when the radiation is completely absorbed at the surface is given as:

$$q_n(t, s) = -k \frac{\partial T}{\partial n}\bigg|_{s,t} = -\sigma F[T_r^4 - T^4(s, t)], \tag{5.133}$$

where $T_r$ is the temperature of the radiation source and $F$ is the combined configuration emissivity factor (i.e., $eF'$, where $F'$ is related to the surface geometry and is called the view factor and $e$ is the emissivity). $\sigma$ is the Stefan-Boltzmann radiation constant given as $5.6697 \times 10^{-8}\,\mathrm{W\,m^{-2\,\circ}K^{-4}}$, or $.1712 \times 10^{-8}\,\mathrm{BTu\,hr^{-1}\,ft^{-2\,\circ}F^{-4}}$. For most polymers the emissivity, $e$, is $0.9 < e < 1.0$. If $T_r \gg T(s, t)$, then we assume a constant heat flux.

The combined shape factors, $F$, take on various forms depending on the geometry of the emitting and absorbing surfaces. Several common situations follow, and other geometries can be found in the books by Holman (1981), Rohsenow and Hartnett (1973), and Siegel and Howell (1981). For parallel flat plates, $F$ is given by:

$$F = (1/e_1 + 1/e_2 - 1)^{-1}, \tag{5.134}$$

where $e_1$ and $e_2$ are the emissivities of the sheet and source, respectively. If one cylindrical surface is enclosed by another such as for a rod, wire, or pipe exposed to a source (e.g., two parallel banks of lights), then

$$F = [1/e_1 + (A_1/A_2)/(1/e_2 - 1)]^{-1}, \tag{5.135}$$

where $A_1$ is the area of the polymer and $A_2$ is the area of the source. As $A_1/A_2 \to 0$, then $F = e_1$.

In some cases in which radiation heating is used, convection may also be employed. This is because in the case of radiation heating, the surface of a polymer may reach temperatures well above the degradation temperature and convection can be used intermittently to cool the surface. The total heat flux at the surface in this case is given as:

$$q_n|_s = (h_c + h_r)(T_1 - T_2), \tag{5.136}$$

where $h_c$ is the convection heat transfer coefficient and $h_r$ is the radiant heat transfer coefficient. The value of $h_r$ is calculated from the following expression:

$$q_n|_s = h_r(T_1 - T_2) = F\sigma(T_1^4 - T_2^4), \tag{5.137}$$

where the subscript 1 pertains to the polymer and the subscript 2 to the heating source. From this equation we find:

$$h_r = F\sigma(T_1^2 + T_2^2)(T_1 + T_2). \tag{5.138}$$

---

**EXAMPLE 5.7**
**Radiation Heating of a Sheet (Thermoforming)**
Prior to being thermoformed, a polymer (PVC) sheet is heated on both sides by radiation from an initial temperature, $T_0$, of 90°F to a final centerline temperature, $T_f$, of 390°F. Determine how long it will take for the centerline of the sheet to reach 390°F for the following set of data.:

Polymer sheet: $k = 0.14\,\mathrm{Btu\,hr^{-1}\,ft^{-1}\,\circ F^{-1}}$; $F = 0.9$; thickness $(2b) = 0.24$ in.; $\rho\bar{C}_p/k = \alpha = 5 \times 10^{-3}\,\mathrm{ft^2\,hr^{-1}}$
Radiation Source: $T_s = 1740$°F and is a bank of heating elements that can be considered as a parallel plate radiation source.

**Solution**

The differential equation to be solved along with the appropriate boundary conditions are:

$$\rho\tilde{C}_p \frac{\partial T}{\partial t} = k \frac{\partial^2 T}{\partial x^2}$$

I.C. at $t = 0$        $T = T_0$

B.C.1 at $x = b$     $q_x|_b = -k\dfrac{\partial T}{\partial x}\Big|_{x=b} = h_r(T_1 - T_2)$

B.C.2 at $x = -b$    $q_x|_b = -k\dfrac{\partial T}{\partial x}\Big|_{x=-b} = h_r(T_2 - T_1)$,

or we can also use:

B.C.2 at $x = 0$     $q_x = 0$.

The solution to this problem can be obtained numerically or by separation of variables if the coefficients are constant. As $\alpha$ is given as constant, we can use the separation of variables solution presented in graphical form in Figures 5.15 and 5.16 for the case when $T_0/T_2 = 0.25$ (based on absolute temperatures). Solutions for other parameters are given in the *Handbook of Heat Transfer* (Rohsenow & Hartnett, 1973).

We assume here that $T_2 \gg T_1$, in which case $h_r = F\sigma T_2^3$. The radiative Biot number, $\mathrm{Bi}_r$, is then defined as:

$$\mathrm{Bi}_r = \frac{h_r b}{k} = \frac{F\sigma T_2^3 b}{k},$$

which comes from the dimensionless temperature gradient. From the data given in this problem we obtain the following quantities:

$$\frac{(T - T_0)}{(T_2 - T_0)} = \frac{390°F - 90°F}{1740°F - 90°F} = 0.182$$

$$\mathrm{Bi}_r = \frac{(0.172 \times 10^{-8})(0.9)(0.12/12) \times (2200)^3}{0.14} = 1.18.$$

From Fig. 5.16, we read the dimensionless time from the abscissa:

$$0.3 = \alpha t/L^2.$$

From this we solve for $t$ and find that $t = 6 \times 10^{-3}$ hr.

The surface temperature of the sheet during the time of heating will reach $(0.7 = (T - T_0)/(T_2 - T_0))$ 824.5°F, which may degrade the sheet surface. In practice we would use forced convection to cool the surface of the sheet. We could then use the combined heat transfer coefficient and resolve the problem. A more accurate solution to this problem can be obtained numerically, and this is done in Pr. 5C.5.

---

In Eq. 5.138 it is assumed that all the radiation from the light source is absorbed at the surface. However, this is not always the case as some materials transmit portions of the incident radiation of certain wavelengths rather than absorb them. Figure 5.17 shows the transmittance curve for a 3.8-mm-thick sample of PET, and here we observe that for wavelengths beyond 2.25 μm no light is transmitted (i.e., for wavelengths of light greater than 2.25 μm PETP of thickness 3.80 mm is opaque). Assuming that the power output of a typical quartz lamp has a specular distribution that can be approximated by a blackbody

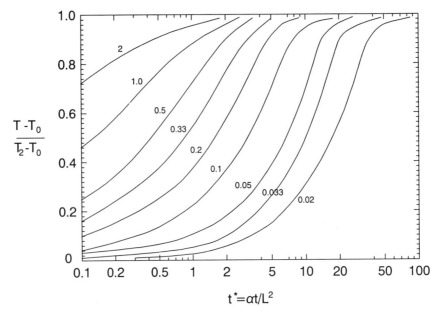

**Figure 5.15** Dimensionless temperature versus dimensionless time for a slab subjected to radiation energy input at the surface ($x/b = 1$). The parameter is the radiative Biot number, $Bi_r = \sigma F T^3 L / k$. The curves are for a specific ratio of $T_0/T_2 = 1/4$ where $T_0$ and $T_2$ are given in absolute units.

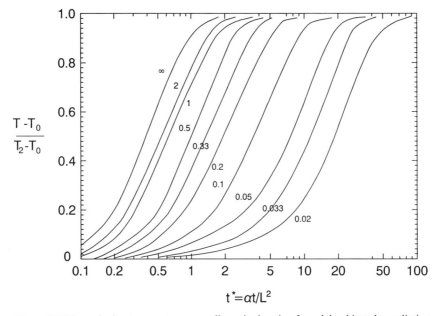

**Figure 5.16** Dimensionless temperature versus dimensionless time for a slab subjected to radiation energy input at the centerline ($x/b = 0$) as in Figure 5.15.

emitter, the spectral emissive power of a blackbody source can be described by Planck's distribution:

$$P(\lambda) = BE(\lambda) = C_1 \lambda^{-6}/[\exp(C_2/\lambda T) - 1], \qquad (5.139)$$

where $B$ is a scaling constant relating the fraction of $E(\lambda)$ that reaches the sample, $\lambda$ is the wavelength in $\mu$m and $C_1$ and $C_2$ are constants with values of $3.742 \times 10^6$ W $\mu$m$^4$/m$^2$ and $1.439 \times 10^4$ $\mu$m $^\circ$K, respectively. The spectral emissive power is superimposed on Fig. 5.17 for various lamp temperatures. Here we see that for the two highest lamp

temperatures most of the incident energy has wavelengths that overlap with the region where PET transmits the radiation rather than absorbs it. Hence, we might expect it difficult to heat a semi-transparent parison such as PET by means of radiation.

Actually there are advantages in having the polymer partially transparent to the incident radiation because the radiation is absorbed internally in the polymer sample, which provides a more uniform heating of the material relative to that produced when the radiation is absorbed only at the surface.

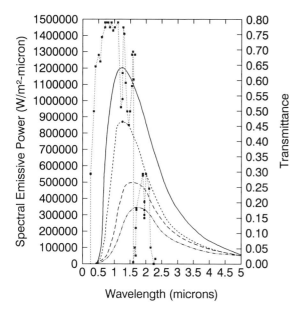

**Figure 5.17** Blackbody radiation spectral distributions superimposed on the transmittance curve for a 3.8-mm-thick PET sample. (Data from Shelby, 1991).

Mathematically the absorbing of radiation is treated as a source term in the equation of energy. The intensity of the incident radiation as a function of depth, $x$, for a planar slab is given by Beer's law (Holman, 1981):

$$\frac{I_x(\lambda)}{I_s(\lambda)} = e^{-k(\lambda)x},$$

$$(5.140)$$

where $I_s(\lambda)$ is the intensity of the radiation at the surface for a given wavelength and $k(\lambda)$ is the absorption coefficient. Noting that $I(\lambda)$ has dimensions of energy per unit area per unit time, we can perform an energy balance to find the energy generated per unit volume per unit time. For a slab of area $A$ and thickness $\Delta x$ the energy balance is (*note: $x = 0$ is taken at the surface here*):

$$(I_x|_x - I_x|_{x+\Delta x})A = \dot{S}A\Delta x.$$

$$(5.141)$$

Dividing by $\Delta x$ and taking the limit as $\Delta x \to 0$, we obtain an expression for the energy generated per unit time and volume:

$$-\frac{dI_x}{dx} = \dot{S} = I_s k(\lambda) e^{-k(\lambda)x}.$$

$$(5.142)$$

For heating cylindrical geometries such as parisons, a similar approach can be used to find $\dot{S}$:

$$\dot{S} = -\frac{1}{r}\frac{d}{dr}(rI_r).$$

$$(5.143)$$

In cylindrical coordinates Eq. 5.140 becomes:

$$\frac{I_r(\lambda)}{I_s(\lambda)} = \exp[-k(\lambda)(R_s - r)],$$

$$(5.144)$$

where $R_s$ is the radius of the outer surface. Substituting Eq. (5.144) into Eq. (5.143), we find $\dot{S}$ for cylindrical coordinates:

$$\dot{S} = I_s(k(\lambda)/r)\exp[-k(\lambda)(R_s - r)].$$

$$(5.145)$$

Provided we know $k(\lambda)$ and $I_s$ we now have the energy generated per unit volume per unit time as the result of absorbed radiation.

### EXAMPLE 5.8
### Radiation Heating of a Semitransparent Sheet

A sheet of (3.8 mm thick, $10 \times 12$ cm) amorphous PET is heated on both sides by a bank of quartz lamps (eight Fostoria T-3 quartz lamps per bank) rated at 500 watts with a peak filament temperature of 2250°C at 120 volts. Formulate the equations and boundary and initial conditions that must be solved to find the time to heat the sheet up to 120°C at the centerline. The following properties are given for PET:

$\rho = 1350\,\mathrm{kg\,m^{-3}}$ $\qquad k = 0.29\,\mathrm{J\,m^{-1}\,s^{-1}\,°C^{-1}}$

$\bar{C}_p = 1250 + 2.0T\,(\mathrm{J\,kg^{-1}\,°C^{-1}})$ $\quad 20 \leqslant T \leqslant 60°C$

$\bar{C}_p = 1370 + (T - 60)^2(0.95)\,(\mathrm{J\,kg^{-1}\,°C^{-1}})\quad 60 \leqslant T \leqslant 80°C$

$\bar{C}_p = 1750\,\mathrm{J\,kg^{-1}\,°C^{-1}}$ $\qquad T \geqslant 80°C.$

At 111 volts the lamp temperature is given as 2200°C. Free convection is assumed to occur at the sheet surfaces with $h$ taken to be $17\,\mathrm{W\,m^{-2}\,°C^{-1}}$. The ambient air temperature is 35°C, and the initial temperature of the preform is 25°C.

### Solution

Taking the coordinates to be at the center of the sheet and the thickness to be $2b$, the energy equation is:

$$\rho\bar{C}_p\frac{\partial T}{\partial t} = k\frac{\partial^2 T}{\partial x^2} + I_s k(\lambda)\,e^{-k(\lambda)x}.$$

The boundary and initial conditions are:

B.C.1 at $x = b$ $\qquad q_x = -k\dfrac{\partial T}{\partial x} = h(T_a - T(b,t))$

B.C.2 at $x = -b$ $\qquad q_x = -k\dfrac{\partial T}{\partial x} = h(T(-b,t) - T_a)$

I.C. at $t = 0$ $\qquad T = T_0 (= 25°C).$

Even if the thermal properties are all taken as constant, one would have to solve the problem numerically because of the source term.

A few comments need to be made regarding the determination of $k(\lambda)$ and $I_s$. $k(\lambda)$ is related to the transmittance, $\tau(\lambda)$, and the thickness of the sample, $L$, by:

$$k(\lambda) = -\ln\tau(\lambda)/L.$$

$$(5.146)$$

**TABLE 5.16 Effective Values of the Absorption Coefficient for PET as a Function of Lamp Temperature**

| Lamp Temperature (°C) | $\bar{k}$ (m$^{-1}$) |
|---|---|
| 2200 | 1300 |
| 2040 | 1460 |
| 1900 | 1610 |
| 1750 | 1790 |

(Data from Shelby, 1991.)

Shelby (1991, p. 1420) obtained an effective value of $k(\lambda)$, $\bar{k}$, for PETP by measuring $\tau(\lambda)$ and then averaging the values over all wavelengths weighted with the blackbody spectral emission curve corresponding to the given lamp temperature. Values of $\bar{k}$ for PETP are given in Table 5.16. This type of data does not appear to be readily available in the literature for other polymers but must be measured for polymers that are semi-transparent. $I_s(\lambda)$ for different lamp temperatures was given for PETP and the lamp arrangement to be:

$$I_s(T_2) = 12{,}000(T_s/2473°K)^4. \tag{5.147}$$

## 5.5. CRYSTALLIZATION, MORPHOLOGY, AND ORIENTATION

The physical properties of a given semi-crystalline polymer depend on the size of the crystallites, the morphology of the crystalline and amorphous regions, and the molecular orientation within the crystalline and amorphous regions. These above factors control the physical properties of polymers and are related to crystallization kinetics, cooling rate (heat transfer), and deformation history. The physical properties of polymers are thought to be associated with the ordering and packing of the molecules that can be affected by processing conditions as illustrated in Table 5.17. In this table it is observed that the modulus of HDPE can be increased from 7 to 70 GPa by the processing technique used, but this is still far from the theoretical limit. Furthermore, the modulus of some organic materials can reach the level of steel, but when compared on a weight basis the organic materials are much stiffer. In this section we first review some qualitative features of crystallization and then the kinetics of crystallization under quiescent conditions. Our goal is to emphasize the fact that the processing conditions can have a significant effect on the properties of a polymer, which in turn affect its end uses.

### 5.5.1. CRYSTALLIZATION IN THE QUIESCENT STATE

Because polymers are only semi-crystalline, which means both crystalline and amorphous phases exist, we must define the degree of crystallinity, $\phi_c$. The degree of crystallinity is the volume fraction of crystallinity and is given by:

$$\phi_c = N\bar{a}^3 f, \tag{5.148}$$

where $N$ is the number of crystalline units (usually they are called spherulites) per unit volume, $\bar{a}$ is the average diameter of the crystallites, and $f$ is the packing factor. $f$ varies between 0.5 and 1.0 depending on the shape of the crystallites. For example, if the crystallites are spherical, the closest packing is when the spheres are arranged in a face centered cubic structure. $f$ for this arrangement is $\pi/3\sqrt{2}$ (0.741).

The manner in which individual molecules crystallize depends on the conformation of the chains. Flexible chain polymers, as shown schematically in Fig. 5.18, fold to form microlamellar plates. Rigid chain polymers form bundles of rodlike structures. The bulk structure or macrostructure of the polymer depends on whether crystallization occurs from dilute solution or from the melt state. When crystallization occurs under quiescent conditions in a dilute solution, then single crystals consisting of folded chains arise. When crystallization from the melt occurs under quiescent conditions, chain folding occurs rapidly in all directions, leading to a spherical structure. These structures are shown schematically in Fig. 5.18. Stress has a significant influence on the macrostructure or morphology. For both dilute solutions and melts of flexible chain polymers the "shish-kebab" structure is observed. Folded chains seem to emanate from rows of highly extended chains, as shown in Fig. 5.18.

The only theory available for describing the conversion of polymer melts to crystalline materials deals with flexible chain polymer melts crystallizing under quiescent (no-deformation) conditions. The rate of crystallization (i.e., the rate of conversion of the volume of amorphous material into crystalline material) consists of two processes as shown in Fig. 5.19: nucleation and growth. *Nucleation* is the initiation of a very small amount of crystalline material emerging from the melt by fluctuation processes. The number of nucleation sites determines the

**TABLE 5.17 Tensile Moduli of Polymers and Other Engineering Materials at 25°C**

| Material | Modulus $E$ (N/m$^2$ × 10$^{-9}$) | Density, $\rho$ (kg/m$^3$ × 10$^{-3}$) | Specific Modulus, $E/\rho$ (× 10$^{-6}$) |
|---|---|---|---|
| Polymers | | | |
| Commonly processed HDPE (Southern & Porter, 1970) | 1–7 | 1 | 1–7 |
| "Extrusion drawn" HDPE fibers (Southern and Porter, 1970) | ~70 | 1 | ~70 |
| Specially cold drawn HDPE fibers (Capaccio & Ward, 1975) | 68 | 1 | 68 |
| DuPont "Kevlar" fibers (rodlike molecules) | 132 | 1.45 | 92 |
| Theoretical limit of HDPE and PVA— fully extended (Sakurada et al., 1966) | 240–250 | 1 | 240–250 |
| Other Materials | | | |
| Aluminum alloys | <70 | | |
| "E" glass fiber | 63 | 2.54 | 35 |
| Steels | ~200 | ~7.0 | ~29 |
| RAE carbon filaments | 420 | 2.0 | 210 |

| TYPE OF MACROMOLECULE | CONFORMATION | MICROSTRUCTURAL UNIT | CONDITIONS | MACROSTRUCTURE | PATTERN (CROSS-SECTION) |
|---|---|---|---|---|---|
| FLEXIBLE MACROMOLECULE IN QUIESCENT CONDITIONS | RANDOM COIL | FOLDED CHAIN LAMELLAE | VERY DILUTE QUIESCENT SOLUTION | SINGLE CRYSTAL | 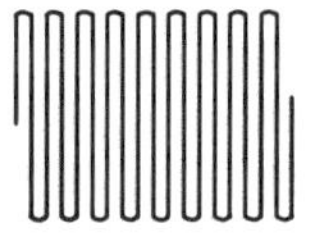 |
| | | | MELT | SPHERULITE | 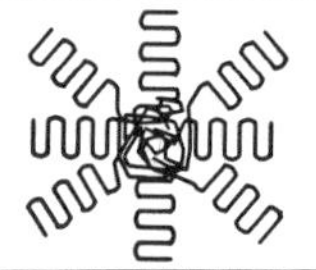 |
| FLEXIBLE MACROMOLECULE IN FIELDS OF FORCE (FLOW) | DEFORMED COIL | CORE OF ALIGNED CRYSTALS (ROW) + LAMELLAR "SIDE PLATES" | DILUTE STIRRED SOLUTION | SHISH-KEBAB |  |
| | | | EXTRUDED "MELT" | ROW NUCLEATED STRUCTURE | |
| RIGID MACROMOLECULE | ROD (LIQUID CRYSTAL) | MICROFIBRIL (RODLET) | SPINNING FROM SOLUTION WITHOUT SPECIAL CONDITIONS | MOZAIC OF RANDOMLY ORIENTED MICROFIBRILS |  |
| | | | SPINNING WITH MAINTENANCE OF FULL ORIENTATION | HIGHLY AND SYMMETRICALLY ORDERED MICROFIBRILS |  |

**Figure 5.18** Morphology of crystallites in polymers. (Reprinted with permission of the publisher from Van Krevelen, 1990.).

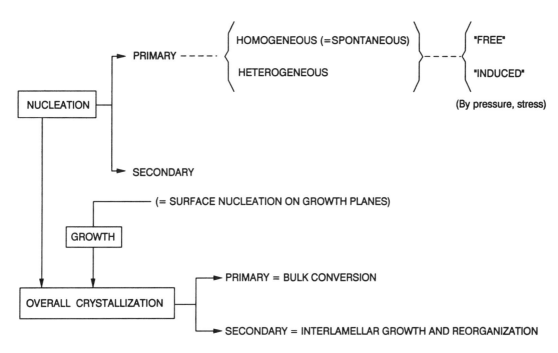

**Figure 5.19** Description of the crystallization process for polymers. (Data from Van Krevelen, 1978.)

morphology of the growing crystallites as a large number of nucleation centers would lead to a large number of small crystallites. Nucleation may be homogeneous or heterogeneous. Homogeneous nucleation occurs in the absence of a second phase, and heterogeneous nucleation is due to the presence of a second phase. In practice heterogeneous nucleation is the rule as most polymer melts contain heterogeneities such as impurities, residues of unmelted polymer, nucleating agents, and so on. *Growth* proceeds also by a nucleation mechanism. Nucleation in this case is a surface process. The whole crystallization process is a continuous interplay between nucleation and diffusive transport of matter to a surface. The rate of overall crystallization depends on the number of available nuclei and on the rate of transport of molecules.

Free crystallization (i.e., no strain or stress) starts from a number of point nuclei and progresses in all directions at equal linear rates (i.e., the rate of increase of the radius of a spherulite, $G$, is linear with time). The rate of growth is very dependent on the temperature of crystallization. In particular $G=0$ at $T_g$ and $T_m$ and passes through a maximum at some intermediate temperature, $T_K$. According to Gandica and Magill (1972), the crystallization process of all the normal polymers follow a master curve. This master curve is a plot of $G/G_{max}$ versus a dimensionless temperature, $\theta$:

$$\theta = (T - T_\infty)/(T'_m - T_\infty),\qquad(5.149)$$

where

$$T_\infty = T_g - 50\qquad(5.150)$$

and $T'_m$ is the thermodynamic equilibrium melting temperature. This master curve is shown in Fig. 5.20. $G/G_{max}$ passes through a maximum when $\theta$ is 0.635, which corresponds approximately ($\approx$) to:

$$T_K \approx 0.5(T'_m + T_g).\qquad(5.151)$$

The quantitative theory of crystallization starts with the linear rate of growth, $G$, which is given by (Lauritzen & Hoffman, 1964):

$$G = G_0 \exp[-E_\text{\DH}/RT]\cdot\exp[-\Delta F_n^*/k_B T].\qquad(5.152)$$

$G_0$ is the molecular jump frequency and is given by $b_0 k_B T/h_p$, where $k_B$ and $h_p$ are the Boltzmann and Planck constants, respectively. $b_0$ is a crystal dimension (see Fig. 5.21), values of which are given in Table 5.18 for PP and in Appendix C for a number of other polymers. The quantity $(-E_\text{\DH}/RT)$ represents the diffusive transport of the molecules in the melt, and $\Delta F_n^*$ is the free energy of a nucleus with $n$-dimensional growth. $E_\text{\DH}$ is the activation energy, $R$ is the gas law constant, and $T$ is the absolute temperature.

Near $T_g$ the term $(-E_\text{\DH}/RT)$ is given by the WLF equation:

$$\frac{E_\text{\DH}}{RT} = \frac{C_1}{R(C_2 + T - T_g)},\qquad(5.153)$$

**TABLE 5.18 Thermal-Physical Properties of Polypropylene**

| Parameter | Value | Units |
|---|---|---|
| $T'_m$ | 447 | °K |
| $T_g$ | 267 | °K |
| $\Delta H_c$ | 45 | cal/gm |
| $\rho_m$ | 0.867 | gm/cm³ |
| $\rho_c$ | 0.938 | gm/cm³ |
| $b_0$ | $5.40 \times 10^{-8}$ | cm |
| $a_0$ | $3.35 \times 10^{-8}$ | cm |
| $\sigma$ | 9.54 | erg/cm² |
| $\sigma_e$ | 47.7 | erg/cm² |
| $\sigma\sigma_e$ | 455 | erg²/cm⁴ |
| $\phi_\infty$ | 0.62 | |

(Data from Bright, B. G., *Quantitative Studies of Polymer Crystallization under Nonisothermal Conditions*, Ph.D. Dissertation, Georgia Institute of Technology, 1975.)

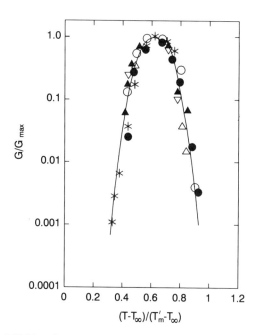

**Figure 5.20** Plot of dimensionless linear growth rate of crystallization versus dimensionless temperature. (Reprinted with permission of the publisher from Gandica & Magill, 1972.)

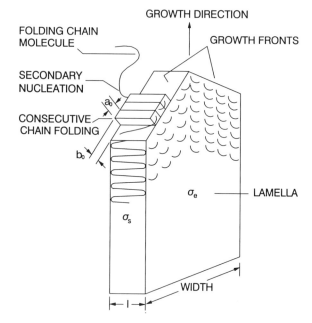

**Figure 5.21** Model of lamellar growth front assuming regular chain folding. (Redrawn from D. G. Bright, "Quantitative Studies of Polymer Crystallization under Non-isothermal Conditions," Ph.D. Thesis, Georgia Tech., 1975.)

where $C_1 = 17.2\,\text{kJ/mol}$ and $C_2 = 51.6°\text{K}$. Near the maximum crystallization temperature the following expression is suggested:

$$E_{\text{Ð}}/R = C_D \frac{T_m'^2}{T_m' - T_g},\qquad (5.154)$$

where $C_D \approx 5.0$. The nucleation factor, $\Delta F_n^*/k_B T$, has the following form:

$$\frac{\Delta F_n^*}{k_B T_x} = \frac{C}{T_x}\left(\frac{T_m'}{T_x}\right)^{n-1}\left(\frac{T_m'}{T_x}\right)^{n-1} \approx \frac{C}{T_x}\left(\frac{T_m'}{\Delta T}\right)^{n-1},\qquad (5.155)$$

where

    $T_x$ = crystallization temperature (°K)
    $\Delta T = T_m' - T_x$ (undercooling)
    $n$ = the dimensionality of the nucleation process, which is usually taken as 2.0
    $C$ = a characteristic constant for every polymer. For a number of polymers $C = 265°\text{K}$
    $T_m'$ = the thermodynamic equilibrium melting point (*note*: this value is slightly different from that of $T_m$)

$C$ can actually be determined theoretically from the following expression:

$$C = \frac{4 b_0 \sigma_s \sigma_e}{k_B \Delta \bar{H}_f \rho_c},\qquad (5.156)$$

where the various parameters are shown in Fig. 5.21. $\Delta \bar{H}_f$ is the heat of fusion and $\rho_c$ is the density of the crystalline plane. $\sigma_s$ is the side surface energy and $\sigma_e$ is the end surface energy associated with lamellar growth as shown in Fig. 5.21.

The linear growth rate is now obtained for two temperature ranges. Using $\xi = T_m'/T_x$ and $\delta = T_g/T_m'$ one obtains for $T_x > T_K$ the following expression

$$\log G = \log G_0 - 2.3 \frac{\xi}{1-\delta} - \frac{115}{T_m'}\frac{\xi^2}{\xi - 1},\qquad (5.157)$$

and for $T_x \ll T_K$:

$$\log G = \log G_0 - \frac{895\xi}{51.63\xi + T_m'(1 - \delta\xi)} - \frac{115}{T_m'}\frac{\xi^2}{\xi - 1},\qquad (5.158)$$

$G_0$ is taken as $10^{12}\,\text{nm/s}$ (*note*: $G_0$ depends on $\bar{M}_w$ and the presence of nucleating agents). Equation 5.157 follows by substituting Eq. 5.154 into Eq. 5.152 while Eq. 5.158 follows by substituting Eq. 5.153 into Eq. 5.152. Equations 5.157 and 5.158 are presented in graphical form in Fig. 5.22. Here $T_m'$ was arbitrarily selected as 473°K so that a two-dimensional plot could be constructed. For every 10°C change in $T_m'$, $G$ will be 0.1 times higher or lower than given in the graph. Furthermore, it is seen that $T_g/T_m'$ has a great influence on the absolute value of $G$.

For symmetrical polymers such as PE, $G$ is high ($T_g/T_m' \approx 0.5$), whereas for asymmetrical polymers such as isotactic polystyrene with $\delta = 0.75$, $G$ is very low. Some numerical values of $G$ for several polymers are presented in Table 5.19. Using these values, we can estimate the length of time for a spherulite to grow to a radius of 10 cm for the various polymers:

| | |
|---|---|
| PE | 1 hour |
| Nylon 6, 6 | 1 day |
| PET | 1 month |
| IPS | 1 year |

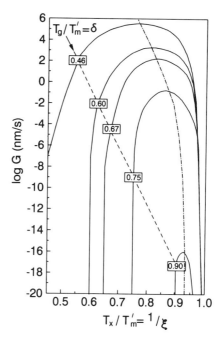

**Figure 5.22** Universal master curve of the rate of growth of spherulites as a function of the dimensionless parameters $T_x/T_m'$ and $T_g/T_m'$ for $T_m' = 473°\text{K}$. (Reprinted with permission of the publisher from Van Krevelen, 1978.)

**TABLE 5.19 Maximum Growth Rates for Some Polymers**

| Polymer | $T_g/T_m'$ | Maximum Growth Rate (nm/s) |
|---|---|---|
| High-density polyethylene | 0.47 | $3 \times 10^4$ |
| Nylon 6,6 | 0.6 | $3 \times 10^3$ |
| Polyester (PET) | 0.64 | $1 \times 10^2$ |
| Isotactic polystyrene (IPS) | 0.73 | $3 \times 10^0$ |

Hence, one can more fully appreciate the relative rates of spherulitic growth.

The overall rate of crystallization under isothermal conditions is given by the Avrami equation (Avrami, 1939, 1940, 1941):

$$\phi_c/\phi_\infty = 1 - \exp(-K t^n).\qquad (5.159)$$

Here $\phi_c$ is the volume fraction of crystallinity and is given by:

$$\phi_c = (\rho - \rho_a)/(\rho_c - \rho_a),\qquad (5.160)$$

where $\rho$ is the density of the sample containing both crystalline and amorphous phases, $\rho_a$ is the density of the amorphous phase, and $\rho_c$ is the density of the spherulitic phase. $\phi_\infty$ is the equilibrium volume fraction of crystallinity (this value is usually less than 1.0). $n$ is a constant that is in the neighborhood of 3.0 but can be determined using DSC for each polymer. $n = 3.0$ for spherulitic growth, but in practice $n$ may be less than 3.0. For PPS $n$ falls between 3.1 and 2.6 for different crystallization temperatures. For PEEK, on the other hand, $n$ is about 2.5 but changes to 1.5 at high values of $\phi$. $K$ is the overall rate constant of crystallization and is determined by fitting DSC data or estimated by the number of nuclei and the linear growth rate as:

$$K = (4/3)\pi N G^3. \tag{5.161}$$

$K$ is related to the half-time for crystallization, $t_{1/2}$, by $t_{1/2} = (\ln 2/K)^{1/n}$.

For nonisothermal crystallization a modified form of Eq. 5.159 is used:

$$\phi_c/\phi_\infty = 1 - \exp\left[ -\int_0^t K(T)nt^{n-1}\,dt \right]. \tag{5.162}$$

Here it is assumed that the contributions to the crystallization process are additive (i.e., at each temperature Eq. 5.159 holds over each temperature interval, $\Delta T$, as the temperature is decreased). One approximation for $K(T)$ is (Velisaris & Seferis, 1986):

$$K(T) = C_1 T \exp[-C_2/(T - T_g + 51.6)]$$

$$\times \exp[C_3/T(T_m' - T)^2]. \tag{5.163}$$

In this equation $C_1$, $C_2$, and $C_3$ are coefficients that are obtained by fitting the function to crystallization data (for example, for PEEK, $C_1$, $C_2$, and $C_3$ are $2.08 \times 10^{10}\,\mathrm{s}^{-n}\,{}^\circ\mathrm{K}^{-1}$, $4050\,{}^\circ\mathrm{K}$, and $1.8 \times 10^7\,{}^\circ\mathrm{K}^3$, respectively). The function given in Eq. 5.163 is based on the assumption that $K(T)$ takes on the same form as $G$.

As $K$ is proportional to the number of nuclei, some estimate of this number is needed. According to Van Krevelen (1978), $N \approx 3 \times 10^4\,\mathrm{cm}^{-3}$ when a polymer is quenched from the melt to $T_x$ and $N \approx 3 \times 10^{11}\,\mathrm{cm}^{-3}$

**TABLE 5.20 Yield stress for Nylon 6,6 as a Function of Spherulite Size**

| Spherulite Size (μm) | Yield Stress, $P$ (psi) | $(Pa \times 10^{-6})$ |
|---|---|---|
| 50 | 10 250 | 72 |
| 10 | 11 800 | 83 |
| 5 | 12 700 | 89 |
| 3 | 14 000 | 98 |

(Data from Van Krevelen, 1978.)

when a sample is heated from the solid state to $T_x$. In other words, there are many more nuclei when starting from the solid state. Since $1\,\mathrm{cm}^3$ contains about $10^{18}$ molecules, the number of nuclei as a fraction of the molecules is extremely small. Starting from the melt there are three nuclei per $10^{13}$ molecules, and starting from the solid state there are three nuclei in $10^6$ molecules.

$N$ determines the maximum size of the spherulites once the whole sample is converted into crystalline material. Since $\phi_c = (4/3)\pi R^3_{max} N$ and if $\phi_c \approx 1$, then $R_{max} N^{1/3} \approx 0.62$. Hence, the more nuclei the smaller the radius of the spherulites.

The properties of a polymer depend on the size of the crystallites. In particular, polymeric materials are brittle when they consist of large spherulites. Hence, it might be better that the spherulites be as small as possible. Although there is very little quantitative data available in the literature in which correlations between spherulite size and mechanical properties are made, some data are presented in Table 5.20 for Nylon 6,6. Here it is observed that the yield stress increases as the spherulite size decreases.

---

**EXAMPLE 5.9**
**Maximum Crystallization Values for Polypropylene**
Determine the maximum crystallization temperature, $T_K$, and the maximum rate of crystallization for polypropylene.

**Solution**

As a first approximation we can use Eq. 5.151, which gives:

$$T_K = 0.5(170°C - 10°C) = 80°C.$$

To obtain a more accurate value we differentiate Eq. 5.157 with respect to $\xi$ (after converting to natural logarithms) and set the derivative equal to zero:

$$\frac{1}{G}\frac{\partial G}{\partial \xi} = -\frac{2.3}{1-\delta} - \frac{115}{T_m'}\left[\frac{\xi^2 - 2\xi}{(\xi-1)^2}\right] = 0.$$

Rearranging this equation we obtain:

$$-\left[\frac{2.3T_m'}{115(1-\delta)} + 1\right]\xi^2 + 2\left[\frac{2.3T_m'}{115(1-\delta)} + 1\right]\xi - \left[\frac{2.3T_m'}{(1-\delta)(115)}\right] = 0.$$

Using the values in Table 5.18 for $T_g$ and $T_m'$. This equation becomes:

$$-23.184\xi^2 + 46.386\xi - 22.184 = 0.$$

The roots of this equation are $\xi = 1.207$ and $0.792$. Because $\xi = (T_m'/T_x)$ we use 1.207 to find that $T_K = 93°C$.

Substituting $T_K$ back into Eq. 5.157 we can now find $G_{max}$:

$$\ln G = \ln(1 \times 10^{12}) - \frac{(2.303)(2.3)(1.207)}{0.403} - \frac{(2.303)(115)(1.207)^2}{447(0.207)}$$

$$= 7.645$$

$$G_{max} = 2.09 \times 10^3\,\mathrm{nm/s}.$$

Finally using Eq. 5.161 we estimate $K_{max}$ to be

$$K_{max} = (4/3)\pi(3 \times 10^4\,\mathrm{cm}^{-3})(2.09 \times 10^{-4}\,\mathrm{cm/s})^3$$

$$= 1.15 \times 10^{-6}\,\mathrm{s}^{-3}.$$

In this calculation we have used the values of $G_0 = 10^{12}\,\mathrm{nm\,s}^{-1}$ and $N = 3 \times 10^4\,\mathrm{cm}^{-3}$, which seem to be average values for a number of polymers. To obtain more accurate values, experimental values of $K$ and $G_0$ are required (these values are usually determined from DSC measurements).

**EXAMPLE 5.10**
**Calculation of the Amount of Crystallization**
Determine the volume fraction of crystallinity in PP crystallized at $T_{max} = 93°C$ for 2 minutes and the size of the spherulites. $\phi_\infty = 0.62$ for PP.

**Solution**

Using Eq. 5.159 with $n = 3$ for spherulitic growth we find:

$$\phi_c = \phi_\infty(1 - \exp[-1.15 \times 10^{-6}(120)^3]) = 0.535.$$

The spherulite size is obtained using Eq. 5.161.

$$G = \frac{dR}{dt} = [K(3/4)/\pi N]^{1/3} = 2.09 \times 10^{-4} \text{ cm s}^{-1}$$

$$R = (2.09 \times 10^{-4} \text{ cm s}^{-1})120 \text{ s} = 2.51 \times 10^{-2} \text{ cm}.$$

**EXAMPLE 5.11**
**Effect of Heat of Crystallization on the Temperature Profile**
A film of PP 0.005 cm thick is cooled from a melt temperature of 220°C to a temperature of 50°C on a casting drum. Determine whether the temperature profile in the film is affected by the heat released during the crystallization process.

**Solution**

The energy equation for this situation is:

$$\rho \bar{C}_p \frac{\partial T}{\partial t} = k \frac{\partial^2 T}{\partial x^2} + \rho_c \Delta \bar{H}_c \frac{d\phi_c}{dt},$$

where the last term in the differential equation is the heat released per unit time per unit volume as a result of crystallization. This comes from the fact that

$$\dot{S} = \frac{\Delta \bar{H}_c}{v}(dw_c/dt),$$

where $w_c$ is the mass of crystalline material and $dw_c/dt$ is the rate of conversion of amorphous to crystalline phase. Using $\phi_c = v_c/v$ and $\rho_c = w_c/v_c$, where $v_c$ is the volume of the crystalline phase, it is now clear how the source term was obtained. Rather than solve the differential equation at this time, we cast it into dimensionless form and evaluate the dimensionless groups that arise. Introducing the following dimensionless groups

$$\theta = \frac{T - T_0}{T_w - T_0} \qquad \xi = x/b \qquad t^* = tK^{\frac{1}{3}},$$

the differential equation becomes

$$\frac{\partial \theta}{\partial t^*} = \left(\frac{kK^{-1/3}}{\rho \bar{C}_p b^2}\right)\frac{\partial^2 \theta}{\partial \xi^2} + \left(\frac{\rho_c \Delta \bar{H}_c}{\rho \bar{C}_p \Delta T}\right)\frac{d\phi_c}{dt^*},$$

where $\Delta T = T_w - T_0$, $2b$ is the thickness and $K$ is the rate constant of crystallization. The dimensionless groups in parentheses can now be used to help us decide whether we should include the source term in the differential equation. Using the values in Table 5.18 and $K_{max}$ calculated in Ex. 5.9 we find the following:

$$\left(\frac{kK^{1/3}}{\rho \bar{C}_p b^2}\right) = \frac{(0.142)(95.4)}{(902)(2.8 \times 10^4)(6.25 \times 10^{-10})} = 86.0$$

$$\left(\frac{\rho_c \Delta \bar{H}_c}{\rho \bar{C}_p \Delta T}\right) = \frac{(938)(18.8 \times 10^4)}{(902)(2.8 \times 10^4)(150)} = 0.46.$$

Hence, we see that the coefficient multiplying $d\phi_c/dt^*$ is about two orders of magnitude smaller than the coefficient multiplying $\partial^2 \theta/\partial \xi^2$, and therefore there should be little contribution to the temperature profile from the heat of crystallization for the conditions given in the problem. We can see that $K$ would have to become extremely large or the polymer sample quite thick (e.g., $b = 0.3$ cm) before the coefficients would be of similar magnitude.

## 5.5.2. OTHER FACTORS AFFECTING CRYSTALLIZATION

Other factors affecting crystallization processes include pressure and stress. Increasing the static pressure has the effect of raising $T_m$ as shown by the Clapeyron equation:

$$\frac{dT_m}{dP} = T_m^0 \frac{\Delta \bar{V}_m}{\Delta \bar{H}_m}, \tag{5.164}$$

where $T_m^0$ is the melting point at atmospheric conditions and $\Delta \bar{V}_m$ and $\Delta \bar{H}_m$ are the volume and enthalpy changes of the melting process. For large pressure variations the change in $T_m$ is given by the Simon equation:

$$P - P_0 = a\left[\left(\frac{T_m}{T_m^0}\right)^c - 1.0\right], \tag{5.165}$$

where $P_0$ and $T_m^0$ are the pressure and melting temperature at the reference conditions and $a$ and $c$ are coefficients. For example, for PE, $T_m^0 = 409°$K (136°C), $a = 3$ kbar, and $c = 4.5$. From this equation, it can be determined that $T_m$ can be raised by 100°C at pressures of 5 kbar. As a result of the increase in $T_m$, the fold length of the chains increases. Unfortunately, the physical properties of polymers crystallized under extreme pressure are poor.

Stress, on the other hand, where deformation is involved, has a considerable effect on crystallization kinetics and morphology. Polymer molecules become oriented during deformation. If the temperature is reduced enough before the molecules relax, then they may crystallize in this oriented state. For polymer melts it is not so easy to create stress-induced crystallization. However, if the product of the relaxation time, $\lambda$, and the time, $t$, for the melt temperature to drop to $T_x$ is approximately equal to 1.0, then we can expect the crystallization of oriented molecules.

Extensional flow is known to have more of an effect on chain extension than shear flow. We illustrate this idea by data obtained in the spinning of PET. Below spinning speeds of 3500 m/min there is very little crystallinity, but above this speed significant crystallinity is observed, as shown in Fig. 5.23. The processing time at these spinning speeds is a few milliseconds, whereas $t^{1/2} = 50$ s. Hence, the crystallization rate must be many decades faster than in the quiescent state.

The stretching of fibers in the neighborhood of $T_g$ is called cold drawing. An example of results for PET is shown in Fig. 5.24. It is noted that at a draw ratio of about 2.5 there is a sudden change in the amount of crystallinity and hence of the density. However, whereas $\rho$ and $\phi_c$ become only slightly dependent, Young's modulus continues to increase with increasing draw ratio. We also note that as $\Delta N$, the birefringence, increases so does Young's modulus, $E$. Crystallinity is not directly correlated with $E$, but $\Delta N$ is. As $\Delta N$ is related to molecular

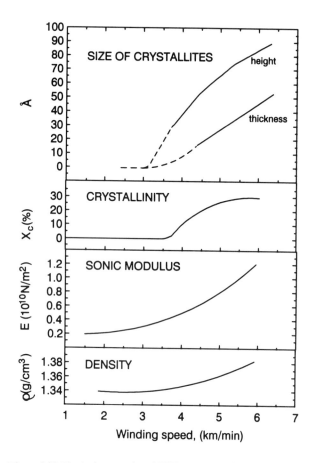

**Figure 5.23** Physical properties of PET yarns wound at various speeds. (Data from Huisman, R. and H. M. Henvel, *J. Appl. Polym. Sci.*, **22**, 943, 1978.)

orientation as discussed in the next section, then $E$ is related to orientation.

One way to incorporate the effect of stress on crystallization is described in Katayama and Yoon (1985). One starts with the generalized Avrami equation:

$$\phi_c/\phi_\infty = 1 - \exp\left[-\left(\int_0^t K(T, \Delta N)dt'\right)^n\right], \qquad (5.166)$$

where $K(T, \Delta N)$ is a crystallization rate constant dependent on $T$ and $\Delta N$. From the stress optic law ($\Delta NC' = \sigma_1 - \sigma_2$), one can relate orientation to stress. According to the theory of Hoffman, the ratio of the oriented ($G_{or}$) to unoriented ($G_{un}$) linear growth rates is:

$$\frac{G_{or}}{G_{un}} = \exp\left[-\frac{4B_0\sigma_s\sigma_e}{KT^{(k)}}\left(\frac{1}{\Delta G_{or}} - \frac{1}{\Delta G_{un}}\right)\right]. \qquad (5-167)$$

$\Delta G_{un}$ is the free energy difference between the amorphous and crystalline states under random orientation and $\Delta G_{or}$ is the free energy difference in the oriented state. $T^{(k)}$ is the absolute temperature. The free energy expressions are:

$$\Delta G_{un}/\rho = \Delta H_{un} - T^{(k)}\Delta S_{un} = \Delta H(\Delta T/T_m^{(k)}) \qquad (5.168)$$

$$\Delta G_{un}/\rho = \Delta H_{or} - T^{(k)}\Delta S_{or}$$
$$= \Delta H(\Delta T/T_m^{(k)}) + T^{(k)}(\Delta S_{or} - \Delta S_{un}), \qquad (5.169)$$

where $\Delta S_{un}$ is the entropy difference between the amorphous and crystalline states under random orientation, $\Delta T$ is the undercooling, $T_m^{(k)}$ is the melting point in $^\circ K$, $\Delta H_{or} = \Delta H_{un} = \Delta H$, and $\Delta S_{or} - \Delta S_{un}$ is the entropy difference between the oriented and the unoriented amorphous states.

The next step is to relate $\Delta S_{or} - \Delta S_{un}$ to some quantity such as orientation. This requires a constitutive equation. Most have used the theory of rubber elasticity, which is not acceptable for polymer melts but which gives:

$$\Delta S_{or} - \Delta S_{un} \propto \upsilon^2 + 2/\upsilon - 3, \qquad (5.170)$$

where $\upsilon$ is the extension ratio. For $\upsilon = 1$ (i.e., small deformations), $\Delta S_{or} - \Delta S_{un} \propto (\Delta N)^2$. Using this relation and Eqs. 5.168 and 5.169, Eq. 5.167 becomes

$$\frac{G_{or}}{G_{un}} = \exp\left\{\frac{C_1}{T^{(k)}}\Delta T\left(1 - \frac{1}{1 + C_2(T^{(k)}/\Delta T)(\Delta N)^2}\right)\right\}, \qquad (5.171)$$

where $C_1$ and $C_2$ are constants. Since $K$ is related to $G$ (see Eq. 5.161), $K(T, \Delta N)$ in Eq. 5.166 is now replaced by Eq. 5.171. There is, however, no significant verification of this approach, and one must use the equation with crystallization data obtained from samples processed under high-stress conditions.

### 5.5.3. POLYMER MOLECULAR ORIENTATION

The physical properties, in particular the modulus, depend on the degree to which the polymer chains lie along a particular direction. In the case of fiber spinning, the degree to which the chains lie along the fiber axis determines the stiffness and strength of the fiber. In this section we define molecular orientation and briefly describe how it is determined.

We first must remember that most polymers are semi-crystalline (although some are just amorphous), and so we must define orientation for both the crystalline and amorphous regions. For the crystalline regions wide-angle x-ray scattering (WAXS) is used to determine orientation. X-rays scatter off the crystallographic planes and are reflected back to the photographic film or a detector. Because polymers are polycrystalline, and if they are unoriented, the reflections are cones having the incident beam as their axis. As the sample is drawn to create orientation, the circles become arcs and then just dots. This is shown in Fig. 5.25.

To quantify the degree of orientation, we define orientation functions. To do this we need to know what the unit cell is. For polyethylene (PE), the unit cell is orthorhombic, in which the three crystallographic axes, $a$, $b$, and $c$, are mutually perpendicular. The orientation functions are:

$$f^a = \frac{3\overline{\cos^2\phi_{a,z}} - 1}{2} \qquad (5.172)$$

$$f^b = \frac{3\overline{\cos^2\phi_{b,z}} - 1}{2}$$

$$f^c = \frac{3\overline{\cos^2\phi_{c,z}} - 1}{2}. \qquad (5.173)$$

The angles $\phi_{i,z}$ are the angles each crystallographic axis makes with

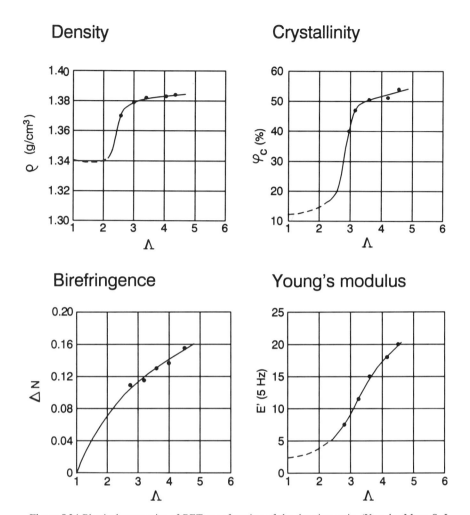

**Figure 5.24** Physical properties of PET as a function of the drawing ratio. (Van der Meer, S. J., Ph.D. Dissertation, Delft University, 1971.)

the $z$ axis (which is the stretch direction). The values $\overline{\cos^2 \phi_{i,z}}$, where $i = a, b,$ or $c$ are evaluated as follows:

$$\overline{\cos^2 \phi_{i,z}} = \frac{\int_0^{\pi/2} I_{hkl}(\phi_{iz}) \cos^2 \phi_{i,z} \sin \phi_{i,z}\, d\phi_{i,z}}{\int_0^{\pi/2} I_{hkl}(\phi_{i,z}) \sin \phi_{i,z}\, d\phi_{i,z}}. \tag{5.174}$$

$I_{hkl}(\phi_{i,z})$ is the intensity diffracted from the $(hkl)$ planes, which are normal to the $i$ crystallographic axis. For orthogonal unit cells the orientation functions are related as follows:

$$f^a + f^b + f^c = 0 \tag{5.175}$$

$$\sum_{i=1}^{3} \cos^2 \phi_{i,z} = 1. \tag{5.176}$$

When a crystallographic axis, for example, $c$, is perpendicular to $z$, $f^c = -0.5$; when it is parallel, $f^c = 1.0$, and when it is randomly oriented

with respect to $z, f^c = 0$. The orientation functions are defined in terms of the cosine squared; otherwise, the integral would always be zero.

For the amorphous regions, we can use sonic waves or birefringence to determine orientation (certainly there are several methods, but our intention is not to review all these or to say which is best). To use birefringence we must know the intrinsic birefringence, $\Delta N_0$, which is either calculated or measured on a perfectly oriented sample. The birefringence, $\Delta N$, is related to the stress field through the stress-optic law. In particular the law reads:

$$\Delta N = C \Delta \sigma, \tag{5.177}$$

where $C$ is the stress optic coefficient and $\Delta \sigma = \sigma_1 - \sigma_2$ is the difference between the principal stresses (*note* that in shear flow $\sigma_1 - \sigma_2 = \sqrt{N_1^2 + 4\tau_{yx}^2}$). For uniaxial stretching flows such as in fiber spinning or drawing, then

$$\Delta N = C \Delta \sigma = C(\tau_{zz} - \tau_{rr}). \tag{5.178}$$

Because the melt is amorphous,

$$\Delta N = \Delta N_{am}^0 f_{am} \tag{5.179}$$

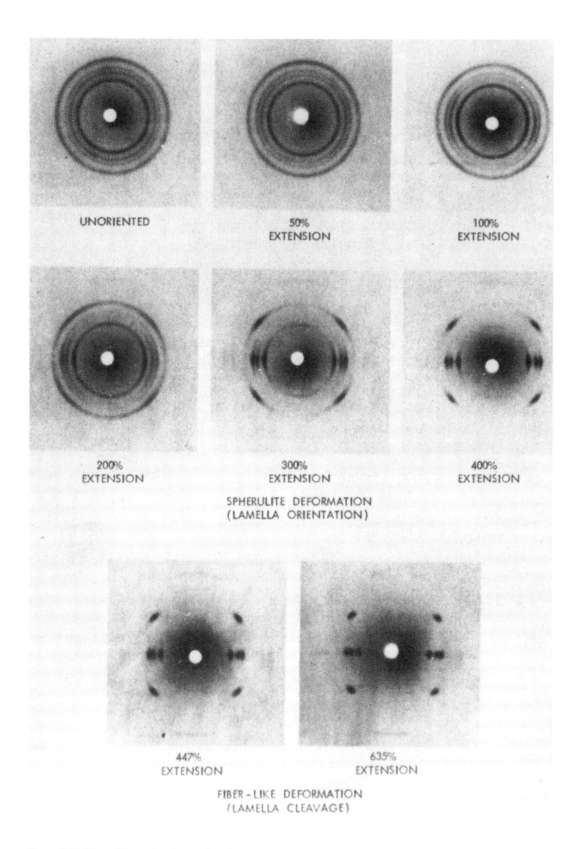

**Figure 5.25** Effect of isotactic polypropylene film extension on the wide-angle x-ray diffraction patterns. (Reprinted with permission of the publisher from Samuels, R. J. *Structured Polymer Properties*, Wiley, New York, p. 27, 1974.)

and

$$f_{am} = (C/\Delta N^0_{am})\Delta\sigma \qquad (5.180)$$

The quantity $(C/\Delta N^0_{am})$ turns out to be a constant for all polymers. Thus, $f_{am}$ is proportional to $\Delta\sigma$ that arises during flow. We must remember that stresses relax when the flow is stopped so that unless the cooling is rapid enough, we will lose a considerable amount of the orientation.

For shear flows the same stress-optic law holds:

$$C\Delta\sigma = C(\sigma_1 - \sigma_2) = \Delta N, \qquad (5.181)$$

where $\sigma_1$ and $\sigma_2$ are the principal stresses, $C$ is the stress optic coefficient, and $\Delta N$ is the birefringence. By a transformation of the coordinates we can relate the stress components in the laboratory coordinates to those in terms of the principal axes. For shear flow the relations are:

$$\tau_{yx} = \tfrac{1}{2}\Delta\sigma \sin 2\chi \qquad (5.182)$$

$$\tau_{xx} - \tau_{yy} = \Delta\sigma \cos 2\chi \qquad (5.183)$$

$$\Delta\sigma = \sigma_1 - \sigma_2. \qquad (5.184)$$

With the assumption that the principal directions and the optical directions coincide,

$$\chi_{opt} = \chi_{stress} = \chi, \qquad (5.185)$$

and introducing Eq. 5.182 into Eqs. 5.183 and 5.184, we obtain the following relations:

$$\tau_{yx} = \frac{1}{2}\frac{\Delta N}{C}\sin 2\chi \qquad (5.186)$$

$$\tau_{xx} - \tau_{yy} = (\Delta N/C)\cos 2\chi. \qquad (5.187)$$

Hence, from the stress field generated during flow we can estimate the degree of molecular orientation. In particular, we find from Eqs. 5.186 and 5.187 that

$$\tan \chi = \frac{(1/2)\tau_{yx}}{\tau_{xx} - \tau_{yy}}. \qquad (5.188)$$

In extensional flow it is assumed that $\chi_{opt} = \chi_{stress} = \chi = 0$, and hence $\tau_{xx} - \tau_{yy} = \sigma_1 - \sigma_2 = \Delta N/C$. So in extensional flow, the molecules are oriented in the flow direction. Values of $C$ for several polymers are listed in Table 5.21.

If we were to measure the birefringence, $\Delta N$, of a polycrystalline material in the solid state, it would not represent the overall orientation. $\Delta N$ consists of contributions from the amorphous and crystalline regions:

$$\Delta N = \phi_c \Delta N^0_{cr} f_{cr} + (1 - \phi_c)\Delta N^0_{am} f_{am}, \qquad (5.189)$$

where $\Delta N^0_{cr}$ and $\Delta N^0_{am}$ are the birefringence values of the perfectly oriented crystalline and amorphous regions and $\phi_c$ is the crystalline volume fraction. (It also should be added that shrinkage measurements on annealing are a widely used method for determining macromolecular orientation in processed articles. The idea is that stretched oriented chains will become randomized on annealing above $T_g$.)

The reason for the discussion of orientation is because physical properties are closely tied to orientation. Our goal is to point out that

**TABLE 5.21 Stress-Optical Coefficients**

| Materials | Temperature (°C) | C ($10^{-9}$ m²/N) |
|---|---|---|
| Polystyrenes | | |
| Styron 666 | 190 | −4.1 |
| Styron 678 | 190 | −4.8 |
| BASF 3 | 178 | −4.5 |
| | 188 | −4.6 |
| | 200 | −4.4 |
| | 214.5 | −4.2 |
| Polyethylenes | | |
| HD | 150 | 2.4 |
| | 190 | 1.8 |
| LD | 150 | 2.0 |
| IUPAC A | 150 | 2.1 |
| Polypropylenes | | |
| PP | 210 | 0.9 |

(Data from H. Janeschitz-Kriegl, "Flow Birefringence of Elastico-Viscous Polymer Systems", *Adv. Polym. Sci.*, **6**, pp. 170–318, 1969, and *Polymer Melt Rheology and Flow Birefringence*, Springer-Verlag, Berlin, 1983.)

molecular orientation and morphology are related to flow and deformation history and that, hence, physical properties are related to processing history. Our discussions of orientation serve only to quantify and define orientation. Likewise they serve to show that at least for amorphous polymers and for polymer melts there is a direct correlation between stress and orientation.

## 5.6. SOLUTION TO DESIGN PROBLEM 4

The solution to Design Problem 4 proceeds as follows. We first develop the form of the energy equation that must be solved. We then determine whether heat transfer from the polymer film to air is important relative to that which occurs at the drum surface. The temperature profile along the film (or in essence as a function of time following a fluid element) will be determined. This can then be substituted into the equation for nonisothermal crystallization to give the percent crystallinity for a given heat transfer coefficient at the drum surface. Assuming spherulitic growth, which is reasonable for PP, we can estimate the spherulite radius from the linear growth rate.

We first determine the form of the energy equation. Following Ex. 5.1 and Ex. 5.11, the energy equation is:

$$\text{Pe}\,\frac{\partial\bar{\theta}}{\partial\zeta} = \frac{-b}{2k}[h_1(T(1) - T_a)/(T_0 - T_a) + h_2(T(-1) - T'_a)/(T_0 - T_a)], \qquad (5.190)$$

where $h_1$ and $h_2$ are the heat transfer coefficients at the film–drum surface and at the film–air interface, respectively, and $T'_a$ is the temperature of the air.

It is not certain whether $h_2$ is significant at this point. Using Eq. 5.123 and 5.124 for forced convection heat transfer involving flow over a flat plate and the conditions given in the problem we estimate $h_2 = 2.8 \, \text{W m}^{-2}\,°\text{K}^{-1}$, which is quite small. For free or natural convection from a vertical flat plate we use Eq. 5.132 to find $h_2 = 3.87 \, \text{W m}^{-2}\,°\text{K}^{-1}$ which again is small. Hence, we consider only the heat transfer at the film–drum surface, and Eq. 5.190 becomes:

$$\text{Pe}\,\frac{\partial\bar{\theta}}{\partial\zeta} = -\frac{bh_1}{2k}\bar{\theta}, \qquad (5.191)$$

where we have introduced the average dimensionless temperature:

$$\bar{\theta} = \frac{\bar{T} - T_a}{T_0 - T_a} = \frac{T(1) - T_a}{T_0 - T_a} \tag{5.192}$$

as there is little variation of $T$ over the film thickness.

We next estimate the magnitude of $h_1$ required to keep the crystallinity below 3.0%. To do this we first estimate what value of $h_1$ is needed to drop the film temperature from 200°C (we expect an insignificant drop in film temperature from the die to the drum based on Ex. 5.1) to 25°C. This temperature is arbitrary, and we may find that we would like to keep the temperature higher as the film is drawn after leaving the chill roll. Eq. 5.191 becomes:

$$\frac{\partial \bar{\theta}}{\partial \zeta} = -\frac{h_1 \bar{\theta}}{2\rho \bar{C}_p V_0}. \tag{5.193}$$

If we follow an element of the film, then we can recast Eq. 5.193 as a transient heat conduction problem. With $\zeta = z/b$ where $b$ is one-half the film thickness and $t = z/V_0$, we are now solving the following differential equation:

$$\frac{\partial \bar{\theta}}{\partial t} = -\frac{h_1 \bar{\theta}}{2b\rho \bar{C}_p} \tag{5.194}$$

with I.C. at $t = 0$, $\bar{\theta} = 1.0$.

Equation 5.194 can be easily solved if $h_1$, $\rho$, and $\bar{C}_p$ are taken to be independent of temperature. Most likely one would use channels carrying a cooling fluid just below the surface, and hence $h_1$ might change slightly across the drum. According to Table 5.6 $\rho$ is not expected to change much, but $\bar{C}_p$ is somewhat temperature-dependent (we could use Eq. C.2 and Table 9 in Appendix C if we want to improve the results). Taking $\bar{C}_p$ to be $2.80 \times 10^3$ kJ kg$^{-1}$°K$^{-1}$ and $\rho$ to be 867 kg m$^{-3}$ we can estimate what $h_1$ should be. Given the line speed and drum diameter, we find that the time the film is in contact with the drum is 1.98 s. With the assumption of constant coefficients, the solution to Eq. 5.194 is:

$$\bar{\theta} = \exp\left[-\frac{h_1 t}{b 2\rho \bar{C}_p}\right]. \tag{5.195}$$

When $\bar{T} = 25$°C, $\bar{\theta} = 0.069$, then $h_1 = 164$ W m$^{-2}$°K$^{-1}$.

Although we have an estimate for $h_1$, we don't know whether this value will keep the amount of crystallinity below the level of 3.0%. To find this out we use Eqs. 5.161 and 5.162:

$$\phi_c / \phi_\infty = 1 - \exp\left[-\int_0^t \left(\frac{4}{3}\right) \pi N G^3 n t'^{(n-1)} dt'\right] \tag{5.196}$$

where $G$ is given by either Eq. 5.157 or Eq. 5.158. $T_k$ has been calculated in Ex. 5.9 to be 93°C. In principle we should use Eq. 5.157 for $93 \leqslant T \leqslant 200$°C and Eq. 5.168 for $25 \leqslant T \leqslant 93$°C, but we will use Eq. 5.157 over the whole temperature interval. For PP Eq. 5.157 becomes:

$$\ln G = \ln G_0 - (2.303)\left[(2.533 \times 10^3/T) + \frac{5.141 \times 10^4}{T(447 - T)}\right] \tag{5.197}$$

or

$$G = G_0 \exp\left[-(2.303)((2.553 \times 10^3/T) + 5.141 \times 10^4/T(447 - T))\right]. \tag{5.198}$$

From Eq. 5.195 we know how T changes with time and hence we can calculate $\phi_c$ by substituting Eq. 5.198 in Eq. 5.196. Before doing this we make a change of variables in Eq. 5.196 from time to temperature by using Eq. 5.195 and we express Eq. 5.196 as:

$$\ln(1 - \phi_c/\phi_\infty) = -\int_0^t (4/3)\pi N G^3 n t'^{(n-1)} dt' \tag{5.199}$$

$$= \int_{T_0}^{T_f} -C \exp\left[-(3)(2.303)((2.553 \times 10^3/T)\right.$$

$$+ 5.141 \times 10^4/T(447 - T))]$$

$$\times (\ln\{(T - T_a)/(T_0 - T_a)\})^2 (1/(T - T_a)) dT, \tag{5.200}$$

where $T_f = 298$°K, $T_0 = 473$°K, and $C = (4/3) \pi N G_0^3 (3)(2b\rho \bar{C}_p/h_1)^3$. To integrate the expression on the RHS of Eq. 5.200, we use the IMSL subroutine QDAGS (see Appendix D.5), and the calling program is given in Table 5.22. We have written the program so that one can read

**TABLE 5.22 Program Listing for Calling Subroutine DQDAGS Used in the Solution to Design Problem 4 (DPRIV.FOR)**

```
C*******************ABSTRACT************************
C  THIS PROGRAM CALLS THE IMSL SUBROUTINE DQDAGS DESCRIBED IN APPENDIX D5
C  TO INTEGRATE THE NONISOTHERMAL FORM OF THE Avrami EQUATION
C  REQUIRED IN THE SOLUTION OF DESIGN PROBLEM 4. THE
C  ESTIMATED HEAT TRANSFER COEFFICIENT, H₁, THE TEMPERATURE OF THE
C  COOLING FLUID, Tₐ, THE INITIAL TEMPERATURE OF THE MELT, A, AND
C  THE ESTIMATED FINAL TEMPERATURE OF THE FILM, B, ARE READ IN FROM
C      THE TERMINAL.
C  •••••••••••••••••••••••••••••••••••••••••••••••••••••
C
      INTEGER NOUT
      DOUBLE PRECISION A,B,C, ABS, ERRABS, ERREST, ERROR, ERRREL, EXACT
   &       ,F, RESULT, H1,TA
      COMMON C,TA,A
      EXTERNAL F, DQDAGS, UMACH
C
C      SET OUTPUT NUMBER
C
```

**TABLE 5.22** *continued*

```
            CALL UMACH(2,NOUT)
C
C   SET LIMITS OF INTEGRATION AND READ IN H1 AND TA
            WRITE(*,*) 'INPUT THE INITIAL TEMPERATURE,A, THE FINAL TEMPERA-
          & TURE, B, THE HEAT TRANSFER COEF., H1, AND TA:'
            READ(*,*) A,B, H1,TA
            C = 6.7417D + 24/H1**3
C
C   SET ERROR TOLERANCES
            ERRABS = 0.0D + 0
            ERRREL = 0.001D + 0
C
C   CALL DQDAGS
            CALL DQDAGS (F,A,B, ERRABS, ERRREL, RESULT, ERREST)
C
C   PRINT RESULTS
            WRITE (NOUT, 100) RESULT, ERREST
100 FORMAT (' COMPUTED =',D22.15,5X,'ERROR ESTIMATE =',1PE10.3)
            END
C
C   FUNCTION SUBROUTINE CONTAINING INTEGRAND
            DOUBLE PRECISION FUNCTION F(X)
            DOUBLE PRECISION X,C,TA,A
            DOUBLE PRECISION DEXP, DLOG
            INTRINSIC DEXP, DLOG
            COMMON C,TA,A
            F = C*DEXP(-1.764D + 04/X - 3.552D + 05/X/(447.-X))*(DLOG((X-
          & TA)/(A-TA)))**2/(X-TA)
            RETURN
            END

INPUT THE INITIAL TEMPERATURE,A, THE FINAL TEMPERATURE, B, THE HEAT
TRANSFER COEF., H1, AND TA:
447.,298.,164.,285.
COMPUTED = -.361433649975024D-08 ERROR ESTIMATE = 2.493E-15
```

in different values of $T_f$, $T_0$, and $h_1$ from the terminal. Furthermore, we find it necessary to use the double precision form of the subroutine because of the magnitude of the terms in Eq. 5.200. For the conditions given and the assumed final temperature of 25°C, $\phi_c \approx 0.0$ for all practical purposes (i.e., $\ln(1 - \phi_c/\phi_\infty) = -0.36 \times 10^{-8}$). Hence, a value of $h_1 = 164\,\mathrm{W\,m^{-2}\,^\circ K^{-1}}$ will keep the crystallinity below 3.0%.

We have only found a desired heat transfer coefficient. Whether we can achieve these conditions is another matter. The design of a cooling system to obtain the desired heat transfer coefficient may be difficult. Two methods that are used presently are the spraying of water on the inside of the drum surface and the circulation of water through channels beneath the drum surface.

The other question of concern in this problem was spherulite size. Because $\phi_c$ is negligible, there is no need to calculate the radius of a spherulite. However, if this was necessary, we would use Eqs. 5.198 and Eq. 5.153 to obtain a differential equation for determing $R$. We would then use the differential equation solver, IVPAG (see Appendix D.7), to find the radius, $R$, as a function of time.

## PROBLEMS
### A. Applications

**5A.1.** Temperature Shifting of Rheological Data
Viscosity versus shear rate data are given in Appendix A, Table A.12, for PPS at various temperatures. Use the shift factors given there as well as the rheological data to obtain values of $\eta$ and $N_1$ at 330°C at higher shear rates than given. Compare the shifted values to the measured values where overlap of the shear rates exists.

**5A.2.** Shift Factor Determined Using the WLF Equation
Calculate the shift factor for PPS at 293°C for a reference temperature of 330°C using the WLF equation (Eq. 5.7), and compare it with the experimental value of 1.708.

**5A.3.** Viscous Heating in a Slit Die
Determine the maximum temperature rise in the slit die shown in Fig. 3.23 (p. 51) for HDPE at 170°C at a wall shear rate of $100\,\mathrm{s^{-1}}$. Use the rheological parameters given in Table 2.3 (p. 14) and Eq. 5.55. Take the die length to be 100 mm and the height to be 2.5 mm.

**5A.4.** Pressure Profile in the Presence Viscous Heating
Determine the effect of viscous heating on the pressure profile in the slit die in Pr. 5A.3.

**5A.5.** Heat Transfer Coefficients in High Speed Wire Coating
A copper wire coated with LDPE leaves a wire-coating die at 165°C. The diameter of the metal wire and coating is 0.09 cm and the coating thickness is 0.02 cm. The line speed is 2000 m/min. The cooling medium is water at 25°C, and the final temperature of the surface of the coating is 25°C. Determine the heat transfer coefficient as the wire enters the cooling bath.

**5A.6.** Heat Transfer Coefficient during the Heating of PET Preforms
A 57-g PET preform is heated from 25 to 120°C using a radiant heat source. The outer diameter of the preform is 2.54 cm, and the wall thickness is 0.32 cm. The air temperature is 25°C. Determine the

convective part of the heat transfer coefficient at the outside surface of the preform assuming that the source of heat transfer is free convection.

**5A.7.** Effect of Carbon Fiber on the Cooling Time of a Composite
A 0.4064-cm-thick laminate (layers of polymer sheet reinforced with long continuous fibers) consisting of 60 volume % carbon fiber and 40 volume % PEEK is cooled from 350 to 100°C in a mold with the wall temperatures set at 75°C. Neglecting crystallization and assuming constant thermal properties, determine how much faster the composite cools down to the final temperature relative to the pure matrix of the same thickness. The properties of the fiber and matrix are given in Table 5.11.

**5A.8.** Crystallization of LDPE
Determine the temperature at which the maximum rate of crystallization occurs and $K(T)_{max}$ for HDPE. Which crystallizes faster, PP or HDPE?

**5A.9.** Orientation of a Polymer Melt in Shear Flow
LDPE (NPE-953) is extruded through a film die at 170°C at a wall shear stress of $1.2 \times 10^5$ Pa. The film die consists of parallel plates with a width of 1.0 m and a height of 0.05 cm. Determine the amorphous orientation function as a function of distance from the center to the die wall in the region where steady shear flow exists.

**5A.10.** Radiation Heating of Carbon Fiber Tow
Before being impregnated with the matrix, carbon fiber (black in color) is heated on both sides by means of a bank of quartz lamps with a filament temperature of 1492°C. The carbon fiber bundles (called tows) are spread to a width of 20 cm and a thickness of 0.25 cm. Treating the bank of lights and the spread carbon fiber as parallel flat plates, determine the time required to heat the centerline of the carbon fibers from 100 to 300°C. What will the surface temperature of the filaments be when the center is 300°C?

**5A.11.** Factors Affecting Spherulitic Crystallization
  (a) The overall rate of crystallization, $r_c = d\phi_c/dt$, is proportional to $(t_{1/2})^{-1}$ where $t_{1/2} = (\ln 2/K)^{1/3}$, for spherulitic crystallization. Show that $r_c \propto 1.5N^{1/3}$ where $N$ is the number of nucleation sites.
  (b) Based on the expression above list four factors that $r_c$ depends on.
  (c) Which two of the these factors can be altered by processing conditions?
  (d) Which two of these factors are determined by the polymer structure?

**B. Principles**

**5B.1.** Viscous Heating in Tangential Annular Flow
Determine the temperature distribution in an incompressible power-law fluid between two coaxial cylinders (see Pr. 2B.6) each assumed to be at the same temperature, $T_w$. Assume that the forced convection assumption holds and that μ, ρ, and $k$ are constant. (*Note*: the equations of motion and energy can be solved independently.)
  (a) Show that the equation of energy is:

$$0 = k\frac{1}{r}\frac{d}{dr}\left(r\frac{dT}{dr}\right) + m\left[r\frac{d}{dr}\left(\frac{v_\theta}{r}\right)\right]^{n+1}. \tag{5.201}$$

Using the velocity profile determined in Pr. 2B.6 show that Eq. 5.201 becomes:

$$0 = k\frac{1}{r}\frac{d}{dr}\left(r\frac{dT}{dr}\right) + m\left[\frac{-2WR^{2/n}k^{2/n}}{n(1-k^{2/n})}\right]^{n+1} r^{-2(1+s)}. \tag{5.202}$$

  (c) Solve Eq. 5.202 for the temperature distribution.
  (d) Under what conditions might we expect a significant temperature rise, say, of the order of 10°C, if the fluid has the properties of HDPE given in Table 2.3 (p. 14).

**5B.2.** Viscous Heating in a Wire-Coating Die
In Example 2.3 in Chapter 2 (p. 18), determine whether any significant rise in temperature will occur for the conditions given. Assume the melt enters the die at 200°C and that the wall and wire temperatures are also 200°C. Repeat the calculations for a wire speed of 2000 m/min.

**5B.3.** Forced Convection Heat Transfer in Tubes (Short Contact Times)
A polymeric fluid whose viscosity function is described by the Ellis model is flowing through the tube as shown in Fig. 5.26. Determine the temperature profile and the wall heat flux for the case of short contact times in which the heat does not penetrate very far into the fluid.
  (a) Obtain the velocity profile.
  (b) Obtain the differential equation for $T(r, z)$ by carrying out an energy balance and by using the equation of thermal energy.
  (c) Set $s = R - r$, and discard any terms that are not important in the vicinity of the wall to obtain the following differential equation:

$$\rho\bar{C}_p[(\tau_R/\eta_0) + (\tau_R^\alpha/\tau_{1/2}^{\alpha-1}\eta_0)]s\frac{\partial T}{\partial z} = k\frac{\partial^2 T}{\partial s^2}, \tag{5.203}$$

in which $\tau_R$ is the momentum flux at $r = R$.
  (d) Under what conditions is it possible to make the substitution suggested in Part c?
  (e) Introduce the following dimensionless variables, and rewrite Eq. 5.203.

$$\theta = \frac{T - T_0}{T_1 - T_0} \qquad \zeta = \frac{z}{R} \qquad \sigma = \frac{s}{R}$$

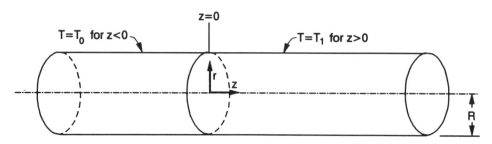

**Figure 5.26** Pipe with sudden change in wall temperature.

$$N = \frac{\rho \bar{C}_p R^2}{k} [(\tau_R/\eta_0) + (\tau_R^{\alpha}/\tau_{1/2}^{\alpha-1}\eta_0)]. \tag{5.204}$$

**(f)** Then show that if a solution of the following form is assumed:

$$\theta = f(\eta), \quad \text{where } \eta = \left(\frac{N\sigma^3}{9\zeta}\right)^{1/3}, \tag{5.205}$$

the partial differential equation in (Part **c**) is transformed into the ordinary differential equation

$$f'' + 3\eta^2 f' = 0, \tag{5.206}$$

in which the prime indicates differentiation with respect to $\eta$.

**(g)** What boundary conditions are required to solve Eq. 5.206?

**(h)** Solve Eq. 5.206, and get

$$f = \left[\Gamma\left(\frac{4}{3}\right)\right]^{-1} \int_{\eta}^{\infty} e^{-\eta^3} d\eta, \tag{5.207}$$

in which $\Gamma(4/3)$ is the gamma function evaluated at 4/3.

**(i)** From the temperature distribution in Part h, evaluate the wall heat flux as a function of the distance down the tube.

**(j)** Integrate the result in Part i to obtain the total heat flow through the pipe surface between $z=0$ and $z=L$, that is, show that:

$$Q = 4\pi R k(T_1 - T_0)\left(\frac{N}{9}\right)^{1/3}\left(\frac{L}{R}\right)^{2/3}\left[\Gamma\left(\frac{7}{3}\right)\right]^{-1}. \tag{5.208}$$

**5B.4.**  Solidification of a Polymer Melt during Injection Molding
A thin rectangular mold is filled with a polymer melt having an initial temperature of $T_0$. The melt is assumed to make perfect thermal contact with the mold walls, which are set at $T_w$. Determine the rate of solidification (i.e., $W_a = \rho_s \, dX_s/dt$) and the time it takes for the centerline to reach the crystallization temperature, $T_c$. Consider the melt to have a distinct crystallization temperature, $T_c$. The heat of crystallization is taken as $\Delta \bar{H}_c$. Assume that there is enough flow into the center of the molten polymer so that the solid polymer always makes contact with the walls of the mold. Use the notation given in Fig. 5.27 and carry out the following steps:

**(a)** Assuming that each region is a finite slab, obtain the differential equations for the temperature distribution and the corresponding initial and boundary conditions for the liquid and solid phases.

**(b)** To relate the temperature distributions in the two phases, carry out an energy balance at the interface to obtain:

$$k_s \frac{\partial T_s}{\partial x_s}\Big|_{x_s} - k_l \frac{\partial T_l}{\partial x_l}\Big|_{x_l} = \rho_s \Delta \bar{H}_c \frac{dX_s}{dt}. \tag{5.209}$$

**(c)** Substitute the series solutions to the differential equations in Part **a** into Eq. 5.209, and explain how this equation can be used to find the solidification rate.

**5B.5.**  Relation of Volume Fraction of Crystallinity to Density
Show that $\phi_c$ is related to $\rho$, $\rho_c$, and $\rho_a$ by the following relation:

$$\phi_c = (\rho - \rho_a)/(\rho_c - \rho_a),$$

where $\rho =$ the density of the semi-crystalline material, $\rho_a$ is the density of the amorphous phase, and $\rho_c$ is the density of the crystalline phase.

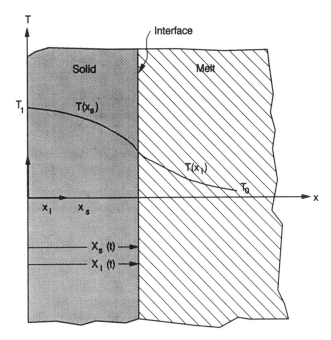

**Figure 5.27**  Crystallization of a semi-infinite melt.

**5B.6.**  Density and Heat Capacity in a Crystallizing Polymer
During crystallization the density and heat capacity change as the polymer is converted from amorphous melt to crystalline solid. Obtain expressions for $\rho$ and $\bar{C}_p$ of the semicrystalline composite material in terms of $\phi$, $\phi_c$, $\phi_\infty$, and the corresponding crystalline and amorphous values of $\rho_c$, $\phi_a$, $\bar{C}_{p,c}$, and $\bar{C}_{p,a}$.

### C.  Numerical Problems

**5C.1.**  Numerical Solution of Heat Conduction in a Slab
Do Ex. 5.5 for the case in which the boundary conditions are given in terms of a heat transfer coefficient and the thermal coefficients are temperature dependent (see Table 5.6). Compare the results to those obtained in Example 5.5 in which constant thermal coefficients were used.

**5C.2.**  Numerical Solution of Heat Conduction in a Slab Using MOLCH
Solve Ex. 5.5 for the case in which the boundary conditions are given in terms of a flux through a heat transfer coefficient. Use the subroutine MOLCH (see Appendix D.8) to obtain the time to cool the centerline to 130°C. To use MOLCH, follow the procedure associated with Eq. 5.112 in the text.

**5C.3.**  Nonisothermal Crystallization of PEEK
The temperature of a PEEK/carbon fiber composite (containing 60% by weight of carbon fiber) is reduced from 350 to 75°C at the rate of 20°C/min. The thickness of the composite is 0.3175 cm. Determine $\phi_c$ using Eqs. 5.162 and 5.163.

**5C.4.**  Film Casting of LDPE
It is desired to use the same film-casting equipment as is used for polypropylene (see Design Problem 4). Using the same conditions as given in Design Problem 4, except for the initial temperature for LDPE, which is 175°C, determine whether $\phi_c$ can be maintained below 3.0%.

**5C.5.  Radiation Heating of a Sheet**
Do Ex. 5.7 numerically using the approach described in Ex. 5.5. Do the problem for both the cases of constant- and temperature-dependent thermal properties (The material is PVC).

**5C.6.  Bonding of Polymer to a Metal Sheet**
PPS sheets (0.018 cm thick) are bonded to both sides of a copper sheet (0.05 cm thick) by heating the materials from 25 to 310°C in a press as shown in Fig. 5.28. The plattens of the press are held at 330°C. Determine the time required to heat the interface between PPS and copper to 310°C. The following properties are given for copper:

$$k_c = 377 \, \text{W m}^{-1} \, ^{\circ}\text{K}^{-1} \qquad \rho_c = 8954 \, \text{kg m}^{-3}$$

$$\bar{C}_{p,c} = 0.381 \, \text{kJ m}^{-1} \, ^{\circ}\text{C}^{-1}.$$

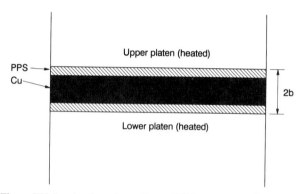

**Figure 5.28** Lamination of two films of PPS to a copper sheet.

**(a)** Formulate the differential equation and boundary and initial conditions required to find the time to heat the interface between PPS and copper up to 310°C.

**(b)** Find an analytical solution to this problem, if possible.

**(c)** Solve the problem numerically using the IMSL subroutine MOLCH (see Appendix D.8).

**D.  Design Problems**

**5D.1.  Design of a Pelletizing Bath**
Based on the conditions given in Ex. 5.6, determine the minimum bath length required to drop the centerline temperature of the strand to 75°C. In your calculations determine the length for three conditions: (1) no thermal resistance at the polymer–water interface; (2) use a heat transfer coefficient based on a mean value calculated at the beginning and end of the bath; and (3) use a variable heat transfer coefficient that depends on temperature.

**5D.2.  Cooling of a Blow Molded Gas Tank**
A coextruded parison consisting of HDPE and nylon 6 are blow-molded to form a gasoline tank for an automobile. The nylon 6 serves as a barrier to gasoline fumes. In order for the tank to exhibit adequate impact properties the spherulite size must be maintained below 10 microns in both resins, and $\phi_c$ for both resins should approach 75% of $\phi_\infty$. Determine the heat transfer coefficient required at the mold wall to provide these conditions. The thickest section of the parsion as it contacts the wall is 0.50 cm with nylon 6 representing one-fifth of the thickness. Treat the composite material as a slab, and assume that there is no heat transfer at the inner surface. The mold is to be cooled by channels with circulating water at 12°C. In your design you must consider that it is necessary to use the shortest time possible to cool the part down.

## REFERENCES

Avrami, M. 1939. *J. Chem. Phys.,* **7**, 1103.

Avrami, M. 1940. *J. Chem. Phys.,* **8**, 212.

Avrami, M. 1941. *J. Chem. Phys.,* **9**, 177.

Bird, R. B., W. E. Stewart, and E. N. Lightfoot. 1960. *Transport Phenomena.* (Wiley, New York).

Brandrup, J., and E. H. Immergut, Eds. 1989. *Polymer Handbook*, 3rd Ed. (Wiley, New York).

Capaccio, G., and I. M. Ward. 1975. "Ultra-High Modulus Linear Polyethylene through Controlled Molecular Weight and Drawing." *Polym. Eng. Sci.,* **15**, 219.

Carslaw, H. S. and J. C. Jaeger. 1973. *Conduction of Heat in Solids*, 2nd Ed. (Oxford University Press, New York).

Fand, R. M., and K. K. Keswani, 1973. "Combined Natural and Forced Convection Heat Transfer from Horizontal Cylinders to Water." *Int. J. Heat Mass Transfer,* **16**, 175.

Gandica, A., and J. H. Magill 1972. "A Universal Relationship for the Crystallization Kinetics of Polymer Materials." *Polymer,* **13**, 595.

Holman, J. P. 1981. *Heat Transfer*, 5th Ed. (McGraw-Hill, New York).

Ke, B. 1960. "Characterization of Polyolefins by Differential Thermal Analysis." *J. Polym. Sci.,* **42**, 15.

Katayama, K., and M. Yoon 1985. An A Ziabicki and H. Kawai, Eds, *High Speed Fiber Spinning* (Wiley, New York).

Lauritzen, I., Jr., and J. D. Hoffman. 1964. *J. Res. Nat'l Bur. Stand. (A)*, **64** 73.

Lopez, L. C., and G. L. Wilkes 1988. "Crystallization Kinetics of Poly(p-phenylene sulphide): Effect of Molecular Weight." *Polymer,* **29**, 106

McAdams, W. H. 1954. *Heat Transmission* (McGraw-Hill, New York).

Nielsen, L. 1962. *Mechanical Properties of Polymers* (Reinhold, New York).

Richardson, M. O. W. 1977. *Polymer Engineering Composites* (Applied Science Publishers Ltd, London).

Riggs, J. B. 1988. *An Introduction to Numerical Methods for Chemical Engineers* (Texas Tech University Press, Lubbock, TX).

Rohsenow, W. M., and J. P. Hartnett, Eds. 1973. *Handbook of Heat Transfer* (McGraw-Hill, New York).

Sakurada, I., T. Ito, and K. Nakamae. 1966. *J. Polym. Sci.,* **C15**, 75.

Shelby, M. D. 1991. "Effects of Infrared Lamp Temperature and Other Variables on the Reheat Rate of PET," *Society of Plastics Engineers Technical Papers,* **37**, 1420.

Siegel, R., and J. R. Howell 1981. *Thermal Radiation Heat Transfer* (McGraw-Hill, New York).

Southern, J. H., and R. S. Porter 1970. "The Properties of PETP Crystallized under the Orientation and Pressure Effects of a Pressure Capillary Rheometer." *J. Appl. Polym. Sci.,* **14**, 2305.

Starkweather, H. W., Jr., and R. H. Boyd 1960. *J. Phys. Chem.,* **64**, 410.

Van der Meer, S. J. 1971. Ph.D. Dissertation, Delft University, Delft, The Netherlands.

Van Krevelen, D. W. 1978. "Crystallization of Polymers and the Means to Influence the Crystallization Process." *Chimia,* **32**(8), 279.

Van Krevelen, D. W. 1990. *Properties of Polymers* (Elsevier, Amsterdam).

Velisaris, N., and J. C. Seferis 1986. "Crystallization Kinetics of Polyetheretherketone (PEEK) Matrices." *Polym. Engng. Sci.,* **26**(22), 1574.

Velisaris, C. N., and J. C. Seferis 1988. "Heat Transfer Effects on the Processing–Structure Relationships of Polyetheretherketone (PEEK) Based Composites." *Polym. Engng. Sci.,* **9**(28), 583.

Whitaker, S. 1977. *Fundamental Principles of Heat Transfer* (Pergamon Press, New York).

Williams, M. L., R. F. Landel, and J. D. Ferry 1955. "The Temperature Dependence of Relaxation Mechanisms in Amorphous Polymers." *J. Am. Chem. Soc.,* **77**, 3701–7.

# 6

# MIXING

DESIGN PROBLEM 5
## Design of a Multilayered Extrusion Die

Coextrusion is an important process for manufacturing layered plastic composites, such as film, sheet, and tubing (Middleman, 1977; Schrenk et al., 1963; Schrenk & Alfrey, 1983). A costly alternative would be the fabrication of individual layers of plastics followed by conventional lamination and coating. An additional advantage of the coextrusion technique is its ability to handle extremely thin films on the order of $10\,\mu m$. One of the most "colorful" applications of this technique is the iridescent film, which consists of hundreds of very thin layers of alternating low and high refractive indices. A typical example of this type of film consists of 116 layers of poly(methyl methacrylate) (PMMA) with refractive index 1.49 and 115 layers of polystyrene (PS) with refractive index 1.59 and total film thickness of about $20\,\mu m$ (Radford et al., 1973).

We seek to produce a multilayered film of two polymers from a coextrusion blown film apparatus. The die is of the annular type with the mandrel rotating (Fig. 6.31.a) and its dimensions are: $\kappa R = 4.8\,cm$, $R = 5.0\,cm$ and $L = 15\,cm$. The two polymers, presented in the schematic by black and white colors, have the same Newtonian viscosity, $\mu = 500\,Pa\cdot s$, at the extrusion temperature. The pressure drop in the die is $\Delta P = 3.45\,MPa$ (500 psi), and the total volumetric flow rate is $Q = 9.43\,cm^3/s$. The feedport system consists of 20 equal-size ports in total, 10 for the first polymer (black) and the other 10 for the second polymer (white). A similar feedport system with 16 ports in total is shown in Fig. 6.31. Calculate the minimum rotational frequency of the mandrel for the maximum layer thickness not to exceed $5\,\mu m$ at the end of the die. To solve this problem, follow both the geometrical and the kinematical approaches. Also, calculate the power to rotate the mandrel and introduce changes necessary for the reduction of the power consumption.

---

In the preceding four chapters momentum, mass, and heat transfers were analyzed as they apply to polymer processing. This is the first chapter among the last five concerned with the specifics of various types of processes. Primarily, mixing in polymer processing addresses two tasks: addition of various ingredients (additives), and production of polymer blends and alloys. The additives are used to alter the properties of the matrix polymer, such as impact strength, flexural modulus, modulus of elasticity, foaming ability and cost. These additives are called modifying additives. Additives are also used to prevent polymer degradation, and they are then called protective additives. The other important task of mixing at present is the blending of other polymers with the given polymer to obtain a desired improvement in the given polymer. For example, the blending of polypropylene (PP) and compatibilizers with nylon 6 leads to a composition lower in price and with improved energy absorbing characteristics (Van Gheluwe et al., 1988). Similarly, blending of rubber with polystyrene (PS) produces a very fine dispersion of rubber particles in the PS matrix, called high-impact PS (HIPS), with great improvements in energy-absorbing characteristics (Bucknall, 1977). Although many would like to think that the properties of a blend are determined by thermodynamics, it

turns out that the method of preparation is of utmost importance. Fig. 6.1 shows the importance of mixing in polymer processing by laying out the major routes for melt processing of polymers to finished products.

Clearly, there are differences between mixing requirements for the additives and the blending of polymers. Blending of polymers usually involves large concentrations (or weight percent) of the additive phase, various degrees of compatibility between the various phases, small density differences, and dispersion of one phase into droplets or fibers with dimensions affected by the physical characteristics of that phase and the hydrodynamics prevailing in the mixing environment. These characteristics of the polymer blending processes contrast with those of polymer-additive systems: low concentrations, large density differences, and solid–liquid distribution and dispersion of agglomerates into particles that cannot be divided any further.

Generally speaking, in polymer processing we are concerned with mixing in three types of systems: liquid–liquid; solid–solid; and liquid–solid, each with different mechanisms and kinetics of mixing. In the liquid–liquid case, mixing is concerned with either low-viscosity monomers or high-viscosity polymer melts. Solid–solid mixing involves blending of either two polymers or resins or an additive and a polymer. In such case, the polymer is granular, pelletized, powdered, or diced form. Finally, solid–liquid mixing involves blending of liquid additives and solid polymers (not melted) or solid additives (below their melting point) and melted polymers.

This chapter is organized as follows. In Section 6.1 we describe mixing with particular emphasis on polymer processing. The characterization of the state of the mixture, statistical analysis, and the various experimental techniques are presented in Section 6.2. Laminar mixing is the only viable mixing mechanism for polymer melts because of their high viscosity, and it is analyzed in Section 6.3 using both the geometrical and kinematical approaches. In Section 6.4 we emphasize the importance of residence time and strain distributions of polymer-processing equipment in understanding the degree of mixing achieved in each process. Dispersive mixing is analyzed in Section 6.5 with focus on both solid aggregates and liquid–liquid dispersions. In Section 6.6 we highlight the importance of thermodynamics in polymer processing, and in Section 6.7 we describe the fundamentals of chaotic mixing as a way to improve the degree of mixing. Finally, Section 6.8 provides the solution to Design Problem 5.

## 6.1. DESCRIPTION OF MIXING

The term *mixing* refers to operations that have a tendency to reduce nonuniformities or gradients in concentration, temperature, size of a dispersed phase, or other properties of materials. Equivalently, a mixing operation increases the configurational entropy of the system, which becomes a maximum as the configuration becomes random. Mixing is considered to be one of the most widespread industrial unit operations, and it is found in the core of many areas in the general industry. This unit operation might be a process in itself or might be a part of a more extended sequence of processes. Typical examples can be taken from the general industry, polymerization processes, and polymer processing.

Here we will restrict ourselves to the study of the mixing of two-component systems. The two components are defined as either *major* or *minor* components by the level of their total concentration.

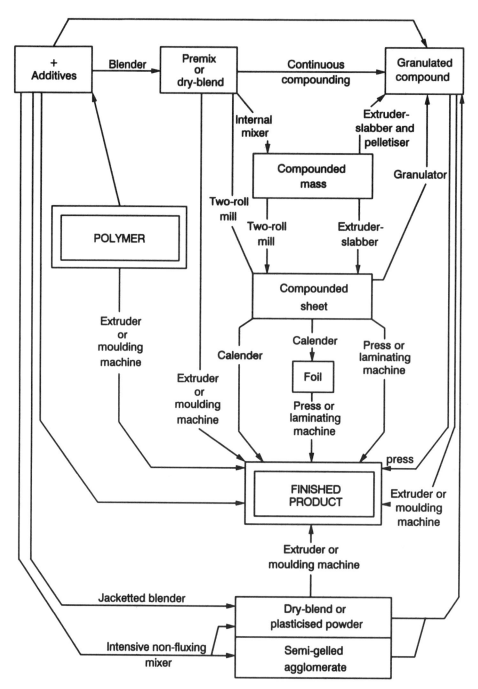

**Figure 6.1** Mixing in the major routes for melt processing of polymers to finished products. (Reprinted by permission of the publisher from Matthews, 1982.)

The goal of the mixing process is usually to achieve a homogeneous dispersion of the minor component into the major one, obtaining the *ultimate particle* or *subdivision* (or *volume element*) level of the minor component (McKelvey, 1962). In the mixing area, that term is used in a restricted sense, because in its general form the ultimate particle is the molecule and ultimate mixing is molecular mixing. However, in typical mixing operations the two parameters that define the size and form of the ultimate particle are the form of the component and the level of satisfaction of the final dispersion. For example, in the mixing of carbon black agglomerates in polyethylene (PE), the ultimate particle

is one particle of carbon black defined by the form of the initial agglomerates (many particles together) as well as by the satisfactory dispersion level of one carbon black particle. In general, typical ultimate particles are molecules and colloidal and microscopic particles.

Mixing is accomplished by movement of material from various parts by the flow field. This movement occurs by a combination of the following mechanisms, two of which are hydrodynamic and one is molecular (Brodkey, 1966). The first is *convective transport*. It is present in both laminar and turbulent regimes, and it can also be called *bulk diffusion*. Generally speaking, a colored pigment being dispersed in a bucket of paint is an example of laminar mixing. In this case, layers of pigment are thinned, lumps are flattened, and threads are elongated by laminar convective flow. Stirring of cream in a cup of coffee is an example of turbulent mixing, in which the mechanism of turbulent bulk flow predominates at the first stages.

The second mechanism is *eddy diffusion*, which is produced by local turbulent mixing. This mechanism prevails at the later stages in the example of stirring of cream into a cup of coffee. The turbulent eddies in the flow field create small-scale mixing, which is sometimes thought to be analogous to molecular diffusion. However, eddy diffusivity is much higher than molecular diffusivity, and it occurs over longer length scales. For gases and low-viscosity liquid systems eddy diffusion becomes the usual mode of mixing.

Finally, there is *molecular diffusion* or interpenetration of molecular species. It is responsible for the ultimate homogenization on a molecular scale (the ultimate particles are the molecules), and it is considered to be true mixing. This form of diffusion is driven by the chemical potential difference due to concentration variation, and it is a very slow process, as its time scale is proportional to the value of the diffusion coefficient. Thus, this mechanism becomes important in gases and low-molecular-weight miscible liquid systems, although there are time scale differences in those two cases.

The major difference between mixing in general and in polymer processing stems from the fact that the viscosity of polymer melts is usually higher than $10^2$ Pa·s, and thus mixing takes place in the laminar regime only (Re < 2000; to achieve such a number the polymer would have to flow down a 1-m-wide channel at a velocity of 20 cm/s). This has a severe consequence, which is the lack of eddy and molecular diffusion that greatly enhances the rate of mixing and reduces the scale of homogenization. Thus, all mixing theories and practices should be adjusted to the laminar regime to find applicability in the polymer processing area. This remark applies also to solid–solid mixing in polymer processing, but it does not to the addition of low-molecular-weight substances into polymers, such as dyes, where molecular diffusion plays a role.

The two basic types of mixing can be identified as *extensive* and *intensive* mixing. *Distributive*, *convective*, *repetitive*, *simple mixing* and *blending* are the main names that extensive mixing is also associated with, whereas *compounding*, *dispersive*, and *dispersing* mixing are the corresponding names associated with intensive mixing. Extensive mixing refers to processes that reduce the nonuniformity of the distribution (viewed on a scale larger than the size of the distributed components) of the minor into the major component without disturbing the initial scale of the minor component. It can be achieved through two mechanisms: rearrangement and deformation in laminar flow. The rearrangement mechanism works in plug-type flows (Tadmor and Gogos, 1979) with absence of deformation, and it can be subdivided into random and ordered types, as shown schematically in Fig. 6.2. Also, deformation achieved in shear, elongation and squeezing flows plays a major role in distributing the minor component.

The term intensive mixing refers to processes that break down the liquid dispersed phase or the initial particle agglomerates, and they decrease the ultimate particle of the dispersion. A typical example is the dispersion of agglomerates of colloidal carbon black particles in PE. In this case the initial ultimate particle is the agglomerate, and the final is the particle itself. Another example is the dispersion of a polymer into another polymer where the minor component should be dispersed into small droplets or elongated fibers (both of them have a length scale of about 10 μm). The analysis of dispersive mixing follows the lines of the analysis of the distributive mixing with the complication that the breakup forces should now be included.

The geometry of the mixing equipment, physical parameters such as viscosity, density, interfacial tension, elasticity, and attractive forces for solids, operating conditions such as temperature, speed of rotating

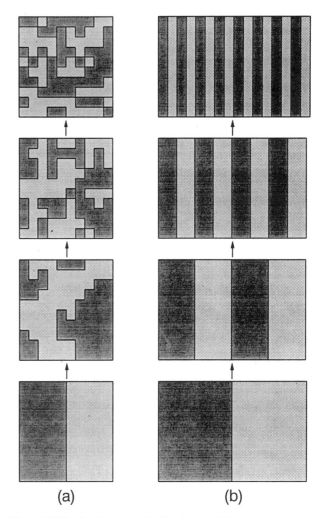

**Figure 6.2** Distributive mixing: (a) Random and (b) ordered rearrangements. (Reprinted by permission of the publisher from Tadmor & Gogos, 1979.)

parts, flowing velocity, and residence time are the important factors that determine the relative strengths of the mixing mechanisms. As a consequence, this relative strength affects the efficiency of mixing and the quality of the product. In almost all cases, both good distribution and good dispersion are required. In some cases, only distributive mixing can be tolerated if the next step offers dispersive characteristics, and dispersive mixing is used when a finely dispersed mixture is required and when the next step does not offer any dispersion characteristics. The distinction between good and poor dispersion and distribution is shown in Fig. 6.3.

Some of the terms mentioned get a specific connotation when referred to polymer processing, and thus we give here some specific definitions (Matthews, 1982). *Compounding* refers to the process of softening, melting, and compaction of the polymer matrix and dispersion of the additive into that matrix. *Blending* refers to all processes in which two or more components are intermingled without significant change of their physical state. Finally, *kneading* refers to mixing achieved by compression and folding of layers over one another; *milling* refers to a combination of smearing, wiping, and possibly grinding; and *mulling* refers to wiping and rolling actions.

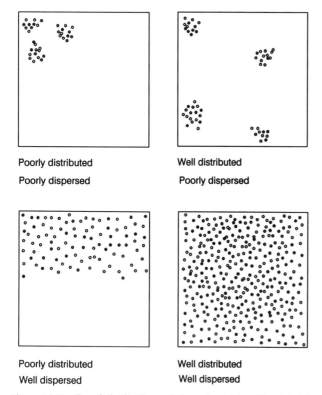

Poorly distributed
Poorly dispersed

Well distributed
Poorly dispersed

Poorly distributed
Well dispersed

Well distributed
Well dispersed

**Figure 6.3** Quality of distributive and dispersive mixing. (Reprinted by permission of the publisher from D. H. Morton-Jones, *Polymer Processing,* Chapman and Hall, London, 1989.)

## 6.2. CHARACTERIZATION OF THE STATE OF MIXTURE

After a certain mixing process is completed, the polymer process engineer faces the questions of the effectiveness of the selected process and of the uniformity of the product. In this section we restrict ourselves to the examination of the product in terms of meeting preset specifications. In its strictest sense, this should include a detailed description of the ultimate particles (aggregates, particles, drops, and fibers) of the minor component (i.e., their number, length scales, shape, orientation, and the statistical description of their spatial distribution). For example, suppose we intend to mix 1% carbon black in PE. After the mixing is carried out, we fracture a number of samples of PE in order to determine whether the initial carbon black aggregates were broken down to either smaller-size aggregates or even to individual particles and to assess the size of the existing carbon black entities and their spatial distribution. Of course, all this information will be integrated into determining the sample concentration statistics and its relation to the expected 1% overall concentration. This procedure can be somewhat altered when measuring the ultimate particles because their number and size might affect some physical property whose measurement can provide a quick way of assessing the success of the mixing step.

Usually, it is sufficient to characterize any mixture by its *gross uniformity, texture,* and *local structure.* The term *gross uniformity* can be also found in the literature as *overall* uniformity or *gross composition* uniformity. In the example of carbon black mixed into PE, the gross uniformity indicates the goodness of the concentration distribution of carbon black, and it can be evaluated by statistical analysis on a number of fractured pellets selected randomly from a large batch of pelletized PE. In the ideal case, all the randomly selected pellets will contain the

same concentration of carbon black, and thus the mixture will be called *perfectly gross uniform.* In practice, the gross uniformity is a measure of the degree of fit of the concentration distribution to the binomial distribution, which was found theoretically and proved experimentally to describe these situations.

The size of the PE pellets determines the *scale of examination* (McKelvey, 1962; Tadmor & Gogos, 1979), which can be expressed by order of magnitude length, area or volume. When a finished product is examined for composition gross uniformity, the scale of examination is determined by the size of the sampling volume. It is obvious that the maximum scale of examination is identical to the scale of the finished product itself and that the minimum is the length scale of the ultimate particles. McKelvey (1962) introduced the ideas of fine- and coarse-grained samples depending on the ratio of the scales of examination and ultimate particles. In particular, if this ratio is very large the sampling volume contains many particles, and its appearance is fine grained, whereas if this ratio is about 10 to 100, then the sampling volume contains a few particles, and it is characterized as coarse grained.

At this point, one should not overlook the importance of the volume fraction of the minor component on the relative sizes of scales of examination and ultimate particles. To be more specific, suppose that we carry out a statistical evaluation of samples of 1 and 10% carbon black in PE and that we would like to have in our microscope eyepiece only about 30 carbon black particles. To meet that requirement we should increase our magnification in the 10% sample more than is required for the 1% sample because of its lower particle concentration.

Is there an ideal scale of examination? Besides the upper and lower limits mentioned, the *texture* of the mixture and its *granularity* pose another lower limit. Tadmor and Gogos (1979) define texture as nonuniformity in the forms of patches, stripes, streaks, and so on. Thus, we can narrow the large range for the scale of examination by raising the lower limit to the scale of the granularity of the texture.

The presence of texture in the samples is important in polymer processing, as: (1) laminar mixing produces texture; (2) the mechanical properties of polymer blends depend on the texture (e.g., skin-core formation, fiber orientation, and distribution of voids along the thickness in structural foams); and (3) the lack or the presence of texture is required in certain products. Samples can exhibit a certain texture and at the same time do or do not exhibit concentration gross uniformity.

The scale of examination for textures might differ from that for gross uniformity. However, the upper and lower limits, in both cases, remain the same. When the scale of examination is about the same as the scale of the ultimate particles, we can probe the *local structure.* In conclusion, gross uniformity, texture, and local structure are the characteristics of the mixture, and their relative importance depends on the specific application. For instance, if carbon black is used for coloring purposes for PE, then its composition gross uniformity and texture are important whereas local structure is not. However, if the purpose of mixing carbon black in PE is for UV protection (obtained by particles themselves not aggregates), then local structure and texture are important. The effect of the various mixing mechanisms on the three parameters of the state of mixture are shown in Fig. 6.4.

### 6.2.1. STATISTICAL DESCRIPTION OF MIXING

The gross uniformity of a mixture can be analyzed by statistical methods that can also be used to define the degree of mixing or index of mixedness. For example, for a mixture of carbon black and PE we assume that there exist no agglomerates, that the carbon black particles are uniform in size (monodisperse sample), and that the PE matrix can be divided into fictitious "particles" having the size of carbon black particles. Thus, the mixture will consist of "white" PE particles and "black" carbon black particles mixed together. Solid–solid, solid–liquid

MIXING MECHANISMS

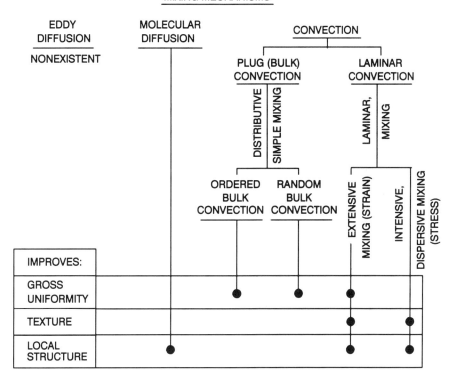

**Figure 6.4** Mixing mechanisms and their effect on the characteristics of mixtures. (Reprinted by permission of the publisher from P. Hold, "Mixing of Polymers—an Overview Part I," *Adv. Polym. Techn.*, **2**(2), 141–51, 1983.)

and liquid–liquid (when the dispersed liquid cannot be dispersed any further) dispersions can be easily visualized by this example. Just before mixing starts, the two components are separated inside the mixer. We assume that the diameter of the particles is very small compared to the length scale of the mixer and that there is a large but finite number of these particles. To test for the kinetics of mixing and the gross uniformity we need to extract on a regular basis small samples from the mixture. Sample size should be large enough to contain a sufficient number of particles for statistical analysis and small enough to leave the mixture undisturbed.

As mixing progresses, the black and white particles intermingle to a greater extent, and, after sufficient time passes, the mixture obtains its random mixing status. At this point the probability of finding a white particle at any point is constant and equal to the overall fraction of the white particles. If a number of samples of equal size are extracted from the mixture, the fraction of the white particles will vary from sample to sample. The mean value over all samples studied should equal the overall fraction of white particles, and the distribution of values should follow the binomial or an equivalent distribution.

Consider a sample, randomly extracted from the mixture that contains $n$ particles, where $n$ is large enough for statistical treatment and small enough compared to the total number of particles in the mixture itself. Let $p$ equal the fraction of black particles in the entire mixture. Then the probability $P$ that this randomly selected sample has exactly $b$ black particles ($x = b/n$) is given by the binomial (or Bernoulli) distribution (Spiegel, 1991):

$$P(b;n,p) = \frac{n!}{b!(n-b)!} p^b (1-p)^{n-b}. \tag{6.1}$$

The variance of that distribution is given by:

$$\sigma_n^2 = \frac{p(1-p)}{n} \tag{6.2}$$

Similarly, if there are $N$ samples tested, where $N$ is a very large number, then the variance of the binomial distribution for these samples is given by the formula:

$$s_N^2 = \frac{p(1-p)}{N}. \tag{6.3}$$

This variance is a measure of how much the concentration differs from the mean value. The procedure to determine whether or not the mixture is grossly uniform is obvious from the discussion: Samples are extracted from the mixture, their average concentration is calculated, and finally the concentration distribution is checked against the binomial distribution. If there is a match between these two distributions, the mixture is considered to be grossly uniform or a random mixture. In the limit of zero variance the mixture attains the uniform state.

The binomial distribution is considered to approach the normal (or Gaussian) distribution if the number of samples is large. More specifically, if the following conditions are met:

$$Np > 5; \qquad p < 0.5, \tag{6.4}$$

then the distribution of concentrations is approximated by the continuous Gaussian form:

$$P = \frac{1}{\sigma\sqrt{2\pi}} \exp\left[-\frac{(x-p)^2}{2\sigma^2}\right] \tag{6.5}$$

Note that if the entire fraction of black particles, $P$, is very small, then a better approximation of the binomial distribution is the Poisson distribution:

$$P = \frac{e^{-np}(np)^b}{b!}. \tag{6.6}$$

In practice, a limited number of samples are examined. Let $N$ be the number of samples. Then, the mean, $\bar{C}$, is defined as:

$$\bar{C} = \frac{1}{N}\sum_{i=1}^{N} C_i, \tag{6.7}$$

where $C_i$ represents the concentration of the $i$th sample. The variance, $s^2$ (or standard deviation $s$), of the measurements is defined as:

$$s^2 = \frac{1}{N-1}\sum_{i=1}^{N}(C_i - \bar{C})^2 = \frac{1}{N-1}\sum_{i=1}^{N} C_i^2 - \frac{1}{N(N-1)}\left[\sum_{i=1}^{N} C_i\right]^2. \tag{6.8}$$

If the sampling procedure has been properly executed, then $\bar{C}$ and $p$ should not be significantly different. The values for the mean and the variance can be used in two ways. First, either with the confidence tests of the statistical theory we can estimate the actual mean and variance of the whole mixture, or, closely to this, with the significance theory we can answer the question, "are the samples taken from the same mixture or not?" Second, the variance can be used in the evaluation of the mixedness or in various kinetic calculations of mixing.

*Confidence intervals* express (in quantitative terms) the percentage of times that the true (yet unknown) values of the mean and standard deviation will lie within a range of specified values. These values are specified based on the statistics of a limited number of samples. The confidence intervals for the mean, $\mu$, are the following:

$$\bar{C} - z\frac{s}{\sqrt{N}} < \mu < \bar{C} + z\frac{s}{\sqrt{N}} \tag{6.9}$$

for $N \geqslant 30$ where the normal distribution applies (*note*: $z$ is the confidence coefficient and is read from the normal distribution table), or:

$$\bar{C} - t\frac{s}{\sqrt{N}} < \mu < \bar{C} + t\frac{s}{\sqrt{N}} \tag{6.10}$$

for $N < 30$, where the $t$ distribution with $N-1$ degrees of freedom applies. Tables for the normal and $t$ distributions are found in standard statistics textbooks (e.g., Spiegel, 1991). For example, if a confidence level of 95% is chosen, the $z$ value ($z_{0.475}$) is 1.96, and the $t$ value ($t_{0.975}$, for 2.5% of the area lying in the "tail") is 2.26 for 9 degrees of freedom.

The confidence intervals for variances are obtained from the $\chi^2$ distribution:

$$\frac{s\sqrt{(N-1)}}{\chi_\alpha} < \sigma < \frac{s\sqrt{(N-1)}}{\chi_{1-\alpha}}. \tag{6.11}$$

For example, for a 95% confidence level with $N-1$ degrees of freedom, the values $\chi_{0.975}$ and $\chi_{0.025}$ are taken from the relevant tables of Spiegel (1991).

---

### EXAMPLE 6.1
**Confidence Intervals for the Mean and Variance**

Suppose that 2% carbon black is mixed with PE in a melt extruder and that the product is pelletized and shipped in 25-kg bags. We randomly select 10 bags, extract a number of pellets, and fracture and analyze them in a microscope for carbon black particle statistics. The weight fraction of carbon black is calculated from the number of particles, their size, and density, and it was found to vary as follows in the 10 bags:

| Bag # | Weight fraction of carbon black |
|-------|-------------------------------|
| 1 | 0.0198 |
| 2 | 0.0185 |
| 3 | 0.0202 |
| 4 | 0.0194 |
| 5 | 0.0211 |
| 6 | 0.0200 |
| 7 | 0.0204 |
| 8 | 0.0189 |
| 9 | 0.0197 |
| 10 | 0.0210 |

Estimate the ranges for the mean and the variance for 95% and 99% confidence.

**Solution**

The mean weight fraction of carbon black is 0.0199, and the standard deviation is 0.00083 (i.e., 4.17% of the mean value). For 95% confidence level and 9 degrees of freedom, $t_{0.975} = 2.26$, $\chi_{0.975} = \sqrt{19} = 4.36$ and $\chi_{0.025} = \sqrt{2.7} = 1.64$. Then Eqs. 6.10 and 6.11 yield:

$$0.01930 < \mu < 0.02050; \qquad 0.00013 < \sigma < 0.00093.$$

Thus, 95% of the time mixed PE will contain carbon black in concentration between 1.93% and 2.05%. Similarly, the standard deviation will vary in the range from 0.00013 to 0.00093. For 99% of the time the corresponding values are:

$$0.01904 < \mu < 0.02076; \qquad 0.00011 < \sigma < 0.00145.$$

The significance tests are interconnected with the concept of hypothesis testing and Type I and II errors. Type I error is made when we reject a hypothesis that should have been accepted, and Type II error is made when we accept a hypothesis that should have been rejected. The *level of significance* of testing a given hypothesis is the maximum probability that a Type I error might occur. In practice a level of significance of 5% or 1% are customary. Thus, if we accept a 5% level of significance, we are 95% confident that the right choice was made or, otherwise stated, there are about 5 out of 100 chances that a Type I error was made.

In polymer processing the significance tests are used to compare the samples with some reference material that might be perfectly mixed, and thus constitutes the goal of the mixing process, might have a certain desirable composition, or might be a material produced in a process considered for scaleup. In all those instances, the mean and the variance of the samples are compared with those of the reference material and a parameter is calculated. That parameter is then compared to tabulated values for the set level of significance. If the value of that parameter is within set limits, then the difference in values of the mean and the variance between the samples and the reference material is considered statistically nonsignificant.

In the significance tests for means, the mean and the variance of the samples are calculated using Eqs. 6.7 and 6.8, and the mean and the variance of the reference material are denoted by $\mu$ and $\sigma^2$, respectively. If the number of samples, $N$, is less than 30, then the Student's $t$ distribution is used for the means:

$$t = \frac{\bar{C} - \mu}{\sigma/\sqrt{N}}. \tag{6.12}$$

This value of $t$ is then compared to the tabulated value of $t$ for $N-1$ degrees of freedom and the set level of significance. The hypothesis that the sample mean is the same as the population mean $\mu$ is not rejected if the calculated $t$ is less than the tabulated value.

The significance test for variances is relevant to cases in which we have to decide whether two samples of sizes $N$ and $M$ and variances $s_N^2$ and $s_M^2$ do or do not come from the same mixture. The hypothesis that the two samples come from the same mixture is tested using the $F$ distribution:

$$F = \frac{s_N^2}{s_M^2}. \tag{6.13}$$

For example, to test the hypothesis at the 10% level, the following relationship should be used:

$$F_{0.05} \leqslant \frac{s_N^2}{s_M^2} \leqslant F_{0.95}, \tag{6.14}$$

where the $F$ values can be found in Spiegel (1991) for $N-1$ and $M-1$ degrees of freedom. Note that the following relationship holds: $F_{0.05} = 1/F_{0.95}$, where the first $F$ value has $M-1$ and $N-1$ degrees of freedom and the other one has $N-1$ and $M-1$ degrees of freedom.

## EXAMPLE 6.2
### Test of the Hypothesis

Test the hypothesis with level of significance 10% that the samples in Example 6.1 truly represent a carbon black/PE mixture with a mean carbon black weight fraction of 0.02 and a standard deviation 0.007 (i.e., 3.5% of the mean). Assume that the standard deviation of the mixture was calculated using 25 samples.

## Solution

To test the hypothesis for the mean we use Eq. 6.12, and thus the $t$ value is equal to $-0.381$. The tabulated value for $t_{0.95}$ and 9 degrees of freedom is 1.83. The hypothesis, with respect to means, is not rejected because $-1.83 < -0.381 < 1.83$. As far as the standard deviation is concerned, Eq. 6.14 applies. The calculated $F$ value is 1.41. The tabulated $F$ values are $F_{0.05} = 0.34$, for 24 and 9 degrees of freedom and $F_{0.95} = 2.30$ for 9 and 24 degrees of freedom. Thus, the hypothesis with respect to standard deviations is not rejected either.

---

These statistics are also useful for the *degree of mixing*, or *mixedness*, or *admixedness*, or *goodness of mixing*. It is obvious that the goal of the mixing process is to produce a mixture where the distribution of the minor component is statistically random. The term *perfect mixing* refers exactly to that state or, equivalently, to the state in which the probability of appearance of the minor component at any point in the mixture is constant. Therefore, the degree of mixing measures the "distance" between the mixing state of our sample and that of the statistically random sample, and it is represented mathematically by mixing indices.

There are more than 30 different criteria of the degree of mixedness (Fan et al., 1970), and they are tied to particular mixing situations. Because the mixing process is a random process, statistical analysis predominates in the calculations for the degree of mixing. More frequently, the variance of a spot sample taken from the mixture is compared to that of the perfectly random sample:

$$M = \frac{\sigma^2}{s^2}, \tag{6.15}$$

where $\sigma^2$ is the limiting minimum value of the variance of the perfectly random sample and $M$ is the degree of mixing.

The index of mixedness can also be expressed as:

$$M = \frac{\sigma}{s} \tag{6.16}$$

if the variances of Eq. 6.15 are substituted by the respective square roots, that is, by the standard deviations. Lacey (1954) introduced the variance of the totally unmixed state, $\sigma_0^2$, in his definition of the mixing index:

$$M = \frac{\sigma_0^2 - s^2}{\sigma_0^2 - \sigma^2}. \tag{6.17}$$

The advantage of using this definition as opposed to that of Eq. 6.15 lies in the range of the index of mixing. Equation 6.15 yields values between $1/n$ and $\infty$, whereas Eq. 6.17 attains values between 0 and 1. The same idea that transformed Eq. 6.15 to Eq. 6.16 can also be applied to Eq. 6.17.

The rate of mixing can be monitored by measuring the index of mixing at various periods. It is obvious that as time progresses (or as the number of revolutions for a rotating mixer increases), mixing is also improving and moving toward its random state. Reference to the "unmixing" or "demixing" that might take place during mixing will not be mentioned. Lacey (1954) developed the idea that all mixing mechanisms for solids, namely, convective, diffusional (see Problem 6B.1) and shear, are consistent with a first-order rate of reaction:

$$M = 1 - e^{-kt}, \tag{6.18}$$

where $k$ is the rate constant (unit of $k$ is s$^{-1}$). The constant $k$ can be calculated from experimental data of the index of mixing as a function of time, and it reflects the quality of the mixer and the suitability of the chosen conditions of mixing. It is not unusual for the variance to reach an asymptotic value different from the corresponding value for the random mixture, because large density differences cause stratification inside the mixer. In such cases, the definition of the index of mixing should be altered to accommodate this difference.

The statistical description of mixing given in this sub-section is restricted to a two-component system for simplicity. Of course, a four-component system can be considered as a two-component system if the second, third, and fourth components are considered as a single entity. If this is not an accurate description of the reality, then the use of Markov chains is inevitable, and the mathematics involved are complex. Such analysis goes beyond the scope of this textbook.

### 6.2.2. SCALE AND INTENSITY OF SEGREGATION

The texture of a sample is characterized by two parameters (Danckwerts, 1952): (1) *scale* and (2) *intensity* of segregation. They can be visualized as shown in Fig. 6.5. We assume that we have mixed two components, $A$, which is included in the circles, and $B$, which is lying in the interstices. The scale of segregation is the measure of the distance between clumps of the same component. It can also be defined as a measure of the size of undistributed parts of the components (Mohr et al., 1957) or as the length scale of the distance between similar interfaces. By following the latter definition we see that the distance between clumps or circles of component $A$ increases as we move toward the right in Fig. 6.5, that is, the scale of segregation increases. In general, mechanical energy, through deformation, is necessary for the reduction of the scale of segregation. Typical mechanisms include: shear, kneading, breakup, and turbulence.

The intensity of segregation refers to the difference in concentration of one component between areas of component $A$ and $B$. The intensity is reduced as component $A$ diffuses out of the circles into the area of component $B$ or as the size of the clumps of $A$ is reduced because of the diffusion of $A$. If the two components do not have any difference in color, the intensity of segregation can be pictured as the difference in the size of clumps $A$ (Fig. 6.5). If there is a difference in color between $A$ (black) and $B$ (white), then the minimum intensity of segregation is achieved when the black has been spread into the white, producing a grey color for the sample. Clearly, as Mohr et al. (1957) pointed out, the reduction in intensity is only affected by diffusion, not by mechanical

mixing. Thus, although diffusion cannot change the shape of the clumps (deformation can do that), it can produce gray color when black and white meet. Note that deformation changes spherical clumps into elongated shapes, breaks them or changes their patterns, but it never causes any diffusional process to take place.

Both reduction of scale and intensity produce uniformity, and so the quality of mixing, as far as texture is concerned, depends on both quantities. Note that the maximum reduction of the scale is achieved when the distance between interfaces approaches the length scale of the ultimate particles (or ultimately molecular scales), and the maximum reduction in intensity is obtained when the minor component is uniformly mixed throughout the major component.

Scale and intensity also have statistical representations. The intensity can be defined mathematically as:

$$I = \frac{s^2}{\sigma_0^2} \tag{6.19}$$

and expresses the difference of concentration of the minor component from the average value normalized by the variance of the completely segregated (totally unmixed; Eq. 6.17) system. Its values range from 1 for a completely unmixed system to 0 for a uniformly mixed system. This definition bears resemblance to the definition of the indices of mixing. Usually the intensity for polymer systems, even in the melt state, is close to 1, because diffusion is extremely slow.

To define the scale of segregation we need to define first the correlation function, $R(\mathbf{r})$. It is very similar to variance in the sense that variance measures the concentration difference of a point and the mean value, whereas the correlation function considers the concentration difference between two points $\mathbf{r}$ distance away. Thus, $R(\mathbf{r})$ is given by:

$$R(\mathbf{r}) = \frac{1}{N} \sum_{i=1}^{N} [(C_i(\mathbf{x}) - \bar{C})(C_i(\mathbf{x}+\mathbf{r}) - \bar{C})]. \tag{6.20}$$

where $C_i$ is the concentration of the $i$-th sample, $\bar{C}$ is the mean concentration, and $N$ is the total number of samples. When $\mathbf{r}$ equals $\mathbf{0}$, then the correlation function becomes approximately (for large $N$) equal to the variance:

$$R(\mathbf{0}) = \frac{N-1}{N} s^2 \simeq s^2. \tag{6.21}$$

The correlation coefficient is obtained by the normalization of the correlation function with the variance:

$$\rho(\mathbf{r}) = \frac{R(\mathbf{r})}{s^2}, \tag{6.22}$$

so that $\rho(\mathbf{0})$ is equal to unity for a sufficiently large $N$. The graph of the correlation coefficient is called a *correlogram*. In isotropic mixtures we can neglect the vector form of $r$ and substitute it with its scalar form. By doing this we presuppose that $\rho(r)$ is the average of $\rho(\mathbf{r})$ over all possible orientations of the vector $\mathbf{r}$.

The values of the correlation coefficient range from $-1$ to $1$. When the origin and the end of the vector $\mathbf{r}$ lie in different components, the correlation coefficient attains a value of $-1$, and the correlation is considered to be perfectly negative. The correlation is considered perfectly positive when both the origin and the end of the $\mathbf{r}$ vector lie in the same component (either minor or major), and consequently the correlation coefficient equals unity. Finally, the correlation coefficient equals zero when the correlation is random or, equivalently, when knowledge of the composition at the origin provides no information about the composition at the end of the vector.

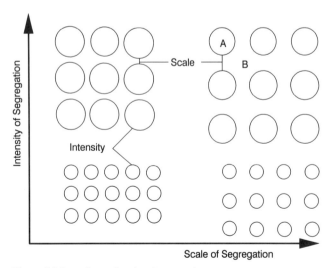

**Figure 6.5** Intensity and scale of segregation.

The correlation coefficient is calculated by the technique of *dipole* or *needle throwing*. Consider a mixture of carbon black in PE with $\bar{C}$ average carbon black concentration. We fracture one pellet, and we take a photograph of the surface with the aid of a high-resolution scanning electron microscope. We take a needle of length $r$ and drop it onto the photograph. Four events can take place: (1) both ends could land in carbon black particles with probability $P_{11}$; (2) both ends could land in the PE matrix with probability $P_{22}$; (3) the first end lands in carbon black and the second in PE with probability $P_{12}$; and (4) the

opposite of the previous event with probability $P_{21}$. Nadav and Tadmor (1973) calculated the correlation coefficient as:

$$\rho(r) = \frac{1-\bar{C}}{\bar{C}} P_{11} + \frac{\bar{C}}{1-\bar{C}} P_{22} - P_{12} - P_{21}. \tag{6.23}$$

A typical example of the application of that formula is given in Example 6.3. Other examples and the derivation of Eq. 6.23 are given in Nadav and Tadmor (1973), Tadmor and Gogos (1979), and Tucker (1981).

---

**EXAMPLE 6.3**
**Calculation of the Correlation Coefficient**
Consider a mixture of carbon black in PE, and assume that the mixture can be represented as a set of unit cells of radius $A$. A carbon black particle of radius $a$ is at the center of the cell, and it is surrounded by PE matrix. The unit cell is shown in Fig. 6.6.a. Approximate the correlation coefficient for this mixture based on Eq. 6.23 for a small $r$ compared to either $a$ or $A$ (i.e., ignore terms of order $r^2$ and higher).

**Solution**

Divide the unit cell into four regions: Region I is inside the particle with radius $a-r$, region II is a spherical shell with outside radius $a$ and inside $a-r$, region III is a spherical shell in the PE matrix with inside radius $a$ and outside radius $a+r$, and finally region IV has inside radius $a+r$ and outside radius $A$. The probabilities of the various ways that a needle of length $r$ might land on this mixture can be calculated. Note that although the cell is pictured in its two-dimensional form in Fig. 6.6.a, the solution will be based on the real three-dimensional picture.

The probability that both needle ends will fall in carbon black depends on the probability that the first end will fall in regions I and II in combination with the probability that the second end will fall in carbon black regions. The probability that the first end will fall in region I is the ratio of the region's volume to the total cell volume:

$$P(I) = \frac{V_I}{V_{\text{cell}}} = \frac{(a-r)^3}{A^3} \simeq \bar{C} - 3\bar{C}\frac{r}{a}. \tag{6.24}$$

The probability that its second end will fall in carbon black is 1. The probability that the first end will fall in region II is similarly equal to the ratio of volumes:

$$P(II) = \frac{V_{II}}{V_{\text{cell}}} = \frac{a^3 - (a-r)^3}{A^3} \simeq 3\bar{C}\frac{r}{a}. \tag{6.25}$$

Now the conditional probability that the second end falls in carbon black region is 3/4. This is calculated by drawing a sphere of radius $r$ centered at a radial distance $x$ from the cell center. That probability is then calculated as the ratio of the area of the sphere inside the carbon black to the total area of the sphere averaged over all possible $x$ (i.e., from $r-a$ to $a$).

The probability that both ends fall in carbon black is:

$$P_{11} = P(I) \times 1 + P(II) \times (3/4) \simeq \bar{C} - \frac{3}{4}\bar{C}\frac{r}{a}. \tag{6.26}$$

The other terms in Eq. 6.23 can be calculated similarly. The correlation coefficient is thus calculated as:

$$\rho(r) \simeq 1 - \frac{3}{4}\frac{r}{a}\frac{1}{1-\bar{C}}. \tag{6.27}$$

The correlogram for various values of $\bar{C}$ is shown in Fig. 6.6.b. It is linear and it intersects the $x$ axis at $4a(1-\bar{C})/3$.

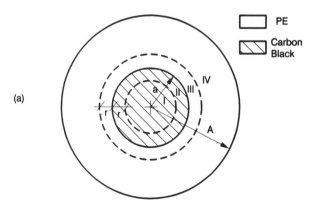

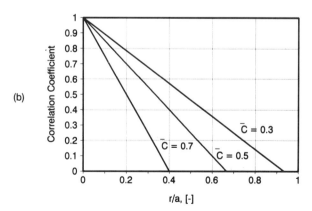

**Figure 6.6** (a) Unit cell of a carbon black particle in a PE matrix. (b) Correlation coefficient as a function of the dimensionless radial distance.

The linear scale of segregation can now be defined in terms of the correlation coefficient:

$$S_L = \int_0^\infty \rho(r)dr, \tag{6.28}$$

and it represents the area under the correlation coefficient curve. Similarly, the volumetric scale of segregation is defined as:

$$S_V = 2\pi \int_0^\infty \rho(r)r^2 dr. \tag{6.29}$$

In the example of a clumpy mixture, the linear scale represents the average size, and the volumetric scale represents the average volume of the clumps. Note that for the same example there is no difficulty in evaluating the integral, because the correlation coefficient becomes zero above some value of $r$. Typical values of the ratio of the linear scale of segregation to the length scale of the structure are: (1) 0.237 for a planar checkered board pattern; (2) 0.250 for an alternating layered structure (one-dimensional) of equal layer thickness; and (3) 0.380 for a collection of spheres.

An alternative description of the texture is its spectral description, or power spectrum, $P(n)$ (Tucker, 1991):

$$P(n) = \int_{-\infty}^\infty R(r)e^{-2\pi nr}dr. \tag{6.30}$$

The importance of the power spectrum stems from the fact that it can be calculated directly from the concentration field and that it can be easily inverted into the correlation function via the fast Fourier transform (FFT) method.

### 6.2.3. MIXING MEASUREMENT TECHNIQUES

The goal of mixing measurement techniques is the acquisition and statistical analysis of data collected from samples to evaluate the quality of the process and of the final product. These techniques are usually time-consuming and laborious. One common problem in measurement techniques that examine a planar section of the product is that two-dimensional information should be transformed into three-dimensional information. This is done with the help of *stereology science* (Underwood, 1977).

The measurement of the variance indicates the quality of the mixing. If the mixture is homogeneous, the variance is small and the average composition is close to $\bar{C}$. To measure the variance we need to withdraw samples from the mixing device or cut samples from the final product. Extraction of samples with a hypodermic needle (Rotz & Suh, 1976), sampling by pumping the mixture through a special sampling device (Tucker & Suh, 1980a), and slicing of the solidified mixture (Hall & Godfrey, 1965) are reported in the literature. For the measurement technique, light transmittance (Nadav & Tadmor, 1973), electrical conductivity (Tucker & Suh, 1980b), titration, and particle counting are the most extensively used techniques.

Lately, computer analysis is used broadly for the evaluation of the state of mixing. Sectioning, acquisition, and analysis are the three steps involved. The acquisition of the digital image is done with a television camera, a video digitizer, and a computer. The camera is attached to the microscope, and its analog picture is translated into digital via the video digitizer. The digital intensity values are now stored into large arrays in a computer. That large array of numbers contains the light intensity of each pixel in terms of its red, green, and blue components. Typical image analysis systems can digitize the screen into $512 \times 512$ or $1024 \times 1024$ pixels. Howland and Erwin (1983) used the Laser Line Scanning technique, which consists of illuminating the picture with a laser light source and measuring the reflectivity of the material with a photodiode.

The analysis of the image requires the knowledge of the correspondence between composition and color (if that is not linear, a calibration curve is required). Furthermore, if diffusion has not altered the composition at the various points, then each pixel is made dark (e.g., minor component) or light (e.g., major component) based on a *threshold* value. The power spectrum is now calculated from the composition data and an FFT algorithm. The correlation function is then calculated by using the inverse Fourier transform of the power spectrum, and the scale and intensity of segregation are then computed using the equations of the preceding subsection. It should be mentioned at this point that the information collected with the image analysis system is limited to distances smaller than the height and the width of the screen and that the resolution is bound by the size of the pixel. However, the great advantage of these systems is their ability to analyze many images fast and with minimum labor.

### 6.3. STRIATION THICKNESS AND LAMINAR MIXING

Mixing of polymer melts cannot be assisted by either diffusion or turbulence. The absence of diffusion makes the two components and their interface easily identifiable. Consequently, the interfacial area per unit volume (or intermaterial area density or specific interfacial area), $A_v$, can be used as a measurement of the extent of mixing, as was suggested by Spencer and Wiley (1951) as well as others. A special case of a polymer mixture is the layered or lamellar mixture shown in Fig. 6.7. This mixture can be produced by either laminar flow (e.g., in a typical single-screw extruder) or ordered rearrangement with plug flow (e.g., motionless mixers, Fig. 6.2). Using the terminology of Section 6.2.1 the lamellar structure comprises the texture of the mixture. The scale of segregation of an equal-thickness alternating structure was mentioned to be one fourth of the thickness, but the scale is not important in this context, because it is a statistical measure and the layered structure is ordered. Of course, the scale of segregation is important for polymer systems with a wide distribution of layer thickness.

The striation thickness concept was first introduced by Mohr et al. (1957) and analyzed extensively by Ranz (1979) and Ottino et al. (1981 and 1979). The striation thickness, $\delta$, is defined as one-half (Ottino et al., 1979) of the thickness of the repeating unit (i.e., one-half of the sum of the thickness of two adjacent layers of components $A$ and $B$).

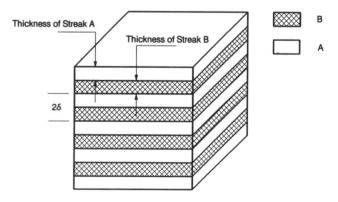

**Figure 6.7** Layered structure and definition of striation thickness $\delta$.

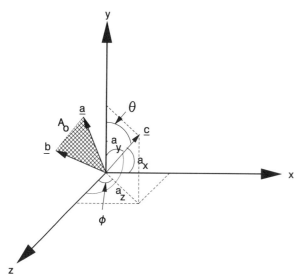

**Figure 6.8** Interfacial element $A_0$ in a Cartesian coordinate system at time $t = 0$.

Note that in some references the striation thickness is defined as the sum of the thickness of two adjacent layers, and thus their striation thickness is twice that we use here. The interfacial area per unit volume, $A_v$, and the striation thickness, $\delta$, are related as follows for a large number of striations:

$$A_v = \frac{1}{\delta}. \tag{6.31}$$

As mixing progresses, the interfacial area per unit volume increases and the striation thickness decreases. If there is a distribution of striation thickness, then not only the mean but also the variance should be taken into consideration for the quality of mixing calculations. In the next subsection the interfacial area growth, or equivalently the striation thickness reduction, is calculated from geometrical arguments.

### 6.3.1. STRIATION THICKNESS REDUCTION FROM GEOMETRICAL ARGUMENTS

The interfacial area per unit volume changes as a function of the strain applied to the system. Consider an interface $A_0$ as shown in Fig. 6.8. This interface is formed by vectors **a** and **b**:

$$\mathbf{a} = x_a \boldsymbol{\delta}_x + y_a \boldsymbol{\delta}_y + z_a \boldsymbol{\delta}_z \tag{6.32}$$

and

$$\mathbf{b} = x_b \boldsymbol{\delta}_x + y_b \boldsymbol{\delta}_y + z_b \boldsymbol{\delta}_z, \tag{6.33}$$

where $\boldsymbol{\delta}_x$, $\boldsymbol{\delta}_y$ and $\boldsymbol{\delta}_z$ are the unit vectors along the $x$, $y$, and $z$ axes, respectively, and $x_a$, $y_a$, $z_a$, $x_b$, $y_b$, and $z_b$ are the components of **a** and **b**. The vector **c** is perpendicular to the plane $A_0$ and equal to:

$$\mathbf{c} = \mathbf{a} \times \mathbf{b}. \tag{6.34}$$

The magnitude of **c** is twice the area of $A_0$:

$$|\mathbf{c}| = 2A_0, \tag{6.35}$$

and it forms $a_x$, $a_y$, and $a_z$ angles with the coordinate axes $x$, $y$, and $z$, respectively.

The area $A_0$ is now subjected to strain with principal directions in the Cartesian coordinates and with principal elongation ratios $\lambda_x$, $\lambda_y$, and $\lambda_z$ (Erwin, 1978a). The area $A_0$ will become $A$, and the vectors defining $A$ are $\mathbf{a}'$, $\mathbf{b}'$ and $\mathbf{c}'$, so that:

$$|\mathbf{a}' \times \mathbf{b}'| = |\mathbf{c}'| = 2A. \tag{6.36}$$

The new vectors $\mathbf{a}'$ and $\mathbf{b}'$ are:

$$\mathbf{a}' = (x_a \lambda_x)\boldsymbol{\delta}_x + (y_a \lambda_y)\boldsymbol{\delta}_y + (z_a \lambda_z)\boldsymbol{\delta}_z$$

$$\mathbf{b}' = (x_b \lambda_x)\boldsymbol{\delta}_x + (y_b \lambda_y)\boldsymbol{\delta}_y + (z_b \lambda_z)\boldsymbol{\delta}_z. \tag{6.37}$$

Substitution of these expressions into Eq. 6.36 yields:

$$A = \frac{1}{2}\{[\lambda_y \lambda_z (y_a z_b - z_a y_b)]^2 + [\lambda_z \lambda_x (z_a x_b - x_a z_b)]^2 +$$
$$[\lambda_x \lambda_y (x_a y_b - y_a x_b)]^2\}^{1/2}. \tag{6.38}$$

This expression can be equivalently written as:

$$A = \frac{1}{2}\sqrt{(\lambda_y \lambda_z x_c)^2 + (\lambda_z \lambda_x y_c)^2 + (\lambda_x \lambda_y z_c)^2}. \tag{6.39}$$

Our primary interest is to calculate the change of the interface $A_0$, and thus we form the ratio of the interface after mixing has started to the initial interface (at time $t = 0$), which is also called the "mixing number" frequently in the literature:

$$\left[\frac{A}{A_0}\right]^2 = \frac{(\lambda_y \lambda_z x_c)^2 + (\lambda_z \lambda_x y_c)^2 + (\lambda_x \lambda_y z_c)^2}{|\mathbf{c}|^2}. \tag{6.40}$$

The angles $a_x$, $a_y$, and $a_z$ can be expressed in terms of the components of the vector **c** as:

$$\cos a_x = \frac{x_c}{|\mathbf{c}|}; \qquad \cos a_y = \frac{y_c}{|\mathbf{c}|}; \qquad \cos a_z = \frac{z_c}{|\mathbf{c}|}, \tag{6.41}$$

and they follow the relationship of the directional cosines, that is:

$$\cos^2 a_x + \cos^2 a_y + \cos^2 a_z = 1. \tag{6.42}$$

Eqs. 6.40 and 6.41 can be combined to give:

$$\left[\frac{A}{A_0}\right]^2 = (\lambda_y \lambda_z \cos a_x)^2 + (\lambda_z \lambda_x \cos a_y)^2 + (\lambda_x \lambda_y \cos a_z)^2. \tag{6.43}$$

If the deformation is considered to preserve the volume of the system (incompressible materials), then:

$$\lambda_x \lambda_y \lambda_z = 1. \tag{6.44}$$

Substituting this equation into Eq. 6.43 yields:

$$\left[\frac{A}{A_0}\right]^2 = \frac{\cos^2 a_x}{\lambda_x^2} + \frac{\cos^2 a_y}{\lambda_y^2} + \frac{\cos^2 a_z}{\lambda_z^2}. \tag{6.45}$$

This is similar to the expression developed by Spencer and Wiley (1951) for unidirectional shear strain. Another form of this expression can be obtained by using the equation of directional cosines, Eqs. 6.42, and 6.43:

$$\left[\frac{A}{A_0}\right]^2 = (\lambda_x \lambda_y)^2 + \lambda_y^2(\lambda_z^2 - \lambda_x^2)\cos^2 a_x + \lambda_x^2(\lambda_z^2 - \lambda_y^2)\cos^2 a_y. \quad (6.46)$$

Next, we apply Eq. 6.46 to simple deformations to show the effect of various types of deformation on mixing.

### 6.3.1.1. Planar Elongation (Pure Shear).

Suppose that we stretch a material surface in the $x$ direction while we constrain it in the $z$ direction. The resulting flow is called planar elongational flow (plane extensional flow; plane strain; see also Subsection 3.1.1). The polymer fluid experiences such flow in the entrance regions to slit dies and nips between rollers. The elongation ratios are:

$$\lambda_x = \lambda_0; \qquad \lambda_y = \frac{1}{\lambda_0}; \qquad \lambda_z = 1. \quad (6.47)$$

Equation 6.46 becomes:

$$\left[\frac{A}{A_0}\right]^2 = 1 + \left(\frac{1}{\lambda_0^2} - 1\right)\cos^2 a_x + (\lambda_0^2 - 1)\cos^2 a_y. \quad (6.48)$$

For large elongation ($\lambda_0 \gg 1$) and for $\cos a_x \neq 0$,

$$\frac{A}{A_0} = \lambda_0 |\cos \alpha_y|. \quad (6.49)$$

The maximum value of the interface growth is achieved when the interface is normal to the minimum principal value that is, $\mathbf{c}$ is in the direction of $y$ ($\cos a_y = 1$). The maximum value is:

$$\left[\frac{A}{A_0}\right]_{\max} = \lambda_0. \quad (6.50)$$

If the orientation of the material interfaces at the entrance of the mixer is random, then the effect of planar elongation on the interfacial area is determined by averaging Eq. 6.49 over all possible orientations. To obtain that, we should evaluate the angles $a_x$ and $a_y$ in terms of spherical coordinates with angles $\phi$ and $\theta$ (see Fig. 6.8):

$$\cos a_x = \sin \phi \sin a_y; \qquad a_y = \theta. \quad (6.51)$$

The average is now obtained by integration:

$$\left[\frac{A}{A_0}\right]_{avg} = \frac{1}{4}\pi$$

$$\int_0^{2\pi}\int_0^{\pi} \sqrt{1 + \sin^2\theta \sin^2\phi\left(\frac{1}{\lambda_0^2} - 1\right) + \cos^2\theta(\lambda_0^2 - 1)}\,\sin\theta\,d\theta\,d\phi. \quad (6.52)$$

This expression is evaluated numerically, and for large values of $\lambda_0$ it follows that:

$$\left[\frac{A}{A_0}\right]_{avg} = \frac{\lambda_0}{2}. \quad (6.53)$$

This means that a mixer that imposes planar elongation on initially randomly oriented fluid interfaces increases the initial interfacial area by a factor of $(\lambda_0/2) - 1$ or, equivalently, decreases the striation thickness by a factor of $(\lambda_0/2) - 1$. If the initial orientation is normal to the minimum principal elongation value, the striation thickness decreases by a factor of $\lambda_0 - 1$.

The strain, $\varepsilon$ (equal to $\dot{\varepsilon}t$, where $\dot{\varepsilon}$ is the strain rate), imposed on the material is related to the increase of linear distance in the $x$ direction by:

$$\frac{dl}{l} = \varepsilon, \quad (6.54)$$

which can be integrated to give: $x = x_0 \exp(\varepsilon)$. The elongation ratio in the $x$ direction, $\lambda_0$, can then be related to strain, $\varepsilon$, by:

$$\lambda_0 = e^{\varepsilon} = e^{\dot{\varepsilon}t}. \quad (6.55)$$

Equation 6.50 can be expressed now in terms of the total strain $\varepsilon$:

$$\left[\frac{A_0}{A}\right]_{\max} = e^{\varepsilon} \quad (6.56)$$

that shows that the maximum interface growth depends on the exponential of the total applied elongation strain. Similarly, the average interface growth equals one-half the exponential of the applied strain.

### 6.3.1.2. Uniaxial (or Pure) Elongation.

Suppose that we stretch an interface in the $x$ direction without imposing any constraints on the $y$ and $z$ directions. The resulting flow is called a *uniaxial elongational flow* (uniaxial extensional flow; a classic example of a mixer creating pure elongational stretch is a taffy puller; see also Subsection 3.1.1). The elongation ratios are now:

$$\lambda_x = \lambda_0; \qquad \lambda_y = \lambda_z = \frac{1}{\sqrt{\lambda_0}}. \quad (6.57)$$

Substitution of these ratios into Eq. 6.46 gives

$$\left[\frac{A}{A_0}\right]^2 = \lambda_0 + \left(\frac{1}{\lambda_0^2} - \lambda_0\right)\cos^2 a_x. \quad (6.58)$$

If the interface is aligned with the $y$ axis, then $\cos a_x = 0$, and thus the maximum value of the interface growth is:

$$\left[\frac{A}{A_0}\right]_{\max} = \sqrt{\lambda_0}. \quad (6.59)$$

Equation 6.58 can be integrated numerically over all possible orientations to give for a large $\lambda_0$:

$$\left[\frac{A}{A_0}\right]_{avg} \simeq \frac{4}{5}\sqrt{\lambda_0}. \quad (6.60)$$

Because this process is not symmetrical about $\lambda_0 = 1$, for deformations $\lambda_0 \ll 1$ it is found that:

$$\left[\frac{A}{A_0}\right]_{avg} \simeq \frac{4}{5}\frac{1}{\lambda_0}. \quad (6.61)$$

In terms of the total applied strain, $\varepsilon$, the arguments are the same as those developed for planar elongation. Thus, the maximum and the average interface growth depend on $\exp(\varepsilon/2)$ for very large values of $\lambda_0$.

### 6.3.1.3. Simple Shear.

The most frequently encountered flow situation in mixing is that of shear. It can be imparted to the fluid by Couette or Poiseuille flow. In principle we can follow the same ideas developed in the preceding two cases, but now the principal axes change as a function of shear, and the analysis is complicated. To simplify the analysis we define the interface $A_0$ as in Fig. 6.9.a, so that **a** lies in the $x$-$y$ plane and **b** lies in the $x$-$z$ plane (McKelvey, 1962). The shear direction is along the $x$ axis so that **b** does not change. A shear strain $\gamma$ is applied.

The new vector, **a**$'$, which defines the interface $A$, can be expressed as (Fig. 6.9.b):

$$\mathbf{a}' = (x_a + \gamma y_a)\boldsymbol{\delta}_x + (y_a)\boldsymbol{\delta}_y, \tag{6.62}$$

so that

$$A = \frac{1}{2}\sqrt{(x_c)^2 + (y_c - \gamma x_c)^2 + (z_c)^2}. \tag{6.63}$$

The interface growth function, $f(\gamma) = A/A_0$, becomes:

$$\left[\frac{A}{A_0}\right]^2 = \frac{(x_c)^2 + (y_c - \gamma x_c)^2 + (z_c)^2}{|\mathbf{c}|^2}. \tag{6.64}$$

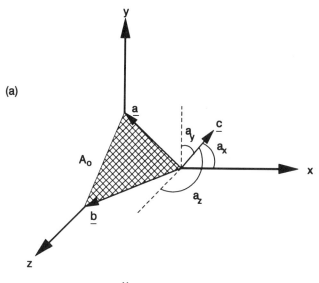

**(a)**

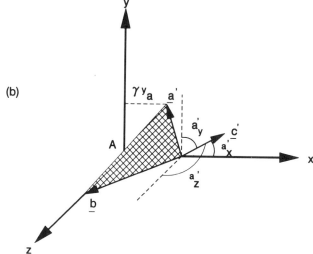

**(b)**

**Figure 6.9** Interfacial element in a shearing experiment. (a) at $t = 0$. (b) After a shear strain, $\gamma$, has been applied to the interface.

Using the expressions of Eq. 6.41 for the directional cosines, we get

$$\left[\frac{A}{A_0}\right]^2 = 1 - 2\gamma \cos a_x \cos a_y + \gamma^2 \cos^2 a_x, \tag{6.65}$$

which is exactly the expression proposed by Spencer and Wiley (1951). Equation 6.65 shows that the interfacial area growth depends on its initial orientation and on the magnitude of the shear strain imposed by the mixer. When $a_x = 90°$, vector **c** lies in a plane parallel to the $y$-$z$ plane and shear strain distorts the interface without changing its area. When $a_y = 90°$, vector **c** lies in the $x$-$z$ plane and shear strain distorts the interface and changes its area. In the latter case the maximum change in area occurs when the interface lies in the $y$-$z$ plane.

For large values of $\gamma$, Equation. 6.65 reduces to:

$$f(\gamma) = \left[\frac{A}{A_0}\right] = \gamma \cos a_x. \tag{6.66}$$

This equation indicates that the interfacial area growth function in a material undergoing large shear proceeds linearly with shear strain. Numerical integration of the equation over all possible orientations following the steps described for elongational strain yields:

$$f(\gamma) = \frac{\gamma}{2}. \tag{6.67}$$

Equation 6.65 can be used to estimate the optimal interface orientation for a maximum in the interface growth function. Differentiation of that expression with respect to $\gamma$, at $\gamma = 0$, yields

$$\frac{dA}{d\gamma} = -A \cos a_x \cos a_y, \tag{6.68}$$

where $A = A_0$. The maximum increase of the interfacial area is achieved at the minimum of the product of the directional cosines $\cos a_x$ and $\cos a_y$ (note the negative sign in Eq. 6.68). The minimum of the product occurs at $a_x = 135°$ and $a_y = 45°$, and it is equal to $-1/2$, so that

$$\left[\frac{A}{A_0}\right]_{max} = e^{\gamma/2}. \tag{6.69}$$

To achieve this maximum interface growth we should orient the interface so that its normal and the $y$ axis form an angle of 45° and its normal and the $x$ axis form an angle of 135°. Then, we shear the interface stepwise with step shear strain equal to a small fraction of the total shear strain $\gamma$ and we reorient the interface before we proceed to the next step. This procedure is shown in the following example.

In Table 6.1 we summarize the values of the average (integrated over all possible orientations) and maximum interface growth function $f(\gamma$ or $\varepsilon)$ for planar and uniaxial elongation as well as for simple shear cases. It is clear that in terms of interface growth (or equivalently striation thickness), elongational flows are much more efficient compared to shear flows. For example, a reduction in striation thickness of three orders of magnitude (from a 3-mm polymer pellet down to 3-μm striations) is achieved by either a total shear strain of 2000 or a total elongation strain of about 8 for random initial orientation. This reduction in striation thickness can be obtained in 20 s in a shear field at an average shear rate of $100\,\text{s}^{-1}$ or in 80 ms in an elongation flow field having an average rate of $100\,\text{s}^{-1}$.

The three strain fields, planar elongation, uniaxial elongation, and simple shear are compared next to determine the flow pattern most

**EXAMPLE 6.4**
**Interface Growth Function with Reorientation for Simple Shear Flow**

Compare the interface growth function for the cases: (**a**) optimum initial orientation and one-step shear strain, and (**b**) initial orientation $a_y = 45°$ and $a_x = 135°$, $N$ steps and reorientation after each step. The total shear strain is 10. Consider 5 and 20 steps for case (**b**).

**Solution**

The optimum initial orientation in case (**a**) is: $a_x = 0°$ and $a_y = 90°$. Equation 6.65 yields:

$$\left[\frac{A}{A_0}\right]_a = \sqrt{1 + \gamma^2}. \tag{6.70}$$

For $\gamma = 10$, the interface growth function is equal to 10.05 in case (**a**). For case (**b**), $\cos a_y = -\cos a_x = 0.707$, and the total shear strain is subdivided into $N$ steps so that in every step the applied strain is $\gamma/N$. Thus, the interface growth function at the end of step $j$ is calculated

from Eq. 6.65 as:

$$\left[\frac{A_j}{A_{0j-1}}\right]_b = \sqrt{1 + 2\frac{\gamma}{N} + \frac{1}{2}\left[\frac{\gamma}{N}\right]^2}. \tag{6.71}$$

At the end of that process the total interface growth function is:

$$\left[\frac{A}{A_0}\right]_b = \left[1 + 2\frac{\gamma}{N} + \frac{1}{2}\left[\frac{\gamma}{N}\right]^2\right]^{N/2}. \tag{6.72}$$

For $N = 10$ and 20 the growth function is equal to 97.66 and 128.39, respectively. Note that its maximum is $e^{10/2} = 148.41$, which is about 14 times larger than case **a**. If we increase the number of steps, the growth function will approach the maximum value, because at a large $N$ the following relationship (keeping only terms of order $(\gamma/N)$ and larger) holds:

$$\left(1 + \frac{\gamma}{N}\right)^{N/2} = \left(1 + \frac{\gamma/2}{N/2}\right)^{N/2} \rightarrow e^{\gamma/2}. \tag{6.73}$$

---

**TABLE 6.1 Interface Growth Function for Various Types of Strain**

| Type of Strain | Interface Growth Function $A/A_0$ | |
|---|---|---|
| | Maximum | Average |
| Planar elongation | $\exp(\varepsilon)$ | $(1/2)\exp(\varepsilon)$ |
| Uniaxial elongation | $\exp(\varepsilon/2)$ | $(4/5)\exp(\varepsilon/2)$, for $\lambda_0 \gg 1$ |
| | | $(4/5)\exp(-\varepsilon)$, for $\lambda_0 \ll 1$ |
| Simple shear | $\gamma$ | $\gamma/2$ |

favorable in terms of power consumption. High-power consumption is unfavorable because of its high cost and requirements for specific equipment. Integration of the mechanical energy equation (Section 2.3) over the volume of the mixer shows that the specific power consumption (power consumption per unit volume), $P_v$, is equal to the viscous dissipation:

$$P_v = -\frac{1}{2}\sum_i\sum_j \tau_{ij}\dot{\gamma}_{ji}, \tag{6.74}$$

where $i, j = x, y$, and $z$; $\tau_{ij}$ is the $ij$ component of the extra stress tensor, $\tau$ and $\dot{\gamma}_{ji}$ is the $ji$ component of the rate-of-strain tensor, $\dot{\gamma}$. Combining Eqs. 6.74, 2.59, and the components of the rate-of-strain tensor from Table 2.10 we get the following expression for the specific power of a Newtonian fluid:

$$\frac{P_v}{\mu} = 2\left[\left(\frac{\partial v_x}{\partial x}\right)^2 + \left(\frac{\partial v_y}{\partial y}\right)^2 + \left(\frac{\partial v_z}{\partial z}\right)^2\right] + \left(\frac{\partial v_x}{\partial y} + \frac{\partial v_y}{\partial x}\right)^2$$
$$+ \left(\frac{\partial v_y}{\partial z} + \frac{\partial v_z}{\partial y}\right)^2 + \left(\frac{\partial v_x}{\partial z} + \frac{\partial v_z}{\partial x}\right)^2, \tag{6.75}$$

where $\mu$ is the fluid viscosity and $v_x$, $v_y$, and $v_z$ are the three velocity components.

The above expression can be applied to idealized systems with three types of deformation: planar (plel) and uniaxial (unel) extensional and simple shear (ss) mixers. For a mixing device dominated by planar

elongation flow, the distances along the $x$ axis are related to strain or strain rate by:

$$x = x_0 e^{\varepsilon} = x_0 e^{\dot{\varepsilon}t}. \tag{6.76}$$

This expression, along with the expressions for the distances along the other axes, on differentiation yield the following velocity profile:

$$v_x = \frac{dx}{dt} = \dot{\varepsilon}x; \qquad v_y = -\dot{\varepsilon}y; \qquad v_z = 0. \tag{6.77}$$

This velocity profile is the same as that of Eq. 3.2 if you account for the difference in axes. The specific power, $P_v$, is calculated by taking appropriate derivatives of the velocity field and substituting them into the power expression, (Eq. 6.75):

$$(P_v)_{plel} = 4\mu\dot{\varepsilon}^2. \tag{6.78}$$

For uniaxial elongational flow we find that:

$$(P_v)_{unel} = 3\mu\dot{\varepsilon}^2. \tag{6.79}$$

For simple shear flow we find that:

$$(P_v)_{ss} = \mu\dot{\gamma}^2. \tag{6.80}$$

We can now make an assessment of the efficiency of the three types of flows to mix materials. Suppose we need to prepare a polymer blend with striation thickness of the minor component in the final product of not more than $3\,\mu m$. The pellets of the minor component fed into the flowing major component have a characteristic length of $3\,mm$. If we allow the components to stay in the flow field no more than $10\,s$, the ratio of the specific power for simple shear flow to that of the uniaxial elongational flow and to that of the planar elongational flow is:

$$(P_v)_{ss} : (P_v)_{unel} : (P_v)_{plel} \simeq 17,309 : 2.64 : 1. \tag{6.81}$$

Hence, shear flow is significantly less efficient than shear-free extensional flows in mixing. Similar results for a power-law fluid can be obtained (see Pr. 6A.10).

## 6.3.2. STRIATION THICKNESS REDUCTION FROM KINEMATICAL ARGUMENTS

A more systematic approach for the calculation of the striation thickness and mixing efficiency was presented by Ottino and co-workers (1979 and 1981). They developed the mathematical formulation for the calculation of the lineal and areal stretch of a material line and area, respectively, subjected to any deformation gradient field. In this text we use the formulation for the lineal stretch because it is easier to apply and correlates to the areal stretch by a simple relation. We should emphasize at this point that the formulation is valid in the absence of interfacial forces between minor and major components. This is true for systems with negligible interfacial tension forces, such as for systems with either negligible interfacial tension or large-length scale or both.

Suppose that a polymer melt is processed in a mixer that deforms the "particles" of the melt. At time $t = 0$ we identify a differential material line at position $\mathbf{x}_0$ of length $|d\mathbf{x}_0|$ and orientation $\mathbf{m}_0$. After a deformation (e.g,. shear, elongation) is applied to the melt by the mixer, we identify the same line with different position, $\mathbf{x}$, length, $|d\mathbf{x}|$, and orientation, $\mathbf{m}$, (Fig. 6.10.a). The lineal stretch, $\lambda$, is then defined as:

$$\lambda \equiv \frac{|d\mathbf{x}|}{|d\mathbf{x}_0|}. \tag{6.82}$$

The deformation applied is characterized by the deformation-gradient tensor, $\mathbf{F}$, with components (for a rectangular coordinate system) equal to:

$$F_{ij} = \sum_i \sum_j \frac{\partial x_i}{\partial x_{0j}}. \tag{6.83}$$

The transpose of the deformation-gradient tensor, $\mathbf{F}^T$, is then given by the components

$$(F^T)_{ij} = \sum_i \sum_j \frac{\partial x_j}{\partial x_{0i}}. \tag{6.84}$$

In these two equations as well as in the rest of equations in this section subscripts 1, 2, 3 indicate the $x$, $y$, and $z$ directions, respectively. The deformation tensor and its transpose can be combined to yield the right relative Cauchy-Green strain tensor, $\mathbf{C}$, with components

$$C_{il} = \sum_j (F^T)_{ij} F_{jl} = \sum_j \frac{\partial x_j}{\partial x_{0i}} \frac{\partial x_j}{\partial x_{0l}}. \tag{6.85}$$

The lineal stretch $\lambda$ is then given by (Ottino et al., 1981):

$$\lambda^2 = \sum_i \sum_l \left( \sum_j \frac{\partial x_j}{\partial x_{0i}} \frac{\partial x_j}{\partial x_{0l}} \right) m_{0l} m_{0i}, \tag{6.86}$$

and the time rate of the lineal stretch per unit of present length is:

$$\frac{\overset{\cdot}{\lambda}}{\lambda} = \frac{1}{2} \sum_i \sum_j \dot{\gamma}_{ij} m_j m_i, \tag{6.87}$$

where $\dot{\gamma}_{ij}$ are the components of the rate-of-strain tensor, $\dot{\gamma}$, and the components of the vector $\mathbf{m}$ are given by:

$$m_i = \frac{1}{\lambda} \sum_j F_{ij} m_{0j}. \tag{6.88}$$

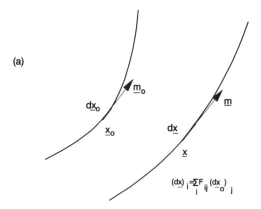

(a)

$$(dx)_i = \sum_j F_{ij} (dx_0)_j$$

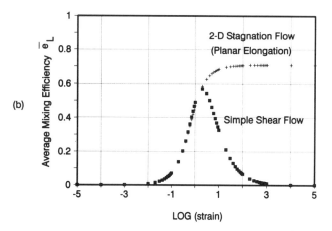

**Figure 6.10** (a) Stretch of a material line at position $\mathbf{x}_0$ of length $|d\mathbf{x}_0|$ and orientation $\mathbf{m}_0$ at time $t = 0$ to a new line at position $\mathbf{x}$ of length $|d\mathbf{x}|$ and orientation $\mathbf{m}$ at present time, produced by a deformation gradient field with components $F_{ij}$. (b) Average mixing efficiency, $\bar{e}_L$, of simple shear flow and planar elongational flow, as a function of strain.

The efficiency of mixing, $e_L$, which relates the rate of lineal stretch and its upper bound, which is proportional to the viscous dissipation, is given by:

$$e_L = 2 \frac{\dot{\lambda}/\lambda}{\sqrt{\sum_k \sum_l \dot{\gamma}_{kl} \dot{\gamma}_{lk}}} \tag{6.89}$$

and the time average efficiency of mixing, $\bar{e}_L$, is given by:

$$\bar{e}_L = \frac{1}{t} \int_0^t e_L(t') dt'. \tag{6.90}$$

Finally, for isochoric deformations (with volume conservation), the product of the areal and lineal stretches is equal to 1, and the sum of the rates of lineal and areal stretches per unit of present length and area (respectively) is equal to zero, for the same orientation vector of the line and the area. The following example illustrates the various steps for calculating the striation thickness reduction function and the time-average mixing efficiency for the simple shear case.

EXAMPLE 6.5
**Striation Thickness Reduction and Efficiency of Mixing for Simple Shear Flow**

Calculate the striation thickness reduction function and the time-average mixing efficiency for simple shear flow, $v_x = \dot{\gamma}y$; $v_y = v_z = 0$, for a line with orientation along the $y$ axis, that is, with initial orientation vector $\mathbf{m}_0 = (0, 1, 0)$.

**Solution**

The kinematics of simple shear flow are:

$$v_x = \frac{dx}{dt} = \dot{\gamma}y; \qquad v_y = \frac{dy}{dt} = 0; \qquad v_z = \frac{dz}{dt} = 0. \qquad (6.91)$$

Integration of this equation with respect to time with initial conditions $x = x_0$, $y = y_0$, and $z = z_0$ yields:

$$x = x_0 + \gamma y_0; \qquad y = y_0; \qquad z = z_0, \qquad (6.92)$$

where $\gamma$ is the shear strain $\dot{\gamma}t$. The only nonzero components of the deformation-gradient tensor are:

$$F_{11} = F_{22} = F_{33} = 1; \qquad F_{12} = \gamma. \qquad (6.93)$$

The nonzero components of the transpose of the deformation-gradient tensor are

$$(F^T)_{11} = (F^T)_{22} = (F^T)_{33} = 1; \qquad (F^T)_{21} = \gamma. \qquad (6.94)$$

The components of the Cauchy-Green strain tensor, $\mathbf{C}$, are calculated from Eq. 6.85. The only nonzero components are:

$$C_{11} = C_{33} = 1; \qquad C_{12} = C_{21} = \gamma; \qquad C_{22} = \gamma^2 + 1. \qquad (6.95)$$

The lineal stretch is then calculated from Eq. 6.86 as:

$$\lambda = \sqrt{1 + \gamma^2}. \qquad (6.96)$$

For long times, the lineal stretch varies linearly with time; that is, $\lambda \propto t$, and so shear flow is considered a *weak* flow in this respect. In extensional flows (see Pr. 6B.4) the lineal stretch grows exponentially with time, that is, $\lambda \propto e^t$, and these flows are considered *strong*. The striation thickness reduction function, $\delta/\delta_0$, which is equal to $\lambda^{-1}$, is then equal to $(1 + \gamma^2)^{-1/2}$. With the proper selection of $a_x$ and $a_y$, note that Eqs. 6.65 and 6.96 are the same, so that the two approaches are equivalent.

To calculate the time-average mixing efficiency of shear flow we need to calculate the components of the rate-of-strain tensor and of the new orientation vector, $\mathbf{m}$. The nonzero components of the rate-of-strain tensor are:

$$\dot{\gamma}_{12} = \dot{\gamma}_{21} = \dot{\gamma}. \qquad (6.97)$$

The new orientation vector, $\mathbf{m}$, is calculated from Eq. 6.88 as:

$$\mathbf{m} = \frac{1}{\lambda}(\gamma, 1, 0). \qquad (6.98)$$

The rate of lineal stretch per unit of length is given by Eq. 6.87 as:

$$\frac{\dot{\lambda}}{\lambda} = \frac{\dot{\gamma}^2 t}{1 + \gamma^2}, \qquad (6.99)$$

and the mixing efficiency is given by Eq. 6.89 as:

$$e_L = \sqrt{2}\frac{\gamma}{1 + \gamma^2} = \sqrt{2}\frac{\dot{\gamma}t}{1 + \dot{\gamma}^2 t^2}. \qquad (6.100)$$

It can be easily proven that the initial and the final value of $e_L$ is 0, its maximum value is $\sqrt{2}/2 = 0.707$ (at $\gamma = 1$), and that at long times it decays as $1/t$. The time-average value of the mixing efficiency is calculated from Eq. 6.90 as:

$$\bar{e}_L = \frac{\sqrt{2}}{2}\frac{\ln(1 + \gamma^2)}{\gamma}, \qquad (6.101)$$

and it is shown graphically in Fig. 6.10.b. Its maximum is about 0.57, and it occurs at $\gamma \simeq 2$. At long times the average efficiency decays, so that simple shear flows are characterized by low efficiencies at long times. The time-average mixing efficiency of a planar elongational flow (also called two-dimensional stagnation flow) is also shown in that figure.

---

Clearly, simple shear is not an effective mixing mechanism, but it is present in all mixing devices. It is difficult to maintain elongational (irrotational) flows between rigid boundaries. Also, the orientation of the interface is very important in the reduction of the striation thickness. Ng and Erwin (1981) experimentally demonstrated that an improvement in the interface growth function can be obtained by stepwise application of shear and reorientation of the interfaces between steps. In a single-screw extruder, the inefficiency of shear can be offset by the incorporation of "mixing sections" in the melt channel. These sections, in the form of vanes, pins, ducts, and so on, increase the pressure drop in the extruder, and so the shear strain is increased. More significantly, they reorient the polymer interfaces, causing a drastic decrease of the striation thickness in the material coming out of the extruder (see Pr. 6B.6; Erwin & Mokhatarian, 1983; Erwin, 1978b).

In conclusion, mixing of highly viscous liquids such as polymer melts is achieved mainly through laminar mixing. In this type of mixing, the interfacial area growth (or striation thickness reduction) function quantifies the degree of mixing. In simple shear laminar mixing the degree of mixing, for large strains, is proportional to the applied shear strain and the orientation of the interface. Consequently, good mixing is obtained by the application of large strains in favorably oriented interfaces. But gross uniformity cannot be achieved through the striation thickness reduction alone. One should also distribute the interfacial elements throughout the system.

### 6.3.3. LAMINAR MIXING IN SIMPLE GEOMETRIES

In this subsection we discuss two examples of mixing in simple geometries: parallel plates and concentric cylinders. The parallel plate geometry is used as the basis of the analysis of the single-screw extruder, and concentric cylinders are used as part of the design of rotational dies. The flow kinematics are calculated, and mixing in terms of striation thickness reduction is evaluated.

### EXAMPLE 6.6
### Mixing in Plane Couette Flow (Parallel Plate Geometry; PCF)

Consider two infinite parallel plates and a polymer blend (with equal viscosity components) filling the space between them. The discussion that follows holds for the case of both Newtonian and non-Newtonian fluids. At time $t=0^-$, the system is at rest. Then, at time $t=0^+$, the upper plate starts moving with velocity $V_z$ in the positive $z$ direction. The movement of the upper plate drags the fluid that starts moving in the same direction. Calculate the striation thickness reduction in this geometry and the effect of reorientation in the case of initially spherical particles of the dispersed phase.

### Solution

Fig. 6.11 shows the parallel plate geometry. Initially, the particles of the minor component can be simulated by idealized geometries such as cubes, spheres, and rectangular parallelepipeds (with the long axis along the $z$ or $y$ directions). The shearing planes are parallel to the plates, and the amount of shear strain is:

$$\gamma = \dot{\gamma}t = \frac{V_z}{H}t, \qquad (6.102)$$

where $H$ is the plate separation distance.

After shearing has started, the edges of the geometries transverse to the shearing planes will rotate through an angle $\theta$. However, the edges that are parallel to the shearing planes will remain parallel and of the same length as before shearing. In the case of the sphere, the diameter transverse to the shearing plane will be transformed into the long axis of an ellipsoid. In all geometries, the initial striation thickness is the thickness $B$ (diameter $2R$ for the sphere). After shearing, the striation thickness becomes $B'$. The cube and the two parallelepipeds will be treated similarly. The treatment for the sphere requires one additional argument, and thus it will be treated separately. The rotation angle $\theta$ and the shear strain are related as follows:

$$\tan\theta = \gamma. \qquad (6.103)$$

The length $L'$, $B'$ and the striation thickness, $\delta$, for all the geometries except the sphere can be calculated from trigonometrical arguments as follows:

$$L' = \frac{L}{\cos\theta} = L\sqrt{1+\gamma^2}$$

$$B' = B$$

$$\delta = B\cos\theta = \delta_0\cos\theta = \frac{\delta_0}{\sqrt{1+\gamma^2}} \qquad (6.104)$$

In the case of a sphere elongated into an ellipsoid, the long axis of the ellipsoid is given by Eq. 6.104, and the short axis is calculated by equating the volume of the initial sphere ($\pi(2R)^3/6$) to the volume of the ellipsoid ($\pi L'(B')^2/6$) as:

$$\delta = B' = \delta_0(1+\gamma^2)^{-1/4} = (2R)(1+\gamma^2)^{-1/4}. \qquad (6.105)$$

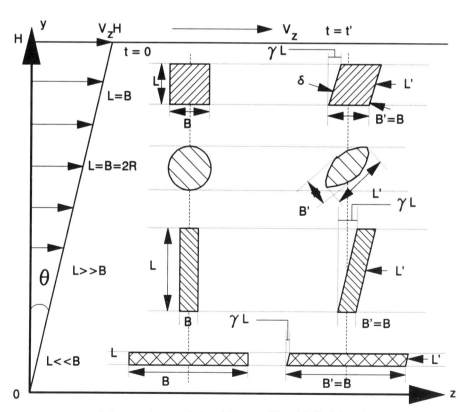

**Figure 6.11** Effect of shear strain on striation thickness of "particles" of the minor component represented by simple geometries in a plane Couette flow (PCF).

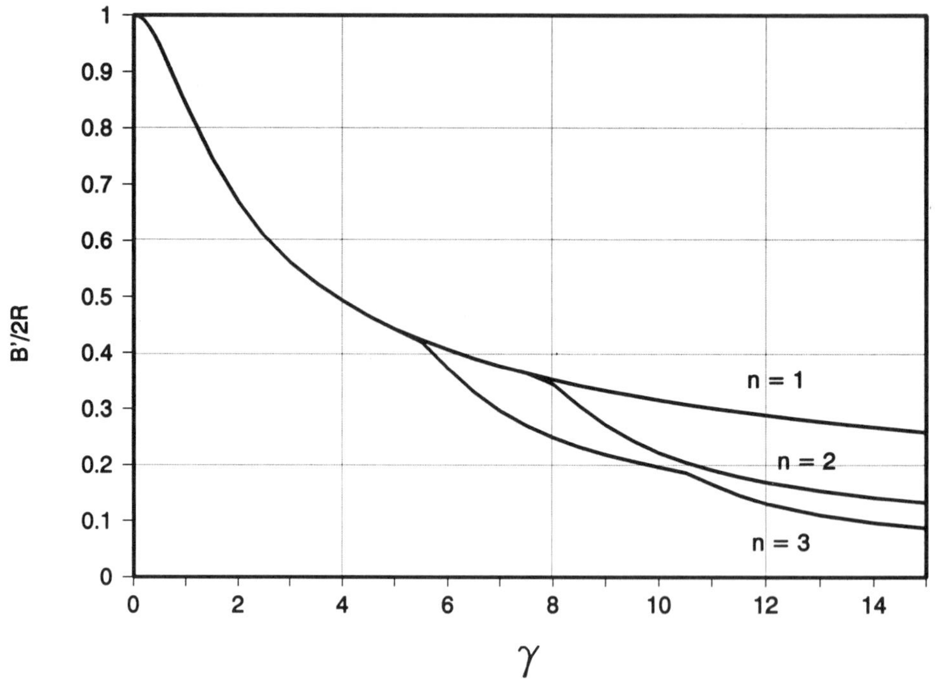

**Figure 6.12** Effect of reorientation on the striation thickness reduction of an initially spherical particle subjected to shear strain, $\gamma$. $n$ is the number of reorientations.

A comparison of Eqs. 6.104 and 6.105 shows that for the same shear strain the sphere offers the maximum reduction in striation thickness.

The generalization of these equations to the case where there is a mismatch of shear viscosity ($p = \mu_d/\mu_c \neq 1$) of the two Newtonian components is straightforward, where $\mu_d$ is the viscosity of the minor (or dispersed) component and $\mu_c$ is the viscosity of the major (or continuous) component. The stress is the same across the gap separation, and at each interface the stresses inside the major and minor components are equal (if interfacial tension is negligible). Thus,

$$\dot{\gamma}_c = p\dot{\gamma}_d, \tag{6.106}$$

where $\dot{\gamma}_c$ and $\dot{\gamma}_d$ are the shear rates in the continuous and dispersed phases, respectively. It is clear that the reduction in striation thickness of the minor component is small if its viscosity is higher than that of the major component.

Finally, the effect of reorientation on the striation thickness reduction can be assessed in the example of an initially spherical particle deformed into an ellipsoid. Equation 6.105 gives the reduction in striation thickness as a function of the shear strain. If the ellipsoid is entering the next step with its long axis perpendicular to the shearing planes, the new striation thickness will be given by the same equation. Thus, after $n$ steps the striation thickness, $\delta$, with reorientation will be:

$$\frac{\delta}{2R} = \prod_{j=1}^{n} (1 + \gamma_j^2)^{-1/4}. \tag{6.107}$$

Fig. 6.12 shows the effect of reorientation on the striation thickness of an initially spherical particle subjected to shear strain $\gamma$.

### EXAMPLE 6.7
### Striation Thickness in Rotational Couette Flow (RCF)
Consider a rotational Couette geometry with the inside cylinder rotating, as shown in Fig. 6.13.a and with the minor component represented by a black line. The position of the black line at time $t = 0$ shows the feedport of the system. After the inner cylinder starts rotating, the black line transforms into a spiral. Calculate the striation thickness as a function of the total number of revolutions of the inner cylinder and the geometry of the system.

### Solution

A typical Couette geometry is shown in Fig. 2.18. Figure 6.13.b shows the geometry of a spiral line that results from the rotation of an initially straight line in a Couette geometry with the inside cylinder rotating with angular velocity $W$, in radians per unit time. For a power-law fluid the tangential velocity $v_\theta$ is given by (see Pr. 2B.6):

$$v_\theta = rW \frac{1 - \left(\dfrac{R}{r}\right)^{2/n}}{1 - \kappa^{-2/n}}, \tag{6.108}$$

where $R$ is the radius of the outside cylinder, $n$ is the power-law index and $\kappa R$ is the radius of the inside cylinder.

Recalling that $v_\theta = r\,d\theta/dt$, that the counterclockwise direction in the $\theta$ coordinate is considered negative, and integrating the previous equation, we get:

$$\theta = -W \frac{1 - \left(\dfrac{R}{r}\right)^{2/n}}{1 - \kappa^{-2/n}} t. \tag{6.109}$$

At time $t = 2\pi/W$ one revolution of the inside cylinder is completed, and the angular displacement of two points that initially were apart by a distance $dr$ is:

$$d\theta = \frac{4\pi}{n(\kappa^{-2/n} - 1)} \frac{R^{2/n}}{r^{(n+2)/n}} dr. \tag{6.110}$$

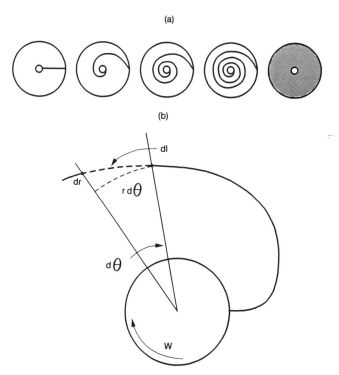

(a)

(b)

**Figure 6.13** (a) Mixing by laminar shear flow in a rotational Couette geometry. (b) Spiral line in the annulus of a rotational Couette geometry.

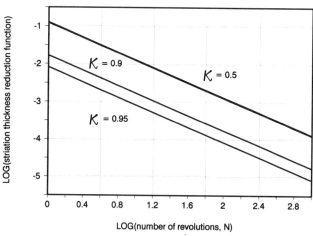

**Figure 6.14** Striation thickness reduction function in rotational Couette flow (RCF) for various ratios of the outside to the inside radius. Newtonian and power-law fluids do not exhibit any significant difference.

This equation was derived with the help of the approximate relation: $(r+dr)^{2/n} - r^{2/n} \simeq (2/n)r^{(2-n)/n} \, dr$.

After $N$ revolutions, the angular displacement will be $N(d\theta)$. Then, the total spiral length is:

$$l_N = \int dl_N = \int [(dr)^2 + r^2(Nd\theta)^2]^{1/2}. \tag{6.111}$$

Equations 6.110 and 6.111 can be combined to give the following:

$$l_N = \int_{\kappa R}^{R} \left[ 1 + \frac{16\pi^2 N^2 R^{4/n}}{n^2(\kappa^{-2/n}-1)^2} r^{-4/n} \right]^{1/2} dr. \tag{6.112}$$

From that, the striation thickness reduction function is calculated as:

$$\left[ \frac{\delta}{\delta_0} \right] = \frac{R - \kappa R}{l_N}. \tag{6.113}$$

For all practical purposes, $(dr)^2 \ll (rNd\theta)^2$, so that the striation thickness reduction function can be calculated from the formula:

$$\left[ \frac{\delta}{\delta_0} \right] = \left[ \frac{2-n}{4\pi N} \right] \left[ \frac{(1-\kappa)(\kappa^{-2/n}-1)}{(\kappa^{(n-2)/n}-1)} \right]. \tag{6.114}$$

Figure 6.14 shows the effect of the total number of revolutions, ratio of inside to outside cylinder radius, and power-law index on the degree of mixing achieved in the Couette rotational device. The lines in this figure represent both the Newtonian ($n=1$) and the power-law solution ($n=0.5$) for $\kappa=0.95$, 0.9, and 0.5. It can be seen from Eq. 6.114 that the power-law index does not play a role in this formula and that $(\delta N)$ is constant depending only on the $\kappa$ value.

## 6.4. RESIDENCE TIME AND STRAIN DISTRIBUTIONS

In Section 6.3 we saw that fluids in simple geometries such as Couette flow in parallel plates or small gap concentric cylinders experience the same shear rate across the gap. Nevertheless, the strain history of each fluid element depends not only on the shear rate but also on the time that the element has been subjected to that shear rate. Consequently, an element at a point close to the moving upper plate in a parallel plate geometry spends less time inside the apparatus than an element at a point close to the stationary lower plate because of different velocities at those two points. At the exit of the apparatus, the fluid collected consists of a set of fluid elements with a distribution of strain histories due to different residence time of each element. Also, because mixing is directly related to strain, at the end of the apparatus the total mixing will exhibit a distribution. The residence time distribution (RTD) is important mainly in the areas of (1) overall mixing efficiency; (2) degradation of temperature-sensitive polymers in processing; (3) foaming and cross-linking of polymers with the aid of temperature-sensitive foaming and cross-linking agents; and (4) backmixing, that is, mixing in the primary flow direction. A knowledge of the RTD is thus necessary for the optimal operation of processing equipment.

### 6.4.1. RESIDENCE TIME DISTRIBUTION

The RTD was first introduced by Danckwerts (1953), and it is defined so that $f(t)dt$ measures the fraction of the exit stream with residence time between $t$ and $t+dt$. The cumulative RTD function or the $F$

function, $F(t)$, is then defined as:

$$F(t) = \int_{t_0}^{t} f(t')dt', \qquad (6.115)$$

where $t_0$ is the minimum residence time. Clearly, $F(t)$ represents the fraction of the exit stream with residence time equal to or less than $t$. If $Q$ denotes the volumetric flow rate at the exit and $dQ$ the fraction of the volumetric exit flow rate with residence time between $t$ and $t + dt$, then the RTD function is given by:

$$f(t)dt = \frac{dQ}{Q}. \qquad (6.116)$$

The mean residence time, $\bar{t}$, is then equal to:

$$\bar{t} = \int_{t_0}^{\infty} tf(t)dt. \qquad (6.117)$$

Extreme cases of RTD are plug flow and perfect mixing. In the plug flow case, all elements of the fluid have exactly the same residence time and thus "appear" at the exit at exactly the same time. In diagrams of the $F(t)$ function, the plug flow mixer is represented by a straight vertical line at $t = \bar{t}$. In reality fluids with no slip at the walls cannot exhibit the characteristics of plug flow, but some fluids approach this behavior. Perfect mixing takes place in a stirred tank, and will be discussed later. Example 6.8 and Pr. 6A.13 illustrate the application of these ideas to specific pressure flows.

---

### EXAMPLE 6.8
### Mean Residence Time and *F(t)* Function for Capillary Flow

Calculate the mean residence time and the cumulative RTD function for pressure (Poiseuille) flow of a power-law fluid through a circular pipe (CPPF).

**Solution:**

The velocity profile is given in Table 2.6. Then the volumetric flow rate, $dQ$, with residence time between $t$ and $t + dt$ is given by the product of the area between the circles with radii $r$ and $r + dr$ and the velocity as:

$$dQ = 2\pi r dr v_z(r), \qquad (6.118)$$

and the total volumetric flow rate is given by the integral Eq. 6.118 as:

$$Q = \int dQ = 2\pi \int_{0}^{R} v_z(r)r dr. \qquad (6.119)$$

By substituting $v_z(r)$ into Eq. 6.119, after some calculations, we get:

$$Q = \frac{\pi n R^3}{3n+1}\left(\frac{R\Delta P}{2mL}\right)^{1/n}, \qquad (6.120)$$

where $L$ is the length of the pipe. Fluid elements at the center line ($r = 0$) stay in the pipe mixer for time $t_0$, which is equal to the minimum residence time:

$$t_0 = \frac{L}{v_z(r=0)} = \frac{(1+n)L}{nR}\left(\frac{2mL}{R\Delta P}\right)^{1/n}. \qquad (6.121)$$

The ratio of the residence times of fluid elements at radial distances $r$ and 0 is:

$$\frac{t}{t_0} = \frac{1}{1 - \left(\dfrac{r}{R}\right)^{\frac{1+n}{n}}}. \qquad (6.122)$$

The RTD function $f(t)$ is given by combining Eqs. 6.118 and 6.120 as:

$$f(r) = \frac{2(1+3n)}{(1+n)R^2}r\left[1 - \left(\frac{r}{R}\right)^{\frac{1+n}{n}}\right]. \qquad (6.123)$$

Equation 6.122 transforms $f(r)$ into $f(t)$:

$$f(t) = \frac{2n(1+3n)}{(1+n)^2}\frac{t_0^2}{t^3}\left(1 - \frac{t_0}{t}\right)^{\frac{n-1}{1+n}}. \qquad (6.124)$$

The mean residence time is calculated from Eq. 6.117 as:

$$\bar{t} = \frac{2n(1+3n)}{(1+n)^2}t_0^2\int_{t_0}^{\infty}\frac{\left(1 - \dfrac{t_0}{t}\right)^{\frac{n-1}{1+n}}}{t^2}dt. \qquad (6.125)$$

After integration by parts the mean residence time is given by:

$$\bar{t} = \frac{1+3n}{1+n}t_0. \qquad (6.126)$$

For Newtonian fluids $n = 1$ and consequently, $\bar{t} = 2t_0$. For power-law fluids $\bar{t} < 2t_0$, so that these fluids approach plug flow.

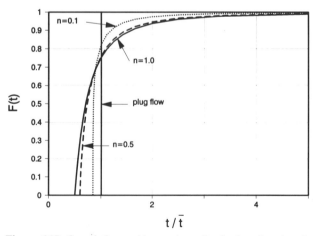

**Figure 6.15** Cumulative residence time distribution function for Poiseuille flow of Newtonian and power-law fluids in a circular pipe (CPPF).

The cumulative RTD function is given by combining Eqs. 6.115 and 6.124 to give:

$$F(t) = \frac{2n(1+3n)}{(1+n)^2} t_0^2 \int\limits_{t_0}^{t} \frac{\left(1 - \frac{t_0}{t'}\right)^{\frac{n-1}{1+n}}}{t'^3} \, dt'. \tag{6.127}$$

It is not difficult to complete the integration in Eq. 6.127 and to combine it with Eq. 6.126 to get:

$$F(t) = \left[1 + \frac{2n}{1+3n}\frac{\bar{t}}{t}\right]\left[1 - \frac{1+n}{1+3n}\frac{\bar{t}}{t}\right]^{\frac{2n}{1+n}}. \tag{6.128}$$

For Newtonian fluids, $F(t) = 1 - (1/4)(\bar{t}/t)^2$. Fig. 6.15 shows the $F$ function for a plug flow mixer and for Poiseuille flow in a long circular pipe for various degrees of pseudoplasticity. The more pseudoplastic the fluid is, the more the flow approaches plug flow. Also, it is evident from the figure that if we introduce a "tagged" material in a CPPF mixer, we expect to see the first trace of that material coming out of the mixer not earlier than 50% of the mean residence time (for the Newtonian case).

---

The $F$ function highly depends on the flow geometry for a given fluid. For Poiseuille flow of a power-law fluid in a parallel-plate geometry (PPF) the mean residence time is given by:

$$\bar{t} = \frac{1+2n}{1+n} t_0 \tag{6.129}$$

and the $F$ function by (see Pr. 6A.13):

$$F(t) = \left[1 + \frac{n}{1+2n}\frac{\bar{t}}{t}\right]\left[1 - \frac{1+n}{1+2n}\frac{\bar{t}}{t}\right]^{\frac{n}{1+n}}. \tag{6.130}$$

The effect of the flow geometry and flow type (Couette and Poiseuille) is shown in Fig. 6.16. Note that the first trace of a "tagged" material does not comes out of a Poiseuille flow in parallel plate geometry earlier than 67% of the mean residence time. Finally, a Poiseuille flow of Newtonian fluids in a circular pipe (CPPF) exhibits the same $F$ function as a Couette flow of Newtonian fluids in the parallel plate geometry (PCF; see Pr. 6A.13).

In these two cases, Poiseuille flow in pipe and parallel plate geometries, the cumulative RTD function was obtained by calculating the velocity profile and the volumetric flow rate. However, in cases in which the velocity profile is not known, other techniques should be used to calculate the distribution of residence times. One technique is to obtain the response of the system under examination to a "step change" in influent concentration of some "tagged" material (tagged with color, pH, radioactivity, etc.). The technique works as follows. Suppose that we want to get the cumulative RTD function of a mixer with an unknown velocity profile. We connect this system with two influent pipes (one for the regular fluid and one for the tagged material with concentration $C_0$) through a three-way valve and we start mixing the regular fluid coming out of one of the influent pipes. After the system attains steady state (at $t=0$), we change the valve position to allow only the tagged material to flow through the mixer, and the concentration of the tagged material in the effluent stream, $C(t)$, is monitored by the appropriate technique (photometry for color as the tagging characteristic; pH measurement for pH as the tagging characteristic, etc.). For the technique to work we assume that the tagged material flow stream exhibits the same characteristics as the main flow stream so that the flow rate is not changed within the system and that the flow enters the system in a plug flow mode.

At time $t$ the tagged material exiting the mixer has been in the mixer for a time less than $t$. The volumetric flow rate of the tagged material at time $t$ is equal to $F(t)C_0 Q$, where $Q$ is the total volumetric flow rate of the influent (or effluent) stream. The tagged material balance yields:

$$QC(t) = F(t)C_0Q, \tag{6.131}$$

and thus the $F$ function is:

$$F(t) = \frac{C(t)}{C_0}. \tag{6.132}$$

In conclusion, the $F$ function for systems with unknown characteristics is obtained through a step-change experiment of a tagged material as the ratio of the tagged material concentrations of the incoming and outgoing streams.

Another technique for measuring the $F$ function includes the instantaneous injection ($\delta$ function or impulse) of a certain amount of the tagged material and the monitoring of the effluent concentration as a function of time. The difference between the two techniques lies in the way the same information is presented: The step-change technique yields the $F(t)$ function whereas the "$\delta$ function change" technique yields the $f(t)$ function.

Danckwerts (1953) distinguishes between the internal RTD function, $g(t)$, and the external RTD function, $f(t)$. The fraction of the material *in the system* with residence time between $t$ and $t+dt$ is $g(t)dt$, whereas $f(t)dt$ refers to the fraction of the material *leaving the system*. The cumulative RTD function corresponding to $g(t)$ is denoted as $G(t)$ and it is equal, to $\int_0^t g(t')dt'$. The relationship between $g(t)$ and $F(t)$ can

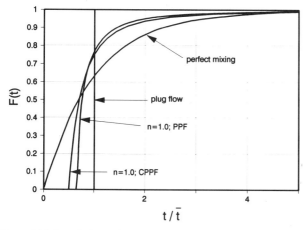

**Figure 6.16** Cumulative residence time distribution function for plug flow, perfect mixing (complete backmixing), and Poiseuille flow of Newtonian fluids in circular pipe (CPPF), and parallel plate geometries (PPF).

be obtained from an imaginary step-change experiment:

$$Q - QF(t) = \frac{d}{dt}[VG(t)], \tag{6.133}$$

where $V$ is the volume of the system occupied by the fluid, and the first, second, and third terms represent the volumetric flow rate into the system, out of the system, and change within the system, respectively. For constant $Q$ and $V$ and for $\bar{t}$ equal to $V/Q$, Eq. 6.133 yields:

$$g(t) = \frac{1 - F(t)}{\bar{t}}. \tag{6.134}$$

The mean residence time of the material inside the mixer is then given by:

$$\bar{t_i} = \int_0^\infty t g(t) dt. \tag{6.135}$$

Further discussion of the internal RTD functions goes beyond the scope of the present textbook.

The first extreme case, as mentioned before, is plug flow, and the other extreme case is that of a "perfect mixer," which is realized in stirred tanks. The latter case corresponds to complete backmixing, that is, mixing along the primary flow direction. Suppose that we perform a step-change experiment in a stirred tank system of volume occupied by the fluid $V$ and volumetric flow rate $Q$. Also suppose that $C_0$ is the tagged material concentration in the influent stream and that the concentration in the effluent stream is $C$. Then, the material balance of the tagged component yields:

$$QC_0 = QC + V\frac{dC}{dt}, \tag{6.136}$$

and the initial condition is $C = 0$ at $t \leqslant 0$. Note that the concentration of the tagged material $C$ in the tank and in the effluent stream is the same because perfect mixing was assumed. The solution to the first-order differential Eq. 6.136 along with the initial condition and Eq. 6.132 is:

$$F(t) = 1 - e^{-t/\bar{t}}, \tag{6.137}$$

where $\bar{t} = V/Q$. Fig. 6.16 shows the $F$ curve for a perfect mixer. Two notes should be made with reference to that curve: (1) The perfect mixer exhibits a broad cumulative RTD curve; and (2) there is no finite time lag, that is, the tagged material exits the mixer immediately after it is introduced to it, although at low concentrations, in contrast to pipe Poiseuille flow with time lag of $0.5\bar{t}$ and plug flow with time lag $\bar{t}$. Also, note that in the perfect mixer there is no distinction between internal and external RTD functions and consequently $f(t) = g(t)$ and $F(t) = G(t)$.

The broad RTD of the perfect mixer seems to contradict the idea of perfect mixing. Caution should be exercised in this respect. The term *mixing* was used in this section to signify the *time distribution of material inside and outside the mixer, not the homogeneity of the effluent stream*. On the other hand, a high degree of turbulence in pipe flow produces a high degree of uniformity in the transverse direction (*transverse mixing*; radial mixing), and poor mixing in the longitudinal direction (primary flow direction; *backmixing*). Thus, although the $F$ curves of turbulent and laminar flows in a pipe are the same, indicating the same degree of backmixing, mixing in the radial direction is greatly improved in the turbulent flow.

Two other qualities characterize the degree of mixing through the

RTD of mixers: *holdback*, $B$, and *segregation*, $S$. Holdback is defined as:

$$B = \frac{1}{\bar{t}} \int_0^{\bar{t}} F(t) dt, \tag{6.138}$$

and it represents the area under the $F$ curve from 0 to $\bar{t}$. $B$ varies from 0 for plug flow to 1 for a mixer full of dead spots, and it measures the deviation from plug flow. To understand the meaning of holdback, we borrow the example given by Danckwerts (1953). Suppose that the inflow stream changes color from white to red in a step-change experiment. Then, $B$ is equal to the fraction of the mixer that is still occupied by white color after a volume of red material equal to the mixer volume has entered the mixer. The holdback of typical flows (see Pr. 6A.14) are:

| | |
|---|---|
| Plug flow | 0 |
| PPF | 0.19 |
| CPPF; PCF | 0.25 |
| Perfect mixing | 0.37 |

Segregation, $S$, can be calculated from the $F$ curves as the area between the perfect mixing and the mixer curves up to the point where these curves cross each other. $S$ varies from $+1/e = 0.37$ for plug flow to $-1$ for a mixer full of dead spots. The value for the CPPF is 0.14 (see Pr. 6A.14).

Finally, the average striation thickness reduction function at the exit of a mixer $\langle\langle\lambda(t)\rangle\rangle$ can be calculated using the idea of the "mixing cup" as follows (Ottino & Chella, 1983):

$$\langle\langle\lambda(t)\rangle\rangle = \left\langle\left\langle\frac{\delta}{\delta_0}(t)\right\rangle\right\rangle = \int_{t_0}^t \lambda(t')f(t')dt'. \tag{6.139}$$

### 6.4.2. STRAIN DISTRIBUTION

It was shown in Section 6.3 that the degree of mixing is proportional to the total shear strain for shear mixing with large applied strains. In other mixing situations the state of mixing is also a function (may be not simple) of the applied strain. However, in a mixer, the fluid particles experience different strain histories, so that the exiting fluid stream consists of particles with different degrees of mixing. The nonuniformity in the state of mixing is measured by the strain distribution function (SDF) in direct analogy with the RTD function. Note that in batch mixers, the SDF depends only on the various paths that the fluid particles have followed, whereas in continuous mixers the RTD and the variation of the path trajectories should be accounted for in the strain distribution. One should also keep track of the sign of the strain, because mixing is enhanced by positive strain (along the flow direction) and reduced by negative (opposite to the flow direction) strain (demixing).

Inside continuous mixers or in batch mixers, in analogy with the RTD discussion (Eq. 6.116), the SDF, $g(\gamma)$, is calculated as the fraction of the fluid that experienced strain from $\gamma$ to $\gamma + d\gamma$. The cumulative SDF, $G(\gamma)$, is then given by the integral of the SDF as in Eq. 6.115. In the exit stream of continuous mixers the SDF, $f(\gamma)d\gamma$, is defined as the fraction of the flow that has experienced strain between $\gamma$ and $\gamma + d\gamma$. The cumulative SDF, $F(\gamma)$, is then calculated as:

$$F(\gamma) = \int_{\gamma_0}^\gamma f(\gamma')d\gamma', \tag{6.140}$$

and it denotes the fraction of the exit flow stream with applied strain less than or equal to $\gamma$, where $\gamma_0$ is the minimum applied strain. The weighted average total strain (WATS) (Pinto & Tadmor, 1970) and the mean total strain, $\bar{\gamma}$, are calculated by weighting the strain with the RTD function and SDF, respectively, as follows:

$$\text{WATS} \equiv \int_{t_0}^{\infty} \gamma(t')f(t')dt'; \qquad \bar{\gamma} \equiv \int_{\gamma_0}^{\infty} \gamma'f(\gamma')d\gamma'. \qquad (6.141)$$

Obviously, the higher the value of the mean strain that is achieved, the better the mixing is. The SDF is only simply related to the RTD for the case when $\dot{\gamma}$ is constant. Example 6.9 illustrates a case of constant $\dot{\gamma}$.

### EXAMPLE 6.9
### Cumulative SDF for PCF

Calculate the cumulative SDF for parallel plate Couette flow (PCF) of a Newtonian fluid, and show its similarity with the cumulative RTD function for Newtonian flow in PCF or CPPF configurations.

### Solution

Referring to Fig. 6.11 for a parallel plate geometry, we see that:

$$\dot{\gamma} = \frac{V_z}{H}; \qquad \gamma = \dot{\gamma}t = \frac{V_z}{H}\frac{L}{v_z} = \frac{L}{y}. \qquad (6.142)$$

The minimum strain, $\gamma_0$, corresponds to $y = H$, so that:

$$\gamma_0 = \frac{L}{H}. \qquad (6.143)$$

The volumetric flow rate across the area $Wdy$ is $dQ$, and it is equal to:

$$dQ = Wdyv_z(y) = W\frac{y}{H}V_zdy. \qquad (6.144)$$

The total flow rate, $Q$, is calculated to be $WHV_z/2$.

According to the definition of the SDF,

$$f(\gamma)d\gamma = \frac{dQ}{Q} = \frac{2}{H^2}ydy. \qquad (6.145)$$

With the aid of Eqs. 6.142 and 6.143, the SDF becomes:

$$f(\gamma)d\gamma = 2\frac{\gamma_0^2}{\gamma^3}d\gamma, \qquad (6.146)$$

which is similar to the RTD function $(2t_0^2dt/t^3)$. Note that to arrive at the Eq. 6.146 we need to take into account the fact that the part of the total flow rate that experiences strain between $\gamma$ and $\gamma + d\gamma$ flows through the cross-sectional area between $W(y-dy)$ and $Wy$. Equivalently, we can use the absolute value of $dQ$, because the SDF is always a positive number. The mean total strain is calculated from Eqs. 6.141 and 6.146 as:

$$\bar{\gamma} = 2\gamma_0 = 2\frac{L}{H}. \qquad (6.147)$$

Finally, the cumulative SDF is calculated with the aid of Eq. 6.140 as:

$$F(\gamma) = 1 - \frac{1}{4}\left(\frac{\bar{\gamma}}{\gamma}\right)^2. \qquad (6.148)$$

This is the same function as the cumulative RTD function of PCF and CPPF with $\gamma_0$ and $\gamma$ replaced by $t_0$ and $t$, respectively.

In geometries and flow configurations where $\dot{\gamma}$. varies spatially, the calculations for the SDF become complicated. Example 6.10 illustrates the complexities involved.

### EXAMPLE 6.10
### Cumulative SDF in CPPF

Show the algorithm for the calculation of the cumulative SDF of a power-law fluid flowing in a CPPF configuration. As special cases, show the final algorithm steps for a Newtonian fluid and a power-law fluid with power-law index $n = 0.5$.

### Solution

The velocity profile and the shear rate are given in Table 2.6. The differential flow rate $dQ$ passing through the "ring" $2\pi rdr$ is equal to $2\pi rv_z(r)dr$, and the total flow rate $Q$ is then calculated as:

$$Q = \int_0^R dQ = \frac{n\pi R^3}{1+3n}\left(\frac{R\Delta P}{2mL}\right)^{1/n}. \qquad (6.149)$$

The SDF can be calculated as a function of the variable $\xi = r/R$ from the ratio $dQ/Q$ as:

$$f(\xi)d\xi = 2\frac{1+3n}{1+n}\xi(1-\xi^{(n+1)/n})d\xi, \qquad (6.150)$$

and the cumulative SDF is calculated as:

$$F(\xi) = \frac{1+3n}{1+n}\xi^2\left(1 - \frac{2n}{1+3n}\xi^{(n+1)/n}\right). \qquad (6.151)$$

The mean total strain is calculated using the expressions for $\dot{\gamma}$, $t = L/v_z(r)$ and the SDF as:

$$\bar{\gamma} = 2\frac{L}{R}\frac{1+3n}{1+2n} \qquad (6.152)$$

It is interesting to note that the mean total strain depends only on the geometry of the system (through $L$ and $R$), not on the flow rate, and also that the smaller the radius or longer the pipe, the higher the strain is. For Newtonian fluids: $\bar{\gamma} = 8L/3R$.

The expressions for $\dot{\gamma}$ and $\bar{\gamma}$ can be combined into the following:

$$\frac{\gamma}{\bar{\gamma}} = \frac{(1+n)(1+2n)}{2n(1+3n)}\frac{\xi^{1/n}}{1-\xi^{(n+1)/n}}. \qquad (6.153)$$

The algorithm now consists of the following steps:

1. Calculation of $\xi$ from Eq. 6.153 for a specified $\gamma/\bar{\gamma}$ ($n$-th-order polynomial; graphically or with the aid of the IMSL subroutine NEQNF; Appendix D.4)

2. Substitution of $\xi$ just calculated into $F(\xi)$

3. Plotting of $F(\gamma/\bar{\gamma})$ versus $\gamma/\bar{\gamma}$

For Newtonian fluids and for power-law fluids with $n = 0.5$ step 1 is simpler because the resulting equations are quadratic and cubic, respectively. So, for Newtonian fluids the equations are:

$$F(\xi) = 2\xi^2\left(1 - \frac{1}{2}\xi^2\right); \qquad \xi^2 + \frac{3\bar{\gamma}}{4\gamma}\xi - 1 = 0, \qquad (6.154)$$

and for a power-law fluid with $n = 0.5$, the corresponding equations are:

$$F(\xi) = \frac{5}{3}\xi^2\left(1 - \frac{2}{5}\xi^2\right); \qquad \xi^3 + \frac{6\bar{\gamma}}{5\gamma}\xi^2 - 1 = 0. \qquad (6.155)$$

The power-law fluid shows a narrower distribution, which is expected because the velocity profile approaches that of plug flow.

---

Similar calculations can be repeated for the case of PPF for power-law fluids in general (see Pr. 6B.9). For $\xi = 2y/H$, where $H$ is the plate separation and $y = 0$ at the center of the separation, the SDF, cumulative SDF, and strain are given by:

$$f(\xi)d\xi = \frac{1+2n}{1+n}(1 - \xi^{(n+1)/n})d\xi \qquad (6.156)$$

$$F(\xi) = \frac{1+2n}{1+n}\xi\left(1 - \frac{n}{1+2n}\xi^{(n+1)/n}\right) \qquad (6.157)$$

$$\frac{\gamma}{\bar{\gamma}} = \frac{(1+n)^2}{n(1+2n)}\frac{\xi^{1/n}}{1 - \xi^{(n+1)/n}}. \qquad (6.158)$$

The mean total strain is equal to:

$$\bar{\gamma} = 2\frac{L}{H}\frac{1+2n}{1+n}. \qquad (6.159)$$

Figure 6.17 compares the cumulative SDF for Newtonian fluids in three flow situations: CPPF, PPF, and PCF.

CPPF and PPF can be compared with each other on the basis of mean total strain applied and flow rate for the same pressure drop. The ratio of the mean total strains for $H = 2R$ is calculated from Eqs.

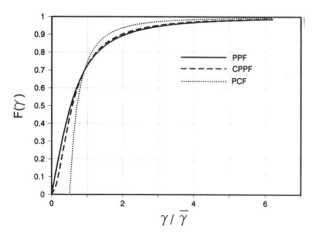

**Figure 6.17** Comparison of the strain distribution function for Poiseuille flow in circular pipe (CPPF) and parallel plate (PPF) and for parallel plate Couette flow (PCF) of Newtonian fluids.

6.152 and 6.159 as:

$$\frac{\bar{\gamma}_{CPPF}}{\bar{\gamma}_{PPF}} = 2\frac{(1+3n)(1+n)}{(1+2n)^2}. \qquad (6.160)$$

That ratio for Newtonian fluids is equal to 16/9, and it approaches 2 for extremely pseudoplastic fluids. Thus, the long circular pipe provides more strain for mixing than in geometrically similar parallel plate configuration. In terms of flow rate, for the same $\Delta P/L$ and for $H = 2R$, the situation is:

$$\frac{Q_{CPPF}}{Q_{PPF}} = \frac{R}{W}\frac{\pi(1+2n)}{2^{\frac{1+n}{n}}(1+3n)}, \qquad (6.161)$$

which for Newtonian fluids becomes $0.59R/W$. For these relationships to hold, $W$ should be very large compared to $R$, so that for $W/R = 20$, the ratio of flow rates becomes 0.0295. In conclusion, CPPF incorporates more strain to the fluid but produces less flow rate than PPF with wide parallel plates. For plates with $W = 2R = H$, a shape factor should be included in the calculation of the flow rate (Middleman, 1977, p. 91), and the ratio of flow rates becomes 4.7 for Newtonian fluids.

Rotational Couette flow (RCF) is analyzed in the next example. Because the flow is confined inside the Couette flow cell and there is no exit stream, we use the nomenclature $G(\gamma)$ and $g(\gamma)d\gamma$ instead of $F(\gamma)$ and $f(\gamma)d\gamma$.

### EXAMPLE 6.11
### SDF in RCF Configuration

Calculate the SDF for RCF, and explain the trends for Newtonian and non-Newtonian fluids.

**Solution**

Consider a rotational geometry with outside radius $R$, inside radius $\kappa R$, and angular velocity of the inside cylinder, $W$. The velocity profile, $v_\theta$, of the fluid in the gap for isothermal laminar flow with no gravitational and centrifugal forces and with no slip at the walls is given in Pr. 2B.6. The shear rate is given (Table 2.10) by:

$$\dot{\gamma} = \dot{\gamma}_{r\theta} = r\frac{\partial}{\partial r}\left(\frac{v_\theta}{r}\right), \qquad (6.162)$$

as the $r$ component of the velocity is zero. Substituting the velocity profile into Eq. 6.162, we obtain the strain:

$$\gamma = \dot{\gamma}t = \frac{2Wt}{n(1 - \kappa^{-2/n})}\left(\frac{R}{r}\right)^{2/n}. \qquad (6.163)$$

The minimum strain is obtained at the outside cylinder and is:

$$\gamma_0 = \frac{2Wt}{n(1 - \kappa^{-2/n})}, \qquad (6.164)$$

and the maximum at the inside cylinder is:

$$\gamma_{max} = \frac{2Wt}{n(1 - \kappa^{-2/n})}\frac{1}{\kappa^{2/n}}. \qquad (6.165)$$

so that the ratio of the maximum to minimum strains is:

$$\frac{\gamma_{max}}{\gamma_0} = \frac{1}{\kappa^{2/n}}. \tag{6.166}$$

For Newtonian fluids and for $\kappa$ equal to 0.9 the maximum shear strain is 23% higher than the minimum, but for $\kappa$ equal to 0.5 the maximum strain is 300% higher than the minimum. These differences cause a striation thickness difference inside the rotational mixer. The SDF can be computed from:

$$g(r)dr = \frac{2\pi r L dr}{\pi L(R^2 - (\kappa R)^2)} \tag{6.167}$$

and Eqs. 6.163 and 6.164 as:

$$g(\gamma)d\gamma = \frac{n}{1-\kappa^2}\frac{\gamma_0^n}{\gamma^{n+1}}d\gamma. \tag{6.168}$$

The mean total strain is calculated as:

$$\bar{\gamma} = \int_{\gamma_0}^{\gamma_{max}} \gamma g(\gamma)d\gamma = \gamma_0 \frac{n}{1-n}\frac{\kappa^{2(n-1)/n}}{1-\kappa^2} \tag{6.169}$$

for power-law fluids, whereas for Newtonian fluids it is given as:

$$\bar{\gamma} = \gamma_0 \frac{2\ln(1/\kappa)}{1-\kappa^2}. \tag{6.170}$$

The cumulative SDF, $G(\gamma)$, is calculated as:

$$G(\gamma) = \frac{1}{1-\kappa^2}\left[1 - \left(\frac{1-n}{n}\frac{1-\kappa^2}{\kappa^{2(n-1)/n}-1}\right)^n\left(\frac{\bar{\gamma}}{\gamma}\right)^n\right] \tag{6.171}$$

for power-law fluids, and it is calculated as:

$$G(\gamma) = \frac{1}{1-\kappa^2}\left[1 - \frac{1-\kappa^2}{2\ln(1/\kappa)}\frac{\bar{\gamma}}{\gamma}\right] \tag{6.172}$$

for Newtonian fluids.

A number of interesting points can be revealed from the SDFs. For Newtonian fluids, only about 52% of the material is subjected to strain less than the mean strain for $\kappa=0.9$, whereas for $\kappa=0.5$ the corresponding value increases to 61%. For comparison with the other mixing flows the values of the percentage of the material subjected to strain less than the mean for Newtonian fluids are for CPPF and PPF 73% and for PCF 75%. The spread of the distribution depends on the gap ratio and the power-law index. The more pseudoplastic the material and the smaller the gap ratio are, the broader the distribution is. A broad distribution suggests nonuniformity in the product, because the degree of mixing will vary substantially for material layers in different parts of the rotational mixer. For power-law fluids with $n=0.5$, 53% of the material is subjected to strain less than the mean for a gap ratio of 0.9 and 67% for a gap ratio of 0.5. The distribution for power-law fluids is also broader than that for Newtonian fluids.

## 6.5. DISPERSIVE MIXING

*Dispersive mixing* is the term used to describe mixing associated with some fundamental change of the physical characteristics of one or more of the components of the mixture. Generally, dispersive mixing is divided into two parts: the incorporation of the additives in terms of agglomerated particles or the second polymer component into the polymer matrix, and the dispersion (or deagglomeration) of the second phase to yield the final product. The microstructures of the blends are determined by rheological, hydrodynamic, and thermodynamic parameters. The rheological parameters are viscosity, elasticity, and yield stress of all components. The hydrodynamic parameters determine the flow fields. The thermodynamic parameters are related to solubility, adhesion, and diffusion of all components. In this section we address the dispersion of agglomerates (additives) and of other polymers (liquid–liquid dispersion) into a polymer matrix.

### 6.5.1. DISPERSION OF AGGLOMERATES

Dispersion of agglomerates has been applied in the polymer processing industry for at least 50 years. It is concerned with the incorporation and deagglomeration of additives in the polymer matrix, the ultimate goal being price reduction or the improvement of the properties of the final product. Of course, if the additive exists in the form of isolated noninteracting particles, then the task of mixing only is to distribute these particles uniformly throughout the final product. However, when the additive exists in the form of clusters of particles (interacting or noninteracting), then dispersive mixing ensures that the agglomerates break into isolated particles that then should be distributed by extensive mixing mechanisms.

The size of the particles and their ability to interact with one another characterize the type of cluster as *cohesionless* or *cohesive*. A *cohesionless* cluster is formed from noninteracting particles or from large particles ($>1$ mm), and its dispersion is determined only by the total deformation of the primary phase. In this case the dispersion is achieved by "peeling off" particles from the surface of the cluster by tangential velocity components close to the particles. On the other hand, a cohesive cluster includes interacting particles or very small particles or particles dispersed in a medium other than the polymer matrix, and its dispersion depends on the applied stresses (or equivalently on the deformation rates).

Usually, additive particles are of maximum size of about $100\,\mu m$, and the cohesive forces cannot be neglected. The significance of the cohesive forces is in the disintegration of the cluster, which requires that the hydrodynamic forces exceed the cohesive forces. Assuming that the aggregates are formed by nontouching spherical particles of like material, the Bradley-Hamaker theory (Elmendorp, 1991) allows one to calculate the attractive Van der Waals force between two particles as:

$$F_w = \frac{2CR_1R_2A}{3}\frac{2xy-x^2-y^2}{x^2y^2}, \tag{6.173}$$

where $A$ is the Hamaker constant ($\simeq 5\times10^{-20}$ to $5\times10^{-19}$ J), $x = C^2 - (R_1+R_2)^2$, $y = C^2 - (R_1-R_2)^2$, and $C = R_1 + R_2 + d$. $R_1$, $R_2$, and $d$ are the radii of the spheres and their distance, respectively. When the two spheres are touching each other, Eq. 6.173 reduces to:

$$F_w = \frac{A}{6z^2}\frac{R_1R_2}{R_1+R_2}, \tag{6.174}$$

where $z$ is the physical adsorption separation distance (equal to about 0.4 nm for adhering spheres). If the adhering spheres are of the same size, then $F_w = (AR)/12z^2$, and for the mean value of $A$ we get $F_w = 0.1R$ in $\mu N$ for $R$ in $\mu m$. Rumpf (1962) showed that the tensile strength of clusters formed by equal-size randomly packed particles is:

$$\sigma_T = \frac{9}{32}\frac{1-\varepsilon}{\varepsilon}\frac{F_w}{R^2}, \tag{6.175}$$

where $\varepsilon$ is the void volume fraction or porosity of the cluster. Combining Eqs. 6.173 and 6.175 we get the cohesive force of the cluster as:

$$F_c = \frac{9}{16} \frac{1-\varepsilon}{\varepsilon} \frac{A}{12z^2R} S, \qquad (6.176)$$

where $S$ is the cross sectional contact area of the rupture plane.

The hydrodynamic forces acting to rupture the cluster can be calculated by assuming simple geometrical shapes. Tadmor and Gogos (1979) assumed that the agglomerate consisted of a dumbbell (two spheres connected with a hypothetical rod to transmit the forces between them) suspended in shear and elongational flow fields. The maximum hydrodynamic force acting on the rod to rupture the dumbbell (of equal-size spheres) was shown to be:

$$F_h = \frac{3}{2} \pi \mu \dot{\gamma} LR \qquad (6.177)$$

for the simple shear case, and

$$F_h = 3\pi\mu\dot{\varepsilon}LR \qquad (6.178)$$

for elongational flow, where $L$ is the distance between the center of the spheres. Note that the maximum force in the shear case is achieved when the dumbbell is oriented 45° to the direction of shear and in the elongation case when the dumbbell is oriented along the direction of the flow. Equations 6.177 and 6.178 show that at the same shear and elongation rates elongational flow creates twice the force created in shear flows. The dumbbell is ruptured whenever the hydrodynamic force exceeds the cohesive force of Eq. 6.173. The better efficiency of the elongational field can be overcome by the shear flow because shear devices can easily produce shear rates of $100 \, s^{-1}$ and those rates cannot be achieved by elongational fields for long times. Thus, in dispersive mixing, shear is the predominant mechanism.

For a doublet of two touching equal-size spheres suspended in a incompressible Newtonian fluid in a shear flow field, Nir and Acrivos (1973) calculated the maximum hydrodynamic force to be:

$$F_h = 6.12\pi\mu\dot{\gamma}R^2. \qquad (6.179)$$

Note that if the spheres are unequal in size, the numerical constant of Eq. 6.179 changes. For instance, the hydrodynamic force for two touching spheres, with the diameter of one of them twice the diameter of the other, becomes equal to $2.57\pi\mu\dot{\gamma}R^2$, and for spheres with a diameter ratio of 10 the force becomes $0.05\pi\mu\dot{\gamma}R^2$. By equating the maximum hydrodynamic force on a doublet to the cohesive force of Eq. 6.174, we get the critical shear rate for break-up as:

$$\dot{\gamma}_{\text{crit}} = 0.004 \frac{A}{z^2} \frac{1}{\mu R}. \qquad (6.180)$$

This equation shows that the smaller the particles and the lower the viscosity of the suspending medium, the harder it is to break them up. The non-Newtonian character of the suspending medium can cause complications in terms of the hydrodynamic force, but this area is still under research.

The maximum hydrodynamic force acting on ellipsoidal, spherical, and highly ellipsoidal clusters of particles can be assumed to be the same as the force acting on single particles of the same shape. For spherical clusters, this maximum hydrodynamic force is given

(Elmendorp, 1991) as:

$$F_h = \frac{5}{2} \pi \mu \dot{\gamma} R_{cl}^2, \qquad (6.181)$$

where $R_{cl}$ is the cluster radius. Equating this force to the cluster cohesive force of Eq. 6.176, we get the critical shear rate for deagglomeration:

$$\dot{\gamma}_{\text{crit}} = \frac{3}{160} \frac{S}{\pi\mu R_{cl}^2} \frac{1-\varepsilon}{\varepsilon} \frac{A}{z^2R}, \qquad (6.182)$$

which leads us to the conclusions that the more viscous the matrix and the larger the individual particles are, the higher the critical shear rate is. In addition, if the agglomerate breaks at an equatorial plane (i.e., $S = \pi R_{cl}^2$), then the critical shear rate does not depend on the initial size of the cluster. The interfacial tension between the particles and the polymer matrix can act as a cohesive force of the cluster, especially when the polymer matrix does not fill completely the voids of the cluster (see Pr. 6A.15).

Equation 6.179 shows the importance of high stresses ($\mu\dot{\gamma}$) in the deagglomeration process. Consequently, it is common practice in the polymer processing industry to make *masterbatches*, or *superconcentrates*, of the dispersed to the continuous phase and thus to increase the applied stresses by increasing the viscosity. $\mu$ in Eq. 6.181 is the viscosity of the medium, which in highly concentrated batches can be orders of magnitude higher than the viscosity of the polymer matrix. For example, the masterbatch of carbon black in PE contains about 50% carbon black, but the final product contains about 2% to 5% only. The deagglomeration takes place in the masterbatch where the viscosity is high, and it is followed by dilution steps in extensive types of mixing.

### 6.5.2. LIQUID–LIQUID DISPERSION

The dispersion of one liquid into another is a very important mixing process for polymer blends. As mentioned in the introduction of this chapter, dispersion of small rubber particles in PS is vital for the improvement of the impact properties of PS. Because the blending takes place in the molten state, understanding of the phenomena in a typical liquid–liquid dispersion is essential for the polymer blend business. However, a lack of agreement between the theoretical predictions and the experimental data in polymer blends should not always be used to discredit the theory, as relaxation processes can alter the structure as the blend goes from the molten to the frozen state.

Liquid–liquid dispersion is characterized by two phases, the dispersed and the continuous. The physical parameters of the two phases affecting a liquid–liquid dispersion are viscosity, elasticity, interfacial tension, solubility, and diffusion rate. For solubility, the system is considered as miscible, immiscible, or partially miscible. Interfacial tension is lowest for miscible systems and highest for immiscible systems. As mentioned in Chapter 4, all high-molecular weight substances have a diffusion coefficient, $Ð$, of about $10^{-12}$ to $10^{-14}$ cm$^2$/s. Consequently, the diffusion rates in molten polymer systems are extremely small, and the relative penetration depths in the time scale of the blending process are extremely small.

Both the dispersed and the continuous phases are fed into the blending or compounding equipment in the form of pellets. The deformation and the dispersion starts after heating both components to temperatures above their melting point. Similarly to the dispersion of agglomerates, the hydrodynamic force is the deforming and disruptive force, and the interfacial tension force is the cohesive force of the dispersed phase. The ratio of these two forces or stresses is called the *capillary* (or *Weber*) number, Ca:

$$Ca = \frac{F_h}{F_c} = \frac{\mu_c \dot{\gamma} R}{\gamma}, \tag{6.183}$$

where $R$ is the characteristic length (radius) of the dispersed phase, $\mu_c$ is the viscosity of the continuous phase and $\gamma$ is the interfacial tension. The initial characteristic length of the dispersed phase is the pellet radius, which is not large enough for interfacial forces, $\gamma/R$, to play any role at that stage. For example, a dispersed system with characteristic length 2 mm, interfacial tension 30 mN/m, continuous phase viscosity 100 Pa·s and subjected to a shear rate of $100\,\text{s}^{-1}$ experiences a viscous disruptive stress of 10,000 Pa whereas the resisting interfacial tension stress is only 15 Pa.

As blending proceeds, the characteristic length of the dispersed phase decreases to the point of equilibrium between the disruptive hydrodynamic and cohesive interfacial tension forces. Of course, during the blending process, dispersed droplets come in contact with each other and may coalesce, so that coalescence and break-up are two competitive mechanisms in polymer blends. In the final blending stages, miscible and immiscible systems behave differently. On the one hand, in miscible systems homogenization is achieved to a very small scale, possibly the molecular scale, if sufficient time is allowed. On the other hand, immiscible systems exhibit a two-phase structure whose characteristics depend on the physical parameters of both polymer phases.

In summary, miscible and immiscible systems show similar behavior in the initial steps of the dispersive mixing process, where hydrodynamic forces deform and disrupt the units of the dispersed phase. In the next stages, interfacial tension forces come into play and induce motion (interfacial tension driven Rayleigh or capillary disturbances). Then, at the final stages, miscible systems are expected to be homogenized at the molecular level (if sufficient time is allowed), and immiscible systems retain the coarser structure of a two-phase system.

Next, we give some fundamental ideas on drop deformation and break-up in shear and extensional fields, which are pertinent to the polymer blend area. We consider first simple shear flow. Figure 6.18.a shows the deformation of an initially spherical droplet in a simple shear field. In a shear field the droplet assumes an ellipsoidal shape with a long axis $L$ and short axis $B$, and it is oriented at an angle $\phi$ with respect to the $y$ axis (see also Fig. 6.11). Theoretical analysis of small deformation, $D$, was carried out by Taylor (1934) and Cox (1969), and it applies to a single Newtonian droplet suspended in a Newtonian medium. The deformation of the droplet is defined as:

$$D \equiv \frac{L - B}{L + B}. \tag{6.184}$$

The above definition is appropriate for mildly deformed droplets. However, for highly deformed droplets (Fig. 6.18.b) the appropriate measure of deformation is $D = L/R$, where $L$ is now the half-length and $R$ is the radius of the undeformed droplet. The steady simple shear solution presented by Cox (1969), which in the limits includes the theory of Taylor (1934), is:

$$D = \frac{5}{4} \frac{19p + 16}{(p+1)\sqrt{(19p)^2 + (20/Ca)^2}}, \tag{6.185}$$

where $p$ is the ratio of the viscosity of the dispersed to the continuous phase, $\mu_d/\mu_c$. The limiting situations of $p \to \infty$ and Ca→0 can be easily determined from Eq. 6.185 as:

$$D = \frac{5}{4p} \quad \text{for} \quad p \to \infty; \qquad Ca = O(1) \tag{6.186}$$

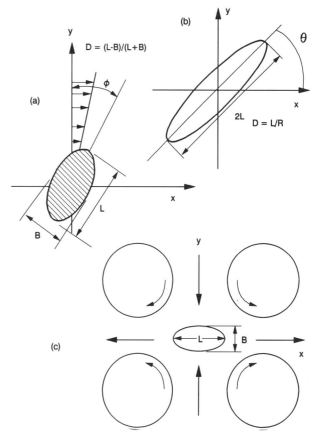

**Figure 6.18** (a) Geometry and definition of the deformation $D$ of a mildly deformed droplet in simple shear. (b) Geometry and definition of the deformation $D$ of a highly extended droplet in simple shear. (c) Four-roll apparatus generating plane hyperbolic (planar extensional) flow of various strengths.

and

$$D = Ca\left(\frac{19p + 16}{16p + 16}\right) = Ca\, f(p) \quad \text{for} \quad Ca \to 0; \qquad p = O(1), \tag{6.187}$$

where $O(1)$ stands for "on the order of magnitude of 1." The orientation of the deformed droplet is

$$\phi = \frac{\pi}{4} + \frac{1}{2}\arctan\left(\frac{19p\,Ca}{20}\right). \tag{6.188}$$

The limitations of this theory are small deformations (much below the bursting limit), isolated droplets, and Newtonian fluids. Figure 6.19 shows the deformation, $D$, as a function of the parameter $\dot{\gamma} R$ for $p = 0.0037$. For small values of $\dot{\gamma} R$, the linear region corresponds to Eq. 6.187. At larger and more realistic values of the shear rate, the curve is no longer linear.

As the deformation of the droplets increases, they assume elongated shapes, and finally at some value of the capillary number, called the critical capillary number, $Ca_c$, the disruptive forces exceed the cohesive forces and the droplets burst. An extensive experimental analysis of large deformations that lead to droplet breakup was

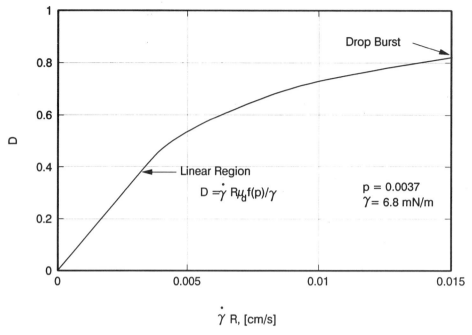

**Figure 6.19** Deformation $D$ of a droplet as a function of the velocity gradient across the droplet, $\dot{\gamma}R$. (Reprinted by permission of the publisher from Grace, 1982.)

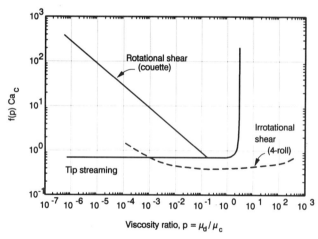

**Figure 6.20** Critical capillary number versus viscosity ratio $p$. (Reprinted by permission of the publisher from Grace, 1982.)

conducted by Grace (1982) and is presented in Fig. 6.20. In this figure, the critical capillary number times a function of the viscosity ratio, $f(p)$, where $f(p)=(19p+16)/(16p+16)$, is plotted against the viscosity ratio for the two modes of bursting, that is, tip streaming (for $p<0.1$) and regular bursting. Tip streaming refers to the situation where droplets assume a sigmoidal shape with tiny droplets shedding off the tips. The important feature shown in Fig. 6.20 is the inability of shear flows to cause droplet breakup at viscosity ratios exceeding 3.5. Note that for this curve to be applicable to other systems, the increase of the shear rate should be stepwise with very small steps. If the shear increases in large steps, the droplets fragment, and the size of the fragments is not correlated by this curve. The left arm of the shear flow curve can be

represented as:

$$Ca_c f(p) = Ca_c \frac{19p+16}{16p+16} = 0.16 p^{-0.6}, \tag{6.189}$$

which seems to agree with some theoretical predictions.

Another important parameter of the droplet break-up process is the time necessary for the interfacial-driven instabilities to cause breakup, $t_b$, when the actual capillary number exceeds the critical capillary number. Grace (1982) provided this information in Fig. 6.21 for Newtonian fluids. Note that the dimensionless burst time is denoted as $t_b^*$, which is equal to $t_b/\tau$, where $\tau$ is the time scale of the bursting process and is equal to $R\mu_c/\gamma$. For example, for a polymer blend with $p=0.1$, $\gamma=10\,\text{mN/m}$, $R=10\,\mu\text{m}$, $\mu_c=1{,}000\,\text{Pa}\cdot\text{s}$ and $Ca/Ca_c=10$, the dimensionless burst time is 11 and the time scale is equal to 1 s. Thus, the burst time, $t_b$, is equal to 11 s.

We consider next planar elongational flow. The experimental setup is a four-roll apparatus (Fig. 6.18.c) used first by Taylor (1934). Cox (1969) developed a theory similar to that for simple shear flow:

$$D(t)=2Ca\left(\frac{19p+16}{16p+16}\right)\left(1-\exp\left(-\frac{19p}{20Ca}t\dot{\varepsilon}_{plel}\right)\right), \tag{6.190}$$

where $\dot{\varepsilon}_{plel}$ is the extension rate in this pure shear case and the capillary number is defined based on the shear viscosity and $\dot{\varepsilon}_{plel}$. This equation shows that even a high-viscosity dispersed phase can be deformed. Data on pure shear are shown in Fig. 6.20 as taken from Grace (1982). Two differences should be pointed out. One is the ability of pure shear flow to break droplets even at high-viscosity ratios, and the other is the lower value (about 1/3) of the minimum capillary number for pure shear flow compared to simple shear flow. Note that for Newtonian fluids the viscosity in pure shear is four times the simple shear viscosity, and this substitution in the capillary number in this figure nearly brings the two curves together.

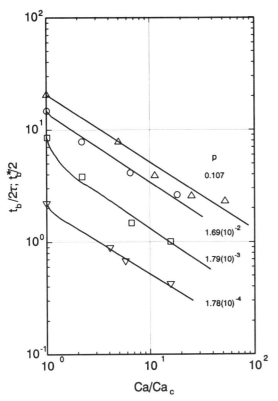

**Figure 6.21** Effect of exceeding the critical capillary number on burst time, $t_b$. $\tau = R\mu_c/\gamma$ is the time scale for burst. (Reprinted by permission of the publisher from Grace, 1982.)

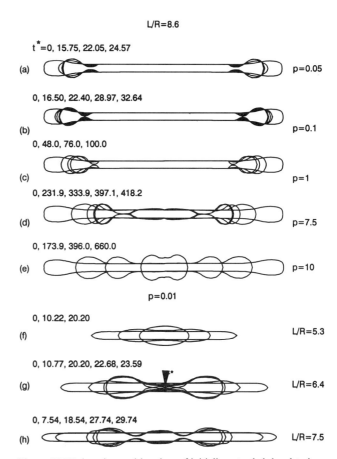

**Figure 6.22** Relaxation and breakup of initially extended droplets in a quiescent fluid. The time scale is $\tau = R\mu_c/\gamma$. The dimensionless time $t^*$, equal to $t/\tau$, is shown above each schematic. (Reprinted by permission of the publisher from Stone & Leal, 1989.)

Finally, we consider uniaxial elongational flow. This flow is more frequently encountered in practice than pure shear flow. Material flowing in a conical die experiences this type of flow. Van der Reijden-Stolk and Sara (1986) studied this flow and found that the deformation grows as:

$$D(t) = \frac{3}{2} \text{Ca} \left( \frac{19p+16}{16p+16} \right) \left( 1 - \exp \left( -\frac{19p}{20\text{Ca}} t\dot{\varepsilon}_{unel} \right) \right), \quad (6.191)$$

where $\dot{\varepsilon}_{unel}$ is the extension rate for uniaxial elongational flow. Comparison of the three flows in terms of the necessary extension rate to achieve the same deformation at long times, that is comparing Eqs. 6.187, 6.190, and 6.191, shows that:

$$\dot{\gamma} : \dot{\varepsilon}_{plel} : \dot{\varepsilon}_{unel} = 6 : 3 : 4. \quad (6.192)$$

The complete picture is formed only when we compare power expended to achieve the same deformation, with the aid of Eqs. 6.78, 6.79, and 6.80:

$$P_{unel} : P_{plel} : P_{ss} = (4/3) : 1 : 1. \quad (6.193)$$

Thus, for Newtonian fluids simple shear and planar elongational flows require the same power to produce the same deformation, whereas uniaxial elongation requires 33% more power.

After the droplet has been extensively deformed (actual capillary number much higher than the critical capillary number) by one of the mechanisms mentioned, it will assume a threadlike shape with a long cylindrical midsection and two bulbous ends when $p > 0.05$, and a long slender shape with nearly pointed ends when $p < 0.05$ (Stone & Leal,

1989). These shapes are unstable, because they can breakup in the flow field because of interfacial tension–driven instabilities (Rayleigh or capillary instabilities; Tomotika, 1935). Moreover, if the flow stops (or the thread moves to a relatively more quiescent environment), the thread can either relax back to its original droplet shape or breakup, depending on the extension $L/R$ and the viscosity ratio. Fig. 6.22.f shows a thread relaxing for $p = 0.01$ and $L/R = 5.3$. Figures 6.22.a to 6.22.e show breakup by shedding off of small droplets from the bulbous ends ("end-pinching" mechanism; Stone & Leal, 1989) for $0.05 < p < 10$ and $L/R = 8.6$. Finally, Fig. 6.23 shows the combined action of instabilities and end-pinching for $p = 1$ and $L/R = 14$. Apparently, the time scale of the end-pinching mechanism is less than the time scale of the instability mechanism in cases where end-pinching prevails.

Interfacial driven disturbances on a Newtonian thread embedded in a Newtonian matrix are briefly discussed next. An initially long cylindrical thread with midsection radius $R_0$ is sinusoidally disturbed by a wave of interfacial tension origin as shown in Fig. 6.24.a. Without going into the details of the analysis we note that the burst time (Elmendorp, 1991) is given as:

$$t_b = \frac{1}{q} \ln \left( \frac{0.82 R_0}{a_0} \right), \quad (6.194)$$

where $1/q$ is the time constant of the distortion process and $a_0$ is the

p=1
L/R=14

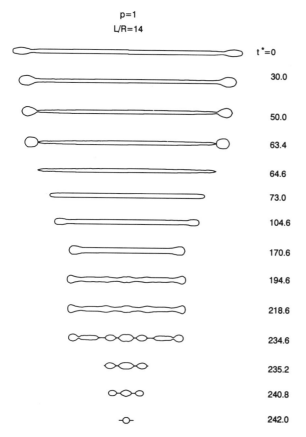

t*=0

30.0

50.0

63.4

64.6

73.0

104.6

170.6

194.6

218.6

234.6

235.2

240.8

242.0

**Figure 6.23** Evolution of end-pinching and capillary waves during the relaxation and breakup of an initially highly extended droplet suspended in a quiescent fluid. The time scale is $\tau = R\mu_c/\gamma$. The dimensionless time $t^*$, equal to $t/\tau$, is shown at the side of each schematic. (Reprinted by permission of the publisher from Stone & Leal, 1989.)

distortion amplitude at time equal to zero. $1/q$ is given by the equation:

$$\frac{1}{q} = \frac{2\mu_c R_0}{\gamma} \frac{1}{\Omega(\lambda, p)} = \frac{2\tau}{\Omega(\lambda, p)}, \tag{6.195}$$

where $\Omega(\lambda, p)$ is a complicated function of the wavelength of the distortion, $\lambda$, and of the viscosity ratio, $p$. This function is shown in Fig. 6.24.b for $\lambda = \lambda_{max}$, which is the wavelength leading to breakup. Then Eq. 6.194 becomes:

$$t_b^* = \frac{t_b}{\tau} = \frac{2}{\Omega(\lambda_{max}, p)} \ln\left(\frac{0.82 R_0}{a_0}\right). \tag{6.196}$$

Although the above equations apply to Newtonian systems, it was shown experimentally that they also apply to situations encountered in polymer blending (Elmendop, 1991). The absence of the effect of viscoelasticity on polymer thread break-up is attributed to the very low deformation rates of the final stages of break-up ($10^{-5}$ to $10^{-3}\,\mathrm{s}^{-1}$). At these rates it can be assumed that most polymers exhibit Newtonian behavior with viscosities equal to their zero-shear values. Examples for the use of these equations follow.

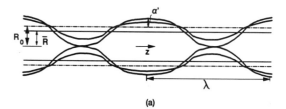

(a)

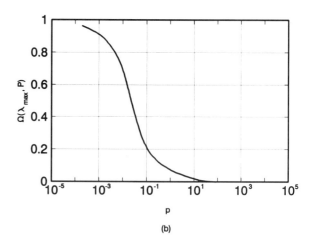

(b)

**Figure 6.24** (a) Capillary waves in an initially uniform-midsection-radius thread. $\lambda$ is the distortion wavelength, $a'$ is the distortion amplitude, $R_0$ is the initial thread radius, $\bar{R}$ is the average thread radius, and $z$ is the axis of the thread. (b) Dominant growth rate $\Omega(\lambda_{max}, p)$ as a function of the viscosity ratio $p$. (Reprinted by permission of the publisher from Elmendorp, 1991.)

**EXAMPLE 6.12**
**Burst Time of a Polymer Cylindrical Thread**
Calculate the dimensionless and the actual burst time for a polymer cylindrical thread of radius 5 μm, viscosity ratio 1, interfacial tension 10 mN/m submerged in a polymeric fluid of zero-shear viscosity 100 Pa·s. What is the dimensionless burst time for $p = 0.1$?

**Solution**

The time scale of the bursting process is $\tau = R_0 \mu_c/\gamma = 0.05$ s. $\Omega = 0.07$ at $p = 1$ from Fig. 6.24.b. The amplitude of the distortion that leads to burst is usually considered to be 0.3% of the initial radius of the midsection. Then Eq. 6.196 gives the dimensionless burst time as:

$$t_b^* = \frac{2}{0.07} \ln\left(\frac{0.82}{0.003}\right) = 160. \tag{6.197}$$

Thus the actual burst time is $160 \times 0.05\,\mathrm{s} = 8\,\mathrm{s}$. Similarly, for a viscosity ratio of 0.1, $t_b^* = 53$.

Figure 6.23 can be used to compare the dimensionless times for end-pinching and instability growth. It is seen from this figure that the dimensionless time for end-pinching is about 64, whereas, the dimensionless time for instability growth is 160, as calculated in Example 6.12. Fig. 6.23 shows also that the dimensionless time for complete breakup is about 240.

In the analysis of drop breakup it is assumed that a single drop exists in an infinite sea of the polymer matrix. In practice, this is not the case. The effect of the volume fraction of the minor component can be double: (1) The presence of drops alters the hydrodynamics around the other drops; and (2) coalescence can take place because of the increased probability of collisions. Mathematically, both phenomena can be grouped into a linear relationship (without rigid foundations; borrowed from experiments in agitated tanks):

$$R = R_T(1 + k\phi), \tag{6.198}$$

where $R_T$ refers to the drop radius according to Taylor's theory (Eq. 6.187), $k$ is a constant and $\phi$ is the volume fraction of the minor component. Typical values of $k$ range from 5 to 200 (Elmendorp, 1991).

### EXAMPLE 6.13
### Blend Morphology

A PP/PS blend is formed in a roll-mill apparatus, and the cooling step to temperatures below the glass transition temperature of PS ($\simeq 100°C$) takes about 30 s. Comment on the possibility that initially in the roll-mill threads were present that later on burst to smaller droplets because of instabilities. The final morphological examination showed PP droplets of 10 μm in size. Assume that $\gamma = 5$ mN/m and that Newtonian viscosities $\mu_c = \mu_d = 1,000$ Pa·s.

### Solution

We test the scenario of small droplets (of radius 10 μm) undergoing end-pinching to form elongated threads with a thread midsection radius of about 4 μm. Then, the extension ratio is $L/R = 6 \times (4/3) \times (R/R_0)^2 = 50$ if we assume that each thread produces six droplets. The calculation of the burst time due to capillary instabilities follows exactly Ex. 6.12, and it is equal to 128 s. But because the extension ratio is large, both end-pinching and capillary instabilities play a role. Fig 6.23 shows that end-pinching precedes burst by capillary instabilities and that the starting time for end-pinching was one-third of the time for capillary break-up. Thus, if the same analogy holds in our system, at about 40 s two droplets will be shed off the ends of the thread. This time is slightly longer than the cooling time. For $R_0 = 3$ μm, $t_b = 95$ s. The conclusion is that the possibility of threads breaking up during the cooling step cannot be totally ruled out.

The following example shows the application of ideas from drop breakup in extruders using a simplified flow theory.

### EXAMPLE 6.14
### Thread Breakup in Extruders

A melt-fed extruder is used for melt blending of two polymers with equal constant viscosities of 100 Pa·s and interfacial tension of 5 mN/m. The maximum shear rate in the screw channel is about 110 s⁻¹. The flight clearance in the extruder, $\delta_f$, is 250 μm, the barrel diameter is $D = 2.54$ cm, the screw rotational frequency is $N = 100$ rpm, the flight width is $e = 0.254$ cm, the mass flow rate is $\dot{m} = 1.4$ g/s, the polymer density is $\rho = 1$ g/cm³, and the mean residence time is $\bar{t} = 138$ s. The molten feed into the extruder consists of a dispersion of 5 μm in radius.

Assuming that the melt passes through the flight clearance once, is it possible that the final morphology consists of an even finer dispersion? Neglect any effect of extensional flows which occur as the fluid passes from the channel through the flight clearance.

### Solution

The shear rate experienced by the blend when it passes through the flight clearance is calculated as (see Subsection 8.3.3):

$$\dot{\gamma}_{\text{clearance}} = \frac{\pi DN}{\delta_f} \simeq 550 \, \text{s}^{-1}. \tag{6.199}$$

Using Fig. 6.20, the critical capillary number for $p = 1$ and shear flow is equal to 0.79. The actual capillary number in the flight clearance is 55, so that the actual value of Ca exceeds the critical value 69 times. The dimensionless burst time for that ratio of capillary numbers is calculated using Fig. 6.21 to be at least 2 (in this figure there is no entry for $p = 1$, so that the corresponding value for viscosity ratio of 0.107 is used as the lower limit). The time scale of the burst process, $\tau$, is equal to $R\mu_c/\gamma = 0.1$ s, and thus the actual burst time is $t_b > 4\tau = 0.4$ s. The residence time in the flight clearance is:

$$\bar{t}_{\text{flight}} = \frac{\pi \rho De \delta_f}{\dot{m}} = 0.03 \, \text{s}. \tag{6.200}$$

Thus, the residence time in the flight clearance is not enough for burst to occur. However, the droplets during the flow in the flight clearance can be deformed into threads, because the shear strain in the clearance is equal to $550 \times 0.03 = 16.5$. According to Eqs. 6.104 and 6.105, the thread will have half-length and half-width to initial radius ratios of:

$$\frac{L}{R} = \sqrt{1 + \gamma^2} = 16.5; \qquad \frac{B}{R} = (1 + \gamma^2)^{-1/4} = 0.25. \tag{6.201}$$

So the total length of the thread will be 160 μm and its radius 1.25 μm. This thread will be subjected to end-pinching and capillary instability. The burst time due to the capillary instability is calculated as in Example 6.12. It is equal to 4 s, and the mean residence time in the extruder is 138 s. Thus, the final morphology will include finer droplets of the dispersed phase.

## 6.6. THERMODYNAMICS OF MIXING

In Section 6.5 polymer–polymer miscibility was considered to be an important parameter in the dispersion of the two polymer phases. In this respect a polymer–polymer system can be considered as miscible, immiscible, or partially miscible depending on the relative solubility of the two polymers. Total solubility characterizes a miscible system, insolubility an immiscible, and partial solubility a partially miscible system. The degree of miscibility has important effects on the mechanical, physical, rheological, and optical properties of the resulting blend.

Measuring the glass transition temperature is the most common technique to detect miscibility. When a polymer blend is immiscible, the two constituent polymers will keep their identity and thus exhibit two distinct glass transition temperatures. However, a miscible blend will consist of one phase and exhibit a single glass transition temperature (in between the glass transition temperatures of the constituent polymers). Finally, a partially miscible system will exhibit two glass transition temperatures, shifted toward each other with respect to the

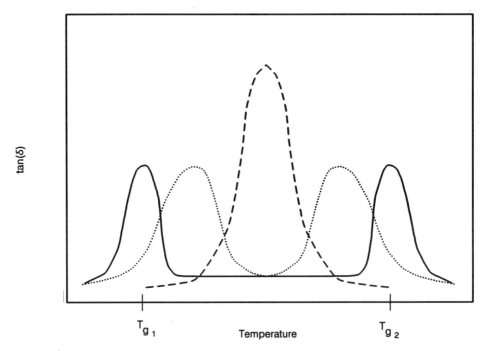

**Figure 6.25** Loss angle, δ, as a function of temperature for various types of polymer blends: (a) miscible (dashed line), (b) immiscible (solid line), and (c) partially miscible (dotted line). The glass transition temperature, $T_g$, is the temperature at the peak of the loss angle.

original glass transition temperatures of the two polymers. Fig. 6.25 shows schematically this behavior in a typical tan δ- (δ is the loss angle) versus-temperature graph.

Thermodynamically, mixing will take place when it is favored energetically, that is, when

$$\Delta G_{\text{mix}} = \Delta H_{\text{mix}} - T \Delta S_{\text{mix}} \leqslant 0, \tag{6.202}$$

where $\Delta G_{\text{mix}}$, $\Delta H_{\text{mix}}$ and $\Delta S_{\text{mix}}$ are the changes of the Gibbs free energy, enthalpy (or heat) and entropy, respectively, of mixing. The heat of mixing, $\Delta H_{\text{mix}}$, is usually a positive quantity (endothermic process). The entropy of mixing is also positive because the randomness of the mixture is higher than the randomness of the components. It should be noted that the condition of Eq. 6.202 is necessary, although not sufficient, for the formation of a stable solution. The heat of mixing is related to a parameter called the *solubility parameter*, δ, through the relationship:

$$\Delta H_{\text{mix}} = V(\delta_1 - \delta_2)^2 \phi_1 \phi_2, \tag{6.203}$$

where $V$ is the volume of the mixture and $\phi_1$ and $\phi_2$ are the volume fractions of the solvent and solute, respectively. If $\delta_1 = \delta_2$, then $\Delta H_{\text{mix}} = 0$ and $\Delta G_{\text{mix}} \leqslant 0$, and so the components will be miscible in all proportions (for equally hydrogen-bonding–capable polymers). Table 6.2 includes some typical values of the solubility parameter for polymers. Finally, the interfacial tension of polymer systems was shown to be very important in the area of dispersive mixing as a determining factor of the final blend morphology. Table 6.3 summarizes interfacial tensions of polymer pairs.

## 6.7. CHAOTIC MIXING

A recent advance in the area of mixing is that of *chaotic mixing*. Mixing was analyzed in Section 6.3 in terms of intermaterial area generation

**TABLE 6.2 Solubility Parameter Values for Various Polymers**

| Polymer | δ (MPa$^{1/2}$) From | To |
|---|---|---|
| PE | 15.8 | 17.1 |
| PP | 16.8 | 18.8 |
| PS | 17.4 | 19.0 |
| PVC | 19.2 | 22.1 |
| PMMA | 18.6 | 26.2 |
| PAN | 25.6 | 31.5 |
| PB | 16.6 | 17.6 |

(Data from D. W. Van Krevelen, *Properties of Polymers*, Elsevier, New York, 1990.)

**TABLE 6.3 Interfacial Tension Values for Typical Polymer Pairs**

| Polymer Pairs | γ, (mN/m) at 140°C | −(dγ/dT) (mN/m·°C) |
|---|---|---|
| PE/PP | 1.1 | — |
| PE/PS | 5.9 | 0.020 |
| PE/PMMA | 9.7 | 0.018 |
| PE/PEO | 9.7 | 0.016 |
| PP/PS | 5.1 | — |

(Data from S. Wu, *Polymer Interfaces and Adhesion*, Marcel Dekker, New York, 1982.)

or reduction of striation thickness. It was shown that simple shear flows, which are extensively present in the area of polymer mixing, have time-decaying average mixing efficiencies (Fig. 6.10). This flow, which is thus considered a weak mixing flow, owes its mixing inefficiency to the eventual orientation of the lineal or intermaterial area along the streamlines. However, simple flows can exhibit chaotic mixing if they become periodic in nature. In that case, the intermaterial area increases exponentially with time, thus providing an effective means of mixing.

Aref (1984) indicated that the equations that describe the particle trajectories in a two-dimensional flow have a *Hamiltonian* structure, that is,

$$v_x = \frac{dx}{dt} = \frac{\partial \psi}{\partial y}; \qquad v_y = \frac{dy}{dt} = -\frac{\partial \psi}{\partial x}, \tag{6.204}$$

where $\psi$ is the stream function. A Hamiltonian system is a physical system of particles whose motions are described by deterministic equations. If $\psi$ in Eq. 6.204 is independent of time, then the velocity field is steady and the system cannot be chaotic. However, if $\psi$ is time-periodic, that is, depends on $x$, $y$, and $t$, then there is a good chance that the system will be chaotic. Aref (1984) applied this idea to the blinking vortex system, which is a system consisting of two alternating corotating vortices. These vortices switch on and off for half the cycle time.

A flow can be termed chaotic if it satisfies any of the following criteria:

1. Positive *Liapunov exponents* in a given region of the flow
2. Presence of *transverse homoclinic* or *heteroclinic* points
3. Presence of *Smale horseshoe maps*

The *Liapunov exponent*, $\sigma$, is related to the long time behavior of the lineal stretch, and it is equal to:

$$\sigma = \lim_{t \to \infty} \frac{\ln(\lambda)}{t}, \tag{6.205}$$

where $\lambda$ is the lineal stretch given by Eq. 6.86. Note that the Liapunov exponent of a simple steady shear flow is 0, because as $t \to \infty$ the term $\ln(\dot{\gamma}t)/t$ goes to zero, and thus simple steady shear flow cannot produce chaos.

The definition of the *homoclinic* and *heteroclinic* points needs first the introduction of *hyperbolic* and *elliptic* points. A two-dimensional flow always consists of hyperbolic and/or elliptic points (Fig. 6.26). At the hyperbolic point the fluid moves toward it in one direction and away from it in another direction. At an elliptic point the fluid moves in closed pathlines. A periodic point is defined as the point at which a particle in a periodic flow returns after a number of periods. The

number of periods defines also the order of the periodic point, as periodic point of period 1, 2 and so on. Note that the periodic elliptic points should be avoided should we want enhanced mixing. A point where the outflow of one hyperbolic point intersects the inflow of another hyperbolic point is called, *transverse heteroclinic* point. When the inflow and outflow refer to the same hyperbolic point, the point is called a *transverse homoclinic* point.

The last identifying feature of chaos is the presence of *Smale horseshoe maps* (Fig. 6.27.a). A typical map involves the stretching and folding of a square with itself. Mixing has been promoted in that sense because the perimeter of the initial square has increased or the striation thickness has decreased. The Smale horseshoe map is similar to the *baker's transformation*, Fig. 6.27.b, which involves stretching, cutting, and stacking of fluid elements (see also Spencer & Wiley, 1951), and it is considered to be the best possible mixing from the mathematical point of view.

The simplest flow that can exhibit chaos is two-dimensional flow. Ottino and co-workers (Chien et al., 1986; Khakhar et al., 1986; Leong & Ottino, 1989, to name a few) produced chaotic mixing in simple prototypical devices such as cavity flow, partitioned-pipe mixer (e.g., a Kenics static mixer; as discussed in Section 8.5), and eccentric helical annular mixer with Newtonian fluids. Of prime interest in the area of polymer processing is, of course, the work in cavity flows. A typical cavity was constructed with the ability of movement of both top and bottom plates. Typical cavity flow, which is described in Chapter 8, corresponds to the steady movement of the top plate only. However, corotational (in the opposite direction) movement of both plates in a periodic fashion induces chaos in the cavity. Leong and Ottino (1989) used two types of movement: discontinuous and continuous in a sinusoidal manner (Fig. 6.28). In the discontinuous corotational flow, the top plate first moves for a half-period, then it stops for 5 s, and the cycle ends with the bottom plate moving for a half-period in the opposite direction. In the continuous type of movement, both plates move sinusoidally at the same time but with a phase difference of $\pi/2$.

In terms of stretching and dispersion, mixing achieved in the discontinuous corotational periodic-type of flow is better than the

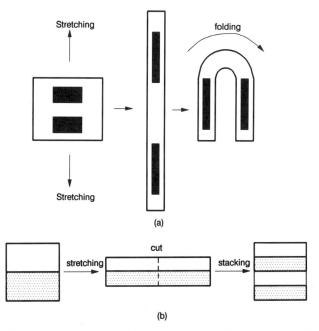

Figure 6.27 (a) Representation of typical Smale horseshoe map. (b) Representation of the baker's transformation.

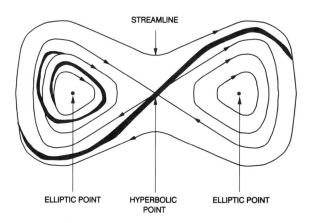

Figure 6.26 Elliptic and hyperbolic points. A blinking vortex system with vortex centers at the elliptic points can produce this streamline pattern.

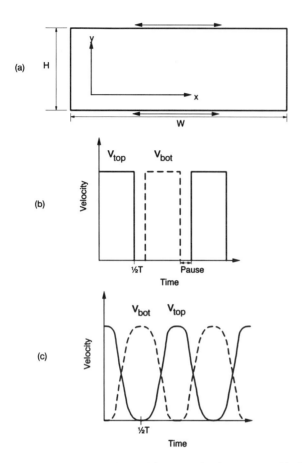

**Figure 6.28** (a) Typical cavity used for corotational movement of both plates. (b) Top and bottom plate motion in discontinuous form. The solid line represents motion of the top plate, whereas the dashed line represents motion of the bottom plate. Pause is for 5 s. (c) Top and bottom plate motion in continuous and sinusoidal form. (Reprinted by permission of the publisher from Leong & Ottino, 1989.)

steady flow, as shown in Fig. 6.29. This figure shows the streamlines of a fluorescent dye injected below the top plate at the center of the cavity at time $t = 0$. The perimeter of the dye regions is indicative of the degree of mixing. Figure 6.30 shows the exponential growth of the degree of mixing for both the discontinuous and the continuous type of periodic flow and the linear growth for steady flow. Furthermore, the efficiency of the periodic flows is higher for the discontinuous case than for the continuous, and the form of the exponential increase is:

$$P = P_0 e^{\beta t}, \tag{6.206}$$

where $P$ is the perimeter and $\beta$ can be considered to be the average Liapunov exponent for the process.

A relevant question at this point is, Can we take advantage of chaos in typical polymer processing conditions? The answer is that we need more understanding of the chaotic processes before we can design better mixers for polymers. This design, when accomplished, will be considered as a major step forward in mixing.

## 6.8. SOLUTION TO DESIGN PROBLEM 5

We finally return to the Design Problem 5 (Fig. 6.31). The flow in the die is helical in nature; that is, it consists of an axial Poiseuille

flow and a drag Couette rotational flow due to the rotation of the mandrel. The analysis of the striation thickness of each layer is based on simple geometrical and kinematical arguments, and it will be shown that the two approaches give the same results. For Newtonian fluids, the axial and angular flow fields are independent and given in Table 2.7 and Example 6.7, respectively.

The volumetric flow rate is calculated (Table 2.7) as:

$$Q = \int_{\kappa R}^{R} v_z(r) 2\pi r dr = \frac{\pi \Delta P R^4}{8\mu L} \left[ 1 - \kappa^4 - \frac{(1-\kappa^2)^2}{\ln(1/\kappa)} \right]. \tag{6.207}$$

The axial velocity can now be written as:

$$v_z = \frac{2Q}{\pi R^2} \frac{1 - \left(\dfrac{r}{R}\right)^2 + \dfrac{1-\kappa^2}{\ln(1/\kappa)} \ln\left(\dfrac{r}{R}\right)}{1 - \kappa^4 - \dfrac{(1-\kappa^2)^2}{\ln(1/\kappa)}}. \tag{6.208}$$

The deformation of a material element as it rotates in the $r\theta$ plane is shown in Fig. 6.32.a, and as it translates down the $z$ axis in Fig. 6.32.b. After time $t$ the material has experienced both deformations. For simplicity we assume that the axial deformation is negligible compared to the rotational deformation:

$$\frac{dv_\theta}{dr} \gg \frac{dv_z}{dr}. \tag{6.209}$$

Thus, the deformation in Fig. 6.32.b is considered negligible. From purely geometrical considerations the striation thickness, $\delta$, and the initial striation thickness, $\delta_0$, are related:

$$\frac{\delta}{\delta_0} = \sin\beta, \tag{6.210}$$

where the angle $\beta$ is related to $d\theta$ through the relation:

$$\tan\beta = \frac{1}{r\dfrac{d\theta}{dr}}. \tag{6.211}$$

But $\sin\beta \simeq \tan\beta$ for small $\beta$ angles, and Eq. 6.210 with the aid of Eq. 6.110 becomes:

$$\frac{\delta}{\delta_0} = \frac{1}{r\dfrac{d\theta}{dr}} = \frac{1-\kappa^2}{2\kappa^2 Wt}\left(\frac{r}{R}\right)^2. \tag{6.212}$$

The residence time $t$ is related to the axial velocity as:

$$t(r) = \frac{L}{v_z(r)}, \tag{6.213}$$

and the combination of Eqs. 6.213 and 6.212 yields:

$$\frac{\delta}{\delta_0} = \frac{Q}{\pi WLR^2} \frac{1-\kappa^2}{\kappa^2}\left(\frac{r}{R}\right)^2 \frac{1 - \left(\dfrac{r}{R}\right)^2 + \dfrac{1-\kappa^2}{\ln(1/\kappa)} \ln\left(\dfrac{r}{R}\right)}{1 - \kappa^4 - \dfrac{(1-\kappa^2)^2}{\ln(1/\kappa)}}. \tag{6.214}$$

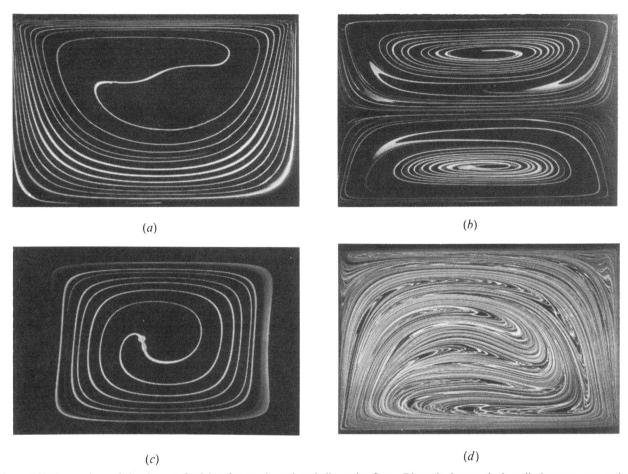

(a)          (b)

(c)          (d)

**Figure 6.29** Comparison of the degree of mixing for steady and periodic cavity flows. Dimensionless total plate displacement per period, $N_d = DP$, where $D$ is the dimensionless plate displacement per period and $P$ is the total number of periods. (a) Steady flow; Re = 1.0; total mixing time is 300 s; $N_d = 55$. (b) Steady flow; both plates moving in the same direction; total mixing time is 300 s; $N_d = 110$. (c) Steady flow; plates moving in the opposite direction; total mixing time is 300 s; $N_d = 110$. (d) Periodic discontinuous flow; total mixing time is 280 s (4 periods); $N_d = 51.4$. (Reprinted by permission of the publisher from Leong & Ottino, 1989.)

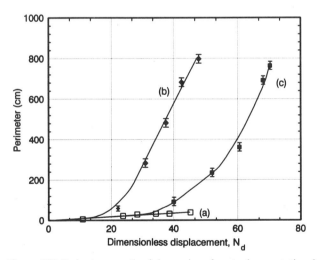

**Figure 6.30** Perimeter growth of dye regions for steady, corotational discontinuous and continuous cavity flows, as a function of the dimensionless displacement, $N_d$. (a) Steady flow. (b) Discontinuous. (c) Continuous. (Reprinted by permission of the publisher from Leong & Ottino, 1989.)

By noting that

$$\bar{t} = \frac{\pi L R^2 (1 - \kappa^2)}{Q}, \tag{6.215}$$

Eq. 6.214 reduces to:

$$\frac{\delta}{\delta_0} = \frac{1}{W\bar{t}} \frac{(1-\kappa^2)^2}{\kappa^2} \left(\frac{r}{R}\right)^2 \frac{1 - \left(\frac{r}{R}\right)^2 + \frac{1-\kappa^2}{\ln(1/\kappa)} \ln\left(\frac{r}{R}\right)}{1 - \kappa^4 - \frac{(1-\kappa^2)^2}{\ln(1/\kappa)}}$$

$$= \frac{f(\kappa, r/R)}{W\bar{t}} \tag{6.216}$$

The radial dependence of the striation thickness reduction function is shown in Fig. 6.33.a for various values of the parameters $\kappa$ and $W\bar{t}$. As is expected, the maximum reduction takes place at the center part of the gap, and the minimum reduction takes place at the cylinders. Also at the gap the higher the parameter $W\bar{t}$, the smaller the striation thickness is. Differentiation of Eq. 6.216 yields the radial position where the maximum reduction in striation thickness takes place, $r_{max}$. The

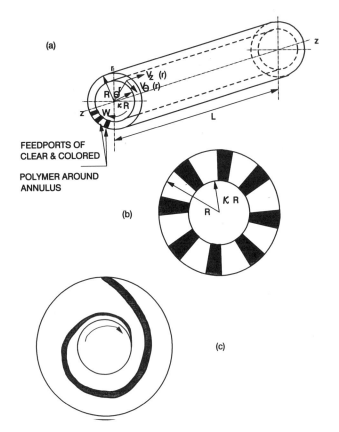

(a)

FEEDPORTS OF
CLEAR & COLORED

POLYMER AROUND
ANNULUS

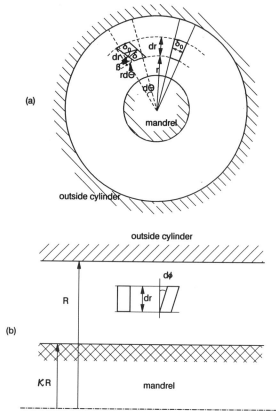

(a)

**Figure 6.31** (a) Geometry of the continuous annular mixer-die. (b) Feed distribution entering the annular die. (c) Mixing pattern generated in the die for a single feedport.

**Figure 6.32** (a) Geometry of the material element deformation at a constant $z$ plane. (b) Geometry of the material element deformation at a constant $\theta$ plane.

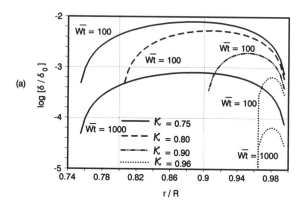

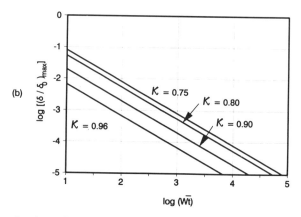

**Figure 6.33** (a) Radial distribution of the striation thickness reduction function for various values of the total angular displacement in radians, $W\bar{t}$. $\kappa$ is the ratio of radii of the mandrel and of the outside cylinder. (b) Maximum reduction of striation thickness as a function of the total angular displacement in radians, $W\bar{t}$.

resulting equation is nonlinear

$$1 - 2\left(\frac{r_{max}}{R}\right)^2 + \frac{1-\kappa^2}{\ln(1/\kappa)}\left[\ln\left(\frac{r_{max}}{R}\right) + \frac{1}{2}\right] = 0, \qquad (6.217)$$

and it can be solved (for example using the IMSL subroutine NEQNF; Appendix D.4) as:

$$r_{max} = \phi(\kappa)R, \qquad (6.218)$$

so that the maximum striation thickness reduction of Eq. 6.216 becomes:

$$\left(\frac{\delta}{\delta_0}\right)_{max} = \frac{f(\kappa, \phi(\kappa))}{W\bar{t}}. \qquad (6.219)$$

Figure 6.33.b shows the maximum striation thickness as a function of the total angular displacement in radians, $Wt$, for various values of the parameter $\kappa$. For specifications that the maximum striation thickness should not exceed a specified value, the minimum rotational frequency, $N_{min}$, is calculated from Eq. 6.219 as:

$$N_{min} = \frac{Q}{2\pi^2 R^2 L (1 - \kappa^2)} f(\kappa, \phi(\kappa)) \left(\frac{\delta}{\delta_0}\right)_{max}^{-1}. \tag{6.220}$$

Using the data of Design Problem 5, for $\delta_0 = (\pi R)/10$, and $\phi(0.96) = 0.98034$ we get:

$$N_{min} = 188 \text{ rev/min}. \tag{6.221}$$

The validity of the above analysis was based on the assumption of Eq. 6.209 that now can be checked to be true ($|dv_\theta/dr| \simeq 16|dv_z/dr|$, at $r = R(\kappa + 1)/2$). The power to rotate the mandrel is calculated as:

$$P_{mandrel} = (2\pi \kappa R L) \tau_{r\theta}(\kappa R) v_\theta(\kappa R), \tag{6.222}$$

which for Design Problem 5 becomes $P_{mandrel} = 3.6 \text{ kW}$. The pumping power is calculated to be: $P_{pumping} = Q \Delta P = 32.5 \text{ kW}$. Suggestions for lowering the power consumptions would be to increase the gap, the temperature, and the rotational frequency. To evaluate the new conditions one should repeat the steps.

Calculations of the striation thickness based on the kinematics developed in Subsection 6.3.2 and outlined in Ex. 6.5 require the use of *scale factors* (Bird et al., 1987). These scale factors are necessary for the calculation of the components of the deformation-gradient tensor in curvilinear coordinates. Note that Eq. 6.83 applies to rectangular coordinates only. Integration of the velocity profiles with respect to time yields:

$$r = r_0; \qquad \theta = \theta_0 + \frac{\kappa^2 Wt}{\kappa^2 - 1} \left[1 - \left(\frac{R}{r}\right)^2\right]$$

$$z = z_0 + Q' \left[1 - \left(\frac{r}{R}\right)^2 + \frac{1 - \kappa^2}{\ln(1/\kappa)} \ln\left(\frac{r}{R}\right)\right] t, \tag{6.223}$$

where $Q'$ is given by:

$$Q' = \frac{2Q}{\pi R^2} \frac{1}{1 - \kappa^4 - \frac{(1 - \kappa^2)^2}{\ln(1/\kappa)}}. \tag{6.224}$$

The only nonzero components of the deformation-gradient tensor are:

$$F_{11} = F_{22} = F_{33} = 1 \tag{6.225}$$

and

$$F_{21} = -\frac{2\kappa^2 WtR^2}{1 - \kappa^2} \frac{1}{r_0^2}$$

$$F_{31} = Q' \left(-2\frac{r_0}{R^2} + \frac{1 - \kappa^2}{\ln(1/\kappa)} \frac{1}{r_0}\right) t. \tag{6.226}$$

Note that the only time the scale factor was used was in the calculation of $F_{21}$:

$$F_{21} = r_0 \frac{\partial \theta}{\partial r_0}. \tag{6.227}$$

Then, the nonzero components of the Cauchy-Green strain tensor are:

$$C_{22} = C_{33} = 1; \qquad C_{12} = C_{21} = F_{21}; \qquad C_{13} = C_{31} = F_{31} \tag{6.228}$$

and

$$C_{11} = 1 + F_{21}^2 + F_{31}^2. \tag{6.229}$$

For a line initially oriented along the $r$ axis, the lineal stretch, $\lambda$, is then given from Eq. 6.86 as:

$$\lambda^2 = 1 + \left[\frac{-2\kappa^2 WR^2 t}{1 - \kappa^2} \frac{1}{r_0^2}\right]^2 + \left[Q' t \left(\frac{-2r_0}{R^2} + \frac{1 - \kappa^2}{\ln(1/\kappa)} \frac{1}{r_0}\right)\right]^2. \tag{6.230}$$

The following two simplifying assumptions will be made. First, the second term on the RHS of Eq. 6.230 is much greater than 1 which is equivalent to angle $\beta$ being small in the geometrical analysis. Second, the second term on the RHS of Eq. 6.230 is much greater than the third term of the same equation, which is equivalent to Eq. 6.209. Then, the analysis based on striation thickness calculations becomes:

$$\frac{\delta}{\delta_0} = \frac{1}{\lambda} \simeq \frac{1 - \kappa^2}{2\kappa^2 Wt} \left(\frac{r}{R}\right)^2, \tag{6.231}$$

which is the same as Eq. 6.212.

## PROBLEMS
### A. Applications

**6A.1.** Probing of Local Structure
Consider a 3% carbon black dispersion in PE. Light is transmitted through a 25-$\mu$m thick sample, and the transmittance is used to characterize the local structure. Determine the required diameter of the light beam to achieve this goal, if the carbon black particle diameter is 1 $\mu$m and the density ratio of carbon black to PE mixture is 1.5.

**6A.2.** Confidence Intervals for Mixtures
A masterbatch of PE and carbon black is used for product formulation. The carbon black weight fraction in the masterbatch is determined by sampling a certain number of pellets from 10 bags. The weight fraction in these 10 bags is: 26.7, 28.0, 33.5, 27.8, 29.3, 31.9, 31.5, 33.6, 30.9, and 34.0. The masterbatch will be mixed with virgin PE to produce a blend with the following specifications: 97.5% of the blended samples should have a carbon black weight fraction of at least 10%. Calculate the additional masterbatch weight fraction that should be blended with the virgin PE due to the variation of the carbon black contained in the masterbatch.

**6A.3.** Statistical Description of Mixtures
Glass fibers are mixed in conventional thermoplastics (e.g., PPS) to enhance the properties of the matrices. Suppose that a requirement for the improvement of the properties is that the number density of the glass fibers is very uniform in a cross section of the final part. Design a statistical analysis that could lead you to assure your client that the properties meet the specifications with a certain level of confidence.

**6A.4.** Mixing Indices
(a) Prove that the values of the mixing index of Eq. 6.15 lie between $1/n$ and $\infty$, where $n$ is the size of the sample.
(b) Calculate the index of mixing for a completely uniform sample, based on Eq. 6.17 as a function of the sample size.

**6A.5.** Correlation Coefficient
Complete all the steps in Ex. 6.3.

**6A.6.** Scale of Segregation
Prove that the ratio of the cube of the linear scale to the volumetric scale of segregation is equal to $3/4\pi$ for the system examined in Ex. 6.3.

**6A.7.** Power Spectrum of Spherical Clumps
Calculate the power spectrum for the clumps shown in Fig. 6.6.a and show that $P(0) = 2 s^2 S_L$. Comment on the relative values of the power spectra.

**6A.8.** Efficiencies of Mixers and Striation Thickness Reduction
Calculate the shear and extension rates required to reduce the striation thickness 1000 times in 10 s in the three mixers: pure shear, uniaxial extensional and simple shear. Then prove Eq. 6.81 assuming Newtonian fluid behavior.

**6A.9.** Interfacial Growth in Simple Shear
A minor component, of volume fraction $\phi$, in the form of cubic pellets is mixed with the major component. Calculate the interface growth function for large shear strains. Furthermore, extend the calculations to the case of different shear viscosities of the minor and major components and of negligible interfacial tension between the components.

**6A.10.** Efficiencies of Mixers for Power-Law Fluids
Calculate the specific power ratios, as in Eq. 6.81, for a power-law fluid with a power-law index of 0.8.

**6A.11.** Effect of Viscosity Ratio on Deformation
Consider a rectangular element of the major component and a smaller rectangular element of the minor component embedded into the first element. Show diagrammatically the shape of both elements after a certain value of shear strain has been imposed for all possible combinations of the viscosities of the two components.

**6A.12.** Rotational Couette Flow for a Power-Law Fluid
Prove that for a rotational Couette geometry with the inside cylinder rotating the striation thickness scales inversely proportional to the shear strain. Furthermore, calculate the proportionality constant for a power-law fluid rotating in a small gap Couette geometry.

**6A.13.** RTD in Poiseuille Flows
Prove that the ratio of the mean residence times in Poiseuille flow in parallel plates to circular pipe is equal to:

$$\frac{\bar{t}_{PPF}}{\bar{t}_{CPPF}} = \frac{1+2n}{1+3n}\left(\frac{1}{2}\right)^{1/n},$$

where $n$ is the power-law index. Assume that the pressure drop per unit length is the same in both cases. Calculate also the $F$ function for the Poiseuille flow of a power-law fluid in a parallel plate geometry.

**6A.14.** Holdback and Segregation in PCF, PPF, and CPPF
   **(a)** Prove that the holdback, $B$, for the Couette flow in the parallel plate geometry is $1/4$.
   **(b)** Prove that the holdback, $B$, for the Poiseuille flow of a Newtonian fluid in the parallel plate geometry is 0.19.
   **(c)** Prove that the segregation, $S$, for the Poiseuille flow in the circular pipe geometry is 0.14.

**6A.15.** Interfacial Tension–Driven Deagglomeration
Determine the critical cluster radius for the attractive Van der Waals forces to become less important than the interfacial tension forces between the cluster and the polymer matrix. Assume that the voids of the cluster are not filled completely with the polymer material and the equal-size particles are randomly packed into the cluster. The physical parameters of the system are: $\gamma = 30$ mN/m, $\varepsilon = 0.9$, Hamaker constant $A = 2 \times 10^{-19}$ J, and particle radius $R = 20$ nm.

**B. Principles**

**6B.1.** "Diffusion" of Particles in Mixers-1
Consider the geometry of the horizontal cylinder-mixer (or single-barrel mixer) of Fig. 6.34. Simulate the random movement of the particles while the cylinder is rotating around its axis by the molecular diffusion in the axial direction (note that diffusion along the radius, due to the rotation of the cylinder, should be much faster than the axial diffusion). The length of the cylinder is $L = 50$ cm, and the black particle concentration is 5%. Initially, the black particles are concentrated in

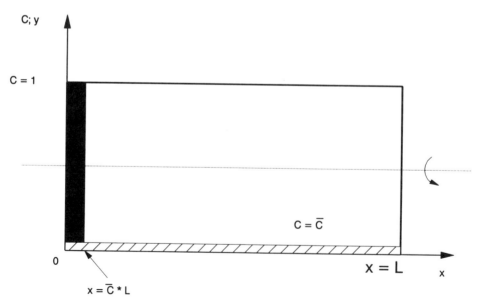

**Figure 6.34** Geometry of a horizontal cylinder-mixer. Initially ($t = 0$), all carbon black particles are at the bottom of the cylinder that then is tumbled horizontally and starts rotating.

the $y$ axis, and finally their concentration is uniformly distributed inside the mixer. If the diffusion coefficient, $\mathcal{D}$, is equal to $10^{-2}$ cm$^2$/s, calculate
(a) the concentration profile $C(x,t)$ at $x = 1.25$, 5, 25, and 40 cm, and
(b) the variance $s^2(t)$ and the index of mixing ($M = 1 - s^2/\sigma_0^2$).
  Also show that for long times
(c) $C(L/4, t) - \bar{C} = s(t)$, and
(d) a graph of $\log(s^2(t))$ versus $t$ has a slope of $2\pi^2\mathcal{D}/L^2$.

**6B.2.**  Diffusion in Layered Structures and Intensity of Segregation—1
Consider a polymer with alternating layered structure. Calculate the time dependence of the concentration variance of the minor component, and evaluate the intensity of segregation after 100 s if the diffusion time constant (ratio of diffusion coefficient to layer thickness squared) is equal to 400 s.

**6B.3.**  Striation Thickness from Kinematical Arguments
Rework Ex. 6.5 for a line oriented at an angle $\phi$ with respect to axis $x$. Also, rework the same example for a two-dimensional point vortex flow with $v_r = 0$ and $v_\theta = \omega/r$.

**6B.4.**  Lineal Stretch Efficiency of Planar Elongational Flow
Prove that the time-average lineal stretch efficiency, $\bar{e}_L$, of a two-dimensional stagnation flow (planar elongational flow; Fig. 6.35) is:

$$\bar{e}_L = \frac{\sqrt{2}}{4\gamma}\left[\ln\left(\frac{e^{4\gamma}+1}{2e^{2\gamma}}\right)\right].$$

The velocity field for that type of flow is given by

$$v_x = \dot{\gamma}y; \qquad v_y = \dot{\gamma}x; \qquad v_z = 0.$$

**6B.5.**  Lineal Stretch Efficiency of Uniaxial Elongational Flow
Prove that the long time lineal stretch efficiency of uniaxial elongational flow is equal to $\sqrt{2/3}$.

**6B.6.**  Improvement of Mixing by Mixing Sections
Consider a single-screw extruder and two types of mixing sections that function as follows: (a) randomize the orientation of the interfaces

entering the section, and (b) orient the incoming interfaces perpendicular to the shearing planes. In both cases the total shear strain imparted to the fluid is equal to $\gamma$. Show that the ratio of the maximum interfacial area growth functions for type (a) and (b) mixing sections, $\lambda$, is equal to:

$$\lambda = \frac{1}{2}\exp\left[\frac{\gamma}{2e}\right].$$

Assume that the mixing sections do not add any shearing to the fluid and that the shear strain between mixing sections is large enough for Eq. 6.66 to apply.

**6B.7.**  Striation Thickness in RCF with Both Cylinders Rotating
A power-law fluid is sheared in a rotational Couette geometry with both cylinders rotating. Prove that the ratio of the striation thickness reduction of this case to the case described in Ex. 6.7 (only the inside cylinder is rotating) is equal to $(\lambda+1)^{-1}$, where $\lambda$ is the ratio of the angular velocities of the inside cylinder to the outside cylinder.

**6B.8.**  Striation Thickness in Axial Annular Couette Geometry
Calculate the striation thickness reduction function for flow in axial annular Couette geometry and for a power-law fluid.

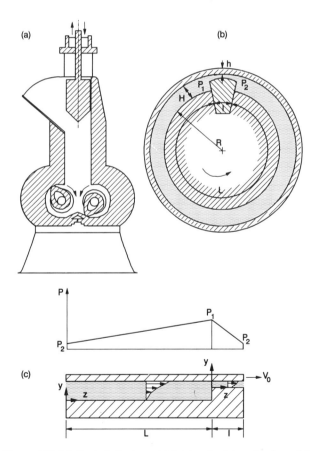

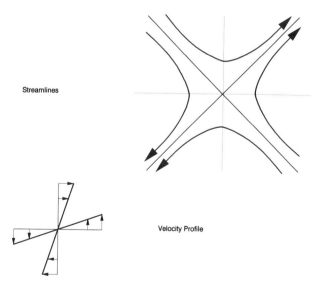

**Figure 6.35** Streamlines and velocity profile of a two-dimensional stagnation (plane hyperbolic; planar extensional) flow.

**Figure 6.36** (a) Typical Banbury high-intensity internal mixer. (b) Idealized chamber of a Banbury mixer with a short low clearance section, consisting of two infinitely long cylinders. The inner cylinder rotates. (c) Idealization of the flow in the clearance by the flow between parallel plates with a step change in channel depth. Pressure distribution as a function of distance.

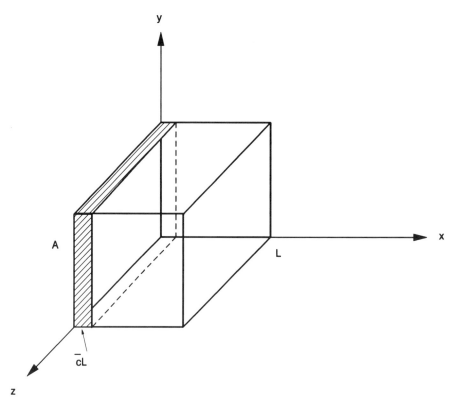

**Figure 6.37** Cubic element of a minor component with simultaneous shear and diffusion

**6B.9.** SDF in PPF

Calculate the algorithm for determining the cumulative SDF versus $\gamma/\bar{\gamma}$ graph for a Newtonian fluid for pressure driven flow through parallel plates. Repeat the exercise for a power-law fluid with power-law index equal to 0.5.

**6B.10.** Internal Banbury Mixer

Consider the high-intensity internal Banbury mixer of Fig. 6.36.a. It consists of a chamber shaped like a figure eight turned 90°, with two rotors counterrotating. The mixture is fed into the chamber through the vertical chute, in which the ram is located. The dispersive mixing takes place in the clearances between the rotors and the chamber walls. Simplify the flow in the chambers by the flow in the clearance of a concentric system of two infinitely long cylinders with a short low clearance section, Fig. 6.36.b. Furthermore, for low clearances, $H/R \ll 1$, assume that the flow is similar to the flow between parallel plates with a step change in the channel depth, Fig. 6.36.c. The pressure profile is given in Fig. 6.36.c. Calculate the shear rate and the maximum shear stress in the lower clearance section. Assume laminar isothermal flow of a Newtonian incompressible fluid, no slip at the walls, and neglibigle gravitational and entrance and exit effects.

## C. Numerical Problems

**6C.1.** "Diffusion" of Particles in Mixers-2
Solve Problem 6B.1 using the IMSL subroutine IVPAG.

**6C.2.** Diffusion in Layered Structures and Intensity of Segregation-2
Solve Problem 6B.2 using the IMSL subroutine IVPAG.

## D. Design Problem

**6D.1.** Simultaneous Laminar Mixing and Diffusion
Consider the case of a cubic element of a minor component (10% by volume) that is simultaneously subjected to laminar mixing and diffusion (Fig. 6.37). The minor component diffuses into the major, and the thickness of the cubic element is reduced as the interface in the $yz$ plane increases. Calculate the time necessary for the minor component to diffuse into the major component so that the concentration standard deviation is 0.11. $Đ = 10^{-10}\,\text{cm}^2/\text{s}$, $L_0 = 3\,\text{mm}$, and $\dot{\gamma} = 50\,\text{s}^{-1}$. Compare this time with the corresponding time without shearing.

## REFERENCES

Aref, H. 1984. "Stirring by Chaotic Advection." *J. Fluid Mech.*, **143**, 1–21.

Bird, R. B., R. C. Armstrong, and O. Hassanger. 1987. *Dynamics of Polymeric Liquids Vol. 1. Fluid Mechanics* (Wiley, New York).

Brodkey, R. S. 1966. "Fluid Motion and Mixing." In V. H. Uhl, and J. B. Gray, Eds., *Mixing,* Vol. 1, Chapter 2 (Academic Press, New York).

Bucknall, C. B. 1977. *Toughened Plastics* (Applied Science Publishers, London).

Chien, W.-L., H. Rising, and J. M. Ottino. 1986. "Laminar Mixing and Chaotic Mixing in Several Cavity Flows." *J. Fluid Mech.*, **170**, 355–77.

Cox, R. G. 1969. "The Deformation of a Drop in a General Time-Dependent Fluid Flow." *J. Fluid. Mech.*, **37**, 601–23.

Danckwerts, P. V. 1952. "The Definition of Measurement of Some Characteristics of Mixtures." *Appl. Sci. Res. A*, **3**, 279–96.

Danckwerts, P. V. 1953. "Continuous Flow Systems-Distribution of Residence Times." *Chem. Eng. Sci.* **2**(1), 1–13.

Elmendorp, J. J. 1991. "Dispersive Mixing in Liquid Systems." In C. Rauwendaal, Ed., *Mixing in Polymer Processing*, Chapter 2 (Dekker, New York).

Erwin, L. 1978a. "Theory of Laminar Mixing." *Polym. Eng. Sci.*, **18**(13), 1044–8.

Erwin, L. 1978b. "Theory of Mixing Sections in Single Screw Extruders." *Polym. Eng. Sci.*, **18**(7), 572–6.

Erwin, L., and F. Mokhatarian. 1983. "Analysis of Mixing in Modified Single Screw Extruders." *Polym. Eng. Sci.*, **23**(2), 49–60.

Fan, L. T., S. J. Chen, and C. A. Watson. 1970. "Solids Mixing." *Ind. Eng. Chem.*, **62**(7), 53–69.

Grace, H. P. 1982. "Dispersion Phenomena in High Viscosity Immiscible Fluid Systems and Application of Static Mixers as Dispersion Devices in Such Systems." *Chem. Eng. Commun.*, **14**, 225–77.

Hall, K. R., and J. C. Godfrey. 1965. "An Experimental and Theoretical Study of Mixing of Highly Viscous Materials." *AIChE J. Chem. E. Symposium Series*, **10**, 71–81.

Howland, C., and L. Erwin. 1983. "Mixing in Counter Rotating Tangential Twin Screw Extruders." 41st SPE Annual Technical Conference, Chicago, IL, **29**, 113–16.

Khakhar, D. V., H. Rising, and J.M. Ottino. 1986. "Analysis of Chaotic Mixing in Two Model Systems." *J. Fluid Mech.*, **172**, 419–51.

Lacey, P. M. C. 1954. "Developments in the Theory of Particle Mixing." *J. Appl. Chem., Lond.*, **4**, 257–68.

Leong, C. W., and J. M. Ottino. 1989. "Experiments on Mixing Due to Chaotic Advection in a Cavity." *J. Fluid Mech.*, **209**, 463–99.

Matthews, G. 1982. *Polymer Mixing Technology* (Applied Science Publishers, London).

McKelvey, J. M. 1962. *Polymer Processing*, Chapter 12 (Wiley, New York).

Middleman, S. 1977. *Fundamentals of Polymer Processing*, Chapter 12 (McGraw-Hill, New York).

Mohr, W. D., R. L. Saxton, and C. H. Jepson. 1957. "Mixing in Laminar-Flow Systems." *Ind. Eng. Chem.*, **49**(11), 1855–6.

Nadav, N., and Z. Tadmor. 1973. "Quantitative Characterization of Extruded Film Texture." *Chem. Eng. Sci.*, **28**, 2115–26.

Ng, K. Y., and L. Erwin. 1981. "Experiments in Extensive Mixing in Laminar Flow. I. Simple Illustrations." *Polym. Eng. Sci.*, **21**(4), 212–7.

Nir, A., and A. Acrivos. 1973. "On the Creeping Motion of Two Arbitrary-Sized Touching Spheres in a Linear Shear Field." *J. Fluid Mech.*, **59**, 209–23.

Ottino, J. M., and R. Chella. 1983. "Laminar Mixing of Polymeric Liquids; A Brief Review and Recent Theoretical Developments." *Polym. Eng. Sci.*, **23**(7), 357–79.

Ottino, J., W. E. Ranz, and C. W. Macosko. 1979. "A Lamellar Model for Analysis of Liquid-Liquid Mixing." *Chem. Eng. Sci.*, **34**, 877–90.

Ottino, J., W. E. Ranz, and C. W. Macosko. 1981. "A Framework for Description of Mechanical Mixing of Fluids." *AIChE J.*, **27**(4), 565–77.

Pinto, G., and Z. Tadmor. 1970. "Mixing and Residence Time Distribution in Melt Screw Extruders." *Polym. Eng. Sci.*, **10**(5), 279–88.

Radford, J. A., T. Alfrey, and W. J. Schrenk. 1973. "Reflectivity of Iridescent Coextruded Multilayered Plastic Films." *Polym. Eng. Sci.*, **13**(3), 216–21.

Ranz, W. E. 1979. "Applications of a Stretch Model to Mixing, Diffusion, and Reaction in Laminar and Turbulent Flows." *AIChE J.*, **25**(1), 41–7.

Rotz, C. A., and N. P. Suh. 1976. "New Techniques for Mixing Viscous Reacting Liquids. Part I: Mechanical Means to Improved Laminar Mixing." *Polym. Eng. Sci.*, **16**, 664–71.

Rumpf, H. 1962. "The Strength of Granules and Agglomerates." In W. A. Knepper, Ed., *Agglomeration*, Chapter 15 (Interscience, New York).

Schrenk, W. J., and T. Alfrey. 1983. "Unmixing in Rotational Laminar Shear Mixers." *Polym. Eng. Rev.*, **2**(4), 363–79.

Schrenk, W. J., K. J. Cleereman, and T. Alfrey. 1963. "Continuous Mixing of Very Viscous Fluids in an Annular Channel." *SPE Trans.*, July, 192–200.

Spencer, R. S., and R. M. Wiley. 1951. "The Mixing of Very Viscous Liquids." *J. Colloid Sci.*, **6**, 133–45.

Spiegel, M. R. 1991. *Probability and Statistics* (McGraw-Hill, New York).

Stone, H. A., and L. G. Leal. 1989. "Relaxation and Breakup of an Initially Extended Drop in an Otherwise Quiescent Fluid." *J. Fluid Mech.*, **198**, 399–427.

Tadmor, Z., and C. G. Gogos. 1979. *Principles of Polymer Processing* (Wiley, New York).

Taylor, G. I. 1934. "The Formation of Emulsions in Definable Fields of Flow." *Proc. R. Soc. Lond. A*, **146**, 501–23.

Tomotika, S. 1935. "On the Instability of a Cylindrical Thread of a Viscous Liquid Surrounded by Another Viscous Fluid." *Proc. R. Soc. Lond. A*, **150**, 322–37.

Tucker, C. L. 1981. "Sample Variance Measurement of Mixing." *Chem. Eng. Sci.*, **36**(11), 1829–39.

Tucker, C. L. 1991. "Principles of Mixing Measurement", In C. Rauwendaal, Ed., *Mixing in Polymer Processing*, Chapter 3 (Dekker, New York).

Tucker, C. L., and N. P. Suh. 1980a. "Mixing for Reaction Injection Molding. I. Impingement Mixing of Liquids." *Polym. Eng. Sci.*, **20**(13), 875–86.

Tucker, C. L., and N. P. Suh. 1980b. "Mixing for Reaction Injection Molding. II. Mixing of Fiber Suspensions." *Polym. Eng. Sci.*, **20**(13), 887–98.

Underwood, E. E. 1977. *Quantitative Stereology* (Addison-Welsley, Reading, MA).

Van der Reijden-Stolk, C., and A. Sara. 1986. "A study on Polymer Blending Microrheology. Part 3: Deformation of Newtonian Drops Submerged in Another Newtonian Fluid Flowing through a Converging Cone." *Polym. Eng. Sci.*, **26**(18), 1229–39.

Van Gheluwe, P., B. D. Favis, and J.-P. Chalifoux. 1988. "Morphological and Mechanical Properties of Extruded Polypropylene/Nylon-6 Blends." *J. Mater. Sci.*, **23**, 3910–20.

# 7

# EXTRUSION DIES

## Design Problem 6
## Coextrusion Blow-Molding Die

The blow molding of gasoline tanks for automobiles from thermoplastics offers the possibility of making more intricate shapes required to fit existing space than is possible with metals. Manufacturers would like to use HDPE, but it has poor barrier properties for gasoline vapors. However, by adding a layer of nylon 6 about 0.1 times the thickness of the tank wall, the material can meet the barrier requirements. The final cross-sectional shape of the gasoline tanks is to be basically as shown in Fig. 7.1. Design a die that has a shape similar to that of the gasoline tank as shown in Fig. 7.2 for coextruding a parison consisting of 2 lb. of nylon 6 and 18 lb. of HDPE in 5 seconds. Based on previous experience it is known that for cylindrically shaped parisons the maximum expansion of the tube should be no more than 2.0 (i.e., the maximum increase in the radius should be no more than a factor of 2). Assume that the density change on cooling and the weight of the parison are enough to offset the increase of the thickness due to extrude swell. The extruders are horizontal, but the die must be mounted vertically. The only rheological data available are the viscosity data given in Table 2.3 for HDPE and in Table 7.3 for nylon 6. Determine the extrusion conditions required to deliver the amount of material and whether one can expect any interfacial stability problems.

Extrusion processes involve the use of extruders that melt and pump polymers and of shaping devices called dies that are placed at the end of the extruder. This chapter is concerned with the design of extrusion dies, and Chapter 8 deals with single- and twin-screw extruders.

Extrusion dies are metal channels that impart a specific cross-sectional shape to a polymer stream. The design difficulty centers on achieving the desired shape within set limits of dimensional uniformity at the highest production rate possible. Because of the viscoelastic nature of polymers and the associated flow behavior, it is no simple matter to design a die that will produce a smooth extrudate with the desired dimensions.

In Section 7.1 we describe briefly the origin of the nonuniformities and the factors that lead to extrudate shapes other than those desired. In Section 7.2 we present flow phenomena associated with the viscoelastic behavior of polymers that affect the design of dies. In Section 7.3 we consider the design of sheet and flat film dies, especially with an emphasis on providing a uniform extrudate. The design of tubular dies presents somewhat different problems, and they are considered in Section 7.4. There are numerous other shapes of extrudates besides flat round, or tubular. Extrusion of irregular-shape extrudates is referred to as profile extrusion, and this is discussed in Section 7.5. Finally, it is now common practice to extrude multiple layers of different polymers through the same die. As this presents even more complications in die design, we introduce the topic in Section 7.6. Finally in Section 7.7, the solution to Design Problem 6 is presented.

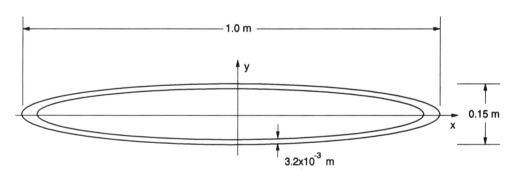

**Figure 7.1** Cross section of an elliptically shaped parison consisting of HDPE and nylon 6 in a 9:1 ratio by weight.

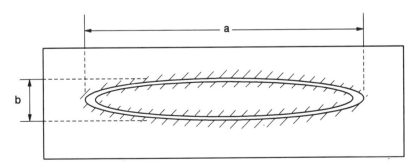

**Figure 7.2** Exit face of the die used to produce the parison in Figure 7.2. The die opening is an elliptically shaped annulus of major axis, $a$, and minor axis, $b$. The thickness, $a$, $b$ are determined in the solution to Design Problem 6.

## 7.1. EXTRUDATE NONUNIFORMITIES

There are basically two types of nonuniformities. Those that occur along the machine direction (MD), or along the extrusion direction, and those that occur in the transverse direction (TD). These non-uniformities are shown in Fig. 7.3 for a planar geometry. The nonuniformities that arise along the MD are usually due to pressure and temperature variations that affect the flow rate, the rheological properties of the melt, and to some degree the die design. The irregularities that occur in the TD are due nearly totally to the die design, but in some cases the rheological properties enter in.

In the case of the MD, variations in the flow rate due to pressure or temperature variations in the pumping device are the main cause of the irregularities. However, flow instabilities associated with the phenomena of melt fracture and draw resonance can lead to variations in the dimensions of the extrudate. These variations are closely connected to the rheological properties of the melt, but die design can at least alleviate the severity of the irregularities.

The TD variations are nearly totally due to die design. The first problem is to design a feed system that will distribute the melt uniformly to the shaping portion of the die (see Fig. 7.4 for definition of parts of a die.) In the event this is not possible, then it must be possible to adjust the die lips in such a way that the fluid will leave the die with a uniform thickness. Part of the thickness variation in the TD is due to the inability to feed the die uniformly from the extruder, and the rest is due to the phenomenon of die swell. Because the degree of swell may vary nonuniformly over the cross section because of variations in the shear rate, then the die lips (main shaping section) may have to be designed to compensate for this.

Before continuing we should note that a lot of the problems concerned with die design are handled empirically. Part of this is due to the lack of the appropriate mathematical tools to simulate the flow of viscoelastic materials through dies. The lack of mathematical tools comes from both inappropriate constitutive equations to describe the rheology of polymer melts and the lack of numerical techniques for handling the nonlinear system of differential equations that must be solved. Although progress is being made to develop finite element techniques for handling die design, these codes are not fully developed or tested at this time. Even once they are developed, they may not be available for every design engineer. For this reason we present in this chapter design considerations that can be handled at this level of the educational process, and at the same time we present material that can be used in a qualitative fashion to improve the design of extrusion dies. It is important to recognize that most die design to date neglects the viscoelastic nature of polymeric fluids.

## 7.2. VISCOELASTIC PHENOMENA

Three phenomena associated with the flow behavior of polymeric fluids must be considered in the design of extrusion dies: pressure drops in contractions (or expansions), die swell, and melt fracture. The latter two bear a direct relation to extrudate uniformity, and the flow behavior in contractions may only be indirectly related to extrudate uniformity. In this section, for illustrative purposes, we present results based primarily on studies in the capillary geometry. One must recognize that the extension of results from a capillary to other geometries may be difficult to make quantitatively.

### 7.2.1. FLOW BEHAVIOR IN CONTRACTIONS

As discussed in Chapter 3 (Section 3.3) the pressure drop across a contraction for a polymeric fluid can be quite large relative to the pressure drop across the die land or lips. The origin of $\Delta P_{ent}$ is thought to be due to the entry flow behavior of the polymer. In some cases, such as for low-density polyethylene (LDPE), the streamlines form natural entry angles as shown in Fig. 7.5. The flow into the die is restricted well into the upstream region, which serves to effectively act as an extension to the capillary length and thereby increase $\Delta P$. It is also observed that large vortices arise in the corners. For polymers such as polystyrene (PS) (see Fig. 7.6) and high-density polyethylene (HDPE), the streamlines only become curved a short distance from the contraction, and the vortices are quite small, as shown in Fig. 7.7. In this case there is very little addition of length to the capillary and $\Delta P_{ent}$ is smaller. This explanation is in line with the results in Fig. 7.7 that shows values of $\Delta P_{ent}$ normalized to the wall shear stress versus $\dot{\gamma}_a$ for various fluids. The normalization with respect to $\tau_w$ removes the difference due to differences in the magnitude of viscosity. LDPE has

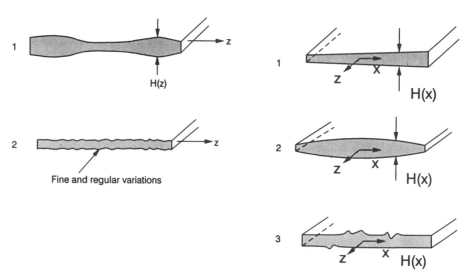

**Figure 7.3** Irregularities in the extrusion of a sheet with those along the machine direction shown on the left and those along the transverse direction shown on the right. The extrusion direction is in the $z$ direction.

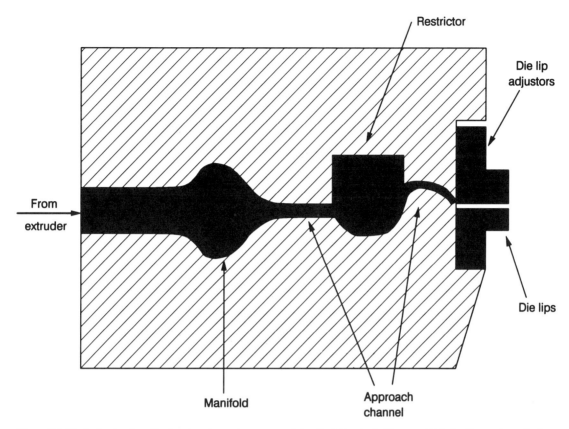

**Figure 7.4** Typical extrusion die showing the various parts of the die (side view). The manifold distributes the melt from the extruder uniformly over the width of the die, and the restrictor can be used to compensate for variations in flow rate across the width of the die. The lips give the melt stream the desired dimensions and shape. (*Note*: because of die swell the dimensions of the extrudate may be considerably different from those of the die lips.)

the highest values of $\Delta P_{ent}/\tau_w$ which is to be expected as it forms large entry vortices. The oil, which is basically Newtonian, exhibits the lowest values. In general, linear polymers exhibit small regions of flow rearrangement and very small regions with vortices and hence lower values of $\Delta P_{ent}/\tau_w$.

The magnitude of $\Delta P_{ent}$ is a function of the rheological properties of the polymer, the contraction ratio, the cross-sectional geometry, and the degree of taper into the entry. Unfortunately, there is no known simple way to translate results from the capillary to other flow geometries, and there have been few attempts to do so. Until more information is available, we will use entrance pressure measurements from capillary geometries to estimate values of $\Delta P_{ent}$ in other geometries provided we maintain at least geometric similarity.

As far as expansions are concerned, there is even less information. We will assume that the pressure rise in an abrupt expansion is the same as the drop in an abrupt contraction. Certainly, this will not be the same because of the viscoelastic nature of polymeric fluids.

### 7.2.2. EXTRUSION INSTABILITIES

The limiting factor in the extrusion rate of polymeric fluids is the onset of a low Reynolds number instability called *melt fracture*. The onset of melt fracture leads to varying degrees of imperfections that may only affect the clarity of a material on one hand but on the other may be so severe as to significantly reduce the physical properties. We first discuss the nature and origin of melt fracture and then what can be done to alleviate or at least mitigate the problem.

There are basically five types of melt fracture: *sharkskin, ripple, bamboo, wavy,* and *severe distortion*. These types of melt fracture are shown in Figs. 7.8, 7.9, and 7.10. Sharkskin melt fracture is shown in Fig. 7.8 for a LLDPE. At the lowest apparent shear rate the extrudate is smooth, but at $\dot{\gamma}_a = 112 \, s^{-1}$, the extrudate exhibits a mild roughness, called sharkskin, that affects the appearance of the surface. This type of fracture is extremely detrimental to the manufacture of packaging films that must meet certain requirements for clarity. As $\dot{\gamma}_a$ is increased, another form of fracture arises. At $\dot{\gamma}_a = 750 \, s^{-1}$, the fracture present is called *bamboo*. Finally, at $\dot{\gamma}_a$ of $2250 \, s^{-1}$, the fracture is *severe*. LLDPE does not seem to exhibit wavy fracture.

HDPE exhibits both sharkskin and bamboo (sometimes referred to as spurt) fracture at lower shear rates as shown in Fig. 7.9. As $\dot{\gamma}_a$ is increased, HDPE is observed to exhibit the wavy form of fracture. LDPE, on the other hand, as shown in Fig. 7.10, does not exhibit sharkskin, but only wavy and severe fracture.

In order to reduce the detrimental effect of melt fracture through die design or polymer modification it is important to know the origin of melt fracture. The major sources for melt fracture are the die entry region, the die land, and the die exit. For a polymer such as LDPE fracture originates at the die entry. As the extrusion rate is increased, the vortices no longer grow in size or intensity. Instead, the flow takes on a spiral motion in the die entry sending sections of the nearly stagnant fluid into the die at regular intervals. This leads to regions of various flow histories passing through the die and leaving the die exit. When this type of fracture occurs, there is no indication of the flow problems in the pressure measured along the die, and hence the wall

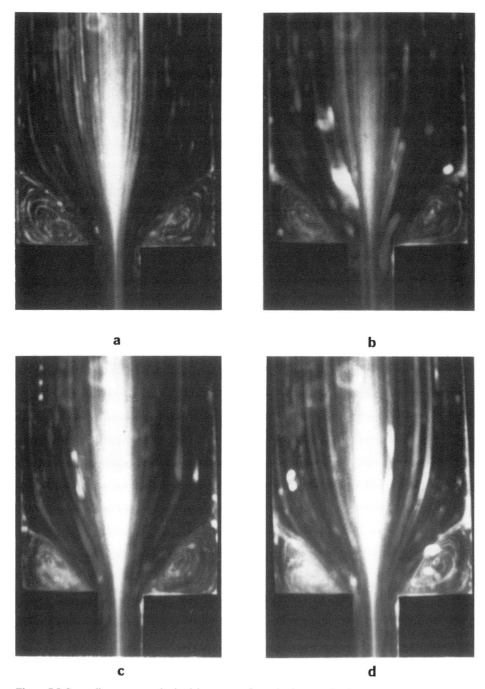

a

b

c

d

**Figure 7.5** Streamline patterns obtained by means of streak photography for LDPE at 150°C in a 4:1 planar contraction. (a) $\dot{\gamma} = 20\,\text{s}^{-1}$, $We = 1.29$. (b) $\dot{\gamma} = 40\,\text{s}^{-1}$, $We = 1.35$. (c) $\dot{\gamma} = 60\,\text{s}^{-1}$, $We = 1.38$. (d) $\dot{\gamma} = 80\,\text{s}^{-1}$, $We = 1.38$ (White, 1988).

shear stress is as shown in Fig. 7.11. By streamlining the die entry or increasing the length of the die land it is possible to reduce the amplitude of the distortion, but the critical shear stress for fracture is unchanged. The critical wall shear stress for the onset of fracture for LDPE is of the order of $10^5$ Pa. A statistical fit of the data for LDPE leads to the following empirical expression for the critical wall shear stress for fracture (Middleman, 1977):

$$\tau_{cr}/T_{abs} = 131.7 + 1.0 \times 10^7/M_w \qquad \text{branched PE,} \qquad (7.1)$$

where $\tau_{cr}$ is in units of Pa and $T_{abs}$ is in °K.

On the other hand, polymers such as HDPE and LLDPE seem to slip in the die land, leading to a "slip-stick" instability. There is a distinct flattening of the flow curve indicating a region where multiple flow rates are possible for the same wall shear stress. This is shown by the data presented in Fig. 7.11. Eventually, the flow curve appears to become normal again at high shear rates. When slip-stick fracture occurs (which results in the ripple and then bamboo types of fracture), increasing the die length just makes the degree of distortion worse.

It has been proposed that polymers such as HDPE, which is

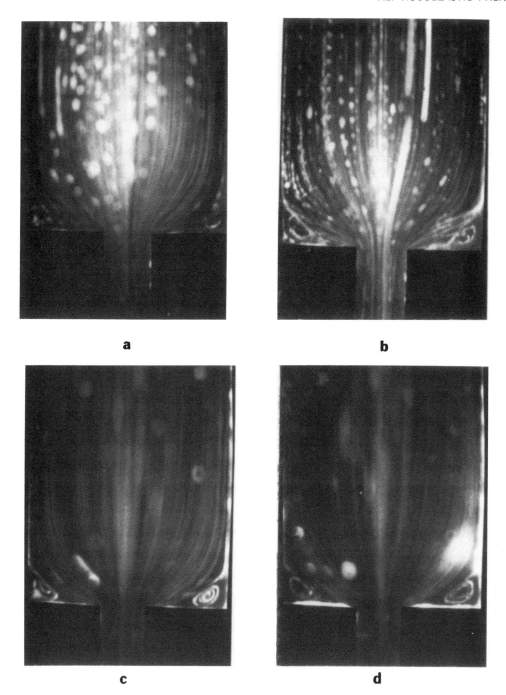

**Figure 7.6** Streamline patterns obtained by means of streak photography for polystyrene at 190°C in a 4:1 contraction. (a) $\dot\gamma = 5\,s^{-1}$, $We = 0.82$. (b) $\dot\gamma = 20\,s^{-1}$, $We = 1.19$. (c) $\dot\gamma = 40\,s^{-1}$, $We = 1.39$. (d) $\dot\gamma = 80\,s^{-1}$, $We = 1.62$.

considered a linear polymer, as well as other linear polymers such as PP and PS have a similar mechanism for fracture. In fact the molecular-weight dependence of the critical shear stress ($\tau_{cr}$) for fracture was found to be similar for linear polymers (Middleman, 1977). A statistical fit of the data for fracture gave the following relation for the critical shear stress for the onset of slip-stick melt fracture, $\tau_{cr}$, for linear polymers:

$$\tau_{cr}/T_{cr} = 171.7 + 2.7 \times 10^{7}/M_{w}. \tag{7.2}$$

It is true that linear polymers do not show large vortices in planar entry flow, although they may occur in axisymmetric flow (White et al., 1987). However, it is known that the origin of fracture for PS is the die entry. In fact, HDPE and LLDPE are the only polymers of the group presented here that readily show any indication of slip-stick in the flow curve. The rest show no indication of fracture in the flow curve, and apparently the origin of fracture is in the die entry. In spite of this, it is interesting to note that $\tau_{cr}$ for the branched polymer, LDPE, falls on a curve separate from that for the linear polymers.

The relations given in Eqs. 7.1 and 7.2 are useful for estimating the onset of gross fracture for a capillary geometry. It does not tell us, however, what will happen in an annular die or some other geometry. However, the equations are at least useful in making an estimate of limiting conditions. There are still many mysteries surrounding the origin of sharkskin melt fracture and the methods proposed to eliminate it. For example, it has been proposed that the metal used in die construction can allow one to alter $\tau_{cr}$, and that the rounding of the

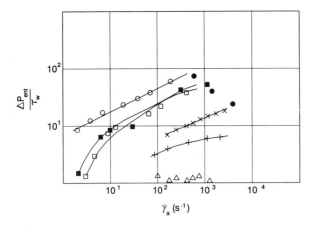

**Figure 7.7** The ratio of entrance pressure drop to wall shear stress versus apparent shear rate, $\dot{\gamma}_a$. (■) PP; (□) PS; (○) LDPE; (+) HDPE; (●) 2.5% PIB in mineral oil; (×) 10% PIB in decalin; (△) NBS-OB oil. (Reprinted with permission of the publisher from J. L. White, *Appl. Polym. Symp., 20,* (155), 1973.).

corners at the die exit is important. Still others recommend processing aids to increase $\tau_{cr}$ for the onset of melt fracture (The addition of fluroelastomers to LLDPE does eliminate both sharkskin and slip-stick fracture.) Once gross distortion occurs, which originates in the die entry, there is very little that can be done to eliminate the problem.

### 7.2.3. DIE SWELL

The phenomenon associated with the increase of the diameter of an extrudate as a polymer leaves a capillary, known as *die swell* or *extrudate swell*, was briefly introduced in Section 3.2. The implication at this point is that die swell is related to unconstrained elastic recovery ($S_\infty$) following shear flow. $S_\infty$ is related to the ratio of the primary normal stress difference to the shear stress through the equation:

$$S_\infty = N_1/2\tau_{yx}. \tag{7.3}$$

Tanner's theory for die swell (see Section 3.2) for flow through a capillary leads to:

$$D_p/D_0 = 0.1 + [1 + \tfrac{1}{2}(S_\infty)^2]^{1/6}. \tag{7.4}$$

In this section we show that die swell is more complicated than indicated by Eq. 7.4 and depends on a number of factors. We then discuss how to deal with die swell in die design.

The first fact that we show is that Tanner's theory does not accurately predict die swell in general. This is illustrated in Fig. 7.12 where values of die swell for four different polymer melts are plotted versus $\tfrac{1}{2}S_\infty$. The solid line represents Eq. 7.4 and is generated for arbitrary values of $S_\infty$. There is as much as a 50% difference between the measured and predicted values of $D_p/D_0$. Furthermore, the values

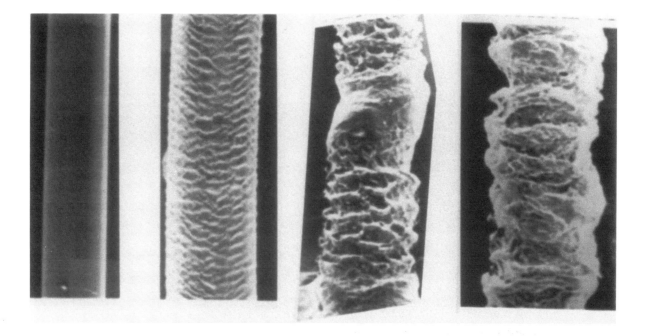

**Figure 7.8** LLDPE extrudates obtained from a capillary at different apparent shear rates, $\dot{\gamma}_a$. From left to right the values of $\dot{\gamma}_a$ are 37, 112, 750, and 2250 s$^{-1}$. (Data from R. H. Moynihan, *The Flow at Polymer and Metal Interfaces.* Ph.D. Thesis, Department of Chemical Engineering, Virginia Tech., Blackburg, VA, 1990.)

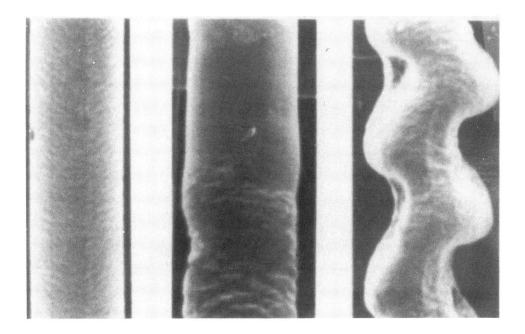

**Figure 7.9** HDPE extrudates obtained from a capillary at different apparent shear rates, $\dot{\gamma}_a$. From left to right the values of $\dot{\gamma}_a$ are 75, 75, 2250 s$^{-1}$, and the corresponding values of $\tau_a$ are 0.20, 0.27, 0.33 MPa. (Data from R. H. Moynihan, *The Flow at Polymer and Metal Interfaces.* Ph.D. Thesis, Department of Chemical Engineering, Virginia Tech., Blackburg, VA, 1990.)

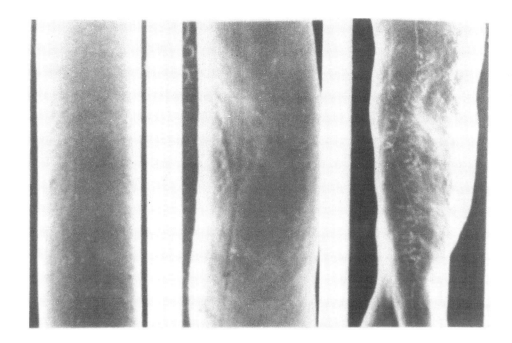

**Figure 7.10** LDPE extrudates obtained from a capillary at different apparent shear rates, $\dot{\gamma}_a$. From left to right the values of $\dot{\gamma}_a$ are 75, 750, 2250 s$^{-1}$ while the corresponding values of $\tau_a$ are 0.1, 0.21, and 0.32 MPa. (Data from R. H. Moynihan, *The Flow at Polymer and Metal Interfaces.* Ph.D. Thesis, Department of Chemical Engineering, Virginia Tech., Blackburg, VA, 1990.)

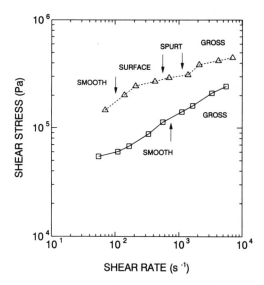

**Figure 7.11** Shear stress versus shear rate for LDPE (*lower curve*) and LLDPE (*upper curve*). The arrows indicate the onset of various types of melt fracture. (Data from R. H. Moynihan, *The Flow at Polymer and Metal Interfaces.* Ph.D. Thesis, Department of Chemical Engineering, Virginia Tech., Blackburg, VA, 1990.)

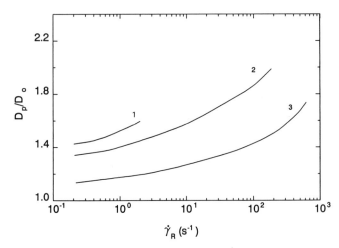

**Figure 7.13** Effect of the method of measurement on the magnitude of die swell for HDPE extruded from a capillary at 180°C at various wall shear rates, $\dot{\gamma}_R$: curve 1, isothermal; curve 2, annealed; curve 3, ambient air. (Data from J. L. White and J. F. Roman, Extrudate Die Swell during the Melt Spinning of Fibers—Influence of Rheological Properties and Take-Up Forces," *J. Appl. Polym. Sci., 20*, 1005, 1976.)

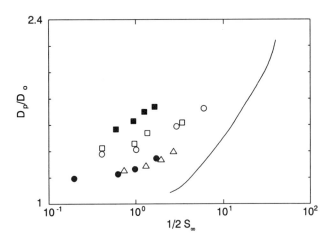

**Figure 7.12** Capillary extrudate swell versus $\frac{1}{2} S_\infty$, the ultimate elastic recovery. (■, □) two HDPEs; (●) PS; (○) LDPE; (△) PP. (Data from J. L. White and J. F. Roman, "Extrudate Die Swell during the Melt Spinning of Fibers—Influence of Rheological Properties and Take-Up Forces," *J. Appl. Polym. Sci., 20*, 1005, 1976.)

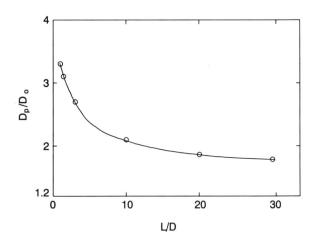

**Figure 7.14** Extrudate swell versus capillary L/D ratio for PP extruded at 219°C at $\dot{\gamma}_R = 700\,\text{s}^{-1}$. (Data from Y. Mori and K. Funatsu, "On Die Swell in Molten Polymers," *J. Appl. Polym. Sci., 20*, 209, 1973.)

of $D_p/D_0$ vary from polymer to polymer. Hence, it does not appear that die swell can be correlated simply to $S_\infty$.

What else does die swell depend on? First it depends on the method used to measure it. This is shown in Fig. 7.13, in which values of $D_p/D_0$ are plotted versus shear rate for three different methods of measurement. The highest values are obtained for polymers that are extruded isothermally into an oil bath. The lowest values are for the extrudate that is extruded into ambient air. In this case the sample is cooled down before die swell is completed. Annealing, as shown by curve 2 in Fig. 7.13, allows the sample to almost reach the values obtained under isothermal conditions.

Die swell depends on the capillary $L/D$ as shown in Fig. 7.14. In this figure we see that $D_p/D_0$ is a function of $L/D$ with the greatest

swell being for the shortest capillary. This behavior has been attributed to the large amount of elastic energy stored during the extensional flow in the entry region.

It is also observed that die swell depends on time after the extrudate leaves the die. This is shown in Fig. 7.15 in which $D_p/D_0$ is plotted versus time for an HDPE melt. Here we observe that a large portion of the swell occurs instantaneously, and that the remainder of the diameter increase can occur over a period of several minutes.

Finally, $D_p/D_0$ (equilibrium swell) is a function of the wall shear stress, $\tau_R$. Data for a commercial polystyrene at three different temperatures when plotted versus shear rate fall on three separate curves. However, when plotted versus $\tau_R$, the data all fall on a single curve as shown in Fig. 7.16. Hence, for a given polymer the equilibrium swell can be correlated with $\tau_R$.

**TABLE 7.1 Description of Resins Studied in Annular Swell**

| McGill University Stock No. | Trade Name/No. | Manufacturer | Density (kg/m³) | Melt Index (dg/min) |
|---|---|---|---|---|
| 22 | Sclair (HDPE) 59C | Dupont Canada | 960 | 0.42 |
| 26 | 40054 (HDPE) | Dow Canada | 954 | 0.40 |
| 27 | DMDJ (HDPE) 5140 | Union Carbide Canada | 962 | 0.72 |
| 28 | PRO-FAX 7723 (PP) | Hercules Canada | 899 | 0.8 |

(Data from Garcia-Rejon & Dealy, 1982.)

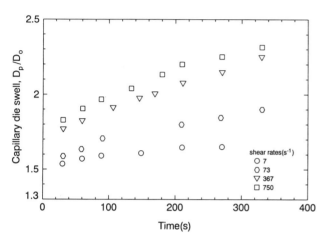

**Figure 7.15** Capillary extrudate swell versus time for four shear rates, $\dot{\gamma}_R$, for HDPE (Resin No. 27). (Data from Garcia-Rejon & Dealy, 1982.)

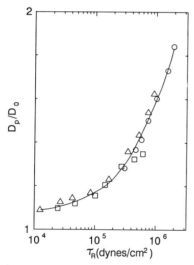

**Figure 7.16** Capillary extrudate swell versus wall shear stress for a commercial polystyrene at three temperatures: (○) 160°C, (△) 180°C, (□) 200°C. (Data from W. W. Graessley, S. D. Glasscock, and R. L. Crawley, "Die Swell in Molten Polymer," *J. Rheol.*, 14, 519, 1970.)

In summary, capillary die swell, $B = D_p/D_0$, is a function of the following variables:

$$B = f(L/D_0, \tau_R, EG, E, t, t_p/\lambda), \qquad (7.5)$$

where $EG$ is the entrance geometry, $E$ is the exit geometry, $t$ is the time after a fluid element leaves the die, $t_p$ is the time required for the melt

to pass through the die, and $\lambda$ is the longest relaxation time for the fluid. The last quantity, $\lambda/t_p$, is referred to as the Deborah number. Certainly the ideas of elastic recovery are involved, but stresses other than those generated in shear flow (e.g., extensional flow at the die exit) must be considered.

The phenomenon of die swell is complex, and the method for incorporating it into die design calculations is unclear. In most cases the processing die geometry is considerably different from that used to make die swell measurements. For example, in the extrusion of a parison used in blow molding there is both swell of the thickness and outer diameter of the parison. How to translate die swell from a capillary to that of a parison is certainly not straightforward.

To illustrate one possible way of translating capillary die swell measurements to some other die geometry we consider the swell of extrudate leaving an annular die. In the swell of polymer extruded from an annular die as shown in Fig. 3.1 (this figure is associated with Design Problem 2), there is swell of the diameter as well as the thickness of the extrudate. The two most common swell parameters are the diameter swell, $B_1$, and the thickness swell, $B_2$, defined, respectively, as:

$$B_1 = D_p/D_0 \qquad (7.6)$$

$$B_2 = H_p/H_0. \qquad (7.7)$$

The thickness swell can be related to the inner (subscript 1) and outer (subscript 2) diameters as follows:

$$B_2 = (D_{p2} - D_{p1})/(D_{02} - D_{01}). \qquad (7.8)$$

Sometimes the weight swell is defined, especially in the case of extruding a parison for blow molding. The weight swell, $S_w$, is:

$$S_w = (D_{p2}^2 - D_{p1}^2)\rho_p/(D_{02}^2 - D_{01}^2)\rho_0, \qquad (7.9)$$

where $\rho_p$ is the density of the polymer extrudate and $\rho_0$ is the density of the melt in the die.

The questions of interest are: what are the relations between $B_1$ and $B_2$ and between $B_1$ and $B$? To answer these questions and to illustrate the complexity of the answer, we consider the following example. Three HDPE samples of different molecular weight characteristics and a polypropylene were used in the study by Garcia-Rejon and Dealy (1982) (see Tables 7.1 and 7.2). The steady shear and dynamic

**TABLE 7.2 Molecular Weight Parameters for Three High Density Polyethylenes**

| Resin Number | $\bar{M}_n (\times 10^{-4})$ | $\bar{M}_w (\times 10^{-4})$ | $\bar{M}_z (\times 10^{-4})$ | $\bar{M}_w/\bar{M}_n$ | $\bar{M}_z/\bar{M}_w$ |
|---|---|---|---|---|---|
| 22 | 1.8 | 18 | 140 | 10.7 | 7.9 |
| 26 | 1.7 | 12 | 88 | 7.6 | 7.5 |
| 27 | 2.1 | 12 | 74 | 5.8 | 7.0 |

(Data from Garcia-Rejon & Dealy, 1982.)

mechanical rheological properties are presented in Figs. 7.17 and 18, and it is seen here that η for resins 22, 26, and 28 are similar, whereas $N_1$ values are different for the three HDPEs and PP. (*Note:* PP is processed at 190°C, whereas the HDPE samples are processed at 170°C.) We would expect resin 28 to have the highest values of $B$ based on the values of η, $\Psi_1$, and Eq. 7.4. From the die swell data obtained from a capillary rheometer presented in Fig. 7.19, we see this is not the case. Resin 28 exhibits the highest values of $B$ while the other resins exhibit similar values.

We next consider how to translate the capillary swell data into values for $B_1$ and $B_2$. To make the comparison we either have to make it at the same wall shear stress or wall shear rate. Because the ratio of $D_{01}/D_{02}$ ($=0.816$ in this case, Fig. 3.1) approaches 1, we can treat the annulus as a thin slit, and hence:

$$\dot{\gamma}_w = 2(2+b)Q/\pi(R_{01}+R_{02})(R_{02}-R_{01})^2, \qquad (7.10)$$

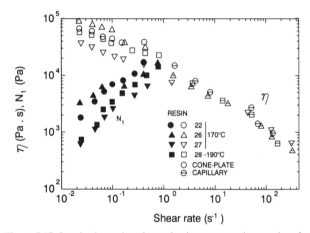

**Figure 7.17** Steady shear viscosity and primary normal stress data for three HDPE and one PP samples described in Tables 7.1 and 7.2. (Data from A. Garcia-Rejon, J. M. Dealy, and M. R. Kamal, *Can. J. Chem. Engng., 59,* 59, 1981.)

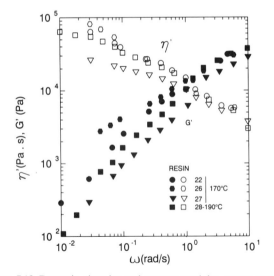

**Figure 7.18** Dynamic viscosity and storage modulus versus angular frequency for the resins described in Tables 7.1 and 7.2. (Data from A. Garcia-Rejon, J. M. Dealy, and M. R. Kamal, *Can. J. Chem. Engng., 59,* 59, 1981.)

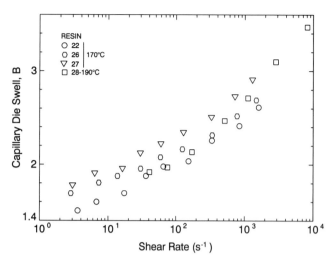

**Figure 7.19** Capillary extrudate swell versus shear rate for the resins described in Tables 7.1 and 7.2. (Data from Garcia-Rejon & Dealy, 1982.)

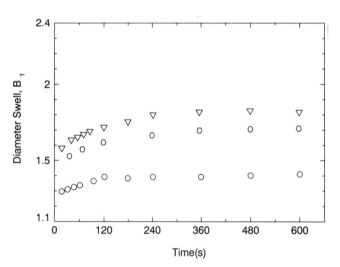

**Figure 7.20** Annular diameter swell versus time for HDPE-22 at 170°C and wall shear rates of (○) 20, (○) 121, and (▽) 270 s$^{-1}$. (Data from Garcia-Rejon & Dealy, 1982.)

where

$$b = d\ln Q/d\ln\tau_w \qquad (7.11)$$

and

$$\tau_w = (R_{02}-R_{01})\Delta P/2L \qquad (7.12)$$

In Figs. 7.20 and 7.21 values of $B_1(t)$ and $B_2(t)$ are presented for sample HDPE-22. Here we observe that the instantaneous swell represents about 85% of the equilibrium swell. For PP (resin 28), values of $B_1(t)$ and $B_2(t)$ are presented in Figs. 7.22 and 7.23. For PP, however, the instantaneous swell is only about 75% to 80% of the equilibrium swell, and it takes much longer to reach the equilibrium swell. This difference cannot be merely accounted for by differences in the longest relaxation times, as they are similar for both polymers based on the viscosity data.

The first relation of interest is that between $B_1$ and $B_2$. These results

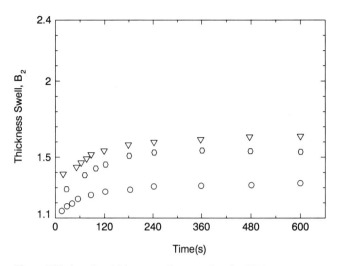

Figure 7.21 Annular thickness swell versus time for HDPE-22 for the same conditions as in Fig. 7.20.

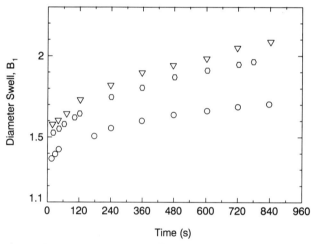

Figure 7.22 Annular diameter swell versus time for PP-28 at 190°C and wall shear rates of (○) 19, (○) 109, and (▽) 231 s$^{-1}$. (Data from Garcia-Rejon & Dealy, 1982.)

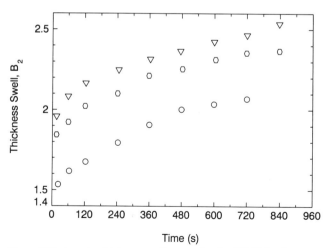

Figure 7.23 Annular thickness swell versus time for PP-28 for the same conditions as in Fig. 7.22.

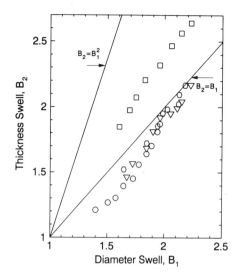

Figure 7.24 Equilibrium thickness swell versus equilibrium diameter swell for HDPE resins (○) 22, (○) 26, (▽) 27, and (□) PP-28. (Data from Garcia-Rejon & Dealy, 1982.)

are shown in Fig. 7.24 in which equilibrium values of $B_2$ are plotted versus $B_1$ for all four polymers. There are several proposed relations between $B_1$ and $B_2$:

$$B_2 = B_1 \tag{7.13}$$

$$B_2 = B_1^2 \tag{7.14}$$

$$B_2 = B_1^3. \tag{7.15}$$

The data are compared with the first two of these relations, and we see that HDPE samples follow Eq. 7.13 more closely, whereas the PP data fall in between Eqs. 7.13 and 7.14. Hence, there is apparently no universal relation between $B_1$ and $B_2$ for all polymers, and it depends on polymer type.

Of greater interest is the relation between $B_1$ and $B_2$ and capillary extrudate swell, $B$. This is shown in Figures 7.25 and 7.26. In the case of diameter swell, all the data fall below the line $B_1 = B$. In the case of thickness swell, most of the data fall below the line $B_2 = B$, except for the PP data that falls more closely to the line. Again, there is no general way to relate the swells.

Finally, Cogswell (1970) proposed a relation for the area swell, $B_1 B_2$, of an annular extrudate and the area swell of the extrudate from a capillary:

$$B_1 B_2 = B_A = 0.25 + 0.73 B^2. \tag{7.16}$$

This equation is compared with data in Fig. 7.27, and we see here that the agreement is good for the HDPEs but not for PP. The cause of the discrepancy is not clear.

The results presented here illustrate the complexity in trying to extend die swell measurements from a capillary to other die geometries. As an initial approximation one can use the relations between $B_1$, $B_2$, and $B$ given in Eqs. 7.13, 7.14, and 7.15. However, one must be aware of the fact that when significant strain hardening arises in the extensional behavior, the data will deviate more dramatically from these relations. As we proceed into more complex geometries, we will find even more problems in trying to make correlations with capillary data.

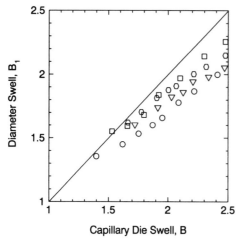

**Figure 7.25** Annular diameter swell versus capillary swell at the same wall shear rate for resins HDPE (○) 22, (○) 26, (▽) 27, and (□) PP-28. (Data from Garcia-Rejon & Dealy, 1982.)

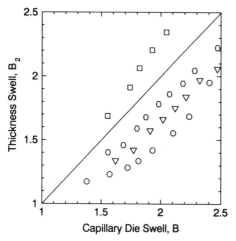

**Figure 7.26** Annular thickness swell versus capillary swell at the same wall shear rate for HDPE-22 (○), −26 (○), −27 (▽), and PP-28 (□). The line is $B_2 = B$. (Data from Garcia-Rejon & Dealy. 1982.)

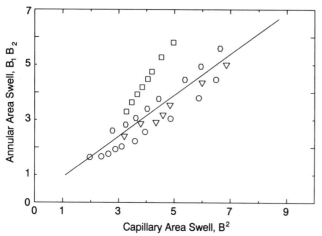

**Figure 7.27** Annular area swell versus capillary area swell at the same wall shear rate for the same resins as in Fig. 7.26. The line is Eq. 7.16. (Data from Garcia-Rejon & Dealy, 1982.)

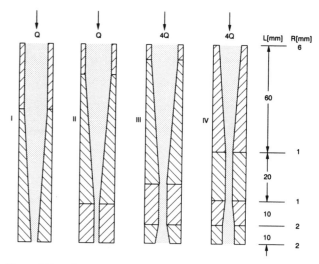

**Figure 7.28** Influence of the die geometry on extrudate swell, $D_p/D_0$, for a LDPE melt: I, $D_p/D_0 = 2.52$; II, $D_p/D_0 = 2.02$; III, $D_p/D_0 = 1.2$; IV, $D_p/D_0 = 1.34$, $\langle v_z \rangle = 7.2 \text{ mm s}^{-1}$. (Data from Laun, 1990.)

Before leaving this topic we would like to further point out the complexity of the phenomenon of extrudate swell. Figure 7.28 shows four different die geometries, and the corresponding extrudate swell values are given in the figure caption. Die geometry I converges to the exit, whereas die geometry II has a capillary of constant radius at the end of the converging section. The final radius in both cases is the same, and hence $\dot{\gamma}_w$ is the same. (*Note*: $\dot{\gamma}_w = (s+3) \langle v_z \rangle / R$ and $\langle v_z \rangle$ is 7.2 mm s$^{-1}$ for all four geometries.) However, $D_p/D_0$ for case I is 2.52, whereas it is 2.02 for case II. In die geometry III a diverging section has been added to the end of the die, and in case IV the die is straight but of the same diameter. $D_p/D_0$ for die geometry IV is 1.34. For die geometries III and IV, $\langle v_z \rangle$ is the same as in cases I and II, but $\dot{\gamma}_w$ is somewhat lower. In case I the additional stresses due to the converging flow lead to an increase in die swell relative to that for the straight tube. In case III the diverging section leads to a relaxation of stresses and a reduction in die swell. Hence, one can see that die design has a significant effect on extrudate swell, but there is no way at present to accurately predict this effect.

## 7.3. SHEET AND FILM DIES

In this section we introduce a few basic ideas behind the design of dies used in the extrusion of flat film and sheet. First, some general statements about the design of these types of dies are made, and then some specific aspects are dealt with. More details about the design of sheet and film dies can be found in the book by Michaeli (1984). The distinction between sheet and flat film rests in the thickness of the extruded product. Flat film is usually less than 0.25 mm thick. Die design considerations are somewhat similar in each case, although different requirements for the properties of film and sheet exist. The main problem is that the die is fed by an extruder with a circular opening. Somehow, the melt must be uniformly distributed over the width of the die so that the extrudate, which may be as wide as 400 cm, leaves the die lips thermally homogeneous and with a uniform stress distribution. A nonuniform flow history can lead to variations in the stress distribution and to extrudate thickness.

The salient features of a film (or sheet) die are shown in Fig. 7.29. A film die consists of four major parts: the manifold, choker bar, the land and the lips. The purpose of the manifold is to distribute the melt

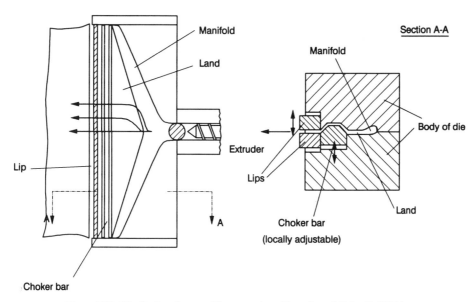

**Figure 7.29** Slit die for sheet or film extrusion. (Data from Michaeli, 1984.)

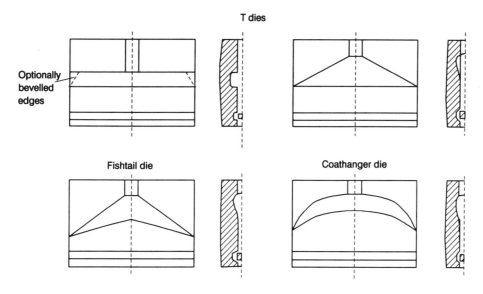

**Figure 7.30** Commonly used manifold designs for film and sheet dies. (Data from Michaeli, 1984.)

uniformly over the width of the die. The land tends to act as a resistance to flow and also promotes better flow uniformity. In the event that the manifold does not quite provide the required uniformity in flow rate across the die width, the choker bar can be used to adjust the flow rate locally. The die lips provide the final film thickness and can also be adjusted locally to account for a nonuniform flow rate or nonuniform die swell.

The design of the manifold is now considered. The most widely used design is that of the coathanger type, shown in Fig. 7.30 along with some other designs. As stated, the purpose of the manifold is to distribute the melt over the width of the die in such a manner that the flow rate is the same everywhere across the width of the exit. The coathanger design is essentially a bent tube of variable radius with a slit in the side wall. A sketch of this design is shown in Fig. 7.31 along with an indication of the pressure variation in the manifold and the land. The basic idea behind the design of a system such as this is that in order to maintain an uniform flow rate across the width of the die the pressure gradient must be the same across the width of the die. To accomplish this the pressure gradient along any length of the manifold must equal the pressure gradient in the land. Because the pressure decreases along the manifold, it must be bent so that the distance from the edge of the manifold decreases in such a way that the pressure gradient is constant.

To carry out the details of the design of a coathanger die, we refer to Fig. 7.32. In order to maintain a uniform flow rate per unit width, $q$, at each distance $x$ across the land, then

$$dp/dy = G = \text{constant}. \tag{7.17}$$

Next, this equation is integrated from $y = 0$ to $y = L(x)$:

$$\int_{P(O)}^{P(L)} dp = \int_{L(O)}^{L(x)} G dz \qquad (7.18)$$

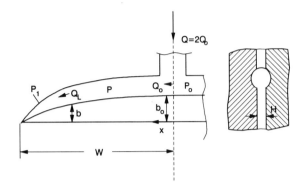

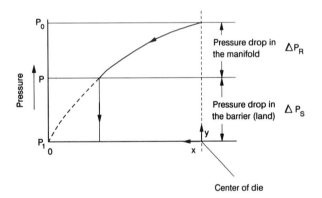

**Figure 7.31** Geometric relationships and pressure variation in the manifold and land sections of a sheet or film die.

to give:

$$P(L) - P(O) = G[L(x) - L(O)]. \qquad (7.19)$$

At $L(x)$, the pressure in the manifold at position l, $P(l)$, is assumed to equal $P(L)$. Substituting $P(l) = P(L)$ into Eq. 7.18 and differentiating with respect to $l$ gives:

$$dP(l)/dl = GdL(x)/dl. \qquad (7.20)$$

This is the general design equation for a coathanger manifold that provides a uniform flow across the slit width. In this analysis, the pressure drop across the contraction has been neglected.

To complete the equation, we must specify $G$ and $dL/dl$. $G$ is determined by assuming that developed flow occurs in the land, and $dL/dl$, which is a geometric variable, is then determined. We first solve for $G$ assuming that the rheological properties are described by the power-law model. For slit flow the volumetric flow rate per unit width, $q$, is given as (see Table 2.5):

$$q = \frac{H^2}{2(s+2)} \left(\frac{H}{2m}\right)^s \left(\frac{dp}{dy}\right)^s. \qquad (7.21)$$

With $G = dp/dy$ we obtain:

$$G = (2^{n+1}(2+s)^n mq^n)/H^{2n+1}. \qquad (7.22)$$

Next, flow inside the manifold is solved for. Although we will assume that the cross section is circular, in practice many times it is drop-shaped. The goal is to determine how $R$ varies with $l$. To do this it is customary to use the lubrication approximation. Starting with the equation for flow of a power-law fluid through a tube we obtain:

$$-dP(l)/dl = \left(\frac{3+s}{\pi}\right)^n 2mQ(l)^n/R(x)^{3n+1}. \qquad (7.23)$$

Employing a mass balance we know that the volumetric flow rate in the manifold at any position $l$, $Q(l)$, must equal the volumetric flow rate

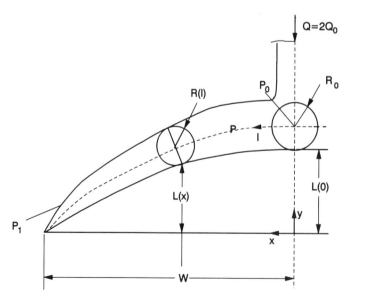

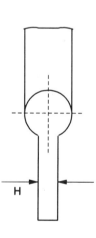

**Figure 7.32** Geometric model of a manifold.

in the slit from that point on until the end of the manifold:

$$Q(l) = q(W - x). \qquad (7.24)$$

Combining Eqs. 7.23 and 7.24, the following expression is obtained:

$$-dP(l)/dl = \left[ \left( \frac{3+s}{\pi} \right)^n 2mq(W-x) \right] / R(x)^{3n+1}. \qquad (7.25)$$

Finally, we replace $dP(l)/dl$ using Eqs. 7.20 and 7.22 to give:

$$(2n(2+s)^n/H^{2n+1})(dL/dl) + ((3+s)/\pi)^n (W-x)^n/R(x)^{3n+1} = 0. \qquad (7.26)$$

For a given fluid and a slit of width $2W$ and height, H, there are two geometric variables: $dL/dl$ and $R(x)$. For example, for a given manifold with curvature, $dL/dl$, there exists a manifold radius profile, $R(x)$, that yields an uniform pressure at any line of constant $y$. On the other hand, one could specify $R(x)$ and then determine $L(l)$ or $L(x)$ such that the pressure would be constant along any line of constant $y$. For instructional purposes one would take $dL/dl$ as constant. However, it is possible to apply the solution to finite segments of width $\Delta W$ and then find values of $dL/dl$ over the segment.

---

**EXAMPLE 7.1**
**Design of a Coathanger Manifold**
For HDPE (rheological data are given in Table 2.3), design a coathanger manifold to feed a film die having a width, $2W = 1.0$ m and height of 0.0508 cm. Assume constant curvature of the manifold.

**Solution**

There are two unknowns in Eq. 7.26, and we must specify one and then solve for the other. We specify $dL/dl$ and then solve for $R(x)$. Taking $dL/dl = -\tan \alpha = -0.087$ (i.e. $\alpha = 5°$), rearranging Eq. 7.26 and using the value for $n$ of 0.56 we find:

$$R(x) = 6.54 \times 10^{-3}(W-x)^{0.209}.$$

This equation gives the manifold radius as a function of $x$.

---

Before leaving this topic a few additional comments should be made about the limitations of Eq. 7.26. First, the degree of taper (i.e., $R(x)$) and the curvature of the manifold (i.e., $dL/dl$) must be small enough that the lubrication approximation is not violated. Furthermore, the solution does not consider viscoelastic effects such as entrance pressure losses or the time dependence of the stresses which might lead to higher values of viscosity than the steady-state values. We also note that the design of a coathanger manifold die is dependent on the rheological properties of the fluid. Hence, in changing from one polymer to another, or changing the melt temperature, the effectiveness of the die design will change. Thus, the choker bar and die lip opening may have to be adjusted to compensate for these changes. Finally, there may be nonuniform die swell across the width of the die as the effective land length changes with position across the die. As has been seen, die swell strongly depends on capillary length. Furthermore, the land length is relatively short, and fully developed flow may be barely achieved. Again, the adjustable die lips can be used to compensate for the nonuniformities associated with the variations in die swell.

## 7.4. ANNULAR DIES

Dies with annular cross sections are used to extrude pipes, tubes, tubular films, and parisons for blow molding. Center-fed dies are commonly used for extruding pipes, and side-fed dies are used for tubular films and parisons. The four basic annular die designs in use at the present time are shown in Fig. 7.33. These include (1) center-fed spider-supported mandrel dies; (2) center-fed screen pack dies; (3) side-fed mandrel dies; and (4) spiral mandrel dies. At the die exit there is usually an outer die ring that forms the die land. Only this part of the die can be analyzed simply as annular flow. The remaining portions must be analyzed through the lubrication approximation. In Section 7.4.1 we discuss the center-fed die and consider in Section 7.4.2 the side-fed and spiral mandrel die designs. Finally, wire-coating dies are discussed in Section 7.4.3. Further details on die design as well as rules of thumb are given by Michaeli (1984).

### 7.4.1. CENTER-FED ANNULAR DIES

Pipes are primarily extruded using center-fed dies of the type shown in Fig. 7.34. The melt stream from the extruder passes from the circular opening to the annular die by means of the mandrel support tip. The melt then passes over the "spider legs" that support the mandrel and through a converging annular region, which for pipes is usually 10° to 15°. The converging region is followed by an annular region with parallel walls that impart the final dimensions to the pipe. The outer diameter of the pipe can range from a few millimeters to approximately 1.6 m. The ratio of the mandrel radius to outer wall radius usually falls

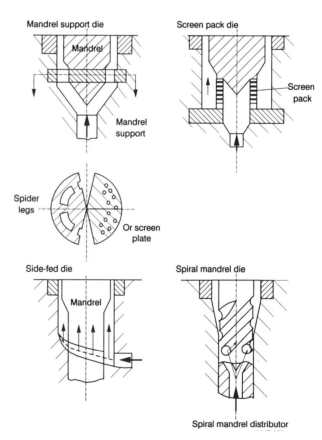

**Figure 7.33** Four common annular die designs. (Data from Michaeli, 1984.)

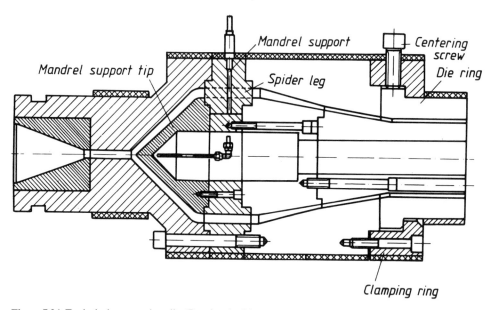

**Figure 7.34** Typical pipe extrusion die. (Reprinted with permission of the publisher from W. Michaeli, 1984.)

in the range of 0.8 to 0.925. This ratio is important as it implies that we can neglect curvature in the analysis of most parts of the die.

The spider legs usually lead to problems, Not only are flow markings visible, but also mechanically weak regions are generated. These flow lines are also referred to as weld lines and are due to a higher degree of molecular orientation caused by the high stresses imparted to the melt and the inability of the molecules to re-entangle during the time the melt spends in the die. The strength of the weld line is most likely related to the degree of re-entanglement of the molecules that occurs as the melt passes through the die at the melt temperature. The re-entanglement time appears to be much longer than the longest relaxation time. It has been estimated from interrupted shear experiments for PS, for example, to be of the order of 300 s (Pissipati, 1983). It is related to self-diffusion of polymer molecules as discussed in Chapter 4.

One way of reducing the effect of the flow lines is through the design of the support system. Some spider leg systems presently used are shown in Fig. 7.35. Instead of arranging the spider legs radially as shown in Fig. 7.35.a, tangential arrangement as shown in Fig. 7.35.b will displace the defect circumferentially over the extrudate. A better way for reducing the flow marks is to use offset spider legs as shown in Fig. 7.35.d. Here the flow marks do not extend all the way through the wall of the extrudate, which offers at least mechanical improvements. Finally, another way to reduce flow marks is to use a screen plate, as shown in Fig. 7.35.c. In this design the melt passes into the annular region by first passing through a plate with many small holes bored in it. In effect the annular die is fed by multiple capillaries.

Design considerations for the center-fed dies consist of the force exerted by the melt on the mandrel support tips, the residence time in the die, the total pressure drop, the extrusion rate, die swell, and the onset of flow instabilities. The significance of the pressure exerted on the mandrel support tip rests in making sure the spider legs are strong enough. Residence time considerations are required to determine whether sufficient time has been allowed for partial healing of the weld lines imparted by the spider legs. The relation between the shear stress and extrusion rate in the land section is also needed, as the onset of melt flow instabilities is the limiting factor in the rate of extrusion. Finally, the diameter of the extrudate and the thickness as the result

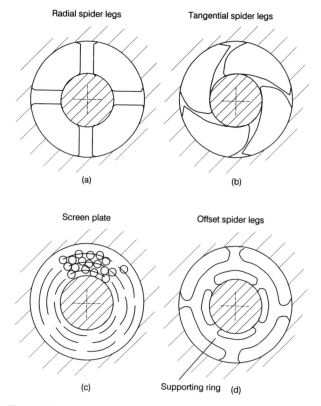

**Figure 7.35** Typical mandrel support systems. (Data from Michaeli, 1984.)

of die swell must be taken into consideration. The complete design of a center-fed die using a viscoelastic constitutive equation is not possible at this time. We can, however, estimate parts of the flow behavior needed in design work by making use of the lubrication approximation.

## 7.4.2. SIDE-FED AND SPIRAL MANDREL DIES

The side-fed mandrel die is used in both blown film and pipe extrusion. The main consideration is to provide a uniform flow rate at the die land. This is done in much the same way as for flat film extrusion by the use of a manifold as shown in Figure 7.36. The main difficulty is in the design of the curved manifold, but one approach is to neglect curvature and consider the manifold to be like that of the flat film die but wrapped around the curved mandrel.

The spiral mandrel die seems to offer the best possibility for providing a uniform flow rate circumferentially at the annular die exit. Two specific designs of this type are shown in Figure 7.37. The die is usually fed from the extruder by means of either a star-shaped or ring-shaped distributing system. The melt then passes into spiral-shaped channels that are machined into the mandrel. The depth of the channels decreases with spiral distance, which ensures that there will be mixing of the melt from channel to channel as the result of leakage. The spiral channels perform in much the same way as the single-screw extruder.

Because of the design of this system there are no mandrel support elements, and hence flow lines are eliminated completely.

## 7.4.3. WIRE-COATING DIES

Wire-coating dies also involve annular cross sections. The basic sections of a wire-coating die are shown in Figure 7.38. The melt usually enters from the side and resembles the side-fed annular dies discussed in Section 7.4.1. The goal in the design of wire-coating dies is to provide a coating of the desired thickness, free of imperfections, and with the wire centered in the insulation. Again, the even distribution of the melt from the die entry is the key design element.

There are two basic types of coating dies used at present: pressure-coating and tube-coating dies. These are shown in Fig. 7.39. In the pressure-coating die the wire is coated under pressure in the die. This technique is used usually for the application of the primary coating where good adhesion is important. In the case of the tube-coating die,

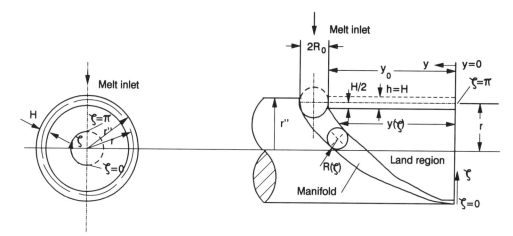

**Figure 7.36** Distribution system for a side-fed mandrel die.

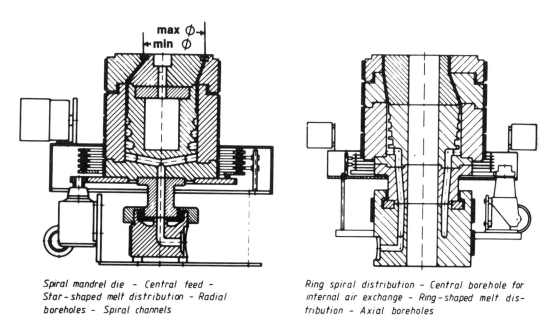

Spiral mandrel die – Central feed –
Star-shaped melt distribution – Radial
boreholes – Spiral channels

Ring spiral distribution – Central borehole for
internal air exchange – Ring-shaped melt distribution – Axial boreholes

**Figure 7.37** Two spiral mandrel die designs. (Reprinted with permission of the publisher from W. Michaeli, 1984.)

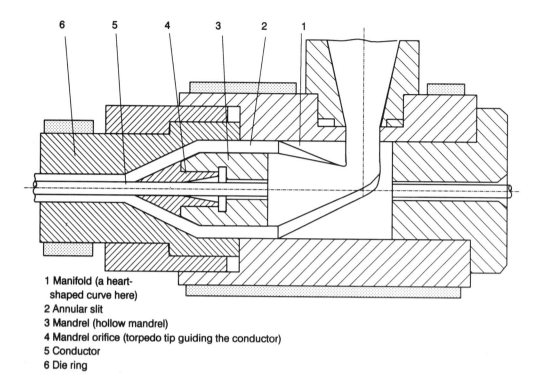

1 Manifold (a heart-
   shaped curve here)
2 Annular slit
3 Mandrel (hollow mandrel)
4 Mandrel orifice (torpedo tip guiding the conductor)
5 Conductor
6 Die ring

**Figure 7.38** Basic features of a wire-coating die. (Data from Michaeli, 1984.)

the polymer coating is applied outside the die and is used for applying a secondary coating.

The main problem in the design of wire-coating dies is to provide the same velocity at each point of the circumference at the die exit. Several widely used ways include the "heart-shaped curve" (see Fig. 7.40) and the circular coathanger designs. The "heart-shaped curve" is shown in Fig. 7.41. The melt stream from the extruder is divided up into two separate melt streams by the bezel that runs parallel to the axis of the mandrel. The flow channels become wider, and their depth decreases. A heart-shaped piece, which is mounted below the bezel and elongates the flow path on the inner arc, functions as a compensating element. The length the flow edge of this heart-shaped piece is dimensioned such that the total path of the melt on the inner arc approximately corresponds to the flow path on the outer arc.

After the melt passes through the distribution system, the die housing narrows down to the wire in order to match the average velocity, $\langle v_z \rangle$, with the wire velocity. If high coating speeds are to be achieved, it is advisable to have the angle between the mandrel and the housing become steadily smaller up to the coating region to suppress vortices.

Further quantitative features of the design of wire-coating dies are discussed in the problems at the end of the chapter. The lubrication approximation is widely used in the design of wire-coating dies.

## 7.5. PROFILE EXTRUSION DIES

Profile extrusion refers to the extrusion of polymer melts through dies of cross sections that are not round, annular, or rectangular with an aspect ratio, $W/H$, greater than 10.0. Because the geometries are quite complex, it is not possible to obtain analytical solutions when the Generalized Newtonian Fluid (GNF) or viscoelastic models are used. The use of finite element methods offers promise in solving problems

associated with the design of profile dies. However, these techniques are not fully developed yet, and the subject matter required to understand these methods is beyond the level of this text. At present, design is carried out by trial-and-error methods. However, some aspects of design can be dealt with, and these are discussed in this section.

We first consider some general aspects of profile die design. Three factors determine the dimensions of a profile die to produce an extrudate of desired dimensions. The first is the degree of die swell, which as discussed earlier (Section 7.2.3) is a function of the flow history in the die as well as the cooling conditions at the die exit. The second involves the shrinkage that occurs as the polymer melt solidifies. Finally, there is the shape change associated with drawing that occurs in the sizing device. Just determining the pressure drop/flow rate relation is complex enough because of irregular boundaries. Only the pressure drop/flow rate relation for some geometries can be quantitatively dealt with at this level.

At present there are three types of profile dies used: *orifice dies*, *multistage dies*, and *tapered profile dies*. An example of an orifice die is shown in Figure 7.42. Basically, the orifice die consists of a die base and a die plate in which the profile is formed. These dies are used for the extrusion of inexpensive profiles where dimensional accuracy is not necessary. Because of the abrupt change in cross-sectional area, there is usually a build up of stagnant material behind the die plate, and high extrusion rates are not possible. These dies are not commonly used for most thermoplastics but are restricted primarily to PVC and rubber.

An example of a multistage die is shown in Fig. 7.43. Multistage dies exhibit step changes in the cross-sectional area of the flow channel. They consist of a series of die plates of similar geometry but of a decreasing cross-sectional area. Certainly, these represent an improvement over the orifice dies, but they still suffer from some of the same deficiencies.

Whenever profiles of high dimensional accuracy are to be produced

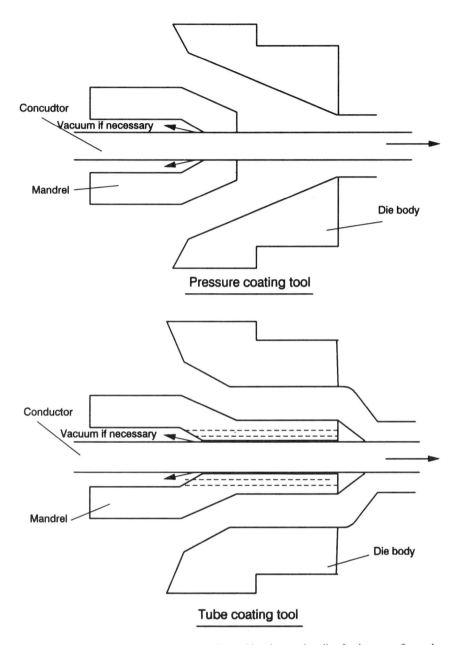

Concudtor

Vacuum if necessary

Mandrel

Die body

**Pressure coating tool**

Conductor

Vacuum if necessary

Mandrel

Die body

**Tube coating tool**

**Figure 7.39** Two mandrel designs commonly used in wire-coating dies. In the upper figure the wire is coated under pressure inside the die, and in the lower figure the wire is coated outside the die.

at high extrusion rates, profile dies with a gradual change of cross-sectional area are required. An example of a profile die for producing highly accurate cross sections is shown in Fig. 7.44. The design of these dies is carried out nearly on an empirical basis at present.

However, if one carries out empirical design work to generate profile dies on the laboratory scale, then one could use dimensional analysis for scaleup. In particular, geometric similarity would require that the following ratios be identical:

$$A_e/A_L \text{ and } L/R_H,\tag{7.27}$$

where $A_L$ is the cross-sectional area of the land and $A_e$ is the cross

sectional area of the entry (this actually represents a contraction ratio) and $R_H$ is the mean hydraulic radius. For dynamic similarity the Deborah (De) numbers would also have to be identical, where De is defined as:

$$\text{De} = \lambda/t_p.\tag{7.28}$$

$\lambda$ is again the longest relaxation time of the melt, and $t_p$ is the process time, which for extrusion is $L/\langle v_z \rangle$, where $\langle v_z \rangle$ is the average velocity in the flow direction. However, because the rheological properties are for the same melt, dynamic similarity requires that:

$$(\langle v_z \rangle/L)_1 = (\langle v_z \rangle/L)_2,\tag{7.29}$$

where the subscript 1 refers to the laboratory system and 2 the scaled-up system.

As one can imagine, it is nearly impossible at this time to carry out quantitative design work for dies such as shown in Figure 7.44. There are, however, a few considerations that can help in doing at least semi-quantitative work. We first consider some ideas concerned with die swell and then shrinkage and sizing.

The complexities associated with die swell are illustrated by the results presented in Fig. 7.45. Here the extrudate shape is presented for two polymers, PVC and LDPE, as a function of the ratio of entry channel design and length to mean hydraulic radius, $R_H$. The mean hydraulic radius, $R_H$, is defined as $A/P$, where $A$ is the cross-sectional area and $P$ is the wetted perimeter. Although the die cross section is square, the extrudate shape barely reflects this. In fact, for low $L/R_H$ ratios, the extrudate reflects the shape of the entry geometry. As the $L/R_H$ ratio increases, the extrudate reflects more closely the shape of the land section. It is also observed that the amount of swell decreases as the ratio increases. Furthermore, the results in Fig. 7.45 illustrate that PVC has considerably less swell than LDPE.

For some shapes it may be possible to carry out estimates of $Q$, $\Delta P$, and extrudate swell by breaking the geometry up into a collection of simple geometries for which the flow can be analyzed. Some examples of profile geometries that can be dealt with by use of the equations derived for thin rectangular slits (Table 2.5) are presented in Fig. 7.46.

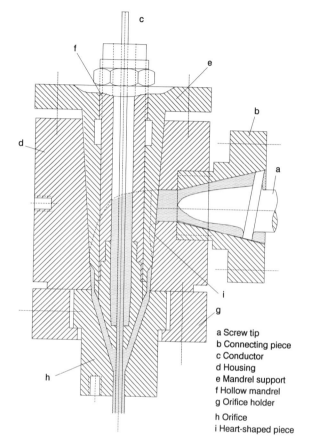

a Screw tip
b Connecting piece
c Conductor
d Housing
e Mandrel support
f Hollow mandrel
g Orifice holder

h Orifice
i Heart-shaped piece

**Figure 7.40** Wire-coating die with a heart-shaped distribution system. (Data from Michaeli, 1984.)

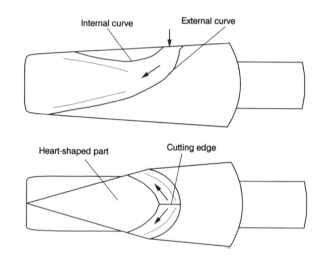

**Figure 7.41** Deflecting distributing system with a heart-shaped piece. (Data from Michaeli, 1984.)

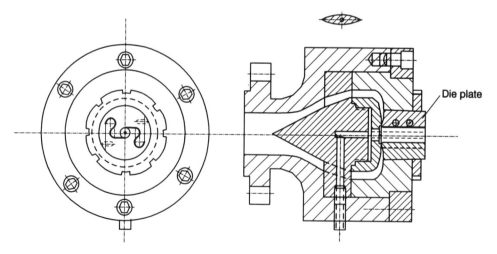

**Figure 7.42** Profile die of the orifice type. The die plate is shown at the left. The extrudate emerges as a tube with two S-shaped flanges. (Data from Michaeli, 1984.)

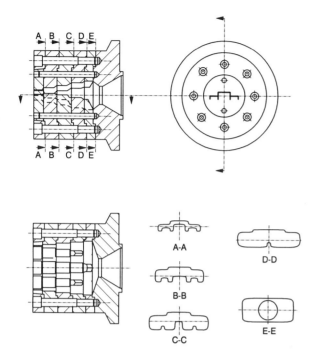

**Figure 7.43** Profile die of the multistage type. The die plate is shown in the upper right-hand figure (the extrudate is used as window sash). In the lower right the segments making up the various stages of the die are shown. (Data from Michaeli, 1984.)

for flow of a power-law fluid through a rectangular slit is given by:

$$Q = \frac{WH^2}{2(s+2)}\left(\frac{H\Delta P}{2mL}\right)^s. \tag{7.30}$$

We can apply Eq. 7.30 to geometry no. 3 by opening up the die to a die that has a width of $4(R_o + R_i)$ and slit height, $H = R_o - R_i$. Likewise, we can estimate die swell from results from a slit die or capillary. Other geometries for which the slit formula can be used are shown in Fig. 7.46 as well as replacements for $W$ and $H$ in Eq. 7.30.

The extruded product shrinks as it cools as a result of density changes. The amount of shrinkage can be estimated by the changes in density from the melt to the solid state. One must also assume that shrinkage occurs uniformly in all directions.

The extrudate is stretched slightly on leaving the die as it is pulled into the sizing device. In the book by Michaeli (1984) there are some suggested amounts by which the die cross-sectional area should be increased to account for the reduction in extrudate cross-sectional area generated by drawing. For example, for LDPE it is recommended that the die cross-sectional area be increased by 15% to 20%. However, these empirical "rules of thumb" are not very dependable.

## 7.6. MULTIPLE LAYER EXTRUSION

It is becoming more common to combine multiple layers of different polymers to form products with properties that take advantage of the best properties of each component. For example, packaging film might be composed of several different types of PE along with a layer of adhesive and a barrier polymer. We first discuss some general aspects of extrusion processes used for generating multi-layered products. Some quantitative aspects of die design are considered along with qualitative considerations for die design.

### 7.6.1. GENERAL CONSIDERATIONS

There are basically three types of multiple layer extrusion techniques: (1) melt streams that flow separately; (2) melt streams that flow separately and then together (3) melt streams that flow together. Examples of type 1 are shown in Fig. 7.47. In this process polymers $A$ and $B$ are extruded through separate flow channels and then joined together outside the die. The advantage in this type of multilayer extrusion is that polymers with widely different processing temperatures and rheological properties can be used. The major problem is that of generating satisfactory adhesion between the components. Usually, the technique is only used for two polymers.

The second technique is shown in Fig. 7.48. Here the streams are brought together inside the die, and then they pass through a common land region. Because the streams are brought together under pressure, adhesion is improved. However, it is not possible to have the streams at widely different temperatures. Likewise, the rheological properties cannot be too widely different, or flow instabilities will arise. Furthermore, at the point where the streams converge, interfacial instability problems may arise.

The third method is not too dissimilar from the second method, as shown in Fig. 7.49. Here the polymer streams are brought together in an adaptor, and then they pass through a common die. In this case the same die as that used for single component extrusion can be used. Again, the melt rheological properties cannot be too dissimilar or an instability will arise that will disrupt the laminar nature of each stream. However, this is one of the simplest and most inexpensive methods for generating multilayer films and sheets.

### 7.6.2. DESIGN EQUATIONS

The most often used die geometries in multilayer extrusion are slits and tubular film dies. However, there are also cases in which profile dies are used. Some quantitative design work can be carried out in the case of flow through flat film and tubular dies. The factors that can be dealt with are flow rates and pressure drops required to provide a given thickness in a multilayer extrudate. The major problem in multilayer extrusion is the instability in the flow, which we discuss in a qualitative fashion in Section 7.6.3. However, the design equations can at least be used to estimate whether the flow will be unstable.

Flow of two polymers through a slit die as shown in Fig. 7.50 is considered first. It is assumed that (1) the rheological properties of the fluids are described by the power-law model; (2) the flow is stable and at steady state; and (3) the flow is isothermal. With these assumptions in hand the goal is to determine a relation between the desired layer thickness, flow rate, and pressure drop. The subscript (1) refers to the polymer in phase 1, and the subscript (2) refers to that in phase 2. The equations of motion are then written for each phase:

$$-\frac{\partial p^{(1)}}{\partial z} - \frac{\partial \tau_{yz}^{(1)}}{\partial y} = 0 \tag{7.31}$$

$$-\frac{\partial p^{(2)}}{\partial z} - \frac{\partial \tau_{yz}^{(2)}}{\partial y} = 0. \tag{7.32}$$

The pressure gradient in each phase must be identical, or there would be flow in the $y$ direction. Equations (7.31) and (7.32) can be integrated to find $\tau_{yz}^{(1)}$ and $\tau_{yz}^{(2)}$:

$$\tau_{yz}^{(1)} = G(y - C_1) \tag{7.33}$$

$$\tau_{yz}^{(2)} = G(y - C_2), \tag{7.34}$$

where $G = -\partial p/\partial z$. At some unknown position, $\alpha$, $\tau_{yz}^{(1)} = 0$, which implies

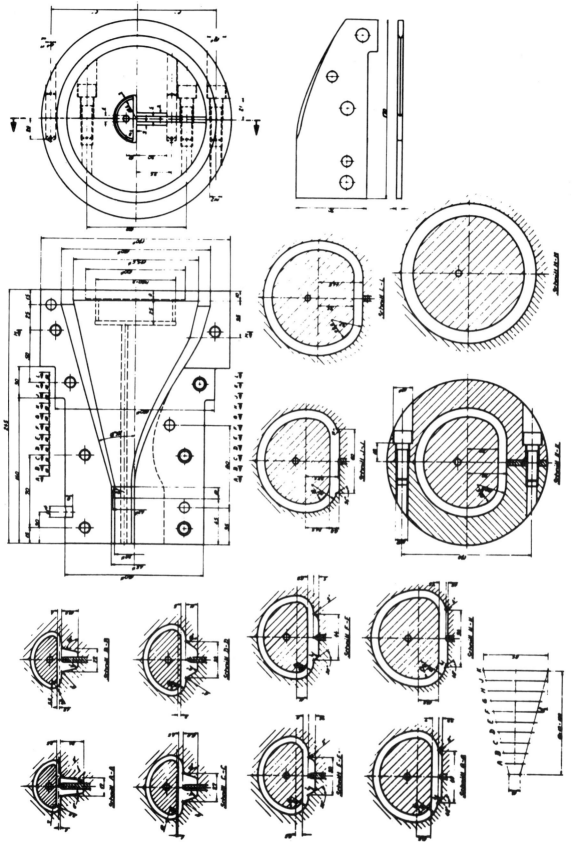

**Figure 7.44** Profile die with a gradual change of cross-sectional area for producing highly accurate shapes. (Reprinted with permission of the publisher from W. Michaeli, 1984.)

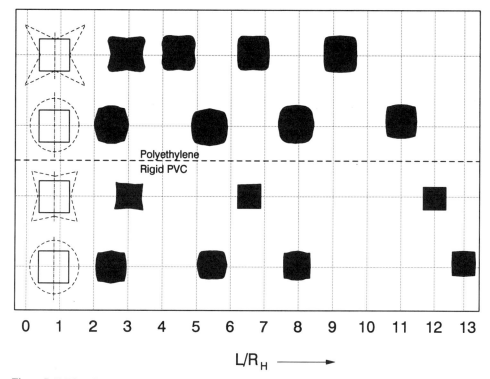

**Figure 7.45** The effect of entrance geometry and land length (here the length is normalized by the mean hydraulic radius) on extrudate swell for LDPE (upper data) and PVC. The broken line represents the shape of the entry while the land is square. (Data from F. Röthemeyer, *Elastic Effects in the Extrusion of Plastic Melts.* Ph.D. Thesis, University of Stuttgart, 1970.)

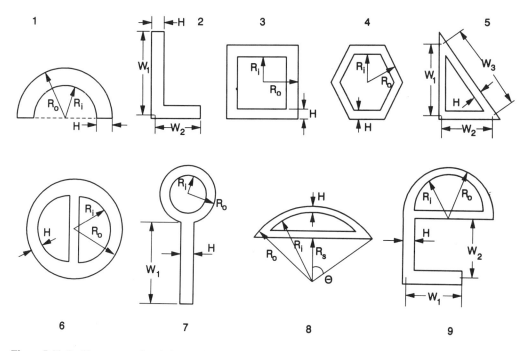

**Figure 7.46** Profile cross-sectional shapes that can be treated as a combination of thin slits. The slit width in Eq. 7.30 is replaced by the following values in each case: 1: $W = \pi/2(R_0 + R_i)$; 2: $W = W_1 + W_2$; 3: $W = 4(R_0 + R_i)$; 4: $W = 3.46(R_0 + R_i)$; 5: $W = W_1 + W_2 + W_3$; 6: $W = \pi R_0 + (2 + \pi)R$; 7: $W = \pi(R_0 + R_i) + W$; 8: $W = (R_0 + R_i)(\theta + \sin \theta)$; $\cos \theta = (2R_s)/(R_0 + R_i)$; 9: $W = ((\pi/2)R_0 + [2 + (\pi/2)]R_i + W_1 + W_2)$. (Data from J. F. Carley, "Problems of Flow in Extrusion Dies." *SPE J.*, 12, 1263, 1963.)

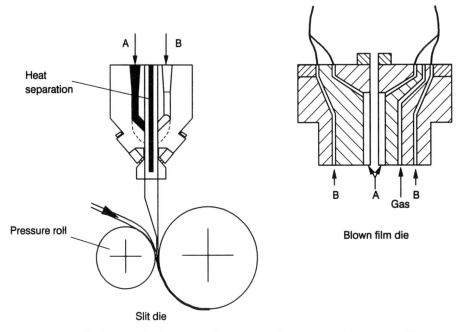

**Figure 7.47** Coextrusion dies in which the two streams of polymer are joined outside the die.

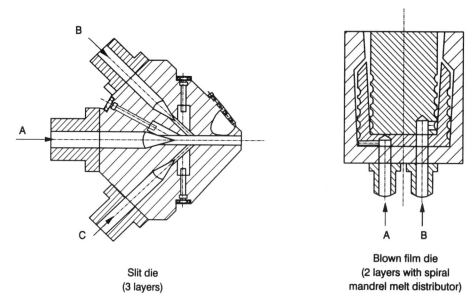

**Figure 7.48** Coextrusion dies in which the streams are combined in the die and then pass through a common land.

that $C_1 = \alpha$. Furthermore, from the assumption that $\tau_{yz}^{(1)} = \tau_{yz}^{(2)}$ at $y = \beta$, it can be shown that $C_2 = \alpha$ also. Therefore, Eqs. 7.33 and 7.34 become:

$$\tau_{yz}^{(1)} = G(y - \alpha) \tag{7.35}$$

$$\tau_{yz}^{(2)} = G(y - \alpha). \tag{7.36}$$

We need four boundary conditions in order to find $v_z^{(1)}$ and $v_z^{(2)}$:

B.C.1 at $y = 0$ $\quad v_z^{(1)} = 0$ (7.37)

B.C.2 at $y = \beta$ $\quad v_z^{(1)} = v_z^{(2)}$ (7.38)

B.C.3 at $y = h$ $\quad v_z^{(2)} = 0$ (7.39)

B.C.4 at $y = \beta$ $\quad \tau_{yz}^{(1)} = \tau_{yz}^{(2)}.$ (7.40)

Here boundary conditions 2 and 4 (B.C.4 was already used to obtain Eq. 7.36) are based on the continuity of stresses and the velocity field at the interface between the two polymer streams. Eventually, as

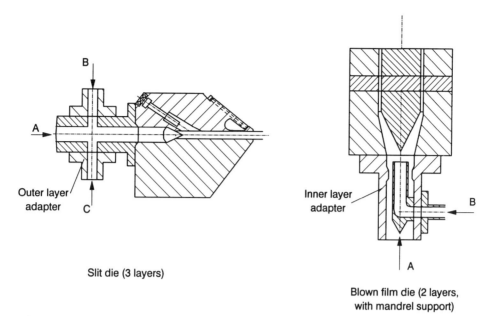

**Figure 7.49** Coextrusion dies in which the streams are brought together in an adaptor and then pass through a common die.

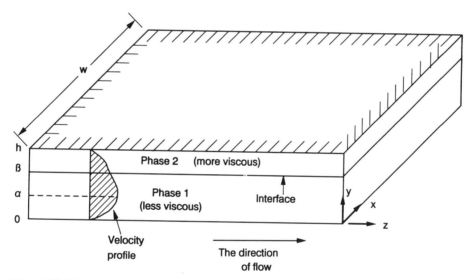

**Figure 7.50** Flow of two polymer melts through a slit die. $\alpha$ is the position of the maximum in the velocity profile, and $\beta$ is the position of the interface between the two melts.

discussed later, the cohesion between the two polymers at the interface may break down, leading to an instability.

We next need expressions for $\tau_{yz}^{(1)}$ and $\tau_{yz}^{(2)}$ in terms of velocity gradients. Because of the changes in sign of the velocity gradient, we must break up the flow region into three parts:

Region (1a):  $0 < y < \alpha$     $dv_z^{(1a)}/dy > 0$       (7.41)

Region (1b):  $\alpha < y < \beta$     $dv_z^{(1b)}/dy < 0$       (7.42)

Region (2):  $\beta < y < h$     $dv_z^{(2)}/dy < 0.$       (7.43)

Substituting the power-law expression for $\tau_{yz}$ into Eqs. 7.35 and 7.36

leads to the following differential equations for $v_z^{(1)}$ and $v_z^{(2)}$:

$$-m_1(dv_z^{(1a)}/dy)^{n_1} = G(y-\alpha) \tag{7.44}$$

$$m_1(-dv_z^{(1b)}/dy)^{n_1} = G(y-\alpha) \tag{7.45}$$

$$m_2(-dv_z^{(2)}/dy)^{n_2} = G(y-\alpha). \tag{7.46}$$

We can now integrate these equations to obtain the velocity field:

$$v_z^{(1a)} = \left(\frac{G}{m_1}\right)^{s_1} \int_0^y (\alpha - y')^{s_1}\,dy' \tag{7.47}$$

$$v_z^{(1b)}(\alpha) - v_z^{(1b)} = \left(\frac{G}{m_1}\right)^{s_1} \int_\alpha^y (\alpha - y')^{s_1} \, dy' \qquad (7.48)$$

$$v_z^{(2)} = \left(\frac{G}{m_2}\right)^{s_2} \int_y^h (\alpha - y')^{s_2} \, dy'. \qquad (7.49)$$

The prime on $y$ indicates that it is a dummy variable of integration. Carrying out the integration and using the condition that $v_z^{(1a)}(\alpha) = v_z^{(1b)}(\alpha)$ at $y = \alpha$, we find the following expressions:

$$v_z^{(1a)} = \left(\frac{G}{m_1}\right)^{s_1} \left(\frac{n_1}{n_1+1}\right)\left[\alpha^{\frac{n_1+1}{n_1}} - (\alpha - y)^{\frac{n_1+1}{n_1}}\right] \qquad (7.50)$$

$$v_z^{(1b)} = \left(\frac{G}{m_1}\right)^{s_1} \left(\frac{n_1}{n_1+1}\right)\left[\alpha^{\frac{n_1+1}{n_1}} - (y - \alpha)^{\frac{n_1+1}{n_1}}\right]. \qquad (7.51)$$

We can write Eqs. 7.50 and 7.51 more succinctly as

$$v_z^{(1)} = \left(\frac{G}{m_1}\right)^{s_1} \left(\frac{n_1}{n_1+1}\right)\left[\alpha^{\frac{n_1+1}{n_1}} - |y - \alpha|^{\frac{n_1+1}{n_1}}\right]. \qquad (7.52)$$

For layer 2 we can complete the integration in Eq. 7.49 to obtain:

$$v_z^{(2)} = \left(\frac{G}{m_2}\right)^{s_2} \frac{n_2}{n_2+1}\left[(h-\alpha)^{\frac{n_2+1}{n_2}} - (y-\alpha)^{\frac{n_2+1}{n_2}}\right]. \qquad (7.53)$$

The goal of this derivation is to obtain an expression for the volumetric flow rate in each layer and pressure drop to produce a desired thickness for each layer. The volumetric flow rate in each layer, $Q_1$ and $Q_2$, is obtained by integrating Eqs. 7.52 and 7.53 over the cross-sectional area:

$$Q^{(1)} = W\left(\frac{G}{m_1}\right)^{s_1}\left(\frac{n_1}{n_1+1}\right)$$

$$\times \left\{\int_0^\alpha \left[\alpha^{\frac{n_1+1}{n_1}} - (\alpha - y)^{\frac{n_1+1}{n_1}}\right]dy + \int_\alpha^\beta \left[\alpha^{\frac{n_1+1}{n_1}} - (y - \alpha)^{\frac{n_1+1}{n_1}}\right]dy\right\} \qquad (7.54)$$

$$= W\left(\frac{G}{m_1}\right)^{s_1}\left(\frac{n_1}{n_1+1}\right)\left[\left(\frac{n_1+1}{2n_1+1}\right)\alpha^{\frac{2n_1+1}{n_1}} + \alpha^{\frac{n_1+1}{n_1}} \cdot \right.$$

$$\left. (\beta-\alpha) - \left[\frac{n_1(\beta-\alpha)}{2n_1+1}\right]^{\frac{2n_1+1}{n_1}}\right] \qquad (7.55)$$

and

$$Q^{(2)} = W\left(\frac{G}{m_2}\right)^{s_2}\left(\frac{n_2}{n_2+1}\right)\int_\beta^h \left[(h-\alpha)^{\frac{n_2+1}{n_2}} - (y-\alpha)^{\frac{n_2+1}{n_2}}\right]dy \qquad (7.56)$$

$$= W\left(\frac{G}{m_2}\right)^{s_2}\left(\frac{n_2}{n_2+1}\right)$$

$$\times \left[(h-\alpha)^{\frac{n_2+1}{n_2}}(h-\beta)\frac{n_2}{2n_2+1}(h-\alpha)^{\frac{2n_2+1}{n_2}} - (\beta-\alpha)^{\frac{2n_2+1}{n_2}}\right]. \qquad (7.57)$$

The total volumetric flow rate, $Q$, is then $Q^{(1)} + Q^{(2)}$.

To solve these equations, one more equation is needed as there are three unknowns, $Q$, $\alpha$, and $\beta$ (for a given $G$). The final equation is obtained from the condition:

$$v_z^{(1)}(\beta) = v_z^{(2)}(\beta) \qquad (7.58)$$

and is

$$\left(\frac{G}{m_1}\right)^{s_1}\left(\frac{n_1}{n_1+1}\right)[\alpha^{1+s_1} - (\beta-\alpha)^{1+s_1}]$$

$$= \left(\frac{G}{m_2}\right)^{s_2}\left(\frac{n_2}{n_2+1}\right)[(h-\alpha)^{1+s_2} - (\beta-\alpha)^{1+s_2}]. \qquad (7.59)$$

Hence, for a given layer thickness (i.e. given $\beta$), one can use Eq. 7.59 to find $\alpha$ (Note: one must guess at $G$.) Then, one can calculate $Q^{(1)}$ and $Q^{(2)}$ from the values of $\alpha$, $\beta$, and $G$. We illustrate the manipulation of these equations in the following example.

---

**EXAMPLE 7.2**
**Coextrusion of Two Polymers through a Film Die**
HDPE (rheological properties are given in Table 2.3) and nylon 6 (rheological properties are given in Table 7.3) are extruded through a film die both at a temperature of 220°C. Determine the pressure gradient, $G$, required to produce an extrudate whose thickness is 1/5 nylon 6 (i.e., $\beta = 0.2h$). Determine the shear rate at the interface for each component. The dimensions of the land of the die are: $W = 76.2$ cm and $h = 0.1016$ cm. The line speed (i.e., linear velocity of the sheet) is 60 m/min.

**TABLE 7.3 Parameters for Nylon 6, Capron™ 8200, Allied Chemical**

| Temperature (°K) | Power Law | | | Carreau Model | | | | Ellis Model | |
| --- | --- | --- | --- | --- | --- | --- | --- | --- | --- |
| | $\dot{\gamma}$ Range (s⁻¹) | $m$ (Pa·sⁿ) | $n$ | $\eta_0$ (Pa·s) | $\dot{\gamma}$ Range (s⁻¹) | $n$ | $\lambda$ (s) | $\alpha$ | $\tau_{1/2}$ (Pa) |
| 498 | 100–2500 | $2.62 \times 10^3$ | 0.63 | $1.6 \times 10^3$ | 100–2000 | 0.63 | 0.27 | 1.64 | $1.06 \times 10^4$ |
| 503 | 100–2000 | $1.95 \times 10^3$ | 0.66 | $1.3 \times 10^3$ | 100–2000 | 0.65 | 0.32 | 1.70 | $1.3 \times 10^4$ |
| 508 | 100–2300 | $1.81 \times 10^3$ | 0.66 | $1.1 \times 10^3$ | 100–2000 | 0.68 | 0.36 | 1.61 | $1.04 \times 10^4$ |

(Data from Z. Tadmor, and C. G. Gogos, *Principles of Polymer Processing*, Wiley, New York, 1979.)

**TABLE 7.4 Program Listing for Calling the Subroutine NEQNF Required to Solve Example 7.2 (COEX1.FOR)**

```
C********************ABSTRACT********************
C   THIS PROGRAM CALLS THE IMSL SUBROUTINE NEQNF TO FIND
C   THE VALUE OF ALPHA(A) IN EQ. 7.59 AS REQUIRED IN THE
C   SOLUTION OF EX.7.2.
C
C               DECLARE VARIABLES
      INTEGER ITMAX, N
      REAL ERRREL
      PARAMETER (N=1)
      INTEGER K, NOUT
      REAL FCN, FNORM, X(N), XGUESS(N)
      COMMON G
      EXTERNAL FCN, NEQNF, UMACH
      WRITE(*,*) 'INPUT THE PRESSURE GRADIENT,G,THE INITIAL
     & GUESS FOR A, ERREL, AND ITMAX'
      READ(*,*) G,XGUESS, ERRREL, ITMAX
      CALL UMACH(2,NOUT)
C
      CALL NEQNF(FCN, ERRREL, N, ITMAX, XGUESS, X, FNORM)
      WRITE (NOUT,100) (X(K),K=1,N), FNORM
100   FORMAT('   THE SOLUTION TO THE SYSTEM IS',/,' X =
      (',2E10.4,
     &   ')',/,' WITH FNORM =',E10.6,//)
      END
C
      SUBROUTINE FCN(X,F,N)
      COMMON G
      INTEGER N
      REAL X(N), F(N)
      F(1) = (G/2.62E+03)**1.587*0.387*(X(1)**2.587-ABS(2.032E-04-X(1))
     &   **2.587)-(G/3.73E+3)**1.639*0.378*(ABS(1.016E-03-X(1))
     &   **1.639-ABS(2.032E-04-X(1))**1.639)
      RETURN
      END
```

**TABLE 7.5 Calculated Values of the Parameter and the Flow Rate for HDPE for Various Guesses of G**

| G (Pa/m) | α | $Q^{(2)}$ (m³/s) |
|---|---|---|
| $1.653 \times 10^9$ | $6.095 \times 10^{-4}$ | $3.022 \times 10^{-4}$ |
| $1.800 \times 10^9$ | $6.095 \times 10^{-4}$ | $3.475 \times 10^{-4}$ |
| $2.200 \times 10^9$ | $6.095 \times 10^{-4}$ | $4.83 \times 10^{-4}$ |
| $2.550 \times 10^9$ | $6.095 \times 10^{-4}$ | $6.194 \times 10^{-4}$ |

**Solution**

The procedure for solving this problem is as follows.

(a) Solve the nonlinear algebraic equation, Eq. 7.59, for α using a guess for G and the given value of β.

(b) Calculate $Q^{(2)}$ using Eq. 7.57 and the values for G, α, and β.

(c) Compare $Q^{(2)}$ against the given value based on the line speed and the thickness, β.

(d) If the difference in values is greater than a specified tolerance, ε, then return to step 1 and repeat the procedure.

An "intelligent guess" for G will accelerate the rate of convergence to the solution. Because the major component of the flow is HDPE, we use Table 2.5 and estimate G for the flow of HDPE as:

$$G = -\frac{\Delta P}{L} = \left[\frac{2Q(s+2)}{WH^2}\right]^n \frac{2m}{H} = 1.653 \times 10^9 \text{ Pa m}^{-1}.$$

We use the IMSL subroutine, NEQNF, for which the calling program is given in Table 7.4. A summary of the guesses for G and the calculated values of α and $Q^{(2)}$ is given in Table 7.5. What we observe is that after the initial guess there is no change in the predicted value of α. In other words α is somewhat insensitive to the magnitude of G. Knowing this and $Q^{(2)} = 6.194 \times 10^{-4}$ m³ s⁻¹, we can calculate G from Eq. 7.57, which we find to be $2.55 \times 10^9$ Pa·m⁻¹.

A more general approach would be to write the program so that values of m, n, and G can be read into the computer. Furthermore, values of $Q^{(2)}$ should be calculated by the computer after each guess using Eq. 7.27 and evaluated to see if they are converging to the value based on the data given in the problem.

Several other cases are commonly encountered. For thin annular dies, Eqs. 7.54 through 7.59 can be adopted directly by making appropriate replacements for W and h. For multilayer extrusion through an annulus in which the gap is too thick to apply the thin slit approximation, the appropriate equations are derived in Pr. 7B.8. For more than two layers the equations for flow through parallel plates can be generalized as follows (Schrenk & Alfrey, 1976). Referring to Fig. 7.51 we can obtain expressions for the volumetric flow rate in each layer. The shear stress in each layer is:

$$\tau_{yz}^{(i)} = G(y-c), \tag{7.60}$$

where c is the position where $\tau_{yz} = 0$. Expressions for the power-law

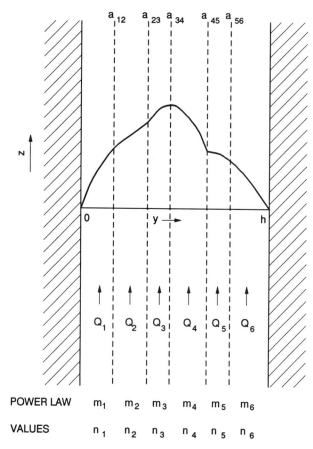

**Figure 7.51** Velocity profile of a non-Newtonian multilayer flow through a wide narrow slit. The parameters $m_i$ and $n_i$ are the power-law parameters for each layer.

model for each layer can now be substituted into Eq. 7.60 and integrated to obtain the velocity field in each layer:

$$v_M(y) = -\text{sign}(G)\left\{\sum_{j=1}^{M-1}\frac{m_j|G|^{n_j}}{n_j+1}[|a_{j-1,j}-\alpha|^{n_j+1} - |a_{j,j+1}-c|^{n_j+1}]\right.$$
$$\left. + \frac{m_M|G|^{n_M}}{n_M+1}[|a_{M-1,M}-\alpha|^{n_M+1} - |y-\alpha|^{n_M+1}]\right\}, \qquad (7.61)$$

where $a_{j-1,j}$ is the interfacial position of layer $j$. Details of the derivation required to obtain Eq. 7.61 are considered in problem 7B.9. The volumetric flow rate in any layer is given as:

$$|Q_M| = \sum_{j=1}^{M-1}\frac{m_j|G|^{n_j}}{n_j+1}$$
$$\times \{|\alpha-a_{j-1,j}|^{n_j+1} - |\alpha-a_{j,j+1}|^{n_j+1}\}(a_{M,M+1}-a_{M-1,M})$$
$$+ m_M\frac{|G|^{n_M}}{n_M+1}\{|\alpha-a_{M-1,M}|^{n_M+1}\}(a_{M,M+1}-a_{M-1,M})$$
$$+ \frac{m_M|G|^{n_M}}{(n_M+1)(n_M+2)}\{\text{sign}(\alpha-a_{M,M+1})|\alpha-a_{M,M+1}|^{n_M+2}$$
$$- \text{sign}(\alpha-a_{M-1,M})|\alpha-a_{M-1,M}|^{n_M+2}\}. \qquad (7.62)$$

These equations are used typically to find $\alpha$ and $a_{ij}$ for a given $Q_M$. By

selecting trial values for $\beta$ and $\alpha$, Eq. 7.61 can be solved sequentially for $a_{ij}$. If $a_{M,M+1}$ does not match the location h and the calculated values of $v_M$ do not match at the interfaces, then new trial values for $a_{ij}$ and $\alpha$ are tried until the correct values are obtained. The values of $Q_M$ can then be calculated. This process is discussed further in Pr. 7C.1.

### 7.6.3. FLOW INSTABILITIES IN MULTILAYER FLOW

There are basically two problems in trying to extrude multiple layers of different fluids through the same die. First, if there are distinct viscosity differences between the fluids, then the lower-viscosity component will try to encapsulate the higher-viscosity component. Second, there are situations when the viscosities of two polymers are closely matched but the interface still becomes wavy and distorted (this type of problem is shown in Fig. 7.52.) Some general comments about each type of instability are made first, followed by some attempt to estimate if an instability is imminent.

The encapsulation of the high-viscosity component by the low viscosity component is illustrated in Fig. 7.53. Viscosity data for the two nylon 6 polymers are shown in Fig. 7.54. The polymers were extruded in a side-by-side configuration through a capillary. The exit angle and the interface shape are shown as a function of the capillary $L/D$. Here it is seen that complete encapsulation does not occur until a $L/D$ of 100 is reached. In many cases the $L/D$ is short enough that not much rearrangement occurs. However, the flow may be on the

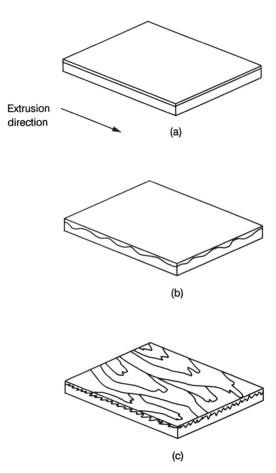

**Figure 7.52** The appearance of two layers of sheet under (a) stable flow conditions, (b) incipient interfacial flow instability, and (c) severe instability.

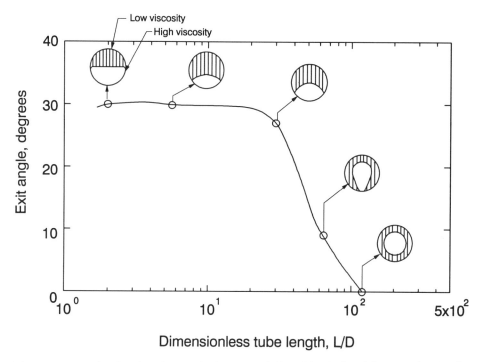

**Figure 7.53** Extrudate interface shape and exit angle variation with tube length for two nylon 6 melts. (Data from A. E. Everage, *J. Rheol., 19*(4), 509–22, 1975.)

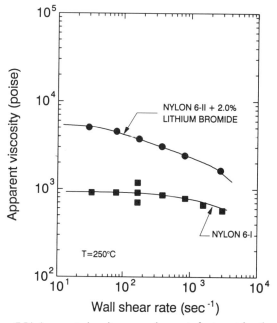

**Figure 7.54** Apparent viscosity versus shear rate for two nylon 6 melts. (Data from A. E. Everage, *J. Rheol., 19*(4), 509–22, 1975.)

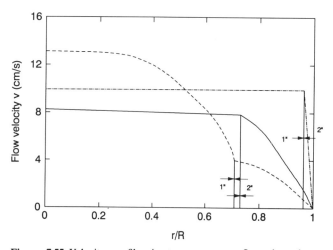

**Figure 7.55** Velocity profiles in two-component flow through a capillary for stable and unstable conditions: *stable* (—) $\eta_{02}/\eta_{01} < 1$ and (–·–) $\eta_{02}/\eta_{01} \ll 1$; *unstable* (– – –) $\eta_{02} > \eta_{01}$.

verge of an instability, and obtaining an extrudate with a straight interface may not be possible.

Although, as mentioned, we are not in a position to determine the interface shape or the distance required for encapsulation, it is at least possible to estimate the conditions when problems are imminent. Following the procedure in Section 7.6.2 one can calculate the velocity profile in tube flow in each layer for various viscosity differences. Figure

7.55 shows velocity profiles for two cases. First, we consider the case when the viscosity of layer 2, $\eta_{02}$, is less than that of layer 1, $\eta_{01}$. (*Note*: This is for tube flow.) When the flow is stable, the velocity profile of layer 1 is flat, whereas in layer 2 there is a large dependence of the velocity profile on the radius of the capillary. However, when $\eta_{02} > \eta_{01}$ the flow is in an unstable condition and the velocity profile varies more strongly with $r$ in layer 1 than in layer 2. The variation of $\dot{\gamma}$ with $r$ is shown in Fig. 7.56. When the flow is in an unstable condition, $\dot{\gamma}(r)$ is larger at the interface than at the wall. Hence, the equations presented in the preceding section can at least tell us whether it is feasible to stably extrude a given set of polymers.

The equations developed in the preceding section can also be used

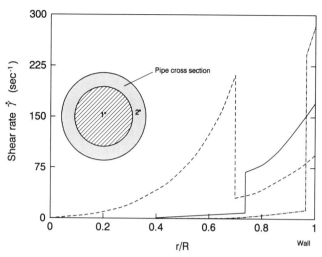

**Figure 7.56** Shear rate versus radial position for the three conditions given in Fig. 7.55.

to estimate the onset of an interfacial flow instability that leads to a rippled interface. According to Schrenk and co-workers (1976, 1978) this instability is due to slip at the interface between the polymers when a critical interfacial shear stress is exceeded. This critical stress varies for various polymer types, but for the system of acrylonitrile-butadiene-styrene (ABS) copolymer, and styron 470, this critical stress was $5.0 \times 10^4$ Pa. The analysis developed in Section 7.6.2 can be used to assess the conditions under which we might expect an interfacial instability to arise.

However, it seems that this instability may be more complicated than just a failure of adhesion between layers in shear flow. In many cases the region of fully developed shear flow is small or nonexistent, and the analysis developed in Section 7.6.2 may not be applicable. The origin of the interfacial instability could be at the die exit where large stresses arise as the velocity profile undergoes a rapid rearrangement. It could also be associated with converging flow upstream of the die lips. In other words, differences in the extensional viscosity of the two polymers as well as differences in the relaxation behavior could lead to interfacial instability.

### 7.7. SOLUTION TO DESIGN PROBLEM 6

The following steps are used in the solution:

1. The volumetric flow rate for each component is determined from the mass and density of each component and the hang time.
2. The dimensions of the die are determined from a mass balance.
3. The curvature in the die cross section is neglected, and it is treated as a parallel plate geometry;
4. The design equations for coextrusion are used to determine the pressure gradient, $G$, and the position where the velocity profile passes through a maximum, $\alpha$.
5. The shear rate and viscosity are calculated at the interface to determine if the flow could be unstable.
6. A manifold system is designed to feed the die.

We first calculate the volumetric flow rate of each component where $Q^{(1)}$ is $Q$ for nylon 6 and $Q^{(2)}$ is $Q$ for HDPE. The density of nylon 6 at 25°C is 1132 kg m$^{-3}$, and that for HDPE at 25°C is 971 kg m$^{-3}$. Given that the mass of the parison is 18 lb. of HDPE and

2 lb. of nylon 6 and that the time for hanging the parison is 5 s we find:

$$Q^{(1)} = 8.02 \times 10^{-4} \text{ m}^3 \text{ s}^{-1}; \qquad Q^{(2)} = 8.42 \times 10^{-3} \text{ m}^3 \text{ s}^{-1}.$$

The cross-sectional area of the gasoline tank is found by using the formula for the area of an ellipse, $A = (\pi/4)ab$, where $a$ is the major axis and $b$ is the minor axis of the ellipse:

$$A_p = (1.0)(0.15)\pi/4 - (9.936 \times 10^{-1})(.1436)\pi/4$$

$$= 5.748 \times 10^{-3} \text{ m}^2.$$

The dimensions of the die are found next. From the information given in the problem, we know that the major axes of the outer wall of the elliptically shaped die are $a_0 = 0.5$ m and $b_0 = 0.075$ m. As the area is assumed to be conserved we can calculate the dimensions of the die (note: extrudate swell must be considered, but the weight of the parison causes some reduction in the thickness, and the density change causes the material to shrink, which offsets some of the extrudate swell):

$$5.748 \times 10^{-3} = (\pi/4)[(0.5)(0.075) - (0.5 - 2H_0)(0.075 - 2H_0)],$$

where $H_0$ is the gap thickness. $H_0$ is determined to be $6.51 \times 10^{-3}$ m. We open the die and treat it as flow between parallel plates having a width, $W_0$, and height, $H_0$. Given $A_p = A_0$ and $H_0$, we find $W_0$ to be 0.883 m.

We can now use the equations derived in Section 7.6 for coextrusion through parallel plates. Equations 7.55, 7.57, and 7.59 represent three equations for finding three unknowns. In our case, $Q^{(1)}$ and $Q^{(2)}$ are known, but $G$, $\alpha$, and $\beta$ are unknown. We could solve the three equations simultaneously using the IMSL subroutine NEQNF (see Appendix D.4). However, to facilitate the understanding of the solution process, we can calculate $\beta$ (i.e., the interfacial position) from the knowledge of $Q^{(1)}$ and $Q^{(2)}$ via the following expression:

$$\beta = \frac{Q^{(1)}H_0}{Q^{(2)} + Q^{(1)}} = 0.087H_0. \tag{7.63}$$

This equation is based on the fact that $Q^{(1)}/A_1 = Q^{(2)}/A_2$, where $A_1$ and $A_2$ are the areas of each stream ($A_1 = \beta H_0 W_0$ and $A_2 = (H_0 - \beta)H_0 W_0$).

Just as in Ex. 7.2 we solve for $\alpha$ using Eq. 7.59 by estimating $G$. We then check to see whether Eq. 7.55 is satisfied. Before doing this, we are faced with a dilemma. The melting temperature of nylon 6 is about 220°C, and hence rheological data are only available at temperatures higher than at those one would normally process HDPE. Based on the discussion of flow instabilities in coextrusion, we must select conditions such that the viscosity of the two polymers is similar at the walls. With this in mind we select the HDPE stream to be at 220°C and the nylon 6 stream to be at 225°C. At this temperature, the $\eta_0$ of nylon 6 is somewhat higher than that of HDPE. However, the temperature mismatch is not so great that we would have to consider the problem to be nonisothermal.

The solution follows the approach in Ex. 7.2 very closely. With the conditions given, we find $G = 5.9 \times 10^6$ Pa/m and $\alpha = 3.536 \times 10^{-3}$ m. From these values, $\beta$, and Eqs. 7.45 and 7.46, we find $\dot{\gamma}_w^{(1)}$ and $\dot{\gamma}_w^{(2)}$:

$$-\dot{\gamma}_w^{(1)} = \left(\frac{G}{m_1}\right)^{s_1}(H_0 - \alpha)^{s_1} = 20.4 \text{ s}^{-1} \qquad -\dot{\gamma}_w^{(2)} = 12.6 \text{ s}^{-1}.$$

The viscosity of nylon 6 and HDPE at these conditions is:

$$\eta_1(\dot{\gamma}_w^{(1)}) = 858 \text{ Pa·s} \qquad \eta_2(\dot{\gamma}_w^{(2)}) = 1386 \text{ Pa·s}.$$

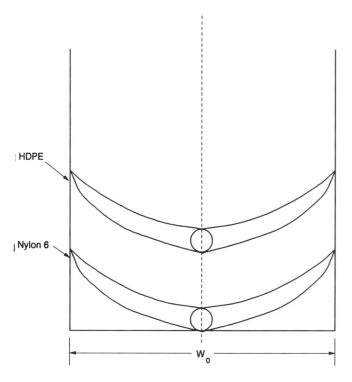

| HDPE

| Nylon 6

$$W_0$$

**Figure 7.57** Planar expanded view of coathanger manifolds for HDPE and nylon 6 for feeding the elliptically shaped annular die in Design Problem 6.

Hence, there is a mismatch in viscosity with nylon 6 having a lower value at the wall. Hence, there would be a tendency for nylon 6 to encapsulate HDPE. However, by keeping the die land fairly short, we can prevent this. The die length will be chosen to be $20H_0 = 6.51 \times 10^{-2}$ m.

In order to adopt the coathanger manifold design equations we must consider that two manifolds are used to feed two streams to the die (see Fig. 7.57). Hence, the direct application of Eq. 7.23 cannot be made. Instead, we use Eq. 7.20 with the value of $|G| = 5.9 \times 10^6$ Pa/m. Furthermore, we use Eq. 7.23 with

$$Q(l)^n = \left[ \frac{Q^{(1)}}{W}(W-x) \right]^n. \tag{7.64}$$

The expression for $R(x)$ for each manifold now becomes:

$$(R(x)^{(i)})^{3n_i+1} = -\left( \frac{3+s_i}{\pi} \right)^{n_i} 2m_i \left[ \frac{Q^{(i)}}{W}(W-x) \right]^{n_i} \left( \frac{1}{G(dL/dl)} \right). \tag{7.65}$$

For nylon 6 the expression for $R(x)$ becomes (using $dL/dl = -0.087$):

$$R(x)^{(1)} = 3.15 \times 10^{-2} [W-x]^{0.223}, \tag{7.66}$$

and for HDPE we find:

$$R(x)^{(2)} = 7.75 \times 10^{-2} [W-x]^{0.211}. \tag{7.67}$$

For example, these equations tell us that the initial openings of the manifolds ($x=0$) should be 2.62 cm for nylon 6 and 6.52 cm for HDPE, which seems reasonable in light of the relative flow rates.

## PROBLEMS

### A. Applications

**7A.1.** Pressure Drop in Segmented Dies: Constant Radius
Calculate the pressure drop across dies II and IV in Fig. 7.28 and the wall shear rate at the die exit. Calculate the die swell based on Tanner's equation, Eq. 7.4, and compare it with the experimental value. Use the rheological data given for NPE 953 in Appendix A, Table A.1, at 170°C. The average velocity at the die exit in all cases is 7.2 mm/s. The dimensions of die II can be obtained by scaling the dimensions from die IV.

**7A.2.** Pressure Drop in Segmented Dies: Tapered Radius
Calculate the pressure drop across dies I and III in Fig. 7.28 and the wall shear rate at the die exit. Estimate the die swell using Eq. 7.4. Use the rheological data given for NPE 953 in Appendix A, Table A.1. The dimensions can be obtained by scaling the dimensions from die IV.

**7A.3.** Scaleup of an Extrusion Die
A die similar to die III in Fig. 7.28 is to be designed so that the final diameter is 10 mm rather than 2 mm. Determine the flow rate and the remaining dimensions of the larger diameter die such that die swell will be the same as for the smaller diameter die.

**7A.4.** Pressure Drop across a Coathanger Die
Determine the pressure drop across the coathanger manifold and land for the sheet die discussed in Ex. 7.1. The extrusion rate is 200 kg/hr. (*Note*: the pressure drop in the whole of the manifold system is calculated by calculating the pressure drop at the center of the die; i.e., $y = L(0)$ and $x = W/2$ in Fig. 7.32).

**7A.5.** Profile Extrusion: Square Duct
A square duct is to be extruded from a profile die, die shape no. 3 in Fig. 7.46. The dimensions of the die are: $R_i = 10.0$ cm and $R_o = 10.5$ cm. The resin to be used is HDPE, resin no. 27 in Fig. 7.17. The material is extruded at 170°C at the rate of 200 kg/hr. Using the rheological data given in Fig. 7.17 and the die swell data given in Fig. 7.19, estimate the dimensions of the duct at 25°C (i.e., include density changes in your calculations). Assume that no sizing of the duct occurs as it leaves the die (by sizing it is meant that no pressure differential is applied to the duct to expand it on leaving the die).

**7A.6.** Profile Extrusion: Tubing with an Internal Wall
Tubing is to be extruded from a profile die, die no. 6 in Fig. 7.46, using HDPE no. 27 (see Fig. 7.17). The dimensions of the die are: $R_i = 10.0$ cm, $R_o = 10.5$ cm, and $H = 0.5$ cm. The rheological properties of the resin are given in Table 2.3. The resin is to be extruded at 170°C at the rate of 200 kg/hr. Die swell data are given in Fig. 7.19. Estimate the dimensions of the tubing including density changes and assuming that no sizing takes place as the extrudate leaves the die, that is, that no pressure differential is generated to expand the tubing to its final dimensions.

### B. Principles

**7B.1.** Flow Distribution in an End-Fed Film Die: Newtonian Case
In some instances end-fed film dies as shown in Fig. 7.58 are used to produce sheet and film. Obtain an expression that can be used to determine the uniformity of flow over the width of the die, where uniformity is determined by the ratio of the maximum to minimum flow rates across the width of the die, by carrying out the following steps.

(a) Show by carrying out a mass balance on a differential element of thickness $\Delta z$ that the pressure variation along the manifold is given

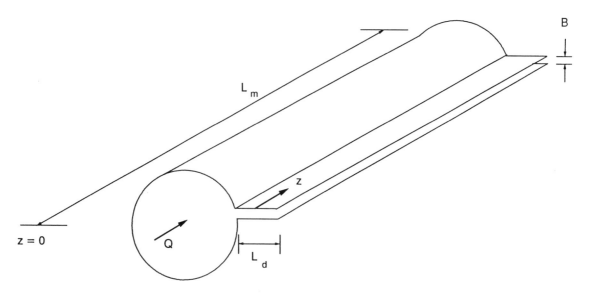

**Figure 7.58** End-fed film die. Melt enters from the side and is distributed over the width of the die.

by the following differential equation:

$$\frac{d^2P}{dz'^2} - \beta^2 P = 0, \tag{7.68}$$

where $\beta^2 = 2B^3 L_m^2 / 3\pi R^4 L_d$ ($R$ is the radius of the manifold) and $z' = z/L_m$.

**(b)** Solve this equation using the boundary conditions that $P = P_0$ at $z' = 0$ and $dP/dz' = 0$ at $z' = 1$, and show that:

$$P/P_0 = e^{-\beta z'} + e^{-\beta} \frac{\sinh \beta z'}{\cosh \beta}. \tag{7.69}$$

**(c)** If the uniformity of flow ($E$) is $q_x(0)/q_x(L_m)$, that is the ratio of the maximum to minimum flow rates per unit width, then show that:

$$E = \frac{P(L_m)}{P(o)} = \frac{1}{\cosh \beta}. \tag{7.70}$$

**(d)** Based on Eq. 7.70, what design stategy should one follow to keep $E$ near unity?

**7B.2.** Analysis of Flow in a *T*-Die
For a straight *T*-die as shown in Fig. 7.30 derive an expression that allows one to evaluate the uniformity of flow across the die similar to that given in Pr. 7B.1 for the end-fed die. **(a)** Do this first for the Newtonian case and **(b)** then for a power-law fluid. Take the radius of the manifold as $R$, the volumetric flow rate as $2Q_0$, the die width as $W$, the die height as $B$, and the die length as $L_d$.

**7B.3.** Pressure Drop along a Flow Path in a Coathanger Die
Show that the pressure drop along any flow path in the coathanger die (see Fig. 7.32) is given by the following expression for a Newtonian fluid:

$$\Delta P = \frac{8Q_0}{\pi L} \int_L^l \frac{l'\mu}{R^4(l')} dl' + \frac{12Q_0\mu}{LH^3} y(l). \tag{7.71}$$

Obtain a similar expression for a power-law fluid.

**7B.4.** Residence Time of a Fluid Particle along a Flow Path
Show that the residence time in a coathanger die of a particle along any flow path is (*note*: the residence time consists of the residence time in the manifold plus that in the land region):

$$t(l) = \frac{\pi L}{Q_0} \int_l^L \frac{R^2(l')}{l'} dl' = \frac{LHy(l)}{Q_0}. \tag{7.72}$$

**7B.5.** Dimensioning of the Distribution System for a Side-Fed Mandrel
Figure 7.36 shows the distribution system for a cylindrically shaped mandrel. Considering the fluid to be Newtonian, carry out the following:

**(a)** Show that the shear rate at the wall of the manifold tube is:

$$\dot{\gamma}_R = \frac{4Q_0\zeta}{\pi R^3(\zeta)}. \tag{7.73}$$

**(b)** Show that the wall shear rate in the slit is:

$$\dot{\gamma}_s = \frac{6Q_0}{\pi r H^2}. \tag{7.74}$$

**(c)** Show that the radius is given as a function of $\zeta$ by:

$$R(\zeta) = R_0(\zeta/\pi)^{\frac{1}{3}}. \tag{7.75}$$

**(d)** Show that y is related to $\zeta$ by:

$$\frac{y(\zeta)}{y_0} = \left(\frac{\zeta}{\pi}\right)^{\frac{2}{3}}. \tag{7.76}$$

**(e)** Show that if the shear rate is to be the same in both the manifold and the slit (this is the design criterion), the maximum land length is:

$$y_0 = \frac{\pi r^2 H^3}{R_0^4}. \tag{7.77}$$

**(f)** Show that the maximum radius of the manifold is:

$$R_0 = (rH^2)^{\frac{1}{3}}. \tag{7.78}$$

**(g)** Show that the total pressure drop is:

$$\Delta P = \frac{12 Q_0 \mu y_0}{\pi r H^3}. \tag{7.79}$$

**(h)** Derive similar expressions as in **a–g** for a power-law fluid.

**7B.6.  Dies with a Triangular Cross Section**
The velocity field for the flow of a Newtonian fluid through a die of triangular cross section is approximately given by the following expression for the equilateral case (Kakovris & Freakley, 1988):

$$v_z = \left[ \frac{3(dp/dz)}{4a\mu} \right] \left( y^2 - \frac{x^2}{3} \right)(x - a), \tag{7.80}$$

where the coordinate system and $a$ are defined in Fig. 7.59.

**(a)** Obtain an expression for the volumetric flow rate, $Q$.

**(b)** Determine expressions for the components of the rate of deformation tensor.

**(c)** Determine where $\dot{\gamma}$ is a maximum.

**7B.7.  Equivalent Newtonian Viscosity**
It has been suggested by Broyer and co-workers (1975) that the solutions to non-Newtonian flow problems can be obtained by using the Newtonian solution with $\mu$ replaced by an equivalent Newtonian viscosity, $\bar{\mu}$. For isothermal flow between parallel plates, carry out the following:

**(a)** Show that the flow rate per unit width is given by:

$$q = \frac{2b^2}{\tau_w^2} \int_0^{\tau_w} \tau \dot{\gamma} d\tau, \tag{7.81}$$

where $b$ is one half the die height.

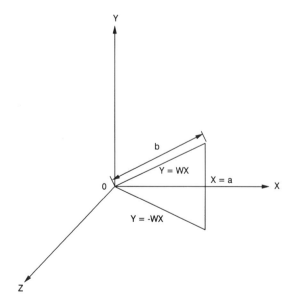

**Figure 7.59** Cross section of a triangularly shaped die.

**(b)** Show that for a Newtonian fluid the flow rate per unit width can be written as:

$$q = \frac{b^2 \tau_w}{3\mu}. \tag{7.82}$$

**(c)** Show that by defining an equivalent Newtonian viscosity,

$$\bar{\mu} = \frac{\tau_w^3}{3 \displaystyle\int_0^{\tau_w} \tau \dot{\gamma} d\tau}, \tag{7.83}$$

the flow rate of a non-Newtonian fluid can be calculated with the Newtonian equation in **(b)** with $\mu$ replaced by $\bar{\mu}$.

**(d)** For a power-law fluid find an expression for $\bar{\mu}$ in terms of $\dot{\gamma}_w$, $m$, $n$, and any geometric factors.

**7B.8.  Multi-layer Flow through an Annulus**
Obtain expressions for $Q^{(1)}$ and $Q^{(2)}$ and for finding $\alpha$ similar to those determined for slit flow in Eqs. 7.55, 7.57 and 7.59, respectively, for flow of two fluids through an annulus. Obtain expressions first for Newtonian fluids and then for power-law fluids.

**7B.9.  Multilayer Flow through a Slit Die**
Obtain expressions for the flow of three fluids through a slit die using the notation in Fig. 7.51. In particular, find the velocity field and the volumetric flow rate for each layer. Confirm your solution by comparing with the expressions in Eqs. 7.60 and 7.61.

**7B.10.  Bicomponent Flow in a Wire-Coating Die**
Derive expressions for the volumetric flow rate in each layer and for determining the position where $\tau_{rz} = 0$ (i.e., $\alpha$) for the flow of two fluids through the annular region in a wire coating die (see Figs. 2.10 and 2.11) for each of the following cases:

**(a)** Newtonian fluids with no imposed pressure gradient.

**(b)** Power-law fluids with no imposed pressure gradient.

**(c)** Power-law fluids with an imposed pressure gradient.

Take the point where $\tau_{rz} = 0$ as $\alpha R$ and the location of the interface between the two layers as $\beta R$

**7B.11.  Flow Distribution in an End-Fed Film Die: Non-Newtonian Case**
Referring to Fig. 7.58 and Pr. 7B.1, derive a similar expression for determining the flow uniformity in an end-fed die for a power-law fluid. In particular, show that the pressure distribution along the manifold is given by the following differential equation:

$$\frac{d^2 p}{dz'^2} - \frac{\beta^2 p^{1/n}}{\left( \dfrac{-dP}{dz'} \right)^{1/n - 1}} = 0, \tag{7.84}$$

where

$$\beta^2 = \frac{n(1 + 3n)B^2 L_m^{s+1}}{2\pi(1 + 2n)R^{3+s}} \left( \frac{B}{L_d} \right)^s. \tag{7.85}$$

**C.  Numerical Problems**

**7C.1.  Multi-layer Sheet Extrusion**
Three layers of polymer are to be extruded through a sheet die having a height 0.15 cm, a width of 20 cm, and a length of 4.5 cm. Following the

notation in Fig. 7.51, layers 1 and 2 are to be 0.3 cm thick and layer 3 is to be 0.9 cm thick. The overall flow rate is to be 100 kg/hr. The power-law parameters for the three fluids are: layer 1, $m = 2.62 \times 10^3$ Pa·s$^n$ and $n = 0.63$; layer 2, $m = 1.55 \times 10^3$ Pa·s$^n$ and $n = 1.0$; layer 3, $m = 3.73 \times 10^3$ Pa·s$^n$ and $n = 0.61$. Determine the pressure gradient required to produce this flow and whether the flow will be stable. Assume the densities of all the fluids are 1000 kg/m$^3$.

**7C.2.** Pressure Distribution in an End-Fed Die
Solve Eq. 7.84 in Pr. 7B.11 numerically using the appropriate IMSL subroutine for various values of $P_0$ (in particular, take values of 500, 1000, and 2000 psi) for a fluid with a power-law index of 0.5. With $B = 0.05$ cm, $L_m = 40$ cm, $R = 5$ cm, and $L_d = 1$ cm, determine the flow uniformity and the volumetric flow rate for these initial pressures.

### D. Design Problems

**7D.1.** Profile Extrusion Coating: Processing Conditions
Profile extrusion coating is a combination of wire-coating and profile extrusion. This technique is used to generate automotive protection molding as shown in Fig. 7.60.a. Polyvinylchloride (PVC) is pumped through the cross-head die while a metal core is fed into the die at a constant velocity. The flow is basically combined pressure and drag flow. The shape and dimensions of the flow channel are shown in Fig. 7.6.b. The melt and the die temperatures are both 150°C. The rheological properties of PVC are described by the power-law model with $m = 5.45 \times 10^4$ Pa·s$^n$ and $n = 0.27$. If the linear speed of the metal core, which is 25 mm in width, is 7 cm/s and the PVC is to be 2.2 mm thick, estimate the pressure at the inlet to the die and the volumetric flow rate required to produced the coating. (This problem is taken from Matsuoka & Takahashi, 1991.)

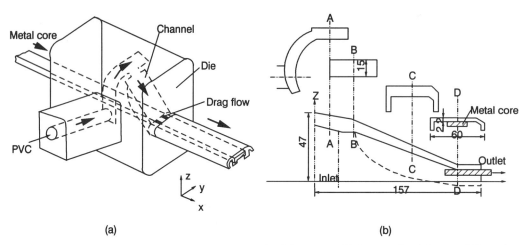

(a)                    (b)

**Figure 7.60** Profile extrusion coating process: (a) Schematic of the process; (b) Cross sectional shape and dimensions of the flow channel. All dimensions are in mm. The distances are: D– Exit = 22 mm B–D along flow channel = 100 mm A–B = 15 mm. Curved inlet channel: average arc length is 100 mm and the channel width is 60 mm and the height is 6 mm.

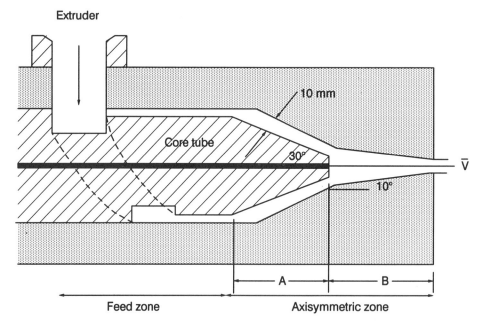

**Figure 7.61** Wire coating die design with possible variations in the dimensions.

**7D.2.** Selection of a Wire-Coating Die Design

Two dies of design similar to that shown in Fig. 7.61 but having the following dimensions:

| | Die no. 1 | Die no. 2 |
|---|---|---|
| A | 1 cm | 1 cm |
| B | 1.5 cm | 3 cm |
| $D_0$ | 0.85 mm | 1.19mm |

are used to coat a 0.5 mm diameter copper wire with HDPE (Note: $D_0$

is the diameter of the die exit). The wire speed is 200 m/min, and the final diameter of the coated wire is to be 0.85 mm. The melt and the die temperatures are set at 200°C. By analyzing the flow in the region starting with the tapered annular section, determine which die design would be best. Make your decision based on the minimum temperature rise due to viscous dissipation, the minimum amount of mechanical degradation due to shear stresses, the minimum pressure drop required across the section, and the smoothest surface. The rheological properties for HDPE are given in Table 2.3.

# REFERENCES

Broyer, E., C. Gutfinger, and Z. Tadmor. 1975. "Evaluating Flows of Non-Newtonian Fluids by the Method of Equivalent Newtonian Viscosity." *AIChE J.*, **21**, 198.

Cogswell, F. N., and R. Lamb. 1970. "Polymer Properties Relevant in Melt Processing." *Plast. Polym.*, **38**, 39.

Garcia-Rejon, A., and J. M. Dealy. 1982. "Swell of Extrudate from an Annular Die." *Polym. Engng. Sci.*, **22**(3), 158.

Kakovris, A. P., and P. K. Freakley 1988. "Flow of a Generalized Power-law Fluid in Triangular Dies for Rubber Extrusion." *Int. Polym. Proc.*, **III**(3), 156–64.

Laun, H. M. 1989. "Transient Elongational Viscosity and Drawability of Polymer Melts." *J. Rheol.*, **33**(1), 119.

Matsuoka, T., and H. Takahashi 1991. "Finite Element Analysis of Polymer Melt Flow in a Profile Extrustion Coating Die." *Int. Polym. Proc.*, **VI**(3),183–7.

Michaeli, W. 1984. *Extrusion Dies* (Hanser, Munich).

Middleman, S. 1977. *Fundamentals of Polymer Processing*, (McGraw-Hill, New York).

Pissipati R. 1983. "A Rheological Characterization of Particulate and Fiber Filled Nylon 6 Melts and Its Application to Weld Line Formation in Molded Parts." Ph.D Thesis, Department of Chemical Engineering, Virginia Polytechnic Institute and State University, Blacksburg, VA.

Schrenk, W. J., and T. Alfrey, 1976. "Coextruded Multilayer Polymer Films and Sheets." In J. A. Mason and L. H. Sperling, Eds, *Polymer Blends and Composites* (Plenum, New York).

Schrenk, W. J., N. L. Bradley, and T. Alfrey 1978. "Interfacial Flow Instability in Multilayer Coextrusion." *Polym. Engng. Sci.*, **18**(8), 620–3.

White, S. A., A. D. Gotsis, and D. G. Baird 1987. "Review of the Entry Flow Problem: Experimental and Numerical." *J. Non-Newt. Fluid Mech.*, **24**, 121–60.

# 8

# EXTRUDERS

## DESIGN PROBLEM 7
### Design of a Devolatilization Section for a Single-Screw Extruder

A self-wiping corotating twin-screw extruder of dimensions shown in Fig. 8.1 and 8.2 has been successfully used to remove residual methymethacrylate (MMA) from polymethylmethacrylate (PMMA). It is desired to use a single-screw extruder to reduce the level of MMA, which is initially 0.65% by weight (6500 ppm), to as much the same level as possible in the twin-screw extruder. Design a devolatilization section for the screw shown in Fig. 8.3. In particular, given that the barrel diameter, $D_b$, is 30 mm and the channel depth in the devolatilization (DV) section is 5.595 mm, determine the length of the DV section, $L_e$, the number of flights, and the screw rpm to reduce the level of MMA to 0.10% by weight. The vacuum pump is capable of providing a mean pressure of 10 torr (133.3 Pa). Use two approaches to obtain your design: Use dimensional analysis and the data given in

Table 8.1 first, and then use the diffusion theory presented in Section 8.5.2. Compare the results of the two approaches, and specify which solution is most accurate.

Thermodynamic data for MMA, PMMA, and the solution are given in Table 8.2 (Biesenberger et al., 1990).

---

Extruders are the heart of the polymer processing industry. They are used at some stage in nearly all polymer processing operations. This chapter is concerned with the basic elements of extruder design. In Section 8.1 we describe some of the technological features of extruders. Section 8.2 is concerned with the design of hoppers, which are often used to feed polymer pellets to the extruder. In Section 8.3 we address the principal features of the design of single-screw extruders. In Section 8.4 we look at some of the most important aspects of the design of

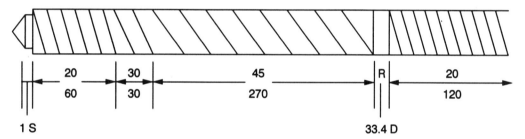

**Figure 8.1** 34-mm diameter screw used in a corotating twin-screw extruder. The lead angle and length (in mm) are given for each selection. (Data from Biesenberger et al., 1990.)

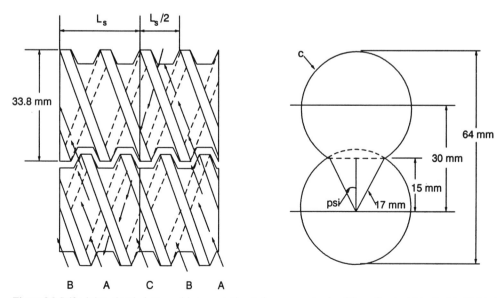

**Figure 8.2** Self-wiping closely intermeshing corotating twin-screw extruder. Flow of material in double-flighted screw elements (left). Cross section of the barrel showing the pertinent dimensions (right).

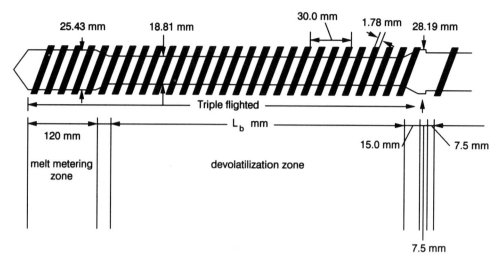

**Figure 8.3** 30-mm diameter screw used in a single-screw extruder. (Data from Biesenberger et al., 1990.)

**TABLE 8.1 Extrusion Data for the Twin-Screw System**

| Temperature ($^\circ$C) | $N$ (rpm) | $PQ$ (kg/hr) | $F_s^*$ | $f^\dagger$ | $\bar{t}$ (s) | $t_p$ (s) | $\bar{t}/t_p$ |
|---|---|---|---|---|---|---|---|
|  | 90 | 3.44 | 0.91 | 0.103 | 13.3 | 0.099 | 134.4 |
| 200 | 30 | 2.82 | 0.78 | 0.172 | 36.0 | 0.297 | 122.0 |
|  | 60 | 4.14 | 0.82 | 0.126 | 18.0 | 0.127 | 142.0 |
|  | 90 | 4.92 | 0.83 | 0.100 | 12.0 | 0.076 | 158.0 |
| 230 | 30 | 2.18 | 0.82 | 0.137 | 36.0 | 0.265 | 135.9 |
|  | 60 | 3.83 | 0.86 | 0.120 | 18.0 | 0.124 | 145.2 |
|  | 90 | 5.80 | 0.84 | 0.120 | 12.0 | 0.084 | 142.9 |
| 250 | 30 | 2.84 | 0.83 | 0.180 | 36.0 | 0.306 | 117.6 |
|  | 60 | 4.15 | 0.88 | 0.133 | 18.0 | 0.127 | 141.7 |
|  | 90 | 5.02 | 0.85 | 0.107 | 12.0 | 0.079 | 152.9 |

*$F_s$: fractional separation.
†$f$: degree of fill channel.

**TABLE 8.2 Thermodynamic Data for Monomer, Polymer, and Solution**

| Temperature ($^\circ$C) | $\rho_m$ (g/cc) | $\rho_p$ (g/cc) | $P_s^0$ (atm) | $S'^\dagger$ (atm) |
|---|---|---|---|---|
| 200 | 0.699 | 1.118 | 8.99 | 64.5 |
| 230 | 0.641 | 1.083 | 14.85 | 113 |
| 250 | 0.592 | 1.061 | 20.30 | 164 |

*$P_s^0$ is a coefficient in the Flory-Huggins theory for calculating $S$.
†$S' = 1/S$ where $S$ is the Henry's law constant.

twin-screw extruders. Extruders have other functions than to melt and pump polymers. In Section 8.5 we present basic elements concerned with mixing, the removal of gases (devolatilization), and reactions in extruders. Finally, in Section 8.6 the solution to Design Problem 7 is presented.

## 8.1. DESCRIPTION OF EXTRUDERS

There are basically three classifications of extruders: screw extruders, disk extruders, and ram extruders. The most common ones are the screw extruders, and for this reason they will be emphasized in this book.

The technology of extruders is extremely vast and has many variations. It is nearly impossible to describe all the technology

associated with extruders within the limits of this book. There is no one reference where all this information can be found, but extruder manufacturers usually can provide many details. The book by Rauwendaal (1986) does contain, in addition to a theoretical description of various aspects of extruders, a significant amount of technological information. Here we describe the most salient features of single- and twin-screw extruders.

### 8.1.1. SINGLE-SCREW EXTRUDERS

The single-screw extruder consists of a metallic barrel and a rotating screw as shown in Fig. 8.4. The screw is a metallic shaft in which a helical channel has been machined. Sometimes parallel channels are machined in the shaft at the same time leading to what are called multi-flighted screws. Typical barrel diameters used in the U.S. are 0.75, 1.0, 1.5, 2.0, 2.5, 3.5, 4.5, 6.0, 8.0, 10.0, 12.0, 14.0, 16.0, 18.0, 20.0, and 24.0 inches. The length to diameter ratios ($L/D$) range from 20 to 30, but the most common ratio is 24.

The main geometrical features of a screw are shown in Fig. 8.5. The diameter of the screw at the tip of the flight (the flight is the metal that remains after machining the channel), $D_s$, is less than the diameter of the barrel, $D_b$, by an amount $2\delta_f$ (i.e., $D_s = D_b - 2\delta_f$). $\delta_f$ is of the order of 0.2 to 0.5 mm. Of course, as the screw and barrel wear, $\delta_f$ increases, and the leakage flow over the flights increases to the point

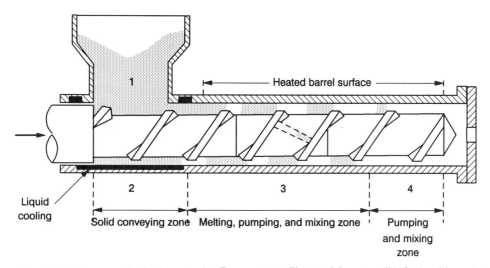

**Figure 8.4** Single-screw plasticating extruder. Four zones are illustrated: hopper, solids feed, melting, and pumping.

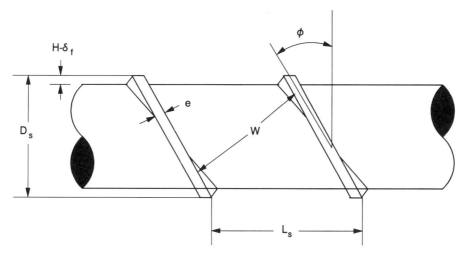

**Figure 8.5** Geometry of a screw.

where the screw loses its pumping efficiency. The lead of the screw, $L_s$, is the axial distance covered in completing one full turn along the flight of the screw. The helix angle, $\phi$, is the angle formed between the flight and the plane normal to the screw axis. The helix angle at the flight tip can be related to the lead and diameter as follows:

$$\tan \phi_s = L_s/\pi D_s. \qquad (8.1)$$

The helix angle is a function of the diameter and hence is different at the base of the flight than at the flight tip. The radial distance between the barrel surface and the root of the screw is the channel depth. The main design variable of screws is the channel depth profile along the helical direction (this is taken as the $z$ direction as discussed later). The width of the channel, $W$, is the perpendicular distance between the flights and is given by:

$$W = L_s \cos \phi - e, \qquad (8.2)$$

where $e$ is the flight width. We note here that $W$ varies with radial position, because $\phi$ does. Finally, the helical distance along the channel, $z$, is related to the axial distance, $l$, by:

$$z = \frac{l}{\sin \phi}, \qquad (8.3)$$

and it is also a function of the distance from the root of the screw.

The most frequently used extruder is a plasticating extruder. Referring to Fig. 8.4 polymer pellets are fed to the extruder by means of a hopper (sometimes the pellets are metered in). The gravitational flow of solids in the hopper is rather complex and will not be covered here. The pellets are compressed in the channel of the screw and then dragged forward by friction between the pellets and the barrel. Heat generated by sliding friction at the barrel surface and transfered from the heated barrel causes the pellets to melt. The melt film is scraped away and collects at one end of the channel. The solid bed width decreases as the solid plug advances along the screw channel until the solid is completely melted. The melt is pressurized by means of a drag flow mechanism. The pressure generated in the extruder and the performance of the extruder are significantly affected by the die geometry.

Although the main function of the single-screw extruder is to melt and pump polymer, there are a number of other applications. Extruders can be used to remove volatiles such as water or trace amounts of monomers. They can be used to generate foamed polymers as the temperature and pressure history can be controlled. They also serve as continuous mixing and compounding devices. Hence, extruders have a wider range of applications than other pumping devices.

### 8.1.2. TWIN SCREW EXTRUDERS

Twin-screw extruders consist of two screws mounted in a barrel having a "figure-eight" cross section. The "figure-eight" cross section comes from the machining of two cylindrical bores whose centers are less than two radii apart. Twin-screw extruders are classified by the degree to which the screws intermesh and the direction of rotation of the screws. Figure 8.6 shows three types of screw arrangements. Part (a) shows an intermeshing counterrotating type; part (b) shows a corotating intermeshing type, part (c) showns a nonintermeshing counterrotating type. In Figure 8.7.a is shown an intermeshing, self-wiping, corotating twin-screw extruder. Not all the elements of a twin-screw extruder are screw elements as shown in Fig. 8.7.b, and kneading elements may also be used. Probably the most frequently used twin screw extruders are the corotating intermeshing and the counterrotating types.

There are two main areas where twin-screw extruders are used. One is for difficult-to-process polymers because they do not flow easily, and they degrade readily. For example, they are used in the profile extrusion of polyvinylchloride (PVC) compounds that are thermally sensitive and do not flow well. The other is for specialty processing operations such as compounding, devolatilization, and chemical reactions. In the case of profile extrusion, counterrotating closely intermeshing extruders are used, because their positive conveying characteristics allow the machine to process hard-to-feed materials (powders, rubber particles, etc.) and yield short residence times and a narrow residence time distribution. In the case of specialty operations high-speed intermeshing corotating extruders are often used, but a wide variety of other designs are also used.

Although the differences between single- and twin-screw extruders will be more apparent by the end of the chapter, we make a few comments now. One of the major differences is the type of transport that takes place in the extruder. Material transport in a single-screw extruder is by drag-induced transport of the solid particles and the molten material. In particular, friction between the barrel walls and the solid pellets advances the polymer in the solids-conveying zone, while viscous drag advances the molten polymer. On the other hand, the transport in an intermeshing twin-screw extruder is to some degree positive displacement. The degree of positive displacement depends on how well the flight of one screw closes the opposing channel of the other screw. Closely intermeshing counterrotating twin-screw extruders provide the most positive displacement. However, some leakage will occur that will reduce the degree of positive conveying that can be achieved.

The flow of material in twin-screw extruders is very complex, and the flow patterns are difficult to predict mathematically. For this reason the simulation of processes in twin-screw extruders is not as well

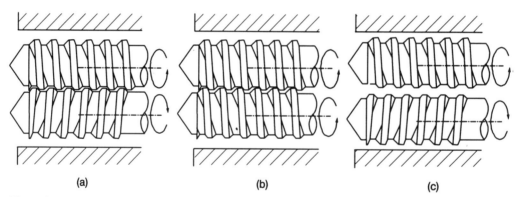

| (a) | (b) | (c) |

**Figure 8.6** Various types of twin-screw extruders. (a) Counterrotating intermeshing elements. (b) Corotating intermeshing elements. (c) Counterrotating nonintermeshing elements.

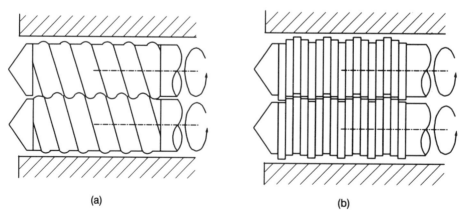

| (a) | (b) |

**Figure 8.7** Corotating screw extruder. (a) Self-wiping intermeshing screw elements. (b) Self-wiping intermeshing kneading block elements.

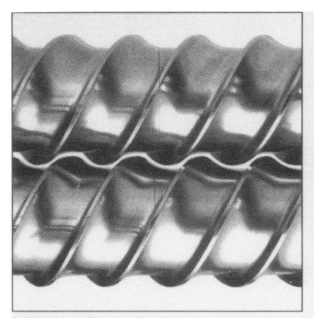
Pushing flight profiles, triple-flighted

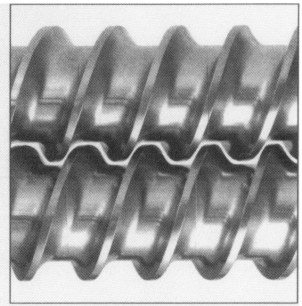
Pushing flight profiles, double-flighted (A-version)

Kneading block

Mixing elements

**Figure 8.8** Various types of screw elements commonly used in corotating twin-screw extruders. (Courtesy of Berstorff Corp., Charlotte, NC.)

developed as it is for single-screw extruders. It is, therefore, difficult to predict the performance of a twin-screw extruder based on geometrical features, polymer properties, and processing conditions. Hence, it is difficult to carry out accurate design calculations. For this reason twin-screw extruders are constructed in modules in which the screw and barrel elements can be changed. The screw design can be changed by changing the sequence of the screw elements. Hence, much of the design of twin-screw extruders is done on an empirical basis. Some of the various types of elements that can be used are shown in Fig. 8.8. One can use a combination of screw elements and kneading blocks to accomplish a given operation.

The sizes of twin-screw extruders range from 25 to 244 mm (this is the diameter of one of the barrels). The barrel-length-to-diameter

ratio, $L/D$, ranges from 39 to 48. The length can be altered as required for most twin-screw extruders because of the modular construction.

## 8.2 HOPPER DESIGN

Most extruders are of the plasticating type in which solid pellets are fed to the extruder where they are converted to melt and pressurized. The extruder is fed by solids that enter the extruder from a hopper (a metallic cylinder with a converging section as shown in Fig. 8.9) or are metered in. The flow patterns in the hopper are complex and are still the subject of research. Our intentions here will be to estimate the pressure at the base of the hopper as this value is needed to calculate the pressure rise in the extruder.

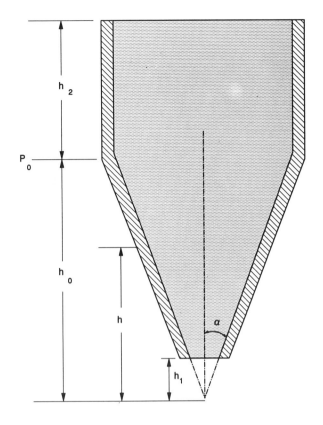

**Figure 8.9** Typical hopper design consisting of cylindrical and conical sections.

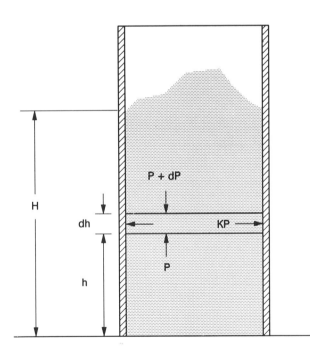

**Figure 8.10** Cylindrically shaped hopper partially filled with granular solids. The nonisotropic pressure distribution is described by the parameter $K$.

To understand how to calculate the pressure at the base of the hopper, we will consider the pressure exerted by solids on the base of a cylindrical container as shown in Fig. 8.10. For a cylinder filled with fluid it is known that the static pressure variation is $P = \rho g(H - h)$ and is the same through the cross section at any position $h$. For granular solids the pressure distribution is not isotropic, because of the ability of the solids to sustain shear stresses. We now perform a force balance on the differential element of thickness, $dh$:

$$A\rho_b g\,dh + (P + dP)A - PA + (C_w + f'_w KP)C\,dh = 0, \tag{8.4}$$

where $\rho_b$ is the bulk density, $A$ is the crosss-sectional area, $C$ is the wetted circumference, and $K$ is the ratio of compressive stress in the horizontal direction to compressive stresss in the vertical direction (note: for a fluid $K = 0$ but because a solid can sustain shear stresses, the pressures are different), $C_w$ is the measure of adhesion of the solids to the wall, and $f'_w$ is the coefficient of friction between the pellets and the wall. The following differential equation is obtained from Eq. 8.4:

$$\frac{dP}{dh} = CC_w + f'_w KPC + A\rho_b g, \tag{8.5}$$

which can be integrated to give:

$$P = P_H \exp\left[\frac{f'_w CK(h - H)}{A}\right]$$
$$+ \frac{(A\rho_b g/C) - C_w}{f'_w K}\left\{1 - \exp\left[\frac{f'_w CK(h - H)}{A}\right]\right\}, \tag{8.6}$$

where $P_H$ is the pressure at $H$ (in this case $p_a$). With $C_w = 0$ and taking the pressures relative to $p_a$ (this is called the gage pressure), the pressure at the base of the cylinder is:

$$P_0 = \frac{\rho_b g D}{4f'_w K}\left\{1 - \exp\left[\frac{4f'_w K(-H)}{D}\right]\right\}, \tag{8.7}$$

where $D$ is the diameter of the cylinder. In the limit as $H$ goes to infinity we obtain

$$P_{0,\max} = \frac{\rho_b g D}{4f'_w K}. \tag{8.8}$$

Hence, most of the weight is supported by friction between the pellets and the metal walls. The maximum pressure is proportional to the bin diameter and inversely proportional to the coefficient of friction at the wall. Whereas the pressure at the base of a cylindrical bin will continue to increase as H increases for a fluid, for solids it will reach a limiting value.

A few comments need to be made about measuring $K$, $C_w$, and $f'_w$. The device used to obtain these parameters is similar to a parallel disk rheometer used to obtain fluid properties except that a large compressive stress can be applied to the solid material. The torque required to turn the upper plate is proportional to $f'_w$. Likewise taking measurements as a function of applied pressure to the upper plate provides $K$ and $C_w$. $K$ is obtained from the effective angle of friction, $\delta$, using the following equation:

$$K = \frac{1 - \sin\delta}{1 + \sin\delta}. \tag{8.9}$$

Most hoppers consist of cylindrical and conical sections as shown in Fig. 8.9. Under mass flow conditions the pressure distribution is given as (Walker 1966):

$$P = (h/h_0)^a P_0 + \frac{\rho_b g h}{a-1} [1 - (h/h_0)^{a-1}], \tag{8.10}$$

where $P_0$ is the pressure at $h_0$, which is the pressure at the base of the vertical section of the hopper, and a is given for conical and wedge shaped hoppers, respectively, as:

$$a = \frac{2B'D^*}{\tan \alpha} \tag{8.11}$$

and

$$a = \frac{B'D^*}{\tan \alpha}. \tag{8.12}$$

$\alpha$ in Eqs. 8.11 and 8.12 is one-half the hopper angle, and $D^*$ is the distribution function taken as 1.0. $B'$ is given by

$$B' = \frac{\sin \delta \sin (2\alpha + \kappa_0)}{1 - \sin \delta \cos (2\alpha + \kappa_0)}, \tag{8.13}$$

where $\kappa_0$ is:

$$\kappa_0 = \beta_w + \arcsin \left(\frac{\sin \beta_w}{\sin \delta}\right) \quad \arcsin > \frac{\pi}{2}, \tag{8.14}$$

where $\beta_w$ is the wall angle of friction (i.e., $\beta_w = \tan^{-1} f'_w$).

The coverage of this topic was brief, to say the least, but it is not within our goals to give a lengthy derivation. Further details can be found in the book by Tadmor and Gogos (1979) and in the original paper by Walker (1966). We will use the following example to illustrate the calculation of the pressure at the base of a hopper required to determine the pressure at the inlet of an extruder.

---

**EXAMPLE 8.1**
**Pressure at the Base of a Hopper**
For a hopper of the design shown in Fig. 8.9 with $h_0 = 0.190$ m, $h_1 = 0.0635$ m, $\alpha = 45°$, and a cylinder diameter of 0.381 m calculate the pressure at the base of the hopper for low-density polyethylene (LDPE) pellets. The bulk density, $\rho_b$, is 595 kg/m³, $f'_w = 0.3$ and $\delta = 33.7°$.

**Solution**

$P_0$ is calculated using Eq. 8.8. To use Eq. 8.8 we must assume that sufficient height of solids are available to give us 99% of the maximum pressure. We first calculate $K$ using Eq. 8.9 and $\delta$:

$$K = \frac{1 - \sin (33.7°)}{1 + \sin (33.7°)} = 0.286.$$

Using this value and those given in the problem we find $P_0$ to be:

$$P_0 = (0.99)(595)(9.806)(0.381)/(4)(0.3)(0.286) = 6.41 \times 10^3 \text{ N/m}^2.$$

Before using Eq. 8.10 to calculate $P$ we must determine $a$ using Eq. 8.11 that contains the parameters $\kappa_0$ and $B'$:

$$\kappa_0 = 16.7 + \arcsin (\sin 16.7/\sin 33.7) = 47.9$$

$$B' = \frac{\sin (33.7) \sin ((2)(45) + (47.9))}{1 - \sin (33.7) \cos ((2)(45) + (47.9))} = 0.2635$$

$$a = \frac{(2)(0.2635)}{\tan (45)} = 0.527.$$

We now calculate the pressure at the base of the hopper, $P_1$:

$$P_1 = (0.0635/0.190)^{0.527}(6.412 \times 10^3)$$
$$+ [(595)(9.806)(0.0635)/(0.527 - 1)]$$
$$\times (1 - (0.0635/0.190)^{-0.473}) = 3.599 \times 10^3 + 532 = 4.131$$
$$\times 10^3 \text{ N/m}^2.$$

---

## 8.3. PLASTICATING SINGLE-SCREW EXTRUDERS

Polymer solids (pellets or powder) enter the throat of the extruder either through the hopper or are metered in by gravimetric or auger-type feeders. From this point they are transported through the extruder first by frictional drag and then by viscous drag. We first describe solids transport, then melting of the compacted solids, and finally the pumping of the melt.

### 8.3.1. SOLIDS TRANSPORT

To help the reader understand how particulate solids are transported through a single-screw extruder we start with a model for drag-induced flow in straight channels. We then summarize the equations for flow of particulate solids in the single-screw extruder. We next add heat transfer to the transport of the particulate solids.

We consider the transport of particulate solids in a rectangular channel as shown in Fig. 8.11. Our goal is to determine the mass flow rate and pressure rise as a function of the plate velocity and friction coefficient between the plate and pellets, $f_{w1}$. Although it would be

desirable to somehow treat this situation in a manner similar to that for fluids in which we solve the equation of motion along with an appropriate constitutive equation, it is uncertain as to what constitutive equation best describes the flow of granular solids. For this reason we consider the particulate solids to be a plug of density $\rho_b$ dragged along by the moving upper plate through Coulomb friction. The upper plate moves with a velocity $V_0$ making an angle $\phi$ with the down-channel direction (i.e., the z direction). The tangential force exerted on the solid plug is in a direction that the plate makes relative to the moving plug as shown in Fig. 8.12. The velocity of the plate relative to the solid bed, $\mathbf{v}_r$, is:

$$\mathbf{v}_r = V_0 \sin \phi \delta_x + V_0 \cos \phi \delta_z - u \delta_z. \tag{8.15}$$

From Eq. 8.15 we find that

$$\tan (\phi + \phi') = \frac{V_0 \sin \phi}{V_0 \cos \phi - u}. \tag{8.16}$$

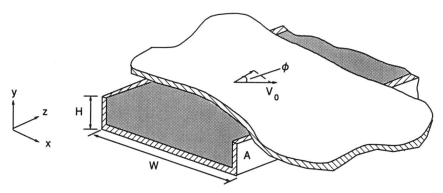

**Figure 8.11** Rectangular channel filled with granular solids with a plate moving at angle φ relative to the down channel direction.

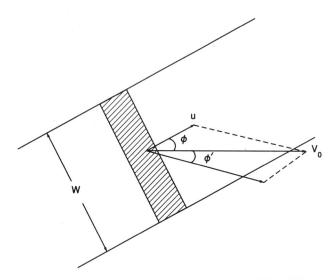

**Figure 8.12** Motion of the plate relative to the motion of the solid bed. The plate moves at an angle $\phi'$ relative to the solid bed.

Using a trigometric identity for tan $(\phi + \phi')$ we obtain the following expression:

$$\tan \phi' = \frac{u \sin \phi}{V_0 - u \cos \phi}.$$ (8.17)

Equation 8.17 contains two unknowns, $u$ and $\phi'$, and hence an additional equation must be derived.

The additional equation is obtained by performing a steady state force balance in the $z$ direction on a differential element of thickness $\Delta z$ (see Fig. 8.12). The contributions to the force balance include the forces due to pressure and the frictional forces at the upper and lower plates. The force balance is:

$$P|_z WH - P|_{z+\Delta z} WH + KPWHf_{w1} \cos(\phi + \phi')\Delta z$$
$$- KPWHf_{w2}\, \Delta z = 0.$$ (8.18)

In this equation $f_{w1}$ and $f_{w2}$ are the friction coefficients between the upper plate and lower plates, respectively, and the solid bed and $K$ is the anisotropy in the stress distribution (see Eq. 8.9). The contribution to the force balance from the side walls has been neglected in this case.

On dividing through by the volume of the element and taking the limit as $\Delta z$ goes to zero, the following differential equation is obtained:

$$-\frac{dP}{dz} + PK(f_{w,1} \cos(\phi + \phi') - f_{w,z}) = 0.$$ (8.19)

This equation is integrated using the initial condition that at $z=0$, $P = P_0$, to give:

$$P = P_0 \exp\{[Kf_{w,1} \cos(\phi + \phi') - Kf_{w,z}]z\}.$$ (8.20)

Equations 8.17 and 8.20 can be solved for the pressure increase and the mass flow rate (i.e., $uWH\rho_b$). Hence, we see that the lower the flow rate the higher the pressure rise, and it will rise exponentially with distance.

We have neglected the resistance to flow offered by the side walls. A force balance in the $x$ direction allows us to determine the normal force exerted by the wall on the solid plug. Hence, there will be an increased frictional force produced by the side wall that will reduce the conveying capacity for a given pressure rise.

We next consider the solids-conveying capacity of a single-screw extruder. The rectangular channel model cannot be used to describe solids conveying in a single-screw extruder because of the presence of deep channels that make curvature effects significant. As we do not intend to rederive the equations for the single-screw extruder, the development for the rectangular channel should serve to facilitate the following discussion.

The model of the feed section of a single-screw extruder is due to Darnell and Mol (1956) and is for the most part similar to the development given for the rectangular channel. The assumptions are:

1. The particulate solid bed is treated as a continuum.
2. The channel depth is constant.
3. The flight clearance is neglected.
4. Plug flow exists.
5. The channel is full so that all surfaces are in contact with the solid (when the channel is not full, this is referred to as starve feeding).
6. The stress distribution in the bed is isotropic.
7. The density is constant.
8. Gravitational forces are neglected.
9. Isothermal conditions hold (this will be relaxed later).

First, we relate the mass flow rate to the angle $\phi'$ which is the angle the relative velocity vector makes with the barrel velocity. A cylindrical plug (or a doughnut with a bite taken out of it) is shown

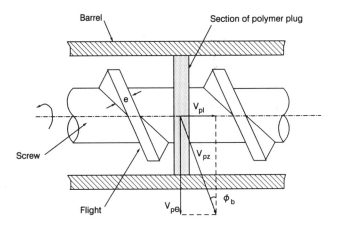

**Figure 8.13** Cylindrically shaped solid plug.

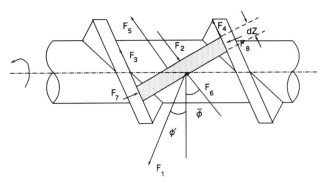

**Figure 8.14** Element of the solid bed showing various forces acting on it.

in Fig. 8.13. The mass flow rate, $G$, is the product of the plug velocity in the axial direction, $V_{pl}$, $\rho_b$, and of the cross sectional area of the plug and is given by:

$$G = V_{pl}\rho_b\left[\frac{\pi}{4}(D_b^2 - D_s^2) - \frac{eH}{\sin\bar{\phi}}\right] \qquad (8.21)$$

where the second term on the right-hand side of the equation is where the flight cuts across the doughnut-shaped plug. $\sin\bar{\phi}$ is the sine of the average helix angle. $D_b$ is the diameter of the barrel opening, and $D_s = D_b - 2H$. In Fig. 8.13 the screw is shown to be rotating in a clockwise direction. It is customary to attach the coordinates to the rotating screw so the velocity vectors are shown relative to the rotating screw. The plug then has velocity components in the tangential direction, $V_{p\theta}$, and in the down-channel direction, $V_{pz}$. Following arguments similar to those used in the previous section we can relate $V_{pl}$, $V_b$, and the angle $\phi'$, which is the angle the relative velocity vector makes with the apparently rotating barrel (remember we have attached the axis to the rotating screw so if you are riding around with the screw, it looks as if the barrel is rotating relative to you), as follows:

$$V_{pl} = V_b\frac{\tan\phi'\tan\phi_b}{\tan\phi' + \tan\phi_b}, \qquad (8.22)$$

where $V_b = \pi N D_b$ and $N$ is the angular velocity of the screw in revolutions per second. We next substitute Eq. 8.22 into Eq. 8.21 to obtain

$$G = \pi^2 N H D_b(D_b - H)\rho_b\frac{\tan\phi'\tan\phi_b}{\tan\phi' + \tan\phi_b}\left[1 - \frac{e}{\pi(D_b - H)\sin\bar{\phi}}\right]. \qquad (8.23)$$

This equation contains two unknowns, $\phi'$ and $G$, and hence we need an additional equation.

The additional equation is obtained by carrying out force and torque balances on an element of the solid bed as shown in Fig. 8.14. For an isotropic stress distribution these forces are:

| | |
|---|---|
| Barrel frictional force | $F_1 = f_b P W_b dz_b$ |
| Force due to pressure difference | $F_6 - F_2 = HW dP$ |
| Normal force at trailing flight | $F_8 = PH dz$ |
| Normal force at leading flight | $F_7 = PH dz + F^*$ |
| Frictional force at lead flight | $F_3 = f_s F_7$ |
| Frictional force at trailing flight | $F_4 = f_s F_8$ |
| Frictional force at screw | $F_5 = f_s P W_s dz_s,$ (8.24) |

where $f_s$ and $f_b$ are the friction coefficients between the polymer and the screw and barrel, respectively. Force and torque balances are then carried out to obtain the following expressions:

$$\cos\phi' = K_s\sin\phi' + M, \qquad (8.25)$$

where

$$K_s = \frac{\bar{D}}{D_b}\frac{\sin\bar{\phi} + f_s\cos\bar{\phi}}{D_b\cos\bar{\phi} - f_s\sin\bar{\phi}} \qquad (8.26)$$

and

$$M = \frac{2H}{W_b}\frac{f_s}{f_b}\sin\phi_b\left(K_s + \frac{\bar{D}}{D_b}\cotan\bar{\phi}\right)$$
$$+ \frac{W_s}{W_b}\frac{f_s}{f_b}\sin\phi_b\left(K_s + \frac{D_s}{D_b}\cotan\phi_s\right)$$
$$+ \frac{\bar{W}}{W_b}\frac{H}{z_b}\frac{1}{f_b}\sin\bar{\phi}\left(K_s + \frac{\bar{D}}{D_b}\cotan\bar{\phi}\right)\ln\left(\frac{P_2}{P_1}\right), \qquad (8.27)$$

where $\bar{D} = (1/2)(D_b + D_s)$, $P_1$ is the initial pressure at $z = 0$ and $P_2$ is the pressure at any down channel distance, $Z_b$. For a given flow rate $\phi'$ is obtained from Eq. 8.23, $M$ is calculated from Eq. 8.25 and the pressure rise is then determined from Eq. 8.27.

We next calculate the power consumed in the solids-conveying section. The power input through the barrel is obtained from

$$P_w = \int \mathbf{F}_b \cdot \mathbf{V}_b dA = \int F_1 V_b\cos\phi' W_b dz_b. \qquad (8.28)$$

Substituting into Eq. 8.28 for $F_1$ using Eq. 8.24 and for $P$ using Eq. 8.27 one obtains (Tadmor & Broyer, 1972):

$$P_w = \pi N D_b W_b Z_b f_b\cos\phi'\frac{P_2 - P_1}{\ln(P_2/P_1)}. \qquad (8.29)$$

Certainly, there are questionable assumptions in the theory of solids conveying. The assumption of an isotropic stress distribution was addressed, but it is not clear that the overall predictions of the model were drastically improved by including an anisotropic stress distribution. In fact, when one realizes how sensitive the predictions of the model are to the value of the friction coefficients and how difficult it is to accurately measure these quantities, then one will understand that improvements of this nature may be of little value.

On the contrary, the assumption of constant temperature must be modified, because the solid plug must increase in temperature as it advances along the screw. Eventually, the surface of the solid bed must reach the melting temperature of the polymer. The heat is supplied through the heated barrel and frictional heating at the solid and barrel interface. The nonisothermal analysis of the solids-conveying section is due to Tadmor and Broyer (1972). The total power introduced through the shaft is partly dissipated into heat at the barrel, flights, and the root of the screw surfaces and is partly used to generate pressure. Most of the power is, however, dissipated into heat at the barrel surface. The heat generated per unit of barrel surface is given by

$$q_b = f_b \pi N D_b \frac{\sin \phi_b}{\sin (\phi_b + \phi')} \frac{P_2 - P_1}{\ln (P_2/P_1)}. \tag{8.30}$$

The heat is conducted into the solid plug and the barrel walls, if the barrel walls are not heated. Tadmor and Broyer (1972) neglected curvature in the system and treated the conveying of the solids as in a rectangular channel, as shown in Fig. 8.11. Assuming that conduction occurs only in the $y$ direction and with bulk flow in the $z$ direction the equation of energy leads to the following differential equation for the temperature distribution:

$$\rho_b \bar{C}_{p,p} V_{pz} \frac{\partial T_p}{\partial z} = k_p \frac{\partial^2 T_p}{\partial y^2}, \tag{8.31}$$

where $T_p$ is the temperature in the plug and $\bar{C}_{p,p}$ and $k_p$ are the heat capacity and the thermal conductivity, respectively, of the plug. Equation 8.31 can be converted to the one-dimensional transient heat conduction equation just as done in Chapter 5 by replacing $dz/V_{pz}$ by $dt$:

$$\frac{\partial T_p}{\partial t} = \alpha_p \frac{\partial^2 T_p}{\partial y^2}, \tag{8.32}$$

where $\alpha_p$ is the thermal diffusivity of the solid bed. $V_{pz}$ is the velocity of the plug in the down-channel direction and is obtained from the equations derived for the isothermal case. The approach here is similar to that used in the forced convection approximation where the velocity field is assumed to be unaffected by changes in temperature. Because heat is assumed to be conducted away through the barrel, the temperature distribution in the barrel, $T_b(y, t)$, is also required. However, because the thermal conductivity of the barrel is at least 100 times higher than the polymer, the temperature distribution is assumed to be linear. The following initial and boundary conditions are used:

$$\text{I.C. at } t = 0 \qquad T_p(y, t) = T_0 \tag{8.33}$$

$$\text{B.C.1 at } y = H \qquad T_p = T_b. \tag{8.34}$$

The additional boundary condition required to solve Eq. 8.32 comes from an energy balance at the interface between the barrel and the solid bed where it is assumed that the frictional heat is conducted away into the barrel and the solid plug:

$$\text{B.C.2 at } y = H \qquad q_b = -K_p \frac{\partial T_p}{\partial y} + K_b \frac{\partial T_b}{\partial y}. \tag{8.35}$$

Tadmor and Broyer (1972) developed a numerical scheme to solve for the temperature profile in the solid bed.

In Pr. 8C.2 the IMSL subroutine IVPAG is used to determine the temperature profile in the solid bed. Two approaches are used in this problem. The first is to assume that the barrel is heated and the solid bed is heated by conduction of heat from the barrel. In the other approach the heat generated by friction is assumed to be conducted into the solid bed, which may only be the case when the barrel is heated to a temperature in the range of the melting point of the polymer.

Properties for a few polymers required in the solids conveying model are presented in Appendix B. Here it is observed that the friction coefficient between the polymer and steel is relatively independent of temperature. As the temperature approaches the melting point, the friction coefficients tend to increase, but over a wide range of temperatures $f_b$ is nearly constant. Hence, it seems justifiable to use the mass flow rate and pressure rise calculated by means of the isothermal model in the nonisothermal model.

---

### EXAMPLE 8.2
### Solids Conveying of LDPE (Tadmor & Gogos, 1979)

LDPE is extruded in a single-screw extruder with a diameter of $6.35 \times 10^{-2}$ m. The square pitched screw (i.e., $L_s = D_b$) is 26.5 turns long with a feed section of 12.5 turns and channel depth of $9.398 \times 10^{-3}$ m, a transition section of 9.5 turns, and a metering section $3.22 \times 10^{-3}$ m deep. The flight width is $6.35 \times 10^{-3}$ m, and the flight clearance is negligible. The screw speed is 60 rpm, and the mass flow rate is 67.1 kg/hr. The pellets enter the extruder at 25°C from the hopper described in Ex. 8.1. The hopper discharge opening is 0.127 m and occupies the first two turns of the screw. The barrel is maintained at 149°C, and melting starts 3 turns from the beginning of the flights. The friction coefficients between the polymer and the barrel and the screw are 0.45 and 0.25, respectively. Calculate (**a**) the pressure at the end of the solids-conveying zone (i.e., over one turn); (**b**) the power consumption in the solids-conveying zone and (**c**) the energy per unit surface area dissipated into heat.

### Solution

First, we compute some geometrical values required in the solids-conveying model:

| | |
|---|---|
| $\phi_b$, helix angle at barrel surface | 17.65° |
| $\bar{\phi}$, mean helix angle | 20.48° |
| $\phi_s$, helix angle at root of screw | 24.33° |
| $\bar{W}$, mean channel width | $5.314 \times 10^{-2}$ m |
| $W_b$, channel width at barrel surface | $5.416 \times 10^{-2}$ m |
| $W_s$, channel width at root of screw | $5.151 \times 10^{-2}$ m |
| $l$, axial length | 10.5 turns, 0.666 m |
| $\bar{z}$, mean helical length | 2.270 m |

We next calculate the pressure rise from Eq. 8.27. To calculate $P_2/P_1$ we determine $\phi'$ from Eq. 8.22, which is in turn is used to calculate $M$ and $K_s$. The axial velocity of the solid plug is:

$$V_{pl} = \frac{(61.7/3600)}{(595)[(\pi/4)[(0.065)^2 - (0.0447)^2] - (0.00635)(0.009398)/\sin(20.48)]}.$$

$$= 0.02195 \text{ m/s}.$$

The velocity of the barrel surface $V_b = \pi N D_b = 0.19995$ m/s, and hence from Eq. 8.22 we find:

$$\tan \phi' = \frac{\tan \phi_b}{(V_b/V_{pl})\tan \phi_b - 1} = 0.1261,$$

and $\phi'$ is 7.57°. From Eq. 8.26 we find $K_s$:

$$K_s = \frac{(0.0541)\sin(20.48) + (0.25)\cos(20.48)}{0.0635\cos(20.48) - (0.25)\sin(20.48)} = 0.5859,$$

and from Eq. 8.25:

$$M = \cos(7.57) - 0.5859\sin(7.57) = 0.9141.$$

The pressure rise, $P_2/P_1$ over one turn of the solids conveying section ($Z_b = 0.0635/\sin 17.6° = 0.209$ m), which starts just after the two flights occupied by the hopper throat and ends with the heating of the barrel, is obtained from:

$$0.9141 = (2)\frac{(0.009398)(0.25)}{(0.05416)(0.45)}\sin(17.65)$$

$$\times \left[(0.5859) + \frac{(0.0541)}{(0.0635)}\cotan(20.48)\right]$$

$$+ \frac{(0.05151)(0.25)}{(0.05416)(0.45)}\sin(17.65)$$

$$\times \left[(0.05859) + \frac{(0.0447)}{(0.0635)}\cotan(24.33)\right]$$

$$+ \frac{(0.05314)(0.009398)\sin(20.48)}{(0.05416)(0.209)(0.45)}$$

$$\times \left[(0.5859) + \frac{(0.0541)}{(0.0635)}\cotan(20.48)\right]\ln\frac{P_2}{P_1}$$

$$= 0.1676 + 0.34328 + 0.09813\ln\frac{P_2}{P_1},$$

and this leads to:

$$P_2/P_1 = 60.9.$$

$P_1$ is the pressure at the base of the hopper and was calculated in Ex. 8.4. Thus $P_2$ is $2.52 \times 10^5$ Pa (36.5 psi). The solids-conveying zone could in fact extend another turn or two until the surface reaches the melting point, $T_m$, of LDPE (135°C). The results indicate that the solids-conveying zone is operating properly and that higher outputs could be obtained before the solid bed no longer fills the channel. The condition in which the channel is only partially full is referred to as *starve feeding* and is dealt with in problem 8B.4. The power input through the barrel is obtained from Eq. 8.29 and is:

$$P_w = (\pi)(1)(0.0635)(0.05416)(0.209)(0.45)\cos(7.570)$$

$$((2.52 \times 10^5) - (4.131 \times 10^3))/\ln(60.9) = 60.8 \text{ W}.$$

The power dissipated into heat per unit of surface area is obtained from Eq. 8.30 and is:

$$q_b = P_w(0.1995)(0.45)[\sin(17.650)]/\sin(17.65 + 7.57)$$

$$= 0.0639P_w.$$

---

## 8.3.2. DELAY AND MELTING ZONES

The basic model for the conversion of the solid bed to melt is due to Tadmor (Tadmor & Klein, 1970) and is based on his observations of the state of material along the screw channel, according to which for most polymer systems a melt film first appears at the barrel surface as the result of heat generation due to friction and heat conducted from the heated barrel. Once the melt film forms, the conveying mechanism changes at the barrel surface where viscous drag is now dominant, but frictional drag is still important at the root of the screw and the flights. The thickness of the melt film continues to increase as the plug proceeds down the channel until it attains a value of several times the flight clearance. At this point, the melt film thickness stays nearly constant and the melt is scraped off and accumulated at the pushing flight. The axial distance from where the melt film first appears until melt begins to accumulate at the pushing flight is referred to as the *delay zone*.

There appears to be no reliable mathematical model for predicting the length of the delay zone. Tadmor and Klein (1970), based on limited experimental data, found an empirical correlation between the number of turns (i.e., length of the delay zone) and a dimensionless parameter $\psi$, where $\psi$ is defined later in Eq. 8.64 and represents the ratio of the melting rate to the mass flux of the solid bed in the channel. This relation is:

$$N' = 0.008\left(\frac{1}{\psi}\right), \tag{8.36}$$

where $N'$ is the number of turns. Although this relation was obtained from data for a limited number of polymers, it is all that is available for estimating the length of this zone at the present time.

Based on visual observations, Tadmor (Tadmor & Klein, 1970) proposed a melting mechanism as described in Fig. 8.15. The melt film is scraped off and accumulates at the pushing flight. The width of the solid bed, $X$, decreases as one proceeds down the channel. The solid is pushed inward, and at the melt and solid interface the bed appears to move upward with a velocity, $V_{sy}$. The melt film thickness, $\delta$, was observed to change along the width of the channel but only slightly, and it did not appear to change significantly along the down-channel direction.

The goal of a model of the melting section is to determine the solid bed width, $X$, as a function of the down-channel distance, $z$. For the situation just described and as shown in Fig. 8.15, the basic idea is to determine the temperature distribution in the melt film and solid bed. Then, an energy balance is performed at what is assumed to be a distinct melt solid interface. The heat flowing into the interface is conducted into the solid bed and used to change the solid bed to melt (enthalpy associated with a phase change). Hence, not only does the melt film remain at nearly the same thickness, but also the temperature does not change along the channel direction. Furthermore, there is also the possibility of a significant contribution to melting from viscous dissipation.

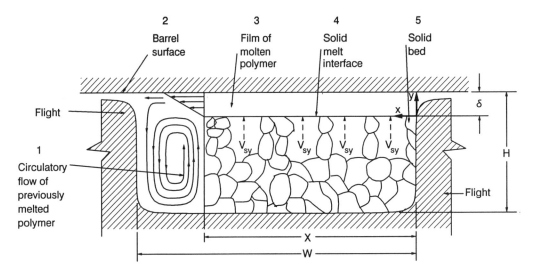

**Figure 8.15** Tadmor melting model.

Based on his observations, Tadmor made the following assumptions:

1. Steady-state conditions are reached in the extruder.
2. Melting takes place only at the barrel surface (in some cases melting has been observed at the root and flights of the screw (Rauwendaal, 1986, p. 271).
3. The solid bed is homogeneous, continuous, and deformable.
4. Physical and thermophysical properties are assumed constant.
5. The solid bed-melt film interface is assumed to be a distinct interface existing at the melting point, $T_m$, of the polymer.

The first step in determining the solid bed profile is to determine the temperature distribution in the melt. Referring to Fig. 8.15 we now develop the model. First, we locate a set of axes at the melt-solid interface at the trailing flight (right side of Fig. 8.15). We next make the following postulates for the velocity and temperature fields in the melt film:

$$v_x = v_x(y) \qquad v_z = v_z(y) \qquad v_y = 0 \qquad T = T(x, y) \qquad (8.37)$$

and the solid bed:

$$T_s = T_s(y). \qquad (8.38)$$

The solid bed is assumed to move as a plug with down-channel velocity of $V_{sz}$ where

$$V_{sz} = G/(\rho_s H W) \qquad (8.39)$$

and is the same as solid bed velocity at the beginning of the melting zone, $V_{pz}$ (the thickness of the melt film has been neglected). For the melt film, the equations of motion and energy are:

$$\frac{\partial \tau_{yx}}{\partial y} = 0 \qquad (8.40)$$

$$\frac{\partial \tau_{yz}}{\partial y} = 0 \qquad (8.41)$$

$$\rho \bar{C}_p v_x \frac{\partial T}{\partial x} = +k \frac{\partial^2 T}{\partial y^2} + k \frac{\partial^2 T}{\partial x^2} - \tau_{yx} \frac{\partial v_x}{\partial y} - \tau_{yz} \frac{\partial v_z}{\partial y}. \qquad (8.42)$$

To solve this set of equations a constitutive relation is needed, and the power-law empiricism for viscosity is used here for convenience. The equations are coupled because of the dependence of viscosity on temperature. However, the velocity field can be obtained independently of the energy equation. The energy equation is coupled to the equations of motion because of the viscous dissipation terms. The velocity field is obtained by integrating Eqs. 8.40 and 8.41 after substituting in the generalized Newtonian fluid (GNF) model and using the following boundary conditions:

B.C.1 at $y = 0$   $v_x = 0$
B.C.2 at $y = 0$   $v_z = V_{sz}$
B.C.3 at $y = \delta$   $v_x = V_{bx}$
B.C.4 at $y = \delta$   $v_z = V_{bz}.$   (8.43)

Because of the homogeneous nature of Eqs. 8.40 and 8.41 the viscosity function drops out, and the velocity field becomes:

$$v_x = \left(\frac{V_{bx}}{\delta}\right) y \qquad (8.44)$$

$$v_z = \left(\frac{V_{bz} - V_{sz}}{\delta}\right) y + V_{sz}. \qquad (8.45)$$

The difficulty comes in solving the energy equation because $\eta$ is a function of temperature and because of the term on the left-hand side of this equation. One should remember that this term is associated with the transport of heat by convection and that if viscous dissipation is large, this term will be important. Tadmor and Klein (1970) assumed a parabolic temperature distribution and initially neglected the conduction and convection terms in the $x$ direction. Rather than assume a temperature distribution, we will use the forced convection assumption. Hence, Eq. 8.42 can be directly integrated (provided we drop the convection term) to obtain the following temperature distribution in the melt:

$$T = \frac{\Phi_v}{2 k_m}(y^2 - y\delta) + y\left(\frac{T_b - T_m}{\delta}\right) + T_m, \qquad (8.46)$$

where $\Phi_v$ is the viscous dissipation term given by:

$$\Phi_v = m\left[\left(\frac{V_{bx}}{\delta}\right)^2 + \left(\frac{V_{bz} - V_{sz}}{\delta}\right)^2\right]^{\frac{n+1}{2}}. \tag{8.47}$$

This temperature profile was obtained using the following boundary conditions:

B.C.1 at $y=0$     $T = T_m$

B.C.2 at $y=\delta$     $T = T_b$, $\tag{8.48}$

where $T_b$ is the barrel temperature.

The temperature distribution in the solid bed is determined next. Assuming that $T_s = T_s(y)$, then the energy equation becomes:

$$\rho_s \bar{C}_{ps} V_{sy} \frac{\partial T_s}{\partial y} = k \frac{\partial^2 T_s}{\partial y^2}. \tag{8.49}$$

The boundary conditions used in solving this equation are

B.C.1 at $y=0$     $T_s = T_m$

B.C.2 at $y=-\infty$     $T_s = T_0$, $\tag{8.50}$

where $T_0$ is the temperature of the bed entering the melting zone. B.C.2 actually presents a problem as there is a temperature gradient in the solid bed and the bed temperature may change as it moves down the channel. Conditions under which one should consider changes in the bed temperature along the channel are discussed by Rawendaal (1986). Using the boundary conditions given in Eq. 8.50, the temperature profile becomes:

$$T_s = (T_m - T_0) \exp\left(\frac{y V_{sy}}{\alpha_s}\right) + T_0. \tag{8.51}$$

The final step in the determination of the melting rate is the energy balance at the melt–solid interface:

$$q_y|_{y-0} - q_{sy}|_{y-0} - \rho_s V_{sy} \Delta \bar{H}_f = 0. \tag{8.52}$$

One now substitutes the temperature distributions given in Eqs. 8.46 and 8.51 into Eq. 8.52 to obtain:

$$\frac{\Phi_v \delta}{2} + k_m \frac{(T_b - T_m)}{\delta} = [\bar{C}_{ps}(T_m - T_0) + \Delta \bar{H}_f]\rho_s V_{sy}. \tag{8.53}$$

Equation 8.53 contains two unknowns, $\delta$ and $V_{sy}$, and hence an additional equation is required. Using the fact that the rate at which the solid is converted to melt at the interface must equal the rate at which it accumulates at the leading flight, one obtains the following equation:

$$w_L(z) = \rho_s V_{sy} X = \rho_m \int_0^\delta \frac{V_{bx}}{\delta} y\, dy = \frac{\rho_m V_{bx} \delta}{2}, \tag{8.54}$$

where $X$ is the bed width at any z distance down the channel and $w_L(z)$ is the rate of melting. For the Newtonian case $\Phi_v$ becomes:

$$\Phi_v = \frac{\mu}{\delta^2}[V_{bx}^2 + (V_{bz} - V_{sz})^2], \tag{8.55}$$

and hence one finds the following expression for $\delta$:

$$\delta = \left[\frac{2k_m(T_b - T_m) + \mu(V_{bx}^2 + (V_{bz} - V_{sz})^2)X}{(\bar{C}_{ps}(T_m - T_0) + \Delta \bar{H}_f)(\rho_m V_{bx})}\right]^{1/2} \tag{8.56}$$

and $w_L(z)$:

$$w_L(z) = \left\{\frac{V_{bx}\rho_m(k_m(T_b - T_m) + \left(\frac{\mu}{2}\right)V_j^2)X}{2(\Delta \bar{H}_f + \bar{C}_{ps}(T_m - T_0))}\right\}^{1/2}, \tag{8.57}$$

where

$$V_j^2 = V_{bx}^2 + (V_{bz} - V_{sz})^2. \tag{8.58}$$

For a GNF with a viscosity function given by the power-law model, Eq. 8.56 represents a nonlinear algebraic equation that must be solved for $\delta$.

Finally, we determine the solid bed profile as a function of distance down the channel. The change in solid bed width is obtained by taking a mass balance on an element of thickness $\Delta z$:

$$\rho_s V_{sz}(H-\delta)X|_z - \rho_s V_{sz}(H-\delta)X|_{z+\Delta z} = w_L(z)\Delta z, \tag{8.59}$$

which on taking the limit as $\Delta z \to 0$ and neglecting the film thickness change in the down channel direction reduces to:

$$\frac{-d(HX)}{dz} = \frac{w_L(z)}{\rho_s V_{sz}}. \tag{8.60}$$

By substituting Eq. 8.57 into Eq. 8.60, we arrive at the following expression:

$$\frac{-d(HX)}{dz} = \frac{\Phi\sqrt{X}}{\rho_s V_{sz}}, \tag{8.61}$$

where

$$\Phi = \left\{\frac{V_{bx}\rho_m[k_m(T_b - T_m) + (\mu/2)V_j^2]}{2(\bar{C}_{ps}(T_m - T_0) + \Delta \bar{H}_f)}\right\}^{1/2}. \tag{8.62}$$

For a constant channel depth, Eq. 8.61 can be integrated to give:

$$\frac{X_2}{W} = \frac{X_1}{W}\left[1 - \frac{\psi(z_2 - z_1)}{2H}\right]^2, \tag{8.63}$$

where $X_1$ and $X_2$ are the widths of the solid bed at locations $z_1$ and $z_2$, respectively, and the dimensionless group is defined as:

$$\psi = \frac{\Phi}{V_{sz}\rho_s\sqrt{X_1}}. \tag{8.64}$$

Hence, for a constant channel depth we can determine the length of the channel required to melt the solid bed from Eq. 8.63.

For a tapered channel of constant taper, which is usually the case, we write Eq. 8.61 as:

$$\frac{d(HX)}{dH} = \frac{\Phi\sqrt{X}}{A\rho_s V_{sz}}, \tag{8.65}$$

where

$$A = -\frac{dH}{dz}. \tag{8.66}$$

Equation 8.65 can be integrated to give:

$$\frac{X_2}{W} = \frac{X_1}{W}\left[\frac{\psi}{A} - \left(\frac{\psi}{A} - 1\right)\sqrt{\frac{H_1}{H_2}}\right]^2, \tag{8.67}$$

where $X_2$ and $X_1$ are the widths of the solid bed at down-channel locations corresponding to $H_2$ and $H_1$, respectively.

Equations 8.63 and 8.67 represent the basic equations for the melting model. The total length of melting for a channel of constant depth is:

$$z_T = \frac{2H}{\psi}, \tag{8.63}$$

and for a tapered channel it is:

$$z_T = \frac{H}{\psi}\left(2 - \frac{A}{\psi}\right). \tag{8.69}$$

For a channel of constant depth, the length of channel required to melt the solid bed is a function of the channel depth and a dimensionless group $\psi$, where $\psi$ expresses the ratio of the local rate of melting per unit solid–melt interface to the local solid mass flux. Thus, the length of melting is proportional to the mass flow rate and inversely proportional to the rate of melting. In the case of tapered channels, the higher the taper, the shorter the melting length, $z_T$. However, if the taper becomes too great, the solid bed can increase in width instead of decreasing, which may lead to plugging of the channel and surging conditions.

## 8.3.3. METERING SECTION

The fully melted polymer now enters the third zone of the extruder where it is pressurized. The buildup of pressure is required in order to pump the melt through the die at the end of the extruder. The pressurization of the melt is based on a viscous drag mechanism. We first illustrate how viscous drag can lead to a pressurization of the melt. This is followed by the development of a nonisothermal non-Newtonian model of the metering section. Because numerical methods are required to solve the equations generated in this model, we end the section by presenting the isothermal Newtonian case where an analytical solution is possible.

The basic principle of operation of the metering section of the single-screw extruder is illustrated by the simple plate model shown in Fig. 8.16. The fluid between the two plates is considered to be Newtonian and under isothermal and steady-flow conditions. Because of a restriction at the end of the channel (which is not shown), the pressure increases along the $z$ direction. $v_z$ is assumed to depend only on $y$, as the aspect ratio of the plates is large (i.e., $W/H > 10$). After substituting in the expression for the shear stress for a Newtonian fluid, the equation of motion becomes:

$$\mu\frac{d^2v_z}{dy^2} - \frac{dp}{dz} = 0. \tag{8.70}$$

This equation is solved using the following boundary conditions:

B.C.1 at $y = 0$    $v_z = 0$

B.C.2 at $y = H$    $v_z = V_0$.      (8.71)

After integrating Eq. 8.70 and using the boundary conditions given above the velocity field is determined to be:

$$v_z = (H^2/2\mu)\left(\frac{dp}{dz}\right)\left[\left(\frac{y}{H}\right)^2 - \frac{y}{H}\right] + \frac{V_0 y}{H}. \tag{8.72}$$

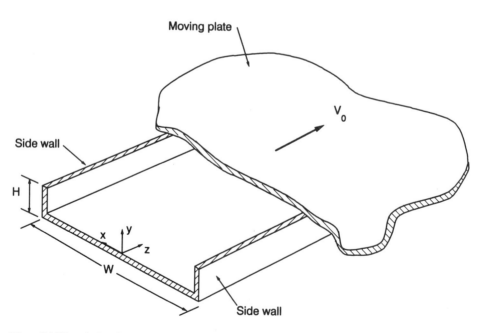

**Figure 8.16** Drag-induced pressurization of a fluid in a rectangular channel having an aspect ratio, $W/H > 10$.

Equation 8.72 is now integrated over the cross-sectional area to obtain the volumetric flow rate:

$$Q = \frac{V_0 WH}{2} - \frac{dp}{dz}\left(\frac{WH^3}{12\mu}\right).$$

(8.73)

$Q$ is seen to consist of two terms: The first is called the drag flow, $Q_d$, and the second is referred to as the pressure flow, $Q_p$. When there is no pressure buildup, then the transport is due entirely to the drag flow term. However, if there is a significant pressure increase, then $Q$ is decreased. In this case the pressure term can dominate to the point where flow can be in the opposite direction (as we will see later, this cannot happen in the extruder). The main point is that, as a result of viscous drag, the fluid can be advanced against resistance because of pressure buildup. This is in essence the principle of operation of the metering section of the single-screw extruder.

As a pump, however, the parallel plate device is not practical in itself. A way is needed to increase the length of the channel and to return the upper plate to the channel after it has transversed the length of the channel. As shown in Fig. 8.17, one way to do this is to machine the channel in a shaft and then give the channel a helical pitch so that the length can be increased. The inner cylinder or outer cylinder could be rotated. In the figure the outer cylinder is rotated. When the channel depth is shallow relative to the radius of the cylinder, then curvature can be neglected, and the flow can be considered as that in flat plates.

We are now in position to develop a model for the metering section of a single-screw extruder. In practice the screw is rotated inside the barrel, and helical or cylindrical coordinates are needed to describe the geometry. However, because the channel depth is usually small relative to the radius of the barrel, we can treat the flow using rectangular Cartesian coordinates, as shown in Fig. 8.18. It is customary to attach the axes to the rotating screw, which makes it appear to the observer that the plate is moving over the channel at an angle $\phi_b$ to the down-channel direction. In principle a three-dimensional model is required to describe the velocity and temperature profiles in the extruder channel. However, this type of detail is usually not necessary in the design of screws, nor is it computationally practical. With the assumptions of constant fluid density and steady-state conditions the following postulates pertaining to the velocity and temperature fields

can be made:

$$v_z = v_z(x, y) \qquad v_x = v_x(x, y)$$

(8.74)

$$T = T(y, z).$$

(8.75)

If $W/H > 10$, then we can simplify matters by making the following postulates for the velocity field:

$$v_z = v_z(y) \qquad v_x = v_x(y).$$

(8.76)

In essence we are neglecting the effect of the side walls of the channel which is probably valid for most single-flighted screw designs. Using these postulates, the equations of motion and energy become, respectively:

$$0 = -\frac{\partial p}{\partial x} - \frac{\partial \tau_{yx}}{\partial y}$$

(8.77)

$$0 = -\frac{\partial p}{\partial z} - \frac{\partial \tau_{yz}}{\partial y}$$

(8.78)

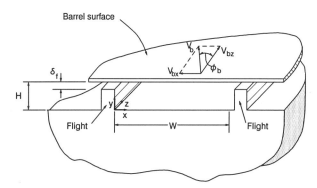

**Figure 8.18** Parallel plate model of flow in the metering section of a single-screw extruder. The axes are attached to the rotating screw that makes the barrel appear to move relative to the screw.

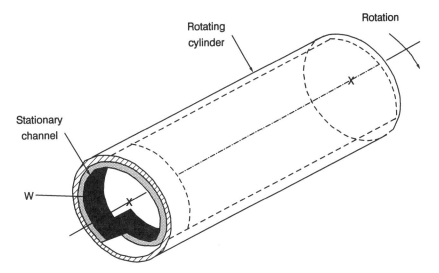

**Figure 8.17** Principle of operation of the metering section of a single-screw extruder. A helical channel is machined in the inner cylinder. The fluid in the channel is dragged forward in the channel by the rotating cylinder.

$$\rho \bar{C}_p v_z \frac{\partial T}{\partial z} = k \frac{\partial^2 T}{\partial y^2} - \tau_{xy} \frac{\partial v_x}{\partial y} - \tau_{yz} \frac{\partial v_z}{\partial y}. \tag{8.79}$$

Using the GNF model, the shear stress components are given by:

$$\tau_{yx} = -\eta(\dot{\gamma}, T) \frac{\partial v_x}{\partial y} \tag{8.80}$$

$$\tau_{yz} = -\eta(\dot{\gamma}, T) \frac{\partial v_z}{\partial y}, \tag{8.81}$$

where $\dot{\gamma}$ is given by:

$$\dot{\gamma} = \sqrt{\left(\frac{\partial v_x}{\partial y}\right)^2 + \left(\frac{\partial v_z}{\partial y}\right)^2}. \tag{8.82}$$

This system of differential equations is nonlinear, and they are coupled through the temperature dependence of viscosity. Numerical techniques are required to solve these equations.

We can obtain reasonable solutions to these equations if we assume that the forced convection assumption holds and that the helix angle, $\phi_b$, is less than 20°. In fact, most screws are designed with a square pitch (i.e., $L_s = D_b$, which means that $\phi_b = 17.7°$). The forced convection assumption allows us to decouple the equations of motion from the equation of energy. Values of $\phi_b < 20°$ allow us to decouple the equations of motion, as $dv_x/dy \ll dv_z/dy$ (to see this, we approximate $dv_x/dy$ as $V_b \sin \phi_b/H$ and $dv_z/dy$ as $V_b \cos \phi_b/H$ and $\dot{\gamma}$ is approximately $dv_z/dy$). The equations of motion become:

$$\frac{d}{dy}\left(m\left|\frac{dv_z}{dy}\right|^{n-1} \frac{dv_z}{dy}\right) = \frac{\partial p}{\partial z} \tag{8.83}$$

$$\frac{d}{dy}\left(m\left|\frac{dv_z}{dy}\right|^{n-1} \frac{dv_x}{dy}\right) = \frac{\partial p}{\partial x}. \tag{8.84}$$

Equation 8.83 can be integrated subject to the following boundary conditions:

B.C.1 at $y=0$ $\quad v_z=0$ $\hspace{2cm}$ (8.85)

B.C.2 at $y=H$ $\quad v_z=V_{bz}=V_b\cos\phi_b.$ $\hspace{0.5cm}$ (8.86)

We also find it convenient to introduce the following dimensionless variables:

$$u_z = v_z/V_{bz} \qquad \xi = y/H.$$

Because $u_z$ passes through a maximum at some point $\beta$, the solution is obtained over two regions:

For $0 \le \xi \le \beta$: $u_z^< = \left(\frac{GH^{n+1}}{mV_{bz}^n}\right)^s \left(\frac{1}{s+1}\right)[(\beta-\xi)^{s+1} - \beta^{s+1}]$ (8.87)

For $\beta \le \xi \le 1$: $u_z^> = \left(\frac{GH^{n+1}}{mV_{bz}^n}\right)^s \left(\frac{1}{s+1}\right)$

$$\times [(\xi-\beta)^{s+1} - (1-\beta)^{s+1}] + 1, \tag{8.88}$$

where $G = \partial p/\partial z$ and $s = 1/n$. The value for $\beta$ is obtained by equating the two velocity fields at $\xi = \beta$ to give:

$$\beta^{s+1} - (1-\beta)^{s+1} + \frac{m^s V_{bz}(1+s)}{G^s H^{1+s}} = 0. \tag{8.89}$$

The volumetric flow rate is obtained by integrating the velocity field over the cross-sectional area and is:

$$\frac{Q}{WHV_{bz}} = \left(\frac{GH^{n+1}}{mV_{bz}^n}\right)^s \left[-\frac{\beta^{s+2}}{s+1} - \frac{(1-\beta)^{s+2}}{s+2}\right] + (1-\beta). \tag{8.90}$$

The expression in Eq. 8.90 is referred to as the screw characteristic and consists of drag flow and pressure flow terms. Finally, the cross-flow term, $v_x$, is obtained by integrating Eq. 8.84 using the velocity field in Eqs. 8.87 and 8.88 and the following boundary conditions:

B.C.1 at $y=0$ $\quad v_x=0$

B.C.2 at $y=H$ $\quad v_x=-V_{bx}=-V_b\sin\phi_b.$ $\hspace{0.5cm}$ (8.91)

The importance of the cross-flow term to mixing will be discussed in Section 8.5.

Using the forced convection assumption and restricting ourselves to small helix angles Eq. 8.79 becomes:

$$\rho \bar{C}_p v_z(y) \frac{\partial T}{\partial z} = k \frac{\partial^2 T}{\partial y^2} + m \left|\frac{dv_z}{dy}\right|^{n+1}. \tag{8.92}$$

We cast the energy equation into dimensionless form by introducing the following variables:

$$z/L = \zeta \qquad v_z/V_{bz} = u \qquad T^* = (T-T_i)/(T_b-T_i), \tag{8.93}$$

where $T_i$ is the initial temperature of the melt entering the metering zone and $T_b$ is the barrel temperature. Equation 8.92 in dimensionless form is:

$$\text{Pe } u_\zeta(\xi) \frac{\partial T^*}{\partial \xi} = \frac{\partial^2 T^*}{\partial \zeta^2} + \text{Br} \left|\frac{du}{d\xi}\right|^{n+1}, \tag{8.94}$$

where $\text{Pe} = H^2\rho\bar{C}_pV_{bz}/kL$ and $\text{Br} = mV_{bz}^{n+1}/H^{n-1}k(T_b-T_i)$. Even with the assumptions made here, numerical methods are needed to solve Eq. 8.94.

The flow patterns in the extruder channel are difficult to visualize for the power-law fluid, and hence we consider the flow for the Newtonian case. The brief development here is mostly for pedagogical purposes, but it does have some applicability to nearly Newtonian fluids under extrusion conditions and for isothermal conditions. For a Newtonian fluid Eqs. 8.83 and 8.84 give the following expressions for $u_x$ and $u_z$, respectively:

$$u_x = -\xi + \xi(\xi-1)(GH^2/2\mu V_{bx}) \tag{8.95}$$

$$u_z = \xi - 3\xi(1-\xi)(H^2G/6\mu V_{bz}). \tag{8.96}$$

Using the fact that there is no net flow in the $x$ direction, we can integrate $u_x$ over $\xi$ to find

$$\frac{\partial p}{\partial x} = -\frac{6\mu V_{bx}}{H^2}. \tag{8.97}$$

Substituting the expression for $\partial p/\partial x$ back into Eq. 8.95, we obtain the cross-channel velocity profile:

$$u_x = -\xi(2-3\xi). \tag{8.98}$$

Equation 8.98 tells us that away from the flights the fluid circulates around a stagnant layer at $y = 2H/3$. Of course, for a pseudoplastic material this position will change. The volumetric flow rate is obtained by integrating Eq. 8.96 over the cross-sectional area of the channel and is:

$$Q = \frac{V_{bz}WH}{2} - \left(\frac{\partial p}{\partial z}\right)\frac{WH^3}{12\mu}.$$ (8.99)

This is the screw characteristic for the Newtonian case, and, as shown in the simple one-dimensional flat plate model described at the start of this section it consists of drag and pressure flow terms. The ratio of pressure to drag flow rates (sometimes called the *throttle ratio*) is:

$$Q_p/Q_d = -\left(\frac{\partial p}{\partial z}\right)\frac{H^2}{6\mu V_{bz}}.$$ (8.100)

One may be under the impression that it is possible to cause the fluid to flow backward toward the hopper. Under no condition is this possible, as we will show. The axial velocity, $v_l$, is obtained by taking the vectorial contributions of $v_x$ and $v_y$ along the axial direction, $l$:

$$v_l = v_x \cos\phi + v_z \sin\phi.$$ (8.101)

Substituting the expressions from Eqs. 8.95, 8.96, and 8.100 into Eq. 8.101 and defining $u_l = v_l/V_b$ we get:

$$u_l = 3\xi(1-\xi)\left(1 + \frac{Q_p}{Q_d}\right)\sin\phi \cos\phi.$$ (8.102)

For the Newtonian case (see Fig. 8.19) it is observed that the cross-flow component is independent of the $Q_p/Q_d$ ratio. On the other hand, $u_z$ and $u_l$ are highly dependent on the $Q_p/Q_d$ ratio, and as the ratio becomes more negative, the flow in the axial direction decreases. When $Q_p/Q_d = -1$, there is no flow out of the extruder (this is known as closed discharge). It is $u_x$ that keeps $v_l > 0$. It should be noted that under closed discharge conditions the residence time of the fluid is infinite and that as one approaches open discharge conditions, the residence time is shortest.

The pressure buildup along the axial direction of the extruder is due to the resistance offered by the die and other elements such as connectors and filtration systems. Hence, the operation of the extruder is directly affected by the design of the die and connecting elements. To illustrate this, we again consider the fluid to be Newtonian, and we consider only the metering section of the extruder. Furthermore, we assume that the die is a simple capillary, and we neglect any pressure losses due to contractions or expansions in the system. Using Eq. 8.99 we write the screw characteristic as

$$Q_s = \frac{1}{2}\pi D_b N \cos\bar{\phi} WH - \left(\frac{\Delta P_s}{L_b}\right)\frac{WH^3 \sin\bar{\phi}}{12\mu},$$ (8.103)

where $Q_s$ is the volumetric flow rate in the extruder, $\Delta P_s$ is the pressure rise in the extruder, $L_b$ is the extruder length, and $\bar{\phi}$ is the average helix angle. Referring to Table 2.6 we can write an expression for the circular die in the following form:

$$Q_D = \left(\frac{\pi R^4}{8L_D}\right)\frac{\Delta P_D}{\mu} = \frac{K_D \Delta P_D}{\mu},$$ (8.104)

where $K_D$ is referred to as the die characteristic and $Q_D$ and $\Delta P_D$ are the volumetric flow rate and pressure drop across the die, respectively. Since $Q_D = Q_s (= Q)$ and $|\Delta P_s| = |\Delta P_D|(=\Delta P)$, then Eqs. 8.103 and 8.104 represent two equations with two unknowns. In this case they can be

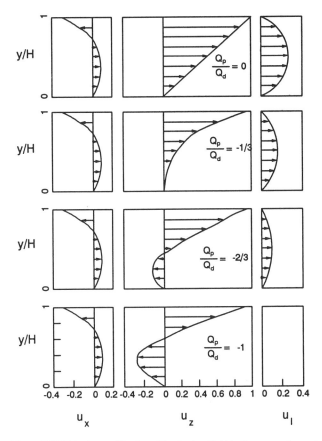

**Figure 8.19** Velocity profiles for a Newtonian fluid in the screw channel for various operating conditions as determined by the ratio of pressure to drag flow. The cross-channel component is unaffected by the ratio $Q_p/Q_d$, and the down channel and axial velocity components change significantly.

solved for $\Delta P$ and $Q$ to give

$$Q = \frac{\frac{1}{2}\pi D_b \cos\bar{\phi}_W H}{1 + (WH^3 \sin\bar{\phi})/(12\mu L_b K_D)}$$ (8.105)

$$\Delta P = \frac{\frac{1}{2}\mu\pi D_b \cos\phi_b WH}{K_D + (WH^3 \sin\bar{\phi}/12L)}.$$ (8.106)

We remind you that Eqs. 8.105 and 8.106 are most likely of little quantitative use, and they serve only to illustrate the effect of the die on the operating conditions of the extruder. In practice one would have to generate values of $\Delta P$ and $Q$ numerically for both the extruder and the die, and then determine under what operating conditions the values are identical.

## 8.4. TWIN-SCREW EXTRUDERS

In this section we consider the essential features of two of the most common types of twin-screw extruders: self-wiping corotating twin-screw extruders (SWCOR) and closely intermeshing counter-rotating twin screw extruders (CICTR). As discussed in Section 8.1.2, there are a vast number of types of twin-screw extruders, and it is not possible to discuss all of these here. Further discussion of the technology and theory for twin-screw extruders is given elsewhere (Rauwendaal 1986; White 1990).

### 8.4.1. SELF-WIPING COROTATING TWIN-SCREW EXTRUDERS

We consider only the melt-conveying section of the SWCOR, because there are no theories available at present for describing the solids transport and melting zones. We first describe some general features of the flow regions in SWCORs and then geometrical features required to understand the more quantitative aspects associated with flow.

SWCORs have a closely matching flight profile, as shown in Fig. 8.20.a. A cross section taken perpendicular to the screw axes (section A-A) as shown in Fig. 8.20.b illustrates how the flights of one screw wipe the surface of the other screw. The cross section shown in this figure is for a double-flighted screw. Because both screws are rotating in the same direction (e.g., counterclockwise in this case), the flight is seen to scrape the surface of the other screw, pushing material over to the other screw. Hence, material travels in a figure-eight pattern in the extruder. By cutting a section B-B that passes through the intermeshing region as shown in Fig. 8.20.c, we see that there is considerable open space between the adjacent channels. Hence, there is little pressure buildup in the intermeshing region. For this reason SWCORs have nonpositive conveying characteristics, and their operation is pressure sensitive in a manner somewhat comparable to single-screw extruders.

To understand the melt-conveying characteristics of SWCORs we must first describe their geometry. We consider the cross section perpendicular to the screw axis as shown in Fig. 8.21 for double-flighted screws. The flight and channel geometry are determined by the screw diameter, $D_b$, (note: this is actually $D_b - 2\delta_s$, but we will neglect flight clearance), the centerline distance, $C_L$, and the number of parallel flights, $p$. From Fig. 8.21 it is easy to see that:

$$C_L = D_b \cos \alpha_i,$$ (8.107)

where $\alpha_i$ is one half the angle of intermesh (i.e., the degree of overlap of the two barrels). $\alpha_i$ is found to be related to the tip angle, $\alpha_t$, by the following relation:

$$\alpha_i = \pi/2p - \alpha_t/2.$$ (8.108)

This relation is not easy to determine from Fig. 8.21, but one must consult the work of Booy (1978). We can now determine the channel depth as a function of the circumferential angle, $\theta$. Referring to Fig.

8.22 we find $H(\theta)$ to be:

$$H(\theta) = \frac{D_b}{2}(1 + \cos \theta) - \left( C_L^2 - \frac{1}{4} D_b^2 \sin^2 \theta \right)^{1/2}.$$ (8.109)

We can now find the cross-channel depth profile along the $x$ direction (note: this is the same as for the single-screw extruder where $z$ is taken along the helical path or down-channel direction and $x$ is taken along the channel width) by substituting in the following coordinate transformation:

$$x = \left( \frac{D_b}{2} \right) \theta \sin \phi$$ (8.110)

to give:

$$H(x) = \frac{D_b}{2}\left[ 1 + \cos \left( \frac{2x}{D_b \sin \phi} \right) \right] - \left[ C_L^2 - \frac{D_b^2}{4} \sin^2 \left( \frac{2x}{D_b \sin \phi} \right) \right]^{1/2}.$$ (8.111)

The cross-channel depth profile is shown in Fig. 8.22.b, and it is seen that the channel is no longer rectangular as is the case for the single-screw extruder.

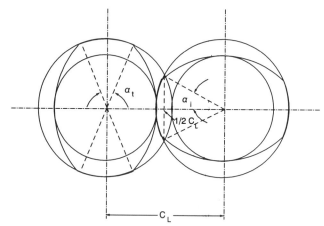

**Figure 8.21** Cross section of a closely intermeshing twin-screw extruder with double-flighted screw elements (or paddle elements). The tip angle, $\alpha_t$, the angle of intermesh, $\alpha_i$, and the centerline distance are shown.

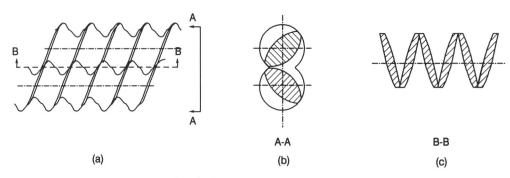

**Figure 8.20** Self-wiping corotating twin-screw extruder. (a) Closely intermeshing double-flighted screw elements. (b) Section A-A showing the barrel and screw cross sections. (c) Section B-B showing the open region between the screw flights.

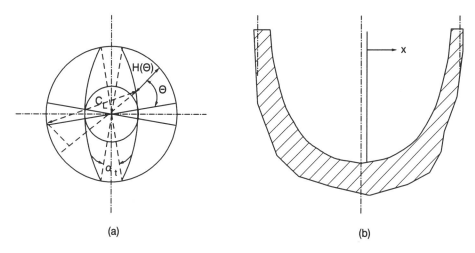

**Figure 8.22** Channel depth for a double-flighted corotating self-wiping twin-screw extruder. (a) Channel depth as a function of circumferential angle. (b) Channel depth as a function of distance $x$ across channel.

We next determine the open cross-sectional area between the barrel and the screw. This is done by finding the cross-sectional area of the barrel and subtracting the cross-sectional area of the two screws. The cross sectional area of the barrel is (Booy, 1978):

$$A_b = \frac{1}{2}(\pi - \alpha_i)D_b^2 + \frac{1}{2}C_L D_B \sin \alpha_i. \tag{8.112}$$

For the double-flighted screw shown in Fig 8.21, this area is easily derivable. The cross-sectional area of one screw is (*note*: this is not easily derived nor seen by refering to Fig. 8.21, but one must consult the paper by Booy, 1978):

$$A_s = p\alpha_i C_L^2 - \frac{1}{2}pC_L D_b \sin \alpha_i + \frac{1}{2}p\alpha_i \left( C_L^2 + \frac{1}{2}D_b^2 - C_L D_b \right). \tag{8.113}$$

The open cross-sectional area between the barrel and screw is just the difference of the two expressions above (*note*: $2A_s$ must be used), which can be written as:

$$A_0 = D_b^2 \left[ \left( p - \frac{1}{2} \right)\alpha_i + \left( p + \frac{1}{2} \right)\sin \alpha_i \cos \alpha_i \right.$$
$$\left. - \pi \cos^2 \alpha_i + (\pi - 2p\alpha_i) \cos \alpha_i \right]. \tag{8.114}$$

This equation tells us that for a fixed barrel diameter, the open area depends on the number of parallel flights and the angle of intermesh, $\alpha_i$.

We are now in a position to consider flow in screw elements. Because of the complicated channel geometry, we will only consider the isothermal flow of Newtonian fluids. The development here will be for pedagogical purposes only and not for quantitative design work. To analyze the flow in SWCORs we use again the flat plate model used for single-screw extruders. The axes are attached to the screws again, and because the channel depth is considered to be small relative to the curvature of the barrel, we use the flat model shown in Fig. 8.23. (*Note*: here we show the situation for a double-flighted screw.) However, the channel length is not the fully unwound channel length but merely the length of one turn. If the flight tip angle is small, then we can consider the flow between channels to be relatively unimpeded and

hence neglect the excess pressure drop in the intermeshing region. (Booy, 1980, and Szydlowski et al., 1987, show how to include the intermeshing region.) Assuming isothermal steady flow of a Newtonian fluid and considering the velocity components, $v_z$ and $v_x$, to depend on $y$ only, the equations of motion are:

$$0 = -\frac{\partial p}{\partial z} + \mu \frac{\partial^2 v_z}{\partial y^2} \tag{8.115}$$

$$0 = -\frac{\partial p}{\partial x} + \mu \frac{\partial^2 v_x}{\partial y^2} \tag{8.116}$$

with boundary conditions

$$v_z(0) = 0 \qquad v_z(H) = V_b \cos \phi$$
$$v_x(0) = 0 \qquad v_x(H) = -V_b \cos \phi. \tag{8.117}$$

Equation 8.115 can be integrated to give the flow rate in one channel:

$$Q = \frac{1}{2}WHV_b \cos \phi - \frac{WH^3}{12\mu}\frac{\partial p}{\partial z}. \tag{8.118}$$

According to Booy (1980), there are $(2p - 1)$ independent channels and hence Eq. 8.118 must be multiplied by $(2p - 1)$ to obtain the flow rate in the screw elements. In principle, the screw characteristic of fully filled screw channels of the SWCOR resembles that of the single-screw extruder. The main differences are in the shape of the channel and when the tip angle becomes large, leading to a large degree of intermeshing.

In most cases the SWCOR is starve-fed, and hence the channel is only partially full. The output from the extruder is determined by means of the device feeding the extruder, not the extruder itself. Actually, the degree of fill changes over the length of the extruder as the final length of the channel must be full if pressure is to be built up to pump the melt through a die. Over most of the section with regular screw elements there will be no pressure buildup. Sometimes, reverse elements or kneading blocks are used as restrictive elements along the extruder.

The fairly simple development that follows is for pedagogical reasons and is due to Werner (1976). A partially filled screw channel

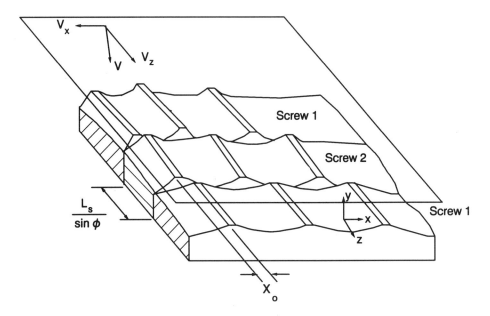

**Figure 8.23** Flat plate model for the closely intermeshing self-wiping twin-screw extruder.

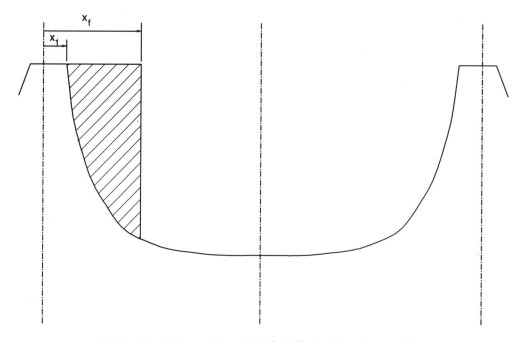

**Figure 8.24** Partially filled screw channel. The fluid fills the channel up to a distance $x_f$.

is shown in Fig. 8.24. The degree of fill is given by the ratio of the filled channel cross-sectional area, $A_f$, to the total cross-sectional area, $A$:

$$f = A_f/A. \tag{8.119}$$

$A$ is found by integrating Eq. 8.111 over the width of the channel and is:

$$A = D_b^2 \left[ \frac{1}{2}\alpha_i - \left( \alpha_i + \frac{1}{2}\alpha_t \right)\cos^2\alpha_i + \left( \frac{1}{2}\sin\alpha_i + \frac{1}{2}\alpha_t \right)\cos\alpha_i \right]\sin\phi. \tag{8.120}$$

$A_f$ is found by integrating Eq. 8.111 from $x_1$ to the filled cross-channel distance, $x_f$:

$$A_f = \int_{x_1}^{x_f} H(x)dx, \tag{8.121}$$

where $x_1 = 0.25\alpha_t D_b \sin\phi$. For isothermal steady-state flow of a Newtonian fluid the equation of motion in the down-channel direction is written as:

$$\frac{\partial^2 v_z}{\partial x^2} + \frac{\partial^2 v_z}{\partial y^2} = 0. \qquad (8.122)$$

By assuming that:

$$\frac{\partial^2 v_z}{\partial x^2} \ll \frac{\partial^2 v_z}{\partial y^2}, \qquad (8.123)$$

then $v_z$ becomes simply:

$$v_z = y V_{bz}/H. \qquad (8.124)$$

The volumetric flow rate is obtained by integrating Eq. 8.124 over the cross section that is filled with fluid, and this is approximately:

$$Q = (1/2)(2p-1)A_f V_{bz}. \qquad (8.125)$$

Equation 8.125 can be written as follows using Eq. 8.119:

$$Q = (1/2)(2p-1)f A\pi D_b N \cos\phi. \qquad (8.126)$$

Thus, $Q$ is directly proportional to the degree of fill and to the screw speed, $N$. The cross-channel velocity profile, $v_x$, is found in much the same way as it was done for the single-screw extruder and is:

$$v_x = \frac{3V_{bx}y^2}{H^2} + \frac{4V_{bx}y}{H} + V_{bx}. \qquad (8.127)$$

Finally, we consider the flow in kneading blocks (see Fig. 8.8) as these are commonly used along with the screw elements. The kneading blocks or paddle elements are the heart of the SWCOR as melting and dispersing of additives occur here. The cross section of a two lobed paddle element looks the same as that of a double-flighted screw element (see Fig. 8.21). White (1990) has attempted to model flow in paddle elements.

### 8.4.2. INTERMESHING COUNTERROTATING EXTRUDERS

The other frequently used twin-screw extruder is the closely intermeshing counterrotating (CICTR) type shown in Fig. 8.25. Looking along the axis of the screws from the right side of the figure we see

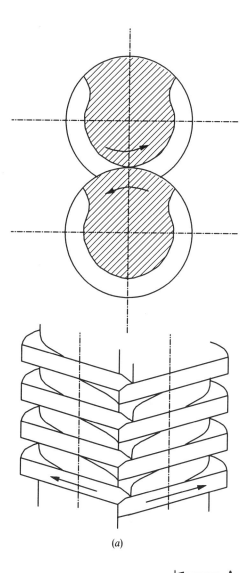

(a)

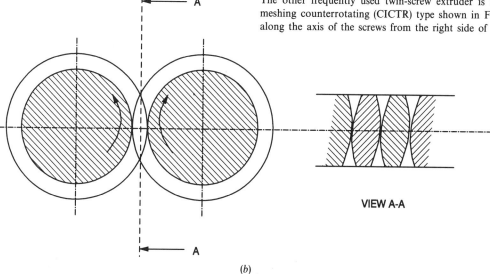

A

A

VIEW A-A

(b)

**Figure 8.25** Closely intermeshing counterrotating twin-screw extruder. (a) Cross-sectional view (end view). (b) Cross-sectional view through flights showing open region.

that the left screw is rotating in a counterclockwise direction and has a right-hand thread, whereas the right screw rotates in a clockwise direction and has a left-hand thread. A cross section taken through the intermeshing region, which is shown in Fig. 8.25.b, reveals that there

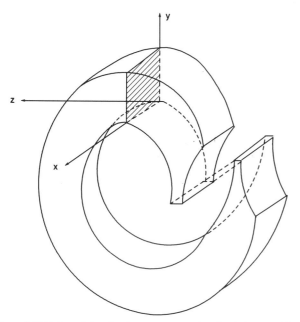

**Figure 8.26** C-shaped chamber in a counterrotating closely intermeshing twin-screw extruder.

is very little opening between the channels of the two screws. Hence, these devices can achieve nearly positive conveying characteristics.

In theory CICTR extruders are positive conveying devices where the maximum possible flow rate is given by:

$$Q_{max} = 2pNV. \tag{8.128}$$

where $p$ is the number of parallel flights, $N$ is the screw speed, and $V$ is the volume of the closed C-shaped chamber as shown in Fig. 8.26. The volume of the C-shaped chamber is approximately:

$$V = \frac{\pi D_b H \bar{W}}{\cos \bar{\phi}} - \frac{\bar{w} D_b^2 (2\alpha_i - \sin 2\alpha_i)}{4 \cos \bar{\phi}}, \tag{8.129}$$

where $\bar{w}$ is the mean flight width given by:

$$\bar{w} = w + H \tan \psi, \tag{8.130}$$

and $\bar{W}$ is the mean channel width given by:

$$\bar{W} = \frac{\pi (D_b - \frac{1}{2} H) \sin \bar{\phi}}{p} - \bar{w}. \tag{8.131}$$

$\psi$ in Eq. 8.130 is the flight flank angle (i.e., the angle the sides of the flight make with the line perpendicular to the channel surface).

In practice the CICTR extruder is not a positive conveying device, and there are a number of places where leakage can occur. Janssen (1978) has identified four places where leakage occurs, as shown in Fig. 8.27:

1. Leakage through the gap between the flight and the barrel wall, called flight leakage, $Q_f$

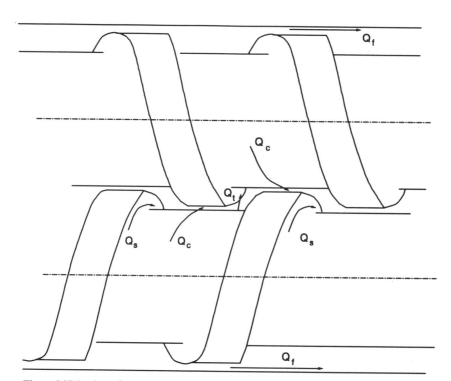

**Figure 8.27** Leakage flows in a counterrotating closely intermeshing twin-screw extruder. $Q_c$, calendaring flow between screw flight and opposite screw root; $Q_f$, leakage flow between screw flight and barrel; $Q_s$, leakage flow between screw flight walls; $Q_t$, leakage flow in tetrahedron region between flights.

2. Leakage between the bottom of the channel of one screw and the flight of the other screw, called calender leakage, $Q_c$.

3. Leakage through the gap that goes from one screw to the other between the flanks of the flights of the two screws, referred to as leakage through the tetrahedron gap, $Q_t$

4. Leakage through the gap between the flanks of the screws perpendicular to the plane through the screw axis, $Q_s$

Quantitative estimates of each type of leakage are given by Janssen (1978) for Newtonian fluids. Because these derivations are rather lengthy and may be of marginal value for quantitative design work involving non-Newtonian fluids, they are not discussed here. The total output from the extruder is given by:

$$Q = 2pNV - 2pQ_s - 2Q_f - Q_t. \tag{8.132}$$

## 8.5. MIXING, DEVOLATILIZATION, AND REACTIONS IN EXTRUDERS

Extruders have other functions than the melting and pumping of polymers. In particular, they are used in mixing operations, in the removal of volatiles, and in the processing of reacting systems. Mixing operations involve primarily the blending of polymers and the dispersion of additives and fillers. Examples of devolatilization include the removal of monomers in the production of polymers, the removal of reaction products during condensation polymerization, and the removal of water from hygroscopic polymers. Applications of extruders as reactors are numerous and include condensation reactions (e.g., the generation of high-molecular-weight PETP), the peroxide degradation of polypropylene, and compatibilization of two polymers through the formation of graft copolymers. The basic elements of these processes are discussed here. Because of the complex nature of these processes, we can only present the most elementary analyses.

### 8.5.1. MIXING

As discussed in Chapter 6, quality of mixing is related to the increase in interfacial area, which is proportional to strain. The calculation of the average strain, $\bar{\gamma}$, requires the residence time distribution function, $f(t)$. We present the analysis for isothermal Newtonian flow in a single-screw extruder only, which is due to Pinto and Tadmor (1970). We give only a descriptive analysis for twin-screw extruders.

The analysis begins with the assumption that curvature is not important and that hence one can use the parallel plate model to describe flow in the screw channel. The axial velocity profile was given in Eq. 8.102 and is reproduced here in the following form:

$$u_l = 3\xi(1-\xi)[1+\Phi'] \sin\phi \cos\phi, \tag{8.133}$$

where $\Phi' = Q_p/Q_d$. The actual path of a fluid particle is quite complex, as shown in Fig. 8.28. The cross-channel velocity component, $v_x$, gives rise to a circulation pattern, and because there is also a $v_z$ component, the actual path of a particle in the channel is spiral in nature. The goal at this point is to determine how long a particle spends in the channel (i.e., its residence time). Because of the circulatory nature of the cross flow being centered at $\xi = 2/3$, a fluid particle located here moves straight down the channel. All other particles move in a helical path. Particles located at $\xi > 2/3$ will turn over on reaching the flight and move in the opposite direction. However, the time spent moving across the channel for the particle located at a position $\xi > 2/3$ is less than the time it takes for the particle to move across the lower portion of the channel. Hence, it is necessary to use an average axial velocity, $\bar{v}_l$, to calculate the residence time of a particle.

First, it is necessary to locate the position of the particle in the

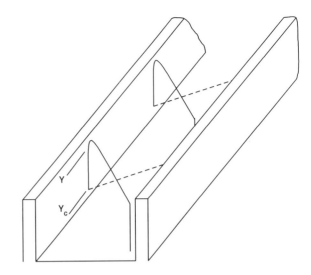

**Figure 8.28** Spiral path of a fluid element in the channel of a single-screw extruder.

lower part of the channel, $\xi_c$, which had a position $\xi$ in the upper portion of the channel (i.e., $\xi > 2/3$). Because the flow is circulatory in the $x$ direction, then the amount of fluid flowing between $\xi$ and 1.0 (the wall) per unit width in the upper part of the channel must be equal to the amount of fluid flowing per unit width in the opposite direction between 0 and $\xi_c$ in the lower portion of the channel. Mathematically, this is given by:

$$\int_0^{\xi_c} u_x d\xi + \int_\xi^1 u_x d\xi = 0, \tag{8.134}$$

which provides the relation between $\xi$ and $\xi_c$. Substituting Eq. 8.98 into Eq. 8.134 gives the following expression:

$$\xi^2 - \xi^3 = \xi_c^2 - \xi_c^3$$

$$0 \leqslant \xi_c \leqslant \frac{2}{3} \qquad \frac{2}{3} \leqslant \xi \leqslant 1. \tag{8.135}$$

A more convenient form for finding $\xi_c$ in terms of $\xi$ is:

$$\xi_c = \frac{1 - \xi + \sqrt{1 + 2\xi - 3\xi^2}}{2}. \tag{8.136}$$

The average axial velocity, $\bar{v}_l$, of a fluid particle that alternates its position between $\xi$ and $\xi_c$ is given by the following equation:

$$\bar{v}_l = v_l(\xi)t_f + v_l(\xi_c)(1 - t_f), \tag{8.137}$$

where $t_f$ is the fraction of time spent by a fluid particle in the upper portion of the channel, which is given by:

$$t_f = \frac{1}{1 + u_x(\xi)/u_x(\xi_c)}. \tag{8.138}$$

The residence time of a particle is equal to the length of the extruder (or metering section), $L$, divided by the $\bar{v}_l$. Substituting $u_x$ from Eq. 8.98 and $u_l$ from Eq. 8.133 into Eqs. 8.137 and 8.138 the residence time, $t$, becomes:

$$t = \left[\frac{L}{3V_b \sin\phi \cos\phi(1+\Phi')}\right]\left(\frac{3\xi - 1 + 3\sqrt{1 + 2\xi - 3\xi^2}}{\xi[1 - \xi + \sqrt{1 + 2\xi - 3\xi^2}]}\right). \quad (8.139)$$

Equation 8.139 represents the distribution of residence times of fluid particles as a function of their initial location in the channel. The minimum residence time occurs for particles located at $\xi = 2/3$, and $t$ increases as one moves toward the barrel or screw surfaces.

We next would like to know what fraction of the fluid leaving the extruder has a certain residence time. This is given by the residence time distribution function, $f(t)$, which was defined in Chapter 6. By definition, $f(t)dt$ is the fraction of material leaving the extruder with a residence time between $t$ and $t + dt$, which is:

$$f(t)dt = \frac{dQ + dQ_c}{Q}, \quad (8.140)$$

where $dQ$ is the fraction of flow between $\xi$ and $\xi + d\xi$ and $dQ_c$ is the fraction of flow btween $\xi_c$ and $\xi_c + d\xi_c$ associated with the residence time $t$. $dQ$ and $dQ_c$ are given, respectively, by:

$$dQ = WHV_{bz}\xi(1 + 3\Phi' - 3\xi\Phi')d\xi \quad (8.141)$$

$$dQ_c = WHV_{bz}\xi_c(1 + 3\Phi' - 3\xi_c\Phi')|d\xi_c|, \quad (8.142)$$

where the absolute value sign in Eq. 8.142 was introduced to account for the change of direction of $d\xi$ with respect to $d\xi_c$. From Eq. 8.136 we find a relation between $d\xi_c$ and $d\xi$:

$$d\xi_c = \frac{1 - 3\xi - \sqrt{1 + 2\xi - 3\xi^2}}{2\sqrt{1 + 2\xi - 3\xi^2}}d\xi. \quad (8.143)$$

Substituting Eqs. 8.136 and 8.143 into Eq. 8.140, and taking $d\xi_c = -d\xi$ leads to the following expression for $f(t)dt$:

$$f(t)dt = \frac{3\xi}{\sqrt{1 + 2\xi - 3\xi^2}}(1 - \xi + \sqrt{1 + 2\xi - 3\xi^2})d\xi. \quad (8.144)$$

Using Eq. 8.139, Eq. 8.144 can be written in the following form:

$$f(t)dt = \frac{9}{2}V_b \sin\phi \cos\phi(1 + \Phi')$$
$$\times \left[\frac{\xi^3(\xi - 1 - \sqrt{1 + 2\xi - 3\xi^2})}{(6\xi^2 - 4\xi - 1)\sqrt{1 + 2\xi - 3\xi^2 + 3\xi - 1}}\right]dt. \quad (8.145)$$

For simple geometries such as pipe flow discussed in Section 6.4 it was possible to solve for $\xi$ in terms of $t$ which is not possible here by using Eq. 8.139. Equations 8.145 and 8.139 must be solved together to calculate $f(t)$. What these two equations do indicate, however, is that $f(t)$ depends only on the group $V_b(1 + \Phi')\sin\phi\cos\phi/L$.

There are two residence time values of specific interest, and these are the mean residence time, $\bar{t}$, and the shortest residence time, $t_0$. The mean residence time was defined in Eq. 6.117 and for the single-screw extruder (i.e., metering section or melt pump) is:

$$\bar{t} = 2L/V_b \sin\phi \cos\phi(1 + \Phi'). \quad (8.146)$$

$\bar{t}$ is also given by the volume of the channel ($WHL/\sin\phi$) divided by the volumetric flow rate. The minimum residence time, $t_0$, is obtained from Eq. 8.139 with $\xi = 2/3$ and is equal to $3\bar{t}/4$.

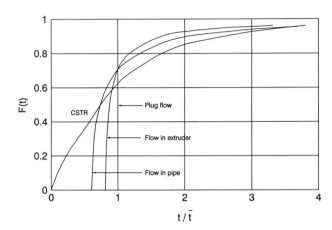

**Figure 8.29** Cumulative residence time distribution function versus reduced time for flow in an extruder, plug flow, flow of a Newtonian fluid in a pipe, and a continuously stirred tank vessel (CST).

The cumulative RTD function, $F(t)$, is found by integrating Eq. 8.144 from $\xi = 2/3$ (this is $t_0$) to $\xi$ (this is $t$):

$$F(t) = F(\xi) = \frac{1}{2}[3\xi^2 - 1 + (\xi - 1)\sqrt{1 + 2\xi - 3\xi^2}]. \quad (8.147)$$

$F(t)$ is plotted versus reduced time, $t/\bar{t}$, in Fig. 8.29, where we also compare $F(t)$ against values for plug flow, laminar flow in a pipe, and a continuous stirred tank reactor. $F(t)$ is seen to lie between plug and laminar flow in pipes. $F(t)$ for the extruder is rather narrow with no long tails.

We next obtain the strain distribution in the metering section of the extruder following the work of Pinto and Tadmor (1970). The shear rate in the upper portion of the channel is:

$$\dot{\gamma} = [\dot{\gamma}_{yx}(\xi) + \dot{\gamma}_{yz}(\xi)]^{1/2}. \quad (8.148)$$

Using Equations 8.95 and 8.96, we find $\dot{\gamma}_{yx}$ and $\dot{\gamma}_{yz}$ to be:

$$\dot{\gamma}_{yx} = 2V_{bx}(1 - 3\xi)/H \quad (8.149)$$

$$\dot{\gamma}_{yz} = V_{bz}[1 + 3\Phi'(1 - 2\xi)]/H. \quad (8.150)$$

In the upper portion of the channel we find:

$$\dot{\gamma} = V_b R(\xi)/H, \quad (8.151)$$

where

$$R(\xi) = [4(1 - 3\xi)^2 \sin^2\phi + (1 + 3\Phi' - 6\xi\Phi')^2 \cos^2\phi]^{1/2}, \quad (8.152)$$

and in the lower portion of the channel it will be:

$$\dot{\gamma}_c = V_b R(\xi_c)/H, \quad (8.153)$$

where

$$R(\xi_c) = [4(1 - 3\xi_c)^2 \sin^2\phi - (1 + 3\Phi' - 6\Phi')^2 \cos^2\phi]^{1/2}. \quad (8.154)$$

The average total strain of a fluid particle alternating between the two portions of the channel is:

$$\bar{\gamma} = \dot{\gamma}t_f t + \dot{\gamma}_c(1 - t_f)t. \quad (8.155)$$

Combining Eqs. 8.139, 8.151, 8.153, and 8.155 leads to:

$$\bar{\gamma} = \frac{1}{3}\left(\frac{L}{H}\right)\left(\frac{1}{1+\Phi'}\right)\left[\frac{2F(\xi,\Phi')}{\cos\phi} + \frac{G(\xi_c,\Phi')}{\sin\phi}\right], \qquad (8.156)$$

where

$$F(\xi,\Phi') = \frac{t_f\left[(1-3\xi)^2 + \dfrac{\cotan^2\phi}{4}(1+3\Phi'-6\xi\Phi')^2\right]^{1/2}}{\xi_c(1-\xi_c)+t_f(\xi-\xi_c)(1-\xi-\xi_c)} \qquad (8.157)$$

and

$$G(\xi,\Phi') = \frac{(1-t_f)[4(1-3\xi_c)^2\tan^2\phi+(1+3\Phi'-6\xi\Phi')^2]^{1/2}}{\xi_c(1-\xi_c)+t_f(\xi-\xi_c)(1-\xi-\xi_c)}. \qquad (8.158)$$

We are now in a position to discuss mixing in the single-screw extruder. The weighted average total strain (WATS) has been proposed by Pinto and Tadmor (1970) to be a measure of mixing in the extruder. The WATS was defined in Eq. 6.141, and its calculation requires both $f(t)$ and the strain distribution (Eq. 8.158). The WATS is a measure of the total deformation experienced by the material leaving the extruder. It is a single number for a given extruder set at specific operating conditions, and it gives a quantitative measure of the quality of mixing.

It is apparent that the WATS depends on three parameters: $L/H$, the helix angle, $\phi$, and the pressure-to-drag-flow ratio, $\Phi'$. Calculations by Pinto and Tadmor (1970) showed that over practical ranges of values for $\phi$ and $\Phi'$, the WATS was not very sensitive to these parameters. Hence, in the metering section of an extruder the most important parameter affecting mixing is the $L/H$ ratio.

Bigg and Midddleman (1974) carried out a similar analysis for non-Newtonian fluids using the power-law model. As there are no analytical results available, we only summarize a few of the most pertinent findings. Fig 8.30 presents the cumulative distribution function, $F(t)$, versus the reduced time for various values of the power-law index, $n$, and for a fixed value of $\Phi' = -0.88$. Here it is seen that for values of $n \geqslant 0.4$ the shape of $F(t)$ is similar (i.e. the $F(t)$ curves for a Newtonian fluid and a power-law fluid with a value of $n \geqslant 0.4$ are similar for $\Phi' = -0.88$). For a highly shear-thinning fluid, $F(t)$ becomes

broader, approaching that of the curve for the well-mixed state. The shape of the $F(t)$ curve was found to be sensitive to the value of $\Phi'$.

The analyses carried out by Pinto and Tadmor (1970) for Newtonian fluids and by Bigg and Middleman (1974) for power-law fluids were for the metering section of a single-screw extruder or for the single-screw melt pump only. More frequently plasticating extruders are used, and mixing probably starts in the melting zone. There are apparently no models available at the time of writing that deal with mixing in the melting zone.

Mixing in twin-screw extruders is much more difficult to estimate because of the complex flow patterns. For intermeshing corotating twin-screw extruders Montes and White (1991) have developed a model to describe mixing. The model of mixing in the screw elements follows the work of Pinto and Tadmor (1970) for a single-screw extruder. The strains generated in each screw element are added together to get the total strain. The major part of the mixing occurs in the kneading blocks. To analyze mixing here the velocity profiles are neeeded, and these can only be calculated numerically for power-law fluids. There is no experimental evidence to support the calculations of Montes and White (1991), and hence it is not known whether their approach is adequate to describe mixing in intermeshing corotating twin-screw extruders. There is apparently no adequate model to describe mixing in intermeshing counterrotating extruders either. Potente and Schultheis (1989) described $F(t)$ in this type of extruder by a distribution function (a double Weibull distribution) that required fitting to experimental data to obtain the constants in the model. The function was based on knowing the minimum and mean residence times. $t_{0i}$ per chamber is given by

$$t_{0i} = \frac{L_i}{V_l} = \frac{L_i}{N\pi D_b \tan\phi_i}, \qquad (8.159)$$

where $L_i$ is the length of the C-shaped chamber and $V_1$ is the axial velocity. $t_0$ results from adding the values of $t_{0i}$ in each chamber. The mean residence time, $\bar{t}$, is just the ratio $V/Q$, where $V$ is the volume of the screw channels. The only certain conclusion is that in general $F(t)$ for twin-screw extruders is narrower than that for single-screw extruders.

Before leaving this section, we offer a few comments about static mixers. To improve distributive mixing and to provide melt streams with more uniform temperature distribution, especially for single-screw extruders, static mixers are placed between the end of the extruder and the die. Static mixers consist of a pipe with nonmoving elements that lead to a rearrangement and distribution of fluid elements. For example, a Kenics static mixer is shown in Fig. 8.31. It consists of helical elements turned 90° relative to each other with each element having alternating pitch. Fluid entering the element near the center is distributed to the wall, and fluid entering near the wall is distributed toward the center.

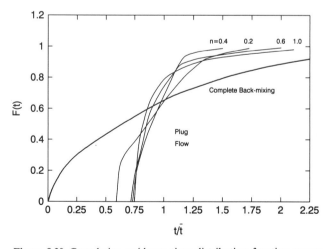

**Figure 8.30** Cumulative residence time distribution function versus reduced time in an extruder for fluids with various values of the power-law index and compared to values for plug flow and complete backmixing continuous stirred tank reactor. (Bigg and Middlemann 1974.)

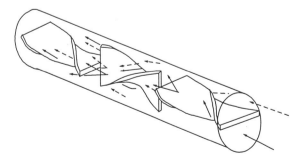

**Figure 8.31** Flow in a Kenics static mixer. The mixer consists of a pipe with helical elements of alternating reverse pitches.

Element no.

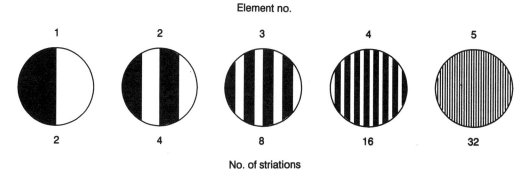

No. of striations

**Figure 8.32** Illustration of how fluid streams are split and separated in a Kenics static mixer. After passing through five elements there are 32 layers of fluid.

A stream entering the first element is split into two streams. The two streams are split into four streams and so on. This is illustrated in Fig. 8.32, where two streams enter the first element, and then they are split into four streams in the second element. In general, the number of striations, $N_\delta$, at the exit of a pipe containing $N$ elements is:

$$N_\delta = 2^N. \tag{8.160}$$

There are quite a few different types of static mixers (see Rauwendaal, 1986), but they all serve the same function.

The flow patterns in static mixers are very complex, and only finite element methods can be used to analyze the details of these flows. The shear rates generated in these devices are rather low (say of the order of $10\,\mathrm{s}^{-1}$), and hence the pressure drop across the static mixer can be estimated from the expression for flow of a Newtonian fluid through a tube with a correction factor for the given geometry as:

$$\Delta P = (4/\pi) N_s \mu Q L / D, \tag{8.161}$$

where $N_s = 220$ for the Kenics static mixer (see Rauwendaal, 1986 for this and other values) and $L$ and $D$ are the total length and diameter of the pipe containing the mixing elements.

### 8.5.2. DEVOLATILIZATION IN EXTRUDERS

The removal of volatiles such as water or residual monomer can be carried out in both single- and twin-screw extruders. In the case of single-screw extruders, specialized screw designs are required, whereas the flexible nature of twin-screw extruders' design allows the appropriate elements to be added. In the case of the single-screw extruder, either an extra long extruder is required with a design as shown in Fig. 8.33 or two extruders in a cascade arrangement are used. The general screw design shown in Fig. 8.33 consists of a standard plasticating screw design followed by a section possessing a deep channel and a vacuum port and then a metering section. Because the pumping capacity of the section with the deep channel is less than that of the last metering section (section 3 in Fig. 8.33), the channel will not be full. For a single-flighted screw as shown in Fig. 8.33 there are a rotating melt pool and a polymer melt film from which volatiles can be removed. The driving force for the diffusion of volatiles from the pool and film is a concentration gradient set up by reducing the surface concentration of the volatile to that of the equilibrium concentration, $C_e$, which is determined by Henry's law. There is also the possibility that additional mass transfer occurs by the generation of bubbles in the rotating melt pool. In the case of twin screws starve-feeding of the screw elements is accomplished by the addition of paddle elements or reverse screw

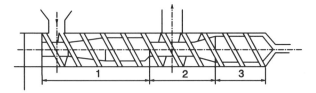

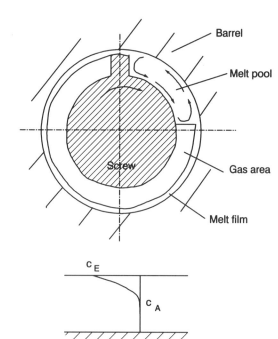

**Figure 8.33** Degassing in a single-screw extruder. Zone 1 consists of a typical plasticating screw section; zone 2 is a deep channel section with a vacuum port; zone 3 is a metering section. The channels in zone 2 are partially full to facilitate degassing.

elements. Because of certain geometric limits, only double-flighted screw elements can be used in twin-screw systems. When this is done, there are three melt pools generated, as shown in Fig. 8.34. To generate a similar number of melt pools in a single-screw extruder a triple-flighted screw is required as shown in Fig. 8.34.b. In principle the process of removing volatiles is similar in both single- and twin-screw extruders.

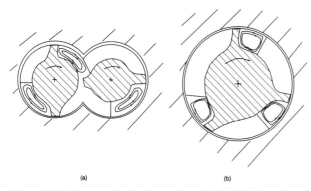

**Figure 8.34** Cross-sectional views of a double-flighted twin-screw extruder (a) and a triple-flighted single-screw extruder (b). Each system exhibits three melt pools that are available for degassing.

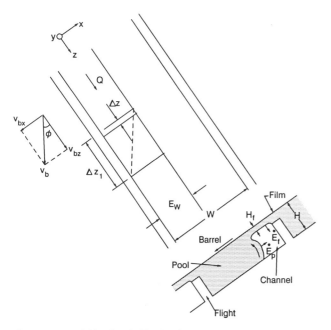

**Figure 8.35** Model for devolatilization in extruders. Gas removal occurs from the melt film on the barrel wall and from the rotating melt pool.

The few models available concerned with devolatilization (DV) in single screw extruders are based on diffusion theory (Biesenberger & Sebastian, 1983). Available experimental data suggest that the process is more rapid than can be accounted for by diffusion alone. However, a summary of the model is useful, because at least the most important variables are brought out.

It is postulated that DV in single-screw extruders occurs by two mechanisms (Roberts, 1970). One is evaporation from the bulk polymer melt flowing in the partially filled screw channel, and the other is evaporation from the film that is generated by the clearance between the flight and barrel as the flight wipes over the barrel surface. The melt in the channel is viewed as an evaporating melt pool that rotates as it flows owing to the angle of drag, $\phi$. These mechanisms are shown in Fig. 8.35.

The goal of a model for DV is to calculate the concentration of the volatile, $C_A$, as a function of distance down the channel. Referring to Fig. 8.35, a mass balance is performed on the element of thickness $\Delta z$. The contributions of mass to this element are from the transport

of $A$ by bulk flow, the loss of $A$ by evaporation at the film surface, $\dot{E}_f(z)$, and by evaporation at the surface of the pool, $\dot{E}_p(z)$, where the dot denotes the rate of evaporation. Mathematically, this mass balance is:

$$0 = fWH[N_A|_z - N_A|_{z+\Delta z}] - \dot{E}_p(z) - \dot{E}_f(z). \tag{8.162}$$

$\dot{E}_p(z)$ is obtained by methods described in Chapter 4 (in particular the derivation is similar to that in Ex. 4.5 but with gas diffusing out of the film) and Eq. 8.162 becomes:

$$0 = fWH[N_A|_z - N_A|_{z+\Delta z}] - V_{bx}H_f\Delta z[C_A - C_A']$$
$$- \left(\frac{4D_A}{\pi t_p}\right)^{1/2}[C_A - C_{Ae}], \tag{8.163}$$

where $C_{Ae}$ is the equilibrium concentration at the vapor–melt interface, $H_f$ is the film thickness, and $t_p$ is the exposure time of the melt–pool surface, which is limited by the circular motion of the pool. Roberts (1970) estimated this exposure time as:

$$t_p = H/V_{bx}. \tag{8.164}$$

The evaporation rate of $A$ from the film is obtained from a mass balance on the film, which is:

$$\dot{E}_f(z) = V_{bx}H_f\Delta z[C_A(z) - C_A'(z)], \tag{8.165}$$

where $C_A'$ is the concentration of $A$ in the melt reentering the melt pool having left the melt pool at a distance of $\Delta z_1$ downstream (see Fig. 8.35). Dividing Eq. 8.163 by $\Delta z$ and taking the limit as $\Delta z \to 0$ we obtain:

$$fWH\frac{dn_{Az}}{dz} = V_{bx}H_f[C_A(z) - C_A'(z)] + \left(\frac{4D_A}{\pi t_p}\right)^{1/2}H[C_A(z) - C_{Ae}]. \tag{8.166}$$

We next must relate $C_A'$ to the concentration at $z + \Delta z_1$, $C_A(z + \Delta z_1)$, where this is the concentration of $A$ in the melt leaving the melt pool at a distance $z + \Delta z_1$ downstream minus the amount of the volatile lost through evaporation during the exposure time $t_f = (1 - f)/N$, where $N$ is the screw speed. This relation is obtained through the staged efficiency of the diffusing film, $X_f$:

$$X_f = \frac{C_A(z + \Delta z_1) - C_A'(z)}{C_A(z + \Delta z_1) - C_{Ae}}, \tag{8.167}$$

Following Roberts (1970), $C_A(z + \Delta z_1)$ can be related to $C_A(z)$ by means of a Taylor series in which only terms through the second order are kept:

$$C_A(z + \Delta z_1) = C_A(z) + \frac{dC_A(z)}{dz}\Delta z_1 + \frac{d^2C_A(z)}{dz^2}\frac{\Delta z_1^2}{2} + \cdots, \tag{8.168}$$

After eliminating $C_A(z + \Delta z_1)$ from Eqs. 8.167 and 8.168, and substituting the result for $C_A'$ into the mass balance (Eq. 8.166), and replacing $N_{Az}$ by $C_A v_z$ we obtain:

$$V_{bx}H_f(1 - X_f)\frac{\Delta z_1^2}{2}\frac{d^2C_A}{dz^2} + [V_{bx}H_f(1 - X_f)\Delta z_1 - Q]\frac{dC_A}{dz}$$
$$- [X_f V_{bx}H_f + k_pH](C_A - C_{Ae}) = 0. \tag{8.169}$$

Equation 8.169 can be written in dimensionless form as:

$$(1/\text{Pe})\frac{d^2 C_A^*}{d\xi^2} - \frac{dC_A^*}{d\xi} - E_x C_A^* = 0, \qquad (8.170)$$

where

$$\xi = z/L_e \qquad \Delta z_1^* = \Delta z_1/L_e \qquad (8.171)$$

and

$$\text{Pe} = \left(\frac{2}{\Delta z_1^*}\right)\left[\frac{1}{n_f(1-X_f)\Delta z_1^*} - 1\right]. \qquad (8.172)$$

$$Ex = \frac{n_f X_f + k_p H L_e/Q}{1 - n_f(1-X_f)\Delta z_1^*}, \qquad (8.173)$$

and

$$n_f = \frac{V_{bx} H_f L_e}{Q} \qquad C_A^* = \frac{C_A - C_{Ae}}{C_{A0} - C_{Ae}}. \qquad (8.174)$$

$L_e$ is the unwound channel length of the devolatilization section.

The Peclet number, Pe, represents the effect of backmixing generated by the film reentering the melt pool. When $\text{Pe} \gg 1$, the effects of backmixing can be neglected. The effect of backmixing is directly determined by the magnitude of $\Delta z_1$. As can be seen in Fig 8.35, $\Delta z_1$ is directly related to the degree of fill, $f$, the channel width, $W$, and the helix angle, $\phi$:

$$\Delta z_1 = \frac{fW}{\tan\phi} = f\pi D_b \cos\phi. \qquad (8.175)$$

The second equality in Eq. 8.175 is correct for single-flighted screws and negligible flight widths. Hence, backmixing increases with the degree of fill and decreasing values of $\phi$. $Ex$ is the extraction number, and it is a measure of the overall devolatilization efficiency.

If the backmixing effect can be neglected, then Eq. 8.169 becomes:

$$\frac{dC_A^*}{d\xi} = Ex C_A^*. \qquad (8.176)$$

With $C_A^*(0) = 1$, the concentration profile becomes:

$$C_A^* = \exp[-Ex\xi], \qquad (8.177)$$

where $Ex$ reduces to $n_f X_f + k_p H L_e/Q$. Hence, the concentration of the volatile is an exponential function of distance.

Finally, the devolatilization efficiency of the machine, $X_T$, is a function of the individual stage efficiency, $X$, and of the extent of surface renewal. In the extruder, the extent of surface renewal is described by $n_f$ in Eq. 8.174. The film stage efficiency, $X_f$, is a function of the surface-to-volume ratio and the exposure time, $t_f$. In general $X_f$ is a function of $n_f$ and of the ratio $t_f/t_D$, where $t_D$ is the residence time in the devolatilization section of the extruder.

The model developed here is based on diffusion. However, in many cases, the length required to reduce the concentration of a volatile below some specified value is overpredicted. Hence, it is believed that the formation of bubbles or foam accelerates the devolatilization process. There is no model available at the time of writing this book, which deals with foam DV.

The process of DV in corotating twin-screw extruders is similar to that in the single-screw extruder. However, no detailed mathematical model for DV in these devices is available.

## 8.5.3. REACTIVE EXTRUSION

There are numerous examples of the use of extruders to carry out chemical reactions in polymeric systems (Xanthos, 1992). In general, two types of reactions are carried out in extruders: polymerization or depolymerization, and polymer modification. In this section we illustrate the approach to modeling reactive extrusion by considering the stepwise polycondensation reaction of two species $A$ and $B$ in a single-screw extruder.

The starting point for the analysis is the flat plate model of the extruder shown in Fig. 8.18. It is assumed that steady-state conditions exist and that material properties such as viscosity, thermal conductivity, and heat capacity do not depend significantly on temperature. It is postulated that:

$$v_z = v_z(y) \qquad v_x = v_x(y) \qquad T = T(y,z) \qquad C_A = C_A(z). \qquad (8.178)$$

The equation of motion becomes:

$$\frac{\partial p}{\partial x} = -\frac{\partial}{\partial y}\left(\eta\frac{\partial v_x}{\partial y}\right) \qquad \frac{\partial p}{\partial z} = -\frac{\partial}{\partial y}\left(\eta\frac{\partial v_z}{\partial y}\right). \qquad (8.179)$$

The energy equation takes on the following form:

$$\rho\bar{C}_p v_z\frac{\partial T}{\partial z} = k\frac{\partial^2 T}{\partial y^2} + \eta\left[\left(\frac{\partial v_x}{\partial y}\right)^2 + \left(\frac{\partial v_z}{\partial y}\right)^2\right] + \dot{S}_r, \qquad (8.180)$$

where $\dot{S}_r$ is the energy generated per unit volume per unit time by the reacting system. The conservation of mass equation for species A is:

$$v_z\frac{dC_A}{dz} = R_A, \qquad (8.181)$$

where $R_A$ is the rate at which species $A$ disappears as the result of the reaction of $A$ and $B$.

The rate of reaction, $R_A$, is assumed to follow $n$th-order reaction kinetics given by:

$$R_A = -k_0\exp[-E/RT]C_A^n, \qquad (8.182)$$

where $k_0$ and $E$ are the reaction rate at the reference temperature and the activation energy, respectively. The heat generated by the reaction is given by:

$$\dot{S}_r = (-\Delta\bar{H}_A)R_A, \qquad (8.183)$$

where $\Delta\bar{H}_A$ is the heat of reaction.

As the two species react, the molecular weight of the product increases, leading to an increase in the viscosity of the bulk material. For the polycondensation reaction of species $A$ and $B$ the weight average molecular weight as a function of the degree of conversion, p, is (Macosko & Miller, 1976):

$$\bar{M}_w = \frac{[r(1+rC^2)M_{A0}^2 + (1+rC^2)M_{B0}^2 + 4rCM_{A0}M_{B0}]}{(rM_{A0} + M_{B0})(1-rC^2)}, \qquad (8.184)$$

where $C = (C_{A0} - C_A)/C_{A0}$, $r$ is the stoichiometric ratio of species $A$, and $B$, $M_{A0}$ and $M_{B0}$ are the initial molecular weights of species $A$ and $B$, respectively. As a result of the increase in $\bar{M}_w$, the viscosity of the melt is changing as it progresses down the channel. The zero shear viscosity increases with increasing $\bar{M}_w$ in the following manner, as described in Chapter 3:

$$\eta_0 = \eta_{CR}\bar{M}_w/M_{CR} \qquad \bar{M}_w \leqslant M_{CR} \tag{8.185}$$

$$\eta_0 = \eta_{CR}(\bar{M}_w/M_{CR})^{3.4} \qquad \bar{M}_w > M_{CR}, \tag{8.186}$$

where $M_{CR}$ is the critical entanglement molecular weight for the given polymer and $\eta_{CR}$ is the zero shear viscosity at $M_{CR}$.

The initial and boundary conditions required to solve Eqs. 8.179, 8.180, and 8.181 are similar to those used in Section 8.3 with the exception that the concentration of species $A$ at $z=0$ is $C_{A0}$. The solution of this set of equations requires numerical methods, but a direct finite difference solution is not possible because of the velocity field changing signs near the surface of the screw. To overcome this problem the space-fixed coordinate system is replaced by one that travels along the streamlines. Neglecting the viscous dissipation terms, Eqs. 8.180 and 8.181 become, respectively:

$$\rho\bar{C}_p\frac{\partial T}{\partial t^*} = k\frac{\partial^2 T}{\partial y^2} + \dot{S}_r \tag{8.187}$$

$$\frac{\partial C_A}{\partial t^*} = R_A, \tag{8.188}$$

where $t^*$ is the residence time of a fluid element along a streamline. The numerical algorithm required to solve these equations is described in a paper by Tzoganakis and co-workers (1988).

## 8.6. SOLUTION TO DESIGN PROBLEM 7

The solution will be carried out by two methods. Because of the modular nature of twin-screw extruders, experiments were carried out in advance to determine what length of devolatilization section was required to reduce the level of MMA to 0.1%, which corresponds to a fractional separation, $F_s$, of about 0.85. Because the single-screw extruder is cheaper to build, it is desirable to determine the length of the DV section and the processing conditions (i.e., $\rho Q$, $N$, and $f$) required to accomplish the same reduction of MMA in PMMA as done in the SWCOR. The first approach is based on dimensional analysis. In the second approach, we use the penetration or diffusion theory summarized in Section 8.5.2.

### 8.6.1. DIMENSIONAL ANALYSIS

The starting point is to make sure that sufficient vacuum is available, so the equilibrium weight fraction, $w_e$ is less than the desired final weight fraction. From Henry's law and the data given in Table 8.2, we calculate $w_e$ at 250°C to be:

$$w_e = P/S = \frac{133.2\,\text{Pa}}{164 \times 10^5\,\text{Pa}} = 81 \times 10^{-6} = 81\,\text{ppm},$$

and at 200°C $w_e = 208\,\text{ppm}$. Hence, sufficient vacuum is available to reduce the level of MMA below 1000 ppm.

Dimensional analysis requires that both extruders be geometrically and dynamically similar. To ensure geometric similarity, it is necessary that the length to diameter of the devolatilization (DV) sections be the same:

$$L_{bl}/D_{bl} = L_{b2}/D_{b2}, \tag{8.189}$$

where the subscripts 1 and 2 refer to the single- and twin-screw systems,

respectively. Furthermore, because double-flighted screw elements are used in the twin-screw extruder, there are three independent flow channels partially filled each with melt pools. Hence, the single screw must be made triple-flighted to produce the same number of melt pools. Dynamic similarity is much more difficult to determine, because it is not clear what mechanisms are important in devolatilization. Based on experimental studies, Biesenberger and coworkers (1990) found that mass transfer in the DV process in extruders occurs primarily by flash evaporation in which foam formation and rupture are enhanced by rotational motion of the melt in the partially filled channel. Based on their observations, they proposed that the pool size and rotational speed are the factors on which dynamic similiarity should be based. With this they proposed that the main dimensionless group should be the ratio:

$$\bar{t}/t_p, \tag{8.190}$$

where $\bar{t}$ is the mean residence time and $t_p$ is the time of rotation of the pool. Hence, the fractional separation, $F_s$, should be expressed as:

$$F_s = F_s(L_b/D_b, \bar{t}/t_p). \tag{8.191}$$

The ratio $\bar{t}/t_p$ represents the number of devolatilization stages ($N_s$) available during the extrusion process.

We first use the condition of geometric similarity to find the length of the DV section:

$$L_{b1} = D_{b1}(L_{b2}/D_{b2}) = 30(300/34) = 265.0\,\text{mm}.$$

The calculation of $\bar{t}_1/t_{p1}$ and $\bar{t}_2/t_{p2}$ requires some additional manipulations for the single-screw extruder. The mean residence time in the DV section of the single-screw extruder is:

$$\bar{t}_1 = L_{e1}/\langle v_{z1}\rangle, \tag{8.192}$$

where $L_{e1}$ is the length of the unwound screw channel and is given by:

$$L_{e1} = L_{b1}/\sin\phi_1. \tag{8.193}$$

$\langle v_{z1}\rangle$ is the average velocity in the down-channel direction and is approximately

$$\langle v_{z1}\rangle \approx (\pi D_b N/2)\cos\phi \tag{8.194}$$

because the velocity profile in the down channel direction is not known exactly in the partially filled channel. $t_p$ is approximated by:

$$t_{p1} = \frac{(f_1 A_1)^{1/2}}{\langle v_{x1}\rangle} = \frac{(f_1 A_1)^{1/2}}{(\pi D_{b1} N_1/2)\sin\phi_1}, \tag{8.195}$$

where $A_1$ is the cross-sectional area of the screw channel and $f$ is the degree of fill. Hence, for the single-screw extruder:

$$N_{s1} = \bar{t}_1/t_{p1} = \frac{L_{b1}}{(f_1 A_1)^{1/2}\cos\phi_1}.$$

For the twin-screw extruder the calculations of $N_{s2}$ are more complicated because of the complex geometry. The flow path for PMMA is shown in Fig. 8.2. For the single-screw extruder to cover a circumferential distance of $\pi D_{b1}$ one must travel a distance of $L_{s1}/\tan\phi_1$ along the helical path (i.e., $\tan\phi_1 \approx L_{s1}/\pi D_{b1}$). For the double-flighted twin screw in a plane orthogonal to the screw axis, one must travel a circumferential distance, $C$, given by:

$$C = 2(2\pi - 2\alpha_i)(D_{b2}/2),$$

where $\alpha_i$ is the angle of intermesh defined in Section 8.4.1. An apparent helix angle, $\phi_2'$, can be defined as:

$$\tan \phi_2' = \left(\frac{3}{2}\right)\frac{L_{s2}}{C},$$

and the unwound channel length, $L_{e2}$, would be:

$$L_{e2} = L_{b2}/\sin \phi_2'.$$

$\alpha_i$ is related to $D_{b2}$ and $C_L$ by means of Eq. 8.107 and is 28.1° (see Fig. 8.21). The mean residence time in the twin-screw extruder, $\bar{t}_2$, is:

$$\bar{t}_2 = \frac{L_{e2}}{\langle v_{z2} \rangle} \approx \frac{L_e}{(\pi D_{b2} N/2)\cos \phi'}, \tag{8.196}$$

where $\langle v_{z2} \rangle$ has been approximated by $(\pi D_{b2} N/2)\cos \phi'$ in the absence of an accurate model for flow in the partially filled screw channels. The time of rotation of the melt pool, $t_{p2}$, is:

$$t_{p2} = \frac{[(f_2 A_0/p_2)\sin \phi_2']^{1/2}}{\langle v_{x2} \rangle} = \frac{[(f_2 A_0/p_2)\sin \phi_2']^{1/2}}{(\pi D_b N/2)\sin \phi_2'}, \tag{8.197}$$

where $A_0$ is defined in Eq. 8.114. Finally, the degree of fill, $f$, is given by:

$$f = Q/Q_D. \tag{8.198}$$

We are now in a position to calculate the remaining unknowns for the single-screw extruder, which are $N$ and $\rho Q$. Using the data for the twin-screw system given in Table 8.1 we can calculate $N_{s2}$ ($= \bar{t}_2/t_{p2}$). The unwound channel lengths are:

$$L_{e1} = \frac{265}{\sin 17.65°} = 874 \text{ mm}$$

$$L_{e2} = (270 \text{ mm}/\sin 20.52°) + (30/\sin 14.01°) = 894 \text{ mm}.$$

At $N = 90$ rpm for the twin-screw extruder,

$$\bar{t}_2 = \frac{770}{(\pi \cdot 34 \cdot 90/60/2)(\cos 20.52°)} + \frac{124}{(80.1)\cos 14.01°} = 11.9 \text{ s}.$$

Using Eqs. 8.197 and 8.114 and the data given in Table 8.1, $t_{p2}$ is found to be 0.079 s. Now using dynamic similarity (i.e., $N_{s1} = N_{s2}$), we find that:

$$\frac{L_{e1}\tan \phi_1}{(f_1 A_1)^{1/2}} = 11.9/0.079 = 150.6.$$

With $A_1 = WH = 58.71 \text{ mm}$, we find $f_1 = 0.058$. We now use Eq. 8.198 to calculate $Q_1$ assuming that $N_1 = 90$ rpm:

$$\begin{aligned}
Q_1 &= f_1 Q_{D1} = (0.058)P_1\langle v_{z1} \rangle A_1 \\
&= 0.058(3)(58.71)(67.4) \\
&= 688.1 \text{ mm}^3/\text{s} = 2.62 \text{ kg/hr}
\end{aligned}$$

Hence, in order to obtain a fractional separation of 0.85 using a single-screw extruder, we must use a triple-flighted screw with a length,

$L_{b1}$, of 265 mm. The processing conditions must be:

| Degree of fill | $f = 0.058$ |
|---|---|
| Screw RPM | $N = 90$ rpm |
| Mass flow rate | $\rho Q = 2.62$ kg/hr |

It should be added that Biesenberger and coworkers (1990) reported the following experimental results for obtaining $F_s = 0.86$ on a similar single-screw extruder with a triple-flighted screw of $D_b = 30$ mm:

| Temperature | $T = 230°C$ (rather than 250°C) |
|---|---|
| Degree of fill | $f = 0.092$ |
| Screw RPM | $N = 90$ |
| Mass flow rate | $\rho Q = 3.15$ kg/hr |
| Screw length | $L_b = 270$ mm |

The dimensional analysis approach with the choice of $\bar{t}/t_p$ as the dimensionless group for determining dynamic similarity seems to be reasonable.

### 8.6.2. DIFFUSION THEORY

Because we may not always have available data for carrying out dimensional analysis, it is worthwhile to see whether the penetration theory (diffusion model) described in Section 8.5.2 provides a reasonable design. In Eq. 8.170 we assume that $Pe > 1$, and hence we can use the expression given in Eq. 8.175, which is:

$$C_A^* = \exp[-Exz/L_e]. \tag{8.199}$$

We also assume that the staged efficiency of the film diffusion process is 1 (i.e., $X_p = 1$) as there is no way to obtain this quantity directly (with $X_f = 1$, $Pe \gg 1$). $Ex$ (the extraction number) is now:

$$Ex = n_f + (k_p H L_e)/Q, \tag{8.200}$$

where

$$n_f = \frac{V_{bx}H_f L_e}{Q} \qquad k_p = \left(\frac{4D_A}{\pi t_p}\right)^{1/2} \tag{8.201}$$

To calculate $n_f$ we must make assumptions about $H_f$ and $f$. We take $H_f$ to be similar to the magnitude of the flight clearance or $2.54 \times 10^{-2}$ mm. The screws are typically run with a degree of fill ($f$) in the range of 0.1 to 0.3. We take $f = 0.1$. $D_A$ for MMA in PMMA is about $1 \times 10^{-10}$ mm²/s (see References in Chapter 4). We now find $n_f$, $k_p$ and $Ex$:

$$n_f = \frac{\tan \phi H_f L_e}{p_f WH} = \frac{(0.318)(2.54 \times 10^{-2})}{(3)(0.1)(58.7)}L_e = 4.59 \times 10^{-4}L_e$$

$$k_p = (4 \times 10^{-10}V_{bx}/\pi H)^{1/2} = 2.21 \times 10^{-5}$$

$$\begin{aligned}
Ex &= 4.59 \times 10^{-4}L_e + \frac{(2.21 \times 10^{-5})(5.595)}{(1186)}L_e \\
&= 4.59 \times 10^{-4}L_e + 1.04 \times 10^{-7}L_e \\
&\approx 4.59 \times 10^{-4}L_e.
\end{aligned}$$

Substituting this value back into Eq. 8.199, we can now solve for the unwound channel length required to reduce the amount of MMA to 0.1%:

$$\ln(0.143) = -E_x z/L_e$$

$$z = \frac{-\ln(0.143)L_e}{E_x} = \frac{1.944 L_e}{4.59 \times 10^{-4} L_e} = 423$$

The barrel length, $L_b$. is:

$$L_b = L_e \sin\phi = (4076)(0.303) = 1236 \text{ mm}.$$

This value is about 4.6 times the value estimated by means of dimensional analysis and reported by Biesenberger and co-workers (1990). Hence, the theory based on diffusion over estimates the length required to reduce the level of MMA to 0.1%. This is true in spite of the fact that in this case most of the volatiles are predicted to be removed from the melt film, not the pool.

## PROBLEMS
### A. Applications

**8A.1.** Solids Conveying of Nylon in a Single-Screw Extruder
The solids-conveying section of a single-screw extruder with a 5.08 cm barrel diameter consists of a screw with a 5.08 cm lead, a screw diameter of 5.06 cm, a 3.49 cm root diameter, and a flight width of 0.508 cm. The bulk density of the nylon pellets is 475 kg/m$^3$, and the friction coeficient between nylon and steel is 0.25 (assume this value for both the barrel and screw). Assuming at first no pressure rise, calculate the solids-conveying rate (kg/rev) for the following conditions: (a) no friction between the screw and the solids, (b) no friction between the solids and screw flights, (c) no friction between the solids and the trailing flight, (d) friction on all contacting surfaces, (e) compare your results with the experimental value of 0.0149 kg/rev reported by Darnell and Mol (1956).

**8A.2.** Solids Conveying with a Pressure Rise
Calculate the pressure at the end of the solids-conveying zone and the power consumption in Pr. 8A.1 when the pressure at the base of the hopper is calculated to be $3.0 \times 10^3$ Pa and the solid-conveying zone is one turn of the screw. Assume that $f_s = 0.8 f_b$ and that only the friction between the solids and the screw surface and the barrel is important. The mass flow rate is 0.0149 kg/rev.

**8A.3.** Scale-Up of Solids-Conveying Section
The optimum channel depth for the solids-conveying zone of a single-screw extruder with a barrel diameter of 5.0 cm was found to be 0.6 cm. At 100 rpm the mass flow rate of nylon was found to be 10.45 g/s and $P_2/P_1 = 100$. $f_s = 0.3$, and $f_b = 0.5$. Determine the screw rpm ($N$), the channel depth ($H$), and the mass flow rate ($G$) in scaling up to a single screw-extruder with $D_b = 11.4$ cm.

**8A.4.** Pressure at the Base of a Silo
A silo 7.0 ft in diameter and 40 ft high contains LDPE pellets with a bulk density of 40 lb/ft$^3$. Assuming the silo is full (i.e., it contains 61,575 lb of pellets), determine the lateral and vertical pressures at the base of the silo if the coefficient of friction, $f'_w$, is 0.2 and the angle of repose, $\delta$ is 45°.

**8A.5.** Delay Zone Length for LDPE
Using Eq. 8.36 and the data in Ex. 8.2, estimate the number of turns in the delay zone for LDPE. Use the rheological data given in Appendix A for NPE 953 (Table A.1).

**8A.6.** Solid Bed Profile in a Single-Screw Plasticating Extruder
Determine the solid bed profile and the length of screw channel required

for melting of LDPE in a 2.5 in. diameter screw extruder with a single-flighted square-pitched screw that has the following screw geometry:

> Feed section: 3.2 turns and a 0.5-in. channel depth
> Compression section: 12 turns with a linear taper
> Metering section: 12 turns and a 0.125-in. channel depth
> Flight width: 0.25 in. and a negligible flight clearance

and the following operating conditions:

> Screw rpm = 82
> Barrel temperature: 150°C
> $G = 120$ lb/hr

Assume that melting starts one turn before the end of the feed section. The physical property data can be found in Chapter 5, and the following relation for the viscosity is to be used:

$$\eta = 5.6 \times 10^4 \exp[-0.01(T-110)]\dot\gamma^{-0.655}, \qquad (8.202)$$

where $\eta$ is given in units of Pa·s and $T$ in °C.

**8A.7.** Shear Rate in the Melting Zone
Using the conditions given in the Pr. 8A.6, calculate the shear rate in the melt film, and compare this with the nominal shear rate in the metering section.

**8A.8.** Operating Conditions for a Single-Screw Extruder
The single-screw extruder described in Pr. 8A.6 is used to pump LDPE at 150°C through a pelletizing die consisting of 10 capillaries each having a diameter of 0.3175 cm and $L/D$ ratio of 20. Considering only the pressure drop across the capillaries (i.e., neglect the pressure drop across the manifold or distribution section and any filtration sections), determine the pressure rise in the metering section and the screw speed in rpm required to extrude 120 lb/hr of polymer (i.e., is 82 rpm sufficient?). Assume isothermal conditions and that there is neglible pressure rise in the melting section.

**8A.9.** Equilibrium Composition by Henry's Law
Molten polystyrene (PS) produced in a bulk polymerization process is flashed to a styrene content of 1.0% but still requires further devolatilization in an extruder at 260°C to a final content of 100 ppm. The Henry's law constant, $K_w$, for styrene in PS at 260°C is 50 atm (note: $P = K_w w_i$). Vacuum at 10 torr (1 torr = 1333.22 microbars) is available. Is this sufficient vacuum and, if not, what final composition can be attained theoretically?

**8A.10.** Geometry of Self-wiping Corotating Screw Elements
Screw elements in a corotating self-wiping twin-screw extruder have the following geometric characteristics: $D = 43$ mm, $L_s = 43$ mm, and $C_L = 39$ mm. Determine the angle of intermesh, the flight tip angle, the flight width (note, $e = \alpha_t L_s/2\pi$), and the open cross-sectional area, $A_0$, for both double- and triple-flighted screw elements. Compare the calculated values for the two cases.

**8A.11** Sizing of Reactive Zones in Extruders
Determine the relative lengths of the reaction zones for a single screw and corotating and counter-rotating twin screw extruders. The extrusion rate of the melt is to be 1000 kg/hr and three minutes of residence time are required for the reaction to occur. The following data is given for all three types of extruders: screw lead, $L_s = 0.3 D_b$; fractional degree of fill, $f = 0.5$; $H = 0.175 D_b$; $\rho = 900$ kg/m$^3$. The single screw is to be triple-flighted while the twin screw devices have

double-flighted screw elements. Assume that the $L_b/D_b$ of the reactive section is in the range of 10 to 30.

## B. Principles

**8B.1** Heat Generation at the Barrel–Solid Interface
Derive Eq. 8.30, which is the heat generated per unit of barrel surface because of friction between the solid bed and the barrel.

**8B.2.** Optimum Channel Depth in the Solids-Conveying Zone
Starting with the expression for the mass flow rate given in Eq. 8.23 find an expression for the optimum channel depth (i.e., maximize $G$ with respect to $H$). Although the value of $H$ must be determined numerically from the expression, specify what parameters $H$ depends on.

**8B.3.** Optimum Helix Angle in the Solids-Conveying Zone
Find an expression from which the optimum helix angle for which the mass flow rate in the solids-conveying section is a maximum can be determined.

**8B.4.** Solids Conveying in a Starve-Fed Single-Screw Extruder
Starve-feeding of a single-screw extruder is a process option whereby solid polymer is metered into the feed throat using a gravimetric or auger-type feeder at a rate less than the solids-conveying capacity of the screw. The resin compacts further downstream in the screw channel as a result of lower pressures, and temperatures are generated which are advantageous for processing thermally sensitive polymers. The degree of fill is given by:

$$f = G_s/G_0, \tag{8.203}$$

where $G_s$ and $G_0$ are the mass flow rates for the starve-fed and flood-fed cases, respectively, at the same screw rotational speed. The average bed width in the starve-fed case is given in terms of the channel width, $W$, as $\overline{W}_s = \overline{W}f$.

(a) If in the case of starve-feeding it is assumed that the forces acting on the trailing flight (i.e., $F_4$ and $F_8$ in Fig. 8.14) are zero, obtain an expression similar to that in Eq. 8.27 for the pressure rise.

(b) Calculate the pressure rise for the conditions given in Pr. 8A.2 when the mass flow rate is only 0.5 of that of the fully flooded screw.

**8B.5.** Maximum Shear Rate in the Metering Section
Obtain an expression for determining the maximum shear rate in the metering section of a single-screw extruder for the Newtonian fluid case.

**8B.6.** Pressure Profile in a Single-Screw Extruder
Show that the pressure profile along the barrel wall of a single-screw extruder for a Newtonian fluid is given by the following expression:

$$P - P_0 = \frac{6\mu\pi D_b N l}{H^2}\left[\frac{-Q_p}{Q_d}\cos^2\phi_b + \sin\phi_b\right], \tag{8.204}$$

where $l$ is the axial distance along the extruder barrel.

**8B.7** Power Input to a Single-Screw Extruder
For the isothermal Newtonian model, show that the power input through the screw is given by the following expression:

$$P_w = \frac{\mu\pi^2 N^2 D_b^2 W}{\sin\bar{\phi}H}\left[4 - 3\cos^2\phi_b\left(\frac{Q}{Q_b}\right)\right]. \tag{8.205}$$

Using this expression, find the helix angle $\phi_b$, which minimizes the power input (Take $\bar{\phi} = \phi_b$)

**8B.8** Optimizing the Design of a Screw
Using the screw characteristic given in Eq. 8.103 (isothermal Newtonian model), find the following: (a) optimum channel depth for maximum pressure rise for a given flow rate; (b) optimum channel depth and helix angle for maximum flow rate at constant screw speed (assume the flow rate through the die is given by $K\Delta P_D/\mu$); (c) channel depth for lowest screw speed for a given flow rate; and (d) determine the $Q/Q_d$ ratio in part (a).

**8B.9.** Scale-Up of a Single-Screw Extruder: Shear Rate Method
The shear rate at the barrel wall in the metering section of a single-screw extruder is approximately given by:

$$\dot{\gamma}_b = \frac{\pi D_b N}{H}. \tag{8.206}$$

If in scaling up from a small extruder ($D_1, L_1, N_1$, etc.) to a large extruder ($D_2, L_2, N_2$, etc.) dynamic similarity is maintained by keeping the shear rate constant (Chung, 1984), and geometric similarity is maintained by keeping the $L/D$ ratio constant, determine the channel width, channel depth, screw speed, flow rate, and power input through the screw for the large extruder in terms of the corresponding quantities of the small extruder and the ratio $D_2/D_1$ (e.g., show that $H_2 = H_1(D_2/D_1)^{0.5}$, $Q_2 = Q_1(D_2/D_1)^2$, etc.). The helix angle is taken to be the same for both extruders.

**8B.10.** Scale-Up of a Single-Screw Extruder: Mixing Method
As discussed in Section 6.4 and 8.5.1, mixing is taken to be a function of strain, which in the metering section of a single-screw extruder is $L/H$ (this follows from $\gamma = \dot{\gamma}t = (\pi D_b N/H)(L/\pi D_b N)$).

(a) Using constant strain as the scale-up criterion, determine $W_2, H_2, N_2, Q_2, P_{w2}$ in terms of the corresponding values for the smaller extruder and the ratio of $D_2/D_1$ (e.g., show that $W_2 = W_1(D_2/D_1)$, $Q_2 = Q_1(D_2/D_1)^3$, $P_{w2} = P_{w1}(D_2/D_1)^2$, etc.).

(b) If the output of a 5.0 cm diameter extruder is 100 kg/hr, determine the mass flow rate in a 15 cm diameter extruder where scale-up is based on a constant strain. Compare this value to that obtained when scale-up is based on a constant shear rate.

**8B.11.** Velocity Profiles in the Curved Channel Model (Pinto & Tadmor. 1970)
Show that the velocity profiles, $v_z(r)$ and $v_\theta(r)$, for flow of a Newtonian fluid in the channel of an extruder where curvature is included are:

$$v_z = \frac{A}{2}\frac{r}{1-\beta^2}\left[\ln\frac{r}{R_b} - \beta^2\ln\frac{r}{R_s} - \left(\frac{R_s}{r}\right)^2\ln\beta\right] + \frac{2\pi N}{1-\beta^2}\frac{r^2 - R_s^2}{r} \tag{8.207}$$

$$v_\theta = \frac{BR_b}{3}\left[\left(\frac{r}{R_b}\right)^2 + (1-\beta^2)\frac{\ln(r/R_b)}{\ln\beta} - 1\right], \tag{8.208}$$

where $\beta = R_s/R_b$ is the ratio between the inner and outer radii and $A$ and $B$ are given as:

$$A = 8\pi N\frac{K(\beta)}{G(\beta)} - \frac{4Q}{\pi R_b^3 G(\beta)\tan\phi_b} \tag{8.209}$$

$$B = \frac{8Q}{\pi R_b^3(\beta)}. \tag{8.210}$$

$K(\beta)$, $G(\beta)$, and $F(\beta)$ are given as:

$$K(\beta) = 1 + \frac{2\beta^2\ln\beta}{1-\beta} \tag{8.211}$$

$$G(\beta) = (1 - \beta^2)\left[ 1 - \left( \frac{2\beta \ln \beta}{1 - \beta^2} \right)^2 \right] \tag{8.212}$$

$$F(\beta) = (\beta^2 - 1)\left[ 1 + \beta^2 + \frac{1 + \beta^2}{\ln \beta} \right]. \tag{8.213}$$

Cylindrical coordinates are used where $z$ is in the axial direction of the extruder.

**8B.12.** Residence Time Distribution: Curved Channel Model (Pinto & Tadmor, 1970)

The residence time distribution (RTD) function for flow of a Newtonian fluid in a rectangular channel was developed in Section 8.5.1. Derive the RTD for the case in which curvature is included. A fluid particle located at position $r$ in the extruder channel will turn over when it hits the screw flight and will start moving in the opposite direction at a position $r_c$.

(a) Show that the analogous equation in cylindrical coordinates to Eq. 8.134 for finding $\xi$ and $\xi_c$ is:

$$\int_{R_s}^{r_c} v_z r dr d\theta + \int_r^{R_b} v_z r dr d\theta = \int_{R_s}^{r_c} v_\theta dr dz + \int_r^{R_b} v_\theta dr dz. \tag{8.214}$$

(b) Show that Eq. 8.214 can be rewritten after changing the integration limits as:

$$\int_r^{R_c} v_z r dr d\theta - \int_r^{r_c} v_\theta dr dz = -\int_{R_s}^{R_b} v_z r dr d\theta + \int_{R_s}^{R_b} v_\theta dr dz. \tag{8.215}$$

(c) Substitute in the velocity profiles for the Newtonian case, and use the relation $dz = R_b \tan \phi_b \, d\theta$ to obtain:

$$C_1(\rho_c^4 - \rho^4) + C_3(\rho_c^2 \ln \rho_c - \rho^2 \ln \rho)$$
$$+ C_4(\rho_c^2 - \rho^2) + C_5(\ln \rho_c - \ln \rho) = 0, \tag{8.216}$$

where $A$ and $B$ are given in Pr. 8B.11 and

$$\rho = r/R_b \qquad \rho_c = r_c/R_b \qquad \beta = R_s/R_b$$

$$C_1 = \frac{(1 - \beta^2)\beta}{2A \tan \phi_b} \qquad C_3 = 2\left[ \frac{C_1(1 - \beta^2)}{\ln \beta} \right]$$

$$C_2 = \frac{4\pi N}{A} \qquad C_4 = 1 - 2C_1 - C_1 \frac{1 - \beta^2}{\ln \beta} - \beta^2 - 2C_2 - 2\beta^2 \ln \beta$$

$$C_5 = 4\beta^2(\ln \beta + C_2).$$

## C. Numerical Problems

**8C.1.** Calculation of the Optimum Channel Depth for Solids Conveying

In problem 8B.2 an expression for the optimum channel depth was obtained. For an 11.4-cm diameter extruder running at 60 rpm with values of $f_b = 0.5$ and $f_s = 0.3$ and $P_2/P_1 = 200$, determine the optimum value of $H$ (i.e., find the value of $H$ that makes $G/\rho$ a maximum). It should be noted that in practice it is difficult to obtain accurate values of the friction coefficients, and hence one must use results of the nature asked for here only as a guideline.

**8C.2.** Nonisothermal Solids-Conveying Model (Tadmor & Broyer, 1972)

As a result of frictional heating at the barrel and plug interface the solid plug can prematurely melt, inhibiting adequate pressure buildup in the solids-conveying zone. The heat generation per unit area of barrel surface is given in Eq. 8.30. Develop a model for predicting the temperature in the solid bed and at the barrel surface, and then use the model to calculate the surface temperature for LDPE under the given conditions. Follow these steps:

(a) Perform an energy balance on the solid plug to obtain the following differential equation:

$$\rho_b \bar{C}_p b v_{pz} \frac{\partial T_p}{\partial z} = k_p \frac{\partial^2 T_p}{\partial y^2}. \tag{8.217}$$

(b) Assuming that $V_{pz}$ can be obtained from the isothermal theory for solids conveying, Eq. 8.21, show that Eq. 8.217 becomes:

$$\frac{\partial T_p}{\partial t} = \alpha_p \frac{\partial^2 T_p}{\partial y^2}, \tag{8.218}$$

where $dt = dz/V_{pz}$.

(c) Carry out an energy balance at the plug and barrel interface, that shows that the heat generated by friction per unit surface area is conducted to the plug and metallic barrel to obtain the following equation:

$$q_b = -k_p \frac{\partial T_p}{\partial y}\Big|_{y=0} + k_b \frac{\partial T_b}{\partial y}\Big|_{y=0}, \tag{8.219}$$

where $q_b$ is the heat generated per unit of barrel surface area and the subscript $b$ refers to the barrel.

(d) As $k_b \gg k_p$, the temperature distribution in the barrel is assumed to be linear. Use this information to specify the boundary condition at the interface where $y = 0$. Specify the remaining boundary condition for solving Eq. 8.217.

(e) Use the numerical approach described in Table 5.14 to obtain the temperature at the surface of a LDPE plug as a function of position along the channel (i.e., represent the spatial derivatives by finite difference expressions and use IVPAG) for the following conditions:

> Extruder geometry: $D_b = 2.5$ in.; Square-pitched screw, $e = 0.25$ in.; $H = 0.375$ in.
> Processing conditions: $G = 150$ lb/hr; $P_1$ (under hopper) $= 0.4$ psi; $T_b$ (barrel temperature) $= 80°F$; bed temperature is initially $80°F$; $N = 80$ rpm
> Material properties: LDPE properties are given in Appendix B.

**8C.3.** Maximum Conveying Rate in a SWCOR

Calculate the maximum conveying rate for LDPE at $150°C$ in a SWCOR twin screw-extruder for the geometry described in Pr. 8A.10 for $p = 2$, except taking $L_s/D_s = 0.71$ and $e = 1.78$ mm. It will be necessary to use the IMSL subroutine QDAGS (Appendix D.5) to find the cross-sectional area. Compare your calculated value to the measured value of 94.3 kg/hr at a screw speed of 43 rpm.

**8C.4.** Effect of Curvature on Fluid Particle Position

An expression for finding the complementary position of a particle as a function of curvature is given in Eq. 8.216 (Pr. B.12). Solve this equation for values of $a = 0.95, 0.9, 0.85$, and $0.82$ for $\beta = 0.8$ and $Q = Q_d$, where $a = (\rho - \beta)/(1 - \beta)$ and $a_c = (\rho_c - \beta)/(1 - \beta)$.

## D. Design Problems

### 8D.1. Design of a Tubing Extrusion Process

A 1-inch diameter garden hose is to be produced at the rate of 500 lb/hr with LDPE. The wall thickness of the hose is to be 0.075 in. Design an extruder and die to accomplish this. Use data given for NPE 953 in various places in the book (e.g., Appendix A, Table A.1, and Chapter 3). In addition to specifying the size of extruder, screw geometry, and die dimensions, list all the assumptions you made in arriving at your design.

### 8D.2. Design of a Coextrusion System

Two resins, acrylonitrile-butadiene-styrene (ABS) and acrylonitrile-EPDM-styrene (AES), are to be coextruded to form a sheet which is 2.8 mm thick, with AES representing 10% of the total thickness of the sheet (i.e., 0.28 mm). The sheet is to be 1.42 m wide, and the overall extrusion rate is to be 455 kg/hr. The density of each polymer is similar (i.e., $\rho = 1100 \, kg/m^3$). ABS and AES have similar viscosity functions at 200°C with values for the power-law model given as:

For ABS, $m = 45,455 \, Pa \cdot s^n$; $n = 0.342$
for AES, $m = 45,000 \, Pa \cdot s^n$, $n = 0.342$

Size and design extruders for providing the desired output with the only requirement being that the $L/D$ ratio of the extruders be 30. Design a feedblock system for joining the fluids and a sheet die for producing the desired sheet dimensions. Because of the close match in viscosity, the two fluids can be joined in the feedblock before entering the sheet die. Use the thermal and frictional properties given for polystyrene, if necessary, in your design calculations.

## REFERENCES

Biesenberger, J. A. and D. H. Sabastian 1983. "Principles of Polymerization Engineering" (Wiley, New York).

Biesenberger, J. A., S. K. Dey, and J. Brizzolara. 1990. "Devotilization of Polymer Melts: Machine Geometry and Scale Factors." *J. Polym. Eng. Sci.*, **30**(23), 1493–9.

Bigg, D., and S. Middleman. 1974. "Mixing in a Screw Extruder. A Model for Residence Time Distribution and Strain." *Ind. Eng. Chem. Fundam.*, **13**(1), 66.

Booy, M. L.. 1978. "Geometry of Fully Wiped Twin-Screw Equipment." *Polym. Eng. Sci.*, **18**, 973–84.

Booy, M. L. 1980. "Isothermal Flow of Viscous Liquids in Corotating Twin Screw Devices." *J. Polym. Eng. Sci.*, **20**(18), 1220.

Broyer, E., and Z. Tadmor. 1972. "Solids Conveying in Screw Extruders. Part I A Modified Isothermal Model." *J. Polym. Eng. Sci.*, **13**, 12–24.

Chung, C. I. "On the Scale-up of Plasticating Extruder Screws." *Polym. Eng. Sci.*, **24**, 626–33.

Darnell, W. H., and E. A. J. Mol. 1956. "Solids Conveying in Extruders." *Soc. Plast. Eng. J.*, **12**, 20-28.

Janssen, L. P. B. M. 1978. *Twin Screw Extrusion* (Elsevier, Amsterdam).

Macosko, C. W., and D. R. Miller. 1976. "A New Derivation, Average Molecular Weights for Newtonian Polymers." *Macromolecules*, **9**, 201.

Montes, S., and J. L. White, 1991, "Fluid Mechanics of Distributive Mixing in a Modular Intermeshing Corotating Twin Screw Extruder." *Int. Polym. Proc.*, **VI**(3), 156.

Pinto, G., and Z. Tadmor 1970. "Mixing and Residence Time Distribution in Melt Screw Extruders." *Polym. Eng. Sci.*, **10**(5), 279–88.

Potente, H. and S. M. Schultheis, 1989. "Investigations of the Residence Time and the Longitudinal Mixing Behavior in Counter-rotating Twin Screw Extruders," *Int. Polym. Proc.*, **4**(4), 255.

Rauwendaal, C. 1986. *Polymer Extrusion* (Hanser, Munich).

Roberts, G. W. 1970. "A Surface Renewal Model for the Drying of Polymers During Screw Extrusion." *AIChE J.*, **16**, 878.

Szydlowski, R., R. Brzoskowski, and J. L. White. 1987. *Int. Polym. Proc.*, **1**, 207.

Tadmor, Z., and E. Broyer. 1972. "Solids Conveying in Screw Extruders. Part II. Non-isothermal Model." *Polym. Engng. Sci.*, **17**, 378–86.

Tadmor, Z., and C. G. Gogos. 1979. *Principles of Polymer Processing* (Wiley, New York).

Tadmor, Z., and I. Klein. 1970. *Engineering Principles of Plasticating Extrusion* (Van Nostrand Reinhold, New York).

Tzoganakis, C., J. Vlachopoulous, and A. E. Hamielec. 1988. "Modeling of the Peroxide Degradation of Polypropylene." *Int. Polym. Proc.*, **3**, 141.

Walker, D. M. 1966. "An Approximate Theory for Pressures and Arching in Hoppers." *Chem. Engng. Sci.*, **21**, 975–99.

Werner, H. 1976. Ph.D. Thesis, University of Munich.

White, J. L. 1990. *Twin Screw Extrusion* (Hanser, Munich).

Xanthos, M., Ed. 1992. *Reactive Extrusion* (Hanser, Munich).

# 9

# POSTDIE PROCESSING

## DESIGN PROBLEM 8
### Design of a Film-Blowing Process for Garbage Bags

Garbage bags are made from low-density polyethylene (LDPE; NPE 953) by the process of film blowing, in which the film leaves the nip rolls as a "lay-flat" film before being cut and sealed to form garbage bags. A typical bag holding 13 gallons of garbage is 25.41 μm (= 1 mil) thick, 61 cm (= 2 ft) wide, and 69.6 cm (= 2 ft 3 3/8″) long. The production line consists of an extruder with an annular die, film-blowing instrumentation (air supply for blowing, air ring, guide rolls, nip rolls, etc.), and it can accommodate 1,500 bags per hour. This design problem consists of two parts: (a) calculation of the dimensions of the annular die from which LDPE is extruded at 170° C without any melt fracture present; and (b) calculation of the blowing air pressure and the drawing force at the nip rolls to obtain the desired strength bags.

The LDPE exiting the die is assumed to exhibit die swell (see Fig. 3.1 in Chapter 3, p. 35), which can be calculated from Eq. 3.89 (i.e., assume that capillary die swell is equal to diameter and thickness swell from an annular die). Melt fracture (Subsection 7.2.2) for LDPE appears when the maximum wall shear stress exceeds $1.13 \times 10^5$ Pa. To secure uninterrupted production, a safety factor of 3 is applied with regard to the maximum wall shear stress. The power-law relationship for LDPE at 170° C and for $\dot{\gamma} \geqslant 0.1\,\mathrm{s}^{-1}$ is:

$$\eta = 5.17 \times 10^3 \dot{\gamma}^{-0.413},$$

where $\eta$ is in Pa·s. The density of the polymer is $0.77\,\mathrm{g/cm^3}$ at the extrusion temperature and $0.92\,\mathrm{g/cm^3}$ at room temperature, and the following relationship correlates the primary normal stress difference, $N_1$, to the shear stress, $\tau_{xy}$:

$$N_1 = 0.119\tau_{xy}^{1.304}$$

for $N_1$ and $\tau_{xy}$ in Pa. Table 1.1 (p. 6) can be used to provide the film thickness at the maximum die swell level as a function of the final thickness.

We then consider the analysis of the film-blowing process, which should be based on the maximum die swell dimensions as the initial dimensions and isothermal conditions are assumed. Calculate the pressure of the blowing air, the drawing force, and the velocity at the nip rolls if the dimensionless frost-line height is confined to 5 and the distance of the nip-roll system from the position where the maximum die swell level occurs is 1.5 m.

---

In spite of the geometry of the final product the processes of fiber spinning, film casting, and film blowing bear an important similarity. There are no constraining surfaces to determine the final dimensions of the product. The surfaces are free, and hence the final dimensions are determined by the rheological properties of the melt as well as processing conditions such as takeup speed, cooling rate, extrusion rate, and die dimensions. Modeling of these processes is complicated not only by the complex rheology of the melt (or solution) but also by the crystallization process. In Section 9.1 we discuss the process of fiber spinning in detail. For the sake of simplicity the analysis starts by

considering a single isothermal filament in which the rheological properties are considered to be Newtonian. These results are then extended to the nonisothermal case in which the processes of crystallization and structure formation are considered. The heat transfer process is analyzed in the context of nonisothermal fiber spinning. We discuss the approach required for viscoelastic fluids next. The solution of these equations is the primary goal, but a knowledge of the assumptions in the equations is necessary. Finally, an analysis of the various instabilities that occur in fiber spinning is presented.

In Section 9.2 the process of film casting is considered along with its associated stability. Both fiber spinning and film casting impart uniaxial stretching. One way to impart biaxial stretching is by film stretching, which is discussed next. The other way of producing biaxially oriented films with high mechanical and physical properties is by film blowing, which is discussed in Section 9.3, along with the associated problems of stability and scaleup. In these last two processes we only present the final equations as the emphasis is on solving the equations.

## 9.1. FIBER SPINNING

A *fiber* is the fundamental unit of textiles, and it is defined as a material unit of axial length scale about 100 times the length scale in the cross direction (width or radius). There are two types of fibers: *natural* and *synthetic* (or *man-made*). The term *spinning* has different meaning for natural and synthetic fibers. Spinning of natural fibers refers to the *twisting of short fibers* into continuous lengths (also called *filaments*). On the other hand, spinning of synthetic fibers refers to production of continuous lengths *by any means*. Finally, the *yarn* is made by twisting many filaments together.

The production of man-made fibers usually includes the following processes (Ziabicki, 1976):

1. Preparation of polymer (polymerization, chemical modification, etc.).
2. Preparation of the spinning fluid (polymer melt or solution).
3. Spinning (extrusion, solidification, and deformation of the spinning line or filament).
4. Drawing (due to higher linear speed at the takeup roll than that at the die; drawing is used to increase the degree of molecular orientation and improve the tensile strength, modulus of elasticity, and elongation of the fibers).
5. Heat treatment.
6. Textile processing (twisting, oiling, dyeing, etc).

Process 3 can be achieved mainly by three procedures: *melt spinning, solution dry-spinning*, and *solution wet-spinning*. Of these three procedures, melt spinning is the simplest and the most economical one. Its simplicity stems from the fact that it involves only heat transfer and extensional deformation, whereas the other methods in addition to these processes also involve mass transfer and diffusion. Melt spinning procedure can be applied to polymers that are thermally stable at the extrusion temperature and that exhibit relatively high fluidity at that temperature. Typical examples of melt-spun polymers are polyamides, polyesters, polystyrene, polyolefins, and inorganic glasses.

In the solution dry-spinning procedure the polymer is dissolved in a volatile solvent, and the solution is extruded. Then the spinning line meets a stream of hot air, and the solvent is evaporated. The recovery of the solvent increases the cost of the whole process. Typical examples of dry-spun polymers are cellulose acetate, acrylonitrile, vinyl chloride, and acetate. In the 1980s, extended-chain PE fibers (ECPE, Spectra fibers) have been made by solution spinning in a typical melt-spinning apparatus. The solution wet-spinning procedure is applied to polymers that meet neither of the criteria of the other two methods (i.e., thermal stability and solubility in a volatile solvent). It involves the extrusion of a polymer solution into a liquid bath of coagulating agents that drive the solvent out of the filament. The basic principles of wet and dry spinning were discussed in Chapter 4 because mass transfer and diffusion are the controlling mechanisms of those types of spinning. There are four other spinning procedures (*phase separation spinning, emulsion spinning, gel spinning,* and *reaction spinning*) which will not be addressed in this textbook.

A schematic (not in scale) of the typical melt-spinning procedure is shown in Fig. 9.1 (see also Fig. 1.5). Polymer is pumped by means of an extruder through a screen pack, in which the polymer is filtered through layers of screens and sand. Then it is divided into many small streams by means of a plate containing many small holes, the *spinneret.* Some spinnerets can have as many as 10,000 holes (rayon spinning from a 15 cm platinum disk spinneret). The extrusion through the spinneret (or die), the subsequent *die swell* (due to the relaxation of the elastic stresses of the polymer; Subsection 7.2.3), the cooling of the filament by cooling air (or water vapor) flowing perpendicularly to the filament axis, the solidification of the polymer, and its cold drawing in the region from the solidification point to the winding at the takeup roll are shown in Fig. 9.1. Part of the time the holes in the spinneret are circular in shape, but mostly they are of other regular shapes.

Other typical shapes include *trilobal,* square, and cross, and they are shown in Fig. 9.2. Note that dry and wet spinning from circular spinneret holes usually results in irregularly shaped fibers.

The key aspects of the modeling of the fiber-spinning process are:

1. Extrusion through a short die (usually of nonround cross section) in which the fluid velocity field must undergo rapid rearrangement.
2. Swell of the liquid leaving the hole.
3. Rapid axisymmetric extension to large strains.
4. Rapid temperature changes and hence large changes in the rheological behavior.
5. Crystallization under conditions of high stress and rapid cooling.

The origin of the coordinate axes is considered to be at the point of maximum cross-sectional area, which occurs at a small distance from the face of the spinneret as a result of die swell. The distance between the spinneret and the point of maximum die swell is only a few die diameters long, and thus it is small compared to the distance between the face of the spinneret and the takeup roll. Therefore, by neglecting it we cause no severe problems in the following analysis.

In terms of the number of the filaments per spinneret plate and of spinning speed, melt spinning is divided into various groups. *Monofilaments* are produced by one-hole spinneret plates, even though most of the time there are numerous filaments extruded through the spinneret (multifilament yarns). *Very-low-speed spinning,* with speeds ranging from 30 to 100 m/min, usually occurs for thick monofilaments spun through liquid baths. *Low-speed spinning* is usually carried out at speeds in the range of 100 to 750 m/min, where the filament tension

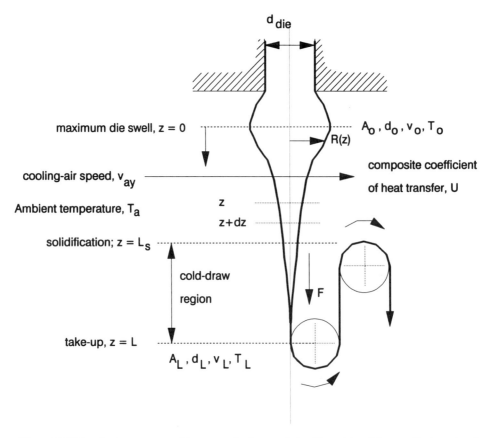

**Figure 9.1** Melt-spinning process and its geometric characteristics. The origin of the coordinate system is located at the point of maximum die swell.

is constant along the entire length. In order to enhance the degree of orientation and crystallinity, and hence physical properties, the yarns are subsequently drawn and annealed, and therefore the melt-spinning process is considered to be a *two-step process* (TSP: first step: spinning; second step: drawing). At *intermediate speeds* of 750 to 3500 m/min, the filament tension is increased because of inertia and air drag. Finally, at *high spinning speeds* of 3500 m/min and above, polymers such as poly(ethylene terephthalate) (PET) undergo stress-induced crystallization. At high spinning speeds (Subsection 9.1.4) the structure, morphology and resulting physical properties are somewhat different from those obtained in conventional low-speed spinning processes.

Typical physical and mechanical properties of melt-spun fibers include the following: density, boil-off shrinkage, birefringence, tensile strength, percent elongation, modulus of elasticity, shrinkage tension, and dyeability. Some of these properties are shown in Table 9.1 for PET spun under conventional conditions of a two-step process and in a one-step process at high takeup speed in the region of 6000 m/min. In the same table, we also provide data for comparison for the top-performing industrial fibers, Spectra (ECPE) and Kevlar (Aramid). In practice, some of the mentioned properties might be found under different names. *Denier per filament* (dpf; unit: denier, d) is usually substituted for density, and *tenacity* (unit: g-force/d or gf/d) is used in place of tensile strength. More about the definitions of these specific terms and units can be found in Subsection 9.1.2.4.

In the following analysis we present first the Newtonian isothermal model, which leads to an analytical solution. Then we discuss the Newtonian nonisothermal model, which gives insight into the complexities of the coupled heat and momentum transfer equations. PET, nylon, and polysiloxanes are three typical polymers that are almost Newtonian at spinning conditions. Finally, we introduce the non-Newtonian isothermal model together with its associated difficulties. High-density polyethylene (HDPE), LDPE, polypropylene (PP) and polystyrene (PS) are all pseudoplastic and viscoelastic and fall into the latter category.

### 9.1.1. ISOTHERMAL NEWTONIAN MODEL

The basic ideas of modeling the fiber-spinning process are best understood by considering the steady-state isothermal Newtonian analysis first. This analysis can be considered to be valid in the case that (1) the drawing take place in a short distance in air and then (2) the fiber is quenched into a water bath. In later steps all the other factors will be added. In the isothermal Newtonian analysis we neglect any interaction between the filament and the surrounding medium; that is, no heat transfer is taking place, and the surface tension and air drag forces are negligible. Schematically, Fig. 9.1 presents the overall melt-spinning picture and Fig. 9.3 presents a section of the filament

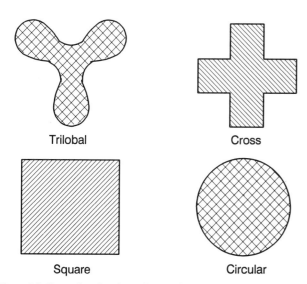

Trilobal    Cross

Square    Circular

**Figure 9.2** Examples of various shapes of spinneret holes.

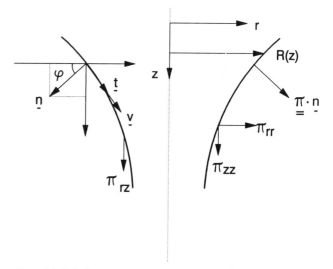

**Figure 9.3** Velocity vectors, stress components, and geometry at a point during melt spinning.

**TABLE 9.1 Typical Properties of Fibers**

| Property | PET | | ECPE | Aramid |
|---|---|---|---|---|
| | OSP (at 6000 m/min) | TSP (conventional) | Spectra 1000 | Kevlar 149 |
| Density (g/cm³) | 1.38 | 1.375 | 0.97 | 1.47 |
| Boil-off shrinkage (%) | 3–4 | 8.0 | | |
| Birefringence, ($10^{-3}$) | 105–115 | 150 | | |
| Tensile strength (MPa) | 460–550 | 610 | 3,000 | 3,450 |
| or tenacity (gf/d) | 3.8–4.5 | 5.0 | | |
| Elongation at break (%) | 45–50 | 35 | 2.7 | 1.5 |
| Modulus of elasticity (GPa) | 13–14 | 18 | 172 | 172 |
| Modulus of elasticity in (gf/d) | 75–80 | 120 | | |

(Data from T. Kawagushi. "Industrial Aspects of High Speed Spinning," in Ziabicki & Kawai, 1985; D. S. Cordova, and D. S. Donnelly, "Reinforced Plastics, Extended-Chain Polyethylene Fibers," in I. I. Rubin, Ed., *Handbook of Plastic Materials and Technology*, Wiley, New York, 1990.)

and all the stress, velocity, and direction vectors associated with the problem.

At a point in the filament boundary the unit outward normal vector is $\mathbf{n}$, and the tangential vector is $\mathbf{t}$. The appropriate coordinate system for the problem is cylindrical with the axis of symmetry coinciding with the $z$ axis (Fig. 9.3). The velocity vector is $\mathbf{v}$, and the total stress tensor is $\boldsymbol{\pi}$. The components of the total stress tensor are related to the components of the extra stress tensor $\boldsymbol{\tau}$ by the following relationship (Eq. 3.7):

$$\pi_{ii} = p + \tau_{ii}; \qquad \pi_{ij} = \tau_{ij} \qquad \text{when} \quad i \neq j, \tag{9.1}$$

where $p$ is the isotropic pressure.

To start our analysis we postulate the following:

$$v_z = v_z(r, z); \qquad v_r = v_r(r, z); \qquad p = p(r, z). \tag{9.2}$$

Because of symmetry and no-rotation condition of the filament, the $\theta$ component of the velocity field is zero and there is no $\theta$ dependence of the velocity and pressure fields. The equation of continuity yields:

$$\dot{m} = \rho A \bar{v}_z = \rho \pi R^2 \bar{v}_z = \text{const.}, \tag{9.3}$$

where $\dot{m}$ is the polymer mass flow rate, $A$ is the cross sectional area of the filament, $R(z)$ is the radius of the filament at the axial position $z$, and $\bar{v}_z$ is the average axial velocity across the filament cross section. The $z$ component of the equation of motion is (Table 2.9, Eq. F, p. 21):

$$\rho \left( v_r \frac{\partial v_z}{\partial r} + v_z \frac{\partial v_z}{\partial z} \right) = -\frac{1}{r} \frac{\partial}{\partial r} (r \pi_{rz}) - \frac{\partial \pi_{zz}}{\partial z}, \tag{9.4}$$

where the gravitational force, $\rho g_z$, has been considered to be negligible. Similarly, the $r$ component of the equation of motion is (Table 2.9, Eq. D):

$$\rho \left( v_r \frac{\partial v_r}{\partial r} \right) = -\frac{1}{r} \frac{\partial}{\partial r} (r \pi_{rr}) - \frac{\partial \pi_{rz}}{\partial z} + \frac{\pi_{\theta\theta}}{r}. \tag{9.5}$$

The Newtonian constitutive equation is used, and thus:

$$\pi_{zz} = p + \tau_{zz} = p - 2\mu \frac{\partial v_z}{\partial z}. \tag{9.6}$$

The boundary conditions for the problem are based on the facts that no fluid crosses the interface between the filament and the surrounding medium and that no stresses are imposed on the free boundary because air drag and surface tension forces are neglected in this analysis. These boundary conditions are stated as:

B.C.1: $\mathbf{v} \cdot \mathbf{n} = 0$;     no flow across the interface     (9.7)

B.C.2: $\boldsymbol{\pi} \cdot \mathbf{n} = 0$;     no air drag and surface tension.     (9.8)

From analytical geometry we get

$$\mathbf{n} = n_z \boldsymbol{\delta}_z + n_r \boldsymbol{\delta}_r, \tag{9.9}$$

where $\boldsymbol{\delta}_i$ represents the unit vector in the $i$ direction. The $z$ and $r$ components of the unit normal vector are related to the angle $\phi$ between the unit normal vector and the $z$ axis and the change of the radius of the filament with respect to $z$ by the following relationships:

$$n_z = \sin \phi = -\frac{dR/dz}{\sqrt{1 + (dR/dz)^2}} = -\frac{R'}{\sqrt{1 + (R')^2}} \tag{9.10}$$

$$n_r = \cos \phi = \frac{1}{\sqrt{1 + (dR/dz)^2}} = \frac{1}{\sqrt{1 + (R')^2}}. \tag{9.11}$$

At the free surface (i.e., at $r = R(z)$), Eq. 9.7 becomes:

$$\mathbf{v} \cdot \mathbf{n} = 0 = v_r n_r + v_z n_z. \tag{9.12}$$

Combining Eqs. 9.10, 9.11 and 9.12, we get $v_r = (dR/dz) v_z = R' v_z$ at $r = R(z)$. Note that if $R' \ll 1$, that is, if the filament radius is slowly changing with $z$, then $v_r \approx 0$, which is true over most of the distance between the spinneret and the takeup roll. Also note that the relationship, $v_r = R' v_z$, indicates that $R(z)$ is a streamline of fluid flow in the filament.

The $z$ and $r$ components of the total stress vector at the free surface are:

$$(\boldsymbol{\pi} \cdot \mathbf{n})_z = \pi_{rz} n_r + \pi_{zz} n_z \tag{9.13}$$

$$(\boldsymbol{\pi} \cdot \mathbf{n})_r = \pi_{rr} n_r + \pi_{rz} n_z, \tag{9.14}$$

and using Eq. 9.8 we get:

$$\pi_{rz}(R) = -\frac{n_z}{n_r} \pi_{zz}(R) = R' \pi_{zz}(R). \tag{9.15}$$

In this relationship $\pi_{zz}$ is the tensile stress due to the drawing, and $\pi_{rz}$ is due to the fact that we do not have a cylindrical geometry (i.e., $R$ is a function of $z$). Thus, we expect that $\pi_{rz}$ is approximately zero for $R$ changing very slowly with $z$ and neglecting air drag on the filament surface. Mathematically the previous argument follows from Eq. 9.15 for $R' \ll 1$.

If we further assume that $v_z = v_z(z) = \bar{v}_z$, then the average of the $z$ component of the equation of motion obtained by multiplication of each term by $2\pi r \, dr$ and integration from 0 to $R(z)$ leads to (Middleman, 1977):

$$\int_0^R \rho v_r \frac{\partial v_z}{\partial r} r \, dr = 0 \qquad \text{since} \quad v_z \neq v_z(r) \tag{9.16}$$

$$\int_0^R \rho v_z \frac{\partial v_z}{\partial z} r \, dr = \frac{1}{2} \rho v_z v_z' R^2 \tag{9.17}$$

$$\int_0^R \frac{1}{r} \left[ \frac{\partial}{\partial r} (r \pi_{rz}) \right] r \, dr = R \pi_{rz}(R) = R R' \pi_{zz}(R) \tag{9.18}$$

$$\int_0^R \frac{\partial \pi_{zz}}{\partial z} r \, dr = \frac{d}{dz} \int_0^R \pi_{zz} r \, dr - \pi_{zz} R R'(R) = \frac{1}{2} \pi_{zz}' R^2, \tag{9.19}$$

where $v_z'$ is the derivative of $v_z$ with respect to $z$. Note that the right equality in Eq. 9.19 is based on Leibnitz's rule of differentiation of an integral. Substitution of Eqs. 9.16 to 9.19 into Eq. 9.4 yields:

$$\rho v_z v_z' = -2 \frac{R'}{R} \pi_{zz} - \pi_{zz}' = -\frac{1}{R^2} \frac{d}{dz} (R^2 \pi_{zz}) \tag{9.20}$$

which is the general equation of the isothermal fiber-spinning problem. Note that we have not yet introduced the constitutive equation given in Eq. 9.6.

Equation 9.20 can be transformed into a differential equation containing only $v_z$ and its derivatives by substituting expressions for $\pi_{zz}$ and $\pi'_{zz}$ obtained from Eq. 9.6. These relations are found from the constitutive equation, Eq. 9.6, by eliminating the pressure term as follows. The isotropic pressure is one-third of the trace of the total stress tensor, that is,

$$p = \frac{1}{3}(\pi_{rr} + \pi_{zz} + \pi_{\theta\theta}). \tag{9.21}$$

Because only the $z$ component of the velocity field exists (see also Eq. 9.14 for $\pi_{rr}$), $\pi_{rr} = \pi_{\theta\theta} = 0$, and thus Eq. 9.21 yields:

$$\pi_{zz} = 3p = -3\mu\frac{dv_z}{dz} = -\bar{\eta}_1\frac{dv_z}{dz}, \tag{9.22}$$

where $\bar{\eta}_1$ is the uniaxial elongational viscosity (Eq. 3.36). For Newtonian fluids, this viscosity is equal to three times the shear viscosity (Trouton's rule, $\bar{\eta}_1 = 3\mu$).

Substitution of Eq. 9.22 and its derivative into the equation of motion, Eq. 9.20, yields:

$$\frac{d}{dz}(v_z)^2 = 12\frac{\mu}{\rho}\frac{R'}{R}\frac{dv_z}{dz} + 6\frac{\mu}{\rho}\frac{d^2v_z}{dz^2}. \tag{9.23}$$

Neglecting the inertial term and with the help of Eq. 9.3 (i.e., $R'/R = -v'_z/2v_z$) we obtain $(v'_z/v_z)' = 0$, which is solved as:

$$v_z = C_1 e^{C_2 z}. \tag{9.24}$$

The melt-spinning problem has the following boundary conditions:

> B.C.1: $v_z = v_0$     at $z = 0$
>
> B.C.2: $v_z = v_L$     at $z = L$. $\qquad(9.25)$

With the help of the above boundary conditions Eq. 9.24 becomes:

$$v_z = v_0 \exp\left[\frac{z\ln D_R}{L}\right] = v_0(D_R)^{z/L}, \tag{9.26}$$

where $D_R$ is the *draw* (or *drawdown*) *ratio* and is equal to:

$$D_R = \frac{v_L}{v_0}. \tag{9.27}$$

Finally, the radius of the filament, $R$, as a function of the axial distance $z$ is found from the continuity equation as:

$$R(z) = R_0 \exp\left[-\frac{1}{2}z\frac{\ln D_R}{L}\right]. \tag{9.28}$$

In summary, the steady-state Newtonian isothermal model is able to provide the axial velocity profile as well as the filament radius profile, and it is based on the following additional assumptions: (1) slowly changing radial profile with axial distance; (2) negligible inertial and gravitational forces; (3) nonexistent radial velocity profile; (4) circular filament; (5) axial velocity profile not dependent on the radial coordinate, and (6) negligible surface tension and air drag forces (see also Pr. 9A.1 for the validity of some of the above assumptions and Schultz (1987) for a challenge of these assumptions).

### EXAMPLE 9.1
### Newtonian and Isothermal Model for Melt-Spun Nylon 6,6

Nylon 6,6 is extruded at 285°C under isothermal conditions (in a temperature-controlled chamber), and it is drawn in such a way that $L = 400$ cm and the draw ratio is equal to 100. If the takeup speed is 1000 m/min, the polymer volumetric flow rate is 0.1 cm³/s, and the die swell diameter is three times the die diameter, carry out the following:

(a) Calculate the maximum stretching rate of the melt.

(b) Compare this stretching with the shear rate inside the die, if the die diameter is 0.16 cm.

(c) Assess the validity of the approximate relation: $v_r \approx 0$

(d) Determine the maximum tensile stress in the melt and the force required to draw the melt.

Assume that nylon 6,6 is Newtonian at the spinning temperature.

### Solution

The stretching rate is calculated by differentiating Eq. 9.26 as follows:

$$\dot{\varepsilon} = v'_z = \frac{dv_z}{dz} = v_0 \exp\left[\frac{z\ln D_R}{L}\right]\frac{\ln D_R}{L}. \tag{9.29}$$

The maximum in $\dot{\varepsilon}$ occurs at $z = L$:

$$\dot{\varepsilon}_{max} = (v'_z)_{max} = v_L\frac{\ln D_R}{L}, \tag{9.30}$$

and it is equal to $19\,\text{s}^{-1}$.

The maximum shear rate (at the walls) inside the die is given by the relation (Table 2.6):

$$\dot{\gamma}_{max} = \frac{4Q}{\pi R_{die}^3}, \tag{9.31}$$

where $Q$ is the polymer volumetric flow rate. Thus, $\dot{\gamma}_{max}$ is equal to about $250\,\text{s}^{-1}$, which is one order of magnitude higher than the maximum stretching rate of $19\,\text{s}^{-1}$.

The validity of the relation $v_r \approx 0$ depends on the value of the slope of the filament radius, which follows from Eq. 9.28:

$$|R'| = \left|\frac{dR}{dz}\right| = \frac{R_0}{2}\frac{\ln D_R}{L}\exp\left[-\frac{1}{2}z\frac{\ln D_R}{L}\right]. \tag{9.32}$$

The maximum of the slope occurs at $z = 0$ (i.e., at the die swell level), and it is equal to:

$$|R'|_{max} = \frac{R_0}{2}\frac{\ln D_R}{L}. \tag{9.33}$$

Numerically the maximum slope is calculated to be $1.4 \times 10^{-3}$, which is much smaller than 1, and so the radial velocity is about equal to zero.

The maximum tensile stress in the melt occurs at the location where the maximum stretching takes place, that is at the takeup roll. It is calculated from Eq. 9.22 as:

$$(\pi_{zz})_{max} = -3\mu(v'_z)_{max}. \tag{9.34}$$

The viscosity of nylon 6,6 at 285°C is taken from Table A.11 in Appendix A as 250 Pa·s, and so the maximum tensile stress is calculated to be equal to $-14.3$ kPa. Finally, the force required to draw the melt is:

$$F = -\pi R_L^2 (\pi_{zz})_{\max} = 9\pi \frac{R_{\text{die}}^2}{D_R} (\pi_{zz})_{\max} \qquad (9.35)$$

which is calculated to be $-2.6$ mN (i.e., 2.6 mN in tension).

### 9.1.2. NONISOTHERMAL NEWTONIAN MODEL

The dynamics of the nonisothermal melt-spinning process has been analyzed since the early 1960s. The analysis that follows is drawn from the work of Kase and Matsuo (1965, 1967) and Kase (1985). These authors presented a model called *thin-filament theory*, and it is based on purely extensional flow field in the filament. The model in its unsteady-state form consists of four partial differential equations based on the continuity, momentum, constitutive, and energy equations. Compared to the previous Newtonian isothermal model, the present model includes the additional complication of nonconstant temperature along the axial distance $z$. The independent variables are time, $t$, and distance, $z$, from the spinneret, and the dependent variables are cross sectional area of the filament, $A(z, t)$, temperature, $T(z, t)$, axial velocity, $v_z(z, t)$, and rheological tensile force, $F(z, t)$. Obviously, the steady-state solution of the four equations gives the dependent variables as a function of the axial distance $z$ only. Figure 9.1 shows a typical melt-spinning process valid in this section.

*9.1.2.1. Assumptions.* The simplifying assumptions of the thin-filament theory are:

1. Circular cross section of the filament; $A = \pi R^2$.
2. Constant polymer density; $\rho = \text{const}$.
3. Constant specific heat of the polymer; $\bar{C}_p = \text{const}$.
4. No die-swell effect.
5. Newtonian viscosity with Arrhenius-type dependence on temperature.
6. No resistance to radial heat conduction within the filament, i.e., $\partial T/\partial r = 0$.
7. No heat conduction within the filament in the axial direction.
8. Vertical filament.
9. Heat transfer at the surface (with composite coefficient $U$) consisting of two parts: convective (with coefficient $h$) and radiant heat transfer with governing equation $\sigma\varepsilon(T^4 - T_a^4)$ with $U = h + \sigma\varepsilon(T^4 - T_a^4)/(T - T_a)$, where $T_a$ is the ambient air temperature, $\sigma$ is the Stefan-Boltzman constant, and $\varepsilon$ is the emissivity).
10. Empirical dependence of the heat transfer coefficient $U$ on the filament velocity, $v_z$, and the cooling air cross-flow velocity, $v_{ay}$.
11. No interactions, either hydrodynamic and thermal, between adjacent filaments.
12. Purely extensional flow field in the filament, that is, $\partial v_z/\partial r = 0$.

Most of the assumptions are valid for industrial melt-spinning conditions. However, assumptions 4 and 5 may not be absolutely valid, because die swell exists and polymers exhibit viscoelastic behavior. To accommodate assumption 4, we consider the origin of the coordinate axis to be at the point of maximum die swell. Furthermore, other developments in numerical schemes (Fisher et al., 1980; Keunings et al., 1983) have verified that the thin-filament theory and finite element calculations give comparable results in a region only a few die diameters downstream from the die swell region. Consequently, for all practical purposes, the thin-filament theory is satisfactorily accurate for most fiber-spinning processes. Finally, industrial and laboratory experience suggests that the neglect of viscoelasticity might not be a serious problem (Ziabicki & Kawai, 1985), except when dealing with instability issues.

*9.1.2.2. Equations.* The continuity equation is the same as Eq. can be easily shown to be of the following form:

$$\frac{\partial A}{\partial t} + v_z \frac{\partial A}{\partial z} + A \frac{\partial v_z}{\partial z} = \frac{\partial A}{\partial t} + \frac{\partial (Av_z)}{\partial z} = 0. \qquad (9.36)$$

The unsteady-state equation of motion (Table 2.9, Eq. F, p. 21), taking also assumptions 8 and 12 into consideration, becomes:

$$\rho\left(\frac{\partial v_z}{\partial t} + v_z \frac{\partial v_z}{\partial z}\right) = \rho g - \frac{1}{r}\frac{\partial}{\partial r}(r\pi_{rz}) - \frac{\partial \pi_{zz}}{\partial z}. \qquad (9.37)$$

The equation of energy (Table 5.1, Eq. B, p. 101) using also assumption 2 becomes:

$$\rho\bar{C}_p\left(\frac{\partial T}{\partial t} + v_z \frac{\partial T}{\partial z}\right) = -\frac{1}{r}\frac{\partial}{\partial r}(rq_r) - \frac{\partial q_z}{\partial z} + \dot{S}, \qquad (9.38)$$

where $\dot{S}$ is the rate of energy production (examples: phase change and chemical reaction). Because there is no significant heat production and no heat conduction in the axial direction (assumption 7), Eq. 9.38 simplifies to:

$$\rho\bar{C}_p\left(\frac{\partial T}{\partial t} + v_z \frac{\partial T}{\partial z}\right) = -\frac{1}{r}\frac{\partial}{\partial r}(rq_r). \qquad (9.39)$$

The heat transfer in the radial direction is given by the following relation (assumption 9):

$$q_r = U(T - T_a) = h(T - T_a) + \sigma\varepsilon(T^4 - T_a^4). \qquad (9.40)$$

By averaging the equation of motion over the cross section (i.e., multiplying each term by $2\pi r dr$ and integrating them from 0 to $R(z)$), we get:

$$\rho A\left(\frac{\partial v_z}{\partial t} + v_z \frac{\partial v_z}{\partial z}\right) = \rho g A - 2\pi(r\pi_{rz}|_{R(z)}) - 2\pi \int_0^{R(z)} \frac{d\pi_{zz}}{dz} r dr. \qquad (9.41)$$

The left-hand-side term in Eq. 9.41 represents the inertia of the filament, and the first term on the right-hand side represents the force due to gravity acting on the filament. The second term on the right-hand side represents the drag force due to air, and the last term represents the tensile stress in the polymer and the surface tension forces. Thus, Eq. 9.41 presents the balance of the acting forces: (inertia) = (gravity) − (air drag) − (surface tension) + (rheological forces).

The air drag force acts along the tangential direction, **t**, as shown in Fig. 9.3 and thus can be decomposed into two components: one in the $r$, and one in the $z$ direction. The summation of the $r$ components of the traction vector over the circumference of the filament at a specific $z$ plane is zero, because they cancel out, and the summation of the $z$ components results in the shear stress $\pi_{rz}$ evaluated at the surface of the filament. When $R' \ll 1$, we are allowed to consider the $z$ component of the air drag to be the same as the air drag itself, for all practical purposes. Thus, the summation of the $z$ components gives: $r\pi_{rz}|_{R(z)} = R\pi_{rz,s} = R(1/2)C_D\rho_a v_z^2 = RD$, where $C_D$ is the hydrodynamic drag (or friction) coefficient, $\rho_a$ is the density of the surrounding medium, that is, the density of air, and the subscript $s$ denotes the surface of the filament.

Using the Leibnitz rule for the last term in Eq. 9.41, we get:

$$\frac{d}{dz}\int_0^{R(z)} \pi_{zz}r\,dr = \pi_{zz}|_R R(z)\frac{dR}{dz} + \int_0^{R(z)} \frac{d}{dz}(\pi_{zz})r\,dr. \qquad (9.42)$$

Based on Eq. 9.42, the equation of motion becomes:

$$\rho A\left(\frac{\partial v_z}{\partial t} + v_z\frac{\partial v_z}{\partial z}\right) = \rho g A - 2D\pi R + 2\pi\pi_{zz}|_R R\frac{dR}{dz} - 2\pi\frac{d}{dz}\int_0^{R(z)} \pi_{zz}r\,dr. \qquad (9.43)$$

A curved surface almost always creates a surface tension term that relates the pressure differential across the interface between the melt and air. The term $\pi_{zz}|_R = \pi_{zz,s}$ is related to the surface tension force via the relationship:

$$\pi_{zz,s} = 3\Delta p = 3(p - p_a) = 3\gamma\left(\frac{1}{R_1} + \frac{1}{R_2}\right), \qquad (9.44)$$

where Eq. 9.22 was used, $R_1$ and $R_2$ are the principal radii of curvature of the fiber surface at position $z$ (Fig. 9.4), $\gamma$ is the surface tension, and $p_a$ is the ambient pressure. The last equality in Eq. 9.44 is the Young–Laplace equation. Note that the principal radii of curvature have opposite signs, as they are on opposite sides of the filament surface. For a slowly changing filament radius (i.e., for $R' \ll 1$) $R_1 \approx R$ and $R_2 \gg R$ so that

$$\pi_{zz,s} \approx 3\gamma/R \qquad (9.45)$$

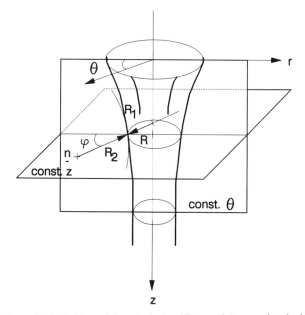

**Figure 9.4** Definition of the principal radii $R_1$ and $R_2$ associated with the surface tension force during melt spinning.

is a good approximation. Based on this the unsteady-state equation of motion attains the following form:

$$\rho A\left(\frac{\partial v_z}{\partial r} + v_z\frac{\partial v_z}{\partial z}\right) = \rho g A - 2D\sqrt{\pi A} + \frac{d}{dz}(H\sqrt{\pi A}) - \frac{dF}{dz}, \qquad (9.46)$$

where $H = 6\gamma$ and $F$ is the tensile force (also called tension), which is equal to $2\pi\int_0^{R(z)}\pi_{zz}r\,dr$.

Similarly, the energy equation is integrated over the cross section to give:

$$2\pi\rho\bar{C}_p\frac{R^2}{2}\left(\frac{\partial T}{\partial t} + v_z\frac{\partial T}{\partial z}\right) = -2\pi R U(T - T_a), \qquad (9.47)$$

which finally can be brought into the following format:

$$\frac{\partial T}{\partial t} + v_z\frac{\partial T}{\partial z} = -\frac{2\sqrt{\pi}U}{\rho\bar{C}_p\sqrt{A}}(T - T_a). \qquad (9.48)$$

Finally, the Newtonian constitutive equation, Eq. 9.6, can be used where the viscosity, $\mu$, follows an Arrhenius-type equation:

$$\mu = \mu_0\exp\left[-\frac{E}{R_g}\left(\frac{1}{T_0} - \frac{1}{T}\right)\right], \qquad (9.49)$$

where $R_g$ is the universal gas constant ($= 8.314\,\text{J/mol}\,^\circ\text{K}$), $E$ is the activation energy, and $\mu_0$ is the viscosity at some reference temperature $T_0$. Note that a WLF relationship (see Section 5.1) or an empirical relationship can also be used to express the effect of temperature on viscosity. In conclusion, Eqs. 9.36, 9.46, 9.48 and 9.49 are the four transient equations that describe the Newtonian nonisothermal unsteady-state melt-spinning problem.

**9.1.2.3. Experimental Correlations.** In order to solve the previous set of equations, we need correlations for the heat transfer coefficient, $U$, and for the hydrodynamic drag coefficient, $C_D$. For typical melt-spinning conditions the values of typical cooling air parameters are: thermal conductivity, $k_a = 0.808 \times 10^{-4}\,\text{cal/cm·s}\,^\circ\text{C}$; kinematic viscosity, $v_a = 0.290\,\text{cm}^2/\text{s}$; and density $\rho_a = 0.815 \times 10^{-3}\,\text{g/cm}^3$.

The literature (Kase & Matsuo, 1967) presents the experimental correlations of the various parameters used in the equations of motion and energy:

$$U = 0.473 \times 10^{-4}\left(\frac{v_z}{A}\right)^{1/3}\left[1 + \left(8\frac{v_{ay}}{v_z}\right)^2\right]^{1/6}, \qquad (9.50)$$

where $v_{ay}$ is the velocity of cross air flow, $v_z$ is in cm/s, $A$ is in cm$^2$, and $U$ is in cal/cm$^2$·s $^\circ$C. Eq. 9.50, which is valid for any polymer, is based on the following formula:

$$\text{Nu} = \frac{(2R)U}{k_a} = \frac{2U\sqrt{A}}{k_a\sqrt{\pi}} = 0.42(\text{Re}^*)^{1/3}\left[1 + \left(8\frac{v_{ay}}{v_z}\right)^2\right]^{1/6}, \qquad (9.51)$$

where Re* is the air-side Reynolds number:

$$\text{Re}^* = \frac{(2R)v_z}{v_a} = \frac{2v_z\sqrt{A}}{v_a\sqrt{\pi}}. \qquad (9.52)$$

Also, note that the term $\{1 + [8(v_{ay}/v_z)]^2\}^{1/6}$ expresses the effect of the direction angle of the air flow measured from the filament axis. Equation 9.50 was developed from data for air moving past a stationary cylinder, and thus it does not take into account the effect of the moving cylinder. George (1982) provides a more accurate relation for moving cylinders, which is similar to Eq. 9.50, but with the two first terms replaced by $1.37 \times 10^{-4}(v_z/A)^{0.259}$.

In terms of the hydrodynamic drag (or friction) coefficient, the following equation is used:

$$C_D = 0.65(\text{Re}^*)^{-0.81}, \tag{9.53}$$

and thus

$$2D\sqrt{\pi A} = 3.12 \times 10^{-4} v_z^{1.19} A^{0.095}. \tag{9.54}$$

Note that researchers (e.g., George, 1982) have used expressions such as $0.44(\text{Re}^*)^{-0.61}$ or $0.37(\text{Re}^*)^{-0.61}$ in Eq. 9.53. Also

$$F = -3\mu \frac{dv_z}{dz} A \tag{9.55}$$

with the following relationship for the viscosity function, which is typical for PET:

$$\mu = \mu_\infty \exp\left(\frac{E'}{T+273}\right) \quad \text{when} \quad T \geqslant 60°\text{C}$$

$$\mu = \infty \quad \text{when} \quad T < 60°\text{C}, \tag{9.56}$$

where $E'$ $(= E/R_g)$ and $\mu_\infty$ are material properties of PET. This form of the viscosity function indicates that crystallization (or solidification) takes place at 60°C. Figure 9.1 shows the solidification point at $z = L_s$, beyond which the polymer is being cold-drawn.

The final form of the equations of continuity, Eq. 9.36, motion, Eq. 9.46, energy, Eq. 9.48, and constitutive, Eq. 9.55, for steady state are (Kase, 1985):

$$\frac{d(Av_z)}{dz} = 0 \tag{9.57}$$

$$\frac{dF}{dz} = \frac{d}{dz}(H\sqrt{\pi A}) + \rho g A - \rho A v_z \frac{dv_z}{dz} - 3.12 \times 10^{-4} A^{0.095} v_z^{1.19} \tag{9.58}$$

$$v_z \frac{dT}{dz} = \frac{1.67 \times 10^{-4} v_z^{1/3}}{\rho \bar{C}_p A^{5/6}} \left[1 + \left(8\frac{v_{ay}}{v_z}\right)^2\right]^{1/6} (T_a - T) \tag{9.59}$$

$$-\frac{dv_z}{dz} = \begin{cases} F/[3A\mu_\infty \exp(E'/(T+273))] & \text{for } T \geqslant 60°\text{C} \\ 0 & \text{for } T < 60°\text{C}. \end{cases} \tag{9.60}$$

In this set of four equations with four unknown $(v_z, A, T,$ and $F)$ the following boundary conditions apply:

B.C.1 at $z=0$: $A = A_0 = \pi R_0^2$; $\quad v_z = v_0$; $\quad T = T_0$; $\quad F = F_0$ (9.61)

B.C.2 at $z=L$ or at $z$ such that $T(z) = 60°\text{C}$: $v_z = v_L$. (9.62)

Equation 9.62 is considered to apply to a constant-take-up-speed filament and the last part of Eq. 9.61 to a constant-tension filament.

### 9.1.2.4. Dimensionless Forms of the Equations.
It is useful to make the governing equations nondimensional, so that the relative importance of the various terms becomes easily identifiable. To this

end we define the following dimensionless variables:

$$\zeta = \frac{z}{L} \quad \xi = \frac{A}{A_0} \quad \psi = \frac{v_z}{v_0} \quad \tau = \frac{v_0 t}{L} \quad \theta = \frac{T}{T_0}$$

$$\lambda = \frac{(-F)L}{3A_0 v_0 \mu_\infty \exp[E'/(T_0+273)]} = \frac{(-F)L}{3A_0 v_0 \mu_{\infty 0}}, \tag{9.63}$$

and the following dimensionless parameters:

$$\psi_{ay} = \frac{v_{ay}}{v_0} \quad \theta_a = \frac{T_a}{T_0} \quad \psi_L = \frac{v_L}{v_0} \quad \lambda_0 = \frac{(-F_0)L}{3A_0 v_0 \mu_{\infty 0}}, \tag{9.64}$$

Substituting these dimensionless variables and parameters into the equations of continuity, motion, and energy and into the constitutive equation yields:

$$\xi\psi = 1 \tag{9.65}$$

$$\frac{d\lambda}{d\zeta} = \frac{d}{d\zeta}\left(\frac{C_1}{\xi} - C_4\sqrt{\xi}\right) - C_3\xi + \frac{C_2}{\xi^{1.095}} \tag{9.66}$$

$$\frac{d\theta}{d\zeta} = (\text{St})\frac{[1 + (8\xi\psi_{ay})^2]^{1/6}}{\xi^{1/6}}(\theta_a - \theta) \tag{9.67}$$

$$\frac{d\xi}{d\zeta} = \begin{cases} \lambda\xi \dfrac{\exp(E'/(T_0+273))}{\exp(E'/(\theta T_0+273))} & \text{when } \theta T_0 \geqslant 60°\text{C} \\ 0 & \text{when } \theta T_0 < 60°\text{C}' \end{cases} \tag{9.68}$$

where the definitions of the various constants are:

$$C_1 = \frac{\rho L v_0}{\mu_{\infty 0}} = \text{Re} \qquad C_2 = \frac{3.12 \times 10^{-4} v_0^{0.19} L^2}{A_0^{0.905} \mu_{\infty 0}}$$

$$C_3 = \frac{\rho g L^2}{v_0 \mu_{\infty 0}} = \frac{\text{Re}}{\text{Fr}} = \left(\frac{\rho L v_0}{\mu_{\infty 0}}\right)\left(\frac{gL}{v_0^2}\right)$$

$$C_4 = \frac{LH\sqrt{\pi}}{v_0 \mu_{\infty 0}\sqrt{A_0}} = \frac{\text{Re}}{\text{Ca}} = \left(\frac{\rho L v_0}{\mu_{\infty 0}}\right)\left(\frac{H\sqrt{\pi}}{\rho v_0^2 \sqrt{A_0}}\right)$$

$$\text{St} = \frac{1.67 \times 10^{-4} L}{\rho \bar{C}_p A_0^{5/6} v_0^{2/3}}, \tag{9.69}$$

and where Re, Fr, Ca, and St are the Reynolds, Froude, capillary (or Weber), and Stanton numbers, respectively. Also, note that in Eqs. 9.66, 9.67, and 9.68 the continuity equation, Eq. 9.65, was used. The boundary conditions, Eqs. 9.61 and 9.62, become:

B.C.1 at $\zeta=0$: $\xi = \psi = \theta = 1$; $\quad \lambda = \lambda_0$ (9.70)

B.C.2 at $\zeta=1$ or at $\zeta$ such that $\theta = 60/T_0$: $\quad \psi = \psi_L$ (9.71)

The solution for the Newtonian and isothermal case developed in Subsection 9.1.1 can be found by solving Eqs. 9.66 to 9.68 (see Pr. 9A.3) using the appropriate simplifications. Furthermore, the Newtonian and isothermal case can be extended to include all the additional forces, that is, to include the forces of air drag, surface tension, gravity, and inertia. The numerical solution of Eqs. 9.66 to 9.68 and its comparison with experimental data were shown by Kase and Matsuo (1967) for low-speed melt-spun polypropylene with the following parameters: $C_1 = C_2 = C_3 = C_4 = 0$, $v_L = 500$ m/min, $v_{ay} = 20$ cm/s,

$T_a = 20°C$, $\rho = 0.83$ g/cm³, $\bar{C}_p = 0.7$ cal/g °C, $E' = 3,500$ °K (i.e., $E \approx 7000$ cal/mol), and take-up denier (dpf) = 8 (see Pr. 9C.1). The denier per filament (dpf) is the weight in grams of a 9000-m-long filament. Thus, for $\dot{m}$ in g/s, $v_L$ in m/s, $\rho$ in g/cm³, and $R_L$ in cm the definition becomes:

$$\text{dpf} = 9000\frac{\dot{m}}{v_L} = 2.83 \times 10^6 \rho R_L^2 \qquad (9.72)$$

The tenacity of a fiber is the ultimate stress expressed as grams-force per denier (gf/d): (tenacity) = (strength)/($9 \times 10^5 \rho$), where the strength is expressed in grams-force/cm². Figure 9.5 shows the data from Kase and Matsuo (1967). The agreement between the theory and experiments is good. Note that similar analyses have been carried out by other groups for polymers such as HDPE, LDPE, PET, and so on (Ziabicki & Kawai, 1985).

The simulation of nonisothermal fiber spinning of PET at intermediate spinning speeds by George (1982) is worth mentioning at this point. His model works well for spinning speeds from 1000 to 3000 m/min. For PET with an intrinsic viscosity (IV) equal to 0.675 dL/g (1 dL = 100 cm³), which is a measure of molecular weight, extrusion temperatures of 290 to 315°C and shear rates of 2500 to 32,000 s⁻¹, the die swell, $B$ (Subsection 7.2.3), is calculated as:

$$B \approx 0.627 \exp\left[-3650\left(\frac{1}{548} - \frac{1}{T+273}\right)\right]\dot{\gamma}^{0.134}, \qquad (9.73)$$

where $\dot{\gamma}$ is the shear rate at the die wall. The elongational viscosity, $\bar{\eta}_1$,

is given by:

$$\bar{\eta}_1 = 3.82 \times 10^{-5} \exp\left[\frac{6802}{T+273}\right], \qquad (9.74)$$

and the momentum equation contains only inertia, air drag, gravitational, and tension forces (note that surface tension force is not included). The air drag force and the heat transfer coefficient are calculated as described in Subsection 9.1.2.3 (George, 1982). Some results from the numerical solution of the four equations (continuity, momentum, heat transfer, and constitutive) and experimental data are shown in Fig. 9.6.

Finally, the relative importance of the various forces in Eq. 9.58 is shown in Fig. 9.7 (from Ziabicki & Kawai, 1985) as a function of the takeup speed. The melt-spun polymer is a polyester, the spinneret radius is 125 µm, the filament radius at takeup is 9.25 µm, the viscosity is 300 Pa·s, the cooling air is stationary, and the mass flow rate is proportional to the spinning speed. It is clear from this figure that for low takeup speeds rheological and inertial forces are significant. As the takeup speed increases, the air drag force increases dramatically, the inertial force increases to a lesser extent, and the rheological force remains constant. Gravity and surface tension forces are negligible at all takeup speeds.

### 9.1.3. ISOTHERMAL VISCOELASTIC MODEL

Analyses of the Newtonian isothermal and nonisothermal models can be even further complicated by the introduction of the viscoelastic

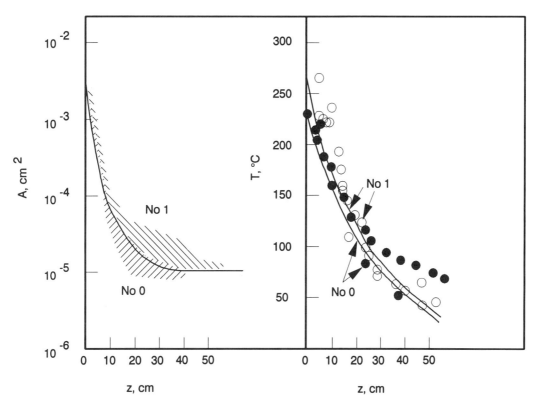

**Figure 9.5** Filament cross sectional area, $A$, and temperature, $T$, at different spinneret temperatures. Data points represent experiments of polypropylene filaments with the following parameters: takeup speed: 500 m/min; cooling cross air flow: 20 cm/s; temperature of cooling air: 20°C; take-up filament denier: 8; diameter of the spinneret nozzle: 0.6 mm; (No 0): spinneret temperature is 240°C, and (No. 1): spinneret temperature is 270°C. Solid lines represent the theoretical calculations. (Reprinted by permission of the publisher from Kase & Matsuo, 1967.)

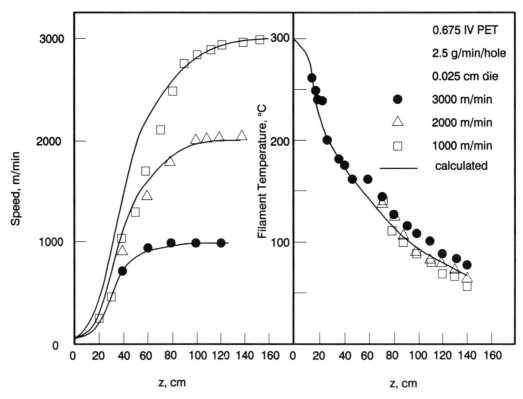

**Figure 9.6** Comparison of calculated and experimental speed and temperature profiles for PET. (Reprinted by permission of the publisher from George, 1982.)

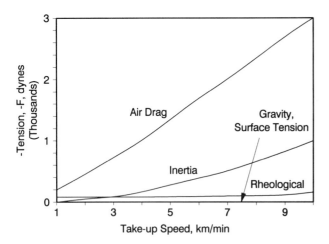

**Figure 9.7** Calculated relative importance of the various forces in the melt-spinning process as a function of the take-up speed. The spinning conditions are: spinneret radius: 125 μm, filament radius at take-up: 9.25 μm, extrusion viscosity: 300 Pa·s, and stationary cooling air. (Reprinted by permission of the publisher from Ziabicki & Kawai, 1985.)

nature of the polymers in the melt state. The viscoelasticity of the polymer is important in cases where the relaxation time, $\lambda$, is of the same order of magnitude or slower than the characteristic time constant of the process, which might be taken to be equal to $v_0/L$. The ratio of these two time constants is called the Deborah number (Eq. 3.90), and

it is equal to:

$$\text{De} = \frac{v_0 \lambda}{L}, \tag{9.75}$$

where $\lambda$ is the relaxation time of the polymer in extensional flows.

The complete analysis should include both the nonisothermal nature of the melt-spinning process and the viscoelastic nature of the polymer. Analytical solutions for this situation do not exist. We outline for pedagogical reasons the solution to the isothermal case but without going into significant details. Denn et al. (1975) and Fisher and Denn (1976) evaluated the effect of the viscoelastic nature of the fluid on melt-spinning.

The momentum equation for the Newtonian case, Eq. 9.20, can be also used in the viscoelastic case and can be written for the case of negligible inertia as:

$$2\frac{R'}{R}\pi_{zz} + \pi'_{zz} = \frac{1}{R^2}\frac{d}{dz}(R^2\pi_{zz}) = 0. \tag{9.76}$$

The force at the takeup point, $F$, is given by:

$$F = \pi R_L^2 \pi_{zz}|_L. \tag{9.77}$$

Combining Eqs. 9.76 and 9.77, together with Eq. 9.3, we get:

$$\pi_{zz} = \frac{\rho F v_z}{\dot{m}} \tag{9.78}$$

A more convenient way of expressing Eq. 9.78 is by considering that $\pi_{rr} \approx 0$ (Eqs. 9.14 and 9.15). Thus,

$$\pi_{zz} - \pi_{rr} = \tau_{zz} - \tau_{rr} = \frac{\rho F v_z}{\dot{m}}. \qquad (9.79)$$

The constitutive equation used by Denn and coworkers (Denn et al., 1975; Fischer & Denn, 1976) was the White-Metzner model (see also Subsection 3.2.1; Eq. 3.42), which was thought to be applicable to high Deborah number processes such as melt spinning. The $zz$ and $rr$ components of the constitutive equation in cylindrical coordinates are:

$$\tau_{zz} + \lambda \left( v_z \frac{d\tau_{zz}}{dz} - 2\tau_{zz} \frac{dv_z}{dz} \right) = -\eta \dot{\gamma}_{zz} = -2\eta \frac{dv_z}{dz}$$

$$\tau_{rr} + \lambda \left( v_z \frac{d\tau_{rr}}{dz} + \tau_{rr} \frac{dv_z}{dz} \right) = -\eta \dot{\gamma}_{rr} = \eta \frac{dv_z}{dz} \qquad (9.80)$$

where the relaxation time, $\lambda$, is related to viscosity, $\eta$, by the relationship:

$$\lambda = \frac{\eta}{G}, \qquad (9.81)$$

and $G$ is the shear modulus of the polymer. Finally, the viscosity function, $\eta$, was considered to obey the power law, so that the model will include both the shear-thinning and the elastic character of the polymer. Equations 9.79 and 9.80 can be combined to eliminate the stress components and yield the following equation (Fisher & Denn, 1976):

$$\psi + (\alpha\psi - 3\varepsilon)(\psi')^n - 2\alpha^2\psi(\psi')^{2n} - n\alpha\psi^2(\psi'')(\psi')^{n-2} = 0, \qquad (9.82)$$

where $\psi$ is the dimensionless velocity, $v_z/v_0$, $\psi'$ denotes differentiation of $\psi$ with respect to $\zeta\ (=z/L)$, $\alpha$ and $\varepsilon$ are dimensionless parameters given by:

$$\alpha = \frac{m(3)^{(n-1)/2}}{G}\left(\frac{v_0}{L}\right)^n \qquad \varepsilon = \frac{m\dot{m}(3)^{(n-1)/2}}{\rho|F|L}\left(\frac{v_0}{L}\right)^{n-1}, \qquad (9.83)$$

and $m$ is the consistency of the polymer melt. Note that the ratio of the two dimensionless parameters is $\varepsilon/\alpha = GA_0/|F|$ and that $\alpha$ is a purely rheological parameter whereas $1/\varepsilon$ is a dimensionless force.

The boundary conditions of the problem are the same as for the Newtonian isothermal case (Eq. 9.25) but with the addition of one more boundary condition:

$$\text{B.C.3 at } z = 0 \qquad \tau_{zz} = \tau_0. \qquad (9.84)$$

It is difficult to make a good estimate of the value of $\tau_0$, but Fisher and Denn (1976) showed that this posed no major problem. The numerical solution of Eq. 9.82 with boundary conditions, Eqs. 9.25 and 9.84, is shown in Fig. 9.8 (from Fisher & Denn, 1976), where also experimental data of melt-spun polystyrene (Zeichner, 1973) are shown. The agreement between theory and experiments is not good because the parameter $\alpha$ for polystyrene has an experimental value of about 0.2 to 0.3 whereas the theory fits the data with $\alpha$ about 0.4 to 0.5. Nevertheless, the viscoelastic theory we described provides the general effect of viscoelasticity on the melt-spinning process.

Better agreement with experiments was achieved by Phan-Thien (1978) who solved the fiber-spinning problem using the PTT viscoelastic model (see Eq. 3.45). In this case the constitutive equation was fitted to data for LDPE and PS, and the solutions to the fiber-spinning problem were compared to experimental data.

**EXAMPLE 9.2**
**Dimensionless Tension, $1/\varepsilon$, for Newtonian and Power-Law Fluids**

Calculate the dimensionless tension, $1/\varepsilon$ (Eq. 9.83), for a power-law fluid. Then find its limit as $n \to 1$ (i.e., for a Newtonian fluid). Use $n = 0.4$ and $D_R = 15$ in both cases.

**Solution**

To calculate $1/\varepsilon$ we need to provide expressions for the tension $F$ and the mass flow rate, $\dot{m}$. The mass flow rate is calculated as:

$$\dot{m} = \rho\pi R_L^2 v_L = \rho\pi R_L^2 v_0 D_R \qquad (9.85)$$

and the force as:

$$F = \pi R_L^2 \pi_{zz}|_L = -\pi R_L^2 m(3)^{(n+1)/2}(v_z')^n|_L. \qquad (9.86)$$

Substitution of Eqs. 9.85 and 9.86 along with the velocity profile obtained in Pr. 9B.1 into Eq. 9.83 yields:

$$\frac{1}{\varepsilon} = 3\left\{ \frac{n}{1-n}\left[1 - (D_r)^{n-(1/n)}\right] \right\}^n. \qquad (9.87)$$

The limit of Eq. 9.87 for $n \to 1$ is easily calculated to be:

$$\frac{1}{\varepsilon} = 3\ln D_R. \qquad (9.88)$$

Finally, we obtain:

$$\left(\frac{1}{\varepsilon}\right)_{n=0.4} = 2.53; \qquad \left(\frac{1}{\varepsilon}\right)_{n=1} = 8.12. \qquad (9.89)$$

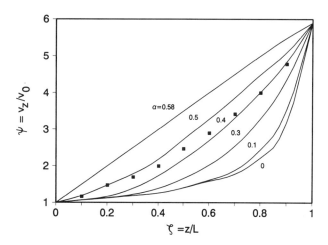

**Figure 9.8** Theoretical dimensionless velocity profiles, $\psi$, as a function of the dimensionless axial distance, $\zeta$, for the following values of the parameters: $n = 1/3$, $D_R = 5.85$, and $\tau_0\dot{m}/\rho v_0 F = 1$. Data shown are for the isothermal spinning of polystyrene at 170°C. (Reprinted by permission of the publisher from Fisher & Denn, 1976.)

### 9.1.4. HIGH-SPEED SPINNING AND STRUCTURE FORMATION

The discussion of fiber spinning in the preceding sections referred to low and moderate spinning speeds in which very little degree of crystallinity can be generated in the spun fibers, especially for slowly crystallizing polymers (e.g., PET; Table 5.19). For example, PET needs to be spun at speeds of 5000 m/min and above in order to exhibit significant crystallinity (see Fig. 5.23). As mentioned at the beginning of this chapter, the lack of crystallinity during the conventional low-speed spinning process forced the industry to use an additional step (drawing) immediately after the spinning process (two-step process, TSP). In this way, the PET fibers exhibit a high degree of orientation, and they are *fully oriented yarns* (FOY).

Nevertheless, the objective of the synthetic fiber industry is to make fibers in a simple, one-step process, in which melt-spun fibers will have a high degree of orientation and crystallinity (structure formation). For slowly crystallizing polymers that simple process is high-speed (3500–5000 m/min) or very-high-speed (> 5000 m/min) spinning, which is a one-step process (OSP) in which a high degree of orientation and crystallinity occur during the spinning process. In this way, relatively lower equipment and energy costs are required than with conventional TSP spinning. Furthermore, increased productivity can be achieved.

High-speed spinning is an attractive process. The mechanical properties of high-speed-spun PET fibers are no better than those produced by TSP (see Table 9.1). The density of the OSP fibers is higher than that of the TSP fibers, which means that the crystallinity is higher, but the birefringence is lower, which means that the orientation is lower. However, the boil-off shrinkage and the elongation at break are better for fibers generated by OSP than by TSP.

Next, we discuss the *formation of structure*, that is, crystallization, crystalline morphology, and orientation during the spinning process. We use PET as a model polymer because data for this polymer are abundant in the literature. PET is a slowly crystallizing polymer ($t_{1/2} \approx 50$ s; Subsection 5.5.1), but under high stress conditions, stress-induced crystallization takes place. Thus, high-speed spinning is necessary for PET crystallization. However, rapidly crystallizing polymers, such as Nylon 6,6 ($t_{1/2} \approx 0.42$ s) and PE (see also Subsection 5.5.1 and Table 5.19), can crystallize even in low- or intermediate-speed spinning processes. Finally, crystallization does not take place in noncrystallizable polymers (e.g., atactic PS), and thus orientation is the only structure formation achieved during the spinning process of these polymers.

Crystallinity is important in fibers, because its degree affects the mechanical, diffusional, and solubility properties, as well the shrinkage of the fibers. In general, crystallization takes place under either quiescent or stress conditions. The effects of both these states on crystallinity were discussed in Section 5.5. When crystallization takes place during the spinning process, the energy equation, Eq. 9.38, should be rewritten as:

$$\rho \bar{C}_p \left( \frac{\partial T}{\partial t} + v_z \frac{\partial T}{\partial z} \right) = -\frac{1}{r} \frac{\partial}{\partial r}(r q_r) - \frac{\partial q_z}{\partial z} + \rho_c \Delta \bar{H}_c \frac{d\phi_c}{dt}, \tag{9.90}$$

where $\Delta \bar{H}_c$ is the (latent) heat of crystallization per unit mass (see Appendix C for tabulated values), $\rho_c$ is the density of the crystalline phase, and $\phi_c$ is the volume fraction of the crystalline phase (Eq. 5.160). Consequently, the steady-state form of Eq. 9.48 becomes (Lin et al., 1992):

$$\frac{dT}{dz} = -\frac{2\pi R U (T - T_a)}{\dot{m} \bar{C}_p} + \frac{\rho_c}{\rho} \frac{\Delta \bar{H}_c}{\bar{C}_p} \frac{d\phi_c}{dz}. \tag{9.91}$$

This equation adds one more unknown, the degree of crystallinity, $\phi_c$,

to the system of equations: continuity, constitutive, momentum, and energy. Thus, one additional equation is needed, which relates the crystallization rate to the spinning and physical characteristics of the system. This additional equation comes from a simplified approach to nonisothermal crystallization under conditions of molecular orientation which, is discussed next.

Nucleation and growth rates, and thus crystallization rates, are very sensitive to temperature. This sensitivity comes from the effect of the temperature on the energy terms in the respective equations of nucleation and growth rate. The experimentally observed crystallization half-time, $t_{1/2}$ (time for $\phi_c/\phi_\infty = 1/2$), or its reciprocal rate constant, $K(T)$, ($t_{1/2} = (\ln 2/K(T))^{1/n}$; see also Eq. 5.159), was shown to obey an empirical relationship of the following form (Ziabicki 1976):

$$K(T) = K_{max} \exp\left[-4 \ln 2 (T - T_{max})^2 / D^2\right], \tag{9.92}$$

where $D$ is the half-width of the $K(T)$ curve (similar to Fig. 5.20), $K_{max}$ is the maximum in the rate–temperature curve, and $T_{max}$ is the temperature at which $K = K_{max}$. Table 9.2 shows some typical values of $K_{max}$ and $D$ for various polymers. Note that Eq. 5.163 is another form of an empirical equation used to fit crystallization-rate-versus-time data.

The area under the rate curve can be shown to be:

$$\int_{T_g}^{T_m} K(T) dT \approx 1.064 K_{max} D = J, \tag{9.93}$$

where $J$ is called *kinetic crystallizability* and characterizes the degree of crystallinity achieved when the material is cooled from the melting temperature, $T_m$, to the glass transition temperature, $T_g$, at unit cooling rate. Typical values of the kinetic crystallizability are shown in Table 9.2. Thus, for the same cooling rate and for $\phi_c \ll \phi_\infty$, the degree of crystallinity of Nylon 6,6 will be 125 times higher than that for PET.

Polymer crystallization is also very sensitive to molecular orientation in the amorphous regions. Orientation affects the entropy and enthalpy of fusion, the nucleation rate, and so on, but the mathematics of the problem goes beyond the scope of the present textbook. Instead, we use Ziabicki's (1976) idea that any function of molecular orientation, $X(f_{am})$, that is, melting temperature $T_m$, crystallization rate $K$, free energy $\Delta F$, and so on can be expanded as a series:

$$X(f_{am}) = X(0) + a_2 f_{am}^2 + \cdots, \tag{9.94}$$

where $f_{am}$ is the orientation of the amorphous region of the polymer before crystallization. Note that for symmetry reasons, Eq. 9.94 does not include a linear term. For low degrees of orientation we can obtain:

$$\ln\left[K(f_{am})/K(0)\right] \approx A(T) f_{am}^2, \tag{9.95}$$

**TABLE 9.2 Kinetic Crystallization Characteristics of Various Polymers**

| Polymer | $K_{max}$ (s$^{-1}$) | $D$ (°C) | $J$ (°C/s) |
|---|---|---|---|
| PP (isotactic) | 0.55 | 60 | 35.0 |
| PET | 0.016 | 64 | 1.1 |
| Nylon 6 | 0.14 | 46 | 6.8 |
| Nylon 6,6 | 1.64 | 80 | 139 |
| PS (isotactic) | 0.0037 | 40 | 0.16 |

(Data from Ziabicki, 1976.)

or, after combination with Eq. 9.92:

$$K(T, f_{am}) = K(T, 0) \exp[A(T) f_{am}^2]$$
$$= K_{max}^0 \exp[-4 \ln 2(T - T_{max}^0)^2/(D^0)^2 + A(T) f_{am}^2], \quad (9.96)$$

where the parameters with superscript 0 refer to the unoriented state. Equation 9.96 shows that the rate of crystallization increases with orientation, $f_{am}$, as $A(T)$ is always positive. Note that Eq. 9.96 is similar to Eq. 5.163. For PET the function $A(T)$ is shown to be equal to:

$$A(T) = \frac{3.09 \times 10^{10} - 1.55 \times 10^8 (284 - T)}{(284 - T)^3}. \quad (9.97)$$

The rate of crystallization can now be assessed using the modified Avrami's equation (Eq. 5.162) as:

$$\frac{d\phi_c}{dz} = \frac{n\phi_\infty K(T, f_{am})}{v_z} \left( \int_0^z \frac{K(T, f_{am})}{v_z} dz' \right)^{n-1}$$
$$\times \exp\left( -\left( \int_0^z \frac{K(T, f_{am})}{v_z} dz' \right)^n \right), \quad (9.98)$$

where $K$ is given by Eq. 9.96. The boundary condition for the above equation is:

$$\phi_c = 0 \quad \text{at} \quad z = 0. \quad (9.99)$$

The relation between the orientation of the amorphous region and the state of stress in the filament is shown next. From Eq. 5.180 we know that the orientation of the amorphous region of a polymer melt is directly related to the stress applied to the polymer. But during the spinning process, temperature, polymer relaxation, and extensional rate might have an effect on the orientation. The equations that relate birefringence to stress in the filament are usually empirical. For polyesters the relevant equation can be either (Katayama & Yoon, in Ziabicki & Kawai, 1985):

$$\Delta N = 0.2 \left[ 1 - \exp\left\{ -\frac{1.65 \times 10^{-6} (F/A)}{T + 273} \right\} \right] \quad (9.100)$$

or (Shimizu et al., in Ziabicki & Kawai, 1985):

$$\frac{d(\Delta N)}{dz} = \frac{C_{opt}}{v_z} \frac{dv_z}{dz} - \frac{G}{\eta} \frac{\Delta N}{v_z}, \quad (9.101)$$

where $C_{opt}$ is equal to 0.53. For LDPE the relevant equation is (Ziabicki, 1976):

$$\frac{\Delta N}{dv_z/dz} = C_{opt} \left[ \frac{F/A}{dv_z/dz} - \eta_1 \right], \quad (9.102)$$

where $\eta_1$ is equal to 6000 Pa·s, and $C_{opt}$ is equal to $1.5 \times 10^{-11}$ Pa.

In conclusion, when crystallization takes place due to either the nature of the polymer or to high-speed spinning (like in PET), the system of steady-state equations that describes the spinning process consists of the following equations: continuity, Eq. 9.57; momentum, Eq. 9.58; energy, Eq. 9.91; constitutive, Eq. 9.60; crystallization, Eq. 9.98; and orientation, Eq. 9.100, 9.101 or 9.102. The relevant boundary conditions are Eqs. 9.61, 9.62, and 9.99. Various numerical solutions of the system of equations relevant to specific polymers are

presented in Ziabicki and Kawai (1985), and they go beyond the scope of this textbook.

Finally, the manner in which individual molecules crystallize depends on the nature of the polymer molecule and the conditions under which crystallization takes place (see Fig. 5.18, p. 121). It is worth noting at this point the work of Dees and Spruiell (1974) concerned with the structure development during melt spinning of HDPE fibers. Figure 9.9 presents their morphological model based on crystallite orientation factors and some other information. At low takeup speeds they assumed spherulitic growth. At higher speeds the morphological model is that of row-nucleated, twisted, ribbonlike lamella, and at even higher speeds that of row-nucleated untwisted lamellae.

## 9.1.5. INSTABILITIES IN FIBER SPINNING

In general, the rate of production in many polymer processing operations is limited by the onset of instabilities. In melt spinning there are two major types of instabilities (Petrie & Denn, 1976). The first type is called *spinnability* and refers to the ability of a polymer melt to be transformed into long fibers (i.e., to be drawn to large elongations) without breaking because of either *capillary waves* and *necking (ductile)* or *cohesive (brittle) fracture*. The spinnability is due to the free boundary flow between the spinneret and the takeup roll. The second type is called *draw resonance*, and it appears as a periodic fluctuation in the takeup cross-sectional area. Besides these two types of instability specific to melt spinning, typical instabilities associated with flow through dies usually referred to as *melt fracture* (Subsection 7.2.2) are also present. We first discuss spinnability and then draw resonance.

Brittle fracture refers to the situation of the tensile stress, $\tau_{zz}$, of a polymer jet exceeding some critical value (tensile strength), $\tau^*$. This type of fracture is possible in viscoelastic materials because these materials store some of the deformational energy, whereas purely viscous materials dissipate all the deformational energy. Figure 9.10 shows a schematic of a polymer filament failing because of cohesive fracture. As the polymer fiber is being drawn, its tensile stress, $\tau_{zz}(z)$, and strength, $\tau^*(z)$, increase with the axial distance $z$. At a certain axial distance, $z_{coh}^*$, both the tensile stress and strength are equal. Beyond that point, the tensile stress exceeds the strength, and the material fails cohesively. For isothermal spinning of Newtonian fluids the maximum length of a polymer fiber is calculated as (Ziabicki, 1976):

$$z_{coh}^* = \frac{1}{\beta} \ln\left( \frac{(2e_{coh}E)^{1/2}}{3\eta v_0 \beta} \right), \quad (9.103)$$

where $e_{coh}$ is the cohesive energy density (CED) of the material, $\beta$ is the deformation gradient defined as $d \ln v/dz$, and $E$ is the modulus of elasticity of the polymer.

The other mechanism responsible for instabilities during melt spinning is referred to as capillary waves or Rayleigh instabilities. Depending on the velocity of the polymer melt in the spinneret hole, three broad regimes can be distinguished: (1) formation of droplets; (2) formation of a liquid jet sustaining waves at its interface, which finally disintegrates into droplets, Fig. 9.11; and (3) complete atomization. For polymer melts, the disintegration step can be described by the following equation:

$$z_{cap}^* = 12d\left( Ca^{1/2} + 3\frac{Ca}{Re} \right), \quad (9.104)$$

where $z_{cap}^*$ is the maximum uninterrupted jet length, $d$ is the diameter of the jet, Ca $(= v^2 d\rho/\gamma)$ is the capillary number of the jet, and Re $(= vd\rho/\eta)$ is the jet Reynolds number. Note that the analysis of the stability of a molten jet is similar to the stability analysis of an extended droplet (mentioned in more detail in Subsection 6.5.2). Finally, note that

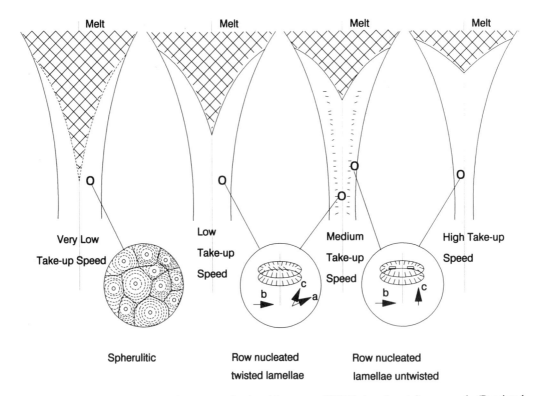

**Figure 9.9** Morphological model of structures developed in as-spun HDPE at various take-up speeds. (Reprinted. by permission of the publisher from Dees & Spruiell, 1974.)

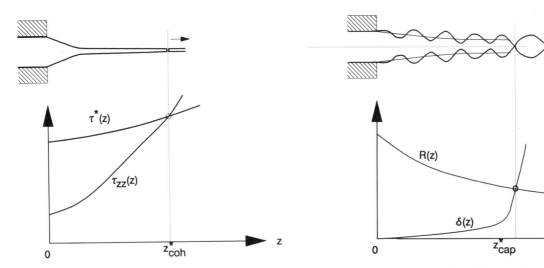

**Figure 9.10** Cohesive (brittle) fracture of a molten thread.

**Figure 9.11** Breakup of a molten thread due to capillary wave instability.

because of their higher viscosity and lower surface tension, polymers compared to metals and glass, can be safely drawn to larger lengths as Eq. 9.104 indicates (for the same extrusion velocity).

Diagrammaticaly, Fig. 9.12 shows the space of all possible conditions of material properties and spinning characteristics. This space is further divided into the various regions of: spinnability $S$, hydrodynamic stability $H$, cohesive fracture $F$, capillary breakup $C$, and hydrodynamic instability $x$-$H$. A system is called hydrodynamically stable if an imposed small perturbation decays with time to either zero

or some small steady value (see about the growth factor $q$ in Eq. 6.194). The cohesive fracture region $F$ is included into the hydrodynamic stability region. Finally, spinnability region $S$ consists of the part of region $H$ where no cohesive fracture takes place and the part of the $x$-$H$ region where the growth is too slow to cause breakage.

The various melt-spun materials can be divided into three groups (Ziabicki, 1976): (1) metals and glasses; (2) linear polycondensates (polyesters and polyamides) with relatively low molecular weights (from 10,000 to 30,000); and (3) linear polyolefins and vinyl polymers (PE,

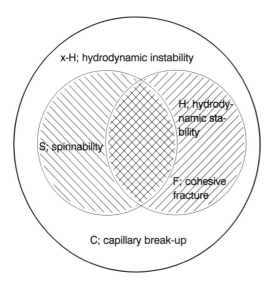

**Figure 9.12** Space of all possible melt-spinning conditions, including regions of: hydrodynamic stability $H$, cohesive fracture $F$, capillary break-up $C$, spinnability $S$, and hydrodynamic instability $x-H$. (Reprinted by permission of the publisher from Ziabicki, 1976.)

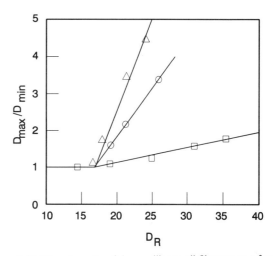

**Figure 9.13** Diameter ratio of drawn silicone oil filament as a function of draw ratio. Polymer volumetric flow rate in mm³/s and spinning length in mm are: △: 64.1, 20; ○: 44.4, 20; □: 30.8, 40. (Reprinted by permission of the publisher from Donnelly & Weinberger, 1975.)

PP, PVC, etc.) with relatively high molecular weights (from 50,000 to 1,000,000). The basic differences between these groups are:

1. Metals and glasses are primarily Newtonian fluids with high surface tension (from 100 to 500 mN/m) and thus a high probability of capillary breakup.
2. Linear polycondensates are also primarily Newtonian (or slightly viscoelastic with short relaxation times) with low shear viscosity (about 100 Pa·s) and high spinnability. The die swell is obviously low (about 1.0 to 1.5) and cohesive fracture is not usually a problem. For example, the critical shear rate for the onset of melt fracture in nylon 6,6 is about $10^5\,\text{s}^{-1}$ at 275°C, and the maximum spinneret shear rate is usually not higher than $10^4\,\text{s}^{-1}$. Usually, takeup velocities of 4000 to 5000 m/min cause no brittle fracture of these materials.
3. Linear polyolefins and vinyl polymers include melts with high shear viscosity (higher than 1000 Pa·s), with strong viscoelastic behavior, and long relaxation times. Usually, the spinning velocities of polyolefins are lower than those of polycondensates because of cohesive fracture. Also, die swell is extremely pronounced in these melts.

Draw resonance appears as a sustained periodic fluctuation with a well-defined and steady period and amplitude of the cross section at the takeup roll, and it occurs even when the flow rate and the takeup speed are constant. This instability should not be confused with the spinnability, as it has nothing to do with breakup of the filament. It appears in both purely viscous and elastic fluids, and two factors reduce its effect: elasticity and nonisothermal conditions of spinning.

The steady-state solution for fiber spinning (Newtonian and isothermal case) was presented in Subsection 9.1.1, and it consists of Eqs. 9.26 and 9.28. Linearized (small disturbances) stability analysis involves (Fisher & Denn, 1976) the study of finite amplitude disturbances, and we do not present it. Rather, we present the results of such an analysis. The value of $D_R = 20.21$ is considered to be the critical draw ratio beyond which the flow becomes unstable. Fig. 9.13 (Donnelly & Weinberger, 1975) shows experimental data that confirm the theory. More specifically, silicone oil (of viscosity equal to 100 Pa·s),

which seems to be Newtonian, was extruded, and the ratio of maximum to minimum filament diameters was plotted against the draw ratio. An instability appears at a draw ratio of about 17, or about 22 if we take into consideration about 14% die swell. The value of the critical draw ratio of 22 compares well with the theoretical value of 20.21. Pearson and Shah (1974) extended the analysis to a power-law fluid and included surface tension, gravitational and inertial forces in the momentum balance, and found that the critical draw ratio is lower than 20.21 for shear-thinning fluids and larger than 20.21 for shear-thickening fluids. For example, the critical draw ratio range is from 3 to 5 for a fluid with a power-law index of 0.4 to 0.5.

The energy equation, Eq. 9.39, should be incorporated into the model to account for temperature variation along the filament axial length. Pearson and Shah (1974) solved the system of equations subjected to a linearized analysis and found that the critical draw ratio depends, besides the power-law index, on the dimensionless number, S:

$$S = k(T_0 - T_a)(\text{St})e^{-\text{St}}, \tag{9.105}$$

where $k$ is the viscosity temperature coefficient of the viscosity function ($\eta_0 \propto \exp[-k(T-T_a)]$) and St is the Stanton number defined as

$$\text{St} = 2\sqrt{\frac{\pi v_0^{1/3}}{\rho \dot{m}}}\frac{\varsigma L}{\bar{C}_p}, \tag{9.106}$$

where $\varsigma$ (units: cal/cm$^{8/3}$·s$^{1/3}$ °C) is the ratio: $U/v_z^{2/3}$. Note that the theoretical analysis of Pearson and Shah is based on the assumption that: $U \propto (v_z)^{2/3}$ (valid if Eq. 9.50 is used). For shear-thinning fluids the critical draw ratio increases very slowly with S, whereas for Newtonian and shear-thickening fluids the increase is dramatic for $S > 0.1$.

Fisher and Denn (1976) presented the linearized stability analysis for the isothermal viscoelastic case as an extension of the steady state case presented in Subsection 9.1.3. The analysis showed (Fig. 9.14) that the critical draw ratio depends on the power-law index, $n$, and on the viscoelastic parameter $\alpha^{1/n}$, where $\alpha$ is defined in Eq. 9.83. Three regions are shown in Fig. 9.14: stable, unstable and unattainable. The lower boundary of the unattainable region is described by the relationship: $D_R = 1 + \alpha^{-1/n}$. At low values of the parameter $\alpha$, the constant viscosity fluids exhibit higher critical draw ratio, and the power-law fluids exhibit a critical draw ratio value as low as 3 to 5 for n from 0.33 to 0.5.

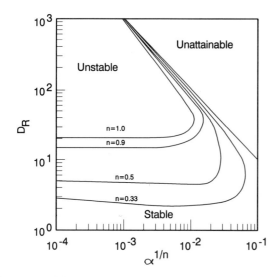

**Figure 9.14** Critical draw ratio, $D_R$, as a function of the viscoelastic parameter $\alpha$ and the power-law index, $n$, for the fiber-spinning process. (Reprinted by permission of the publisher from Fisher & Denn, 1976.)

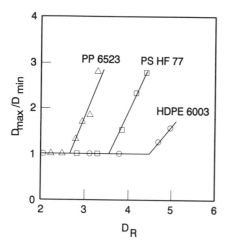

**Figure 9.15** Diameter ratio of drawn polymer filaments as a function of draw ratio for isothermal spinning at 218°C. Other conditions are: $v_0 = 5$ cm/s; $L = 16.5$ cm; and $R_0 = 0.08$ cm. (Reprinted by permission of the publisher from Weinberger et al., 1976.)

Finally, as the viscoelastic parameter $\alpha$ increases, the critical draw ratio can increase dramatically, and it can be extended up to the unattainable region.

Experimentally, the critical draw ratio for various polymers is measured as the draw ratio at which the ratio of maximum to minimum filament diameters increases above 1. Figure 9.15 shows experimental data for PP, HDPE, and PS. The corresponding critical draw ratios are: 2.7, 4.6, and 3.6 (Weinberger et al., 1976). The power-law index of PP and PS is about 0.5, and thus the agreement between experimental data and theory is generally good.

Finally, finite (large) amplitude stability analysis, which can give information about the large disturbances and macroscopic diameter variations, supports the findings of the linearized stability analysis. Thus, for draw ratios less than 20.21 a Newtonian system is stable to finite amplitude disturbances, and for draw ratios larger than 20.21 disturbances grow and reach a sustained oscillation (draw resonance).

### EXAMPLE 9.3
### Critical Draw Ratio and Spinning Length

A viscoelastic material is being melt-spun at $D_R = 10$. Estimate the possibility of draw resonance if the spinning length is 3 m. The various properties of the material are: $n = 1/3$, $m = 3000$ Pa·s$^{1/3}$, $G = 2000$ Pa, $R_0 = 200$ μm, $\rho = 1$ g/cm$^3$, and $\dot{m} = 0.02$ g/s. Apply the analysis of Fisher and Denn (1976). Also indicate the maximum spinning length for stable operation at the same draw ratio.

### Solution

At a draw ratio of 10 and power-law index of 1/3, Fig. 9.14 shows that the minimum and the maximum attainable values of the viscoelastic parameter $\alpha^{1/n}$ are 0.043 and 0.1, respectively. The velocity at zero axial distance is given as 7.09 cm/s. Thus, the parameter $\alpha^{1/n}$ is calculated as 0.027, which is outside the limits of 0.043 and 0.1. Melt spinning with these conditions is expected to result in draw resonance. To get rid of this problem we need to decrease the spinning length to about 180 cm.

## 9.2. FILM CASTING AND STRETCHING

A large activity of the polymer processing industry is the production of films and sheets of thermoplastic polymers. By definition, the term *film* is used for thicknesses less than 250 μm (equal to about 0.010″), and the term *sheet* is used for thicker films. Note that in this section we will use the term film generically, and we will occasionally mention the term sheet when confusion might occur. These products are primarily used in the packaging industry for either foodstuffs (groceries, dairy produce, etc.) or other consumer products. Quite frequently, the properties of various polymers need to be combined by coating, lamination, or coextrusion. The major properties of the films and sheets are transparency, toughness, flexibility, and a very large aspect ratio (width or length to thickness) of about 10$^3$. Typical thickness values range from about 10 to 2500 μm, whereas the other two dimensions can vary from 40 to 320 cm.

Flat-film production consists mainly of the following three processes: *extrusion*, *casting*, and *stabilization*. Depending on the film thickness there are three major groups: fine film, with thickness of 10 to 50 μm, thicker cast film and sheet, with thickness of 100 to 400 μm, and thermoformable sheet, with thickness of 200 to 2500 μm. The first two groups are produced on chrome-plated chill-roll or water-bath lines (see Fig. 9.16), whereas the third one is produced with a special roll. All film types, after the chill roll or water bath, are trimmed at the edges (some curling might occur there) and either wound or undergo stretching (uniaxial or biaxial) or thermoforming. PP, PE, polyester, and polyamide are the four most frequently used polymers on chill-roll lines.

As an example, film casting is used for the production of polymer films for video and magnetic tapes (d'Halewyn et al., 1990). The film emerging from the die is first stretched between the die and the roll. Then, it is stretched in both transverse and machine directions and it assumes a thickness of about 12 μm. Finally, the film is stabilized at a temperature of around 100°C, cooled and rolled up.

### 9.2.1. FILM CASTING

Fig. 9.16 shows two typical configurations for film casting: the film freezes on contact with a chill roll or on submersion in a water bath. In both cases freezing occurs at a fixed point that moves at a known velocity. Figure 9.17 shows two views of the film between the extrusion die and the chill roll. In the following sections we describe first the

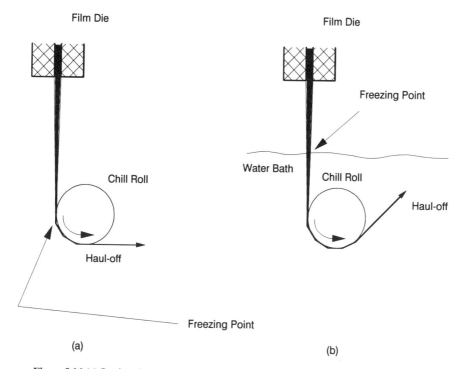

**Figure 9.16** (a) Sectional view of chill-roll casting. (b) Sectional view of waterbath casting.

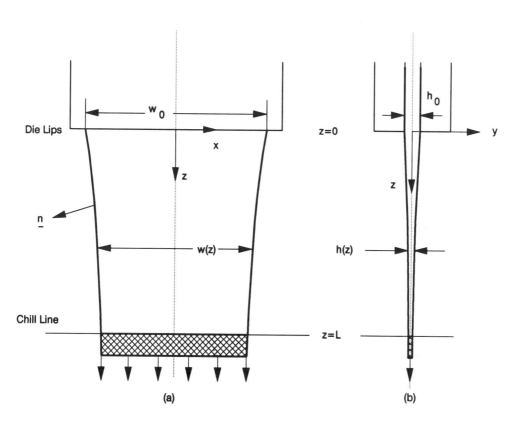

**Figure 9.17** (a) Geometry of the front view of the film-casting process. The polymer film between the die and the roll is called the web. (b) Side view of the film-casting process.

general form of equations for very thin films ($|\nabla h| \ll 1$), and then we present the isothermal Newtonian case for thicker films.

Consider the origin of the Cartesian coordinate system, $xyz$, at the intersection of the line of symmetry and the die lips (Fig. 9.17). The original width of the polymer film (in the $x$ direction and at the die lips) is $w_0$, and the extension length is $L$. The width of the film decreases along the $z$ direction because the film is being drawn in that direction by the chill roll. Its width at the roll is $w_L$. Similar drawing takes place in the $y$ direction as well, where the thickness changes from $h_0$ at the die lips to $h_L$ at the chill roll. The deformation in the $y$ and $x$ directions distinguishes the film casting flow from a typical plane flow.

The die swell effect will be neglected here as was done in the fiber-spinning analysis. In film casting no significant cooling is obtained between the face of the die and the chill roll (see Section 5.2), whereas in the melt spinning of fibers cooling was very important. If the $z$ component of the film velocity at the die lips is $v_{z0}$ and at the freeze line is $v_{zL}$, then the draw ratio, $D_R$, is defined as (see Eq. 9.27):

$$D_R = \frac{v_{zL}}{v_{z0}}. \tag{9.107}$$

Typical values of the draw ratio range from 2 to 20.

The objective of the analysis is to calculate the thickness and temperature at the chill roll as a function of the takeup speed, rheology, distance, $L$, and conditions at the die lips. In the case of $w_0 > L \gg h_0$, then changes in the $y$ direction can be considered insignificant for the rest of the analysis. Then, we postulate the following surface velocity field for the steady-state film-casting problem:

$$v_x = v_x(x, z); \qquad v_z = v_z(x, z) \tag{9.108}$$

with the associated thickness, $h$, being $h(x, z)$. The continuity equation at steady state can be written as:

$$\frac{\partial}{\partial x}(\rho h v_x) + \frac{\partial}{\partial z}(\rho h v_z) = 0. \tag{9.109}$$

The $x$ and $z$ components of the equation of motion are (Table 2.9, p. 21):

$$\frac{\partial}{\partial x}(\rho h v_x^2) + \frac{\partial}{\partial z}(\rho h v_x v_z) = -\frac{\partial(h\pi_{xx})}{\partial x} - \frac{\partial(h\pi_{xz})}{\partial z} \tag{9.110}$$

$$\frac{\partial}{\partial x}(\rho h v_x v_z) + \frac{\partial}{\partial z}(\rho h v_z^2) = -\frac{\partial(h\pi_{xz})}{\partial x} - \frac{\partial(h\pi_{zz})}{\partial z} + \rho g h, \tag{9.111}$$

where the surface tension and air drag forces are neglected because they are really unimportant. Also, note that we assume that the $z$ coordinate is vertical. If the film is drawn away from the vertical, the equations need to be modified. Finally, the energy equation can be written as follows. Since the thickness of the film is considered small, the average temperature (Pearson, 1985) over the film thickness is:

$$\bar{T} = \bar{T}(x, z) = \int_{-h/2}^{+h/2} T(x, y, z) \frac{dy}{h}. \tag{9.112}$$

The energy equation can now be based on the average temperature as:

$$\frac{\partial}{\partial x}(\rho h \bar{C}_p \bar{T} v_x) + \frac{\partial}{\partial z}(\rho h \bar{C}_p \bar{T} v_z) = -2h_a(\bar{T} - T_a), \tag{9.113}$$

where $T_a$ is the temperature of the surrounding medium (air), and $h_a$

is the air-side heat transfer coefficient. The heat generation by viscous dissipation has been neglected in Eq. 9.113.

The boundary conditions are:

B.C. 1 at $z = 0$    $v_x = 0$;    $v_z = v_{z0}$;    $h = h_0$;     $\bar{T} = \bar{T}_0$

B.C. 2 at $z = L$    $v_x = 0$;    $v_z = v_{zL}$

B.C. 3 at $x = \pm 1/2w(z)$    $\pi \cdot \mathbf{n} = 0$.      (9.114)

Note that B.C.3 is valid in cases where symmetry along the $z$ axis is preserved, and surface tension is insignificant (see also Eq. 9.8). One additional condition specifies the edge:

$$\frac{\partial w}{\partial z} = \frac{v_x}{v_z}. \tag{9.115}$$

Finally, the set of equations is complete with the incorporation of the rheological constitutive equation. The complicated set of equations, Eqs. 9.109, 9.110, 9.111, and 9.113 along with the constitutive equation can be simplified in the following special cases for Newtonian and isothermal conditions: (1) $L \gg w_0$, and (2) $L \ll w_0$ (see Pr. 9B.5).

In the case of thicker films the flow of the polymer will be considered as two-dimensional along the $z$ (machine) and $y$ (transverse) axes instead of the $z$ and $x$ axes as in the preceding section (Fig. 9.17.b). Thus, the width of the film will be considered constant:

$$w(z) = w_0. \tag{9.116}$$

The steady-state continuity equation now becomes:

$$\frac{\partial v_z}{\partial z} + \frac{\partial v_y}{\partial y} = 0, \tag{9.117}$$

and because $v_z = v_z(z)$ Eq. 9.117 yields:

$$v_y = -\frac{dv_z}{dz} y. \tag{9.118}$$

The mass flow rate is calculated as:

$$\dot{m} = w_0 \int_{-h/2}^{h/2} \rho v_z \, dy = \rho v_z h w_0, \tag{9.119}$$

and because it is constant over time, it follows that:

$$\frac{1}{h} \frac{dh}{dz} = -\frac{1}{v_z} \frac{dv_z}{dz} \tag{9.120}$$

for constant density.

The steady-state momentum equation in the $z$ direction (for vertical casting) after neglecting inertia and gravity becomes:

$$\frac{\partial \tau_{yz}}{\partial y} + \frac{\partial \pi_{zz}}{\partial z} = 0. \tag{9.121}$$

Note that the shear stress in Eq. 9.121 comes from the geometry of the system (as in the fiber-spinning case). Integration of Eq. 9.121 from $-h/2$ to $+h/2$ yields:

$$w_0 h \pi_{zz} = F, \tag{9.122}$$

where $F$ is the force necessary to draw the film. $\pi_{zz}$ is now correlated to the velocity gradient through an equation similar to Eq. 9.22 in the fiber-spinning case as:

$$\pi_{zz} = -4\mu \frac{dv_z}{dz} = -\bar{\eta}_p \frac{dv_z}{dz}, \tag{9.123}$$

where $\bar{\eta}_p$ is the planar elongational viscosity (equal to $4\mu$).

Upon combining Eqs. 9.119, 9.120, 9.122, and 9.123, we get the following differential equation:

$$\frac{1}{h}\frac{dh}{dz} = \frac{\rho F}{\bar{\eta}_p \dot{m}}, \tag{9.124}$$

which is solved as:

$$\frac{h}{h_0} = \exp\left[-z\frac{\rho|F|}{\bar{\eta}_p \dot{m}}\right] = \exp\left[z\frac{\rho F}{\bar{\eta}_p \dot{m}}\right] = (D_R)^{-z/L}, \tag{9.125}$$

and consequently:

$$v_z = v_{z0}\exp\left[z\frac{\rho|F|}{\bar{\eta}_p \dot{m}}\right] = v_{z0}\exp\left[-z\frac{\rho F}{\bar{\eta}_p \dot{m}}\right] = v_{z0}(D_R)^{z/L}. \tag{9.126}$$

Note that the draw ratio, $D_R$, in this case can be written as:

$$D_R = \exp\left[-\frac{\rho F}{\bar{\eta}_p \dot{m}}L\right] = \exp\left[\frac{\rho|F|}{\bar{\eta}_p \dot{m}}L\right]. \tag{9.127}$$

Eq. 9.126 is similar to Eq. 9.26 of the fiber-spinning case, and in general the film-casting process can be considered the two-dimensional counterpart of the fiber-spinning process (see also Pr. 9A.4).

## 9.2.2. STABILITY OF FILM CASTING

Theoretical analysis of the stability problem for film casting shows the similarities between fiber spinning and film casting. Yeow (1974) showed that the critical draw ratio for draw resonance of Newtonian fluids is 20.21, which is exactly the same as that in the fiber-spinning case (Subsection 9.1.5). For non-Newtonian fluids the similarity is not necessarily preserved, but the qualitative effects of the non-Newtonian viscosity, viscoelasticity, and cooling are expected to be the same. For power-law fluids the theory suggests that the critical draw ratio increases with increasing power-law index $n$, being 40 and 91 for $n = 1.2$ and 1.5, respectively. Kase (1974a) and Bergonzoni and DiCresce (1966) studied the problem experimentally and showed that the ratio of maximum to minimum film thickness ranges from 1.8 to about 4.5. Furthermore, data for PS and PP exhibited a critical draw ratio of about 20, which is in agreement with the theory.

Anturkar and Co (1988) studied the effects of viscoelasticity on the stability of the film-casting process in a similar fashion to that of Fisher and Denn (1976). The constitutive equation used in their theoretical analysis is the UCM (Eqs. 3.40 and 3.41) with the Carreau viscosity function (Eq. 2.8; with $\eta_\infty = 0$) $\eta(\dot{\gamma}) = \eta_0(1 + \lambda_v^2\dot{\gamma}^2)^{(n-1)/2}$, which is the White-Metzner model, and a similar function for the characteristic fluid time, $\lambda(\dot{\gamma}) = \lambda_0(1 + \lambda_t^2\dot{\gamma}^2)^{(n'-1)/2}$. The critical draw ratio, $D_R$, as a function of the viscoelastic parameter defined below:

$$\Lambda_0(2\Lambda_t)^{n'-1} = \lambda_0(2\lambda_t)^{n'-1}/(L/v_{z0}) \tag{9.128}$$

is shown in Fig. 9.18. Note the similarity between Figs. 9.14 and 9.18 for the fiber spinning and film casting processes.

Finally, one problem associated with film casting is the presence

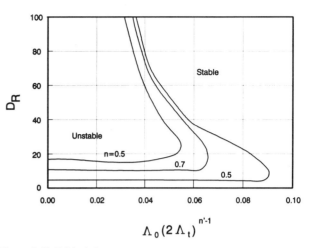

**Figure 9.18** Critical draw ratio, $D_R$, as a function of the viscoelastic parameter $\Lambda_0(2\Lambda_t)^{n'-1}$ and the power-law index $n$, for the film-casting process. (Reprinted by permission of the publisher from Anturkar & Co, 1988.)

of edge beads (Dobroth & Erwin, 1985; d'Halewyn et al., 1990). These edge beads in the form of thick edges appear in the web, and they cause problems in other downstream processes such as winding. The most common way to get rid of the edge beads is to cut them and either recycle or scrap them, and the associated cost is about 12% of the production cost. The main cause of this instability is the difference in stress that the center and the edges of the web experience (plane strain at the center; Subsection 6.3.1.1, and uniaxial elongational stress at the edges, Subsection 6.3.1.2). The experimentally observed ratio of the thicknesses of the edge to the centerline, also called *bead ratio*, is about 2 to 4. The theory based on the plain strain at the centerline and plane stress at the edges predicts a bead ratio equal to the square root of the draw ratio, and experiments confirm such a relationship.

## 9.2.3. FILM STRETCHING AND PROPERTIES

The term *film stretching* can also be encountered in the literature as *cold drawing* of the film. The objective of this process is to impart biaxial orientation in the film and to increase its modulus. These improvements are achieved by stretching the film in two directions ($x$ and $z$; transverse [TD] and machine [MD] directions, respectively) simultaneously. The operation temperature should be below the temperature of maximum crystallization rate and above the glass transition temperature. PP and PET have been successfully used in this process. Industrially, the process is carried out in the *stender* (or *tender*; from the textile industry). This device grips the edges of the film and extends them to larger widths as the film moves from the inlet to the exit roller. As soon as the film reaches the exit roller, the stender clamps release the grips of the film. During this stretching the polymer experiences an extensional flow.

The biaxially stretched films are characterized by the amounts of crystallinity and orientation, and their mechanical properties are correlated to the extension rates in both the machine and transverse directions. As an example we present the results for poly(phenylene sulfide) (PPS) (Maemura et al., 1989), which is an important engineering thermoplastic because of its high dimensional stability, solvent resistance, and temperature stability. PPS film 300 μm in thickness was biaxially stretched at temperatures in the range of 90 to 115°C and at 600 to 15,000%/min stretching rates. PPS crystallizes under stress, and the increase of stretching rate increases birefringence and crystalline orientation. As far as the tensile properties of the stretched film are

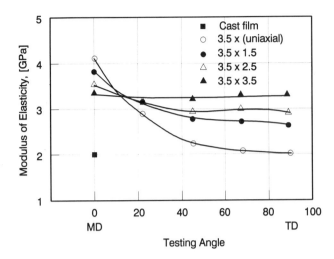

**Figure 9.19** Modulus of elasticity of stretched and unannealed PPS films as a function of testing angles, for various stretch ratios. (Reprinted by permission of the publisher from Maemura et al., 1989.)

concerned, the modulus of elasticity and tensile strength increase, and the elongation at break decreases with stretching. Fig. 9.19 shows the effect of the stretching and the testing angle on the modulus of elasticity. The testing angle refers to the angle between the MD and the direction of the sample cut for testing. The most anisotropic material is that which is uniaxially drawn, and the most isotropic is that which is equally biaxially drawn. All of the samples have modulus of elasticity greater than that of the cast film, which is 2.0 GPa.

## 9.3. FILM BLOWING

One method to produce film with a good balance of mechanical properties is by extruding a polymer through a film die and then subsequently stretching the film in two directions, as described in the preceding section. The other technique involves extrusion through an annular die. Then, the moving tubular film is stretched and inflated by an air stream flowing from inside the annular die (the pressure is slightly higher than atmospheric pressure, Fig. 9.20) creating a 'bubble' (see also Section 1.2). This bubble is cooled by an air jet flowing from an air ring toward its outside surface. The cooling results in crystallization and solidification that start at the crystallization line. Beyond this line the polymer melt is transformed into a two-phase mixture consisting of molten and solidified polymer. Beyond the crystallization line there is the freeze line, where the bubble boundaries become parallel to the centerline. Finally, the frost line (Campbell & Cao, 1987) is the other boundary of the region that starts with the freeze line. Beyond the frost line the deformation of the bubble is practically zero, and the bubble consists of one-phase material only, the solidified polymer.

The bubble is then flattened by a set of guide rolls and taken up by a set of rubber nip rolls that form an air-tight seal at the upper end of the bubble. The takeoff at the nip rolls may be either of constant speed or constant torque. Finally, the film is wound onto cylinders and sold as "lay-flat" tubing or trimmed at the edges and wound into two rolls of flat film. Figure 9.20 (or Fig. 1.6) shows a schematic of the film-blowing process called *tubular film blowing* (or simply film blowing). In terms of direction, most frequently this process takes place vertically upward and less frequently vertically downward and horizontally. The major advantages of this method over the first method are the economics and the speed of production.

Film blowing and fiber spinning have general similarities. Both processes have free boundaries, and the flows are predominantly

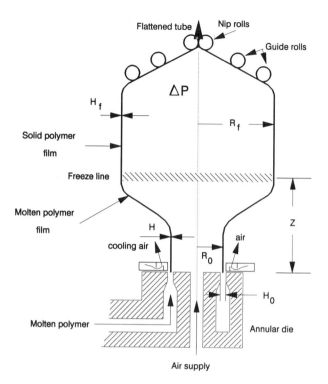

**Figure 9.20** Film-blowing process.

elongational. They differ with respect to the orientation generated. The fiber-spinning process imparts orientation in the axial direction only, whereas the film-blowing process imparts unequal (in general) biaxial orientation. The two axes of orientation are the axial (*machine*; MD) direction due to the drawing of the tube and the circumferential (*non-machine*, or *transverse*; TD) direction due to the blowup of the tube. The mechanical properties of blown film are nearly uniform in both directions as a result of biaxial orientation, and this is the reason why flat film is produced by the film-blowing process.

The two main parameters of this process are the *blow ratio* (or *blow-up ratio*), $B_R$ (or BUR), and the machine-direction *draw* (or *drawdown*) ratio, $D_R$. The blow ratio is defined as the ratio of the final tube radius, $R_f$, to the initial tube outer radius just downstream of the annular die, $R_0$ (see also Figs. 9.20 or 9.21 and Section 1.2):

$$B_R = \frac{R_f}{R_0}. \tag{9.129}$$

Similarly, the draw ratio is defined as:

$$D_r = \frac{V}{v_0}, \tag{9.130}$$

where $V$ is the takeup speed, and $v_0$ is the die extrusion speed. The final film thickness, $H_f$, can be calculated from the blow and draw ratios and the mass conservation equation as follows:

$$H_f = \frac{H_0}{B_R D_R}, \tag{9.131}$$

where $H_0$ is the initial film thickness, or equivalently the die gap thickness. Typical parameters in the film-blowing process are: $H_0 = 1$–$2$ mm; $R_0 = 2.5$–$25$ cm; $v_0 = 1$–$5$ cm/s; $B_R = 1.5$–$5$; $D_R = 5$–$25$; $\Delta P = 50$ Pa, that

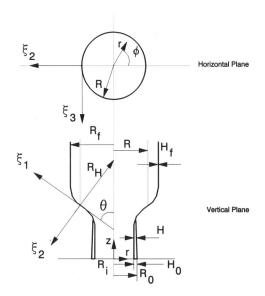

**Figure 9.21** Moving Cartesian coordinate system and cylindrical polar coordinate system, for the film-blowing process.

is, the internal pressure is about 0.05% of the atmospheric pressure; and freeze line height $Z = 0.25$–5 m. An average value of the blow and draw ratios and of the initial film thickness yields a final film thickness on the order of 50 μm (i.e., about 2 mils in English units). In terms of nomenclature, the final film is considered to be thick gauge-blown film whenever its thickness exceeds 75 μm or equivalently 3 mils. In terms of applications, thick gauge-blown film is used in the production of dunnage bags, heavy duty shrink film, greenhouse film, lawn and garden bags, and resin and chemical packaging.

As far as the mechanical properties are concerned, the tear (test name: Elmendorf tear), impact (test name: dart drop), and tensile strengths give an indication of the mechanical strength of the tubular film. Some discussion of these properties and the effect the processing parameters have on them were discussed in Section 1.2. The amorphous as well as the crystalline orientation development during the blowing process depend on the stretching imparted in the machine and transverse directions. Finally, besides orientation, the amount of crystallinity as well as the size of the crystallites may play a significant role in the mechanical and physical properties of the blown film.

This process is not as fast as fiber spinning, which results in a more uniform temperature distribution in the film relative to that in the fiber. Usually, cooling is achieved by blowing an air stream from an axisymmetric air ring toward the external film surface (see Fig. 9.20). In some cases, in addition to the external air ring, an internal air cooling system is provided. Finally, in some other cases, especially in thick tube and large bag production, cooling is achieved by a water spray or ring. Note that in the latter cases the film must be dried before winding up, which leads to an additional step.

Commercially, the film-blowing process is extensively used for the production of polyolefin (LDPE, HDPE, and PP) wrapping film. Mechanical strength, optical clarity, which depends on the degree and type of crystallinity for crystallizable polymers, and the uniformity of thickness (variations of about 5% for films with a length scale of 10 mm to 10 m is acceptable) are the three most important and general properties of the film.

Before we start analyzing the equations describing the film-blowing process in detail, it is worth mentioning a variation of the main technique referred to as that of the double bubble tubular film extrusion process. It has been applied by White and coworkers (Kang et al., 1990;

Kang & White, 1990) to PPS and PET. It consists of producing two bubbles. The first one is generated under conditions of moderate blow and draw ratios, and the second is produced by reheating and blowing the first bubble. The process is usually employed for polymers with low melt strength and slow crystallization rates. By quenching the bubble in the first step very little crystallization occurs. The material is then heated above $T_g$ and blown again.

As in the fiber-spinning section, we first gain extensive understanding of the film-blowing process by analyzing the simplest case, that of isothermal Newtonian film blowing. Then, we discuss some points about nonisothermal film blowing for Newtonian and viscoelastic materials. The stability analysis of this process is introduced, and some overall remarks are presented at the end of this section.

### 9.3.1. ISOTHERMAL NEWTONIAN MODEL

Figure 9.21 shows a schematic of the geometry of the film-blowing process. The analysis that follows comes from the work of Pearson and Petrie (1970a, b, and c). Besides the assumptions of isothermal conditions and Newtonian (homogeneous and incompressible) behavior of the polymer, the following assumptions are also incorporated:

1. Inertial, gravitational, air drag, and surface tension forces are neglected.
2. The film thickness is small with respect to other characteristic dimensions (i.e., $H \ll R$ which is the *thin-sheet approximation*).
3. Die swell is neglected, or the origin of the fixed coordinate system is assumed to be just beyond the maximum die swell.
4. Steady-state conditions exist.
5. The fluid bubble is axisymmetric.
6. The region between the freeze and frost lines is collapsed into a single line demarking a sharp transition between liquid and solid phases.

The significance of the second assumption is that the deformation field in the film is essentially elongational. Shear stresses are not present if the film thickness is very small. The first assumption makes the final equations simpler without losing any significant information. Finally, overall these assumptions resemble those used in Subsection 9.1.2 in the thin filament theory for the fiber-spinning process.

Cylindrical polar coordinates $(r, \phi, z)$ are taken with $z$ in the direction of flow (or axial direction). Also, symmetry around the $z$ axis is assumed. Although the cylindrical system is space-fixed, for the calculation of stresses, a moving Cartesian coordinate system, $\xi_1$, $\xi_2$, $\xi_3$, embedded in the inner surface of the bubble is used, as shown in Fig. 9.21 (where both coordinate systems are shown). The $\xi_2$ direction is normal to the film (i.e., thickness direction); the $\xi_1$ direction is in the direction of flow (i.e., MD); and the $\xi_3$ direction is perpendicular to $\xi_2$ and $\xi_1$ and tangent to the circumferential directions.

We will first calculate the strain field, then the stress field in the bubble, and finally the force balance at the frost line will be written in terms of the radius and thickness of the bubble. The principal rates of strain are:

$$\dot{\gamma}_{ii} = 2 \frac{\partial v_i}{\partial \xi_i}, \tag{9.132}$$

where $i = 1, 2, 3$, and $v_i$ is the $i$th component of the velocity vector in the moving coordinate system. Note that the $\dot{\gamma}_{ij}$'s are equal to zero for $i \neq j$. Also note that for an incompressible fluid:

$$\sum_{i=1}^{3} \dot{\gamma}_{ii} = 0. \tag{9.133}$$

The goal for the subsequent analysis is to relate the various strain rates to the geometry of the film.

The $v_2$ velocity component is zero at the inner surface of the bubble ($\xi_2 = 0$), and it is equal to $dH/dt$ at the outer surface of the bubble ($\xi_2 = H$). Then, the velocity gradient across the thickness of the bubble for very thin-walled bubbles is:

$$\dot{\gamma}_{22} \approx 2\frac{v_2}{H} = \frac{2}{H}\frac{dH}{dt} = \frac{2}{H}\frac{dH}{d\xi_1}\frac{d\xi_1}{dt} = \frac{2v_1}{H}\frac{dH}{d\xi_1} = \frac{2v_1\cos\theta}{H}\frac{dH}{dz}, \quad (9.134)$$

where use of the following correspondence between the space-fixed cylindrical and the moving Cartesian systems was made:

$$d\xi_1 = \frac{1}{\cos\theta}dz. \quad (9.135)$$

Similarly, the extension rate in the $\xi_3$ direction is calculated using the circumferential velocity, $v_3$, which in turn is related to the rate of bubble expansion. Thus,

$$\dot{\gamma}_{33} = \frac{2v_1\cos\theta}{R}\frac{dR}{dz}, \quad (9.136)$$

and from Eq. 9.133 we get the following expression for the last extension rate:

$$\dot{\gamma}_{11} = -2v_1\cos\theta\left(\frac{1}{H}\frac{dH}{dz} + \frac{1}{R}\frac{dR}{dz}\right). \quad (9.137)$$

The continuity equation relates the mass polymer flow rate, $\dot{m}$, to $v_1$ as follows:

$$\dot{m} = 2\rho\pi RHv_1. \quad (9.138)$$

Equation 9.138 can be incorporated into Eqs. 9.134, 9.136, and 9.137 to yield the following rate-of-deformation tensor:

$$\dot{\gamma} = \frac{\dot{m}\cos\theta}{\rho\pi RH}\begin{bmatrix} -\dfrac{1}{H}\dfrac{dH}{dz} - \dfrac{1}{R}\dfrac{dR}{dz} & 0 & 0 \\[2ex] 0 & \dfrac{1}{H}\dfrac{dH}{dz} & 0 \\[2ex] 0 & 0 & \dfrac{1}{R}\dfrac{dR}{dz} \end{bmatrix}. \quad (9.139)$$

Now, we can calculate the stress field in the bubble. For a Newtonian fluid:

$$\pi_{ij} = p\delta_{ij} - \mu\dot{\gamma}_{ij}, \quad (9.140)$$

where the components of $\pi$ refer to the $\xi$ coordinate system. Because no external forces act on the bubble, $\pi_{22} = 0$ (for $\Delta P \ll p$), and thus the isotropic pressure in the fluid is:

$$p = \mu\dot{\gamma}_{22} = \frac{\mu\dot{m}\cos\theta}{\rho\pi RH}\frac{1}{H}\frac{dH}{dz}. \quad (9.141)$$

Thus, the rest of the nonzero total stress components are equal to:

$$\pi_{11} = \frac{\mu\dot{m}\cos\theta}{\rho\pi RH}\left(\frac{2}{H}\frac{dH}{dz} + \frac{1}{R}\frac{dR}{dz}\right) \quad (9.142)$$

$$\pi_{33} = \frac{\mu\dot{m}\cos\theta}{\rho\pi RH}\left(\frac{1}{H}\frac{dH}{dz} - \frac{1}{R}\frac{dR}{dz}\right). \quad (9.143)$$

In order to be able to determine the thickness and the radius of the bubble, we must apply force balances on the bubble. Taking a fluid element with dimensions $2\pi R$, $H$, and $d\xi_1$, then the viscous forces in the axial ($L$) and transverse ($H$) directions are, respectively:

$$F_L = 2\pi RH\pi_{11}$$

$$dF_H = Hd\xi_1\pi_{33}. \quad (9.144)$$

This equation can also be written in terms of viscous forces per unit length as:

$$P_L = H\pi_{11} = \frac{F_L}{2\pi R} = \frac{\mu\dot{m}\cos\theta}{\rho\pi R}\left(\frac{2}{H}\frac{dH}{dz} + \frac{1}{R}\frac{dR}{dz}\right)$$

$$P_H = H\pi_{33} = \frac{dF_H}{d\xi_1} = \frac{\mu\dot{m}\cos\theta}{\rho\pi R}\left(\frac{1}{H}\frac{dH}{dz} - \frac{1}{R}\frac{dR}{dz}\right). \quad (9.145)$$

For thin shells the pressure difference between the inside and outside of the film, $\Delta P$ (also called internal overpressure), is related to the forces per unit length in the film as follows:

$$\Delta P = \frac{(-P_L)}{R_L} + \frac{(-P_H)}{R_H}, \quad (9.146)$$

which is similar to the Young–Laplace equation (Eq. 9.44) for the pressure inside a bubble. In the above equation, $R_L$ and $R_H$ are the principal radii of curvature of the bubble surface at the point of interest (Fig. 9.21). These radii are calculated by geometrical arguments as:

$$R_H = R\sec\theta = R(1 + (dR/dz)^2)^{1/2} \quad (9.147)$$

$$R_L = -\frac{\sec^3\theta}{d^2R/dz^2} = -\frac{(1 + (dR/dz)^2)^{3/2}}{d^2R/dz^2}. \quad (9.148)$$

Finally, the draw (or drawdown) force at the frost line, $F_Z$, is related to the pressure difference and the viscous force in the axial direction as:

$$2\pi RP_L\cos\theta - \pi(R_f^2 - R^2)\Delta P = F_z = -|F_z| \quad (9.149)$$

where $R_f$ is the constant bubble radius beyond the frost line and $F_z$ is considered to be constant beyond the frost line. Equations 9.145–9.149 can be combined together to yield the following two differential equations:

$$2r^2(\hat{T} + r^2B)r'' = 6r' + r\sec^2\theta(\hat{T} - 3r^2B), \quad (9.150)$$

and

$$\frac{w'}{w} = -\frac{r'}{2r} - \sec^2\theta\frac{\hat{T} + r^2B}{4}, \quad (9.151)$$

where the various dimensionless variables and parameters are defined as follows:

$$r = \frac{R}{R_0}; \quad w = \frac{H}{R_0}; \quad x = \frac{z}{R_0}; \quad \sec^2\theta = 1 + (r')^2 \quad (9.152)$$

$$B = \frac{\rho\pi R_0^3\Delta P}{\mu\dot{m}}; \quad T_z = \frac{\rho R_0|F_z|}{\mu\dot{m}}; \quad \hat{T} = T_z - B_R^2B \quad (9.153)$$

and ( )' means differentiation with respect to $x$ (i.e., $\equiv d/dx$). Thus, $B$ is the dimensionless internal overpressure, and $\hat{T}$ is the dimensionless stress. Typical values of these parameters are: $0.075 \leqslant B \leqslant 0.4$; $0.5 \leqslant T_z \leqslant 2.5$. Equation 9.150 is a second-order differential equation for the bubble radius, and Eq. 9.151 is a first-order differential equation for the bubble wall thickness. The boundary conditions for these equations are:

B.C.1 at $x = 0$     $r = 1$                        (9.154)

B.C.2 at $x = X = \dfrac{Z}{R_0}$     $r' = 0$          (9.155)

B.C.3 at $x = 0$     $w = w_0 = \dfrac{H_0}{R_0}$.      (9.156)

Typical values of $X$ range from 5 to 20. Equation 9.150 with boundary conditions Eqs. 9.154 and 9.155 and Eq. 9.151 with boundary condition Eq. 9.156 constitute the set of differential equations and boundary conditions that describe the isothermal Newtonian model. Note that to solve this model we need to specify $Z$ and $H_0$, which in general are not known a priori. Pearson and Petrie (1970c) solved this model numerically with three parameters: $B$, $T_z$, and $X$ (see also Problem 9C.2). Their results are shown in parametric graphs such as Fig. 9.22. Similarly, Fig. 9.23 shows the blow ratio as a function of the thickness reduction for $X = 5$. Furthermore, if one specifies one of these parameters, the other two must be adjusted to give the desired product dimensions (see Table 9.3). Finally, note that in the following relationship the thickness ratio is calculated as a function of the draw and blow ratios (see also Eq. 9.131):

$$\frac{H_0}{H_f} = B_R D_R. \qquad (9.157)$$

**TABLE 9.3**   $X$, $B$ and $T_z$ Required to Give $B_R = 3$ and $D_R = 20/3$

| $X$ | 8 | 9 | 10 | 20 | 23 |
|-----|------|-------|-------|------|------|
| $B$ | 0.2 | 0.175 | 0.165 | 0.1 | 0.09 |
| $T_z$ | 2.3 | 2.0 | 1.85 | 1.15 | 1.0 |

(Data from Pearson & Petrie, 1970c)

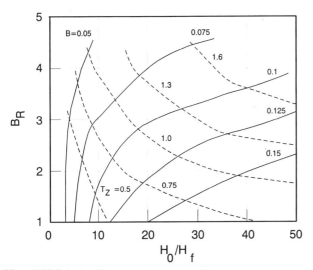

**Figure 9.22** Calculated isoparametric curves of blow-up versus thickness ratio, for the Newtonian isothermal model and for $X = 20$. (Reprinted by permission of the publisher from Pearson & Petrie, 1970c.)

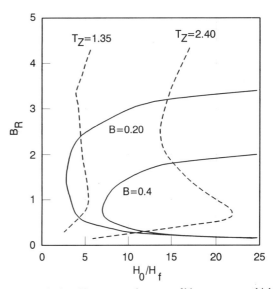

**Figure 9.23** Calculated isoparametric curves of blow-up versus thickness ratio, for the Newtonian isothermal model and for $X = 5$. (Reprinted by permission of the publisher from Cain & Denn, 1988.)

### EXAMPLE 9.4
### Blown LLDPE Film

Tubular linear low-density polyethylene (LLDPE) film of thickness equal to $34.1\,\mu m$ is produced by a blown film operation with draw ratio equal to 4. The annular die has an inner diameter of 1.387 cm and outer diameter of 1.496 cm. Calculate the pressure of the air to blow a bubble (i.e., the internal overpressure) of the given specifications and the axial tension to draw such a bubble. Consider that LLDPE is Newtonian with a viscosity of $720\,Pa\cdot s$, the process is isothermal at $180°C$, and that LLDPE freezes at an axial distance of 15 cm. The mass flow rate of the polymer is 0.21 g/s, and its density is $0.919\,g/cm^3$.

### Solution

$Z = 15$ cm and $R_0 = 1.496/2$ cm $= 0.748$ cm. Thus,

$$X = \frac{Z}{R_0} = \frac{15}{0.748} = 20. \qquad (9.158)$$

Consequently, Fig. 9.22 applies to this blown-film case. The thickness ratio is easily calculated to be:

$$\frac{H_0}{H_f} = \frac{\left(\dfrac{1.496 - 1.387}{2}\right)}{34.1 \times 10^{-4}} \approx 16, \qquad (9.159)$$

and the blow ratio is then:

$$B_R = 16/4 = 4, \qquad (9.160)$$

From Fig. 9.22 and for a blow ratio of 4 and thickness ratio of 16 the corresponding approximate values of the parameters $B$ and $T_z$ are 0.0675 and 1.15, respectively. Finally, Eq. 9.153 yields the following values for the pressure and tension:

$$\Delta P = 8.45\,Pa; \qquad F_z = -25\,mN. \qquad (9.161)$$

### 9.3.2. NONISOTHERMAL NEWTONIAN MODEL

The isothermal Newtonian model is a useful model, because it reveals most of the characteristics of the tubular film-blowing process. Nevertheless, it suffers from two disadvantages: The actual film-blowing process is basically a nonisothermal process, and the polymer melt is non-Newtonian in character. In this section we address the nonisothermal case and in the next section the matter of the non-Newtonian character of the polymer melt.

In industrial practice, cooling of the blown film is enhanced from the outside by an annular air jet, which flows from an air ring (see Fig. 9.20) attached very close to the die and/or from the inside by a process called *internal bubble cooling* (IBC). Note that the air jet usually "hits" the film bubble at a point slightly above the die, so that the heat transfer that occurs in the region between the die and that point is rather low. There are two ways to approach the nonisothermal nature of the film-blowing process. The first one includes the temperature dependence of viscosity and density only but does not employ the energy equation, whereas the second approach includes the use of the energy equation.

Petrie (1975) incorporated the effect of temperature on Newtonian viscosity and density through exponential and linear relationships respectively (for the viscosity function see also Eq. 9.105):

$$\mu = \mu_0 \exp\left[-k(T - T_0)\right] \tag{9.162}$$

$$\rho = \frac{\rho_0}{1 + c(T - T_0)}, \tag{9.163}$$

where the various constants have the following values for LDPE: $k = 0.03°C^{-1}$; $c = 0.00069°C^{-1}$; $\rho_0 = 0.801$ g/cm$^3$; and $T_0 = 115°C$. Note that the actual value of $\mu_0$ is not needed until the pressure difference across the film is calculated. In this model, $\dot{\gamma}_{11}$ differs from that of Eq. 9.137, and it is equal to:

$$\dot{\gamma}_{11} = -2v_1 \cos\theta\left(\frac{1}{H}\frac{dH}{dz} + \frac{1}{R}\frac{dR}{dz} + \frac{1}{\rho}\frac{d\rho}{dT}\frac{dT}{dz}\right). \tag{9.164}$$

The temperature profile was taken from experimental data. This theory underestimates the actual bubble size obtained from experimental data.

The next approach incorporates the energy equation (Wagner, 1976). By assuming no heat transfer at the inner surface the steady-state energy equation (Eq. 9.38) takes the following form (see also Eq. 9.47):

$$\bar{C}_p \frac{\dot{m}\cos\theta}{2\pi R}\frac{dT}{dz} = h(T_a - T) + \sigma\varepsilon(T_a^4 - T^4), \tag{9.165}$$

where all the variables are defined as in Eq. 9.40. The boundary condition required to solve Eq. 9.165 is:

B.C.1 at $x = 0$    $T = T_0$. \tag{9.166}

Moreover, the temperature at the frost line $X$ is equal to the solidification temperature of the polymer, which is the crystallization temperature for semi-crystalline polymers (about 116°C for LDPE) and the glass transition temperature for amorphous polymers. Wagner (1976) assumed a constant heat transfer coefficient, $h$, equal to $8.33 \times 10^{-4}$ cal/cm$^2$·s·°C, relative emissivity, $\varepsilon$, equal to 0.3, exit temperature, $T_0$, equal to 160°C, $k = 0.033°C^{-1}$, $c = 0.00073°C^{-1}$, $\rho_0 = 0.782$ g/cm$^3$, and

$$\bar{C}_p = 0.478 + 8.46 \times 10^{-4} T \tag{9.167}$$

in cal/g °C (see also Table 5.6 for equivalent relationships). These values of the parameters are for LDPE. Note that general relations for the

heat transfer coefficient can be found in Kanai and White (1984). The prediction of the model agreed well with experimental data only when an average Newtonian viscosity was assumed for the process, which also depended on the draw ratio.

Kanai and White (1985) combined the dynamics of the process, which is expressed in Eqs. 9.138, 9.150, 9.151, and the following equations for the energy balance and the Newtonian viscosity:

$$\bar{C}_p \frac{\dot{m}\cos\theta}{2\pi R}\frac{dT}{dz} = h(T_a - T) + \sigma\varepsilon(T_a^4 - T^4) + \frac{\rho_c}{\rho}\frac{\dot{m}\cos\theta}{2\pi R}\Delta\bar{H}_c\frac{d\phi_c}{dz} \tag{9.168}$$

$$\mu = \mu_0 \exp\left[\frac{E}{R_g}\left(\frac{1}{T} - \frac{1}{T_0}\right)\right]\exp\left[C\phi_c\right], \tag{9.169}$$

where $C$ is a constant obtainable from experimental results of viscosity versus percent crystallinity. The resulting set of four differential equations was solved numerically, and it was shown that the predicted and experimental data agree well, at least qualitatively.

### 9.3.3. NONISOTHERMAL NON-NEWTONIAN MODEL

A purely viscous non-Newtonian approach was followed by Han and Park (1975b). They used the power-law model and the energy equation, assuming that the effects of crystallization were insignificant. The agreement of this model with experimental data in terms of the bubble radius and thickness as a function of the axial distance for LDPE and HDPE was reported to be reasonable. In terms of viscoelastic models, Luo and Tanner (1985) considered the Leonov model, Cain and Denn (1988) considered the upper convected Maxwell (UCM) and Marrucci models, and Alaie and Papanastasiou (1993) considered an integral constitutive model in nonisothermal cases of film blowing. In some of the cases analyzed, multiple steady-state solutions were present (see also Pr. 9C.2).

Campbell and Cao (1987) presented a different model that incorporates the interaction of crystallinity, viscoelasticity and the two phases, liquid and solid, on the bubble shape. Equations 9.139 and 9.146–9.149 remain the same, whereas Eq. 9.145 is now given by:

$$P_L = H^l\pi_{11}^l + H^s\pi_{11}^s$$

$$P_H = H^l\pi_{33}^l + H^s\pi_{33}^s, \tag{9.170}$$

where the superscripts $l$ and $s$ denote liquid and solid phase, respectively. The mass conservation equation is:

$$\dot{m} = 2\rho\pi R(H^l\rho^l + H^s\rho^s)v_1. \tag{9.171}$$

The rheological modeling of the solid phase was based on a mechanical model of a dashpot connected in parallel with a spring and a slide. For small stresses, the model is equivalent to a dashpot and a spring (Kelvin-Voigt two-element model). For large stresses, the model is equivalent to the Bingham model. The rheology of the liquid phase was described by a truncated power-law model. The overall model is essentially a UCM model with altered parameters. By combining all these equations as well as the energy equation (similar to Eq. 9.91) Campbell and coworkers solved the tubular film-blowing problem numerically, and their results are shown in Figs. 9.24 and 9.25. Figure 9.24 shows the bubble radius as a function of the axial distance. The agreement between theory and experiments (HDPE, from Kanai & White, 1984) is quite good, especially near the freeze and frost lines. The agreement is also better near the freeze and frost lines for the temperature as a function of the axial distance (Fig. 9.25). Cao and

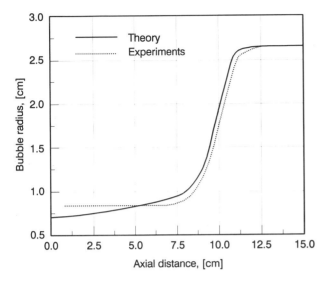

**Figure 9.24** Comparison of model and experimental bubble shape. (Reprinted by permission of the publisher from Campbell & Cao, 1987.)

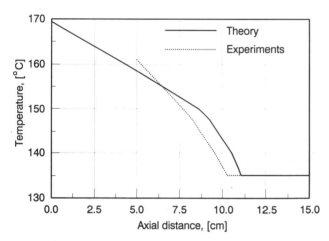

**Figure 9.25** Comparison of model and experimental temperature profiles. (Reprinted by permission of the publisher from Campbell & Cao, 1987.)

Campbell (1990) also found similar agreement between experimental data for PS and theoretical results.

It is clear from all this that the film-blowing process still remains difficult to analyze theoretically. In practice, the operator of the film-blowing tower adjusts the various parameters by trial and error. Consequently, one major consideration is the scaleup (Subsection 9.3.6) of a small experimental setup to an industrial-scale film-blowing tower, especially for polymers tested for the first time in the experimental setup.

### 9.3.4. BIAXIAL STRETCHING AND MECHANICAL PROPERTIES

Biaxial stretching is part of the tubular blown-film process. This type of stretching was also encountered in the second step of a film-casting process (see Subsection 9.2.3). Biaxially stretched PPS from a film-casting process was analyzed in Subsection 9.2.3, where it was shown that the modulus of elasticity in the MD decreases with increasing TD

stretching. On the other hand, the TD modulus of elasticity increases with increasing stretching in the same direction. The same trend was shown to be followed by PET in tubular film-blowing experiments (Ma & Han 1988). As far as the effect of MD stretching (which in the film-blowing nomenclature is called draw ratio) on the modulus of elasticity is concerned, Ma and Han (1988) showed experimentally that the MD modulus increases and the TD modulus decreases with increasing draw ratio.

### 9.3.5. STABILITY OF FILM BLOWING

Unstable flow in film blowing, fiber spinning, film casting is unacceptable for two reasons: (1) It limits the production rate of the facility; and (2) it lowers the quality of the product. As was mentioned in Subsection 9.1.5 for melt spinning, besides melt fracture, which is associated with die flow, *draw resonance* is present as a periodic diameter fluctuation when the draw ratio exceeds a critical value. Draw resonance as it applies to the uniaxial extension cases of fiber spinning and film casting was analyzed in Subsection 9.1.5. On the other hand, film blowing in the general sense imparts biaxial orientation in the film, and the equivalence of draw resonance appears now as a sequence of surface waves (Han & Park, 1975c; Fig. 9.26). Furthermore, if the blow and draw ratios are such that uniaxial extension (see Pr. 9B.7) is dominant in the film-blowing process, then draw resonance appears as in the fiber-spinning case.

Han and Park (1975c) experimentally studied flow instabilities in film blowing by introducing disturbances in the air overpressure or the take up speed. They found that HDPE and LDPE are more sensitive to take up speed disturbances than to air overpressure disturbances and that a decrease in melt temperature tends to stabilize the bubble after it has been disturbed. Finally, the disturbed bubble stabilizes itself when the size of the disturbance is below a critical value.

Yeow (1976) theoretically analyzed the instabilities due to axisymmetric disturbances in an isothermal Newtonian fluid and presented neutral stability curves in the space $w_f$ ($= H_f/R_0$) and $B_R$ and for various values of the parameter $X$ ($X = Z/R_0$ which is the dimensionless freeze-line). Kanai and White (1984) experimentally studied the stability of nonisothermal film blowing of viscoelastic melts such as LLDPE, LDPE, and HDPE, and their results are shown in Figs. 9.27 and 9.28. LDPE is more stable than LLDPE and HDPE, which is in accord with LDPE's strain hardening extensional behavior. In addition, for LLDPE as the frost-line height increases, the bubble becomes more unstable.

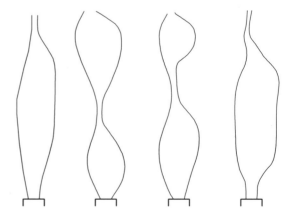

**Figure 9.26** Typical bubble instability shapes. (Reprinted by permission of the publisher from Kanai & White, 1984.)

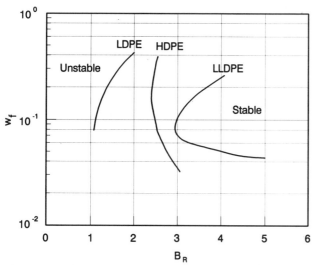

**Figure 9.27** Stability and instability regions in the space: $w_f$ and $B_R$, for $X = 16$. (Reprinted by permission of the publisher from Kanai & White, 1984.)

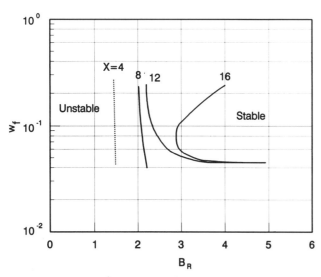

**Figure 9.28** Stability and instability regions in the space: $w_f$ and $B_R$, for various $X$, and for LLDPE. (Reprinted by permission of the publisher from Kanai & White, 1984.)

### 9.3.6. SCALEUP

Scaleup is a process that is best investigated through the use of dimensionless numbers and scale up criteria. The latter determine the variables that are expected to be the same in the laboratory and industrial scale, and the equality of the various dimensionless numbers and groups ensures that the solution of the governing equations remains the same in both scales. For example, if the ratios of various geometrical lengths in both cases are the same, then *geometrical similarity* is observed. Similarly, both scales are considered to be *dynamically similar* if the Reynolds number is the same. Weissenberg, Deborah, Reynolds, Froude, and Weber numbers are some of the most frequently encountered dimensionless numbers in the scale up process. The best approach for scaleup is discussed in Pr. 9D.3. See also Pearson (1985) for a detailed discussion of film-blowing scaleup and the relevant dimensionless groups for a viscoelastic polymer melt.

## 9.4. SOLUTION TO DESIGN PROBLEM 8

In this design problem we need to analyze the process in two sections: (**a**) from the die to the die swell level, and (**b**) from the die swell level to the nip rolls. In the first section, the critical wall shear rate for melt fracture, taking into consideration the safety factor, dictates the dimensions of the die. The volume of material in each bag is calculated as $2 \times (61 \times 69.6 \times 0.002541) = 21.6 \, cm^3$. The mass flow rate is then calculated from the requirement of production of 1500 bags/hr as:

$$\dot{m} = 1500 \frac{bags}{3600 \, s} \times 0.92 \frac{g}{cm^3} \times 21.6 \frac{cm^3}{bag} = 8.3 \, g/s. \quad (9.172)$$

From the statement of the problem we know that:

$$R_f = \frac{2 \times 61}{2\pi} \, cm = 19.4 \, cm; \qquad H_f = 0.002541 \, cm. \quad (9.173)$$

Assume that $R_0$ is the die outer radius, and that $\kappa R_0$ is the die inner radius. As the final thickness of the film is 25.41 μm, it is reasonable to expect that the die gap is small, and consequently that the annular die can be approximated by a slit of thickness, $H_0$, and width, $W_0$, where

$$H_0 = R_0 (1 - \kappa) \quad (9.174)$$

$$W_0 = \pi R_0 (1 + \kappa). \quad (9.175)$$

Die swell is present in this problem, and the radius and thickness at the maximum die swell level will be noted as $R_p$ and $H_p$, respectively (see Fig. 3.1, p. 35). Note that the dimensions at this die swell level will be considered as the initial dimensions for the film-blowing process. From Table 1.1 (p. 6) we get that $H_p = 0.018'' = 0.0457 \, cm$ for a die land length of 3.81 cm. Thus, the reduction in film thickness becomes:

$$\frac{H_p}{H_f} = B_R D_R = 18. \quad (9.176)$$

The dimensions of the die will be determined by means of a mass balance and the thickness increase in the film due to die swell as follows. Equation 3.89 is used as:

$$B = \frac{R_p}{R_0} = \frac{H_p}{H_0} = 0.1 + \left[ 1.0 + \frac{1}{2} \left( \frac{N_{1,w}}{2\tau_{xy,w}} \right)^2 \right]^{1/6}. \quad (9.177)$$

The wall shear stress inside the die is one-third of the maximum wall shear associated with the onset of melt fracture, that is, it is equal to $(1.13 \times 10^5/3) \, Pa = 37,667 \, Pa$. Thus, the allowed ratio in parentheses in Eq. 9.177 (primary normal stress difference over wall shear stress) is equal to:

$$\frac{N_{1,w}}{2\tau_{xy,w}} = \frac{0.119}{2} \tau_{xy,w}^{0.304} = 1.464. \quad (9.178)$$

Consequently, the die swell ratio (Eq. 9.177) becomes:

$$\frac{R_p}{R_0} = \frac{H_p}{H_0} = 1.23. \quad (9.179)$$

Substituting the known die swell film thickness into Eq. 9.179, we get the first equation for the unknowns $R_0$ and $\kappa$:

$$R_0(1 - \kappa) = \frac{0.0457}{1.23} \, cm = 0.03721 \, cm. \quad (9.180)$$

The second equation comes from the mass balance and is:

$$\dot{m} = \frac{\rho \pi R_0^3 (1-\kappa^2)}{2(s+2)} \left(\frac{\tau_{xy,w}}{m}\right)^s,$$ (9.181)

where use of Table 2.5 (p. 19) for flow in a slit die was made. Thus, Eq. 9.181 yields:

$$R_0^3(1-\kappa^2) = 0.8641 \text{ cm}^3.$$ (9.182)

Equations 9.180 and 9.182 can be recast into a single quadratic equation with the following solution:

$$R_0 = 3.41 \text{ cm}; \qquad \kappa = 0.99.$$ (9.183)

The value of $\kappa$, which is very close to 1, justifies the approximation of the annular die with a slit die made at the beginning of this section. The die gap is thus calculated to be $0.03721 \text{ cm} = 372.1 \,\mu\text{m}$ or 14.65 mils. Equations 9.179 and 9.183 can be combined to yield $R_p = 4.2 \text{ cm}$ or 1.65 in.

The blow and draw ratios of the film-blowing process can be calculated from the known values of the radii of the die due to die swell and the final bubble, and from Eq. 9.176 are:

$$B_R = \frac{R_f}{R_p} = \frac{19.4}{4.2} = 4.62$$ (9.184)

$$D_R = \frac{V}{v_p} = \frac{18}{B_R} = 3.89.$$ (9.185)

To calculate the drawing velocity we need to calculate first the velocity at the die and the die swell level. This is done through mass balances as follows:

$$v_0 = \frac{\dot{m}}{\rho \pi R_0^2 (1-\kappa^2)} = 13.6 \text{ cm/s},$$ (9.186)

and

$$v_p = v_0 \frac{R_0^2 - (R_0 - H_0)^2}{R_p^2 - (R_p - H_p)^2} = 9.0 \text{ cm/s}.$$ (9.187)

Thus, the drawing velocity is calculated as

$$V = D_R v_p = 35.0 \text{ cm/s},$$ (9.188)

which is equivalent to 68.8 fpm (feet per min).

The film-blowing process in this case is assumed to be isothermal, and we consider the viscosity to be Newtonian as well. For $X = 5$, $B_R = 4.62$, and $D_R = 3.89$ solution of Eqs. 9.150 and 9.151 with boundary conditions given in Eqs. 9.154 through 9.156 (with the use of the IMSL subroutine IVPRK; see also Pr. 9C.2 or Fig. 9.23) yields:

$$B = 0.11 \quad \Rightarrow \quad \Delta P = 22.9 \text{ Pa},$$ (9.189)

and

$$T_Z = 2.23 \quad \Rightarrow \quad |F_Z| = 2.6 \text{ N}$$ (9.190)

The viscosity required to get the above values is calculated as follows. The total strain in the $\xi_1$ direction is calculated from Eq. 9.132 as:

$$\dot{\gamma}_{11}(t_z - t_p) = \ln \frac{v_z}{v_p},$$ (9.191)

where $v_z$ is the velocity at the frost line, and $t_z - t_p$ represents the travel time from the die swell level to the frost line level. Based on the mass balance, $v_z$ is calculated to be equal to 29.1 cm/s, and thus the total strain is equal to 1.2 units. Similarly, the total strain in the transverse direction $\xi_3$ is equal to 1.5 units. The average strain rates in those directions are about 1.2 and $1.5 \text{ s}^{-1}$, respectively, as the travel time is approximated to be 1 s. Then, the viscosity is calculated to be equal to 4500 Pa·s.

Finally, the force at the nip rolls is calculated by applying a force balance on the film from the frost line to the nip rolls to give:

$$F_L = F_Z - \rho g (L-Z) 2\pi R_f H_f.$$ (9.192)

Based on Eq. 9.192, the force at the nip rolls is calculated as: $F_L = -3.0 \text{ N}$ (i.e., 3.0 N in tension).

## PROBLEMS
### A. Applications

**9A.1.** Significance of Various Terms in Newtonian Isothermal Fiber Spinning

Calculate the relative importance of the inertial terms to viscous terms in the Newtonian and isothermal fiber-spinning process. Use the data of Ex. 9.1 to assess this importance in the isothermal fiber spinning of nylon 6,6 at 285°C. Furthermore, estimate the relative importance of (1) gravitational forces, and (2) shear rate.

**9A.2.** Heat Transfer Coefficient and the Melt-Spinning Process

(a) Prove that the heat transfer coefficient for cross flow is twice that for parallel flow.

(b) Show that in the upper part of the spinning chamber (i.e., close to the spinneret) the heat transfer coefficient depends on the cooling air velocity only, whereas in the lower part it depends on the filament speed only.

**9A.3.** The Newtonian and Isothermal Model as a Special Case of the Newtonian and Nonisothermal Model

Prove that the solution of the Newtonian isothermal model, Eq. 9.26, can be reduced from Eqs. 9.66 to 9.68 with the appropriate simplifications.

**9A.4.** Draw Ratios in Fiber Spinning and Film Casting

Prove that the following relationship holds for isothermal and Newtonian fiber spinning and film casting:

$$\frac{[\ln D_R]_{\text{spin.}}}{[\ln D_R]_{\text{cast.}}} = \frac{4}{3}.$$

**9A.5.** Principal Radii of Curvature

Prove Eqs. 9.147 and 9.148.

### B. Principles

**9B.1.** Isothermal Spinning of a Power-Law Fluid

Show that the axial velocity profile of a power-law fluid in the fiber spinning process is given by the following equation:

$$v_z = v_0 \left[ 1 + ((D_R)^{1-(1/n)} - 1) \frac{z}{L} \right]^{n/(n-1)},$$

where $n$ is the power-law index. To show that, use all the assumptions of the Newtonian case. Also, show diagramatically the effect of the power-law index on the axial velocity profile.

**9B.2.** Centerline Temperature in Intermediate Speed PET Spinning
Melt-spun PET is cooled by an air cross flow of 20 cm/s. The extrusion
temperature is 300°C, the take up velocity is 2000 m/min, the polymer
mass flow rate is 2.5 g/min, and the filament surface temperature and
speed are shown in Fig. 9.6. Estimate the centerline filament temperature
60 cm below the spinneret plate. The physical parameters of PET
are: $\rho = 1.37\ \text{g/cm}^3$, $k = 6.9 \times 10^{-4}\ \text{cal/cm·s °C}$, and $\bar{C}_p = 0.3\ \text{cal/g °C}$. The
cooling air temperature is 25°C. Also, comment on the validity of the
assumption 6 in Subsection 9.1.2.1, i.e., there is no resistence to radial
heat conduction within the filament.

**9B.3.** Draw Resonance of a Power-Law Nonisothermal Melt Spinning
Estimate the critical draw ratio for draw resonance for a melt-
spun polymer. The properties of the polymer are: $\bar{C}_p = 0.7\ \text{cal/g °C}$,
$k = 0.02°\text{C}^{-1}$ (see Eq. 9.105), $\rho = 0.83\ \text{g/cm}^3$, and $T_0 = 270°\text{C}$, and the
characteristics of the spinning system are: $L = 50\ \text{cm}$, dpf (at takeup) $= 8$,
$v_L = 500\ \text{m/min}$, $T_a = 20°\text{C}$, $v_{ay} = 20\ \text{cm/s}$, and $R_0 = 0.3\ \text{cm}$. The power-
law index of the polymer is about 0.4. Base your calculations on
Subsection 9.1.5.

**9B.4.** Spinning Length for Stable Melt Spinning
Using the data of Ex. 9.3 sketch the spinning length as a function of
the draw ratio for stable and attainable operation. Apply the results
of the theoretical analysis by Fisher and Denn (1976). Propose changes
that will decrease the spinning length.

**9B.5.** Special Cases of Very Thin Film Casting
Prove that the following relationship holds for two special cases of
very thin film casting:

$$[h/h_0]_{L \ll w_0} = ([h/h_0]_{L \gg w_0})^2.$$

Assume Newtonian and isothermal flow in the case of $L \ll w_0$ and the
same draw ratio in both cases.

**9B.6.** Straight Tubular Film Extrusion
Prove that when there is no blowing of the polymer film extruded
from an annular die, the film thickness is given by the following
relationship:

$$H = H_0 \exp\left[-B \frac{z}{R_0}\right],$$

where $B$ is given by Eq. 9.153. Compare this case to the isothermal
fiber spinning of a Newtonian fluid.

**9B.7.** Various Forms of Extension in Film Blowing
Prove that the following relationships hold for the film-blowing
operation:

1. For uniaxial extension: $D_R B_R^2 = 1$
2. For equibiaxial extension: $D_R = B_R$
3. For planar extension: $B_R = 1$

Then, for $D_R = 3$ calculate the relative values of the air pressure and the
axial tension at the nip-roll in each extensional case.

**9B.8.** Production of Straight Tube
Calculate the air pressure and the nip-roll tension for the production
of a straight tube of LLDPE from the extruder of Ex. 9.4.

**9B.9.** Draw Resonance in Coextrusion Fiber Spinning
Coextrusion fiber spinning is an industrial process used for the
production of plastic optical fibers. Based on the rheological differences
between LDPE and LLDPE (i.e., LDPE has a longer relaxation time

than LLDPE, and it exhibits strain hardening), provide qualitative
arguments about the expectation of draw resonance for a bicomponent
fiber-spun with LLDPE as the core material. (*Hint:* Read the paper by
Lee & Park, 1992.)

**C. Numerical Problems**

**9C.1.** Newtonian and Nonisothermal Fiber-Spinning
Calculate the diameter of the filament at the takeup roll for the
Newtonian and non-isothermal model using Eqs. 9.66 to 9.68 and the
IMSL subroutine IVPAG (Appendix D.7). Use the same parameters
as Kase and Matsuo (1967) which are given in the figure caption under
Fig. 9.5. (a) Neglect air drag, surface tension, inertial and gravitational
forces. (b) Include only the air drag forces ($\mu_{\infty 0} = 200\ \text{Pa·s}$).

**9C.2.** Isothermal Newtonian Film Blowing
Consider the isothermal Newtonian film-blowing process described by
Eqs. 9.150 and 9.151 and boundary conditions and Eqs. 9.154, 9.155
and 9.156. Using the IMSL subroutine IVPRK show that the following
two sets of conditions constitute steady state solutions of the model:
(1) $B = 0.2$, $X = 5$, $T_Z = 1.35$, $B_R = 2.4$, $H_0/H_f = 4.4$ and (2) $B = 0.2$, $X = 5$
$T_Z = 1.35$, $B_R = 0.5$, $H_0/H_f = 4.2$, and thus multiple solutions might exist
for a certain set of conditions (Cain & Denn, 1988). More specifically,
plot the dimensionless bubble radius, $r$, and the dimensionless film
thickness, $w$, versus the dimensionless axial distance, $x$.

**D. Design Problems**

**9D.1.** Elongational Viscosity Measurements and the Fiber-Spinning
and Film-Blowing Processes
Design a set of experiments, based on fiber-spinning and film-blowing
processes, which allow one to calculate extensional viscosity (uniaxial,
biaxial, and planar) as a function of extension rate. (*Hint:* Read the
paper by Han & Park, 1975a.)

**9D.2.** Spinning of Hollow Fibers
Consider the fiber-spinning process of a hollow fiber (Freeman et al.,
1986). Formulate the continuity, momentum, energy, and rheological
equations for this process along the same principles as in Subsection
9.1.1 and Subsection 9.1.2. For the special case of isothermal Newtonian
low-speed (no inertia, air drag, and gravity) spinning, and using the
thin-filament theory prove that the average velocity profile is given by
the relation

$$v_z = v_0 \exp\left[\frac{z}{R_{oo} - R_{io}}\left(\frac{|F| - \pi R_{io}^2 p_i}{3\pi\mu v_0(R_{oo} + R_{io})}\right)\right],$$

where $R_{oo}$ and $R_{io}$ are the initial outside and inside radius, respectively,
at $z = 0$, $p_i$ is the internal pressure and $F$ is the axial tension needed to
draw the hollow fiber.

**9D.3.** Scaleup of the Film-Blowing Process
It is very useful to be able to predict industrial-scale film processability
and resulting physical and mechanical properties from laboratory-scale
film-blowing experiments. Based on the nonisothermal Newtonian
model analyzed in Subsection 9.3.2, discuss the scale up principles and
dimensionless numbers for the film-blowing process. Then assume that
the scale up criterion requires equal stresses at the freeze line for machine
and transverse directions. Propose a scale up model for the radii of
the annular die (multiplication factor is $\alpha$) and the final (beyond freezing)
bubble radius (multiplication factor is $\kappa$). Show that the scale up
criterion is satisfied as long as the reciprocal of the internal
overpressure and the freeze-line tension, both at the industrial scale,
are $\kappa$ times the corresponding variables in the laboratory scale, for the
same temperature profiles and final film thickness in both scales.

## REFERENCES

Alaie, S. M., and T. C. Papanastasiou. 1993. "Modeling of Non-isothermal Film Blowing with Integral Constitutive Equations." *Intern. Polym. Proc.*, **8**(1), 51–65.

Anturkar, N. R., and A. Co. 1988. "Draw Resonance in Film Casting of Viscoelastic Fluids: A Linear Stability Analysis," *J. Non-Newton. Fluid Mech.*, **28**, 287–307.

Bergonzoni, A., and A. J. DiCresce. 1966. "The Phenomenon of Draw Resonance in Polymeric Melts." *Polym. Eng. Sci.*, **6**, 45–59.

Cain, J. J., and M. M. Denn. 1988. "Multiplicities and Instabilities in Film Blowing." *Polym. Eng. Sci.*, **28**(23), 1527–41.

Campbell, G. A., and B. Cao. 1987. "The Interaction of Crystallinity, Elastico-plasticity, and a Two-Phase Model on Blown Film Bubble Shape." *J. Plast. Film Sheet.*, **3**, 158–70.

Cao, B., and G. A. Campbell. 1990. "Viscoplastic-Elastic Modeling of Tubular Blown Film Processing." *AIChE J.*, **36**(3), 420–30.

Dees, J. R., and J. E. Spruiell. 1974. "Structure Developmemt During Melt Spinning of Linear Polyethylene Fibers." *J. Appl. Polym. Sci.*, **18**, 1053–78.

Denn, M. M., C. J. S. Petrie, and P. Avenas. 1975. "Mechanics of Steady Spinning of a Viscoelastic Liquid." *AIChE J.*, **21**(4), 791–9.

Dobroth, T., and L. Erwin. 1985. "Causes of Edge Beads in Cast Films." 43rd SPE Annual Technical Conference, Washington, DC, **31**, 89–92.

Donnelly, G. J., and C. B. Weinberger. 1975. "Stability of Isothermal Spinning of a Newtonian Fluid." *Ind. Eng. Chem. Fundam.*, **14**(4), 334–7.

Fisher, R. J., and M. M. Denn. 1976. "A Theory of Isothermal Melt Spinning and Draw Resonance." *AIChE J.*, **22**(2), 236–46.

Fisher, R. J., M. M. Denn, and R.I. Tanner. 1980. "Initial Profile Development in Melt Spinning." *Ind. Eng. Chem. Fundam.*, **19**, 195–7.

Freeman, B. D., M. M. Denn, R. Keunings, G. E. Molau, and J. Ramos. 1986. "Profile Development in Drawn Hollow Tubes." *J. Polym. Eng.*, **6**(1–4), 171–86.

George, H. H. 1982. "Model of Steady-State Melt Spinning at Intermediate Take-up Speeds." *Polym. Eng. Sci.*, **22**(5), 292–9.

d'Halewyn, S., J. F. Agassant, and Y. Denouy. 1990. "Numerical Simulation of the Cast Film Process." *Polym. Eng. Sci.*, **30**(6), 335–40.

Han, C. D., and J. Y. Park. 1975a. "Studies on Blown Film Extrusion. I. Experimental Determination of Elongational Viscosity." *J. Appl. Polym. Sci.*, **19**, 3257–76.

Han, C. D., and J. Y. Park. 1975b. "Studies on Blown Film Extrusion. II. Analysis of the Deformation and Heat Transfer Processes." *J. Appl. Polym. Sci.*, **19**, 3277–90.

Han, C. D., and J. Y. Park. 1975c. "Studies on Blown Film Extrusion. III. Bubble Instability." *J. Appl. Polym. Sci.*, **19**, 3291–7.

Kanai, T., and J. L. White. 1984. "Kinematics, Dynamics and Stability of the Tubular Film Extrusion of Various Polyethylenes." *Polym. Eng. Sci.*, **24**(15), 1185–1201.

Kanai, T., and J. L. White. 1985. "Dynamics, Heat Transfer and Structure Development in Tubular Film Extrusion of Polymer Melts: A Mathematical Model and Predictions." *J. Polym. Eng.*, **5**(2), 135–57.

Kang, H. J., and J. L. White. 1990. "A Double Bubble Tubular Film Extrusion Process of Poly *p*-Phenylene Sulfide (PPS)." 48th SPE Annual Technical Conference, Dallas, TX, **36**, 104–9.

Kang, H. J., J. L. White, and M. Cakmak. 1990. "Single and Double Bubble Tubular Film Extrusion of Polyethylene Terephthalate." *Intern. Polym. Proc.*, **5**(1), 62–73.

Kase, S. 1974a. "Studies on Melt Spinning. IV. On the Stability of Melt Spinning." *J. Appl. Polym. Sci.*, **18**, 3279–304.

Kase, S. 1974b. "Studies on Melt Spinning. III. Velocity Field Within the Thread." *J. Appl. Polym. Sci.*, **18**, 3267–78.

Kase, S. 1985. "Mathematical Simulation of Melt Spinning Dynamics." In A. Ziabicki, and H. Kawai, Eds., *High-Speed Fiber Spinning* (Wiley, New York).

Kase, S., and T. Matsuo. 1965. "Studies on Melt Spinning. I. Fundamental Equations on the Dynamics of Melt Spinning." *J. Polym. Sci. A*, **3**, 2541–54.

Kase, S., and T. Matsuo. 1967. "Studies on Melt Spinning. II. Steady-State and Transient Solutions of Fundamental Equations Compared with Experimental Results." *J. Appl. Polym. Sci.*, **11**, 251–87.

Keunings, R., M. J. Crochet, and M. M. Denn. 1983. "Profile Development in Continuous Drawing of Viscoelastic Liquids. *Ind. Eng. Chem. Fundam.*, **22**, 347–55.

Lee, W. S., and C.-W. Park. 1992. "Draw Resonance Instability in Coextrusion Fiber Spinning." 50th SPE Annual Technical Conference, Detroit, MI, **38**, 2181–3.

Lin, C.-Y., P. A. Tucker, and J. A. Cuculo. 1992. "Poly(Ethylene Terepthalate) Melt Spinning Via Controlled Threadline Dynamics." *J. Appl. Polym. Sci.*, **46**, 531–52.

Luo, X.-L. and R. I. Tanner. 1985. "A Computer Study of Film Blowing." *Polym. Eng. Sci.*, **25**, 620–9.

Ma, T. C., and C. D. Han. 1988. "Processing-Structure-Property Relationships in Poly(ethylene Terepthalate) Blown Film." *J. Appl. Polym. Sci.*, **35**, 1725–57.

Maemura, E., M. Cakmak, and J. L. White. 1989. "Characterization of Crystallinity, Orientation, and Mechanical Properties in Biaxially Stretched Poly(p-Phenylene Sulfide) Films." *Polym. Eng. Sci.*, **29**(2), 140–50.

Middleman, S. 1977. *Fundamentals of Polymer Processing*, Chapter 9 (McGraw-Hill, New York).

Pearson, J. R. A. 1985. *Mechanics of Polymer Processing* (Elsevier, New York).

Pearson, J. R. A., and C. J. S. Petrie. 1970a. "The Flow of Tubular Film. Part 1. Formal Mathematical Representation." *J. Fluid Mech.*, **40**(1) 1–19.

Pearson, J. R. A., and C. J. S. Petrie. 1970b. "The Flow of Tubular Film. Part 2. Interpretation of the Model and Discussion of Solutions." *J. Fluid Mech.*, **42**(3), 609–25.

Pearson, J. R. A., and C. J. S. Petrie. 1970c. "A Fluid-Mechanical Analysis of the Film Blowing Process." *Plastics & Polymers*, **38**, 85–94.

Pearson, J. R. A., and Y. T. Shah. 1974. "On the Stability of Isothermal and Nonisothermal Fiber Spinning of Power-Law Fluids." *Ind. Eng. Chem. Fundam.*, **13**(2), 134–8.

Petrie, C. J. S. 1975. "A Comparison of Theoretical Predictions with Published Experimental Measurements on the Blown Film Process." *AIChE J.*, **21**(2), 275–82.

Petrie, C. J. S., and M. M. Denn. 1976. "Instabilities in Polymer Processing." *AIChE J.*, **22**(2) 209–36.

Phan-Thien, N. 1978. "A Nonlinear Network Viscoelastic Model." *J. Rheol.*, **22**(3) 259–83.

Schultz, W. W. 1987. "Slender Viscoelastic Fiber Flow." *J. Rheol.*, **31**(8), 733–50.

Wagner, M. H. 1976. "Das Folienblasverfahren als rheologisch-thermo-dynamischer Prozeß." *Rheol. Acta*, **15**, 40–51.

Weinberger, C. B., G. F. Cruz-Saenz, and G. J. Donnelly. 1976. "Onset of Draw Resonance during Isothermal Melt Spinning: A Comparison between Measurements and Predictions." *AIChE J.*, **22**(3), 441–8.

Yeow, Y. L. 1974. "On the Stability of Extending Films: A Model for the Film Casting Process." *J. Fluid Mech.*, **66**(3), 613–22.

Yeow, Y. L. 1976. "Stability of Tubular Film Flow: A Model of the Film-Blowing Process." *J. Fluid Mech.*, **75**(3), 577–91.

Zeichner, G. R. 1973. "Spinnability of Viscoelastic Fluids." M.Ch.E. thesis, Univ. Delaware, Newark, DE.

Ziabicki, A. 1976. *Fundamentals of Fibre Formation* (Wiley, London).

Ziabicki, A., and H. Kawai, Eds. 1988. *High-Speed Fiber Spinning* (Wiley, New York).

# 10

# MOLDING AND FORMING

**Design of a Compression-Molding Process**

Polypropylene (PP) containing 30 wt% glass fiber mat is to be compression-molded to form a panel with curved ends, as shown in Fig. 10.1. The PP/glass composite, referred to commercially as Azdel PM 10300 (Giles & Reinhard, 1991), comes in the form of sheets 3.6 mm thick by 19.1 cm by 21.6 cm. The final part is to be 3.2 mm thick, have a breadth of 64.8 cm, and the remaining dimensions as shown in Fig. 10.1. The sheets, also referred to as blanks, are to be heated to the processing temperature by means of an infrared oven through which the blanks are able to pass on a continuous basis. The infrared heaters are located on both sides of the conveyor system and are considered to be parallel plate sources with a maximum surface temperature of 427°C. Based on dynamic mechanical thermal analysis, in which a strip of the composite is tested in the dynamic oscillatory mode as a function of temperature (see Fig. 10.21), the material begins to flow at about 160°C, which is about the melting point of PP. Because of the complex nature of these composite materials, the only rheological data available are the complex viscosity versus frequency at three temperatures as shown in Fig. 10.22. The upper processing temperature limit for PP is known to be 230°C. Typical presses can be operated in a speed range of 4.23 to 33.9 mm/s. Determine the arrangement of the blanks in the mold and the number of blanks required, the minimum time required to heat the blanks to a temperature where the material will flow into the remainder of the mold, rate of closing of the press and the required operating force, and the temperature of the mold plattens that will minimize the time required for the part to remain in the mold.

In this chapter we are concerned with processes in which a discrete mass of polymer, which is either above its melting temperature in the case of semi-crystalline polymers or above its glass transition temperature in the case of amorphous polymers, is forced to take the shape of a cavity by means of applied pressure. The processes vary somewhat in the details but in general involve heating the polymer mass,

pressurizing it, and then cooling the formed sample. Processes of this nature include injection molding, compression molding, thermoforming, and blow molding. In this chapter we discuss these various processes in the order just mentioned. Our intentions are to describe the most important features as well as point out where design and analysis can be carried out.

## 10.1. INJECTION MOLDING

Injection molding is probably the most widely used cyclic process for manufacturing parts from thermoplastics. In this section we discuss some of the general aspects of the process, the fluid mechanics of mold filling, the method by which structuring occurs, and the basis for computer-aided design.

### 10.1.1. GENERAL ASPECTS

A general description of injection molding was given in Chapter 1. In essence polymer pellets are plasticated in a single-screw extruder, and the molten polymer accumulates at the tip of the screw in a reservoir. As this part of the process has been described in Chapter 8, we will concentrate on what happens to the melt from this point on. Figure 10.2 shows the end of the extruder and its connection to the mold. The melt, which has accumulated in the reservoir, is pushed forward by the screw whose displacement is controlled by hydraulic pressure. The melt flows through the nozzle, which connects the extruder to the mold, passes through the sprue, along the runner, through the gate, and into the mold cavity. The sprue is designed to offer as little resistance to flow as possible while minimizing the amount of wasted polymer. The runner is designed to carry melt to the mold cavity, and when multiple cavities are involved, it must be designed to ensure that each cavity fills at the same time. The gate represents the entrance to the mold. Its location is of utmost importance to the appearance of the part. Furthermore, it is desirable to make the gate as small as possible not only for cosmetic reasons but also to facilitate the separation of the part from the rest of the material solidified in the runner. The melt enters the mold cavity where it begins to solidify as it touches the mold wall. As semicrystalline polymers solidify, they shrink as a result

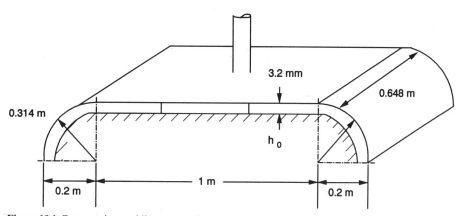

**Figure 10.1** Compression-molding process for producing a panel from polypropylene and glass fiber mat composite sheets.

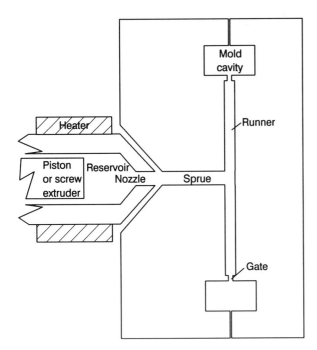

**Figure 10.2** Injection-molding tooling showing the tip of the injection system and its connection to the mold. Melt passes from the reservoir through the nozzle, the sprue, and runner system, entering the mold cavities through the gate.

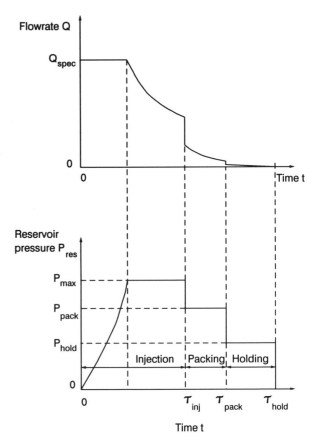

**Figure 10.3** Flow rate and pressure in the reservoir of an injection-molding unit as a function of time.

of increases in density. Pressure is maintained during the cooling process to ensure that melt continues to flow into the mold. Once solidification is complete, the mold plates open and the part is ejected. Although the screw is being pulled back, it rotates, plasticating more polymer.

The flow rate and pressure in the reservoir are shown schematically in Fig. 10.3. The injection pressure, which is the hydraulic pressure applied to the screw, is one of the variables that can be selected. Because there is little resistance to flow in the beginning, the flow rate through the nozzle is constant. However, as the melt advances through the sprue and runner, the resistance to flow increases, and the pressure increases. As the cavity fills, the set pressure reaches a constant pressure (this is the injection pressure, and it is a machine setting), but the resistance to flow continues to increase. The flow rate through the nozzle as well as the flow rate into the cavity must decrease. If the resistance to flow is too great, either as a result of the polymer solidifying or of the melt viscosity being too high, the polymer will fail to fill the mold, leading to what is known as a "short-shot." Once the mold fills, the hydraulic pressure applied to the screw is reduced (this is called the holding pressure) to a value that maintains enough flow of material into the mold to compensate for the volume changes due to shrinkage. Some pressure is maintained during the complete cooling cycle.

Mold filling involves both high deformation and high cooling rates. For this reason a considerable amount of orientation and structure or morphology can be developed in an injection-molded part. The shrinkage distribution in an injection molded sample of polystyrene (PS), an amorphous polymer, is shown in Fig. 10.4. Shrinkage is used here as a measure of molecular orientation, with the highest amount of shrinkage representing material with the highest degree of orientation. In part (a) of this figure the shrinkage distribution is shown as a function of thickness for two different fill rates. For the highest fill rate the highest shrinkage is at the mold wall, and there is a local maximum at a distance of about 0.22 mm from the mold wall. The shrinkage then decreases as one approaches the centerline of the mold. For the lowest

fill rate the shrinkage is still highest at the mold wall, but it is lower in value than that for the higher fill rate. There is also a local maximum, but it is now at a distance of about 0.44 mm from the wall. In Fig. 10.4.b the shrinkage distributions along and transverse to the flow direction are compared. The shrinkage along the transverse direction is negligible over most of the part thickness except near the wall. It is always lower along the transverse direction than along the flow direction, which suggests that a significant amount of anisotropy will exist in the properties.

To account for this distribution of shrinkage (orientation) Tadmor (1974) proposed the fountain flow mechanism, which occurs in the advancing front. In Fig. 10.5 the flow patterns in normal mold filling are shown schematically. As the melt leaves the gate, the front is found to occupy various positions in the mold at different times. The velocity profiles in the fully developed flow behind the front are shown in part (b) of Fig. 10.5. The flow well behind the front is primarily shear flow, and that at the front involves stagnation flow. In essence fluid passes through the center of the cavity to the front where it turns and then is laid up on the wall of the mold where it solidifies. In Fig. 10.6 the velocity profile in the fully developed region as seen by an observer moving with the average velocity of the flow is shown. The velocity gradient, and hence stress, is seen to pass through a maximum at an interior point in the flow. Because orientation is directly related to stress (see Section 5.5), one can see why there would be at least a local maximum in shrinkage at this point. A fluid element near the centerline will decelerate as it approaches the front and become compressed along the $x$ direction and stretched along the $y$ direction. The element is then stretched further at the front and laid up on the wall where it is rapidly

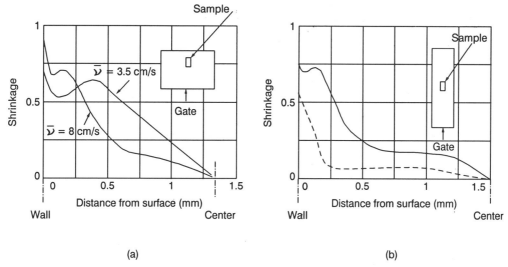

**Figure 10.4** Shrinkage distribution as a function of distance from the surface of an injection-molded part consisting of amorphous polystyrene. (a) Shrinkage along the flow direction for two different fill rates. (b) Comparison of the shrinkage distribution along the flow direction (solid line) and transverse to the flow direction (broken line). (Reprinted by permission of the publisher from G. Menges and W. Wübken, "Influence of Processing Conditions on Molecular Orientation in Injection Molds," *Society of Plastics Engineers Technical Papers*, **31**, 519, 1973.)

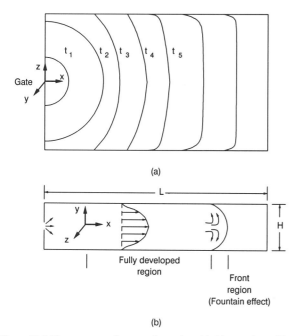

**Figure 10.5** Flow patterns in an end-gated mold. (a) top view of fronts as a function of time. (b) side view of velocity profiles in the fully developed region and frontal region. (Reprinted by permission of the publisher from Tadmor & Gogos, 1979.)

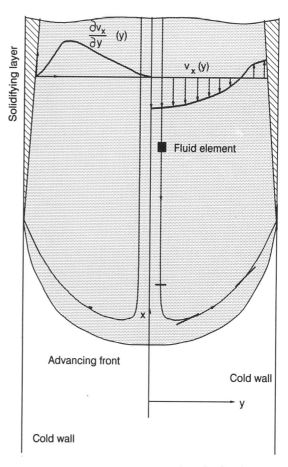

**Figure 10.6** Flow pattern in the advancing front for flow between two parallel plates as observed relative to the average velocity. A fluid element approaching the front is compressed along the flow direction and stretched along the $y$ direction before being laid up on the cold wall. (Reprinted by permission of the publisher from Tadmor, 1974.)

solidified in a highly oriented state. Hence, the extensional flow at the front stretches the fluid element and leads to a higher degree of orientation in the material at the mold wall than at the interior of the material.

The fountain flow associated with the advancing front is extremely important to the properties of materials generated by means of injection molding. In the case of homogeneous polymer systems we have already seen how the molecular orientation is affected. For fiber-filled systems the flow at the front can lead to highly oriented fibers at the surface of an injection-molded part, which certainly will affect the flexural properties of the part. In the case of blends, extensional flow at the front leads to a morphology in which the minor component exists as fibrils. Hence, in simulating the injection-molding process it is important that frontal flow eventually be included.

On occasion the opening at the gate is smaller than the cavity thickness, and the melt no longer fills the mold by an advancing front mechanism. Rather, it "snakes" its way into the mold, leading to a material with a poor surface appearance and reduced physical properties. Snaking does not seem to be a common method found in the filling of molds, but it can occur.

One of the major problems in injection molding of parts is the formation of *weld lines* that lead to surface imperfections and weak spots in the part. Weld lines arise from the presence of obstructions in the flow and from the impingement of advancing fronts from different gates. The former type of weld line is referred to as a *hot weld*, and the latter is referred to as a *cold weld*. In the case of the hot weld the melt as it flows into the cavity is split by an obstruction such as a pin, for example, and then the streams are brought back together. Usually, the temperature at the interface does not change much, and hence the streams are brought back together at the processing temperature. On the other hand, when two fronts impinge on each other, as shown in Fig. 10.7, the temperature of the free surfaces has dropped somewhat,

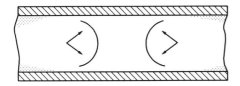

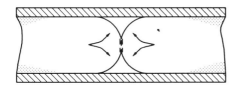

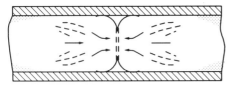

Intrusion by hot core

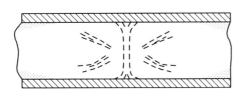

**Figure 10.7** Impingement of two fronts during injection molding to form a weld line. (Reprinted by permission of the publisher from S. W. Hobbs. *Polym. Engng. Sci.*, **14**, 621, 1974.)

leading to the formation of what are called cold welds. Healing of these weld lines is controlled by self-diffusion, which was discussed in Section 4.2.5. According to the theory of self-diffusion one must either use higher melt temperatures and longer hold times in the mold or use lower-molecular-weight polymers to accelerate the healing process.

### 10.1.2. SIMULATION

The design of injection-molding tooling is a complex process. In a commercial process the tooling usually contains multiple cavities, and the cavities are fed by several gates. The simulation of the injection-molding process must at least contain the capability to predict cavity layout and runner design, the fill rate as a function of injection pressure, gate location, weld line positions, and cooling time. In addition, it would be desirable to predict molecular orientation, morphology, residual stresses, warpage, and shrinkage. In this section we discuss the approach presently taken in simulating injection molding and its role in computer-aided design. Because quite large computer programs are required to simulate injection molding, our goal is to explain the approach and its capabilities. Those aspects of design which can be handled at this level, such as heat transfer and the cooling of parts, will be discussed.

The flow geometries frequently encountered in injection molding can be reduced to the three geometrically simple units shown in Fig. 10.8 (Pearson & Richardson, 1983). Flow through runners and the sprue can be handled by the first geometric unit (a), although many times the runners are semi-circular in cross-sectional shape. The cavities can be considered to be a combination of thin planar geometries as shown in part (b) or as center-gated disk geometries as shown in Fig. 10.8(c).

We consider the flow of polymer melt into thin rectangular cavities as a means of illustrating the approach taken. Hieber & Shen (1980) model the flow of polymer melt in a thin cavity using classical Hele-Shaw flow. In this approach the velocity field is considered to consist of two components, $v_x$ and $v_y$, that depend primarily on $z$ but not on $x$ or $y$ (i.e., $\partial v_x/\partial x \ll \partial v_x/\partial z$). The components of the equation of motion are then taken as:

$$0 = \frac{\partial}{\partial z}\left(\eta \frac{\partial v_x}{\partial z}\right) - \frac{\partial p}{\partial x} \tag{10.1}$$

$$0 = \frac{\partial}{\partial z}\left(\eta \frac{\partial v_y}{\partial z}\right) - \frac{\partial p}{\partial y}, \tag{10.2}$$

where $\eta$ is the viscosity taken to be of the form:

$$\eta = \eta(\dot{\gamma}, T), \tag{10.3}$$

and $\dot{\gamma}$ is:

$$\dot{\gamma} = \sqrt{\left(\frac{\partial v_x}{\partial z}\right)^2 + \left(\frac{\partial v_y}{\partial z}\right)^2}. \tag{10.4}$$

The continuity equation is of the following form for this flow:

$$0 = \frac{\partial}{\partial x}(b\bar{v}_x) + \frac{\partial}{\partial y}(b\bar{v}_y), \tag{10.5}$$

where $b$ is the half gap width that may depend on $x$ and $y$ and the bar denotes an average over the gapwise coordinate, $z$. The energy equation takes the following form:

$$\rho \bar{C}_p\left(\frac{\partial T}{\partial t} + v_x \frac{\partial T}{\partial x} + v_y \frac{\partial T}{\partial y}\right) = k \frac{\partial^2 T}{\partial z^2} + \eta \dot{\gamma}^2. \tag{10.6}$$

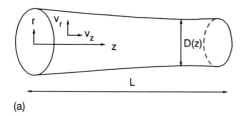

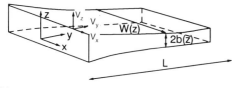

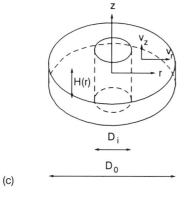

(a)

(b)

(c)

**Figure 10.8** Basic geometric elements to which runners and mold cavities can be reduced: (a) tubular element; (b) planar element; (c) center-gated disk.

In the development of these equations the viscoelastic nature of the fluid and the fountain flow at the advancing front have been neglected. This set of equations is solved using the following boundary conditions:

At $z=b$: $v_x=v_y=0$;     $T=T_w$           (10.7)

At $z=0$: $\dfrac{\partial v_x}{\partial z}=\dfrac{\partial v_y}{\partial z}=0=\dfrac{\partial T}{\partial z}$.       (10.8)

The solution of this set of equations is facilitated by the following procedure. Because $p$ is independent of $z$, Eqs. 10.1 and 10.2 can be integrated to give:

$$\eta\left(\frac{\partial v_x}{\partial z}\right)=\left(\frac{\partial p}{\partial x}\right)z \tag{10.9}$$

$$\eta\left(\frac{\partial v_y}{\partial z}\right)=\left(\frac{\partial p}{\partial y}\right)z, \tag{10.10}$$

where Eq. 10.8 has been used. These equations can be integrated again

using Eq. 10.7 to give:

$$v_x=\left(-\frac{\partial p}{\partial x}\right)\int_z^b \frac{z'dz'}{\eta} \tag{10.11}$$

$$v_y=\left(-\frac{\partial p}{\partial y}\right)\int_z^b \frac{z'dz'}{\eta}. \tag{10.12}$$

The gapwise-averaged velocities are obtained by integration of Eqs. 10.11 and 10.12:

$$\bar v_x=\left(-\frac{\partial p}{\partial x}\right)\frac{S}{b}; \qquad \bar v_y=\left(-\frac{\partial p}{\partial y}\right)\frac{S}{b}, \tag{10.13}$$

where

$$S=\int_0^b \frac{z^2dz}{\eta}. \tag{10.14}$$

Substituting Eq. 10.13 into Eq. 10.5 gives:

$$\frac{\partial}{\partial x}\left(S\frac{\partial p}{\partial x}\right)+\frac{\partial}{\partial y}\left(S\frac{\partial p}{\partial y}\right)=0. \tag{10.15}$$

Equation 10.15 is solved subject to the following boundary conditions:

Along the advancing front: $p=0$        (10.16)

At the entry to the mold: $p=p_e(x,y,t)$     (10.17)

At the mold wall: $\partial p/\partial n=0$.        (10.18)

Finally, when two melt fronts coalesce, forming a weld line, the boundary conditions are that the pressure and normal velocity be continuous, that is:

$$p^+=p^- \tag{10.19}$$

and

$$(S\partial p/\partial n)^+=-(S\partial p/\partial n)^-, \tag{10.20}$$

where the $+$ and $-$ signs denote values on either side of the weld line.

In summary the primary governing equations are given by Eqs. 10.15 and 10.6 for $p$ and $T$, respectively, while $S$, $v_x$, $v_y$, $\dot\gamma$, and $\eta$ are calculated by means of Eqs. 10.14, 10.11, 10.12, 10.4 and 10.3, respectively. To complete the solution of these equations an empiricism for viscosity must be used. The power-law and the Carreau models are often used. The solution of these equations requires numerical methods as discussed by Hieber & Shen (1980) and Wang and coworkers (1986).

In Ex 10.1 the solution to Eq. 10.15 is illustrated for the isothermal Newtowian case.

Once the cavity is filled, additional material is forced into the cavity to compensate for an increasing polymer density arising from crystallization and compressibility of the melt. Wang and coworkers (1986) have included compressibility throughout the entire filling, packing, and cooling phases for amorphous polymers. However, for semi-crystalline polymers their approach may not be satisfactory.

If one is to predict the orientation distribution in a part, then it is necessary to include fountain flow at the advancing front. Mavridis

## EXAMPLE 10.1
### Isothermal Newtonian Flow in a Rectangular Cavity

Consider the isothermal flow of a Newtonian fluid in a rectangular mold, as shown in Fig. 10.8.b but with $W$ and $2b$ constant and $W/2b > 10$. The gate at the entrance to the mold is a fan gate, so that the flow is considered to occur only in the $y$ direction: that is, $v_y = v_y(z)$ and $p = p(y)$. The location of the front at any instant is $L(t)$. The pressure at the gate is constant and is $P_0$. Determine $Q(t)$ and the time to fill a cavity of length $L$ using the approach of Hieber and Shen (1980).

### Solution

The pertinent equations are:

$$\mu \frac{\partial^2 v_y}{\partial z^2} - \frac{\partial p}{\partial y} = 0 \tag{10.21}$$

$$\frac{\partial}{\partial y}\left(S \frac{\partial p}{\partial y}\right) = 0; \qquad S = \frac{b^3}{3\mu}. \tag{10.22}$$

Equation 10.22 can be integrated to find the pressure distribution, which is:

$$P = \frac{-P_0}{L(t)} y + P_0, \tag{10.23}$$

where $P_0$ is the pressure at $y = 0$ and $P = 0$ at $y = L(t)$, which is the position of the front. $v_y$ is obtained from Eq. 10.12:

$$v_y = \frac{b^2 P_0}{2L(t)\mu}\left[1 - \left(\frac{z}{b}\right)^2\right]. \tag{10.24}$$

The position of the front is found from a mass balance:

$$Q = \frac{b^3 P_0}{3L(t)\mu} = 2bW \frac{dL}{dt}, \tag{10.25}$$

where $Q$ is obtained by integrating Eq. 10.24 over the cross-sectional area. Equation 10.25 can be solved to find the position of the front as a function of time:

$$L(t) = \sqrt{\frac{b^2 P_0 t}{3\mu W}}. \tag{10.26}$$

The time for filling the mold is determined by setting $L(t) = L$. One can also see by substituting Eq. 10.26 back into Eq. 10.25 that $Q$ decreases with increasing time as follows:

$$Q = b^2 \sqrt{\frac{P_0 W}{3\mu t}}. \tag{10.27}$$

and coworkers (1988) have presented an approach for modeling fountain flow under isothermal conditions. Referring to Fig. 10.6 axes are attached to the moving front that move with the average velocity of the flow, $\langle v_x \rangle$. At the upstream boundary the flow is modeled as a one-dimensional shear flow. The equation of motion and the boundary conditions for this flow are:

$$\frac{dp}{dx} = \frac{d}{dy}\left(\eta \frac{dv_x}{dy}\right) \tag{10.28}$$

$$\text{B.C.1: } \frac{dv_x}{dy} = 0 \qquad \text{at } y = 0 \tag{10.29}$$

$$\text{B.C.2: } v_x = B\left(\eta \frac{dv_x}{dy}\right) \qquad \text{at } y = 1, \tag{10.30}$$

where $B$ is the slip coefficient. The flow in the rest of the region is such that the following postulates are made:

$$v_x = v_x(x, y); \qquad v_y = v_y(x, y); \qquad p = p(x, y). \tag{10.31}$$

The boundary conditions are:

$$\text{At } x = 0: \ v_x = v_x(y) \tag{10.32}$$

$$\text{At } y = b: \ v_y = 0 \qquad v_x = B\tau_{yx} \tag{10.33}$$

$$\text{At the front: } \mathbf{v} \cdot \mathbf{n} = 0 \qquad \boldsymbol{\pi} \cdot \mathbf{n} = 0, \tag{10.34}$$

where $\mathbf{v} \cdot \mathbf{n}$ means there is no flow across the front and $\boldsymbol{\pi} \cdot \mathbf{n} = 0$ means there are no forces acting on the front. The parameter $B$ is a measure of slip, and when $B \to 0$, there is no slip, and when $B \to \infty$, there is perfect slip. The postulates in Eq. 10.31 along with the equation of motion and a constitutive equation lead to a set of nonlinear differential equations that must be solved by means of numerical methods, preferably finite element methods. At present most simulations of injection molding do not include fountain flow.

The cooling of the melt can be considered as a transient one-dimensional heat transfer problem. The approach used in Chapter 5 can be employed to calculate temperature distributions and cooling times as well as estimate the degree of crystallinity and the morphology. The main difficulty is to establish the appropriate boundary conditions at the mold walls. In a commercial process where a large number of parts are made per hour, it may take several hours before the mold surfaces reach an equilibrium temperature. This temperature can be somewhat higher than what is expected based on the temperature settings of cartridge heaters inserted in the mold base or fluids circulated through channels machined in the mold base. The most accurate way to determine the temperature distribution in the melt is to use a heat transfer coefficient. The overall heat transfer coefficient per unit length of coolant line, $U$, is given by:

$$\frac{1}{U} = \frac{1}{Sk_m} + \frac{1}{\pi Dh}, \tag{10.35}$$

where $k_m$ is thermal conductivity of the mold material (e.g., $43.3\ \text{W m}^{-1}\,^\circ\text{K}^{-1}$ for steel), $h$ is the convective heat transfer coefficient for flow through tubes (see Chapter 5 for determining this value), $D$ is the diameter of the coolant line, and $S$ is the shape factor, which is defined in Fig. 10.9 and given (Throne, 1979) as:

$$S = \frac{2\pi}{\ln\left[(2P/\pi d)\sinh(2\pi D/P)\right]}. \tag{10.36}$$

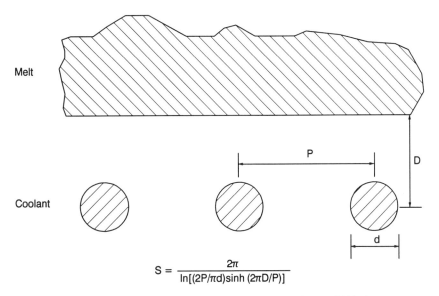

Melt

Coolant

$$S = \frac{2\pi}{\ln[(2P/\pi d)\sinh(2\pi D/P)]}$$

**Figure 10.9** Shape factor geometry for coolant lines in a mold block.

## 10.2. COMPRESSION MOLDING

Compression molding is primarily used to process thermosetting systems and difficult-to-process thermoplastics, such as fiber-filled systems or thermoplastic elastomers. We first describe some of the basic features of the process and the areas where design is needed. This is followed by a discussion of the elementary ideas in the modeling of compression molding.

### 10.2.1. GENERAL ASPECTS

The essential features of the compression molding process are illustrated in Fig. 10.10. In the case of thermoplastics a preheated mass of polymer, which may either be a sheet or a pile of pellets or powder, is placed in the mold. The temperature of the mold is set low enough to cause the polymer to solidify but not so rapidly that it will not flow. Hydraulic pressure is applied to the top or bottom plate, pushing the platens together. The molds are designed to prevent the top part of the mold from touching the bottom part, which would squeeze the resin from the mold.

The design of a compression-molding process consists of four aspects. The first is the selection of the proper amount of material to fill the cavity when the mold halves are closed. The second is the determination of the minimum time required to heat the blank to the desired processing temperature and the selection of the appropriate heating technique (radiation heating, forced convection, etc.). It is necessary to make sure that the center reaches the desired processing temperature without the surface being held at too high a temperature for too long time. The third is the prediction of the force required to fill out the mold. Finally, the temperature of the mold must be determined keeping in mind that one wants to cool the part down as rapidly as possible but that too rapid a cooling rate will prevent the polymer from filling the mold.

Compression molding is used for processing thermoplastics that do not flow readily such as highly filled systems and granular materials that do not really melt but only fuse under pressure. In the case of thermoplastic composite systems reinforced with long fibers, the process is referred to as stamping. It is a process that is being used more frequently for processing thermoplastic composites used in the manufacture of panels such as would be found in car trunks and hoods.

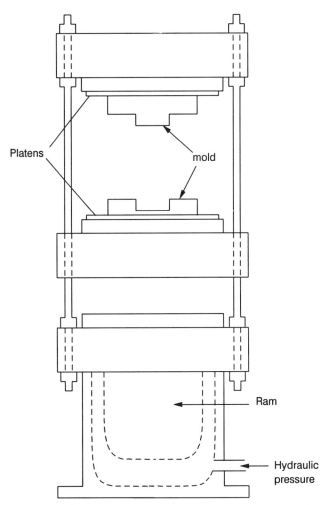

**Figure 10.10** Compression-molding process showing the hydraulic system and mold.

Typically, the process involves much less flow than found in injection molding or other forming processes discussed in the remainder of the chapter.

### 10.2.2. SIMULATION

Compression molding of thermoplastics typically involves very little flow. A discrete mass of material is placed in the mold whose volume is adequate to fill the mold when the plates are pushed together. The preliminary material may be in the form of a sheet, referred to as a *blank*, or stacked sheets reinforced with long continuous fibers, called *prepreg*. Initially, the surface area of the plates may be completely covered or only partially covered with material. When the plates of the mold are brought together, the material is forced to cover the rest of the mold surface. For example, blanks may cover initially from 20 to 80% of the mold surface. Thermoplastic prepregs are placed over the entire mold surface, and hence the application of pressure leads only to shaping, compaction, and bonding of the stacked sheets.

Compression molding is a highly nonisothermal process. In the processing of blanks consisting of a thermoplastic reinforced with long fibers, the blanks are preheated and then placed in the mold, which is at a temperature lower than that of the blanks. In the case of thermoplastic prepregs, the stacked sheets are heated and cooled in the same mold. In either case there is a temperature distribution in the sample before molding, and it is enhanced during molding. The accurate simulation of the compression-molding process requires solving of the coupled momentum and energy equations.

When there is significant flow, the deformation of the blank can be complex. For example, the blank may be initially square but during compression be required to deform more in one direction than the other. In any event, the modeling of the deformation may not be easily handled without the use of numerical techniques such as the finite element method.

To illustrate the modeling of the compression-molding process we consider the compression molding of a blank as shown in Fig. 10.11. The blank is constrained on the sides so that it can only flow on squeezing along the $x$ direction. In principle, one would like to develop the nonisothermal viscoelastic model of the process, but this disguises some of the essential features of the modeling process. For this reason we first develop the isothermal Newtonian model of the process and then look at the difficulties involved with the development of a complete nonisothermal viscoelastic model.

In the process shown in Fig. 10.11 a blank of polymeric material is compressed in a mold that allows the material to deform in the $x$-direction only. The mold is of width $W$, and the blank is of width $W$, initial thickness $h_0$, and initial length $2X_0$. The blank is assumed to cover 50% of the surface area of the plates, and hence only low deformation of the blank occurs. We will assume that the blank rheology is Newtonian and that the flow is isothermal and quasi-steady. With these assumptions we postulate the following for the velocity and pressure fields:

$$v_x = v_x(x, z); \qquad v_z = v_z(z); \qquad p = p(x, z). \tag{10.37}$$

The continuity equation for this situation is:

$$\frac{\partial v_x}{\partial x} + \frac{\partial v_z}{\partial z} = 0. \tag{10.38}$$

The equations of motion are:

$$-\frac{\partial p}{\partial x} - \frac{\partial \tau_{zx}}{\partial z} - \frac{\partial \tau_{xx}}{\partial x} = 0 \tag{10.39}$$

$$-\frac{\partial p}{\partial z} - \frac{\partial \tau_{zz}}{\partial z} = 0. \tag{10.40}$$

For the Newtonian fluid the stress components are:

$$\tau_{xx} = -2\mu \frac{\partial v_x}{\partial x}; \qquad \tau_{zz} = -2\mu \frac{\partial v_z}{\partial z}; \qquad \tau_{zx} = -\mu \frac{\partial v_x}{\partial z}. \tag{10.41}$$

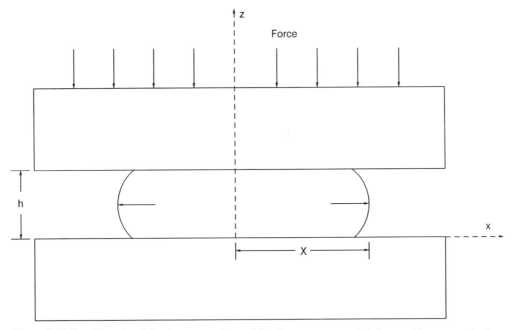

**Figure 10.11** Two-dimensional flow in compression molding between two parallel plattens. Flow occurs in the $z$ and $x$ directions only.

From the conditions given in the problem it is possible to show:

$$\frac{\partial v_x}{\partial z} \gg \frac{\partial v_x}{\partial x} \quad \text{and} \quad \frac{\partial v_z}{\partial z}, \tag{10.42}$$

and hence the dominant stress component is $\tau_{zx}$. The equations of motion become on introducing the expression for $\tau_{zx}$ in Eq. 10.41 and conditions in Eq. 10.42

$$\mu \frac{d^2 v_x}{dz^2} = \frac{dp}{dx}. \tag{10.43}$$

Equation 10.43 is integrated using the following boundary conditions:

At $z = 0$: $v_x = 0$

At $z = h$: $v_x = 0$ (10.44)

to give

$$v_x = (Gh^2/2\mu)\left[\left(\frac{z}{h}\right)^2 - \frac{z}{h}\right], \tag{10.45}$$

where $G = dp/dx$.

The goal is to determine a relation between the rate of plate closing ($dh/dt$), which is usually constant in most commercial presses, the force required to maintain this rate, the plate geometry, and the viscosity of the fluid. It is also desired to know how long it takes to completely fill out the mold. The first step is to relate the rate of flow out through a plane perpendicular to the $x$ direction to the rate of closing of the plate, which is given by:

$$(dh/dt)WX = W\int_0^h v_x dz, \tag{10.46}$$

or on substituting Eq. 10.45 into Eq. 10.46 for $v_x$ we obtain:

$$\left(\frac{dh}{dt}\right)WX = \frac{Gh^3 W}{\mu}\int_0^1 (\xi^2 - \xi)d\xi = \frac{-Gh^3 W}{12\mu}, \tag{10.47}$$

where $G = dp/dx$, $X$ is the half width of the sheet, and $\xi = z/h$. From Eq. 10.47, we obtain the pressure distribution in the sample:

$$p = \left(\frac{dh}{dt}\right)\left(\frac{G\mu}{h^3}\right)(X^2 - x^2). \tag{10.48}$$

Finally, we carry out a force balance on the upper plate:

$$-F + 2W\int_0^X p(x)dx = 0 \tag{10.49}$$

to obtain the following relation (*note*: Eq. 10.48 is substituted into Eq. 10.49, and the integration is carried out):

$$F = \frac{8W\mu\dot{h}X^3}{h^3}. \tag{10.50}$$

Equation 10.50 can be used to determine the force as a function of plate position if $dh/dt = \dot{h}$ is constant, or if $F$ is constant, a nonlinear

ordinary differential equation arises that can be solved for $h(t)$. In the following example we use the equations just presented to estimate conditions in a compression-molding process.

---

**EXAMPLE 10.2**
**Compression Molding of a Newtonian Fluid**

Consider the compression molding of a blank of PP reinforced with glass fiber as shown in Fig. 10.11. For the time being, consider the material to be Newtonian with a viscosity of $5 \times 10^2$ Pa·s. The thickness of the blank is 3.6 mm, and it is to be molded to fill a mold of length of 1 m, width of 0.25 m, and thickness of 3.2 mm. Determine the initial dimensions of the sheet required to fill out the mold (take the width as 0.25 m), the time for filling the mold, and the force required to complete the filling of the mold. The press can exert a maximum force of 20 MN and operate at rates between 4.23 and 33.9 mm/s.

**Solution**

Starting with the lowest closing rate of 4.23 mm/s, we obtain:

$$h(t) = -4.23 \times 10^{-3}t + 3.6 \times 10^{-3}.$$

The time for the gap to reach 3.2 mm is then 0.0946 s. Using Eq. 10.50, the force required to fill out the mold is:

$$F = \frac{8(0.25)(5.0 \times 10^2)(4.23 \times 10^{-3})(0.5)^3}{(3.2 \times 10^{-3})^3} = 16.14 \text{ MN}.$$

Hence, one could operate at a slightly higher closing rate if necessary, because the force required at the lowest rate is less than the maximum force available.

---

Although we have considered a highly idealized case of isothermal Newtonian flow, the model and calculations give an idea of the approach one must take in modeling and designing an actual process. In practice the time for heating the blank up to the processing temperature and the time for cooling the sample in the mold are by far the longest times. The time for filling out the mold may be of the order of seconds. Hence, although there will be an initial temperature distribution in the sample, very little change in temperature will occur during the deformation process. As a first approximation, provided we can handle the initial temperature variation in the sample, we can decouple the equation of motion from the energy equation. Once the mold is filled out, the time for cooling the part can be determined by treating the situation as a one-dimensional transient heat conduction problem.

There are still several parts of the problem that cannot be dealt with at this level. First, the viscoelastic nature of the flow must be considered as the deformation occurs over short times and may involve a significant extensional component. Second, the rheological properties of materials that are usually processed by means of compression molding are difficult to obtain. Materials such as PP filled with long glass fibers are extremely difficult to characterize rheologically. In fact, the best approach may be to use lubricated squeezing flow, which was discussed in Chapter 3. Finally, the shape changes that the blank may undergo may require the use of finite element methods.

## 10.3. THERMOFORMING

Thermoforming is used primarily for the manufacturing of packaging and disposable containers. However, it is also becoming a useful technique in the processing of engineering thermoplastics to produce parts used in the transportation industry. Polymers processed by this technique must have sufficient melt strength so that on heating they do not sag significantly under their own weight, yet can be deformed under pressure to take the shape of a mold. Hence, highly crystalline polymers with high melting temperatures and low molecular weight cannot be readily thermoformed. For example, Nylon 6, 6 ($T_m = 265°C$ and $\bar{M}_w = 30,000$) is not usually processed by means of thermoforming, whereas low-density polyethylene (LDPE) ($T_m = 110°C$ and $\bar{M}_w = 200,000$) is. In this section we first describe the general aspects of thermoforming and then some basic features of modeling this process as required in design.

### 10.3.1. GENERAL ASPECTS

The essential features of the thermoforming process are shown in Fig. 10.12. A thermoplastic sheet is heated usually by means of radiation but sometimes in conjunction with convection cooling to temperatures just above either $T_g$ in the case of amorphous polymers or $T_m$ in the case of semicrystalline polymers. The exact temperature depends on the degree of sag exhibited by the material under its own weight, which is determined by its rheological properties. The sample is then removed from the heating system and brought into position over the mold. The sample is forced to take the shape of the mold by applying pressure to the top of the sheet or by generating a vacuum on the underside of the sheet, as shown in Fig. 10.12. The forming step occurs in the matter of a second. The sample is maintained in the mold until it is rigid enough to be removed from the mold without altering its shape.

There are a number of variations on the basic process that are described in more detail by Throne (1986). For example, in plug-assisted vacuum forming the heated sheet is forced by a plug into the mold with the remainder of the shape being produced by the application of vacuum to the underside of the sheet. This method is used to help maintain a more uniform wall thickness throughout the part. Another example is matched die molding, which is shown in Fig. 10.13. The heated sheet is forced to take the shape of the female portion of the mold by the male part. This process is characterized by the formation of parts with more intricate shapes and uniform wall thickness. Finally, one last technique is that of twin-sheet thermoforming as shown in Fig. 10.14. Here two sheets are heated and then forced to take the shape of the mold by applying air pressure on the inside of the sheets and possibly vacuum on the outside. This process resembles somewhat that of blow molding, which is discussed in Section 10.4. However, the sheets can be forced to take on different shapes as each half of the mold can have a different shape. Furthermore, different polymers can be used for each half. The sheets must be held in the mold long enough for bonding to occur.

Thermoforming can be divided into four sections: (1) sheet heating without deformation; (2) sheet stretching without significant heat transfer; (3) part cooling in the mold; and (4) postmolding operations such as trimming. The time to make a part is primarily determined by steps 1 and 3 as these are of the order of minutes. However, the successful functioning of the part is determined by step 2 as the distribution of wall thickness is determined in this step. In the next section we consider the modeling of each of these steps.

### 10.3.2. MODELING

Of the four subdivisions we discussed the first three, heating, forming, and cooling, lend themselves to simulation and modeling, which in turn

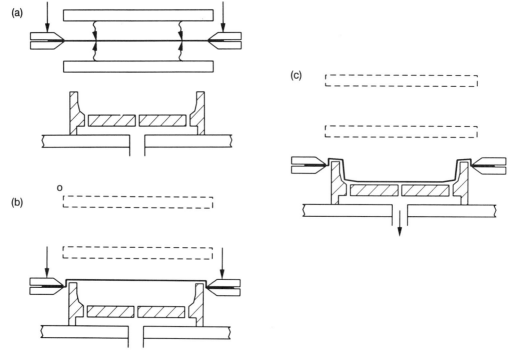

**Figure 10.12** Typical pressure thermoforming process. (a) The clamped sheet is heated to the processing temperature by means of infrared heaters. (b) The heated sheet is removed from the radiation heating source and placed into position for forming. (c) Vacuum is applied to the underside of the sheet, or air pressure is applied to the topside of the sheet, forcing the heated sheet into the mold.

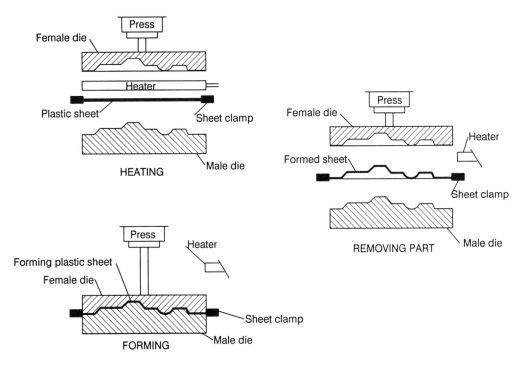

**Figure 10.13** Forming in a matched metal die system. After heating, the sheet is formed by the application of mechanical pressure to the upper part of the mold.

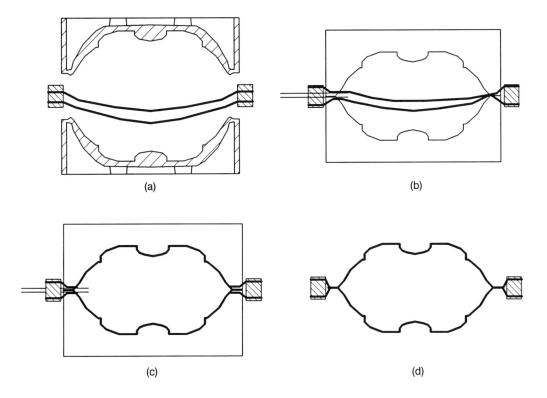

**Figure 10.14** Twin-sheet thermoforming process. (a) Heated sheets are placed between the mold halves. (b) The mold is closed sealing the two sheets at the edges and around an inflation tube. (c) Air pressure is supplied through the tube, causing the sheets to inflate against the mold walls. (d) The mold is opened once the sheets are cooled, and the part is removed.

leads to the possibility of carrying out effective design considerations. We first discuss the heating and cooling aspects followed by ideas on the forming step.

The success of a thermoforming process depends primarily on the rate at which a part can be produced. This is controlled by the time required to heat the sheet to the processing temperature and the time to cool the part down once it has been formed. The time required to heat a sheet to the processing temperature depends on the heat transfer conditions and, of course, the temperature at which the sheet becomes sufficiently formable but not so low in melt strength that significant sagging occurs. In the case of amorphous polymers this is usually 30 to 60°C above $T_g$, and for semicrystalline polymers it is usually just above the temperature where melting is just completed. In the case of semicrystalline polymers the molecular weight must be sufficiently high to impart a viscosity that is high enough to resist sagging.

At present there seems to have been no quantitative attempts to estimate the temperature at which a polymer is suitably formable. There are some empirical guidelines based on experience with given polymers. For example, Gruenwald (1987) gives the thermoforming temperature for PP to be between 150 to 199°C and for PS to be between 143 and 177°C (*note*: $T_g$ for PS is about 100°C). Dynamic mechanical properties obtained as a function of temperature may provide a way to estimate the processing temperatures. For example, the storage modulus ($G'$) versus temperature at an angular frequency of 1.0 rad/s is shown in Fig. 10.15 for PP and glass-filled PP. Based on experience when $G'$ reaches about $1 \times 10^7$ Pa, the material is formable. This appears to occur at a temperature above 150°C for PP. For glass-filled PP this temperature appears to be above 170°C. Experience coupled with dynamic mechanical properties may provide the best approach for estimating the processing temperature.

Once the processing temperature has been determined, the time required to heat the sheet to this temperature can be calculated by methods given in Chapter 5. Basically, the problem becomes one of one-dimensional transient heat conduction. Heating is usually accomplished by radiation combined with convection. The radiation heat transfer coefficient is given in Eq. 5.138, and the view factor for two parallel planes (Eq. 5.134) is most frequently used. The emissivities of the heating source fall in the range of 0.9 to 0.95, and that of the polymer sheet is about 0.9. The rating of the heating source is available from the manufacturer of the radiant heating source. For example,

resistance heated strip elements usually found in commercial thermoforming machines have a surface temperature of 427°C at the highest power input. If the distance of the heaters from the sheet is greater than about 30 cm, then free convection cooling of the sheet surface can occur. Throne (1986) has estimated the free convection heat transfer coefficient to be of the order of 11.3 W/m² °K, but this can be calculated via methods given in Chapter 5. Sometimes, forced convection is used cyclically to cool the sheet surface. Again, the methods described in Chapter 5 can be used to estimate the heat transfer coefficient. The numerical procedure presented in the code in Table 5.14 (p. 113) can be used for determining the heating time and temperature distribution in the sheet.

Although the time for producing a part is determined by the heating and cooling times, the utility of the part is determined by the wall thickness distribution. If during the forming process the wall becomes too thin in certain regions, then the part may fail under loading or have reduced barrier properties. The ability to estimate the wall thickness distribution is crucial to the design of a thermoforming process.

Estimating the wall thickness distribution is extremely difficult, especially for irregularly shaped parts. Numerical methods, such as the finite element method, are required. Furthermore, an appropriate nonlinear constitutive equation for the polymer is required, and this may be difficult to obtain for materials that exist in the near rubbery state at the forming temperature. Obtaining rheological data at the forming temperature is very difficult. Finally, the solution of the coupled equations of motion and the nonlinear constitutive equation requires sophisticated numerical codes that are not readily available at this time.

In principle it is desirable to be able to predict wall thickness distribution for a given polymer being formed into a given shape. Then, given the various processing variables, including temperature, differential pressure, and sheet rheological properties, one wishes to determine how to move the material around in order to meet the necessary part design criteria while using the minimum initial sheet thickness. We are not in position at this point to be able to do this accurately. However, there are at least two things we can do. First, we can estimate whether sufficient material is available to maintain a desired average wall thickness, and second we can see how for simple geometries the wall thickness distribution is obtained using geometric arguments and a mass balance.

As a first estimate we must be able to determine on the average whether sufficient material is available in the sheet for producing a certain part. This is in essence done by carrying out a mass balance on the sheet. For example, consider Fig. 10.16 in which a sheet of arbitrary dimensions is to be formed into a female mold. Because the original volume of material available for forming in this configuration is $CDt_i$ (*note*: $AB - CD$ is the area that contacts the mold surface first and this will deform very little leaving only the area $CD$) where $t_i$ is the initial thickness of the sheet. After forming, the volume of the material is

$$V_f = (2DE + 2CE + CD)t_f, \tag{10.51}$$

and hence using the dimensions given in arbitrary units $t_f = t_i/4$. Obviously if the average thickness does not meet the design requirements, then sufficient material is not available.

However, because there will be a distribution of wall thickness, there may be regions that do not satisfy the design requirements. This is in spite of sufficient material being available to provide the wall thickness on the average. It is desirable, therefore, to be able to predict the wall thickness distribution as a function of processing conditions and rheological properties of the polymer. This has been done for several simple geometries (Rosenzweig & coworkers, 1979) and we illustrate the approach in the following example.

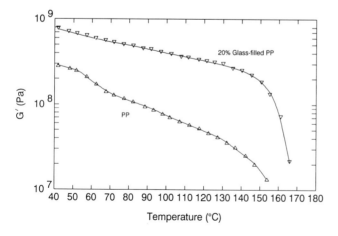

**Figure 10.15** Dynamic mechanical thermal analysis ($G'$ versus temperature at an angular frequency of 1.0 rad/s) of polypropylene (PP) and PP filled with 20 wt% glass fiber determined on a Rheometrics Mechanical Spectrometer using the torsional mode.

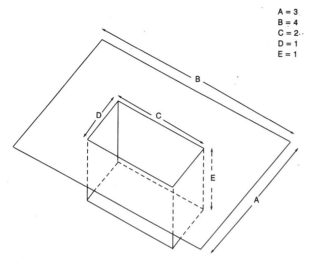

A = 3
B = 4
C = 2
D = 1
E = 1

**Figure 10.16** Relative dimensions of a container generated from a thermoformed sheet using a female mold. Before forming, the area of the section to be formed is 2 units. After forming, the surface area of the box is 8 units.

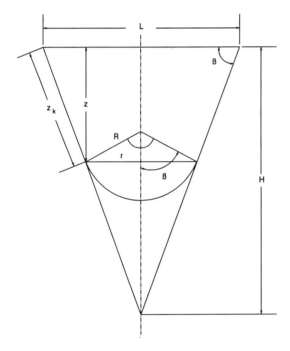

**Figure 10.17** Geometry of a conical mold used in the thermoforming of a sheet. The sheet is assumed to expand as a spherically shaped bubble freezing as it contacts the walls.

---

### EXAMPLE 10.3
### Wall Thickness Distribution in a Conical Mold

Consider the thermoforming of a sheet in which the mold is conical in shape as shown in Fig. 10.17. The sheet is forced into the mold by means of an applied pressure differential (e.g., air pressure is applied to the surface). The sheet is initially of uniform thickness, $h_0$, and the process is isothermal. Obtain an expression for the thickness distribution as a function of the initial sheet thickness and mold geometry.

### Solution

The following assumptions are made:

1. The polymer is incompressible.
2. Polymer deformation occurs only in the bubble.
3. The free surface is uniform in thickness and has a spherical shape.
4. The sheet solidifies on touching the mold walls, but the spherical surface remains isothermal.
5. The polymer does not slip on the walls.
6. The spherical surface, or bubble, thickness is small compared to its radius.

Referring to Fig. 10.17 a differential mass balance is made to obtain the thickness distribution:

$$2\pi R^2(1-\cos\beta)h|_{z_k} - 2\pi R^2(1-\cos\beta)h|_{z_k+\Delta z_k} = 2\pi r h\Delta z_k, \qquad (10.52)$$

where the first term is the volume of the spherical bubble at position $z_k$ along the wall and the second term is the volume of the bubble at a position $z_k + \Delta z_k$. The volume of material deposited on the wall as the bubble moves from $z_k$ to $z_k + \Delta z_k$ is represented on the right side of Eq. 10.52. Dividing through by $\Delta z_k$ and taking the limit as $\Delta z_k$ goes to zero gives the following differential equation:

$$-\frac{d}{dz_k}(R^2h) = \frac{rh}{(1-\cos\beta)}. \qquad (10.53)$$

From geometrical considerations we can replace $r$ by $R\sin\beta$ and $R$ by:

$$R = \frac{H - z_k\sin\beta}{\sin\beta\tan\beta} \qquad (10.54)$$

to give the following differential equation for $h$:

$$\frac{dh}{h} = \left(2 - \frac{\tan\beta\sin\beta}{1-\cos\beta}\right)\sin\beta\frac{dz_k}{H - z_k\sin\beta}. \qquad (10.55)$$

Equation 10.55 can be integrated using the initial condition that $h(0)=h_1$, where $h_1$ is the initial thickness of the bubble tangent to the cone at $z_k=0$, to give:

$$\frac{h}{h_1} = \left(1 - \frac{z_k}{H}\sin\beta\right)^{\sec\beta-1}. \qquad (10.56)$$

Finally, it is possible to relate the initial thickness, $h_1$, to the original sheet thickness, $h_0$, by the following relation:

$$\frac{\pi L^2 h_0}{4} = \frac{\pi L^2(1-\cos\beta)h_1}{2\sin^2\beta}, \qquad (10.57)$$

where the left side of this equation is just the volume the circular piece of sheet above the mold opening before forming, and the right side is the spherical surface when $z_k=0$. On replacing $h_1$ in Eq. 10.56 with the value for $h_1$ in terms of $h_0$ in Eq. 10.57, the thickness distribution is thus given by:

$$\frac{h}{h_0} = \frac{1+\cos\beta}{2}\left(1 - \frac{z_k}{H}\sin\beta\right)^{\sec\beta-1}. \qquad (10.58)$$

The approach described in Ex. 10.3 is totally based on conservation of mass with no mention of the role played by the rheological properties of the polymer. In spite of this, it seems to work reasonably well for shallow geometries that lend themselves to geometric analysis. Where, then, is the constitutive equation required in the analysis of thermoforming? In principle, the shape of the bubble is not known but is related to both processing conditions and the rheology of the polymer. Furthermore, the pressure required to deform the sheet and the time to fill the mold are determined by the rheology of the polymer. In the following example we illustrate how the constitutive equation is used in simulating thermoforming.

**EXAMPLE 10.4**
**Inflation Pressure of the Bubble**

Establish the equations required to determine the pressure and time to form a sheet in the conical mold shown in Fig. 10.17 for two cases: (1) a Newtonian fluid; and (2) a viscoelastic fluid.

**Solution**

In addition to the assumptions used in Ex. 10.3 we assume that the deformation of the spherical film is occurring under quasi-steady-state conditions, and hence we use an unsteady-state mass balance, but the steady-state equation of motion. The bubble thickness is also assumed to be small relative to the radius. Using spherical coordinates, the following postulates are made:

$$v_r = v_r(r); \qquad p = p(r), \tag{10.59}$$

where $r$ is the spherical coordinate taken along $R$ in Fig. 10.17, and not the radius of the conical cross section as indicated in this figure. The continuity equation becomes:

$$\frac{\partial}{\partial r}(r^2 v_r) = 0, \tag{10.60}$$

which can be integrated to give

$$v_r = A(t)/r^2, \tag{10.61}$$

where $A(t)$ is an arbitrary function of time. At the inside surface of the bubble, which is located at $R$ the velocity of the fluid just equals the velocity of the surface, which is $dR/dt$ (we denote this as $\dot{R}$), that is,

$$v_r(R) = \dot{R}. \tag{10.62}$$

From Eq. 10.62 we find that $A(t) = \dot{R}R^2$, and $v_r$ becomes:

$$v_r = \frac{\dot{R}R^2}{r^2}. \tag{10.63}$$

Hence, an expression for the velocity field has been obtained directly from the continuity equation without using the equation of motion.

In order to relate the rate at which the mold is filled to the applied pressure, it is necessary to use the equation of motion. Independent of the choice of constitutive equation the equation of motion becomes:

$$0 = -\frac{\partial p}{\partial r} - \frac{1}{r^2}\frac{\partial}{\partial r}(r^2 \tau_{rr}) + \frac{\tau_{\theta\theta} + \tau_{\phi\phi}}{r} \tag{10.64}$$

Equation 10.64 can be rewritten in the following form:

$$\frac{\partial \pi_{rr}}{\partial r} = \frac{\tau_{\theta\theta} + \tau_{\phi\phi} - 2\tau_{rr}}{r}, \tag{10.65}$$

where $\pi_{rr}$ is the total stress (see Chapter 3). At the inside surface (i.e., at $R$) it can be shown that $\pi_{rr}(R) = -P(R)$, and at the outside surface $\pi_{rr}(R+h) = -P(R+h)$. Therefore, the pressure differential across the film is given by:

$$P(R) - P(R+h) = \Delta P = \int \frac{\tau_{\theta\theta} + \tau_{\phi\phi} - 2\tau_{rr}}{r}dr. \tag{10.66}$$

In our case it is assumed that the film is thin and the stresses do not vary significantly over the film thickness. Hence Eq. 10.66 becomes:

$$\Delta P = \frac{(\tau_{\theta\theta} + \tau_{\phi\phi} - 2\tau_{rr})|_R h}{R}. \tag{10.67}$$

To complete our goal we now must assume a constitutive equation for the polymer. We first assume the Newtonian model, as this allows one to obtain an analytical solution. This is also an advisable procedure to follow before embarking on a solution using a viscoelastic model that will more than likely require numerical methods to obtain an answer. For the Newtonian model (see Chapter 2) the stress components are:

$$\tau_{\theta\theta} = -2\mu \frac{v_r}{r} = \frac{-2\mu\dot{R}R^2}{r^3}$$

$$\tau_{\phi\phi} = -2\mu \frac{v_r}{r} = \frac{-2\mu\dot{R}R^2}{r^3}$$

$$\tau_{rr} = -2\mu \frac{\partial v_r}{\partial r} = \frac{4\mu\dot{R}R^2}{r^3}. \tag{10.68}$$

These quantities are now substituted back into Eq. 10.67 to obtain an expression for the pressure difference:

$$\Delta P = \frac{-8\mu\dot{R}h}{R^2}. \tag{10.69}$$

By using Eqs. 10.58 and 10.54 we can express Eq. 10.69 as:

$$\Delta P = \frac{-8\mu\dot{R}h_0}{R^2}\left[\frac{R\sin\beta\tan\beta}{H}\right]^{\sec\beta - 1}. \tag{10.70}$$

Equation 10.70 represents a nonlinear ordinary differential equation for finding $R(t)$ for a given pressure differential across the film or bubble. The solution of this equation is obtained using the IMSL subroutine IVPAG (see Pr. 10C.3).

We next consider the formulation of the problem for the viscoelastic case. We select the Phan Thien-Tanner (PTT) model again (Eq. 3.45), which in spherical coordinates (Bird et al., 1987) leads to the following equations for the stress components:

$$\exp(-\varepsilon\lambda tr\tau/\mu)\tau_{rr}+\lambda\frac{\partial}{\partial t}\tau_{rr}+\frac{4\lambda\dot{R}}{R}(1-\xi)\tau_{rr}=\frac{4\mu\dot{R}}{R} \tag{10.71}$$

$$\exp(-\varepsilon\lambda tr\tau/\mu)\tau_{\theta\theta}+\lambda\frac{\partial\tau_{\theta\theta}}{\partial t}-\frac{2\lambda\dot{R}}{R}(1-\xi)\tau_{\theta\theta}=\frac{-2\mu\dot{R}}{R} \tag{10.72}$$

$$\exp(-\varepsilon\lambda tr\tau/\mu)\tau_{\phi\phi}+\lambda\frac{\partial\tau_{\phi\phi}}{\partial t}-\frac{2\lambda\dot{R}}{R}(1-\xi)\tau_{\phi\phi}=\frac{-2\mu\dot{R}}{R}. \tag{10.73}$$

This set of equations represents three coupled nonlinear ordinary

differential equations for finding $\tau_{rr}$, $\tau_{\theta\theta}$, and $\tau_{\phi\phi}$. Because they cannot be solved explicitly for the stresses, as was the case for the Newtonian fluid, it is not possible to obtain a differential equation such as Eq. 10.70 which can be solved to find $\dot{R}/R$ directly. To solve these equations, one must guess at values of $\dot{R}/R$ first, solve the set of coupled differential equations numerically (IVPAG), and then determine whether the stress values satisfy Eq. 10.67. One must repeat this process until Eq. 10.67 is satisfied for a given pressure differential (this approach is used in Pr. 10C.3). Because the solution is obtained numerically, the guess for $\dot{R}/R$ is made for a small time step over which it is assumed that $\dot{R}/R$ is constant, and hence $R(t)=R_0\exp(Ct)$. The thickness distribution is then determined using the calculated value of $R(t)$ and Eqs. 10.58 and 10.54.

Before leaving this section we make a few comments about solving problems using nonlinear rheological equations of state. The solution of the nonlinear equations is facilitated by obtaining the Newtonian solution first. In the example just presented, the solution of the Newtonian case serves to provide an estimate of the time required to fill the mold, the magnitude of the pressure difference, and $\dot{R}/R$. Because the solution must be obtained numerically using an iterative procedure in the viscoelastic case, the rate of convergence of the solution is greatly enhanced by the initial guesses for $\dot{R}/R$ and the time for filling the mold provided by the solution for the Newtonian case. Hence, it is advisable to obtain a solution to the Newtonian case before embarking on the solution to the viscoelastic case.

## 10.4. BLOW MOLDING

Blow molding is a process for generating hollow plastic articles such as bottles and containers. It was initially used by the packaging industry, but more recently it has been used by the automotive industry to produce parts such as fuel tanks, bumpers, dashboards, and seatbacks. In other words, plastic parts are being manufactured for applications where some structural integrity is required.

In this section we first review some of the more salient technological features of blow molding. We then consider the parts of the process that can be subjected to quantitative analysis and design.

### 10.4.1. TECHNOLOGICAL ASPECTS

Although there are a number of variations in the way in which blow molding is carried out, there are a number of common steps. First, conventional melt processes are used to make a cylindrical tube (*note*: the preformed sample may be of other shapes). When extrusion is used, this tube is referred to as a *parison*, and when injection molding is used, it is referred to as a *preform*. The softened preformed tube is transferred to a mold consisting of two halves, where it is sealed and inflated to assume the internal contours of the mold. The part is cooled in the mold until it reaches a temperature where it will maintain the shape of the mold after the mold is opened.

*Extrusion blow molding* is used frequently for polymers that exhibit high melt strength such as polyolefins. The process is shown schematically in Fig. 10.18. In this figure the tubular parison is continuously extruded from a die into position between the two mold halves and then separated from the main stream by a knife. The mold closes, sealing the end of the parison, and air pressure is applied, inflating the parison against the walls of the mold. The time for inflation is very short, usually in the range of a second depending on the size of the part. The longest step in the process is the cooling of the part. The time for cooling depends on the temperature to which the part must be lowered in order for it to maintain the shape of the mold and the

rate of heat transfer between the mold wall and the part. The mold finally opens, and the part is ejected.

In some cases injection molding is used to generate preforms rather than extrusion, but otherwise the process is nearly the same. Injection molding is used primarily when the screw-thread dimensions must be precise and to avoid flash, weld lines, and material waste at the base of the container. Just as in the case of extrusion blow molding, it is

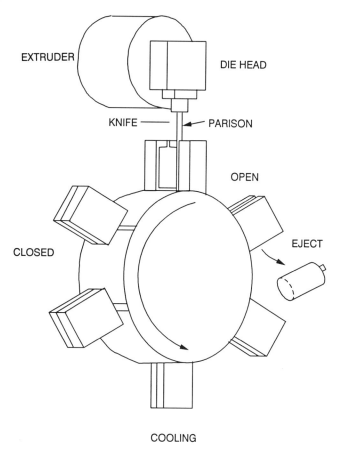

**Figure 10.18** Commercial extrusion blow-molding process. The tubular parison is extruded into position between the walls of the open mold. The mold walls close, pinching the ends of the parison, and simultaneously the parison is cut from the continuously extruded tube. The parison is inflated against the mold walls, where the part cools and is removed when the mold is opened.

possible to generate preforms with multiple layers for situations where barrier properties are required.

In continuous processes the parison or preform must have sufficient melt strength so that it does not sag under its own weight. Sagging leads to unacceptable variations in the wall thickness of the part. For polymers such as PET it is usually not possible to use extrusion blow molding because of severe sagging problems. To overcome sagging problems, preforms are injection-molded in a separate step where they can be rapidly quenched to inhibit crystallization and hence remain clear. The preforms are then heated by radiation to a temperature about 30°C above glass transition temperature, $T_g$, where cyrstallization kinetics are slow but the material is deformable. The heated preforms are then transferred to the mold, where they are inflated by air pressure. This type of two-step process can be used for a resin such as PPS that also has slow crystallization kinetics and low melt strength.

Resistance to sagging is due to the rheological properties of the melt. It is preferable to modify the polymer so that the extensional behavior of the melt is altered rather than raise $\eta_0$ by increasing the molecular weight. In Fig. 10.19 the length of a parison as function of time is presented for PET and a modified form of PET that presumably contains branching (Birley et al., 1991). The boken lines in Fig. 10.19 represent the length of the parison in the absence of any swell or sagging (i.e., $\langle v_z \rangle t$). One can see that for short times the parison length is below that given by the broken line (this is for extrusion times up to about 5 s, which corresponds to a parison length of about 12 cm), whereas after about 5 s there is considerable parison sagging. For the modified PET, referred to as PETG, parisons of length of about 20 cm can be extruded before sagging starts, but even then there is less sagging than for PET. Although no rheological data were given, it is assumed that PETG contains branching, and that the material exhibits extensional strain hardening.

The process of inflating the parison is primarily one of planar extensional flow, especially away from the ends of the parsion. Because the ends of the parison are constrained as the parison expands, the thickness of the wall decreases as the diameter expands, leading to primarily planar extensional deformation. For this reason the blow-molded part contains primarily orientation along the circumferential or hoop direction and hence will exhibit mechanical anisotropy.

To generate a better balance of mechanical properties, it is necessary to create biaxial orientation in the part. *Stretch blow molding* is used to accomplish this. In essence the parison is stretched along the axial direction before being inflated. Biaxial orientation is specifically required in large containers for fluids. For example, bottles for carbonated beverages are typically processed by means of stretch blow molding.

Our coverage of blow molding is by no means thorough, as it only serves to provide enough background for the following section, which is concerned with more quantitative aspects of blow molding as required in design and analysis. More details about the technology of blow molding can be found in Lee (1990). We now turn our attention to the aspects of blow molding that can be handled in a quantitative fashion.

### 10.4.2. SIMULATION

Figure 10.20 schematically shows the basic parts of the blow-molding process. There are basically four steps that must be considered in the analysis of a blow-molding process (these are in addition to the extrusion or injection molding steps that have already been considered): (1) the cooling of the hanging parison or preform by free convection before the mold closes or the heating of the preform in a two-step process; (2) the sagging of the parison in the case of extrusion processes; (3) the expansion of the parison or preform against the mold walls; and (4) the cooling of the part by means of forced convection. Although we will make a few comments about the heat transfer parts of the process, most of this analysis falls under the area of transient heat transfer discussed in Chapter 5. We emphasize topics such as parison sagging and expansion.

During the hanging of the parison in the case of extrusion blow molding there is some cooling of the parison as the hang times may be of the order of 10 to 20 s for large parts. There is probably little heat transfer at the inner surface, which, hence, can be considered as insulated. At the exterior surface heat transfer occurs by free convection. The heat transfer coefficient can be estimated using the material given in Section 5.3. The problem can be considered as one of one-dimensional

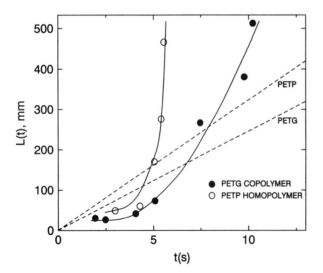

**Figure 10.19** Length of an extruded parison of PETP and PETG (modified PETP) versus time. The lines represent the length of the parison as a function of time under conditions of no extrudate swell or sagging. (Reprinted by permission of the publisher from Birley et al., 1991.)

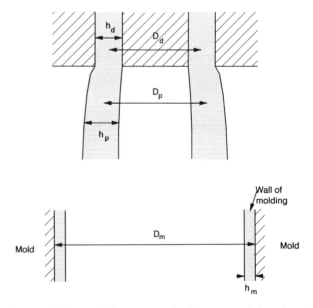

**Figure 10.20** Blow-molding process showing an extruded parison of cylindrical shape leaving an annular die and the walls of the mold.

transient heat conduction with the axes attached to the moving parison. There is some thickness variation along the length of the hanging parison that may present some problem in the analysis. Certainly, if there is a significant drop in temperature, then the temperature of the melt leaving the die will have to be adjusted accordingly.

In two-step processes, such as used for blow molding PET, the preform must be heated up to a temperature suitable for inflation. This is usually accomplished by means of radiation heating as discussed in Section 5.4. We do not discuss this topic further here.

Once the parison has been inflated, it is cooled primarily by heat transfer at the mold surface. Channels through which water can be circulated are machined in the mold walls in a way similar to that used in injection molding. Approximately 50 to 95% of the total processing time is involved in cooling the part. At the inner surface very little heat transfer occurs with the heat transfer coefficient because of free convection being estimated in the range of 5.7 to 57 $W\,m^{-2}\,°K^{-1}$. Hence, the cooling time of the part can be shortened by increasing the heat transfer at the inner surface. Using chilled air in the blowing process certainly helps, but other methods, such as injecting liquid nitrogen (Gibbs, 1989) or a high-pressure mixture of air and water in which the water freezes as the air expands to form fine ice crystals (Hunkar, 1973), may provide more rapid cooling. Again the cooling of the side walls is described by the one-dimensional transient heat conduction equation.

Sagging of the parison, as discussed, leads to a nonuniform distribution of the wall thickness. If the wall becomes too thin, it will fail during the inflation process. In the next example, we consider how to approach the modeling of sagging.

---

### EXAMPLE 10.5
### Sagging of a Cylindrical Parison

Determine the equations required to calculate the length of the parison shown in Fig. 10.20 as a function of time and the thickness of the parison as a function of position. Assume that the melt is Newtonian and that the thickness of the parison is small relative to the diameter, so that curvature can be neglected. Take the initial thickness as the maximum thickness due to extrudate swell.

### Solution

For most cases it can be assumed that the thickness of the parison is small relative to the diameter, and hence we can treat the parison as a flat sheet. Furthermore, we will assume that the deformation occurs along the axis of the parison and in the thickness direction but that the diameter does not change significantly. With these assumptions, we can follow the development for thick films given in Section 9.1 and refer to Fig. 9.17.b (p. 263). We will assume quasi-steady-state conditions in which the mass balance is unsteady but the steady-state equation of motion holds. With these assumptions we postulate that

$$v_z = v_z(z); \qquad v_y = v_y(y). \tag{10.74}$$

Following the development given in Eqs. 9.116 through 9.119, we can determine a relation between $v_z$ and $h$ (Eq. 9.120)

$$\frac{1}{h}\frac{dh}{dz} = -\frac{1}{v_z}\frac{dv_z}{dz}. \tag{10.75}$$

The equation of motion in the z direction including the gravitational term is

$$\frac{\partial \tau_{yz}}{\partial y} + \frac{\partial \pi_{zz}}{\partial z} - \rho g = 0. \tag{10.76}$$

Neglecting the effect of surface tension and air drag, which means that the surface is stress free, we can relate $\tau_{rz}$ to $\pi_{zz}$ by the following relation:

$$\tau_{yz} = \frac{1}{2}\frac{dh}{dz}\pi_{zz}, \tag{10.77}$$

where $h$ is the wall thickness at any z position. Because $v_z$ is assumed to be nearly constant over the cross section of the parison, we can integrate Eq. 10.76 across the thickness to obtain the following form of the equation of motion:

$$\frac{h'}{h}\pi_{zz} + \frac{d\pi_{zz}}{dz} = \rho g, \tag{10.78}$$

where $h' = dh/dz$. At this point we will assume that the fluid can be described as Newtonian, but a viscoelastic constitutive equation could be used as well. Using Eq. 9.123, Eq. 10.78 becomes

$$-4\mu\frac{d^2v_z}{dz} + \frac{4\mu}{v_z}\left(\frac{dv_z}{dz}\right)^2 - \rho g = 0. \tag{10.79}$$

Equation 10.79 can be integrated once by first replacing $dv_z/dz$ by $p$, separating variables and then integrating to give

$$\frac{dv_z}{dz} = \sqrt{v_z}\left\{\frac{1+\exp\left[2\sqrt{\rho g/4\mu v_z}(z-L)\right]}{1-\exp\left[2\sqrt{\rho g/4\mu v_z}(z-L)\right]}\right\}. \tag{10.80}$$

To obtain Eq. 10.80 it was assumed that $dv_z/dz = 0$ at $z = L$ where $L$, is the instantaneous length of the parison. The solution to Eq. 10.80 must be obtained numerically using the initial condition that $v_z = \langle v_z \rangle$ at $z = 0$ (See Pr. 10C.6). The solution is obtained for arbitrary values of $L$. The thickness profile can be obtained by means of Eq. 10.75 using the initial thickness estimated from die swell data. One must assume that die swell is unaffected by the weight of the hanging parison. The complete solution of the problem can only be obtained using numerical techniques.

---

Finally, we consider one more example, the pressure required to inflate a cylindrical parison. In the following example we consider the inflation process up to the point where the cylinder just touches the walls of the mold. In the central region of the parison the deformation is primarily that of planar extensional flow. The complete filling of the mold requires the use of finite element methods, as once the parison touches the wall, the deformation of the remaining parison is very complex.

**EXAMPLE 10.6**
**Inflation of a Cylindrical Parison**

Referring to Fig. 10.20, determine the time required at a given inflation pressure for a cylindrically shaped parison of length $L$ to contact the mold wall. Assume that as the ends of the parison are clamped, the process is primarily one of planar extensional flow. Furthermore, assume that surface tension is negligible, $\rho$ is constant, inertial effects are negligible, $h \ll R$, and the process is isothermal. Obtain solutions for first the Newtonian case and then the viscoelastic case.

**Solution**

The solution follows closely that used in Ex. 10.3 for the inflation of a spherical bubble, and hence some steps will be carried out without detailed justification. With the postulate that $v_r = v_r(r, t)$, the continuity equation plus the boundary condition that $v_r(R, t) = dR/dt$ lead to the following velocity field:

$$v_r = \dot{R}R/r, \tag{10.81}$$

where $\dot{R} = dR/dt$. Assuming that the fluid can be described as Newtonian, then the stresses are:

$$\tau_{rr} = -2\mu \frac{\partial v_r}{\partial r} = 2\mu \dot{R}R/r^2$$

$$\tau_{\theta\theta} = -2\mu \frac{v_r}{r} = -2\mu \dot{R}R/r^2. \tag{10.82}$$

The equation of motion for this type of flow is:

$$0 = -\frac{\partial p}{\partial r} - \frac{1}{r}\frac{\partial}{\partial r}(r\tau_{rr}) + \frac{\tau_{\theta\theta}}{r}. \tag{10.83}$$

Using the definition for the total stress Eq. 10.83 can be rewritten as:

$$-\frac{\partial \pi_{rr}}{\partial r} = \frac{\tau_{rr} - \tau_{\theta\theta}}{r}. \tag{10.84}$$

With the assumptions that $h \ll R$ and that the stresses are constant over the thickness of the parison, Eq. 10.84 is integrated to give:

$$P - p_a = (\tau_{rr} - \tau_{\theta\theta})(h/R), \tag{10.85}$$

where $P$ is the applied pressure and $p_a$ is the pressure on the outside of the parison. For the Newtonian case the expressions for the stresses given in Eq. 10.82 are substituted into Eq. 10.85 to give:

$$P - p_a = 4\mu \dot{R}h/R^2. \tag{10.86}$$

In order to solve Eq. 10.86 another expression that relates $h$ and $R$ must be obtained from the conservation of mass. Because $\rho$ is constant and the length of the parison is constant, then one finds that $\dot{h}/h = -\dot{R}/R$ (*note*: this is obtained by differentiating $V = 2\pi RhL$ with respect to time). Equation 10.86 plus this last equation must be solved simultaneously to find $R(t)$ and $h(t)$ using the initial conditions that at $t = 0$, $R = R_p$ and $h = h_p$. The solution of these equations is obtained numerically in Pr. 10C.4.

We next obtain the equations required to find $R(t)$ and $h(t)$ for the case of a viscoelastic fluid. Using the PTT model and the velocity field given in Eq. 10.81, the equations for determining the stress components are:

$$\lambda \frac{\partial \tau_{rr}}{\partial t} + \exp[-(\varepsilon\lambda/\mu)(\tau_{rr} + \tau_{\theta\theta})]\tau_{rr} + (2\lambda \dot{R}R)(1 - \xi)\tau_{rr} = 2\mu\dot{R}/R \tag{10.87}$$

$$\lambda \frac{\partial \tau_{\theta\theta}}{\partial t} + \exp[-(\varepsilon\lambda/\mu)(\tau_{rr} + \tau_{\theta\theta})]\tau_{\theta\theta} + (2\lambda \dot{R}R)(1 - \xi)\tau_{\theta\theta} = -2\mu\dot{R}/R \tag{10.88}$$

The details of determining Eqs. 10.87 and 10.88 are considered in Pr. 10B.6. As in the situation in Ex. 10.4, Eqs. 10.87 and 10.88 represent two nonlinear ordinary differential equations that must be solved numerically using guesses for $\dot{R}/R$ based on the Newtonian solution. The stresses must satisfy Eq. 10.85 as well as the expression relating $h$ and $R$ obtained from the mass balance. The solution requires an iterative approach, as discussed in Pr. 10C.5.

---

The discussed approach for dealing with the inflation of a parison certainly represents a way to estimate the required inflation pressure and inflation time. However, the complete filling of the mold cavity is more complicated than that represented by planar extensional flow, and hence finite element methods are required to more accurately handle the simulation of the inflation of the parison. The capability of accurately predicting the wall thickness distribution in the part is crucial to the successful design of a blow-molding process. Furthermore, it would be desirable to be able to predict molecular orientation and associated physical properties as a function of processing conditions. The ultimate goal would be to model the complete extrusion blow-molding process, including extrudate swell, parison sag, and the blowing process.

## 10.5. SOLUTION TO DESIGN PROBLEM 9

The solution to Design Problem 9 basically consists of four parts: (1) determining the number of sheets required to fill the mold when the

mold is closed; (2) determining the temperature setting of the infared heaters in order to heat the blanks to the processing temperature as fast as possible without exceeding the upper processing temperature of PP at the surfaces; (3) determining the rate of closing of the press and the maximum operating force required to fill out the mold; and (4) specifying the cooling conditions at the mold wall to minimize the time required for the part to remain in the mold.

We first determine the arrangement of the sheets in the mold and the number of sheets required. The total length of the mold including the curved sections is 1.628 m. The breadth of the mold as given is 64.8 cm, and the final thickness is given as 3.2 mm. Hence, the total volume of the final part will be $3.3758 \times 10^{-3}$ m³. Because the mold is 64.8 cm in breadth, we can place three sheets across this dimension. If we fill the flat section of the mold with 5.23 sheets that are 19.1 cm in width, they will not fill the curved section of the mold when it is closed. Therefore, we must stack the sheets. If we stack the sheets two deep, then covering 72.39 cm of the flat section of the mold will provide enough material to fill the mold when it is closed. Hence, we need two

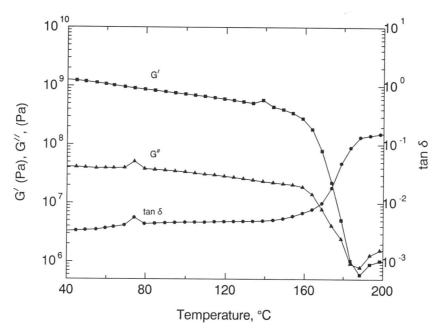

**Figure 10.21** Dynamic mechanical thermal analysis of polypropylene reinforced with 30 wt% glass fiber mat (Azdel PM-10300). Measurements were made at an angular frequency of 1.0 rad/s on rectangular strips in a Rheometrics Mechanical Spectrometer (RMS 800).

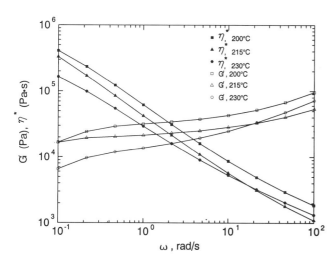

**Figure 10.22** Dynamic oscillatory shear properties of Azdel PM-10300 obtained at three temperatures.

layers of sheet in which three are needed to cover the breadth of the mold, and 3.79 are needed along the width of the mold. A total of 12 sheets are needed in which three of the sheets must be cut to a width of 15.1 cm.

We next consider at what temperature the infared heaters should be set in order to heat the sheets to an acceptable processing temperature in the shortest time possible. The sheets are heated in the oven from both sides and the maximum temperature of the surface of the infared heaters is given as 427°C. The problem is that if the infared heaters are used at the highest temperature, then the surface will come up to temperature in a much shorter time than required to heat the centerline to temperature. Hence, the PP will degrade at the surface leading to a

reduction in properties. The other problem is to determine the temperature to which the sheets must be heated in order for them to flow under a reasonable force level. It is suggested by the dynamic mechanical properties of the composite sheet presented in Fig. 10.21 that at a temperature between 160°C and 168°C a sheet begins to soften (*note*: the melting point of PP is about 165°C). The storage modulus decreases rapidly with increasing temperature at this point. Above 184°C the method of testing changes, and the parallel plate fixtures must be used. Dynamic mechanical properties ($\eta^*$ and $G'$) at three temperatures are presented in Fig. 10.22 for this material. Based on past experience it is known that the complex viscosity at low values of $\omega$(1.0 rad/s) should be no more than about $2 \times 10^5$ Pa·s in order for the material to flow with the application of reasonable pressure levels. Hence, it is desirable to keep the sheet or blank between about 230 and 215°C during compression molding.

We next formulate the solution to the heat transfer problem. Although the blanks will be stacked two deep in the compression mold, they will pass through the oven as single blanks. The problem to be solved is that of one-dimensional transient heat conduction in which the top and bottom surfaces are subjected to radiation heating. Following the development in Sections 5.3 and 5.4 the differential equation for the heat transfer process becomes:

$$\frac{\partial \theta}{\partial t} = (k/\rho \bar{C}_p b^2)\frac{\partial^2 \theta}{\partial \xi^2}, \qquad (10.89)$$

where the dimensionless variables $\theta$ and $\xi$ are defined, respectively, as:

$$\theta = \frac{T - T_i}{T_R - T_i}; \qquad \xi = \frac{x}{b}, \qquad (10.90)$$

where $T_R$ is the surface temperature of the infared heater, $T_i$ is the initial temperature of the blank, and $b$ is the half thickness of the blank $(1.8 \times 10^{-3}\,\text{m})$. The initial and boundary conditions in dimensionless form are:

I.C. at $t = 0$; $\quad \theta(0, \xi) = 0$

B.C.1 at $\xi = 0$; $\quad \dfrac{\partial \theta}{\partial \xi}(t, 0) = 0$

B.C.2 at $\xi = 1$, $\quad \dfrac{\partial \theta}{\partial \xi}(t, 1) = -(bh_r/k) \left[ \dfrac{\theta(t, 1)(T_R - T_i) + T_i}{T_R - T_i} \right]$,

$$(10.91)$$

where $h_r$ is the radiation heat transfer coefficient defined in Eq. 5.137 and in terms of $\theta$ is given as:

$$h_r = \frac{\sigma F[(\theta(t, 1)(T_R - T_i) + T_i)^4 - T_R^4]}{[(T_i - T_R)(1 - \theta(t, 1))]}. \qquad (10.92)$$

For parallel flat plates $F$ is given in Eq. 5.134 and taking values of $e_1$ and $e_2$ to be about 0.9, $F$ becomes 0.82.

Before solving Eq. 10.89 a few comments regarding the material properties and energy absorbed on melting of PP must be made. In order to solve Eq. 10.89 values of $\rho$, $\bar{C}_p$, and $k$ for both the glass and PP are required. For PP the values at 180°C are given in Table 5.5 (p. 104). For glass, values are given only at 22°C (Harper, 1992), and these are 2500 kg m$^{-3}$, 2584.3 J kg$^{-1}$ °C$^{-1}$, and 1.021 W m$^{-1}$ °C$^{-1}$ for $\rho$, $\bar{C}_p$, and $k$, respectively. Blending rules for calculating $k$ and $\bar{C}_p$ were given in Section 5.3 where Eq. 5.73 was used for $k$ and Eq. 5.34 for $\bar{C}_p$. The values for the composite sample are:

$\rho_b = 1084$ kg m$^{-3}$; $\qquad \bar{C}_{pb} = 2207$ J kg$^{-1}$ °K$^{-1}$;

$k_b = 0.162$ W m$^{-1}$ °K$^{-1}$.

The melting point of PP is around 165°C, and as the polymer reaches this temperature during the heatup cycle, additional energy is absorbed delaying the time for the blank to reach the desired processing temperature. Although as a first approximation we will neglect the latent heat of fusion, this should be included as a source term in the energy equation (see Ex. 5.11).

To solve Eq. 10.89 we use the numerical approach discussed in Section 5.4. We can adopt the computer program given in Table 5.14 that uses the method of lines to solve an equation such as Eq. 10.89. The main differences are in the form of the heat transfer coefficient, $h_r$, which is given in Eq. 10.92, the expression for the surface node temperature changes to:

$$\theta_{NEQ+1} = \frac{4\theta_{NEQ} - \theta_{NEQ-1} + \left( \dfrac{2hb\Delta\xi}{k} \right)(T_R - T_i)}{3 + (2hb\Delta\xi(T_R - T_i)/k)}, \qquad (10.93)$$

and the ordinary differential equation for the nodal temperature next to the last node at the surface changes to:

$$\frac{\partial \theta_{NEQ}}{\partial t} = \left( \frac{k}{\rho \bar{C}_p b^2 (\Delta\xi)^2} \right)(\theta_{NEQ+1} - 2.0\theta_{NEQ} + \theta_{NEQ-1}). \qquad (10.94)$$

The computer code for solving Eq. 10.89 with a heat flux due to radiation at the slab surface is given in Table 10.1. The code given here is not particulary user-friendly in that one must make changes in the various parameters internally in the program.

Some experimentation with the numerical technique was needed before a stable numerical solution was obtained. Because the size of the time steps was relatively large ($1 \times 10^{-4}$ s), then it was necessary to refine the spacing of the nodal points until a stable solution was obtained. Division of the half thickness of the slab into 10 segments was not sufficient to obtain a stable solution. However, 20 segments or 21 nodal points gave a stable solution that improved only slightly with further refinement of the nodal spacings.

Results for two different heater settings are plotted in Fig. 10.23. Using the highest temperature of the heating elements of 427°C resulted in the surface of a blank reaching the upper processing temperature (UPT) of 230°C in less than 10 s. Meanwhile, it took about 30 s for the centerline to reach the suggested lowest processing temperature (LPT) limit. Severe degradation of the blank will occur under these conditions. Taking the heater temperature at 230°C, which is the UPT limit, it is observed that it takes about 70 s before the centerline reaches the LPT limit and 90 s before it reaches 215°C. However, the surface of the blank remains below the UPT limit. Certainly, one could optimize the time for heating the centerline to the LPT limit by trying different values of the heating element surface temperature. The actual time for heating the blank to the processing temperature will be longer than this as there will be some cooling of the blank by convection and energy will be absorbed at the melting temperature. Furthermore, it has been observed that when the temperature of the blank reaches the melting point of PP, then there is a swelling of the blank to almost twice its original thickness due to the recovery of elastic stresses imparted to the material during the fabrication of the composite sheet. These calculations at least give an estimate of the time required to heat the blank to the processing temperature and more importantly an approach for more accurately calculating this temperature.

We next estimate the pressure required to fill out the mold. A relation relating the force required to close the mold platens to the viscosity, geometric factors, and the rate of closing the platens was given in Eq. 10.50. In the derivation of Eq. 10.50 it was assumed that the fluid was Newtonian and isothermal conditions prevailed. Some cooling of the blanks will occur during the time of transfer from the oven to the mold. Giles and Reinhard (1991) have reported this time to be in the range of 15 s to a minute. There will be some drop in surface temperature due to free convection cooling. It can be estimated using the numerical approach presented in Section 5.4 that the average temperature of the blanks will be about 215°C when the compression process starts for transfer times less than 30 s. At the lowest closing rate available of $\dot{h} = 4.23$ mm/s it takes only about 0.76 s to close the mold. Hence, there should be little drop in the temperature of the blanks during the compression process, and the assumption of isothermal conditions should be valid. The maximum force required is that which occurs just as the mold is filled with the composite blank. In calculating this quantity by means of Eq. 10.50, the viscosity is required. The only data available are the complex viscosity as a function of angular frequency as shown in Fig. 10.22. The rheology of the composite blanks consisting of PP and random long glass fiber mat is unknown and difficult to obtain (lubricated squeezing flow discussed in Section 3.5 may be a way to obtain these data). We must therefore estimate the viscosity of the blank. At the point of closing the mold, the average rate of deformation (based on planar extensional deformation only) is $\dot{h}/h = 1.32$ s$^{-1}$. From the data in Fig. 10.22, $\eta^* = 500$ Pa·s. Using this value for the viscosity, the force is given by:

$$F = \frac{8(0.648)(1.2 \times 10^4)(2.23 \times 10^{-3})(0.814)^3}{(3.2 \times 10^{-3})^3} = 5.160 \times 10^7 \text{ N}$$

$$= 1.16 \times 10^7 \text{ lb}_f = 5.80 \times 10^3 \text{ tons}. \qquad (10.95)$$

Hence, to manufacture a part of this size, a high-force hydraulic press is required.

Finally, we make a few comments about the cooling of the part

**TABLE 10.1 Program Listing for Solving the Heat Transfer Problem Associated with Design Problem 9 (DPRIX.FOR)**

```
C**************ABSTRACT******************************
C
C  THIS PROGRAM CALLS THE SUBROUTINE DIVPAG DESCRIBED IN APPENDIX D
C  TO SOLVE THE ONE-DIMENSIONAL TRANSIENT HEAT CONDUCTION EQUATION
C  WITH A FLUX DEFINED AT THE WALL BOUNDARY. THE METHOD OF LINES
C  IS USED TO TRANSFORM THE PARTIAL DIFFERENTIAL EQUATION INTO A
C  SYSTEM OF FIRST-ORDER ORDINARY DIFFERENTIAL EQUATIONS.
C  THIS CODE IS USED IN THE SOLUTION OF DESIGN PROBLEM 9.
C
C*********************************************************
C
C  SPECIFICATION OF LOCAL VARIABLES, PARAMETERS, AND FILES
C
      INTEGER NEQ, NPARAM
      PARAMETER (NEQ=19, NPARAM=50)
      INTEGER IDO, IEND, IMETH, INORM, NOUT
      DOUBLE PRECISION A(1,1),FCN,FCNJ,HINIT,PARAM(NPARAM),
     &     TOL,X,XEND,Y(NEQ),T,DX,H,AL,T0,T10,RK
      EXTERNAL FCN,DIVPAG, SSET, UMACH
      COMMON T20,H
      OPEN(6,FILE='DATAIX.R3',STATUS='NEW')
C
C SPECIFICATION OF DIVPAG PARAMETERS
C
      HINIT=1.0D-5
      INORM=2
      IMETH=2
      PARAM(1)=HINIT
      PARAM(10)=INORM
      PARAM(12)=IMETH
      PARAM(4)=50000
      IDO=1
      TOL=1.0D-4
C
C  SPECIFICATION OF INITIAL CONDITIONS
C
      DO 1 II=1,19
         Y(II)=0.0
    1 CONTINUE
      T0=0.0
      T20=0.0
C
C  MATERIAL PROPERTIES
C
      RK=0.162
      AL=RK/(2553.0*1084.0)
      DX=9.0D-5
C
C  T=TIME(SEC)
C
      T=0.0
      DO 10 IEND=10,100,10
           WRITE(6,20)T,(N,N=0,NEQ+1),T0,(Y(I),I=1,19),T20
   20      FORMAT(25X,'THE SOLUTION AT TIME T = ',F5.1,
     &         /,21I6,/,21F6.3,//)
           T=DFLOAT(IEND)
           XEND=T
           H = 4.643D-08*((T20*203.0 + 304.0)**4 - 507.0**4)/(304.0*
     &         (1.0-T20) + 507.0*(T20-1.0))
           CALL DIVPAG (IDO,NEQ,FCN,FCNJ,A,X,XEND,TOL,
     &         PARAM,Y)
      T0=(1.0/3.0)*(4.0*Y(1)-Y(2))
      T20=(4*Y(19) - Y(18) + H*DX*406.0/RK)/(3.0+ DX*H*406.0/RK)
   10 CONTINUE
C
C  FINAL CALL TO RELEASE WORKSPACE
C
      IDO=3
```

**TABLE 10.1** *continued*

```
                CALL DIVPAG (IDO,NEQ,FCN,FCNJ,A,X,XEND,TOL,PARAM,Y)
C
          STOP
          END
C
C
C
C   SUBROUTINE CALLED BY DIVPAG TO CALCULATE TIME
C   DERIVATIVES
C
          SUBROUTINE FCN (NEQ,X,Y,YPRIME)
          INTEGER NEQ, JJ
          DOUBLE PRECISION X,Y(NEQ),YPRIME(NEQ),C,D,RK,DX,H,AL,TB
          COMMON T20,H
          RK = 0.162
          AL = RK/(2553.0*1084.0)
          DX = 9.0D-05
C    H = 6.917D-08*((T20*400.0 + 304.0)**4 - 704.0**4)/(304.0*
C &      (1.0-T20) + 704.0*(T20-1.0))
C
C   YPRIME FOR INTERIOR NODES
          DO 2 JJ = 1,17
              YPRIME(JJ + 1) = (AL/DX**2)*(Y(JJ)-2.0*Y(JJ + 1) + Y(JJ + 2))
          2 CONTINUE
C
C   YPRIME FOR NODAL POINT NEXT TO WALL WITH ZERO ENERGY
C   FLUX
C
          YPRIME(1) = (AL/DX**2)*(-(2.0/3.0)*Y(1) + (2.0/3.0)*
       &          Y(2))
C
C   YPRIME FOR NODAL POINT NEXT TO WALL WITH FINITE ENERGY
C   FLUX
C
              YPRIME(19) = (AL/DX**2)*(T20-2.0*Y(19) + Y(18))
C
C   YPRIME FOR CONSTANT WALL TEMPERATURE
C
C             YPRIME(9) = AL*(25.0-2.0*Y(9) + Y(8))/(DX**2)
C
          RETURN
          END
C
C   DUMMY SUBROUTINE FOR DIVPAG (JACOBIAN)
C
          SUBROUTINE FCNJ (NEQ,X,Y,DYPDY)
          INTEGER NEQ
          DOUBLE PRECISION X,Y(NEQ),DYPDY(*)
          RETURN
          END
```

in the mold. If the mold wall temperature is set too low, then the blanks will cool too rapidly and will not flow well enough to fill out the mold. On the other hand, if the mold temperature is set too high, then the part will take too long to cool. The maximum rate of crystallization for PP occurs at about 90°C. By setting the mold temperature at this value it may be possible to keep the material at a high enough temperature so that it will flow without the application of excessively high pressures and yet crystallize rapidly enough so that the part can be removed from the mold in the shortest time possible. At this point one can use the numerical approach described in Section 5.4 to calculate the time for cooling. One may find that variations of the mold temperature around 90°C will lead to the optimum time for cooling. However, because these calculations are straightforward as already described, there is little to be gained in repeating them. The main point

is to provide some insight for selecting an initial guess for the mold temperature.

## PROBLEMS
### A. Applications

**10A.1.** Heat Removal from an Injection Mold Using Coolant Lines The overall heat transfer coefficient per unit length of coolant line is given in Eq. 10.35. Using this equation, answer the following questions if $k_m$ and $h$ are held fixed:

(a) If the coolant line diameter is doubled from initial values of $P/d = 2$ and $D/d = 2$ how does $U$ change?

(b) If a second row of coolant lines is added, decreasing $P/d$ from 4 to 2, how is $U$ changed (remember $h$ is held constant)?

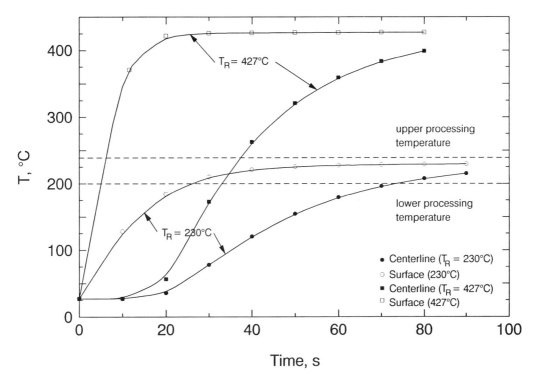

**Figure 10.23** Calculated surface and centerline temperatures as a function of time for Azdel sheets 3.6 mm thick subjected to two different infared heater surface temperatures.

**(c)** If $D/P$ is decreased from 1 to 0.5 by decreasing $D$, how is $U$ changed?

**10A.2.** Equilibration Temperature of a Mold Base

PP containing 30 wt% glass is injection-molded into a rectangular cavity 8.9 × 8.9 cm by 0.15 cm thick. The mold base consists of two rectangular stainless steel plates 15 cm x 30 cm by 2.5 cm thick. One-half the cavity thickness is machined in each of the bottom and top plates of the mold. The polymer enters the cavity at 230°C with a fill time of 1.0 s. The mold temperature is originally 25°C. The mold is held closed for 25 s while the polymer cools and is in the open position for 5 s. During the time that the mold is opened, it is subjected to free convection cooling by air at 25°C. The following properties are given for stainless steel:

$$\rho = 7750 \, \text{kg/m}^3; \qquad k_m = 2.30 \, \text{Wm}^{-1}\text{°K}^{-1};$$

$$C_{pm} = 460.2 \, \text{Jkg}^{-1}\text{°C}^{-1}.$$

Show the procedure for determining the time required for the mold base to come to an equilibrium temperature listing all your assumptions, and carry the calculations out for two cycles of the process.

**10A.3.** Flow Rate of Coolant Through a Mold Base

Coolant lines of 1.27 cm in diameter are machined in the base of a mold with a spacing of 3.81 cm (this is the centerline distance) and a distance of 2.54 cm from the mold surface. The total length of the line is 1.83 m. Tap water is used as the coolant and enters the line at 12°C with a line pressure of $1 \times 10^8$ Pa. The coolant must remove heat at the rate of 1758.3 J/s. The mold base is made of stainless steel with the thermal properties given in Pr. 10A.2. Determine the convection heat transfer coefficient in the coolant lines, the increase in temperature of

the water (i.e., the temperature of water at the exit of the coolant line), and the flow rate of water required to produce the heat transfer coefficient.

**10A.4** Time for Healing a Weld Line

LDPE (NPE 953) is injection-molded into a cavity from two gates. The fronts meet at the center of a mold cavity, which is 0.3175 cm thick, with a melt temperature of 180°C. Determine the mold temperature and the length of time required for the melt to remain in the mold in order to obtain a part with adequate weld line strength. Treat the geometry as that of a slab, and use the conditions given in Section 4.2.5 for determining weld line strength.

**10A.5.** Effect of Calcium Carbonate on the Thermal Properties of Polypropylene

Calcium carbonate ($CaCO_3$) is added to polymers as a filler with the intent of lowering material costs. Calculate the thermal conductivity, the heat capacity, and the density of a PP composite containing 40 wt% $CaCO_3$. Take the following properties for PP:

$$\rho = 900 \, \text{kg/m}^3; \qquad k = 0.20 \, \text{W/m°C}; \qquad \bar{C}_p = 1.8 \, \text{kJ/kg°C},$$

while the following properties are reported for $CaCO_3$:

$$\rho = 3000 \, \text{kg/m}^3; \qquad k = 2.7 \, \text{W/m°C}; \qquad \bar{C}_p = 0.86 \, \text{kJ/kg°C}.$$

Compare your calculated values to the experimental values, which are given as:

$$\rho = 1250 \, \text{kg/m}^3; \quad k = 0.56 \, \text{W/m°C}; \quad \bar{C}_p = 1.34 \, \text{kJ/kg°C}.$$

A = 3
B = 4
C = 2
D = 1
E = 1

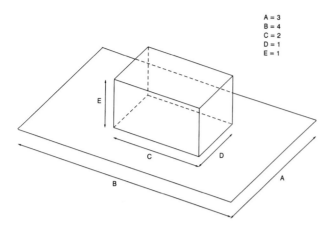

**Figure 10.24** Drape forming over a positive or male mold.

Under similar conditions of heating, which material will come to temperature sooner; PP or PP/CaCO$_3$?

**10A.6 Average Wall Thickness in a Thermoformed Sheet**
Consider the sheet of arbitrary dimensions shown in Fig. 10.24, which is formed by draping over a positive or male mold. Calculate the average reduction in wall thickness assuming that the hot sheet first contacts area CD that retains its original thickness.

### B. Principles

**10B.1.** Extension Rate at the Advancing Front
Tadmor (1974) estimated the extension rate at the advancing front in the filling of a rectangular cavity to be:

$$\dot{\varepsilon} = \frac{dv_y}{dy} = -\frac{dv_x}{dx} = \frac{\langle v_x \rangle - v_{max}}{2b}, \tag{10.96}$$

where $2b$ is the thickness of the mold, $\langle v_x \rangle$ is the average velocity in the $x$ direction and $v_{max}$ is the maximum velocity. For a power-law fluid show that Eq. 10.96 becomes

$$-\dot{\varepsilon} = \left(\frac{n}{n+1}\right)\frac{\langle v_x \rangle}{2b}. \tag{10.97}$$

For LDPE (NPE 953) at 170°C calculate $\dot{\varepsilon}$ for the filling of a square cavity, 8.89 × 8.89 cm and 0.16 cm thick in the times of 0.5 and 1.0 s.

**10B.2.** Lubricated Compression Molding
For the compression-molding process shown in Fig. 10.11 the plates are lubricated either with mold-release agent or by using Teflon sheets. Calculate an expression for the force required to close the plates similar to Eq. 10.50 when the polymer is assumed to exhibit complete slip. Do this first for a Newtonian fluid and then a polymer melt with rheological properties described by the PTT model.

**10B.3.** Compression Molding in a Cup-Shaped Cavity
Consider the compression molding of a polymeric material in the cup-shaped mold shown in Fig. 10.25. For isothermal flow of a power-law fluid carry out the following steps to obtain an expression for the compression force.
  **(a)** Show that the velocity field for the radial flow portion of the

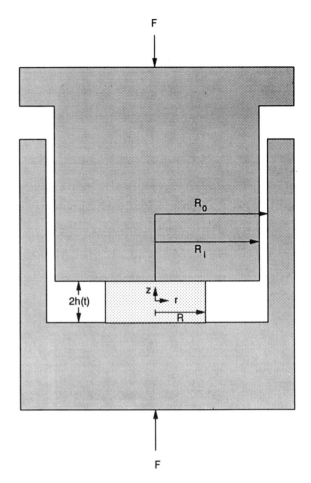

**Figure 10.25** Cup-shaped cavity used in compression molding of a thermoplastic.

flow (i.e., the flow between the two parallel disks up to $R_i$) is:

$$v_r(z, r, t) = \frac{h^{1+s}}{1+s}\left(-\frac{1}{m}\frac{\partial p}{\partial r}\right)^s\left[1 - \left(\frac{z}{h}\right)^{1+s}\right]. \tag{10.98}$$

What assumptions are made in obtaining Eq. 10.98?
  **(b)** Show that the pressure gradient is:

$$P - p_a = \frac{m(2+s)^n(-\dot{h})R^{n+1}}{2^n(n+1)h^{2n+1}}\left[1 - \left(\frac{r}{R}\right)^{n+1}\right]. \tag{10.99}$$

  **(c)** Show that the applied force is:

$$F = \frac{m\pi(2+s)^n(-\dot{h})^nR^{n+1}}{2^n(n+3)h^{2n+1}}. \tag{10.100}$$

  **(d)** When the flowing material reaches the outer wall at $r = R_o$, it is forced to flow in the annular space. Show that for a constant squeeze rate the rate of increase of material along the axial distance in the annulus is:

$$-\pi R_i^2 \dot{h}(t) = \pi(R_o^2 - R_i^2)\dot{l}, \tag{10.101}$$

where $l$ is the axial coordinate.

**(e)** For a thin annulus (i.e., $\Delta R = (R_o - R_i) \ll R_o$), show that the volumetric rate of flow, $Q$, is:

$$Q = \pi \bar{R}^2 \dot{h}; \tag{10.102}$$

where $\bar{R} = 0.5 \, (R_i + R_o)$.

**(f)** Since $\dot{h} \ll \dot{l}$ for a thin annulus (i.e., the plunger travel rate, $\dot{h}$, is small compared to the rate of advancement of the fluid in the annulus, $\dot{l}$), the flow can be considered to be one of pressure flow only, not a combined drag and pressure flow. Show that the pressure drop across the annular flow region for the case of a small gap is:

$$P(R_i) - p_a = \left( \frac{2 \, m \dot{l}}{R_o^{3n+1}} \right) \frac{[2(s+z)\bar{R}\Delta R \dot{l}]^n}{(1-\kappa)^{1+2n}}. \tag{10.103}$$

**(g)** Obtain a final expression for the plunger force when flow exists in the annular space.

**10B.4.** Wall Thickness Distribution in a Deep Truncated Conical Mold

The problem of determining the wall thickness distribution in a deep truncated conical mold as shown in Fig. 10.26, along with several other geometries, was analyzed by Rosenzweig (1983). Until the spherical bubble touches the bottom of the mold, the thickness distribution was assumed to be that given in Eq. 10.58. Carry out the following steps to show how Rosenzweig arrived at an expression for the thickness distribution once the bubble touched the bottom of the mold.

**(a)** When the spherical bubble just touches the bottom of the mold, show that the position of detachment along the wall, $z_T$, is:

$$z_T = \frac{H(1 + \cos \beta)}{\sin \beta} - \frac{L}{2}. \tag{10.104}$$

**(b)** Show that the initial area of the free bubble just as it makes contact with the bottom of the mold is:

$$S_I = 2\pi(1 + \cos \beta)(H/\sin \beta - z_T)^2. \tag{10.105}$$

**(c)** Show that the area of the free bubble at any instant is:

$$S = (2\pi/(1 - \cos \beta))(H - z \sin \beta)[(z - z_r)\beta + (H - z \sin \beta)]. \tag{10.106}$$

**(d)** Finally, show that the thickness distribution in this part of the forming process is:

$$h/h_I = S_I/SF, \tag{10.107}$$

where $h_I$ is the initial thickness calculated by substituting $z_T$ into Eq. 10.68 and $F$ is given by:

$$F = \left| \frac{aZ^2 + bZ + c}{aZ_T^2 + bZ_T + c} \right|^{\frac{f \sin \beta}{2a(2-f)}}$$

$$+ \left| \frac{(2aZ + b - g)(2aZ_T + b + g)}{(2aZ_T + b - g)(2aZ + b + g)} \right|^{\frac{\sin \beta[d - (fb/2a)]}{g(2-f)}}, \tag{10.108}$$

where

$$a = \sin \beta - \beta$$

$$b = \beta(H/\sin \beta + Z_T) - 2H$$

$$c = (H - Z_T \beta)H/\sin \beta$$

$$d = \frac{W}{2} - Z_T - H\left( \frac{1}{\tan \beta} - \cos \beta \right)$$

$$f = 1 - \cos \beta$$

$$g = \sqrt{b^2 - 4ac} = \beta\left( \frac{H}{\tan \beta} - \frac{W}{2} \right). \tag{10.109}$$

**10B.5.** Wall Thickness Distribution in a Long Triangular Prism Mold (Rosenzweig, 1983)

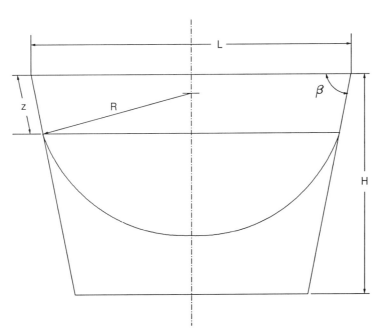

**Figure 10.26** Truncated conical mold.

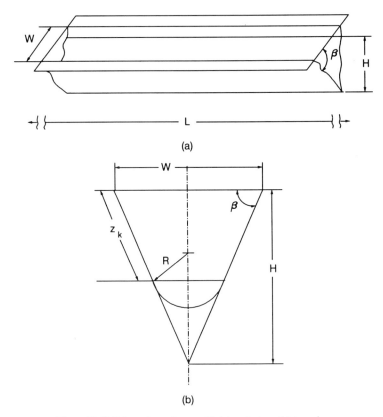

**Figure 10.27** Triangular prism mold: (a) end view; (b) top view.

Following the development used in Ex. 10.3 obtain an expression for the wall thickness distribution in the long triangular prism mold shown in Fig. 10.27 by carrying out the following steps.

**(a)** Neglecting the deformation at the ends of the mold, perform a mass or volume balance to obtain the following differential equation for the thickness, $h$:

$$\frac{-dh}{h} = \frac{dR}{R} + \frac{dz}{\beta R}. \tag{10.110}$$

**(b)** Show that the radius of the bubble as a function of contact position along the wall, $z_k$, is:

$$R = \frac{H - z_k \sin \beta}{\sin \beta \tan \beta}. \tag{10.111}$$

**(c)** On integrating, Eq. 10.110 becomes:

$$h/h_1 = \left[ \frac{H - z_k \sin \beta}{H} \right]^{\frac{1}{\beta} \tan \beta - 1)}, \tag{10.112}$$

where $h_1$ is related to the initial sheet thickness by $h_1 = (\sin \beta / \beta) h_0$. The thickness distribution is obtained by substituting the expression for $h_1$ into Eq. 10.112.

**10B.6.** Inflation of a Viscoelastic Cylindrical Parison
In determining $R(t)$ for the expansion of a cylindrically shaped parison the components of stress for the PTT model were given in Eqs. 10.87 and 10.88. Show how to obtain these two equations. (See Bird and coworkers, 1987, for convected time derivatives in cylindrical and spherical coordinates.)

**10B.7.** Inflation of a Bicomponent Parison
A cylindrically shaped parison consists of two layers of polymer: an inner layer of initial thickness $h_{01}$ and viscosity $\mu_1$ and an outer layer of initial thickness $h_{02}$ and viscosity $\mu_2$. Assuming that the fluids are Newtonian, derive expressions for determining $R_i(t)$ and $h_i(t)$ where the subscript $i$ is 1 or 2.

**C. Numerical Problems**

**10C.1.** Effect of Filler on the Thermoforming Time of
Polypropylene
In the thermoforming of pen barrels, $CaCO_3$ is added to PP to reduce material costs. The pen barrel is formed from a cylindrical tube having an O.D. of 5.28 mm and an I.D. of 2.82 mm. The tube is heated by conduction heating in which the tube is in contact with a metal ring having a temperature of 143°C. The inside of the tube is not heated, and no heat transfer is considered to occur there. Determine the time required for the inner wall of the tube to reach a temperature of 143°C for both pure PP and PP/40% $CaCO_3$. The thermal properties are given in Pr. 10A.5. By how much does the presence of $CaCO_3$ reduce the heating time?

**10C.2.** Infrared Heating of a Transparent Sample of Polyurethane
A clear disk of PUR, 25.4 cm in diameter and 3.5 mm thick, is to be heated by means of an infared heater with a surface temperature rating of 2400°K. Menezes and Watt (1992) obtained by means of a bolometric detector the following expression for the transmitted intensity

$$I = 6.43 \times 10^6 \, e^{-536x}. \tag{10.113}$$

Calculate the time to heat the sheet from an initial temperature of 27°C to 180°C for the following two cases: **(a)** the convective heat transfer coefficient at both surfaces is 100 W/m² °K; **(b)** the convective heat transfer coefficient at the top surface is 12 W/m² °K and 5 W/m² °K at the bottom surface. The thermal properties of polyurethane are: $\rho = 1250\,\mathrm{kg/m^3}$, $k = 0.31\,\mathrm{J/ms\,°K}$, $\bar{C}_p = 1.88 \times 10^3\,\mathrm{J/kg\,°K}$. Consider the problem to be one of one-dimensional transient heat conduction.

**10C.3** Pressure Forming of a Polymeric Sheet in a Conical Mold
A sheet of LDPE (NPE 953) of thickness 0.3175 cm is thermoformed in the conical cavity shown in Fig. 10.17 at a temperature of 170°C. The dimensions of the mold are $H = 30.48$ cm and $L = 20.0$ cm. The applied pressure differential is $6.9 \times 10^5$ Pa. Calculate the wall thickness distribution and the time to fill the mold, using first the Newtonian constitutive equation and then the PTT model. Compare the predictions of the two models.

**10C.4.** Expansion of a Cylindrical Parison: Newtonian Case
A cylindrical parison consisting of LDPE (NPE 953) at 170°C is inflated with a pressure differential of $3.5 \times 10^5$ Pa. Initially, the diameter of the parison is 2.60 cm and the wall thickness is 0.13 cm. Determine the time for the parison to contact the mold walls and the thickness of the inflated parison at this point, if the diameter of the mold is 13 cm (see Fig. 10.20). Assume that melt can be considered as a Newtonian fluid, and use the rheological properties given for LDPE (NPE 953) in Appendix A, Table A.1. The equations to be solved are Eq. 10.86 and the expression based on conservation of mass (i.e., $\dot{h}/h = -\dot{R}/R$).

**10C.5.** Expansion of a Cylindrical Parison: Viscoelastic Case
Do Pr. 10C.4 again, but use the PTT model. The solution requires the solving of Eqs. 10.85, 10.87, and 10.88. Compare the solution against that obtained for the Newtonian case.

**10C.6.** Sagging of a Cylindrical Parsion
Solve Eqs. 10.75 and 10.80 to find the thickness distribution and velocity as a function of position for a thin-walled parison. Also determine the length of the parison as a function of time, and compare the results

against the ideal case (i.e., $L = \langle v_z \rangle t$). The viscosity of the melt is 500 Pa·s, and its density is 900 kg/m³. The wall shear rate in the die is 200 s⁻¹. The outer diameter of the parison is 2.54 cm, and its thickness is 0.127 cm. The parison thickness and diameter swell are 1.5.

## D. Design Problems

**10D.1.** Design of a Cooling Line System for Injection Molding
Nylon 6,6 containing 30 wt% glass is injection-molded to form a manifold cover for an engine block. The manifold is basically a box 0.76 m long by 0.20 m wide by 0.10 m high with a wall thickness of 0.635 cm. Design a cooling line system that uses tap water at 10°C that will minimize the time the part must remain in the mold. The melt enters the mold at 285°C. The mold base consists of stainless steel. The coolant lines are to be continuous with a single inlet and outlet. In your design specify the dimensions of the coolant lines, their spacing, and their location relative to the melt (i.e., $D$ in Fig. 10.9). Furthermore, specify the velocity of the water, the pressure drop, and the temperature of the water on leaving the mold. The only restriction is that the coolant lines should not exceed a diameter of 1.27 cm. Also assume that the cavity is machined in one side of the mold base.

**10D.2.** Design of a Multiple Cavity Runner System (Tadmor & Gogos, 1979)
In many commercial situations molds are designed with multiple cavities. In order that each part have identical physical properties it is necessary that the cavities fill simultaneously. For the runner-cavity system shown in Fig. 10.28 (here only one-half of the cavity and runner system are shown), it is desired to have the cavities fill at the same time. First assuming the fluid is Newtonian and then nonNewtonian with the viscosity described by the power-law model, carry out the following for isothermal conditions:

**(a)** Design the runner branches for each of the four cavities with the gates taken identical in order to get simultaneous filling;

**(b)** Design the gates so that there is simultaneous filling in each of the cavities with the runner branches taken to be identical. The runners and gates are taken to be cylindrical. Neglect pressure losses across the contractions and the expansions.

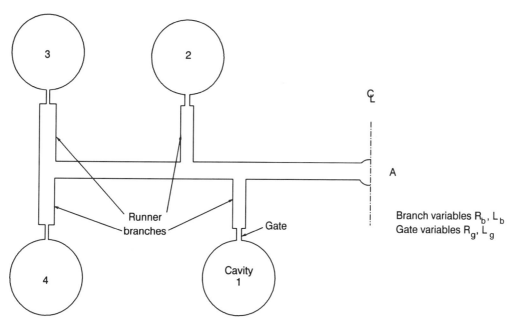

**Figure 10.28** Schematic of a multi-cavity mold. Only half of the symmetric arrangement of the cavities is shown.

**10D.3.** Cooling Conditions for a Two-Layer Blow-Molded Tank
A gasoline tank consisting of an inner layer of nylon 6 and an outer layer of HDPE is to be produced by blow molding. Design a cooling process that will provide the minimum amount of cooling time. Consider both the cooling of the interior and exterior walls of the blown parison. Tap water at 12°C is available for cooling the mold, and air as cool as 12°C or as hot as 100°C can be used on the interior. The temperature of the parison as it contacts the mold wall is 220°C. The thickness of the parison as it contacts the mold is 0.635 cm with one-fifth of the thickness being nylon 6. In order for the nylon to act as an effective barrier to gasoline, it must reach 75% of its maximum degree of crystallinity, and the HDPE should reach 100% of its maximum degree of crystillinity to have adequate impact properties.

## REFERENCES

Bird, R. B., R. C. Armstrong, and O. Hassager. 1987. *Dynamics of Polymeric Liquids: Vol. I, Fluid Mechanics* (Wiley, New York).

Birley, A. W., B. Haworth, and J. Batchelor. 1991. *Physics of Plastics* (Hanser, Munich).

Gibbs, M. L. 1989. "Liquid Nitrogen Cooling for Blow Molded Parts: Effect on Cycle Time and Bottle Performance." *Society of Plastics Engineers Technical Papers*, **35**, 918.

Giles, H. F., and D. L. Reinhard. 1991. "Compression Molding of Polypropylene Glass Composites." *SAMPE*, **36**, 556.

Gruenwald, G. 1987. *Thermoforming* (Technomatic Lancaster, PA).

Harper, C. A., Ed. 1992. *Handbook of Plastics, Elastomers, and Composites*, 2nd Ed. (McGraw-Hill, New York).

Hieber, C. A., and S. F. Shen. 1980. "A Finite Element/Finite Difference Simulation of the Injection Molding Filling Process." *J. Non-Newtonian Fluid Mech.*, **7**, 1–32.

Hunkar, D. B. 1973. "Internal Surface Cooling of Blow Molded Articles." *Society of Plastics Engineers Technical Papers*, **19**, 448.

Lee, N., Ed. 1990. *Plastic Blow Molding Handbook* (Van Nostrand Reinhold, New York).

Mavridis, H., A. N. Hrymak, and J. Vlachopoulos. 1988. "The Effect of Fountain Flow in Molecular Orientation in Injection Molding." *J. Rheol.*, **32**(6), 639.

Menezes, M., and D. F. Watt 1992. "Modeling of Infrared Heating of Transparent Polyurethane." *Society of Plastics Engineers Technical Papers*, **38**, 109–13.

Pearson, J. R. A., and S. M. Richardson. 1983. *Computational Analysis of Polymer Processing* (Applied Science Publishers, London).

Rosenzweig, N. 1983. "Wall Thickness Distribution in Thermoforming." *SPE Technical Papers*, **29**, 478–82.

Rosenzweig, N., M. Narkis, and Z. Tadmor. 1979. "Wall Thickness Distribution in Thermoforming." *Polym. Engng. Sci.*, **19**, 946.

Tadmor, Z. 1974. "Molecular Orientation in Injection Molding." *J. Appl. Polym. Sci.*, **18**, 1753.

Tadmor, Z., and C. G. Gogos 1979. *Principles of Polymer Processing* (Wiley, New York).

Throne, J. L. 1979. *Plastics Process Engineering* (Marcel Dekker, New York).

Throne, J. L. 1986. *Thermoforming* (Hanser, Munich).

Wang, V. W., C. A. Hieber, and K. K. Wang. 1986. "Dynamic Simulation and Graphics for the Injection Molding of Three-Dimensional Thin Parts." *J. Polym. Engng.*, **7**, 21.

# NOMENCLATURE

| | | | | | |
|---|---|---|---|---|---|
| $A$ | Area of the plates | Eq. 2.1 | $C_{Ae}$ | Equilibrium concentration | Eq. 8.163 |
| $A$ | Hamaker constant | Eq. 6.173 | $C_{Ai}$ | Concentration of $A$ at interface | Eq. 4.96 |
| $A$ | Radius of a set of unit cells | Eq. 6.24 | $C_D$ | Constant | Eq. 5.154 |
| $A$ | Interfacial area | Eq. 6.36 | $C_D$ | Hydrodynamic drag coefficient | Eq. 9.53 |
| $A$ | Cross-sectional area | Eq. 8.4 | $C_i$ | Molar concentration of $i$th species | Eq. 4.1 |
| $A$ | Filament cross-sectional area | Eq. 9.3 | $C_L$ | Centerline distance | Eq. 8.107 |
| $A_L$ | Cross-sectional area of land | Eq. 7.27 | $C_w$ | Measure of adhesion of solids to wall | Eq. 8.4 |
| $A_e$ | Cross-sectional area of entry | Eq. 7.27 | | | Eq. 9.69 |
| $A_f$ | Filled-channel cross-sectional area | Eq. 8.119 | $C_1, C_2, C_3,$ | | |
| $A_s$ | Area across which mass transfer takes place | Eq. 4.99 | $C_4$ | Constants | |
| $A_v$ | Interfacial area per unit volume | Eq. 6.31 | $C'_A$ | Concentration of $A$ in melt reentering melt pool | Eq. 8.165 |
| $A_0$ | Interfacial area | Eq. 6.35 | $\bar{C}$ | Mean concentration | Eq. 6.7 |
| $A_1$ | Area of polymer | Eq. 5.135 | $\bar{C}_P$ | Constant pressure heat capacity per unit mass | Eq. 5.16 |
| $A_2$ | Area of source | Eq. 5.135 | | | |
| $a$ | Acceleration | Eq. 2.68 | $\bar{C}_{p,p}$ | Heat capacity of plug | Eq. 8.31 |
| $a$ | coefficient | Eq. 5.165 | $\bar{C}_v$ | Constant volume heat capacity per unit mass | |
| $a$ | crystallographic axis | Eq. 5.172 | | | |
| $a$ | Radius of a particle | Eq. 6.24 | Ca | Capillary (or Weber) number | Eq. 6.183 |
| $\mathbf{a}$ | Vector that forms interface $A_0$ | Eq. 6.32 | $c$ | coefficient | Eq. 5.165 |
| $\mathbf{a}'$ | Vector that forms interface $A$ | Eq. 6.62 | $c$ | crystallographic axis | Eq. 5.174 |
| $\bar{a}$ | Average diameter of crystallites | Eq. 5.148 | $c$ | Density coefficient | Eq. 9.163 |
| $a_T$ | Shifting factor | Eq. 5.3 | $\mathbf{c}$ | Vector perpendicular to $A_0$ | Eq. 6.34 |
| $a_x$ | Angle | Eq. 6.41 | $\mathbf{c}'$ | Vector perpendicular to $A$ | Eq. 6.36 |
| $a_y$ | Angle | Eq. 6.41 | $D$ | Diameter of extrudate | Eq. 3.89 |
| $a_z$ | Angle | Eq. 6.41 | $D$ | Characteristic length | |
| $a_0$ | Distortion amplitude of time 0 | Eq. 6.194 | $D$ | Deformation | Eq. 6.184 |
| $B$ | Scaling constant relating the fraction of $E(\lambda)$ that reaches the sample | Eq. 5.139 | $D$ | Tube diameter | Eq. 5.127 |
| | | | $D$ | Film thickness | |
| $B$ | Holdback | Eq. 6.138 | $D$ | Half-width of $K(T)$ curve | Eq. 9.92 |
| $B$ | Short axis | Eq. 6.184 | $D$ | Air drag force | Eq. 9.46 |
| $B$ | Capillary die swell | Eq. 7.5 | $D/Dt$ | Material time derivative or time derivative following the fluid motion | Eq. 5.58 |
| $B$ | Dimensionless internal overpressure | Eq. 9.150 | | | |
| $B$ | Slip coefficient | Eq. 10.30 | $D_b$ | Barrel diameter | Eq. 8.21 |
| $\mathrm{Bi}^{-1}$ | Reciprocal of Biot number, $k/hb$ | | De | Deborah number | Eq. 3.90 |
| $\mathrm{Bi}_r$ | Radiative Biot number | Eq. 5.117 | $Dp$ | Diameter of polymer | |
| Br | Dimensionless number | Eq. 8.94 | $D_R$ | Draw (or draw down) ratio | Eq. 9.26 |
| $B_R$ | Blow (blowup) ratio | Eq. 9.129 | $D_s$ | Screw diameter | Eq. 8.1 |
| $B'$ | Length | Eq. 6.104 | $D^*$ | Distribution function | Eq. 8.11 |
| $B_1$ | Diameter swell | Eq. 7.6 | $D^*$ | Tracer diffusion coefficient | |
| $B_2$ | Thickness swell | Eq. 7.7 | $\bar{D}$ | $1/2(D_B + D_s)$ | Eq. 8.26 |
| $b$ | Shear-free flow constant | Eq. 3.2 | $Ð_{AB}$ | Mutual diffusion coefficient | Eq. 4.12 |
| $b$ | crystallographic axis | Eq. 5.173 | $Ð_s$ | Self-diffusion coefficient | Eq. 4.87 |
| $b$ | Number of black particles | Eq. 6.1 | $Ð_0$ | Preexponential factor | Eq. 4.53 |
| $b$ | Half thickness of slab | Eq. 4.33 | $Ð_w$ | Diffusivity of moisture | Eq. 4.83 |
| $b_0$ | Crystal dimension | Fig. 5.21 | $Ð_{0w}$ | Preexponential factor | Eq. 4.84 |
| $\mathbf{b}$ | Vector that forms interface $A_0$ | Eq. 6.33 | $Ð_\alpha$ | Diffusivity in amorphous material | Eq. 4.58 |
| $\mathbf{b}'$ | Vector that forms interface $A$ | Eq. 6.36 | $Ð_{sc}$ | Diffusivity in semi-crystalline material | Eq. 4.58 |
| $C$ | Characteristic constant for every polymer; for a number of polymers, $C = 265\,\mathrm{K}$ | Eq. 5.155 | $d$ | distance between spheres | |
| | | | $d$ | Diameter of jet | Eq. 9.104 |
| $C$ | Stress optic coefficient | Eq. 5.177 | $d_A$ | Density of species $A$ | Eq. 4.61 |
| $C$ | Wetted circumference | Eq. 8.4 | dpf | Takeup denier (denier per filament) | Eq. 9.72 |
| $C$ | Constant | Eq. 9.169 | $E$ | Young's modulus of elasticity | |
| $C$ | Total molar concentration | Eq. 4.1 | $E$ | Exit geometry | Eq. 7.5 |
| $\mathbf{C}$ | Cauchy-Green strain tensor | Eq. 6.85 | $E$ | Activation energy | Eq. 9.49 |
| $C_A$ | Concentration of the volatile | Eq. 8.163 | $EG$ | Entrance geometry | Eq. 7.5 |

| | | |
|---|---|---|
| $E_D$ | Activation energy of diffusion | Eq. 4.53 |
| $E_{\bar p}$ | Activation energy for permeation | Eq. 4.75 |
| $E_D\mathbf{RT}$ | Diffusive transport of molecules in melt | Eq. 5.152 |
| $E_f$ | Evaporation at film surface | Eq. 8.162 |
| $E_p$ | Evaporation at surface of pool | Eq. 8.162 |
| $E_v$ | Friction loss | Eq. 2.69 |
| $E_x$ | Extraction number | Eq. 8.173 |
| $E'$ | Material property | Eq. 9.56 |
| $e$ | Emissivity | |
| $e$ | Flight width | Eq. 8.2 |
| erf | Error function | Eq. 4.51 |
| erfc | Complementary error function | Eq. 4.92 |
| $e_{coh}$ | Cohesive energy density | Eq. 9.103 |
| $e_L$ | Efficiency of mixing | Eq. 6.89 |
| $\bar e_L$ | Time average efficiency of mixing | Eq. 6.90 |
| $e_1$ | Emissivity of sheet | Eq. 5.134 |
| $e_2$ | Emissivity of source | Eq. 5.134 |
| $F$ | Force | Eq. 2.1 |
| $F$ | Tensile force | Eq. 3.106 |
| $F$ | Combined configuration emissivity factor | Eq. 5.133 |
| $F$ | Distribution | Eq. 6.13 |
| $F$ | Force at takeup point | Eq. 9.77 |
| $F$ | Cohesive fracture | |
| $F$ | Drawing force acting on film | Eq. 9.16 |
| $F_L$ | Viscous forces in axial direction | Eq. 9.144 |
| $F_H$ | Viscous forces in transverse direction | Eq. 9.144 |
| Fr | Froude number | Eq. 9.69 |
| $F'$ | View factor | Eq. 5.133 |
| $F(t)$ | Cumulative RTD function, or $F$ function | Eq. 6.115 |
| $\mathbf{F}$ | Gradient-deformation tensor | Eq. 6.83 |
| $\mathbf{F}^T$ | Transpose of deformation-gradient tensor | Eq. 6.84 |
| $F_Z$ | Force at frost line | Eq. 9.149 |
| $F_w$ | Attractive Van der Waals force | Eq. 6.173 |
| $F_c$ | Cohesive force of cluster | Eq. 6.176 |
| $F_h$ | Hydrodynamic force | Eq. 6.177 |
| $\Delta F_n^*$ | Free energy of the nucleus with an $n$-dimensional growth | Eq. 5.152 |
| $\Delta F_n^*/k_B T$ | Nucleation factor | Eq. 5.155 |
| $f$ | Packing factor | Eq. 5.148 |
| $f$ | Degree of fill | Eq. 8.119 |
| $f_{am}$ | Amorphous orientation function | Eq. 5.179 |
| $f_w'$ | Coefficient of friction between pellets and wall | Eq. 8.4 |
| $f(t)$ | External RTD function | Eq. 6.116 |
| $f(p)$ | Function of $p$ | Eq. 6.189 |
| $G$ | Shear modulus | Eq. 3.51 |
| $G$ | Linear rate of growth | Eq. 5.152 |
| $G$ | Pressure gradient | Eq. 7.17 |
| $G$ | Mass flow rate | Eq. 5.127 |
| $G$ | Shear modulus | Eq. 9.81 |
| $G(t)$ | Cumulative internal RTD function | Eq. 6.133 |
| $Gr_D$ | Grashof number for diffusion | Eq. 4.103 |
| $Gr_f$ | Grashof number evaluated at film temperature | Eq. 5.130 |
| $G_{or}$ | Oriented linear growth rate | Eq. 5.167 |
| $G_{un}$ | Unoriented linear growth rate | Eq. 5.167 |
| $\Delta G_{un}$ | Free energy difference between amorphous and crystalline states under random orientation | Eq. 5.167 |
| $\Delta G_{mix}$ | Change of Gibbs free energy of mixing | Eq. 6.202 |
| $G_0$ | Molecular jump frequency | Eq. 5.152 |
| $G^*$ | Complex shear modulus | Eq. 3.24 |
| $G'$ | Storage modulus | Eq. 3.24 |
| $G''$ | Loss modulus | Eq. 3.24 |

| | | |
|---|---|---|
| $g_c$ | Mass of crystalline material | |
| $g(t)$ | Internal RTD function | Eq. 6.134 |
| $H$ | Separation distance | Eq. 2.1 |
| $H$ | Distance of the capillary to the wheel | Eq. 3.105 |
| $H$ | Slit height | Eq. 7.30 |
| $H$ | Hydrodynamic stability | |
| $H$ | Function of surface tension | Eq. 9.46 |
| $H(t)$ | Unit step function | Eq. 3.25 |
| $H_f$ | Final film thickness | Eq. 9.131 |
| $H_0$ | Initial film thickness | Eq. 9.131 |
| $H^l$ | Film thickness in liquid phase | Eq. 9.170 |
| $H^s$ | Film thickness in solid phase | Eq. 9.170 |
| $\Delta \bar H_A$ | Heat of reaction | Eq. 8.183 |
| $\Delta \bar H_c$ | Heat of crystallization | |
| $\Delta \bar H_c$ | Latent heat of crystallization per unit mass | Eq. 9.90 |
| $\Delta \bar H_f$ | Heat of fusion | |
| $\Delta \bar H_m$ | Enthalpy change of melting process | Eq. 5.164 |
| $\Delta H_{mix}$ | Change in enthalpy of mixing | Eq. 6.202 |
| $\Delta \bar H_s$ | Molar heat of sorption | Eq. 4.62 |
| $h$ | Heat transfer coefficient | Eq. 5.122 |
| $h$ | Wall thickness | Eq. 10.75 |
| $h_a$ | Air-side heat transfer coefficient | Eq. 9.113 |
| $h_L$ | Thickness of film at chill roll | |
| $h_p$ | Planck constant | |
| $h_r$ | Radiation heat transfer coefficient | Eq. 5.138 |
| $\bar h$ | Average heat transfer coefficient | Eq. 5.124 |
| $I$ | Intensity of segregation | Eq. 6.19 |
| $I_{hkl}(\Phi_{i,z})$ | Intensity diffracted from ($hkl$) planes that are normal to the $i$ crystallographic axis | Eq. 5.168 |
| $I_s(\lambda)$ | Intensity of radiation at surface for a given $\lambda$ | Eq. 5.140 |
| $I_1$ | Invariant flow 1 | Eq. 2.60 |
| $I_2$ | Invariant flow 2 | Eq. 2.61 |
| $I_3$ | Invariant flow 3 | Eq. 2.62 |
| $J$ | Kinetic crystallizability | Eq. 9.93 |
| $\mathbf{J}_i^*$ | Molar flux of species $i$ | Eq. 4.11 |
| $j_D$ | Mass transfer Chilton-Colburn $j$ factor | Eq. 4.106 |
| $\mathbf{j}_A$ | Mass flux of species $A$ | Eq. 4.10 |
| $\mathbf{j}_i^*$ | Mass flux of species $i$ | Eq. 4.11 |
| $K$ | Related to half-time for crystallization; volume rate of crystallization | Eq. 5.161 |
| $K$ | Partition Coefficient | Pr. 4C.1 |
| $K$ | Ratio of compressive stress in horizontal direction to compressive stress in vertical direction | Eq. 8.4 |
| $K(T)$ | Rate constant | Eq. 9.92 |
| $K(\lambda)$ | Adsorption coefficient | Eq. 5.130 |
| $K_{max}$ | Maximum of rate–temperature curve | Eq. 9.92 |
| $K_{tot}$ | Total kinetic energy | Eq. 2.69 |
| $k$ | Viscosity temperature coefficient | Eq. 9.105 |
| $k$ | Rate constant for mixing | Eq. 6.18 |
| $k_B$ | Boltzmann constant | Eq. 5.152 |
| $k_c$ | Convective mass transfer coefficient | Eq. 4.98 |
| $k_m$ | Thermal conductivity | Eq. 10.35 |
| $k_p$ | Thermal conductivity of plug | Eq. 8.31 |
| $k_x$ | Convective mass transfer coefficient | Eq. 4.98 |
| $k_0$ | Reaction rate | Eq. 8.182 |
| $\bar k_c$ | Average mass transfer coefficient | Eq. 4.99 |
| $k_c^\bullet$ | Local mass transfer coefficient | Eq. 4.96 |
| $\bar k_x$ | Average mass transfer coefficient | Eq. 4.100 |
| $k_x^\bullet$ | Local mass transfer coefficient | |
| $L$ | Length | Eq. 2.15 |
| $L$ | Long axis | Eq. 6.184 |
| $L_e$ | Unwound length of devolatilation section | Eq. 8.174 |
| $L_n$ | Horizontal length | |

| | | |
|---|---|---|
| $L_s$ | Lead of screw | Eq. 8.1 |
| $L_v$ | Thickness of plate | |
| $l_N$ | Total spiral length | Eq. 6.111 |
| $M$ | Degree of mixing | Eq. 6.15 |
| $M_i$ | Molecular weight of component $i$ | Eq. 4.4 |
| $M_t$ | Amount sorbed or desorbed at time $t$ | Eq. 4.91 |
| $M_\infty$ | Total amount sorbed or desorbed | Eq. 4.91 |
| $\bar{M}_W$ | Weight average molecular weight | Eq. 3.112 |
| $M$ | Total possible amount absorbed or desorbed | |
| $M_{A0}, M_{B0}$ | Initial molecular weight | Eq. 8.184 |
| $m$ | Mass | Eq. 2.68 |
| $m$ | Consistency, power-law parameter | Eq. 2.26 |
| $\mathbf{m}$ | Orientation vector | Eq. 6.88 |
| $m'$ | Power-law parameter in Hershel-Bulkley model | Eq. 2.11 |
| $m^0$ | Value of parameter at reference temperature | Eq. 5.26 |
| $\dot{m}$ | Polymer mass flow rate | Eq. 9.3 |
| $\dot{m}_p$ | Dope output | Eq. 4.124 |
| $N$ | Number of crystalline units | Eq. 5.148 |
| $N$ | Number of samples tested | Eq. 6.3 |
| $N$ | Number of steps | Eq. 6.71 |
| $N$ | Number of revolutions | Eq. 6.111 |
| $N$ | Angular velocity of screw, rad/s | Eq. 8.23 |
| $N'$ | Number of turns | Eq. 8.36 |
| $\mathbf{N}_i$ | Molar flux of species $i$ | Eq. 4.9 |
| $N(d\theta)$ | Displacement after $N$ revolutions | Eq. 6.111 |
| Nu | Nusselt number | Eq. 5.23 |
| $N_{min}$ | Minimal rotational frequency | Eq. 6.220 |
| $N_\delta$ | Number of striations | Eq. 8.160 |
| $\mathbb{N}_A$ | Molar rate over entire interface | Eq. 4.99 |
| $\mathbb{N}_B$ | Molar rate over entire interface | Eq. 4.99 |
| $\Delta N$ | Birefringence | Eq. 5.177 |
| $\Delta N_0$ | Intrinsic birefringence | |
| $\Delta N_{cr}^0$ | Birefringence value of perfectly oriented crystalline region | Eq. 5.189 |
| $\Delta N_{am}^0$ | Birefringence value of perfectly oriented amorphous region | Eq. 5.179 |
| $N_1$ | Primary normal stress difference | Eq. 3.11 |
| $Nu_m$ | Nusselt number | Eq. 5.129 |
| $n$ | Power-law index | Eq. 2.6 |
| $n$ | Dimensionality of nucleation process, which is usually taken as 2.0 | Eq. 5.155 |
| $n$ | Number of species in the system | Eq. 4.2 |
| $n$ | Total number of individual groups per structural unit of the macromolecule | Eq. 4.79 |
| $n$ | Number of particles | Eq. 6.1 |
| $n'$ | Power-law parameter in Hershel-Bulkley model | Eq. 2.11 |
| $n^0$ | Value of power-law index at reference | Eq. 5.13 |
| $\mathbf{n}$ | Normal vector | Fig. 9.3 |
| $\mathbf{n}_i$ | Mass flux of species $i$ | Eq. 4.9 |
| $P$ | Partial pressure | Eq. 4.61 |
| $P$ | Probability | Eq. 6.1 |
| $p$ | Ratio of viscosity of dispersed to continuous phase $\mu_d/\mu_c$ | Eq. 6.185 |
| $P$ | Perimeter | Eq. 6.206 |
| $P$ | Hydrostatic pressure | Eq. 9.1 |
| $P(n)$ | Spectral description or power spectrum | Eq. 6.30 |
| $P_H$ | Pressure at $H$ | Eq. 8.6 |
| $P_v$ | Power consumption per unit volume | Eq. 6.74 |
| $P_w$ | Power input through barrel | Eq. 8.28 |
| $P_0$ | Pressure at $h_0$ | Eq. 8.10 |
| $P_0$ | Pure solvent vapor pressure | Eq. 4.86 |
| $P_1$ | Initial pressure at $z=0$ | Eq. 8.27 |
| $P_2$ | Pressure at any down channel distance | Eq. 8.27 |
| $P_0$ | Pure solvent vapor pressure | Eq. 4.86 |
| $\bar{P}$ | Permeability | Eq. 4.71 |
| $\bar{P}_0$ | Preexponential factor | Eq. 4.75 |
| $P_{sc}$ | Permeability in semicrystalline materials | Eq. 4.81 |
| $\bar{P}\star$ | Preexponential permeability factor | Eq. 4.78 |
| $Pe_D$ | Peclet number for diffusion | Eq. 4.105 |
| Pr | Prandtl number | Eq. 5.123 |
| $Pr_f$ | Prandtl number evaluated at film temperature | Eq. 5.130 |
| $P_{end}$ | End pressure | Eq. 3.92 |
| $P_{ent}$ | Entrance pressure | Eq. 3.92 |
| $P_{ex}$ | Exit pressure | Eq. 3.92 |
| $P_H$ | Hole pressure | Eq. 3.100 |
| $P_H$ | Viscous forces per unit length | Eq. 9.145 |
| $P_L$ | Viscous forces per unit length | Eq. 9.145 |
| $P'_L$ | Combined dynamic and hydrostatic pressure at exit | Eq. 2.17 |
| $P_{tot}$ | Total pressure | Eq. 3.93 |
| $P'_0$ | Combined dynamic and hydrostatic pressure | Eq. 2.17 |
| $\Delta P$ | Internal overpressure | Eq. 9.146 |
| $p$ | Pressure | Eq. 2.71 |
| $p$ | Fraction of black particles in entire mixture | Eq. 6.1 |
| $p$ | Number of parallel flights | Eq. 8.128 |
| $p$ | Degree of conversion | |
| $p$ | Viscosity ratio | Eq. 6.106 |
| $p_a$ | Atmospheric pressure | Eq. 3.102 |
| $p_0$ | Pressure at capillary entrance | Eq. 2.16 |
| $p_L$ | Pressure at capillary exit | Eq. 2.16 |
| $Q$ | Volumetric flow rate | Eq. 2.31 |
| $Q_c$ | Calender leakage | |
| $Q_d$ | Drag flow | Eq. 8.100 |
| $Q_f$ | Flight leakage | Eq. 8.132 |
| $Q_p$ | Pressure flow | Eq. 8.100 |
| $Q_s$ | Volumetric flow in the extruder | Eq. 8.103 |
| $Q_t$ | Tetrahedron gap | Eq. 8.132 |
| $Q_t$ | Accumulated amount of diffusant | Eq. 4.88 |
| $q$ | Flow rate per unit width | Eq. 7.21 |
| $q$ | Rate of distortion process | Eq. 6.194 |
| $q_{n|s}$ | Heat flux in the direction normal to the surface and evaluated at the surface | Eq. 5.122 |
| $q_r$ | Heat flux due to conduction in the $r$ direction | Eq. 5.43 |
| $q_z$ | Heat flux due to conduction in the $z$ direction | Eq. 5.43 |
| $R$ | Radius of capillary | Eq. 3.105 |
| $R$ | Function | Eq. 9.15 |
| $R$ | Radius of outer cylinder | Eq. 2.19 |
| $R$ | Radius of roll | |
| $R(\mathbf{r})$ | Correlation function | Eq. 6.20 |
| $R(z)$ | Radius of spinning line at every $z$ distance | Eq. 4.117 |
| Re | Reynolds Number | |
| $\dot{R}_A$ | Net rate of mole $A$ production by chemical reaction | Eq. 4.21 |
| $R_A$ | Rate of reaction of species $A$ | Eq. 8.181 |
| $R_{die}$ | Radius of die | Eq. 9.31 |
| $R_{ex}$ | Local Reynolds number | Eq. 5.123 |
| $R_f$ | Initial tube outside radius | Eq. 9.129 |
| $R_f$ | Constant bubble radius beyond frost line | Eq. 9.149 |
| $R_g$ | Universal gas constant | Eq. 4.53 |
| $R_H$ | Mean hydraulic radius | Eq. 7.27 |
| $R_H$ | Principal radius of curvature of bubble | Eq. 9.146 |
| $R_L$ | Radius of die at exit | Eq. 2.111 |

| | | |
|---|---|---|
| $R_L$ | Principal radius of curvature of bubble surface at point of interest | Eq. 9.146 |
| $R_1, R_2$ | Principal radii of curvature of fiber | Eq. 9.44 |
| $R_s$ | Radius of outer surface | Eq. 5.144 |
| $R_T$ | Droplet radius according to Taylor's theory | Eq. 6.198 |
| $R_0$ | Die radius at entrance | Eq. 2.11 |
| $R_w$ | Wheel radius | Eq. 3.105 |
| Re* | Air-side Reynolds number | Eq. 9.51 |
| $R_{cl}$ | Cluster radius | Eq. 6.181 |
| $\kappa R$ | Radius of inside cylinder | |
| $r$ | Thickness of a thin cylindrical shell | Eq. 2.16 |
| $r$ | Distance between molecules | Eq. 4.59 |
| $r$ | Stoichiometric ratio | Eq. 8.184 |
| $r$ | Spherical coordinate | Eq. 10.59 |
| $\mathbf{r}$ | Separation vector | Eq. 6.20 |
| $\dot{r}_A$ | Net rate of mass $A$ production by chemical reaction | Eq. 4.18 |
| $S$ | Solubility | Eq. 4.61 |
| $S$ | Segregation | |
| $S$ | Spinnability | |
| $S$ | Shape factor | Eq. 10.36 |
| $S$ | Dimensionless number | Eq. 9.105 |
| Sc | Schmidt number | Eq. 4.102 |
| Sh | Sherwood number | Eq. 4.101 |
| St | Stanton number | Eq. 9.69 |
| $St_Ð$ | Stanton number for diffusion | Eq. 4.104 |
| $S_L$ | Linear scale of segregation | Eq. 6.28 |
| $S_V$ | Volumetric scale of segregation | Eq. 6.29 |
| $S_w$ | Weight swell | Eq. 7.9 |
| $S_\infty$ | Unconstrained elastic recovery | Eq. 7.3 |
| $\dot{S}$ | Source term | Eq. 5.43 |
| $\dot{S}$ | Rate of energy production | Eq. 9.38 |
| $\dot{S}_r$ | Energy generated per unit volume per unit time by a reacting system | Eq. 8.180 |
| $\Delta S_{un}$ | Entropy difference between amorphous and crystalline states under random orientation | Eq. 5.168 |
| $\Delta S_{mix}$ | Change in entropy of mixing | Eq. 6.202 |
| $S_0$ | Preexponential factor | Eq. 4.62 |
| $S_a$ | Solubility in amorphous regions | Eq. 4.67 |
| $S_{sc}$ | Solubility in semicrystalline polymers | Eq. 4.67 |
| $S$ | Cross-sectional contact area of rupture plate | Eq. 6.176 |
| $s$ | Constant | Eq. 2.42 |
| $s$ | Standard deviation | Eq. 6.8 |
| $s$ | Scaling factor | Eq. 4.78 |
| $s_N^2$ | Variance of the binomial distribution | Eq. 6.3 |
| $\hat{T}$ | Dimensionless stress | Eq. 9.150 |
| $T_a$ | Temperature of the cooling fluid | Eq. 5.81 |
| $T_a$ | Ambient air temperature | |
| $T_{abs}$ | Absolute temperature | Eq. 7.1 |
| $T_f$ | Film temperature | |
| $T_g$ | Glass transition temperature | Eq. 4.54 |
| $T^K$ | Absolute temperature | |
| $T_m$ | Melting temperature | |
| $T_{max}$ | Temperature at $K_{max}$ | Eq. 9.92 |
| $T_m^{(K)}$ | Melting point in Kelvin | |
| $T_m'$ | Thermodynamic equilibrium melting point | Eq. 5.155 |
| $T_m^\circ$ | Melting point at atmospheric conditions | Eq. 5.164 |
| $T_R$ | Reference temperature | Eq. 5.16 |
| $T_r$ | Temperature of radiation source | Eq. 5.133 |
| $T_x$ | Crystallization temperature | Eq. 5.155 |
| $\Delta T$ | $T_0 - T_\infty$ | |
| $\Delta T$ | $T_m' - T_x$ (undercooling) | Eq. 5.155 |
| $T_b$ | Barrel temperature | Eq. 8.46 |
| $T_p$ | Temperature in plug | Eq. 8.31 |

| | | |
|---|---|---|
| $T_0$ | Temperature of bed entering melting zone | Eq. 8.50 |
| $t$ | Time after a fluid element leaves the die | Eq. 7.5 |
| $t$ | Residence time | Eq. 8.139 |
| $t$ | Confidence coefficient | Eq. 6.10 |
| $t_{av}$ | Process time | Eq. 3.90 |
| $t_b$ | Burst time | Eq. 6.194 |
| $t_f$ | Fraction of time spent in upper portion of channel | Eq. 8.138 |
| $t_f$ | Exposure time | |
| $t_H$ | Hang time | Eq. 2.127 |
| $t_{lag}$ | Time lag | Eq. 4.90 |
| $t_p$ | Time required for melt to pass through the land | Eq. 7.5 |
| $t_p$ | Exposure time of melt pool surface | |
| $t_p$ | Time of rotation of pool | Eq. 8.190 |
| $t_0$ | Shortest residence time | Eq. 6.121 |
| $\mathbf{t}$ | Tangential vector | Fig. 9.3 |
| $t^*$ | Dimensionless time | Eq. 5.107 |
| $t^*$ | Residence time of fluid element along streamline | |
| $t_b^*$ | Dimensionless burst time | Eq. 6.196 |
| $\bar{t}$ | Mean residence time | Eq. 6.117 |
| $t_{1/2}$ | Half-time | Eq. 4.94 |
| $t_i$ | Mean residence time inside mixer | Eq. 6.135 |
| $U$ | Composite heat transfer coefficient | Eq. 9.40 |
| $u$ | Dimensionless velocity | Eq. 8.87 |
| $V$ | Velocity | Eq. 2.1 |
| $V$ | Air speed | |
| $V_T$ | Length of a needle | |
| $V$ | Takeup speed | Eq. 9.130 |
| $V$ | Volume of closed C-shaped chamber | Eq. 8.128 |
| $\bar{V}$ | Velocity of filament | Eq. 4.123 |
| $V_A$ | Volume of gas at STP dissolved into polymer per unit volume of solution | Eq. 4.61 |
| $V_b$ | Velocity of barrel surface | |
| $V_{cell}$ | Volume of a cell | Eq. 6.24 |
| $V_P$ | Volume of polymer per unit volume of solution | Eq. 4.61 |
| $V_r$ | Velocity of plate relative to solid bed | Eq. 8.15 |
| $V_I$ | Volume of region I | Eq. 6.24 |
| $V_0$ | Velocity of plate | Eq. 8.15 |
| $V_0$ | Film velocity | Eq. 5.16 |
| $\bar{V}_i$ | Molar volume of component $i$ | Eq. 4.4 |
| $\Delta \bar{V}_m$ | Volume change of melting process | Eq. 5.164 |
| $v_{ay}$ | Cooling air cross-flow velocity | Eq. 9.64 |
| $v_x$ | Velocity component | Eq. 6.75 |
| $v_y$ | Velocity component | Eq. 6.75 |
| $v_z$ | Axial velocity | Eq. 6.207 |
| $v_z$ | Filament velocity | |
| $v_z$ | Circumferential velocity | |
| $v_\theta$ | Tangential velocity | Eq. 6.108 |
| $v_l$ | Axial velocity | Eq. 8.101 |
| $\bar{v}$ | Time-averaged velocity | Eq. 2.69 |
| $\bar{v}_l$ | Average axial velocity | Eq. 8.137 |
| $\bar{v}_z$ | Average velocity | Eq. 9.3 |
| $\mathbf{v}$ | Mass average bulk velocity | Eq. 4.2 |
| $\mathbf{v}_i$ | Velocity of the $i$th species | Eq. 4.2 |
| $\mathbf{v}_p$ | Velocity of membrane with respect to a fixed reference system | Eq. 4.14 |
| $\langle v \rangle$ | Average velocity | Eq. 2.31 |
| $\mathbf{v}^*$ | Molar average bulk velocity | Eq. 4.3 |
| $\mathbf{v}^\blacksquare$ | Volume average bulk velocity | Eq. 4.4 |
| $v_{ay}$ | Velocity of air crossflow | Eq. 9.50 |
| $W$ | Width | Eq. 6.144 |

| | | |
|---|---|---|
| $W$ | Work input to the system | Eq. 2.69 |
| $W$ | Film width | Eq. 4.40 |
| $W$ | Width of channel | Eq. 8.2 |
| $W$ | Angular velocity | Eq. 6.108 |
| $\bar{W}$ | Mean channel width | |
| $W_b$ | Channel width at barrel surface | |
| $W_L$ | Final width | Eq. 2.114 |
| $W_s$ | Channel width at root of screw | |
| $W_0$ | Initial width | Eq. 2.114 |
| WATS | Weighted average total strain | Eq. 6.141 |
| $w$ | Mass flow rate | Eq. 2.69 |
| $w$ | Water content | Eq. 4.83 |
| $w$ | Dimensionless thickness | Eq. 9.152 |
| $w_e$ | Equilibrium weight fraction | |
| $w_L$ | Width at roll | |
| $w_L(z)$ | Rate of melting | Eq. 8.54 |
| $\bar{w}$ | Mean flight width | Eq. 8.130 |
| $X$ | Width of solid bed | Eq. 8.54 |
| $X$ | Individual stage efficiency | |
| $X$ | Dimensionless freezeline height | Eq. 9.155 |
| $X_c$ | Crystalline volume fraction | Eq. 5.182 |
| $X_f$ | Staged efficiency of diffusing film | Eq. 8.167 |
| $X_T$ | Devolatilization efficiency of machine | |
| $x$ | Depth | Eq. 5.140 |
| $x_a$ | Component of **a** | Eq. 6.32 |
| $x_b$ | Component of **b** | Eq. 6.33 |
| $x_i$ | Mole fraction of $i$th species | Eq. 4.1 |
| $y_a$ | Component of **a** | Eq. 6.32 |
| $y_b$ | Component of **b** | Eq. 6.33 |
| $Z$ | Freezeline height | Eq. 9.155 |
| $Z_b$ | Down channel distance | Eq. 8.28 |
| $z$ | Length | Eq. 2.128 |
| $z$ | Confidence coefficient | Eq. 6.9 |
| $z$ | Machine axis | Eq. 9.115 |
| $z$ | Frost line height | |
| $z$ | Helical distance along channel | Eq. 8.3 |
| $\bar{z}$ | Mean helical length | |
| $z_a$ | Component of **a** | Eq. 6.32 |
| $z_b$ | Component of **b** | Eq. 6.33 |
| $z_T$ | Total length of melting | Eq. 8.68 |
| $z_{\mathrm{coh}}^{*}$ | Axial distance | Eq. 9.103 |
| $z_{\mathrm{cap}}^{*}$ | Maximum uninterrupted jet length | Eq. 9.104 |

**Greek Symbols**

| | | |
|---|---|---|
| $\alpha$ | Angle | Eq. 2.66 |
| $\alpha$ | Thermal diffusivity | |
| $\alpha$ | Rheological parameter | Eq. 9.83 |
| $\alpha$ | One-half the hopper angle | Eq. 8.11 |
| $\alpha$ | Dimensionless parameter | Eq. 9.83 |
| $\alpha_i$ | One-half the angle intermesh | Eq. 8.107 |
| $\alpha_p$ | Thermal diffusivity of solid bed | Eq. 8.32 |
| $\alpha_t$ | Tip angle | Eq. 8.108 |
| $\beta$ | Function of geometry, $r/R$ | Eq. 2.19 |
| $\beta$ | Volume coefficient of expansion | |
| $\beta$ | Angle | Eq. 6.210 |
| $\beta$ | Average Liapunov exponent | Eq. 6.206 |
| $\beta$ | Deformation gradient | Eq. 9.103 |
| $\beta_n$ | Eigenvalues | Eq. 5.92 |
| $\beta_w$ | Wall angle of friction | Eq. 8.14 |
| $\Gamma_b$ | Bulk composite property ($\bar{C}_p$ or $k$) | Eq. 5.73 |
| $\Gamma_1$ | Matrix property ($\bar{C}_p$ or $k$) | Eq. 5.73 |
| $\Gamma_2$ | Second component property ($\bar{C}_p$ or $k$) | Eq. 5.73 |
| $\gamma$ | Shear strain | Eq. 6.62 |

| | | |
|---|---|---|
| $\gamma$ | Surface tension | Eq. 9.45 |
| $\dot{\gamma}$ | Shear rate | Eq. 2.8 |
| $\dot{\gamma}_r$ | Reduced shear rate | Eq. 5.2 |
| $\dot{\gamma}_w$ | Wall shear rate | Eq. 3.97 |
| $\gamma$ | Rate of strain tensor | Eq. 2.59 |
| $\dot{\gamma}_c$ | Shear rate in continuous phase | Eq. 6.106 |
| $\dot{\gamma}_d$ | Shear rate in dispersed phase | Eq. 6.106 |
| $\bar{\gamma}$ | Mean total strain | Eq. 6.141 |
| $\gamma_0$ | Strain amplitude | Eq. 3.22 |
| $\gamma_0$ | Minimum strain | Eq. 6.140 |
| $\dot{\gamma}_0$ | Amplitude | Eq. 3.18 |
| $\dot{\gamma}_{\mathrm{crit}}$ | Critical shear rate for breakup | Eq. 6.180 |
| $\delta$ | Coating thickness | Eq. 2.53 |
| $\delta$ | $T_g/T'_m$ | Eq. 5.157 |
| $\delta$ | Thickness over which concentration changes from $C_{Ai}$ to $C_{A0}$ | |
| $\delta$ | Penetration of dye into the film | |
| $\delta$ | Striation thickness | Eq. 6.31 |
| $\delta$ | Solubility parameter | Eq. 6.203 |
| $\delta$ | Effective angle of friction | Eq. 8.9 |
| $\delta_{ij}$ | Kronecker delta | Eq. 3.7 |
| $\delta_s$ | $D_b - D_s = 2\delta_s$ | |
| $\boldsymbol{\delta}_i$ | Unit vector in the $i$ direction | Eq. 9.9 |
| $\boldsymbol{\delta}_x$ | Unit vector along $x$ direction | Eq. 6.32 |
| $\boldsymbol{\delta}_y$ | Unit vector along $y$ direction | Eq. 6.32 |
| $\boldsymbol{\delta}_z$ | Unit vector along $z$ direction | Eq. 6.32 |
| $\Delta$ | Difference | |
| $\varepsilon$ | Emissivity | |
| $\varepsilon$ | Potential energy constant | Eq. 4.52 |
| $\varepsilon$ | Strain | Eq. 6.54 |
| $\varepsilon$ | Void volume fraction or porosity of cluster | Eq. 6.175 |
| $\varepsilon$ | Dimensionless parameter | Eq. 9.83 |
| $\dot{\varepsilon}$ | Shear-free flow constant | Eq. 3.2 |
| $\dot{\varepsilon}$ | Strain rate | Eq. 6.55 |
| $1/\varepsilon$ | Dimensionless force | |
| $\varepsilon/k$ | Lennard-Jones temperature | Table 4.8 |
| $\zeta$ | Dimensionless coordinate | Eq. 5.21 |
| $\zeta$ | Dimensionless variable | Eq. 9.63 |
| $\eta$ | Shear viscosity | Eq. 2.5 |
| $\eta$ | $x/\sqrt{(4Ð_{AP}t)}$ | Eq. 4.51 |
| $\eta_p$ | Planar elongational viscosity | Eq. 9.123 |
| $\eta_r$ | Reduced viscosity | Eq. 5.1 |
| $\eta_0$ | Zero shear viscosity | Eq. 2.17 |
| $\eta_0(T)$ | Zero shear viscosity at temperature $T$ | Eq. 5.3 |
| $\eta_0(T_0)$ | Zero shear viscosity at temperature $T_0$ | |
| $\eta_\infty$ | Viscosity as shear rate $\to\infty$ | Eq. 2.8 |
| $\eta^{*}$ | Complex viscosity | Eq. 3.21 |
| $\eta'$ | Viscous contribution to dynamic viscosity | Eq. 3.21 |
| $\eta''$ | Elastic contribution to dynamic viscosity | Eq. 3.21 |
| $\theta$ | Dimensionless variable | Eq. 9.63 |
| $\theta$ | Dimensionless temperature | Eq. 5.112 |
| $\theta$ | Spherical coordinate angle | Eq. 6.51 |
| $\theta$ | Circumferential angle | Eq. 8.109 |
| $\theta_a$ | Dimensionless parameter | Eq. 9.64 |
| $\theta$ | Mean temperature | Eq. 5.25 |
| $\theta_a$ | Dimensionless parameter | Eq. 9.64 |
| $\kappa$ | Limit of integration | Eq. 2.25 |
| $\kappa$ | Modified friction angle | Eq. 8.14 |
| $\kappa$ | Ratio of inside to outside radius | Eq. 6.108 |
| $\kappa R_0$ | Die inner radius | Eq. 9.180 |
| $\lambda$ | Dimensionless variable | Eq. 9.63 |
| $\lambda$ | $\sim 1/\dot{\gamma}$ for the onset of shear thinning | Eq. 2.8 |
| $\lambda$ | Wavelength | Eq. 5.139 |
| $\lambda$ | Relaxation time | Eq. 3.39 |

| | | | | | |
|---|---|---|---|---|---|
| $\lambda$ | Lineal stretch | Eq. 6.82 | $\tau_R$ | Wall shear stress | Eq. 7.5 |
| $\lambda_x$ | Principal elongation ratio | Eq. 6.37 | $\tau_w$ | Wall shear stress | Eq. 3.98 |
| $\lambda_y$ | Principal elongation ratio | Eq. 6.37 | $\tau_0$ | Yield stress | Eq. 2.9 |
| $\lambda_z$ | Principal elongation ratio | Eq. 6.37 | $\tau_0$ | Tortuosity | Eq. 4.81 |
| $\lambda_0$ | Elongation ratio | Eq. 6.47 | $\tau_0$ | Initial stress | Eq. 9.84 |
| $\lambda_0$ | Dimensionless parameter | Eq. 9.64 | $\tau_{1/2}$ | $\tau_{yx}$ when $\eta = (1/2)\eta_o$ | Eq. 2.7 |
| $\lambda_0$ | Characteristic time | Eq. 9.128 | $\tau_{xy}^*$ | Shear stress (mechanics convention) | Eq. 2.4 |
| $\dot{\lambda}$ | Rate of linear stretch | Eq. 6.87 | $\tau_{xx}$ | Extra normal stress component | Eq. 3.6 |
| $\Lambda_O$ | Characteristic time | Eq. 9.128 | $\tau_{yx}$ | Shear stress, viscous flux of $x$ momentum in the $y$ direction | Eq. 2.3 |
| $\Lambda_t$ | Characteristic time | Eq. 9.128 | | | |
| $\mu$ | Newtonian viscosity | Eq. 2.1 | $\tau_{yy}$ | Extra normal stress component | |
| $\mu$ | Mean value | Eq. 6.9 | $\upsilon$ | Angular velocity | Eq. 3.105 |
| $\mu_c$ | Viscosity of continuous phase | | $\upsilon$ | Kinematic viscosity | |
| $\mu_d$ | Viscosity of minor (dispersed) phase | | $\upsilon_0$ | Die extrusion speed | Eq. 9.130 |
| $\mu_k$ | Roots of Bessel function $J_0(\mu_k)=0$ | Eq. 4.121 | $\Phi$ | Viscous dissipation term | Eq. 8.47 |
| $\mu_0$ | Bingham model parameter | Eq. 2.9 | $\Phi$ | Phase shift | Eq. 3.19 |
| $\mu_\infty$ | Material property | Eq. 9.56 | $\phi_c$ | Volume fraction of crystallinity | Eq. 5.159 |
| $\mu_{\infty 0}$ | Material property | Eq. 9.63 | $\phi_{i,z}$ | Angles each orthographic axis makes with the $z$ axis | Eq. 5.174 |
| $\xi$ | Limit of integration | Eq. 2.11 | | | |
| $\xi$ | $T_m'/T_x$ | Eq. 5.157 | $\phi_1$ | Volume fraction matrix | Eq. 5.73 |
| $\xi$ | Dimensionless variable | Eq. 8.87 | $\phi_2$ | Volume fraction second component | Eq. 5.73 |
| $\xi_c$ | Particle position in lower part of channel | Eq. 8.134 | $\phi_\infty$ | Equilibrium volume fraction of crystallinity | Eq. 5.159 |
| $\xi'$ | Dummy variable of integration | Eq. 2.11 | $\phi$ | Spherical coordinate angle | Eq. 6.51 |
| $\pi$ | Permachor | Eq. 4.78 | $\phi$ | Angle between unit normal vector and $z$ axis | Eq. 9.10 |
| $\pi_a$ | Permachor for amorphous materials | Eq. 4.80 | $\phi$ | Helix angle | Eq. 8.16 |
| $\pi_{ii}$ | Normal component of stress tensor | Eq. 9.1 | $\phi$ | Angle of drag | |
| $\pi_{ij}$ | $ij$th component of the total stress tensor | Eq. 3.7 | $\phi$ | Orientation of deformed droplet | Eq. 6.188 |
| $\pi_{rr}$ | $r$ component of total stress | Eq. 10.65 | $\phi_A$ | Solvent volume fraction | Eq. 4.86 |
| $\pi_{sc}$ | Permachor for semicrystalline materials | Eq. 4.80 | $\phi_b$ | Helix angle of barrel surface | |
| $\rho$ | Density | Eq. 2.15 | $\phi_c$ | Crystallinity of material | Eq. 4.58 |
| $\rho$ | Total mass concentration | Eq. 4.1 | $\phi(r)$ | Intermolecular energy | Eq. 4.52 |
| $\rho$ | Density of surrounding medium | | $\phi_s$ | Helix angle of screw | Eq. 8.1 |
| $\rho_a$ | Density of amorphous phase | Eq. 5.160 | $\phi_{tot}$ | Total potential energy | Eq. 2.69 |
| $\rho_b$ | Bulk density | Eq. 8.4 | $\phi_1$ | Volume fraction of mixture | Eq. 6.203 |
| $\rho_c$ | Density of spherulitic phase | Eq. 5.160 | $\phi_2$ | Volume fraction of mixture | Eq. 6.203 |
| $\rho_c$ | Density of crystalline plane | Eq. 5.156 | $\hat{\phi}$ | Potential energy per unit mass | Eq. 2.69 |
| $\rho_i$ | Mass concentration of $i$th species | Eq. 4.1 | $\chi$ | Relaxation time | Eq. 9.73 |
| $\rho(\mathbf{r})$ | Correlation coefficient | Eq. 6.22 | $\chi$ | Interaction parameter of solvent-polymer system | Eq. 4.86 |
| $\sigma$ | Stefan-Boltzmann radiation constant | Eq. 5.133 | | | |
| $\sigma$ | Liapunov exponent | Eq. 6.205 | $\chi$ | Confidence coefficient | Eq. 6.11 |
| $\sigma$ | Potential length, or collision diameter | Eq. 4.52 | $\chi_{opt}$ | Principal optical direction | Eq. 5.185 |
| $\Delta\sigma$ | Difference between the principle stresses | Eq. 5.177 | $\chi_{stress}$ | Principal stress direction | Eq. 5.185 |
| $\sigma_e$ | End surface energy associated with lamellar growth | Eq. 5.156 | $\psi$ | Stream function | Eq. 6.204 |
| | | | $\Psi_1$ | Primary normal stress difference coefficient | Eq. 3.16 |
| $\sigma_n^2$ | Variance of the distribution | Eq. 6.2 | $\Psi_2$ | Secondary normal stress difference coefficient | Eq. 3.17 |
| $\sigma_o^2$ | Variance of the totally unmixed state | Eq. 6.17 | $\psi$ | Dimensionless variable in melting model | Eq. 8.64 |
| $\sigma_s$ | Side surface energy | Eq. 5.156 | $\psi$ | Flight flank angle | Eq. 8.130 |
| $\sigma_x$ | Collision diameter of molecule $x$ | Eq. 4.54 | $\psi$ | Dimensionless variable | Eq. 9.63 |
| $\sigma_T$ | Tensile strength of cluster | Eq. 6.175 | $\psi$ | Stream function | Eq. 6.204 |
| $\varsigma$ | Constant | Eq. 9.106 | $\psi_{ay}$ | Dimensionless parameter | Eq. 9.64 |
| $\tau$ | Dimensionless variable | Eq. 9.63 | $\psi_L$ | Dimensionless parameter | Eq. 9.64 |
| $\tau$ | Stress tensor | Eq. 2.59 | $\omega$ | Angular velocity | Eq. 3.18 |
| $\tau$ | Time scale of burst process | Fig. 6.21 | $\omega_A$ | Dope solvent mass fraction | Eq. 4.119 |
| $\tau_{cr}$ | Critical wall shear stress | Eq. 7.1 | $\omega_i$ | Mass fraction of $i$th species | Eq. 4.1 |
| $\tau_{ii}$ | Normal component of extra stress tensor | Eq. 9.1 | $\omega$ | Reduced angular frequency | |
| $\tau_{ij}$ | A component of the extra stress tensor | Eq. 3.7 | | | |

# A

# RHEOLOGICAL DATA FOR SEVERAL POLYMER MELTS

This appendix contains rheological data for a few polymer melts (LDPE, HDPE, LLDPE, mineral-filled nylon 6, 6, and PPS).

## A.1. LDPE DATA

**TABLE A.1 Steady Shear Viscosity and Primary Normal Stress Difference Data for LDPE (NPE 953, Quantum Chemicals)**

| $\dot{\gamma}\,(s^{-1})$ | $\eta\,(Pa \cdot s)$ 170°C | $\eta\,(Pa \cdot s)$ 180°C | $\eta\,(Pa \cdot s)$ 190°C |
|---|---|---|---|
| 0.010 | 2.310E+04 | 1.530E+04 | 1.112E+04 |
| 0.0215 | 2.215E+04 | 1.526E+04 | 1.147E+04 |
| 0.0464 | 2.013E+04 | 1.446E+04 | 1.129E+04 |
| 0.100 | 1.693E+04 | 1.309E+04 | 1.088E+04 |
| 0.215 | 1.437E+04 | 1.124E+04 | 9.237E+03 |
| 0.464 | 1.122E+04 | 9.224E+03 | 7.538E+03 |
| 1.00 | 8.192E+03 | 7.023E+03 | 5.845E+03 |

| $\dot{\gamma}\,(s^{-1})$ | $N_1\,(Pa)$ 170°C | $N_1\,(Pa)$ 180°C | $N_1\,(Pa)$ 190°C |
|---|---|---|---|
| 0.0100 | | | |
| 0.0215 | | | |
| 0.0464 | 6.474E+02 | 2.486E+02 | 1.498E+02 |
| 0.100 | 1.216E+03 | 6.852E+02 | 5.568E+02 |
| 0.215 | 3.717E+03 | 2.184E+03 | 1.765E+03 |
| 0.464 | 1.071E+04 | 5.051E+03 | 5.104E+03 |
| 1.00 | 2.652E+04 | 1.328E+04 | 1.306E+04 |

**TABLE A.2 Dynamic Oscillatory Shear Data for LDPE (NPE 953, Quantum Chemicals) at 170°C and 190°C**

| Frequency (rad/s) | $G'$ (Pa) | $G''$ (Pa) | $\eta^*$ (Pa·s) |
|---|---|---|---|
| | 170°C | | |
| 1.000E−01 | 5.851E+02 | 1.339E+03 | 1.461E+04 |
| 2.154E−01 | 1.189E+03 | 2.127E+03 | 1.131E+04 |
| 4.642E−01 | 2.204E+03 | 3.326E+03 | 8.597E+03 |
| 1.000E+00 | 3.895E+03 | 5.013E+03 | 6.349E+03 |
| 2.154E+00 | 6.477E+03 | 7.336E+03 | 4.542E+03 |
| 4.641E+00 | 1.023E+04 | 1.027E+04 | 3.123E+03 |
| 1.000E+01 | 1.556E+04 | 1.422E+04 | 2.108E+03 |
| 2.154E+01 | 2.297E+04 | 1.935E+04 | 1.394E+03 |
| 4.641E+01 | 3.285E+04 | 2.585E+04 | 9.008E+02 |
| 1.000E+02 | 4.539E+04 | 3.366E+04 | 5.651E+02 |
| | 190°C | | |
| 1.000E−01 | 3.291E+02 | 9.073E+02 | 9.651E+03 |
| 2.154E−01 | 7.283E+02 | 1.535E+03 | 7.888E+03 |
| 4.642E−01 | 1.456E+03 | 2.525E+03 | 6.280E+03 |
| 1.000E+00 | 2.644E+03 | 3.877E+03 | 4.693E+03 |
| 2.154E+00 | 4.589E+03 | 5.785E+03 | 3.428E+03 |
| 4.641E+00 | 7.565E+03 | 8.476E+03 | 2.448E+03 |
| 1.000E+01 | 1.190E+04 | 1.193E+04 | 1.685E+03 |
| 2.154E+01 | 1.807E+04 | 1.665E+04 | 1.141E+03 |
| 4.641E+01 | 2.661E+04 | 2.260E+04 | 7.523E+02 |
| 1.000E+02 | 3.765E+04 | 2.984E+04 | 4.804E+02 |

**TABLE A.3 Capillary Rheometer Data for LDPE (NPE 953, Quantum Chemicals) at 170°C**

| $\dot{\gamma}_a\,(s^{-1})$ | $\tau_a\,(Pa)$ | $\dot{\gamma}_c\,(s^{-1})$ | $\tau_c\,(Pa)$ |
|---|---|---|---|
| | $L/D=12.5$ | | |
| 0.37504E+02 | 0.66010E+05 | 0.56246E+02 | 0.64136E+05 |
| 0.75007E+02 | 0.88013E+05 | 0.11249E+03 | 0.65933E+05 |
| 0.11251E+03 | 0.10617E+06 | 0.16874E+03 | 0.74074E+05 |
| 0.22502E+03 | 0.14082E+06 | 0.33747E+03 | 0.96463E+05 |
| 0.37504E+03 | 0.17603E+06 | 0.56246E+03 | 0.13049E+06 |
| 0.75007E+03 | 0.24204E+06 | 0.11249E+04 | 0.14130E+06 |
| 0.11251E+04 | 0.26844E+06 | 0.16874E+04 | 0.17624E+06 |
| 0.22502E+04 | 0.39166E+06 | 0.33747E+04 | 0.23477E+06 |
| 0.37504E+04 | 0.44006E+06 | 0.56246E+04 | 0.25862E+06 |
| | $L/D=37.1$ | | |
| 0.37504E+02 | 0.48188E+05 | 0.54467E+02 | 0.47556E+05 |
| 0.75007E+02 | 0.63756E+05 | 0.10893E+03 | 0.56317E+05 |
| 0.11251E+03 | 0.74135E+05 | 0.16340E+03 | 0.63322E+05 |
| 0.22502E+03 | 0.96375E+05 | 0.32680E+03 | 0.81430E+05 |
| 0.37504E+03 | 0.11584E+06 | 0.54467E+03 | 0.10049E+06 |
| 0.75007E+03 | 0.17422E+06 | 0.10893E+04 | 0.14028E+06 |
| 0.11251E+04 | 0.17978E+06 | 0.16340E+04 | 0.14871E+06 |
| 0.22502E+04 | 0.24464E+06 | 0.32680E+04 | 0.19179E+06 |
| 0.37504E+04 | 0.29191E+06 | 0.54467E+04 | 0.23077E+06 |
| | $L/D=75.1$ | | |
| 0.37504E+02 | 0.54935E+05 | 0.55277E+02 | 0.54623E+05 |
| 0.75007E+02 | 0.64091E+05 | 0.11055E+03 | 0.60416E+05 |
| 0.11251E+03 | 0.73246E+05 | 0.16583E+03 | 0.67905E+05 |
| 0.22502E+03 | 0.95220E+05 | 0.33166E+03 | 0.87837E+05 |
| 0.37504E+03 | 0.12086E+06 | 0.55277E+03 | 0.11328E+06 |
| 0.75007E+03 | 0.15748E+06 | 0.11055E+04 | 0.14071E+06 |
| 0.11251E+04 | 0.17579E+06 | 0.16583E+04 | 0.16045E+06 |
| 0.22502E+04 | 0.23622E+06 | 0.33166E+04 | 0.21011E+06 |
| 0.37504E+04 | 0.27284E+06 | 0.55277E+04 | 0.24264E+06 |

(Data from R. H. Moynihan, "Flow Stability of Linear Low Density Polyethylene at Polymer and Metal Interfaces," Ph.D. Thesis, Virginia Tech., Blacksburg, VA, 1990.)

**TABLE A.4 Extensional Data for LDPE (NPE 953, Quantum Chemicals) at 170°C**

| Time (s) | Strain | $\bar{\eta}$ (Pa·s) | $\sigma$ (Pa) |
|---|---|---|---|
| | $\dot{\varepsilon}=0.020\,s^{-1}$ | | |
| 0.00000E+00 | 0.00000E+00 | 0.00000E+00 | 0.00000E+00 |
| 0.50000E+01 | 0.10000E+00 | 0.39900E+05 | 0.79800E+03 |
| 0.12500E+02 | 0.25000E+00 | 0.53850E+05 | 0.10770E+04 |
| 0.20000E+02 | 0.40000E+00 | 0.61050E+05 | 0.12210E+04 |
| 0.27500E+02 | 0.55000E+00 | 0.65200E+05 | 0.13040E+04 |
| 0.35500E+02 | 0.71000E+00 | 0.72450E+05 | 0.14490E+04 |
| 0.43000E+02 | 0.86000E+00 | 0.76500E+05 | 0.15300E+04 |
| 0.50500E+02 | 0.10100E+01 | 0.78200E+05 | 0.15640E+04 |
| 0.58000E+02 | 0.11600E+01 | 0.81550E+05 | 0.16310E+04 |
| 0.65500E+02 | 0.13100E+01 | 0.83600E+05 | 0.16720E+04 |
| 0.73000E+02 | 0.14600E+01 | 0.85350E+05 | 0.17070E+04 |
| 0.80500E+02 | 0.16100E+01 | 0.93650E+05 | 0.18730E+04 |
| 0.88500E+02 | 0.17700E+01 | 0.98050E+05 | 0.19610E+04 |
| 0.96000E+02 | 0.19200E+01 | 0.94550E+05 | 0.18910E+04 |

**TABLE A.4** *continued*

| Time (s) | Strain | $\bar{\eta}$ (Pa·s) | $\sigma$ (Pa) |
|---|---|---|---|
| | | $\dot{\varepsilon} = 0.053\,\mathrm{s}^{-1}$ | |
| 0.00000E+00 | 0.00000E+00 | 0.00000E+00 | 0.00000E+00 |
| 0.28302E+01 | 0.15000E+00 | 0.37679E+05 | 0.19970E+04 |
| 0.58491E+01 | 0.31000E+00 | 0.50811E+05 | 0.26930E+04 |
| 0.88679E+01 | 0.47000E+00 | 0.57604E+05 | 0.30530E+04 |
| 0.11887E+02 | 0.63000E+00 | 0.65792E+05 | 0.34870E+04 |
| 0.14906E+02 | 0.79000E+00 | 0.72755E+05 | 0.38560E+04 |
| 0.17925E+02 | 0.95000E+00 | 0.76774E+05 | 0.40690E+04 |
| 0.20943E+02 | 0.11100E+01 | 0.83415E+05 | 0.44210E+04 |
| 0.23962E+02 | 0.12700E+01 | 0.91453E+05 | 0.48470E+04 |
| 0.26981E+02 | 0.14300E+01 | 0.96755E+05 | 0.51280E+04 |
| 0.30000E+02 | 0.15900E+01 | 0.10666E+06 | 0.56530E+04 |
| 0.33019E+02 | 0.17500E+01 | 0.11323E+06 | 0.60010E+04 |
| 0.36038E+02 | 0.19100E+01 | 0.12523E+06 | 0.66370E+04 |
| 0.39057E+02 | 0.20700E+01 | 0.14026E+06 | 0.74340E+04 |
| 0.42075E+02 | 0.22300E+01 | 0.15343E+06 | 0.81320E+04 |
| | | $\dot{\varepsilon} = 0.200\,\mathrm{s}^{-1}$ | |
| 0.00000E+00 | 0.00000E+00 | 0.00000E+00 | 0.00000E+00 |
| 0.70000E+00 | 0.14000E+00 | 0.20565E+05 | 0.41130E+04 |
| 0.14500E+01 | 0.29000E+00 | 0.30115E+05 | 0.60230E+04 |
| 0.22000E+01 | 0.44000E+00 | 0.37800E+05 | 0.75600E+04 |
| 0.29500E+01 | 0.59000E+00 | 0.44105E+05 | 0.88210E+04 |
| 0.37000E+01 | 0.74000E+00 | 0.52050E+05 | 0.10410E+05 |
| 0.44500E+01 | 0.89000E+00 | 0.59800E+05 | 0.11960E+05 |
| 0.52000E+01 | 0.10400E+01 | 0.68650E+05 | 0.13730E+05 |
| 0.59500E+01 | 0.11900E+01 | 0.79300E+05 | 0.15860E+05 |
| 0.67000E+01 | 0.13400E+01 | 0.89700E+05 | 0.17940E+05 |
| 0.74500E+01 | 0.14900E+01 | 0.10190E+06 | 0.20380E+05 |
| 0.82000E+01 | 0.16400E+01 | 0.11495E+06 | 0.22990E+05 |
| 0.89500E+01 | 0.17900E+01 | 0.12975E+06 | 0.25950E+05 |
| 0.97500E+01 | 0.19500E+01 | 0.14495E+06 | 0.28990E+05 |

(Data from S. A. White, "The Planar Entry Flow Behavior of Polymer Melts: An Experimental and Numerical Analysis," Ph.D. Thesis, Dept. of Chem. Eng,, Virginia Tech, Blacksburg, VA, 1987.)

## A.2. HDPE DATA

**TABLE A.5 Steady Shear Data for HDPE (EMN 885, Philips Petroleum Co.)**

| $\dot{\gamma}$ (s$^{-1}$) | $\eta$ (Pa·s) 170°C | $\eta$ (Pa·s) 180°C | $\eta$ (Pa·s) 190°C |
|---|---|---|---|
| 0.100 | 7.801E+02 | 6.649E+02 | 5.575E+02 |
| 0.2154 | 7.608E+02 | 6.657E+02 | 5.390E+02 |
| 0.4641 | 7.049E+02 | 6.111E+02 | 4.846E+02 |
| 1.00 | 6.396E+02 | 5.689E+02 | 4.561E+02 |
| 2.154 | 5.908E+02 | 5.138E+02 | 4.139E+02 |
| 4.641 | 5.207E+02 | 4.535E+02 | 4.267E+02 |
| 10.0 | 4.336E+02 | 3.922E+02 | 3.705E+02 |

| $\dot{\gamma}$ (s$^{-1}$) | $N_1$ (Pa) 170°C | $N_1$ (Pa) 180°C | $N_1$ (Pa) 190°C |
|---|---|---|---|
| 0.464 | 1.161E+02 | 9.379E+01 | 8.240E+01 |
| 1.00 | 2.917E+02 | 2.165E+02 | 2.075E+02 |
| 2.15 | 7.430E+02 | 5.089E+02 | 4.743E+02 |
| 4.64 | 1.608E+03 | 1.423E+03 | 1.098E+03 |
| 10.0 | 3.658E+03 | 3.052E+03 | 2.640E+03 |

**TABLE A.6 Dynamic Oscillatory Shear Data for HDPE (EMN 885, Phillips Petroleum Co.)**

| Frequency (rad/s) | $G'$ (Pa) | $G''$ (Pa) | $\eta^*$ (Pa·s) |
|---|---|---|---|
| | | 170°C | |
| 1.000E−01 | 1.171E+01 | 1.044E+02 | 1.050E+03 |
| 2.154E−01 | 4.193E+01 | 1.523E+02 | 7.330E+02 |
| 4.642E−01 | 5.541E+01 | 3.120E+02 | 6.826E+02 |
| 1.000E+00 | 1.267E+02 | 6.364E+02 | 6.489E+02 |
| 2.154E+00 | 2.819E+02 | 1.165E+03 | 5.566E+02 |
| 4.641E+00 | 5.850E+02 | 2.238E+03 | 4.984E+02 |
| 1.000E+01 | 1.290E+03 | 4.173E+03 | 4.368E+02 |
| 2.154E+01 | 2.746E+03 | 7.680E+03 | 3.786E+02 |
| 4.641E+01 | 5.586E+03 | 1.370E+04 | 3.189E+02 |
| 1.000E+02 | 1.095E+04 | 2.348E+04 | 2.591E+02 |
| | | 180°C | |
| 1.000E−01 | 1.359E+01 | 6.945E+01 | 7.077E+02 |
| 2.154E−01 | 2.167E+01 | 1.366E+02 | 6.420E+02 |
| 4.642E−01 | 2.567E+01 | 2.642E+02 | 5.718E+02 |
| 1.000E+00 | 1.146E+02 | 5.296E+02 | 5.418E+02 |
| 2.154E+00 | 2.196E+02 | 1.044E+03 | 4.951E+02 |
| 4.641E+00 | 5.022E+02 | 1.986E+03 | 4.413E+02 |
| 1.000E+01 | 1.122E+03 | 3.696E+03 | 3.862E+02 |
| 2.154E+01 | 2.367E+03 | 6.841E+03 | 3.360E+02 |
| 4.641E+01 | 4.859E+03 | 1.227E+04 | 2.843E+02 |
| 1.000E+02 | 9.585E+03 | 2.107E+04 | 2.314E+02 |
| | | 190°C | |
| 1.000E−01 | 8.319E+00 | 4.105E+01 | 4.189E+02 |
| 2.154E−01 | 1.635E+01 | 1.141E+02 | 5.351E+02 |
| 4.642E−01 | 3.233E+01 | 2.151E+02 | 4.687E+02 |
| 1.000E+00 | 7.965E+01 | 4.210E+02 | 4.284E+02 |
| 2.154E+00 | 1.711E+02 | 8.179E+02 | 3.879E+02 |
| 4.641E+00 | 3.893E+02 | 1.581E+03 | 3.508E+02 |
| 1.000E+01 | 8.810E+02 | 3.017E+03 | 3.143E+02 |
| 2.154E+01 | 1.872E+03 | 5.602E+03 | 2.742E+02 |
| 4.641E+01 | 3.854E+03 | 1.013E+04 | 2.335E+02 |
| 1.000E+02 | 7.679E+03 | 1.759E+04 | 1.920E+02 |

## A.3. LLDPE DATA

**TABLE A.7 Steady Shear Data for LLDPE (NTA 101, Mobil) at 170°C**

| $\dot{\gamma}$ (s$^{-1}$) | $\eta$ (Pa·s) | $N_1$ (Pa) |
|---|---|---|
| 0.1000E−01 | 0.1334E+05 | 0.7010E+00 |
| 0.2154E−01 | 0.1321E+05 | 0.4717E+01 |
| 0.4641E+01 | 0.1296E+05 | 0.9272E+02 |
| 0.1000E+00 | 0.1250E+05 | 0.4520E+03 |
| 0.2154E+00 | 0.1168E+05 | 0.1444E+04 |
| 0.4641E+00 | 0.1061E+05 | 0.3925E+04 |
| 0.1000E+01 | 0.8871E+04 | 0.9759E+04 |
| 0.2154E+01 | 0.6962E+04 | 0.2043E+05 |

(Data from R. H. Moynihan, "Flow Stability of Linear Low Density Polyethylene at Polymer and Metal Interfaces," Ph.D. Thesis, Chem. Eng. Dept., Virginia Tech., Blacksburg, VA, 1990.)

**TABLE A.8 Dynamic Shear Data for LLDPE (NTA 101, Mobil) at 170°C**

| $\omega$ (rad/s) | $\eta^*$ (Pa·s) | $G'$ (Pa) |
|---|---|---|
| 0.1000E+00 | 0.1213E+05 | 0.1709E+03 |
| 0.1585E+00 | 0.1155E+05 | 0.3031E+03 |
| 0.2512E+00 | 0.1102E+05 | 0.5204E+03 |
| 0.3981E+00 | 0.1042E+05 | 0.8915E+03 |
| 0.6310E+00 | 0.9781E+04 | 0.1500E+04 |
| 0.1000E+01 | 0.9109E+04 | 0.2514E+04 |
| 0.1585E+01 | 0.8374E+04 | 0.4137E+04 |
| 0.2512E+01 | 0.7613E+04 | 0.6707E+04 |
| 0.3981E+01 | 0.6833E+04 | 0.1069E+05 |
| 0.6310E+01 | 0.6046E+04 | 0.1666E+05 |
| 0.1000E+02 | 0.5265E+04 | 0.2535E+05 |
| 0.1585E+02 | 0.4508E+04 | 0.3760E+05 |
| 0.2512E+02 | 0.3794E+04 | 0.5433E+05 |
| 0.3981E+02 | 0.3140E+04 | 0.7653E+05 |
| 0.6310E+02 | 0.2551E+04 | 0.1049E+06 |
| 0.1000E+03 | 0.2026E+04 | 0.1394E+06 |

(Data from R. H. Moynihan, "Flow Stability of Linear Low Density Polyethylene at Polymer and Metal Interfaces," Ph.D. Thesis, Virginia Tech., Blacksburg, VA, 1990.)

**TABLE A.9 Capillary Data for LLDPE (NTA 101, Mobil) at 170°C**

| $\dot{\gamma}_a$ (s$^{-1}$) | $\tau_a$ (Pa) | $\dot{\gamma}_c$ (s$^{-1}$) | $\tau_c$ (Pa) |
|---|---|---|---|
| | | $L/D=12.5$ | |
| 0.37504E+02 | 0.14082E+06 | 0.77567E+02 | 0.15929E+06 |
| 0.75007E+02 | 0.19803E+06 | 0.15513E+03 | 0.23295E+06 |
| 0.11251E+03 | 0.23983E+06 | 0.23270E+03 | 0.28144E+06 |
| 0.22502E+03 | 0.31685E+06 | 0.46540E+03 | 0.28939E+06 |
| 0.37504E+03 | 0.37846E+06 | 0.77567E+03 | 0.29303E+06 |
| 0.75007E+03 | 0.45327E+06 | 0.15513E+04 | 0.29107E+06 |
| 0.11251E+04 | 0.48847E+06 | 0.23270E+04 | 0.39236E+06 |
| 0.22502E+04 | 0.55888E+06 | 0.46540E+04 | 0.40904E+06 |
| 0.37504E+04 | 0.73491E+06 | 0.77567E+04 | 0.45584E+06 |
| | | $L/D=37.1$ | |
| 0.37504E+02 | 0.12974E+06 | 0.65226E+02 | 0.13596E+06 |
| 0.75007E+02 | 0.16680E+06 | 0.13045E+03 | 0.17857E+06 |
| 0.11251E+03 | 0.20387E+06 | 0.19568E+03 | 0.21789E+06 |
| 0.22502E+03 | 0.26503E+06 | 0.39136E+03 | 0.25578E+06 |
| 0.37504E+03 | 0.32249E+06 | 0.65226E+03 | 0.29370E+06 |
| 0.75007E+03 | 0.38179E+06 | 0.13045E+04 | 0.32714E+06 |
| 0.11251E+04 | 0.41515E+06 | 0.19568E+04 | 0.38277E+06 |
| 0.22502E+04 | 0.48188E+06 | 0.39136E+04 | 0.43139E+06 |
| 0.37504E+04 | 0.53933E+06 | 0.65226E+04 | 0.44530E+06 |
| | | $L/D=75.1$ | |
| 0.37504E+02 | 0.14283E+06 | 0.69768E+02 | 0.14591E+06 |
| 0.75007E+02 | 0.19593E+06 | 0.13954E+03 | 0.20175E+06 |
| 0.11251E+03 | 0.23805E+06 | 0.20930E+03 | 0.24498E+06 |
| 0.22502E+03 | 0.27467E+06 | 0.41861E+03 | 0.27010E+06 |
| 0.37504E+03 | 0.30763E+06 | 0.69768E+03 | 0.29342E+06 |
| 0.75007E+03 | 0.33876E+06 | 0.13954E+04 | 0.31177E+06 |
| 0.11251E+04 | 0.40285E+06 | 0.20930E+04 | 0.38686E+06 |
| 0.22502E+04 | 0.44680E+06 | 0.41861E+04 | 0.42186E+06 |
| 0.37504E+04 | 0.49624E+06 | 0.69768E+04 | 0.44979E+06 |

(Data from R. H. Moynihan, ((Flow Stability of Linear Law Density Polyethylene at Polymer and Metal Interfaces," Ph.D. Thesis, Virginia Tech., Blacksburg, VA, 1990.)

**TABLE A.10 Extensional Data for LLDPE (NTA 101 Mobil) at 170°C**

| Time (s) | Strain | $\bar{\eta}$ (Pa·s) | $\sigma$ (Pa) |
|---|---|---|---|
| | | $\dot{\varepsilon}=0.200\,\text{s}^{-1}$ | |
| 0.00000E+00 | 0.00000E+00 | 0.00000E+00 | 0.00000E+00 |
| 0.80000E+00 | 0.16000E+00 | 0.33755E+05 | 0.67510E+04 |
| 0.15500E+01 | 0.31000E+00 | 0.36615E+05 | 0.73230E+04 |
| 0.24000E+01 | 0.48000E+00 | 0.38480E+05 | 0.76960E+04 |
| 0.31500E+01 | 0.63000E+00 | 0.39485E+05 | 0.78970E+04 |
| 0.39000E+01 | 0.78000E+00 | 0.38780E+05 | 0.77560E+04 |
| 0.47000E+01 | 0.94000E+00 | 0.40495E+05 | 0.80990E+04 |
| 0.54500E+01 | 0.10900E+01 | 0.39770E+05 | 0.79540E+04 |
| 0.62500E+01 | 0.12500E+01 | 0.40085E+05 | 0.80170E+04 |
| 0.70000E+01 | 0.14000E+01 | 0.40045E+05 | 0.80090E+04 |
| 0.77500E+01 | 0.15500E+01 | 0.40805E+05 | 0.81610E+04 |
| 0.85500E+01 | 0.17100E+01 | 0.40295E+05 | 0.80590E+04 |
| 0.93000E+01 | 0.18600E+01 | 0.39560E+05 | 0.79120E+04 |
| 0.10050E+02 | 0.20100E+01 | 0.39670E+05 | 0.79340E+04 |
| | | $\dot{\varepsilon}=0.053\,\text{s}^{-1}$ | |
| 0.00000E+00 | 0.00000E+00 | 0.00000E+00 | 0.00000E+00 |
| 0.30189E+01 | 0.16000E+00 | 0.37962E+05 | 0.20120E+04 |
| 0.60377E+01 | 0.32000E+00 | 0.33245E+05 | 0.17620E+04 |
| 0.90566E+01 | 0.48000E+00 | 0.38943E+05 | 0.20640E+04 |
| 0.12075E+02 | 0.64000E+00 | 0.39321E+05 | 0.20840E+04 |
| 0.15094E+02 | 0.80000E+00 | 0.39000E+05 | 0.20670E+04 |
| 0.18113E+02 | 0.96000E+00 | 0.38019E+05 | 0.20150E+04 |
| 0.21132E+02 | 0.11200E+01 | 0.37925E+05 | 0.20100E+04 |
| 0.24151E+02 | 0.12800E+01 | 0.39642E+05 | 0.21010E+04 |
| 0.27170E+02 | 0.14400E+01 | 0.37189E+05 | 0.19710E+04 |
| 0.30189E+02 | 0.16000E+01 | 0.35925E+05 | 0.19040E+04 |
| 0.33208E+02 | 0.17600E+01 | 0.35226E+05 | 0.18670E+04 |
| 0.36226E+02 | 0.19200E+01 | 0.34358E+05 | 0.18210E+04 |
| | | $\dot{\varepsilon}=0.020\,\text{s}^{-1}$ | |
| 0.00000E+00 | 0.00000E+00 | 0.00000E+00 | 0.00000E+00 |
| 0.75000E+01 | 0.15000E+00 | 0.31550E+05 | 0.63100E+03 |
| 0.15000E+02 | 0.30000E+00 | 0.32750E+05 | 0.65500E+03 |
| 0.22500E+02 | 0.45000E+00 | 0.33900E+05 | 0.67800E+03 |
| 0.30000E+02 | 0.60000E+00 | 0.38600E+05 | 0.77200E+03 |
| 0.38000E+02 | 0.76000E+00 | 0.40300E+05 | 0.80600E+03 |
| 0.45500E+02 | 0.91000E+00 | 0.39550E+05 | 0.79100E+03 |
| 0.53000E+02 | 0.10600E+01 | 0.39400E+05 | 0.78800E+03 |
| 0.60500E+02 | 0.12100E+01 | 0.36950E+05 | 0.73900E+03 |
| 0.68000E+02 | 0.13600E+01 | 0.42550E+05 | 0.85100E+03 |
| 0.75500E+02 | 0.15100E+01 | 0.42450E+05 | 0.84900E+03 |
| 0.83000E+02 | 0.16600E+01 | 0.38500E+05 | 0.77000E+03 |
| 0.90500E+02 | 0.18100E+01 | 0.39500E+05 | 0.79000E+03 |
| 0.98500E+02 | 0.19700E+01 | 0.37900E+05 | 0.75800E+03 |

*Tables continue*

## A4. Nylon 6,6 Data

### TABLE A.11 Steady Shear Cone-and-Plate Data for Mineral-Filled Nylon 6,6 at 285°C

| $\dot{\gamma}$ (s$^{-1}$) | $\eta$ (Pa·s) | $N_1$ (Pa) |
|---|---|---|
| 3.981E−02 | 2.978E+03 | |
| 6.310E−02 | 2.642E+03 | |
| 1.000E−01 | 2.046E+03 | |
| 1.585E−01 | 1.680E+03 | |
| 2.512E−01 | 1.350E+03 | |
| 3.981E−01 | 1.014E+03 | |
| 6.310E−01 | 8.116E+02 | 9.482E+01 |
| 1.000E+00 | 6.369E+02 | 9.311E+01 |
| 1.585E+00 | 5.232E+02 | 8.648E+01 |
| 2.512E+00 | 4.359E+02 | 7.696E+01 |
| 3.981E+00 | 3.861E+02 | 7.101E+01 |
| 6.310E+00 | 3.444E+03 | 8.464E+01 |
| 1.000E+01 | 3.105E+02 | 1.257E+02 |
| 1.585E+01 | 2.703E+03 | 3.937E+02 |
| 2.512E+01 | 2.067E+02 | 5.958E+02 |
| 3.982E+01 | 1.505E+02 | 1.437E+03 |
| 6.310E+01 | 1.557E+02 | 3.284E+03 |

(Data from R. Pisipati, "A Rheological Characterization of Particulate and Fiber filled Nylon 6,6 Melts and its Application to Weldline Formation in Molded Parts," Ph.D. Thesis, Chem. Eng. Dept., Virginia, Tech., Blacksburg, VA, 1983.)

## A5. PPS Data

### TABLE A.12 Dynamic Oscillatory Shear Data for PPS (Ryton, Phillips Petroleum Co.)

| Frequency (rad/s) | $G'$ (Pa) | $G''$ (Pa) | $\eta^*$ (Pa·s) |
|---|---|---|---|
| | **293°C** | | |
| 0.1 | 45 | 170 | 1759 |
| 0.32 | 89 | 481 | 1529 |
| 1.00 | 248 | 1406 | 1428 |
| 3.16 | 1008 | 3508 | 1155 |
| 10.00 | 2277 | 8373 | 868 |
| 31.62 | 10100 | 18250 | 660 |
| 100.00 | 24660 | 35740 | 434 |
| | **330°C** | | |
| 0.1 | 13 | 89 | 899 |
| 0.32 | 39 | 238 | 754 |
| 0.46 | 50 | 360 | 790 |
| 1.00 | 89 | 790 | 795 |
| 2.15 | 230 | 1654 | 777 |
| 3.16 | 394 | 1974 | 637 |
| 6.81 | 1105 | 4366 | 661 |
| 10.00 | 1500 | 5150 | 536 |
| 21.54 | 4055 | 10680 | 530 |
| 31.62 | 5267 | 12330 | 424 |
| 68.13 | 12060 | 23310 | 385 |
| 100.00 | 14690 | 25930 | 298 |

### TABLE A.13 Shift Factors for PPS for a Reference Temperature of 330°C

| Temperature (°C) | Shift Factor |
|---|---|
| 330 | 1.000 |
| 312 | 1.307 |
| 293 | 1.708 |
| 273 | 2.918 |
| 253 | 4.546 |

# B

# PHYSICAL PROPERTIES AND FRICTION COEFFICIENTS FOR SOME COMMON POLYMERS IN THE BULK STATE

TABLE B.1

| Polymer | $\rho$ (kg/m$^3$) | $k_b$ (Wm$^{-1}$ °K$^{-1}$) | $\alpha_b$ (m$^2$/s) | Friction Coefficient on Steel Temperature (°C) | | |
|---|---|---|---|---|---|---|
| | | | | 20 | 60 | 100 |
| LDPE | 500 | 0.346 | $9.29 \times 10^{-8}$ | 0.34 | 0.40 | 0.32 |
| PVC | 620 | 0.156 | $9.55 \times 10^{-8}$ | 0.43 | 0.46 | 0.76 |
| Nylon 6,6 | | | | 0.25 | | |
| Teflon | 1000 | 0.208 | $9.55 \times 10^{-8}$ | 0.04 | 0.04 | 0.04 |
| PS | | | | 0.45 | | |

# C

# THERMAL PROPERTIES OF MATERIALS

**TABLE C.1 Thermal-Physical Properties of Polyethyleneterephthalate**

| Parameter | Value | Units |
|---|---|---|
| $T'_m$ | 540 | °K |
| $T_g$ | 353 | °K |
| $\Delta \bar{H}_c$ | 30 | cal/g |
| $\rho_m$ | 1.335 | g/cm$^3$ |
| $\rho_c$ | 1.455 | g/cm$^3$ |
| $b_0$ | $4.04 \times 10^{-8}$ | cm |
| $a_0$ | $5.76 \times 10^{-8}$ | cm |
| $\sigma$ | 5 | erg/cm$^2$ |
| $\sigma_e$ | 40 | erg/cm$^2$ |
| $\sigma\sigma_e$ | 200 | erg$^2$/cm$^4$ |
| $\phi_\infty$ | 0.342 | |

(Data from D. G. Bright, *Quantitative Studies of Polymer Crystallization under Non-Isothermal Conditions*, Ph.D. Thesis, Georgia Institute of Technology, 1975.)

**TABLE C.3 Thermal-Physical Properties of Polycaprolactam, Nylon 6**

| Parameter | Value | Units |
|---|---|---|
| $T'_m$ | 505 | °K |
| $T_g$ | 323 | °K |
| $\Delta \bar{H}_c$ | 45.3 | cal/g |
| $\rho_m$ | 1.0840 | g/cm$^3$ |
| $\rho_c$ | 1.2255 | g/cm$^3$ |
| $b_0$ | $8.62 \times 10^{-8}$ | cm |
| $a_0$ | $8.83 \times 10^{-8}$ | cm |
| $\sigma$ | 8 | erg/cm$^2$ |
| $\sigma_e$ | 60 | erg/cm$^2$ |
| $\sigma\sigma_e$ | 480 | erg$^2$/cm$^4$ |
| $\phi_\infty$ | 0.312 | |

(Data from D. G. Bright, *Quantitative Studies of Polymer Crystallization under Non-Isothermal Conditions*, Ph.D. Thesis, Georgia Institute of Technology, 1975.)

**TABLE C.2 Thermal-Physical Properties of High-Density Polyethylene**

| Parameter | Value | Units |
|---|---|---|
| $T'_m$ | 415 | °K |
| $T_g$ | 231 | °K |
| $\Delta \bar{H}_c$ | 68.4 | cal/g |
| $\rho_m$ | 0.8838 | g/cm$^3$ |
| $\rho_c$ | 1.0075 | g/cm$^3$ |
| $b_0$ | $4.13 \times 10^{-8}$ | cm |
| $a_0$ | $4.46 \times 10^{-8}$ | cm |
| $\sigma$ | 10.25 | erg/cm$^2$ |
| $\sigma_e$ | 57.0 | erg/cm$^2$ |
| $\sigma\sigma_e$ | 584 | erg$^2$/cm$^4$ |
| $\phi_\infty$ | 0.722 | |

(Data from D. G. Bright, *Quantitative Studies of Polymer Crystallization under Non-Isothermal Conditions*, Ph.D. Thesis, Georgia Institute of Technology, 1975.)

**TABLE C.4 Thermal-Physical Properties of Poly (hexamethylene adipamide), Nylon 6,6**

| Parameter | Value | Units |
|---|---|---|
| $T'_m$ | 545 | °K |
| $T_g$ | 330 | °K |
| $\Delta \bar{H}_c$ | 46.6 | cal/g |
| $\rho_a$ | 1.07 | g/cm$^3$ |
| $\rho_c$ | 1.266 | g/cm$^3$ |
| $b_0$ | $4.77 \times 10^{-8}$ | cm |
| $a_0$ | $4.04 \times 10^{-8}$ | cm |
| $\sigma$ | 8.5 | erg/cm$^2$ |
| $\sigma_e$ | 42.35 | erg/cm$^2$ |
| $\sigma\sigma_e$ | 360 | erg$^2$/cm$^4$ |
| $\phi_\infty$ | 0.32 | |

(Data from D. G. Bright, *Quantitative Studies of Polymer Crystallization under Non-Isothermal Conditions*, Ph.D. Thesis, Georgia Institute of Technology, 1975.)

**TABLE C.5 Crystallization Parameters for Polyetheretherketone (PEEK)**

| | |
|---|---|
| Crystal dimension | $b_0 = 4.7$ Å |
| Side surface energy | $\sigma_s = 38$ erg/cm$^2$ |
| End surface energy | $\sigma_e = 49$ erg/cm$^2$ |
| Heat of fusion | $\Delta \bar{H}_f = 130$ J/g |
| Thermodynamic melting point | $T'_m = 395$°C |
| Glass transition temperature | $T_g = 144$°C |
| Activation energy | $E_D = 2000$ cal/mol |
| Crystal density | $\rho_c = 1.40$ g/cm$^2$ |
| Amorphous density | $\rho_a = 1.263$ g/cm$^2$ |
| $\phi_\infty$ | 0.33 |

(Data from D. J. Blundell and B. N. Osborn, "The Morphology of Poly(aryl-ether-ether-ketone)." *Polymer, 24*, 753, 1983.)

**TABLE C.6 Properties of Water (Saturated Liquid)**

| $T$ (°C) | $\bar{c}_p$ (kJ/kg·°C) | $\rho$ (kg/m³) | $\mu$ (kg/m/s) | $k$ (W/m·°C) | $Pr$ | $\dfrac{g\beta\rho^2\bar{c}_p}{\mu k}$ (1/m³·°C) |
|---|---|---|---|---|---|---|
| 0.0 | 4.225 | 999.8 | $1.79 \times 10^{-3}$ | 0.566 | 13.25 | $1.91 \times 10^{9}$ |
| 4.44 | 4.208 | 999.8 | 1.55 | 0.575 | 11.35 | $6.34 \times 10^{9}$ |
| 10.0 | 4.195 | 999.2 | 1.31 | 0.585 | 9.40 | $1.08 \times 10^{10}$ |
| 15.56 | 4.186 | 998.6 | 1.12 | 0.595 | 7.88 | $1.46 \times 10^{10}$ |
| 21.11 | 4.179 | 997.4 | $9.8 \times 10^{-4}$ | 0.604 | 6.78 | $1.91 \times 10^{10}$ |
| 26.67 | 4.179 | 995.8 | 8.60 | 0.614 | 5.85 | $2.48 \times 10^{10}$ |
| 32.22 | 4.174 | 994.9 | 7.65 | 0.623 | 5.12 | $3.3 \times 10^{10}$ |
| 37.78 | 4.174 | 993.0 | 6.82 | 0.630 | 4.53 | $4.19 \times 10^{10}$ |
| 43.33 | 4.174 | 990.6 | 6.16 | 0.637 | 4.04 | $4.89 \times 10^{10}$ |
| 48.89 | 4.174 | 988.8 | 5.62 | 0.644 | 3.64 | $5.66 \times 10^{10}$ |
| 54.44 | 4.179 | 985.7 | 5.13 | 0.649 | 3.30 | $6.48 \times 10^{10}$ |
| 60.0 | 4.179 | 983.3 | 4.71 | 0.654 | 3.01 | $7.62 \times 10^{10}$ |
| 65.55 | 4.183 | 980.3 | 4.30 | 0.659 | 2.73 | $8.84 \times 10^{10}$ |
| 71.11 | 4.186 | 977.3 | 4.01 | 0.665 | 2.53 | $9.85 \times 10^{10}$ |
| 76.67 | 4.191 | 973.7 | 3.72 | 0.668 | 2.33 | $1.09 \times 10^{11}$ |
| 82.22 | 4.195 | 970.2 | 3.47 | 0.673 | 2.16 | |
| 87.78 | 4.199 | 966.7 | 3.27 | 0.675 | 2.03 | |
| 93.33 | 4.204 | 963.2 | 3.06 | 0.678 | 1.90 | $1.23 \times 10^{11}$ |
| 104.4 | 4.216 | 955.1 | 2.67 | 0.684 | 1.66 | |
| 115.6 | 4.229 | 946.7 | 2.44 | 0.685 | 1.51 | |
| 126.7 | 4.250 | 937.2 | 2.19 | 0.685 | 1.36 | |
| 137.8 | 4.271 | 928.1 | 1.98 | 0.685 | 1.24 | |
| 148.9 | 4.296 | 918.0 | 1.86 | 0.684 | 1.17 | $2.81 \times 10^{11}$ |
| 176.7 | 4.371 | 890.4 | 1.57 | 0.677 | 1.02 | |
| 204.4 | 4.467 | 859.4 | 1.36 | 0.665 | 1.00 | $5.02 \times 10^{11}$ |
| 232.2 | 4.585 | 825.7 | 1.20 | 0.646 | 0.85 | |
| 260.0 | 4.731 | 785.2 | 1.07 | 0.616 | 0.83 | $8.59 \times 10^{11}$ |
| 287.7 | 5.024 | 735.5 | $9.51 \times 10^{-5}$ | | | |
| 315.6 | 5.703 | 678.7 | 8.68 | | | |

**TABLE C.7 Properties of Saturated Liquids**

| $T$ (°C) | $\rho$ (kg/m³) | $\bar{c}_p$ (kJ/kg·°C) | $\nu$ (m²/s) | $k$ (W/m·°C) | $\alpha$ (m²/s) | $Pr$ | $\beta$ (°K⁻¹) |
|---|---|---|---|---|---|---|---|
| | | | Ethylene Glycol, $C_2H_4(OH)_2$ | | | | |
| 0 | 1,130.75 | 2.294 | $57.53 \times 10^{-6}$ | 0.242 | $0.934 \times 10^{-7}$ | 615 | |
| 20 | 1,116.65 | 2.382 | 19.18 | 0.249 | 0.939 | 204 | $0.65 \times 10^{-3}$ |
| 40 | 1,101.43 | 2.474 | 8.69 | 0.256 | 0.939 | 93 | |
| 60 | 1,087.66 | 2.562 | 4.75 | 0.260 | 0.932 | 51 | |
| 80 | 1,077.56 | 2.650 | 2.98 | 0.261 | 0.921 | 32.4 | |
| 100 | 1,058.50 | 2.742 | 2.03 | 0.263 | 0.908 | 22.4 | |
| | | | Engine Oil (Unused) | | | | |
| 0 | 899.12 | 1.796 | 0.00428 | 0.147 | $0.911 \times 10^{-7}$ | 47,100 | |
| 20 | 888.23 | 1.880 | 0.00090 | 0.145 | 0.872 | 10,400 | $0.70 \times 10^{-3}$ |
| 40 | 876.05 | 1.964 | 0.00024 | 0.144 | 0.834 | 2,870 | |
| 60 | 864.04 | 2.047 | $0.839 \times 10^{-4}$ | 0.140 | 0.800 | 1,050 | |
| 80 | 852.02 | 2.131 | 0.375 | 0.138 | 0.769 | 490 | |
| 100 | 840.01 | 2.219 | 0.203 | 0.137 | 0.738 | 276 | |
| 120 | 828.96 | 2.307 | 0.124 | 0.135 | 0.710 | 175 | |
| 140 | 816.94 | 2.395 | 0.080 | 0.133 | 0.686 | 116 | |
| 160 | 805.89 | 2.483 | 0.056 | 0.132 | 0.663 | 84 | |

**TABLE C.8 Properties of air at atmospheric pressure (Holman, 1980)**

| $T$ (°K) | $\rho$ (kg/m³) | $\bar{C}_p$ (kJ/kg °C) | $\mu$ (kg/m/s × 10⁵) | $\nu$ (m²/s × 10⁶) | $k$ (W/m °C) | $\alpha$ (m²/s × 10⁴) | $Pr$ |
|---|---|---|---|---|---|---|---|
| 100 | 3.6010 | 1.0266 | 0.6924 | 1.923 | 0.009246 | 0.02501 | 0.770 |
| 150 | 2.3675 | 1.0099 | 1.0283 | 4.343 | 0.013735 | 0.05745 | 0.753 |
| 200 | 1.7684 | 1.0061 | 1.3289 | 7.490 | 0.018090 | 0.10165 | 0.739 |
| 250 | 1.4128 | 1.0053 | 1.4880 | 9.49 | 0.022270 | 0.13161 | 0.722 |
| 300 | 1.1774 | 1.0057 | 1.9830 | 16.840 | 0.026240 | 0.22160 | 0.708 |
| 350 | 0.9980 | 1.0090 | 2.0750 | 20.760 | 0.030030 | 0.29830 | 0.697 |
| 400 | 0.8826 | 1.0140 | 2.2860 | 25.900 | 0.033650 | 0.37600 | 0.689 |
| 450 | 0.7833 | 1.0207 | 2.4840 | 31.710 | 0.037070 | 0.42220 | 0.683 |
| 500 | 0.7048 | 1.0295 | 2.6710 | 37.900 | 0.040380 | 0.55640 | 0.680 |
| 550 | 0.6423 | 1.0392 | 2.8480 | 44.340 | 0.043600 | 0.65320 | 0.680 |
| 600 | 0.5879 | 1.0551 | 3.0180 | 51.340 | 0.046590 | 0.75120 | 0.680 |
| 650 | 0.5430 | 1.0635 | 3.1770 | 58.510 | 0.049530 | 0.85780 | 0.682 |
| 700 | 0.5030 | 1.0752 | 3.3320 | 66.250 | 0.052300 | 0.96720 | 0.684 |
| 750 | 0.4709 | 1.0856 | 3.4810 | 73.910 | 0.055090 | 1.07740 | 0.686 |
| 800 | 0.4405 | 1.0978 | 3.6250 | 82.290 | 0.057790 | 1.19510 | 0.689 |
| 850 | 0.4149 | 1.1095 | 3.7650 | 90.750 | 0.060280 | 1.30970 | 0.692 |
| 900 | 0.3925 | 1.1212 | 3.8990 | 99.300 | 0.062790 | 1.42710 | 0.696 |
| 950 | 0.3716 | 1.1321 | 4.0230 | 108.200 | 0.065250 | 1.55100 | 0.699 |
| 1000 | 0.3524 | 1.1417 | 4.1520 | 117.800 | 0.067520 | 1.67790 | 0.702 |
| 1100 | 0.3204 | 1.1600 | 4.4400 | 138.600 | 0.07320 | 1.9690 | 0.704 |
| 1200 | 0.2947 | 1.1790 | 4.6900 | 159.100 | 0.07820 | 2.2510 | 0.707 |
| 1300 | 0.2707 | 1.1970 | 4.9300 | 182.100 | 0.08370 | 2.5830 | 0.705 |
| 1400 | 0.2515 | 1.2140 | 5.1700 | 205.500 | 0.08910 | 2.9200 | 0.705 |
| 1500 | 0.2355 | 1.2300 | 5.4000 | 229.100 | 0.09460 | 3.2620 | 0.705 |
| 1600 | 0.2211 | 1.2480 | 5.6300 | 254.500 | 0.10000 | 3.6090 | 0.705 |
| 1700 | 0.2082 | 1.2670 | 5.8500 | 280.500 | 0.10500 | 3.9770 | 0.705 |
| 1800 | 0.1970 | 1.2870 | 6.0700 | 308.100 | 0.11100 | 4.3790 | 0.704 |
| 1900 | 0.1858 | 1.3090 | 6.2900 | 338.500 | 0.11700 | 4.8110 | 0.704 |
| 2000 | 0.1762 | 1.3380 | 6.5000 | 369.000 | 0.12400 | 5.2600 | 0.702 |
| 2100 | 0.1682 | 1.3720 | 6.7200 | 399.600 | 0.13100 | 5.7150 | 0.700 |
| 2200 | 0.1602 | 1.4190 | 6.9300 | 432.600 | 0.13900 | 6.1200 | 0.707 |
| 2300 | 0.1538 | 1.4820 | 7.1400 | 464.000 | 0.14900 | 6.5400 | 0.710 |
| 2400 | 0.1458 | 1.5740 | 7.3500 | 504.000 | 0.16100 | 7.0200 | 0.718 |
| 2500 | 0.1394 | 1.6880 | 7.5700 | 543.500 | 0.17500 | 7.4410 | 0.730 |

(Data from J. P. Holman, *Heat Transfer*, McGraw-Hill, New York, 1981.)

**TABLE C.9 Experimental Heat Capacities of Polymers**

| Polymer | $\bar{C}_p^s$ (298°K) (J/kg °K) | $\bar{C}_p^m$ (J/kg °K) |
|---|---|---|
| Polyethylene | 1550/1760 | 2260 |
| Polypropylene | 1630/1760 | 2140 |
| Polybutene | 1550/1760 | 2140 |
| Polyvinylchloride (PVC) | 960/1090 | 1220 |
| Nylon 6 | 1470 | 2140/2470 |
| Nylon 6,6 | 1470 | |
| Polystyrene | 1220 | 1720 |
| PET | 1130 | 1550 |

(Data from D. W. Van Krevelen, *Properties of Polymers*, Elsevier, Amsterdam, 1992.)

In the event experimental data are not available for $\bar{C}_p(T)$ the following expressions can be used for estimating the heat capacity in the solid state, $\bar{C}_p^s$, and in the melt state, $\bar{C}_p^m$ (Van Krevelen, 1990):

$$\bar{C}_p^s(T) = \bar{C}_p^s(298°K)(0.106 + 3 \times 10^{-3}T) \tag{C.1}$$

$$\bar{C}_p^m(T) = \bar{C}_p^m(298°K)(0.64 + 1.2 \times 10^{-3}T), \tag{C.2}$$

where $T$ is in °K and $\bar{C}_p^s$ (298°K) and $\bar{C}_p^m$ (298°K) are given in Table C.9 for a number of common polymers. Equation C.1 is valid for $T_g \leqslant T \leqslant T_m$ and Eq. C.2 for $T > T_m$ for semi-crystalline polymers. For amorphous polymers, Eq. C.2 is used for $T > T_g$.

# D

# IMSL DOCUMENTATION

This appendix contains documentation describing how to call the various IMSL (Visual Numerics Inc., Houston, TX) subroutines used throughout the book and in the problem sets. These subroutines represent just a few of the many available through Visual Numerics, Inc. Additional information about other products and services are available by calling (713) 782-6060.

The recommended procedure for using the IMSL subroutines is to copy the included disk onto the hard drive of your PC (locate them in either a subdirectory or the main directory). At least the 4.0 version of Microsoft Fortran is needed to compile, link, and execute the source program that calls the given subroutine. When using the 5.1 version of Microsoft Fortran the "Gb" option must be used (see p. 370 in the Fortran Reference Manual). The calling program should be in the disk drive, although it is possible to have it on the hard drive (C drive). With the calling program in the A drive, one should type the following command to compile, link, and create an executable file:

A:>fl /Gb filename.for /link IMSL1.lib+IMSL2.lib+mathcore.lib

where IMSL1 and IMSL2 are the IMSL libraries containing the required subroutines and mathcore is a library which must always be used. If there are no Fortran errors or linking errors, then after the cursor reappears one types the filename:

A:>filename

The program will execute, and if the program executes properly the following statement will appear:

stop-program terminated

A:>

The subroutines are stored in the following libraries:

| Subroutine | Library |
|------------|---------|
| 1. UMACH | MATHCORE |
| 2. WRRRN | SMATH10 |
| 3. RNLIN | SSTAT2B |
| 4. NEQNF | SMATH7 |
| 5. QDAGS/DQDAGS | SMATH4A/DMATH4A |
| 6. TWODQ/DTWODQ | SMATH4A/DMATH4A |
| 7. IVPAG/DIVPAG | SMATH5A/DMATH5A |
| 8. MOLCH | SMATH5B |
| 9. BVPFD | SMATH5A |
| 10. FPS2H | SMATH5B |
| 11. SSET | BLAS.LIB |

The subroutine WRRRN prints out arrays without the user having to write format statements. UMACH sets or retrieves the input or output device unit numbers. It is not necessary to use either of these subroutines.

As an example of the command for compiling, linking, and creating an executable file for the Fortran code found in Table 10.1, one would

type the following:

A:>fl /Gb dprix.for /link dmath5a.lib+mathcore.lib

Once the program is error-free one would type the following to execute the program:

A:>dprix

Finally, many universities and companies have the IMSL subroutines on the mainframes and workstations. With the addition of the appropriate job control lines (JCL) the programs given in the text that call the IMSL subroutines can be run directly on the mainframe. One of the reasons for selecting the IMSL subroutines is their general availability.

## Appendix D.1.   Subroutine UMACH

SUBROUTINE UMACH(N, NUNIT) (MATHCORE.LIB)

Routine UMACH sets or retrieves the input or output device unit numbers. UMACH is set automatically so that the default Fortran unit numbers for standard input and output are used. These unit numbers can be changed by inserting a call to UMACH at the beginning of the main program that calls IMSL MATH/LIBRARY routines. If the input or output numbers are changed from the standard values, the user should insert an appropriate OPEN statement in the calling program.

The calling sequence for UMACH is

CALL UMACH (N, NUNIT),

where NUNIT is the input or output unit number that is either retrieved or set, depending on which value of N is selected.

The arguments are summarized by the following:

| N | Effect |
|---|--------|
| 1 | Retrieves input unit number in NUNIT. |
| 2 | Retrieves output unit number in NUNIT. |
| -1 | Sets the input unit number to NUNIT. |
| -2 | Sets the output unit number to NUNIT. |

If the value of $N$ is negative, the input or output unit number is reset to NUNIT. If the value of $N$ is positive, the input or output unit number is returned in NUNIT.

In the following example, an input value of IYEAR=0 causes IMSL routine IDYWK to issue a terminal error. With a call to UMACH, the error message will be written to a file named CHECKERR.

```
        INTEGER     IDAY, IMONTH, IYEAR
        EXTERNAL    IDYWK, UMACH6
C
```

```
      IDAY = 7
      IMONTH = 12
      IYEAR = 0
C
      CALL UMACH (-2, 9)
      OPEN (UNIT = 9,FILE = 'CHECKERR')
C
      CALL IDYWK (IDAY, IMONTH, IYEAR)
C
      END
```

The output from this example, written to CHECKERR is:

*** TERMINAL ERROR from IDYWK. IYEAR = 0. IYEAR must not be equal to 0
***      since the year 0 does not exist.

### REFERENCE

Stevenson, D. 1982, *A Proposal Standard for Binary Floating-Point Arithmetic, Draft 10.0.* IEEE Floating-point subcommittee working document P754/82-8.6.

### Appendix D.2.   Subroutine WRRRN

WRRRN/DWRRRN (Single/double precision)
(SMATH10/DMATH10.LIB)

Purpose
Print a real rectangular matrix with integer row and column labels.

Usage
CALL WRRRN (TITLE, NRA, NCA, A, LDA, ITRING)^

**Arguments**

| | |
|---|---|
| TITLE | Character string specifying the title. (Input) TITLE = ' ' suppresses printing of the title. Use '%/' within the title to create a new line. Long titles are automatically wrapped. |
| NRA | Number of rows (input). |
| NCA | Number of columns (input). |
| A | NRA by NCA matrix to be printed (input). |
| LDA | Leading dimension of A exactly as specified in the dimension statement in the calling program (input). |
| ITRING | Triangle option (input). |

| ITRING | Action |
|---|---|
| 0 | Full matrix is printed. |
| 1 | Upper triangle of A is printed. |
| 2 | Upper triangle of A excluding the diagonal of A is printed. |
| −1 | Lower triangle of A is printed. |
| −2 | Lower triangle of A excluding the diagonal of A is printed. |

**Remarks**

1. A single format is chosen automatically in order to print pretty output. A field width of 10 is used. If all of the elements of A have no fractional part and are less than $10**9$ in absolute value, an I10 format is used. Otherwise, a D, E, or F format is used to print four significant digits for the largest element of A in absolute value. IMSL routine WROPT can be used to change the default format.

2. Horizontal centering, method for printing large matrices, paging, method for printing NaN (not a number), and printing a title on each page can be selected by invoking IMSL routine WROPT.
3. A page width of 78 characters is used. Page width and page length can be reset by invoking IMSL routine PAGE.
4. Output is written to the unit specified by IMSL routine UMACH.

**Algorithm**

Subroutine WRRRN prints a real rectangular matrix with the rows and columns labeled 1,2,3.... The printing parameters described in Remarks 1, 2, 3, and 4 can all be changed via IMSL routines PAGE (page 1063), WROPT (page 1059), and UMACH (section "Machine-Dependent Constants" in Reference Material). More information about these topics can be obtained by consulting the manual document for each of these routines.

In addition, subroutine WRRRN can print only the elements of the upper or lower triangles of matrices via the ITRING option. Generally, the ITRING option is used with symmetric matrices.

**Example**

The following example prints all of a $3 \times 4$ matrix A, where $a_{ij} = i + j/10$.

```
      INTEGER       (ITRING, LDA, NCA, NRA
      PARAMETER     (ITRING = 0, LDA = 10, NCA = 4, NRA = 3)
C
      INTEGER       I, J
      REAL          A(LDA,NCA)
      EXTERNAL      WRRRN
C
      DO 20 I = 1, NRA
         DO 10 J = 1, NCA
            A(I,J) = I + J*0.1
 10      CONTINUE
 20   CONTINUE
C                   Write A matrix.
      CALL WRRRN ('A', NRA, NCA, A, LDA, ITRING)
      END
```

Output

|   | A |   |   |   |
|---|---|---|---|---|
|   | 1 | 2 | 3 | 4 |
| 1 | 1.1 | 1.2 | 1.3 | 1.4 |
| 2 | 2.1 | 2.2 | 2.3 | 2.4 |
| 3 | 3.1 | 3.2 | 3.3 | 3.4 |

### Appendix D.3.   Subroutine RNLIN

RNLIN/DRNLIN (Single/double precision)
(SSTAT2B/DSTAT2B.LIB)

Purpose
Fit a nonlinear regression model.

Usage
CALL RNLIN(FUNC, NPARM, IDERIV, THETA, R, LDR, IRANK, DFE, SSE)

**Arguments**

| | |
|---|---|
| FUNC | User-supplied SUBROUTINE to return the weight, frequency, residual, and optionally the derivative of the |

residual at the given parameter vector THETA for a single observation. The usage is

CALL FUNC (NPARM, THETA, IOPT, IOBS, FRQ, WT, E, DE, IEND)

where

| | |
|---|---|
| NPARM | Number of unknown parameters in the regression function (input). |
| THETA | Vector of length NPARM containing parameter values (input). |
| IOPT | Function/derivative evaluation option (input). |

| IOPT | Meaning |
|---|---|
| 0 | Evaluate the function. |
| 1 | Evaluate the derivative. |

| | |
|---|---|
| IOBS | Observation number (input) |
| | The function is evaluated at the IOBS-th observation. |
| FRQ | Frequency for the observation (output). |
| WT | Weight for the observation (output). |
| | Use WT = 1.0 for equal weighting (unweighted least squares). |
| E | Error (residual) for the IOBS-th observation (output, if IOPT = 0). |
| DE | Vector of length NPARM containing the partial derivatives of the residual for the IOBS-th observation (output, if IOPT = 1). |
| | IF IDERIV = 0, DE is not referenced and can be a vector of length 1. |
| IEND | Completion indicator (output). |

| IEND | Meaning |
|---|---|
| 0 | IOBS is less than or equal to the number of observations. |
| 1 | IOBS is greater than the number of observations. WT, FRQ, E, and DE are not output. |
| | FUNC must be declared EXTERNAL in the calling program. |

| | |
|---|---|
| NPARM | Number of unknown parameters in the regression function (input). |
| IDERIV | Derivative option (input). |

| IDERIV | Meaning |
|---|---|
| 0 | Derivatives are obtained by finite differences. |
| 1 | Derivatives are supplied by FUNC. |

| | |
|---|---|
| THETA | Vector of length NPARM containing parameter values (input/output). |
| | On input, THETA must contain the initial estimate. |
| | On output, THETA contains the final estimate. |
| R | NPARM by NPARM upper triangular matrix containing the R matrix from a QR decomposition of the Jacobian (output). |
| LDR | Leading dimension of R exactly as specified in the dimension statement in the calling program (input). |
| IRANK | Rank of R (output). |
| | IRANK less than NPARM may indicate the model is overparameterized. |

| | |
|---|---|
| DFE | Degrees of freedom for error (output). |
| SSE | Sums of squares for error (output). |

**Remarks**

1. Automatic workspace usage is

| RNLIN | 13*NPARM + 17 units, or |
|---|---|
| DRNLIN | 25*NPARM + 28 units. |

Workspace may be explicitly provided, if desired, by use of R2LIN/DR2LIN. The reference is

CALL R2LIN (FUNC, NPARM, IDERIV, THETA, R, LDR, IRANK, DFE, SSE, IPARAM, RPARAM, SCALE, IWK, WK)

The additional arguments are as follows:

| | |
|---|---|
| IPARAM | Vector of length 6 containing convergence parameters (input/output). |
| | On input, set IPARAM(1) = 0 for default convergence parameter settings. If IPARAM(1) = 0, the remaining elements of IPARAM, and the arguments RPARAM and SCALE need not be initialized. |

| I | (IPARAM(I) |
|---|---|
| 1 | (Initialization flag (input) |
| | 0 means use default settings for IPARAM, RPARAM, and SCALE. |
| | 1 means use the input IPARAM and RPARAM settings. |
| 2 | Number of good digits in the residuals (input, if IPARAM(1) = 1). |
| 3 | Number of iterations (input, if IPARAM(1) = 1; output). |
| | On input, this is the maximum number of iterations allowed. The default is 100. On output, it is the actual number of iterations taken. |
| 4 | Number of SSE evaluations (input/output, if IPARAM(1) = 1; output if IPARAM(1) = 0). On input, this is the maximum number of evaluations allowed. The default is 400. On output, it is the actual number of evaluations taken. |
| 5 | Number of Jacobian evaluations (input, if IPARAM(1) = 1 and IDERIV = 1; output, if IDERIV = 1). On input, this is the maximum number of Jacobian evaluations allowed. The default is 100. On output, it is the number of iterations taken. |
| 6 | Scaling option (input, IF IPARAM(1) = 1). If IPARAM(6) = 1 the values for SCALE are set internally. The default is 1. Otherwise, SCALE must be input. |

| | |
|---|---|
| RPARAM | Vector of length 7 containing convergence parameters (input, if IPARAM(1) = 1). |
| | In the following table, the default settings are given in parentheses. For single precision EPS = AMACH(4), and for double precision EPS = DMACH(4). (See the documentation for IMSL routines AMACH and DMACH.) |

| I | (RPARAM(I). |
|---|---|
| 1 | (Scaled gradient tolerance (SQRT(EPS) for single precision; EPS**(1.0/3.0) for double precision). |
| 2 | (Scaled step tolerance (EPS**(2.0/3.0)). |
| 3 | (Relative function tolerance MAX(1.0E-10,EPS**(2.0/3.0)) for single precison; MAX(1.0E-20,EPS**(2.0/3.0)) for double precision). |

4  Absolute function tolerance
MAX(1.0E-20,EPS**2.0) for single precision;
MAX(1.0D-40,EPS**2.0) for double precision.

5  False convergence tolerance (100.0*EPS).

6  Maximum allowable step size
(1000*MAX(TOL1,TOL2) where
TOL1 = SNRM2(NPARM,SCXTH,1)
TOL2 = SNRM2(NPARM,SCALE,1)
and SCXTH is the elementwise product of
SCALE and THETA, i.e.,
SCXTH(I) = SCALE(I)*THETA(I)).

7  (Size of initial trust region radius (based on the
initial scaled Cauchy step).

SCALE  Vector of length NPARM (input, if IPARAM(1) = 1
and IPARAM(6) = 0).
A common choice is to set all elements of SCALE to 1.0.

IWK  Work vector of NPARM.

WK  Work vector of length 11*NPARM + 4.

2.  Informational errors

| Type | Code | |
| --- | --- | --- |
| 3 | 1 | Both the scaled actual and predicted reductions in the function are less than or equal to the relative function convergence tolerance. |
| 3 | 2 | The iterates appear to be converging to a noncritical point. Incorrect gradient information, a discontinuous function, or stopping tolerances being too tight may be the cause. |
| 4 | 3 | Maximum number of iterations is exceeded. |
| 4 | 4 | Maximum number of function evaluations is exceeded. |
| 3 | 6 | Five consecutive steps of the same size have been taken. Either the function is unbounded below, has a finite asymptote in some direction, or the stepsize is too small. |

3.  The first stopping criterion for RNLIN occurs when SSE is less
than the absolute function tolerance. The second stopping criterion
occurs when the norm of the scaled gradient is less than the given
gradient tolerance. The third stopping criterion occurs when the
scaled distance between the last two steps is less than the step
tolerance.

4.  To use some nondefault convergence parameters, first call R8LIN,
then reset the corresponding convergence parameters to the desired
value and call R2LIN. For example, the following code could be
used if nondefault convergence parameters are to be used:

```
C
      CALL R8LIN (IPARAM, RPARAM)
C   R8LIN outputs IPARAM(1) = 1 to indicate some
C   nondefault convergence parameters are to be set.
C   R8LIN outputs the remaining elements of IPARAM and
C   RPARAM as their default values.
C
C   Set some nondefault convergence parameters.
      IPARAM(3) = 20
      IPARAM(6) = 0
      SCALE(1) = 0.1
      SCALE(2) = 10.0
C
      CALL R2LIN (FUNC, NPARM, IDERIV, THETA, R,
                  LDR, IRANK, DFE, SSE, IPARAM,
                  RPARAM, SCALE, IWK, WK)
```

If double precision is being used, then DR8LIN and DR2LIN
are called, and RPARAM is declared double precision.

*Keywords*:  Levenberg-Marquardt; nonlinear least squares; approximation

**Algorithm**

Subroutine RNLIN fits a nonlinear regression model using least
squares. The nonlinear regression model is:

$$y_i = f(x_i, \theta) + \varepsilon_i \qquad i = 1, 2,,,,, n,$$

where the observed values of the $y_i$'s constitute the responses or values
of the dependent variable, the known $x_i$'s are the vectors of the values
of the independent (explanatory) variables, $\theta$ is the vector of $p$ regression
parameters, and the $\varepsilon_i$'s are independently distributed normal errors
with mean zero and variance $\sigma^2$. For this model, a least-squares estimate
of $\theta$ is also a maximum likelihood estimate of $\theta$.

The residuals for the model are

$$e_i(\theta) = y_i - f(x_i, \theta) \qquad i = 1, 2, \ldots, n.$$

A value of $\theta$ that minimizes $\sum_{i=1}^{n}[e_i(\theta)]^2$ is a least squares estimate of
$\theta$. RNLIN is designed so that these residuals are input one at a time
from a user-supplied subroutine. This permits RNLIN to handle the
case when $n$ is large and the data cannot reside in an array but must
reside on some secondary storage device.

RNLIN is based on MINPACK routines LMDIF and LMDER
by Moré et al. (1980). RNLIN uses a modified Levenberg-Marquardt
method to generate a sequence of approximations to a minimum point.
Let $\hat{\theta}$ be the current estimate of $\theta$. A new estimate is given by $\hat{\theta}_c + s_c$,
where $s_c$ is a solution to

$$(J(\hat{\theta}_c)^T J(\hat{\theta}_c) + \mu_c I)s_c = J(\hat{\theta}_c)^T e(\hat{\theta}_c).$$

Here $J(\hat{\theta}_c)$ is the Jacobian evaluated at $\hat{\theta}_c$. The algorithm uses a "trust
region" approach with a step bound of $\delta_c$. A solution of the equations
is first obtained for $\mu_c = 0$. If $\|s_c\|_2 < \delta_c$, this update is accepted.
Otherwise, $\mu_c$ is set to a positive value, and another solution is obtained.
The method is discussed by Levenberg (1944), Marquardt (1963), and
Dennis and Schnabel (1983, Chapter 10).

If IDERIV = 0, forward finite differences are used to estimate
the Jacobian numerically. If IDERIV = 1, the Jacobian is computed
analytically via the user-supplied subroutine. With IDERIV = 0 and
single precision arithmetic, the estimate of the Jacobian may be so poor
that the algorithm terminates at a noncritical point. In such instances,
IDERIV = 1 or double precision arithmetic are recommended.

RNLIN does not actually store the Jacobian but uses fast Givens
transformations to construct an orthogonal reduction of the Jacobian
to upper triangular form (stored in R). The reduction is based on fast
Givens transformations (see IMSL routines SROTMG and SROTM
in Chapter 20, "Mathematical Support," Golub & Van Loan, 1983,
pp. 156–62; Gentleman, 1974). This method has two main advantages:
(1) the loss of accuracy resulting from forming the cross-product matrix
used in the equations for $s_c$ is avoided, and (2) the $n \times p$ Jacobian need
not be stored, saving space when $n > p$.

A weighted least-squares fit can also be performed. This is
appropriate when the variance of $\varepsilon_i$ in the nonlinear regression model
is not constant but instead is $\sigma^2/w_i$. Here the $w_i$'s are weights input via
the user-supplied subroutine. For the weighted case, RNLIN computes
a minimum weighted sum of squares for error (stored in SSE).

## Example 1

This example uses data discussed by Neter, et al. (1983, pp. 475–78). A nonlinear model

$$y_i = \theta_1 e^{\theta_2 x_i} + \varepsilon_i \qquad i = 1, 2, \ldots, 15$$

is fitted. The option IDERIV = 0 is used.

```
      INTEGER      LDR, NOBS, NPARM
      PARAMETER    (NOBS = 15, NPARM = 2, LDR = NPARM)
C
      INTEGER      IDERIV, IRANK, NOUT
      REAL         DFE, R(LDR,NPARM), SSE,
                   THETA(NPARM)
      EXTERNAL     EXAMPL, RNLIN, UMACH, WRRRN
C
      DATA THETA/60.0, -0.03/
C
      CALL UMACH (2, NOUT)
C
      IDERIV = 0
      CALL RNLIN (EXAMPL, NPARM, IDERIV, THETA, R,
     &            LDR, IRANK, DFE, SSE)
      WRITE (NOUT,*) 'THETA = ', THETA
      WRITE (NOUT,*) 'IRANK = ', IRANK, ' DFE = ', DFE, '
     &            SSE = ', SSE
      CALL WRRRN ('R', NPARM, NPARM, R, LDR, O)
      END
C
      SUBROUTINE EXAMPL (NPARM, THETA, IOPT, IOBS,
     &            FRQ, WT, E, DE, IEND)
      INTEGER      NPARM, IOPT, IOBS, IEND
      REAL         THETA(NPARM), FRQ, WT, E, DE(1)
C
      INTEGER      NOBS
      PARAMETER    (NOBS = 15)
C
      REAL         EXP, XDATA(NOBS), YDATA(NOBS)
      INTRINSIC    EXP
C
      DATA YDATA/54.0, 50.0, 45.0, 37.0, 35.0, 25.0, 20.0, 16.0,
     &     18.0, 13.0, 8.0, 11.0, 8.0, 4.0, 6.0/
      DATA XDATA/2.0, 5.0, 7.0, 10.0, 14.0, 19.0, 26.0, 31.0, 34.0,
     &     38.0, 45.0, 52.0, 53.0, 60.0, 65.0/
C
      IF (IOBS .LE. NOBS) THEN
         WT   = 1.0E0
         FRQ  = 1.0E0
         IEND = 0
         E    = YDATA(IOBS) - THETA(1)*EXP(THETA(2)*
                XDATA(IOBS))
      ELSE
         IEND = 1
      END IF
      RETURN
      END
```

### Output

```
      THETA = 58.60771   -3.95888E-02
      IRANK = 2   DFE = 13.00000   SSE = 49.45930
```

```
      R
      1       2
   1  2.      1136.
   2  0.      1131.
```

## Example 2

This example fits the model in Ex. 1 with the option IDERIV = 1.

```
      INTEGER      LDR, NOBS, NPARM
      PARAMETER    (NOBS = 15, NPARM = 2, LDR-NPARM)
C
      INTEGER      IDERIV, IRANK, NOUT
      REAL         DFE, R(LDR,NPARM), SSE,
                   THETA(NPARM)
      EXTERNAL     EXAMPL, RNLIN, UMACH, WRRRN
C
      DATA THETA/60.0, -0.03/
C
      CALL UMACH (2, NOUT)
C
      IDERIV = 1
      CALL RNLIN (EXAMPL, NPARM, IDERIV, THETA, R,
     &            LDR, IRANK, DFE, SSE)
      WRITE (NOUT,*) 'THETA = ', THETA
      WRITE (NOUT,*) 'IRANK = ', IRAN, ' DFE - ', DFE, '
     &            (SSE = ',SSE
      CALL WRRRN ('R', NPARM, NPARM, R, LDR, O)
      END
C
      SUBROUTINE EXAMPL (NPARM, THETA, IOPT, IOBS,
     &            FRQ, WT, E, DE, IEND)
      INTEGER      NPARM, IOPT, IOBS, IEND
      REAL         THETA(NPARM), FRQ, WT, E,
                   DE(NPARM)
C
      INTEGER      NOBS
      PARAMETER    (NOBS = 15)
C
      REAL         EXP, XDATA(NOBS), YDATA(NOBS)
      INTRINSIC    EXP
C
      DATA YDATA/54.0, 50.0, 45.0, 37.0, 35.0, 25.0, 20.0, 16.0,
     &     18.0, 13.0, 8.0, 11.0, 8.0, 4.0, 6.0/
      DATA XDATA/2.0, 5.0, 7.0, 10.0, 14.0, 19.0, 26.0, 31.0, 34.0,
     &     38.0, 45.0, 52.0, 53.0, 60.0, 65.0/
C
      IF (IOBS .LE. NOBS) THEN
         WT   = 1.0E0
         FRQ  = 1.0E0
         IEND = 0
         IF (IOPT .EQ. 0) THEN
            E = YDATA(IOBS) -
                THETA(1)*EXP(THETA(2)*XDATA (IOBS))
         ELSE
            DE(1) = -EXP(THETA(2)*XDATA(IOBS)
            DE(2) = -THETA(1)*XDATA(IOBS)*EXP(THETA(2)*
                    XDATA(IOBS)
         END IF
      ELSE
         IEND = 1
      END IF
      RETURN
      END
```

**Output**

THETA = 58.606468  -3.95863E-02
IRANK = 20  DFE = 13.00000  SSE = 49.45930

|   | R |   |
|---|---|---|
|   | 1 | 2 |
| 1 | 2. | 1140. |
| 2 | 0. | 1140. |

**References**

Dennis, J. E., Jr., and R. B. Schnabel. 1983. *Numerical Methods for Unconstrained Optimization and Nonlinear Equations*, Prentice-Hall, Englewood Cliffs, NJ.

Gentleman, W. M. (1974). "Basic Procedures for Large, Sparse or Weighted Linear Least Squares Problems." *Appl. Stat.*, **23**, 448–54.

Golub, G. H., and C. F. Van Loan. 1983. *Matrix Computations* (Johns Hopkins University Press, Baltimore, MD.

Levenberg, K. 1944. "A Method for the Solution of Certain Problems in Least Squares." *Quarterly of Applied Mathematics*, **2**, 164–8.

Marquardt, D. 1963. "An Algorithm for Least-Squares Estimation of Nonlinear Parameters." *SIAM J. Appl. Math.*, **11**, 431–41.

Moré, J. J., B. S. Garbow, and K. E. Hillstrom. 1980. *User Guide for MINPACK-1* (Argonne National Labs Report ANL-80-74, Argonne, IL).

Neter, J., W. Wasserman, and M. H. Kutner (1983), *Applied Linear Regression Models* (Irwin, Homewood, IL).

## Appendix D.4.  Subroutine NEQNF

NEQNF/DNEQNF (Single/double precision)
(SMATH7/DMATH7.LIB)

Purpose
Solve a system of nonlinear equations using the Levenberg-Marquardt algorithm and a finite-difference approximation to the Jacobian.

Usage
CALL NEQNF (FCN, ERRREL, N, ITMAX, XGUESS, X, FNORM)

**Arguments**

FCN  User-supplied SUBROUTINE to evaluate the system of equations to be solved. The usage is
CALL FCN (X, F, N)
where
 X  The point at which the functions are evaluated (input).
  X should not be changed by FCN.
 F  The computed function values at the point X (output).
 N  Length of X and F (input).
  FCN must be declared EXTERNAL in the calling program.

ERRREL  Stopping criterion (input).
The root is accepted if the relative error between two successive approximations to this root is less than ERRREL.

N  The number of equations to be solved and the number of unknowns (input).

ITMAX  The maximum allowable number of iterations (input).
The maximum number of calls to FCN is ITMAX*(N + 1).
Suggested value = 200.

XGUESS  A vector of length N (input).
XGUESS contains the initial estimate of the root.

A  vector of length N (output).

X  contains the best estimate of the root found by NEQNF.

FNORM  A scalar that has the following value:
F(1)**2 + ... + F(N)**2 at the point X (output).

**Remarks**

1. Automatic workspace usage is

 NEQNF  1.5*N**2 + 7.5*N  units, or
 DNEQNF  3*N**2 + 15*N  units.

Workspace may be explicitly provided, if desired, by use of N2QNF/DN2QNF. The reference is
CALL N2QNF (FCN, ERRREL, N, ITMAX, XGUESS, X, FNORM, FVEC, FJAC, R, QTF, WK)
The additional arguments are as follows:

FVEC  A vector of length N. FVEC contains the functions evaluated at point X.

FJAC  An N by N matrix. FJAC contains the orthogonal matrix Q produced by the QR factorization of the final approximate Jacobian.

R  A vector of length N*(N + 1)/2. R contains the upper triangular matrix produced by the QR factorization of the final approximation Jacobian. R is stored row-wise.

QTF  A vector of length N. QTF contains the vector TRANS(Q)*FVEC.

WK  A work vector of length 5*N.

2. Informational errors

| Type | Code | |
|---|---|---|
| 4 | 1 | The number of calls to FCN has exceeded ITMAX*(N + 1). A new initial guess may be tried. |
| 3 | 2 | ERRREL is too small. No further improvement in the approximate solution is possible. |
| 4 | 3 | The iteration has not made good progress. A new initial guess may be tried. |

*Keywords*:  Powell hybrid method; Nonlinear equations; Roots; Zeros, MINPACK; System of equations.

**Algorithm**

NEQNF is based on the MINPACK subroutine HYBRD1, which uses a modification of M.J.D. Powell's hybrid algorithm. This algorithm is a variation of Newton's method, which uses a finite-difference approximation to the Jacobian and takes precautions to avoid large step sizes or increasing residuals. For further description, see Moré et al. (1980).

Since a finite-difference method is used to estimate the Jacobian, for single precision calculation the Jacobian may be so incorrect that the algorithm terminates far from a root. In such cases high-precision arithmetic is recommended. Also, whenever the exact Jacobian can be easily provided, IMSL routine NEQNJ should be used instead.

**Example**

The following $3 \times 3$ system of nonlinear equations

$$f_1(x) = x_1 + e^{x_1 - 1} + (x_2 + x_3)^2 - 27$$

$$f_2(x) = e^{x_2 - 2}/x_1 + x_3^2 - 10$$

$$f_3(x) = x_3 + \sin(x_2 - 2) + x_2^2 - 7$$

is solved with the initial guess (4.0, 4.0, 4.0).

```
C                           Declare variables
      INTEGER    ITMAX, N
      REAL       ERRREL
      PARAMETER  (N=3)
C
      INTEGER    K, NOUT
      REAL       FCN, FNORM, X(N), XGUESS(N)
      EXTERNAL   FCN, NEQNF, UMACH
C                           Set values of initial guess
C                           XGUESS = ( 4.0 4.0 4.0)
C
      DATA XGUESS/4.0, 4.0, 4.0/
C
      ERRREL = 0.0001
      ITMAX = 100
C
      CALL UMACH (2, NOUT)
C                           Find the solution
      CALL NEQNF (FCN, ERRREL, N, ITMAX, XGUESS, X,
     &            FNORM)
C                           Output
      WRITE (NOUT,99999)   (X(K),K-1,N), FNORM
99999 FORMAT (' The solution to the system is', /, ' X = (', 3F5.1,
     &        ')', /,' with FNORM =', F5.4, //)
C
      END
C                           User-defined subroutine
      SUBROUTINE FCN (X, F, N)
      INTEGER    N
      REAL       X(N), F(N)
C
      REAL       EXP, SIN
      INTRINSIC  EXP, SIN
C
      F(1) = X(1) + EXP(X(1)-1.0) + (X(2)+X(3))*(X(2)+X(3)) - 27.0
      F(2) = EXP(X(2)-2.0)/X(1) + X(3)*X(3) - 10.0
      F(3) = X(3) + SIN(X(2)-2.0) + X(2)*X(2) = 7.0
      RETURN
      END
```

**Output**

The solution to the system is
X = ( 1.0 2.0 3.0 )
with FNORM = 0.0000

**Reference**

Moré, J., B. Garvow, and K. Hillstrom. 1980. *User guide for MINPACK-1* (Argonne National Labs Report ANL-80-74, Argonne, IL).

## Appendix D.5  Subroutine QDAGS

QDAGS/DQDAGS (Single/double precision)
(SMATH4A/DMATH4A.LIB)

Purpose
Integrate a function (which may have endpoint singularities).

Usage
CALL QDAGS (F, A, B, ERRABS, ERRREL, RESULT, ERREST)

**Arguments**

| | |
|---|---|
| F | User-supplied FUNCTION to be integrated. The form is F(X), where |
| | X  Independent variable (input). |
| | F  The function value (output). |
| | F must be declared EXTERNAL in the calling program. |
| A | Lower limit of integration (input). |
| B | Upper limit of integration (input). |
| ERRABS | Absolute accuracy desired (input). |
| ERRREL | Relative accuracy desired (input). |
| RESULT | Estimate of the integral from A to B of F (output) |
| ERREST | Estimate of the absolute value of the error (output) |

**Remarks**

1. Automatic workspace usage is
   QDAGS   2500 units, or
   DQDAGS  4500 units.
   Workspace may be explicitly provided, if desired, by use of Q2AGS/DQ2AGS. The reference is
   CALL Q2AGS (F, A, B, ERRABS, ERRREL, RESULT, ERREST MAXSUB, NEVAL, NSUBIN, ALIST, BLIST, RLIST, ELIST, IORD)
   The additional arguments are as follows:

| | |
|---|---|
| MAXSUB | Number of subintervals allowed (input). A value of 500 is used by QDAGS. |
| NEVAL | Number of evaluations of F (output). |
| NSUBIN | Number of subintervals generated (output). |
| ALIST | Array of length MAXSUB containing a list of the NSUBIN left endpoints (output). |
| BLIST | Array of length MAXSUB containing a list of the NSUBIN right endpoints (output). |
| RLIST | Array of length MAXSUB containing approximations to the NSUBIN integrals over the intervals defined by ALIST, BLIST (output). |
| ELIST | Array of length MAXSUB containing the error estimates of the NSUBIN values in RLIST (output). |
| IORD | Array of length MAXSUB (output). Let K be |
| | NSUBIN  if NSUBIN .LE. (MAXSUB/2 + 2) |
| | MAXSUB + 1-NSUBIN otherwise. |
| | The first K locations contain pointers to the error estimates over the subintervals, such that ELIST(IORD(1)),..., ELIST(IORD(K)) form a decreasing sequence. |

2. Informational errors

| Type | Code | |
|---|---|---|
| 4 | 1 | The maximum number of subintervals allowed has been reached. |
| 3 | 2 | Roundoff error, preventing the requested tolerance from being achieved, has been detected. |
| 3 | 3 | A degradation in precision has been detected. |
| 3 | 4 | Roundoff error in the extrapolation table, preventing the requested tolerance from being achieved, has been detected. |
| 4 | 5 | Integral is probably divergent or slowly convergent. |

3. If EXACT is the exact value, QDAGS attempts to find RESULT such that ABS(EXACT-RESULT) .LE. MAX(ERRABS,ERRREL *ABS(EXACT)). To specify only a relative error, set ERRABS to zero. Similarly, to specify only an absolute error, set ERRREL to zero.

**Algorithm**

QDAGS is a general-purpose integrator that uses a globally adaptive scheme to reduce the absolute error. It subdivides the interval (A, B) and uses a 21-point Gauss-Kronrod rule to estimate the integral over each subinterval. The error for each subinterval is estimated by comparison with the 10-point Gauss quadrature rule. This subroutine is designed to handle functions with endpoint singularities. However, the performance on functions that are well-behaved at the endpoints is quite good also. In addition to the general strategy described in QDAG, this subroutine uses an extrapolation procedure known as the ε-algorithm. QDAGS is an implementation of the subroutine QAGS that is fully documented by Piessens et al. (1983). Should QDAGS fail to produce acceptable results, then either IMSL routines QDAG or QDAG* may be appropriate.

**Example**

The value of

$$\int_0^1 \ln(x)x^{-1/2}dx = -4$$

is estimated. The values of the actual and estimated error are machine dependent.

```
      INTEGER    NOUT
      REAL       A, ABS, B, ERRABS, ERREST, ERROR,
     &           ERRREL, EXACT, F, RESULT
      INTRINSIC  ABS
      EXTERNAL   F, QDAGS, UMACH
C                         Get output unit number
      CALL UMACH (2, NOUT)
C                         Set limits of integration
      A = 0.0
      B = 1.0
C                         Set error tolerances
      ERRABS = 0.0
      ERRREL = 0.001
      CALL QDAGS (F, A, B, ERRABS, ERRREL, RESULT,
     &            ERREST)
C                         Print results
      EXACT = -4.0
      ERROR = ABS(RESULT-EXACT)
      WRITE (NOUT,99999) RESULT, EXACT, ERREST,
     &            ERROR
99999 FORMAT ('Computed =', F8.3, 13X, ' Exact =', F8.3, /,/,
     &         Error estimate =', 1PE10.3, 6X, 'Error =', 1PE10.3)
      END
C
      REAL FUNCTION F (X)
      REAL       X
      REAL       ALOG, SQRT
      INTRINSIC  ALOG, SQRT
      F = ALOG(X)/SQRT(X)
      RETURN
      END
```

**Output**

Computed = -4.000;    Exact = -4.000
Error estimate = 3.110E-03;    Error = 3.300E-04

**Reference**

Piessens, R., E. deDoncker-Kapenga, C. W. Überhuber, and D. K. Kahaner. 1983. *QUADPACK* (Springer-Verlag, New York).

**Appendix D.6.  Subroutine TWODQ**

TWODQ/DTWODQ (Single/Double precision)
(SMATH4A/DMATH4A.LIB)

Purpose
Compute a two-dimensional iterated integral.

Usage
CALL TWODQ (F, A, B, G, H, ERRABS, ERRREL, IRULE, RESULT, ERREST)

**Arguments**
**Arguments**

F          User-supplied FUNCTION to be integrated. The form is F(X, Y), where
           X     First argument of F (input).
           Y     Second argument of F (input).
           F     The function value (output).
           F must be declared EXTERNAL in the calling program.
A          Lower limit of outer integral (input).
B          Upper limit of outer integral (input).
G          User-supplied FUNCTION to evaluate the lower limits of the inner integral. The form is G(X), where
           X     Only argument of G (input).
           G     The function value (output).
           G must be declared EXTERNAL in the calling program.
H          User-supplied FUNCTION to evaluate the upper limits of the inner integral. The form is H(X), where
           X     Only argument of H (input)
           H     The function value (output)
           H must be declared EXTERNAL in the calling program.
ERRABS     Absolute accuracy desired (input)
ERRREL     Relative accuracy desired (input)
IRULE      Choice of quadrature rule. (input)
           A Gauss-Kronrod rule is used with
               7–15 points if IRULE = 1
               10–21 points if IRULE = 2
               15–31 points if IRULE = 3
               20–41 points if IRULE = 4
               25–51 points if IRULE = 5
               30–61 points if IRULE = 6
           If the function has a peak singularity use IRULE = 1.
           If the function is oscillatory use IRULE = 6.
RESULT     Estimate of the integral from A to B of F (output).
ERREST     Estimate of the absolute value of the error (output).

### Remarks

1. Automatic workspace usage is
   TWODQ      2500      units, or
   DTWODQ    4500     units.
   Workspace may be explicitly provided, if desired, by use of
   T2ODQ/DT2ODQ. The reference is
   CALL T2ODQ (F, A, B, G, H, ERRABS, ERRREL, IRULE,
                  RESULT, ERREST, MAXSUB, NEVAL,
                  NSUBIN, ALIST, BLIST, RLIST, ELIST,
                  IORD, WK, IWK)
   The additional arguments are as follows:

| | |
|---|---|
| MAXSUB | Number of subintervals allowed (input). A value of 250 is used by TWODQ. |
| NEVAL | Number of evaluations of F (output). |
| NSUBIN | Number of subintervals generated in the outer integral (output). |
| ALIST | Array of length MAXSUB containing a list of the NSUBIN left endpoints for the outer integral (output). |
| BLIST | Array of length MAXSUB containing a list of the NSUBIN right endpoints for the outer integral (output). |
| RLIST | Array of length MAXSUB containing approximations to the NSUBIN integrals over the intervals defined by ALIST, BLIST, pertaining only to the outer integral (output). |
| ELIST | Array of length MAXSUB containing the error estimates of the NSUBIN values in RLIST (output). |
| IORD | Array of length MAXSUB (output). Let K be    NSUBIN.   if NSUBIN .LE. (MAXSUB/2 + 2)    MAXSUB + 1-NSUBIN otherwise. Then the first K locations contain pointers to the error estimates over the corresponding subintervals, such that ELIST(IORD(1)),..., ELIST(IORD(K)) form a decreasing sequence. |
| WK | Work array of length 4*MAXSUB, needed to evaluate the inner integral. |
| IWK | Work array of length MAXSUB, needed to evaluate the inner integral. |

2. Informational errors

| Type | Code | |
|---|---|---|
| 4 | 1 | The maximum number of subintervals allowed has been reached. |
| 3 | 2 | Roundoff error, preventing the requested tolerance from being achieved, has been detected. |
| 3 | 3 | A degradation in precision has been detected. |

3. If EXACT is the exact value, TWODQ attempts to find RESULT such that ABS(EXACT-RESULT) .LE. MAX(ERRABS,ERRREL *ABS(EXACT)). To specify only a relative error, set ERRABS to zero. Similarly, to specify only an absolute error, set ERRREL to zero.

### Algorithm

TWODQ approximates the two-dimensional iterated integral

$$\int_a^B \int_{g(x)}^{h(x)} f(x,y)\,dy\,dx,$$

with the approximation returned in RESULT. An estimate of the error is returned in ERREST. The approximation is achieved by iterated calls to QDAG. Thus, this algorithm will share many of the characteristics of the IMSL routine QDAG. As in QDAG, several options are available. The absolute and relative error must be specified, and in addition, the Gauss-Kronrod pair must be specified (IRULE). The lower-numbered rules are used for less smooth integrands while the higher-order rules are more efficient for smooth (oscillatory) integrands.

### Example 1

In this example we approximate the integral

$$\int_0^1 \int_1^3 y\cos(x+y^2)\,dy\,dx.$$

The value of the error estimate is machine dependent.

```
      INTEGER      IRULE, NOUT
      REAL         A, B, ERRABS, ERREST, ERRREL, F, G, H,
                   RESULT
      EXTERNAL     F, G, H, TWODQ, UMACH
C                           Get output unit number
      CALL UMACH (2, NOUT)
C                           Set limits of integration
      A = 0.0
      B = 1.0
C                           Set error tolerance
      ERRABS = 0.0
      ERRREL = 0.01
C                           Parameter for oscillatory function
      IRULE = 6
      CALL TWODQ (F, A, B, G, H, ERRABS, ERRREL, IRULE,
                  RESULT, ERREST)
C                           Print results
      WRITE (NOUT,99999) RESULT, ERREST
99999 FORMAT (' Result = ', F8.3, 13X, ' Error estimate = ', 1PE9.3)
      END
C
      REAL FUNCTION F (X, Y)
      REAL    X, Y
      REAL    COS
      INTRINSIC COS
      F = Y*COS(X + Y*Y)
      RETURN
      END
C
      REAL FUNCTION G (X)
      REAL    X
      G = 1.0
      RETURN
      END
C
      REAL FUNCTION H (X)
      REAL    X
      H = 3.0
      RETURN
      END
```

### Output

Result = −0.514; Error estimate = 2.453E−05.

**Example 2**

We modify the preceding example by assuming that the limits for the inner integral depend on $x$ and in particular are $g(x) = -2x$ and $h(x) = 5x$. The integral now becomes

$$\int_0^1 \int_{-2x}^{5x} y \cos(x + y^2)\, dy\, dx.$$

The value of the error estimate is machine dependent.

```
C                       Declare F, G, H
      INTEGER    IRULE, NOUT
      REAL       A, B, ERRABS, ERREST, ERRREL, F, G,
                 H, RESULT
      EXTERNAL   F, G, H, TWODQ, UMACH
C
      CALL UMACH (2, NOUT)
C                       Set limits of integration
      A = 0.0
      B = 1.0
C                       Set error tolerance
      ERRABS = 0.001
      ERRREL = 0.0
C                       Parameter for oscillatory function
      IRULE = 6
      CALL TWODQ (F, A, B, G, H, ERRABS, ERRREL, IRULE,
                 RESULT, ERREST)
C                       Print results
      WRITE (NOUT,99999) RESULT, ERREST
99999 FORMAT (' Computed =', F8.3, 13X, ' Error estimate = ',
             1PE9.3)
      END
      REAL FUNCTION F (X, Y)
      REAL    X, Y
C
      REAL    COS
      INTRINSIC COS
C
      F = Y*COS(X + Y*Y)
      RETURN
      END
      REAL FUNCTION G (X)
      REAL    X
C
      G = -2.0*X
      RETURN
      END
      REAL FUNCTION H (X)
      REAL    X
C
      H = 5.0*X
      RETURN
      END
```

**Output**

Computed $= -0.083$; Error estimate $= 1.677E - 05$.

**Reference**

Piessens, R., E. deDoncker-Kapenga, C. W. Überhuber, and D. K. Kahaner (1983), *QUADPACK*, Springer-Verlag, New York.

**Appendix D.7.    Subroutine IVPAG**

IVPAG/DIVPAG (Single/double precision)
(SMATH5A/DMATH5B.LIB)

Purpose
Solve an initial-value problem for ordinary differential equations using an Adams-Moulton or Gear method.

Usage
CALL IVPAG (IDO, NEQ, FCN, FCNJ, A, X, XEND, TOL, PARAM, Y)

**Arguments**

IDO    Flag indicating the state of the computation (input/output).
1    Initial entry.
2    Normal reentry.
3    Final call to release workspace.
4    Return because of interrupt 1.
5    Return because of interrupt 2 with step accepted.
6    Return because of interrupt 2 with step rejected.
7    Return for new value of A. The matrix A at X must be recomputed and IVPAG/DIVPAG called again. No other argument (including IDO) should be changed. This value of IDO is returned only if PARAM(19) = 2. Normally, the initial call is made with IDO = 1. The outine then sets IDO = 2, and this value is then used for all but the last call, which is made with IDO = 3. This final call is only used to release workspace, which was automatically allocated by the initial call with IDO = 1. See Remark 5 for a description of the interrupts.

NEQ    Number of differential equations (input).

FCN    User-supplied SUBROUTINE to evaluate functions. The usage is
CALL FCN (NEQ, X, Y, YPRIME)
where
NEQ        Number of equations (input).
X          Independent variable (input).
Y          Array of length NEQ containing the dependent variable values (input).
YPRIME     Array of length NEQ containing the values of the right-hand side of the system of of equations    (output). See Remark 3.
FCN must be declared EXTERNAL in the calling program.

FCNJ    User-supplied SUBROUTINE to compute the Jacobian. The usage is
CALL FCNJ (NEQ, X, Y, DYPDY)
where
NEQ        Number of equations (input).
X          Independent variable (input).
Y          Array of length NEQ containing the dependent variable values (input).
DYPDY      Matrix (whose type is determined by PARAM(14) = MTYPE), containing the derivatives dYPRIME(i)/dY(j) (output). YPRIME is the right-hand side of the ODE.
FCNJ must be declared EXTERNAL in the calling program. If PARAM(19) = IATYPE is nonzero, then FCNJ should compute the Jacobian of the right-hand side of the equation $Ay' = f(x, y)$. FCNJ is used only if PARAM(13) = MITER = 1.

| A | Matrix used when ODE system is implicit (input). |
|---|---|
| | A is referenced only if PARAM(19) = IATYPE is nonzero. Its data structure is determined by PARAM(14) = MTYPE. A must be nonsingular and MITER must be 1 or 2. See Remark 3. |
| X | Independent variable (input/output). |
| | On input, X supplies the initial value. |
| | On output, X is replaced by XEND unless error conditions arise. See IDO for details. |
| XEND | Value of X at which the solution is desired (input). |
| | XEND may be less than the initial value of X. |
| TOL | Tolerance for error control (input). |
| | An attempt is made to control the norm of the local error such that the global error is proportional to TOL. More than one run, with different values of TOL, can be used to estimate the global error. Generally, it should not be greater than 0.001. |
| PARAM | Vector of length 50 containing optional parameters (input/output). |
| | If a parameter is zero then the default value is used. The following parameters must be set by the user. |

| PARAM | | Meaning |
|---|---|---|
| 1 | HINIT | Initial value of the step size H. Always nonnegative. Default: ABS(.001*(XEND-X)). |
| 2 | HMIN | Minimum value of the step size H. Default: 0.0. |
| 3 | HMAX | Maximum value of the step size H. Default: No limit is imposed on the step size. |
| 4 | MXSTEP | Maximum number of steps allowed. Default: 500. |
| 5 | MXFCN | Maximum number of function evaluations allowed. Default: No limit. |
| 6 | MAXORD | Maximum order of the method. Default: If Adams' method, 12. If Gear's method, 5. The defaults are also the maximum values allowed. |
| 7 | INTRP1 | If nonzero then the subroutine will return, with IDO = 4, before every step. See Remark 5. Default: 0. |
| 8 | INTRP2 | If nonzero then the subroutine will return, with IDO = 5, after every successful step and with IDO = 6 after every unsuccessful step. See Remark 5. Default: 0. |
| 9 | SCALE | A measure of the scale of the problem, such as an approximation to the average value of a norm of the Jacobian along the trajectory. Default: 1.0. |
| 10 | INORM | Switch determining error norm. In the following Ei is the absolute value of an estimate of the error in Y(i), called Yi here. |
| | | 0 min(absolute error, relative error) = max(Ei/Wi), |

| | | |
|---|---|---|
| | | $i = 1.2, \ldots,$ NEQ, where Wi = max(abs(Yi,1.0), |
| | 1 | absolute error = max(Ei), $i = 1, 2, \ldots$ |
| | 2 | max(Ei/Wi), $i = 1,2,\ldots$, where Wi = max(abs(Yi), FLOOR), and FLOOR is PARAM(11). |
| | 3 | Euclidian norm scaled by YMAX = sqrt(sum (Ei**2/Wi**2)), where Wi = max(abs(Yi),1.0); for YMAX, see Remark 1. |
| 11 | FLOOR | Used in the norm computation. Default: 1.0. |
| 12 | METH | Method indicator. METH = 1 selects Adams' method METH = 2 selects Gear's backward difference method. Default: 1 |
| 13 | MITER | Iteration method indicator. MITER = 0 selects functional iteration. IATYPE must be set to zero with this option. MITER = 1 selects the chord method with an analytic, user-supplied Jacobian. MITER = 2 selects the chord method with a finite-difference Jacobian. MITER = 3 selects the chord method with the Jacobian replaced by a diagonal approximation based on a directional derivative. IATYPE must be set to zero with this option. Default: 0 |
| 14 | MTYPE | Matrix type for A (if used) and the Jacobian (if MITER is 1 or 2). These must have the same type. MTYPE = 0 selects full matrices. MTYPE = 1 selects band matrices. MTYPE = 2 selects symmetric positive definite matrices. MTYPE = 3 selects band symmetric positive definite matrices. Default: 0 |
| 15 | NLC | Number of lower codiagonals, used if MTYPE = 1. Default: 0 |
| 16 | NUC | Number of upper codiagonals, used if MTYPE is 1 or 3. Default: 0. |
| 17 | | Not used. |
| 18 | EPSJ | Epsilon used in computing finite-difference Jacobians. Default: SQRT(eps), where eps is the machine precision. |
| 19 | IATYPE | Type of the matrix A. |

|  |  |  |
|---|---|---|
|  |  | IATYPE = 0 implies A is not used (the ODE system is explicit).<br>IATYPE = 1 implies A is constant.<br>IATYPE = 2 implies A depends on X.<br>Default: 0 |
| 20 | LDA | Leading dimension of A exactly as specified in the dimension statement in the calling program. Used if IATYPE is not zero.<br>Default: N, if MTYPE = 0 or 2<br>NUC + NLC + 1, if MTYPE = 1<br>NUC + 1, if MTYPE = 3 |
| 21–30 |  | Not used. |

The following entries in PARAM are set by the program.

| 31 | HTRIAL | Current trial step size. |
|---|---|---|
| 32 | HMINC | Computed minimum step size. |
| 33 | HMAXC | Computed maximum step size. |
| 34 | NSTEP | Number of steps taken. |
| 35 | NFCN | Number of function evaluations used. |
| 36 | NJE | Number of Jacobian evaluations. |
| 35–50 |  | Not used. |

Y    Vector of length NEQ of dependent variables (input/output). On input, Y contains the initial values. On output, Y contains the approximate solution.

**Remarks**

1. Automatic workspace usage is

   IVPAG    4*NEQ + NMETH + NPW + NIPVT, or
   DIVPAG   8*NEQ + 2*NMETH + 2*NPW + NIPVT units.

   Here
   NMETH = 13*NEQ if METH is 1,
   NMETH = 6*NEQ if METH is 2.
   NPW = 2*NEQ + NPWM + NPWA
   where
   NPWM = 0 if MITER is 0 or 3,
   NPWM = NEQ**2 if MITER is 1 or 2, and if MTYPE is 0 or 2.
   NPWM = NEQ*(2*NLC + NUC + 1) if MITER is 1 or 2 and MTYPE = 1.
   NPWM = NEQ*(NLC + 1) if MITER is 1 or 2 and if MTYPE = 3.
   NPWA = 0 if IATYPE is 0.
   NPWA = NEQ**2 if IATYPE is nonzero and MTYPE = 0,
   NPWA = NEQ*(2*NLC + NUC + 1) if IATYPE is nonzero and MTYPE = 1
   NIPVT = 1, otherwise

   Workspace may be explicitly provided, if desired, by use of

   I2PAG/DI2PAG. The reference is
   CALL I2PAGE (IDO, NEQ, FCN, FCNJ, A, X, XEND, TOL, PARAM, Y, YTEMP, YMAX, ERROR, SAVE1, SAVE2, PW, IPVT, VNORM)

None of the additional array arguments should be changed from the first call with IDO = 1 until after the final call with IDO = 3.
The additional arguments are as follows:

| YTEMP | Vector of length NMETH (workspace). |
|---|---|
| YMAX | Vector of length NEQ containing the maximum Y values computed so far (output). |
| ERROR | Vector of length NEQ containing error estimates for each component of Y (output). |
| SAVE1 | Vector of length NEQ (workspace). |
| SAVE2 | Vector of length NEQ (workspace). |
| PW | Vector of length NPW. PW is used both to store the Jacobian and as workspace (workspace). |
| IPVT | Vector of length NEQ (workspace). |
| VNORM | User-supplied SUBROUTINE to compute the norm of the error (input). The routine may be provided by the user, or the IMSL routine I3PRK/DI3PRK may be used. |

   The usage is
   CALL VNORM (NEQ, V, Y, YMAX, ENORM)
   where

| NEQ | Number of equations (input). |
|---|---|
| V | Vector of length NEQ containing the vector whose norm is to be computed (input). |
| Y | Vector of length NEQ containing the values of the dependent variable (input). |
| YMAX | Vector of length NEQ containing the maximum Y values computed so far (input). |
| ENORM | Norm of the vector V (output). |

   VNORM must be declared EXTERNAL in the calling program.

2. Informational errors

| Type | Code |  |
|---|---|---|
| 4 | 1 | After some initial success, the integration was halted by repeated error-test failures. |
| 4 | 2 | The maximum number of function evaluations have been used. |
| 4 | 3 | The maximum number of steps allowed have been used. The problem may be stiff. |
| 4 | 4 | On the next step X + H will equal X. Either TOL is too small or the problem is stiff. |
| 4 | 5 | After some initial success, the integration was halted by a test on TOL. |
| 4 | 6 | Integration was halted after failing to pass the error test, even after reducing the step size by a factor of $1.0E + 10$. TOL may be too small. |
| 4 | 7 | Integration was halted after failing to achieve corrector convergence, even after reducing the step size by a factor of $1.0E + 10$. TOL may be too small. |
| 4 | 8 | IATYPE is nonzero and the input matrix A is singular. |

3. Both explicit ODE systems, of the form $y' = f(x, y)$, and implicit ODE systems, of the form $Ay' = f(x, y)$ can be solved. If the system is explicit then PARAM(19) = 0 and the matrix A is not referenced. If the system is implicit, then PARAM(14) determines the data structure of the matrix A. If PARAM(19) = 1, then A is assumed to be a constant matrix (not depending on x or y). The value of A used on the first call (with IDO = 1) is used until after a call with IDO = 3. The value of A must not be changed between these calls. IF PARAM(19) = 2, then the matrix is assumed to be a function of x.

4.  If MTYPE is greater than zero, then MITER must equal 1 or 2.

5.  If PARAM(7) is nonzero, the subroutine returns with IDO = 4 and will resume calculation at the point of interruption if reentered with IDO = 4. If PARAM(8) is nonzero, the subroutine will interrupt the calculations immediately after it decides whether or not to accept the result of the most recent trial step. IDO = 5 if the routine plans to accept, or IDO = 6 if it plans to reject. IDO may be changed by the user in order to force acceptance of a step (by changing IDO from 6 to 5) that would otherwise be rejected, or vice versa. Relevant parameters to observe after return from an interrupt are IDO, HTRIAL, NSTEP, NFCN, NJE, and Y. Y is the newly computed trial value, accepted or not.

### Algorithm

IVPAG solves a system of first-order ODEs of the form $y' = f(x, y)$ or $Ay' = f(x, y)$ with initial conditions, where A is a matrix of order $N$. Two classes of implicit linear multistep methods are available. The first is the implicit Adams' method (up to order 12); the second is the backward differentiation formula (up to order 5), also called Gear's stiff method. In both cases, because the basic formula is implicit, an algebraic system of equations must be solved at each step. The matrix in this system has the form $L = A + \eta J$, where $\eta$ is a small number and $J$ is the Jacobian (from either the user-supplied routine FCNJ, or computed internally). If A is not user specified, then the identity matrix is used for A. The matrix L may be declared to be real general, real band, symmetric positive definite, or band symmetric positive definite.

### Example 1

Euler's equation for the motion of a rigid body not subject to external forces is

$$y'_1 = y_2 y_3 \qquad y_1(0) = 0$$

$$y'_2 = -y_1 y_3 \qquad y_2(0) = 1$$

$$y'_3 = -0.51 y_1 y_2 \qquad y_3(0) = 1.$$

Its solution is, in terms of Jacobi elliptic functions, $y_1(x) = \text{sn}(x, k)$, $y_2(x) = \text{cn}(x, k)$, $y_3(x) = \text{dn}(x, k)$, where $k^2 = 0.51$. The Adams method option of IVPAG is used to solve this system.

The last call to IVPAG with IDO = 3 releases IMSL workspace, which was reserved on the first call to IVPAG. It is not necessary to release the workspace in this example, because the program ends after solving a single problem. The call to release workspace is made as a model of what would be needed if the program included further calls to IMSL routines.

Because PARAM(13) = MITER = 0, functional iteration is used and so subroutine FCNJ is never called. It is included only because the calling sequence for IVPAG requires it.

```
      INTEGER      NEQ, NPARAM
      PARAMETER    (NEQ = 3, NPARAM = 50)
C
      INTEGER      IDO, IEND, IMETH, INORM, NOUT
      REAL         A(1,1), FCN, FCNJ, HINIT,
                   PARAM(NPARAM),
     &             TOL, X, XEND, Y(NEQ)
      EXTERNAL     FCN, IVPAG, SSET, UMACH, FCNJ
C                           Initialize
      HINIT = 1.0E-3
      INORM = 2
```

```
      IMETH = 1
      CALL SSET (NPARAM, 0.0, PARAM, 1)
      PARAM(1) = HINIT
      PARAM(10) = INORM
      PARAM(12) = IMETH
C
      IDO = 1
      X = 0.0
      Y(1) = 0.0
      Y(2) = 1.0
      Y(3) = 1.0
      TOL = 1.0E-6
C                           Write title
      CALL UMACH (2, NOUT)
      WRITE (NOUT,99998)
C                           Integrate ODE
      DO 10 IEND = 1, 10
         XEND = IEND
         CALL IVPAG (IDO, NEQ, FCN, FCNJ, A, X, XEND, TOL,
                     PARAM, Y)
         WRITE (NOUT,99999) X, Y
   10 CONTINUE
C                           Finish up
      IDO = 3
      CALL IVPAG (IDO, NEQ, FCN, FCNJ, A, X, XEND, TOL,
                  PARAM, Y)
C
99998 FORMAT (11X, 'X', 14X, 'Y(1)', 11X, 'Y(2)', 11X, 'Y(3)')
99999 FORMAT (4F15.5)
      END
C
      SUBROUTINE   FCN (NEQ, X, Y, YPRIME)
      INTEGER      NEQ
      REAL         X, Y(NEQ), YPRIME(NEQ)
C
      YPRIME(1) = Y(2)*Y(3)
      YPRIME(2) = -Y(1)*Y(3)
      YPRIME(3) = -.51*Y(1)*Y(2)
      RETURN
      END
C
      SUBROUTINE   FCNJ (NEQ, X, Y, DYPDY)
      INTEGER      NEQ
      REAL         X, Y(NEQ), DYPDY(*)
C                           This subroutine is never called
      RETURN
      END
```

### Output

| X | Y(1) | Y(2) | Y(3) |
|---|------|------|------|
| 1.00000 | 0.80220 | 0.59705 | 0.81963 |
| 2.00000 | 0.99537 | −0.09615 | 0.70336 |
| 3.00000 | 0.64141 | −0.76720 | 0.88892 |
| 4.00000 | −0.26961 | −0.96296 | 0.98129 |
| 5.00000 | −0.91172 | −0.41079 | 0.75899 |
| 6.00000 | −0.95750 | −0.28841 | 0.72967 |
| 7.00000 | −0.42876 | 0.90341 | 0.95197 |
| 8.00000 | 0.51092 | 0.85962 | 0.93106 |
| 9.00000 | 0.97566 | 0.21925 | 0.71730 |
| 10.00000 | 0.87789 | −0.47886 | 0.77907 |

**Example 2**

Solve the partial differential equation

$$e^{-t}\frac{\partial u}{\partial t} = \frac{\partial^2 u}{\partial x^2}$$

with the initial condition

$$u(t=0, x) = \sin x$$

and the boundary condition

$$u(t, x=0) = u(t, x=\pi) = 0$$

on the interval $|0, \pi|$ using the method of lines with a piecewise linear Galerkin discretization. The solution is known to be $u(t, x) = \exp(1-e^t)\sin x$. The interval $|0, \pi|$ is divided into equal intervals by the breakpoints $x_k = k\pi/(N+1)$ for $k = 0, \ldots, N+1$. The unknown function $u(t, x)$ is approximated by $\sum_{k=1}^{N} c_k(t)\phi_k(x)$, where $\phi_k(x)$ is a piecewise linear function that equals 1 at $x_k$ and is zero at all of the other breakpoints. We can now approximate the partial differential equation by a system of $N$ ODEs, $Adc/dt = Rc$, where $A$ and $R$ are matrices of order $N$. The matrix $A$ is given by

$$A_{ij} = e^{-t}\int \phi_i(x)\phi_j(x)dx = \begin{cases} e^{-t}2h/3 & \text{if } i=j \\ e^{-t}h/6 & \text{if } i=j\pm 1 \\ 0 & \text{otherwise,} \end{cases}$$

where $h = 1/(N+1)$ is the mesh spacing. The matrix $R$ is given by

$$R_{ij} = \int \phi_i''(x)\phi_j(x)dx = -\int \phi_i'(x)\phi_j'(x)dx = \begin{cases} -2/h & \text{if } i=j \\ 1/h & \text{if } i=j\pm 1 \\ 0 & \text{otherwise.} \end{cases}$$

The first integral above is interpreted in the distributional sense.

Because this system may be stiff, Gear's method is used.

In the following program the vector $Y$ corresponds to $c$. Note that $Y$ contains $N+2$ elements; $Y(0)$ and $Y(N+1)$ are used to store the boundary values. The matrix $A$ depends on $t$, so we set PARAM(19)=2 and evaluate $A$ when IVPAG returns with IDO=7. The subroutine FCN computes the vector $Rc$ and the subroutine FCNJ computes $R$. The matrices $A$ and $R$ are stored as band-symmetric, positive-definite matrices with one upper codiagonal.

```
      INTEGER    LDA, NEQ, NPARAM, NSTEP, NUC
      PARAMETER  (NEQ=9, NPARAM=50, NSTEP=4,
                 NUC=1, LDA=NUC+1)
C
      INTEGER    I, IATYPE, IDO, IMETH, INORM,
                 ISTEP, MITER, MTYPE
      REAL       A(LDA,NEQ), C, CONST, EXP, FCN,
                 FCNJ,
     &           FLOAT, HINIT, HX, PARAM(NPARAM),
                 PI, SIN,
     &           T, TEND, TMAX, TOL,
                 XPOINT(0:NEQ+1), Y(0:NEQ+1)
      CHARACTER  TITLE*10
      COMMON     /COMHX/ HX
      INTRINSIC  EXP, FLOAT, SIN
      EXTERNAL   CONST, FCN, FCNJ, IVPAG, SSET,
                 WRRRN
C                             Initialize PARAM
      HINIT = 1.0E-3
      INORM = 1
      IMETH = 2
      MITER = 1
      MTYPE = 3
      IATYPE = 3
      CALL SSET (NPARAM, 0.0, PARAM, 1)
      PARAM(1) = HINIT
      PARAM(10) = INORM
      PARAM(12) = IMETH
      PARAM(13) = MITER
      PARAM(14) = MTYPE
      PARAM(16) = NUC
      PARAM(19) = IATYPE
C                             Initialize other arguments
      PI = CONST('PI')
      HX = PI/FLOAT(NEQ+1)
      CALL SSET (NEQ-1, HX/6., A(1,2), LDA)
      CALL SSET (NEQ, 2.*HX/3., A(2,1), LDA)
      DO 10 I=0, NEQ + 1
         XPOINT(I) = I*HX
         Y(I) = SIN(XPOINT(I))
   10 CONTINUE
      T0L = 1.0E-6
      T = 0.0
      TMAX = 1.0
C                             Integrate ODE
      IDO = 1
      DO 30 ISTEP=1, NSTEP
      TEND = TMAX*FLOAT(ISTEP)/FLOAT(NSTEP)
   20 CALL IVPAG (IDO, NEQ, FCN, FCNJ, A, T, TEND, TOL,
                 PARAM, Y(1))
C                             Set matrix A
      IF (IDO .EQ. 7) THEN
         C = EXP(-T)
         CALL SSET (NEQ-1, C*HX/6., A(1,2), LDA)
         CALL SSET (NEQ, 2.*C*HX/3., A(2,1), LDA)
         GO TO 20
      END IF
C                             Print solution
      WRITE (TITLE, '(A,F5.3,A)') 'U(T-', T, ')'
      CALL WRRRN (TITLE, 1, NEQ+2, Y, 1, 0)
   30 CONTINUE
C                             Final call to release workspace
      IDO = 3
      CALL IVPAG (IDO, NEQ, FCN, FCNJ, A, T, TEND, TOL,
                 PARAM, Y(1))
      END
C
      SUBROUTINE FCN (NEQ, T, Y, YPRIME)
      INTEGER    NEQ
      REAL       T, Y(NEQ), YPRIME(NEQ)
C
      INTEGER    I
      REAL       HX
      COMMON     /COMHX/ HX
      EXTERNAL   SSCAL
C
      YPRIME(1) = -2.0*Y(1) + Y(2)
      DO 10  I = 2, NEQ - 1
         YPRIME(I) = -2.0*Y(I) + Y(I-1) + Y(I+1)
   10 CONTINUE
      YPRIME(NEQ) = -2.0*Y(NEQ) + Y(NEQ-1)
      CALL SSCAL (NEQ, 1.0/HX, YPRIME, 1)
      RETURN
      END
C
```

```
SUBROUTINE FCNJ (NEQ, T, Y, PD)
INTEGER    NEQ
REAL       T, Y(NEQ), PD(*)
C
REAL       HX
COMMON     /COMHX/ HX
EXTERNAL   SSET
C
CALL SSET (NEQ-1, 1.0/HX, PD(3), 2)
CALL SSET (NEQ, -2.0/HX, PD(2), 2)
RETURN
END
```

## Output

|  |  | $U(t=0.250)$ |  |  |  |
|---|---|---|---|---|---|
| 1 | 2 | 3 | 4 | 5 | 6 |
| 0.0000 | 0.2321 | 0.4414 | 0.6076 | 0.7142 | 0.7510 |
| 7 | 8 | 9 | 10 | 11 | |
| 0.7142 | 0.6076 | 0.4414 | 0.2321 | 0.0000 | |

|  |  | $U(t=0.500)$ |  |  |  |
|---|---|---|---|---|---|
| 1 | 2 | 3 | 4 | 5 | 6 |
| 0.0000 | 0.1607 | 0.3056 | 0.4206 | 0.4945 | 0.5199 |
| 7 | 8 | 9 | 10 | 11 | |
| 0.4945 | 0.4206 | 0.3056 | 0.1607 | 0.0000 | |

|  |  | $U(t=0.750)$ |  |  |  |
|---|---|---|---|---|---|
| 1 | 2 | 3 | 4 | 5 | 6 |
| 0.0000 | 0.1002 | 0.1906 | 0.2623 | 0.3084 | 0.3243 |
| 7 | 8 | 9 | 10 | 11 | |
| 0.3084 | 0.2623 | 0.1906 | 0.1002 | 0.0000 | |

|  |  | $U(t=1.000)$ |  |  |  |
|---|---|---|---|---|---|
| 1 | 2 | 3 | 4 | 5 | 6 |
| 0.0000 | 0.0546 | 0.1039 | 0.1431 | 0.1682 | 0.1768 |
| 7 | 8 | 9 | 10 | 11 | |
| 0.1682 | 0.1431 | 00.103 | 0.0546 | 0.0000 | |

## References

Hindmarsh, A. C. (1974), *GEAR: Ordinary Differential Equation System Solver* (Lawrence Livermore Laboratory Report UCID-30001, Revision 3,      ).

## Appendix D.8.   Subroutine MOLCH

MOLCH/DMOLCH (Single/double precision)
(SMATH5B/DMATH5B.LIB)

Purpose
Solve a system of partial differential equations of the form UT = FCN(X, T, U, UX, UXX) using the method of lines with cubic Hermite polynomials.

Usage
CALL MOLCH (IDO, FCNUT, FCNBC, NPDES, T, TEND, NX, XBREAK, TOL, HINIT, Y, LDY)

**Arguments**

IDO      Flag indicating the state of the computation (input/output).
     1    Initial entry.
     2    Normal reentry.
     3    Final call, release workspace.

Normally, the initial call is made with IDO = 1. The routine then sets IDO = 2, and this value is then used for all but the last call, which is made with IDO = 3. Usually, the final call is made with X = XEND, so that the only task of the final call is to release workspace.

FCNUT    User-supplied SUBROUTINE to evaluate the function. The usage is

CALL FCNUT (NPDES, X, T, U, UX, UXX, UT)

where
NPDES    Number of equations (input).
X    Space variable (input).
T    Time variable (input).
U    Array of length NPDES containing the dependent variable values (input).
UX    Array of length NPDES containing the derivative of U with respect to X (input).
UX    Array of length NPDES containing the second derivative of U with respect to X (input).
UT    Array of length NPDES containing the derivative of U with respect to T (output).

FCNUT must be declared EXTERNAL in the calling program.

FCNBC    User-supplied SUBROUTINE to evaluate the boundary conditions. The boundary conditions are ALPHA(i)*U(i) + BETA(i)*UX(i) = GAMMA(i). The usage is

CALL FCNBC (NPDES, X, T, ALPHA, BETA, GAMMAP)

where
NPDES    Number of equations (input).
X    Space variable (input).
T    Time variable (input).
ALPHA    Array of length NPDES containing the ALPHA values (output).
BETA    Array of length NPDES containing the BETA values (output).
GAMMA    Array of length NPDES containing the vaues of the derivative of GAMMA(i) with respect to T (output).

FCNBC must be declared EXTERNAL in the calling program.

NPDES    Number of differential equations (input).
T    Independent variable (input/output).
On input, T supplies the initial time. On output, T is set to the value to which the integration has been completed.
TEND    Value of T at which the solution is desired (input).
NX    Number of mesh points (input).
XBREAK    Array of length NX containing the break points for the cubic Hermite splines used in the spatial discretization (input).
The points in X must be strictly increasing with X(1) and X(NX) the endpoints of the interval.
TOL    Differential equation error tolerance (input).
An attempt is made to control the local error in such a way that the global relative error is proportional to TOL.

HINIT      Initial step size (input).

It must be nonnegative. If HINIT is zero, an initial step size of ABS(.001*(XEND-X)) will be used.

Y      Array of size NPDES by NX containing the solution (input/output).

Y contains the solution as Y(k,i) = U(k) at X(i).

On input, Y supplies the initial solution. It MUST satisfy the boundary conditions.

On output, Y contains the computed solution.

LDY      Leading dimension of Y exactly as specified in the dimension statement of the calling program (input).

**Remarks**

1. Automatic workspace usage is

MOLCH    2*NX*NPDES*(12*NPDES**2 + 21*NPDES + 10)
DMOLCH   2*NX*NPDES*(24*NPDES**2 + 42*NPDES + 19)

Workspace may be explicitly provided, if desired, by use of M2LCH/DM2LCH. The reference is

CALL M2LCH (IDO, FCNUT, FCNBC, NPDES, T, TEND, NX,
               XBREAK, TOL, HINIT, Y, LDY, WK, IWK)

The additional arguments are as follows:

WK     Work array of length
        2*NX*NPDES*(12*NPDES**2 + 21*NPDES + 9).
        WK should not be changed between calls to M2LCH.

IWK    Work array of length 2*NX*NPDES.
        IWK should not be changed between calls to M2LCH.

2. Informational errors

| Type | Code | |
|---|---|---|
| 4 | 1 | After some initial success, the integration was halted by repeated error test failures. |
| 4 | 2 | On the next step X+H will equal X. Either TOL is too small or the problem is stiff. |
| 4 | 3 | After some initial success, the integration was halted by a test on TOL. |
| 4 | 4 | Integration was halted after failing to pass the error test, even after reducing the step size by a factor of 1.0E+10. TOL may be too small. |
| 4 | 5 | Integration was halted after failing to achieve corrector convergence, even after reducing the step size by a factor of 1.0E+10. TOL may be too small. |

**Algorithm**

Let M = NPDES, N = NX and $x_i$ = XBREAK(I).
MOLCH uses the method of lines to solve the partial differential equation system:

$$\frac{\partial u^k}{\partial t} = f^k \left( x, t, u^1, \ldots, u^M, \frac{\partial u^1}{\partial x}, \ldots, \frac{\partial u^M}{\partial x}, \frac{\partial^2 u^1}{\partial x^2}, \ldots, \frac{\partial^2 u^M}{\partial x^2} \right),$$

with the initial conditions

$$u^k = u_0^k(x) \quad \text{at} \quad t = t_0$$

and the boundary conditions

$$\alpha_1^k u^k + \beta_1^k \frac{\partial u^k}{\partial x} = \gamma_1^k \quad \text{at} \quad x = x_1$$

$$\alpha_M^k u^k + \beta_M^k \frac{\partial u^k}{\partial x} = \gamma_M^k \quad \text{at} \quad x = x_M,$$

for $k = 1, \ldots, M$. Cubic Hermite polynomials are used in the spatial approximation, so that the trial solution is expanded in the series

$$\hat{u}^k(x, t) = \sum_{i=1}^{M} (a_i^k(t)\phi_i(x) + b_i^k(t)\psi_i(x)),$$

where $\phi_i(x)$ and $\psi_i(x)$ are the standard basis functions for the cubic Hermite polynomials with the knots $x_1 < x_2 < \cdots < x_N$. These are piecewise cubic polynomials with continuous first derivatives. At the breakpoints they satisfy

$$\phi_i(x_l) = \delta_{il} \qquad \psi_i(x_l) = 0$$

$$\frac{d\phi_i}{dx}(x_l) = 0 \qquad \frac{d\psi_i}{dx}(x_l) = \delta_{il}.$$

According to the collocation method, the coefficients of the approximation are obtained, so that the approximation satisfies the differential equation at the two Gaussian points in each subinterval,

$$p_{2j-1} = x_j + \frac{3 - \sqrt{3}}{6}(x_{j+1} - x_j)$$

$$p_{2j} = x_j + \frac{3 - \sqrt{3}}{6}(x_{j+1} + x_j)$$

for $j = 1, \ldots, N$.
The collocation approximation to the differential equation is

$$\frac{da_i^k}{dt}\phi_i(p_j) + \frac{db_i^k}{dt}\psi_i(p_j)$$

$$= f^k(p_j, t, \hat{u}^1(p_j), \ldots, \hat{u}^M(p_j), \hat{u}_x^1(p_j), \ldots, \hat{u}_x^M(p_j), \hat{u}_{xx}^1(p_j), \ldots, \hat{u}_{xx}^M(p_j)),$$

for $k = 1, \ldots, M$ and $j = 1, \ldots, 2(N-1)$. This is a system of $2M(N-1)$ ODEs in $2MN$ unknown coefficient functions, $a_i^k$ and $b_i^k$. The last $2M$ equations are obtained by differentiating the boundary conditions

$$\alpha_1^k \frac{da_1^k}{dt} + \beta_1^k \frac{db_1^k}{dt} = \frac{d\gamma_1^k}{dt}$$

$$\alpha_M^k \frac{da_M^k}{dt} + \beta_M^k \frac{db_M^k}{dt} = \frac{d\gamma_M^k}{dt}$$

for $k = 1, \ldots, M$.
The initial conditions $u_0^k(x)$ must satisfy the boundary conditions; also $\gamma_1^k(t)$ and $\gamma_1^N(t)$ must be continuous or the boundary conditions will not be properly imposed for $t > t_0$.
If $\alpha_1^k - \beta_1^k = 0$, it is assumed that no boundary condition is desired for the $k$th unknown at the left endpoint (similarly for the right endpoint), and so collocation is done at the endpoint. This is generally useful for first-order partial differential equations.
This system can be written in the form $Adc/dt = F(t, y)$ with $c(t_0) = c_0$, where $c$ is a vector of coefficients of length $2MN$ and $c_0$ holds the initial values of the coefficients.

If the number of partial differential equations is $M = 1$, and the number of breakpoints is $N = 4$, then

$$
A \quad
\begin{array}{cccccccc}
\alpha_1 & \beta_1 & & & & & & \\
\phi_1(p_1) & \psi_1(p_1) & \phi_2(p_1) & \psi_2(p_1) & & & & \\
\phi_1(p_2) & \psi_1(p_2) & \phi_2(p_2) & \psi_2(p_2) & & & & \\
& & \phi_3(p_3) & \psi_3(p_3) & \phi_4(p_3) & \psi_4(p_3) & & \\
& & \phi_3(p_4) & \psi_3(p_4) & \phi_4(p_4) & \psi_4(p_4) & & \\
& & & & \phi_5(p_5) & \psi_5(p_5) & \phi_6(p_5) & \psi_6(p_5) \\
& & & & \phi_5(p_6) & \psi_5(p_6) & \phi_6(p_6) & \psi_6(p_6) \\
& & & & & & \alpha_4 & \beta_4
\end{array}
$$

The vector $c$ is

$$c = [a_1, b_1, a_2, b_2, a_3, b_3, a_4, b_4]^T,$$

and the right-side $F$ is

$$F = [\gamma'(x_1), f(p_1), f(p_2), f(p_3), f(p_4), f(p_5), f(p_6), \gamma'(x_4)]^T.$$

If the number of partial differential equations, $M$, is greater than one, then each entry in the above matrix is replaced by an $M \times M$ diagonal matrix. The element $\alpha_1$ is replaced by $\mathrm{diag}(\alpha_1^1, \ldots, \alpha_1^M)$. The elements $\alpha_M$, $\beta_1$, and $\beta_M$ are handled in the same manner. The $\phi_i(p_j)$ and $\psi_i(p_j)$ elements are replaced by $\phi_i(p_j)I_M$ and $\psi_i(p_j)I_M$, respectively, where $I_M$ is the identity matrix of order $M$.

The input/output vector y contains the values of the $a_i^k$. The inital values of the $b_i^k$ are obtained by using the IMSL cubic spline routine CSINT to construct functions $\hat{u}_0^k$ such that $\hat{u}_0^k(x_i) = a_i^k$. The IMSL routine CSDER is then used to compute $\hat{u}_0'^k(x_i) = b_i^k$.

The order of matix $A$ is $2MN$ and its maximum bandwidth is $6M - 1$. The band structure of the Jacobian of $F$ is the same as the band structure of $A$. This system is solved using a modified version of IVPAG, page 328. The modifications are due to Fortran limitations; the algorithm is unchanged. Gear's method is used because the system is often stiff.

## Example

MOLCH is used to solve the partial differential equation system

$$\frac{\partial u}{\partial t} = \frac{\partial^2 u}{\partial x^2} - 6xe^t + v$$

$$\frac{\partial v}{\partial t} = x \frac{\partial^2 v}{\partial x^2} \Big/ 2 - \frac{\partial u}{\partial x} + v,$$

subject to the boundary conditions

$$\frac{\partial u}{\partial x} = 0, \qquad v = 0 \qquad \text{at } x = 0$$

$$u = e^t, \qquad -2v + v_x = e^t \qquad \text{at } x = 1$$

and to the initial conditions

$$u = x^3, \qquad v = x^3 \qquad \text{at } t = 0.$$

A nonuniform grid of six points is used. The exact solution is $u = v = x^3 e^t$.

The last call to MOLCH with IDO $= 3$ releases IMSL workspace, which was reserved on the first call to MOLCH. It is not necessary to release the workspace in this example, because the program ends after solving a single problem. The call to release workspace is made as a model of what would be needed if the program included further calls to IMSL routines.

The routine that generates the plot is now shown.

```
C                  SPECIFICATIONS FOR LOCAL VARIABLES
      INTEGER    LDY, NPDES, NX
      PARAMETER  (NPDES = 2, NX = 6, LDY = NPDES)
C
      INTEGER    I, IDO, J, NOUT, NSTEP
      REAL       FCNBC, FCNUT, FLOAT, HINIT, T,
     &           TEND, TOL, XBREAK(NX), Y(LDY, NX)
      CHARACTER  TITLE*19
      INTRINSIC  FLOAT
      EXTERNAL FCNBC, FCNUT, MOLCH, UMACH, WRRRN
C                  Set breakpoints and initial
C                  conditions
      DO 10 I = 1, NX
         XBREAK(I) = (FLOAT(I-1)/FLOAT(NX-1))**1.1
         Y(1,I)    = XBREAK(I)**3
         Y(2,I)    = XBREAK(I)**3
   10 CONTINUE
C                  Set parameters for MOLCH
      TOL = 1.E-4
      HINIT = 0.01
      T = 0.0
      IDO = 1
      NSTEP = 5
      CALL UMACH (2, NOUT)
      DO 20 J = 1, NSTEP
         TEND = FLOAT(J)/FLOAT(NSTEP)
C                  Solve the problem
         CALL MOLCH (IDO, FCNUT, FCNBC, NPDES, T,
     &        TEND, NX, XBREAK, TOL, HINIT, Y, LDY)
C                  Print results
         WRITE (TITLE,'(A,F4.2)')881 'Solution at T =', T
         CALL WRRRN (TITLE, NPDES, NX, Y, LDY, O)
   20 CONTINUE
C                  Final call to release workspace
      IDO = 3
      CALL MOLCH (IDO, FCNUT, FCNBC, NPDES, T, TEND,
     &        NX, XBREAK, TOL, HINIT, Y, LDY)
      END
      SUBROUTINE FCNUT (NPDES, X, T, U, UX, UXX, UT)
      INTEGER    NPDES
      REAL       X, T, U(2), UX(2), UXX(2), UT(2)
C
      REAL       EXP
      INTRINSIC  EXP
C                  Define the PDE
      UT(1) = UXX(1) - 6.*X*EXP(T) + U(2)
      UT(2) = X*UXX(2)/2.0 - UX(1) + U(2)
      RETURN
      END
      SUBROUTINE FCNBC (NPDES, X, T, ALPHA, BETA,
     &        GAMP)
      INTEGER    NPDES
      REAL       X, T, ALPHA(2), BETA(2), GAMP(2)
C
      REAL       EXP
      INTRINSIC  EXP
```

```
C                        Define the boundary conditions
      IF (X .LT. 0.5) THEN
         ALPHA(1) = 0.0
         BETA(1) = 1.0
         GAMP(1) = 0.0
         ALPHA(2) = 1.0
         BETA(2) = 0.0
         GAMP(2) = 0.0
      ELSE
         ALPHA(1) = 1.0
         BETA(1) = 0.0
         GAMP(1) = EXP(T)
         ALPHA(2) = -2.0
         BETA(2) = 1.0
         GAMP(2) = EXP(T)
      END IF
      RETURN
      END
```

**Output**

Solution at $t = 0.20$

|   | 1 | 2 | 3 | 4 | 5 | 6 |
|---|---|---|---|---|---|---|
| 1 | 0.000 | 0.006 | 0.059 | 0.226 | 0.585 | 1.222 |
| 2 | 0.000 | 0.006 | 0.059 | 0.226 | 0.585 | 1.222 |

Solution at $t = 0.40$

|   | 1 | 2 | 3 | 4 | 5 | 6 |
|---|---|---|---|---|---|---|
| 1 | 0.000 | 0.007 | 0.073 | 0.277 | 0.715 | 1.492 |
| 2 | 0.000 | 0.007 | 0.073 | 0.277 | 0.715 | 1.492 |

Solution at $t = 0.60$

|   | 1 | 2 | 3 | 4 | 5 | 6 |
|---|---|---|---|---|---|---|
| 1 | 0.000 | 0.009 | 0.089 | 0.338 | 0.873 | 1.823 |
| 2 | 0.000 | 0.009 | 0.089 | 0.338 | 0.873 | 1.823 |

Solution at $t = 0.80$

|   | 1 | 2 | 3 | 4 | 5 | 6 |
|---|---|---|---|---|---|---|
| 1 | 0.001 | 0.012 | 0.109 | 0.413 | 1.067 | 2.227 |
| 2 | 0.000 | 0.011 | 0.108 | 0.413 | 1.067 | 2.228 |

Solution at $t = 1.00$

|   | 1 | 2 | 3 | 4 | 5 | 6 |
|---|---|---|---|---|---|---|
| 1 | 0.001 | 0.014 | 0.133 | 0.505 | 1.303 | 2.720 |
| 2 | 0.000 | 0.013 | 0.133 | 0.505 | 1.304 | 2.723 |

### Appendix D.9.  Subroutine BVPFD

BVPFD/DBVPFD (Single/double precision) (SMATH5A.LIB)

Purpose
Solve a (parameterized) system of differential equations with boundary conditions at two points, using a variable order, variable step size finite-difference method with deferred corrections.

Usage
CALL BVPFD (FCNEQN, FCNJAC, FCNBC, FCNPEQ, FCNPBC, NEQNS, NLEFT, NCUPBC, XLEFT, XRIGHT, PISTEP, TOL, NINIT, XINIT, VINIT, LDYINI, LINEAR, PRINT, MXGRID, NFINAL, XFINAL, YFINAL, LDYFIN, ERREST)

**Arguments**

FCNEQN    User-supplied SUBROUTINE to evaluate derivatives.

The usage is

CALL FCNEQN (NEQNS, X, Y, P, DYDX)

where
NEQNS    Number of differential equations (input).
X        Dependent variable (input).
Y        Array of length NEQNS containing the independent variable values (input).
P        Continuation parameter (input). See Remark 3.
DYDX     Array of length NEQNS containing the derivative of Y(I) with respect to X at X,Y (output).
FCNEQN must be declared EXTERNAL in the calling program.

FCNJAC    User-supplied SUBROUTINE to evaluate the Jacobian. The usage is

CALL FCNJAC (NEQNS, X, Y, P, DYPDY)

where
NEQNS    Number of differential equations (input).
X        Dependent variable (input).
Y        Array of length NEQNS containing the independent variable values (input).
P        Continuation parameter (input). See Remark 3.
DYPDY    NEQNS by NEQNS array containing the derivative of DYDX(I) with respect to Y(J) at X,Y (output).
FCNJAC must be declared EXTERNAL in the calling program.

FCNBC    User-supplied SUBROUTINE to evaluate the boundary conditions. The usage is

CALL FCNBC (NEQNS, YLEFT, YRIGHT, P, F)

where
NEQNS    Number of differential equations (input).
YLEFT    Array of length NEQNS containing the values of the independent variable at the left endpoint (input).
YRIGHT   Array of length NEQNS containing the values of the independent variable at the right endpoint (input).
P        Continuation parameter (input). See Remark 3.
F        Array of length NEQNS containing the boundary conditions (output).
         The boundary conditions are defined by F(I) = 0, for I = 1,..., NEQNS. The left endpoint conditions must be defined first, then the conditions involving both endpoints, and finally the right endpoint conditions.
FCNBC must be declared EXTERNAL in the calling program.

FCNPEQ    User-supplied SUBROUTINE to evaluate the derivative of Y' with respect to the parameter P. The usage is

CALL FCNPEQ (NEQNS, X, Y, P, DYPDP)

where

| | |
|---|---|
| NEQNS | Number of differential equations (input). |
| X | Dependent variable (input). |
| Y | Array of length NEQNS containing the independent variable values (input). |
| P | Continuation parameter (input). See Remark 3. |
| DYPDP | Array of length NEQNS containing the derivative of Y'(I) with respect to P at X,Y (output). |

FCNPEQ must be declared EXTERNAL in the calling program.

FCNPBC  User-supplied SUBROUTINE to evaluate the derivative of the boundary conditions with respect to the parameter P. The usage is

CALL FCNPBC (NEQNS, YLEFT, YRIGHT, P, F)

where

| | |
|---|---|
| NEQNS | Number of differential equations (input). |
| YLEFT | Array of length NEQNS containing the values of the independent variable at the left endpoint (input). |
| YRIGHT | Array of length NEQNS containing the values of the independent variable at the right endpoint (input). |
| P | Continuation parameter (input). See Remark 3. |
| F | Array of length NEQNS containing the derivative of F(I) with respect to P (output). |

FCNPBC must be declared EXTERNAL in the calling program.

| | |
|---|---|
| NEQNS | Number of differential equations (input). |
| NLEFT | Number of initial conditions (input). NLEFT must be greater than or equal to zero and less than NEQNS. |
| NCUPBC | Number of coupled boundary conditions (input). NLEFT + NCUPBC must be greater than zero and less than or equal to NEQNS. |
| XLEFT | The left endpoint (input). |
| XRIGHT | The right endpoint (input). |
| PISTEP | Initial step size for P (input). If zero, then the problem is assumed not to be parameterized, and the routine FCNPEQ and FCNPBC will not be called. |
| TOL | Relative error control parameter (input). The computations stop when ABS(ERROR(J,I))/MAX(ABS(Y(J,I)),1.0) .LT. TOL for all J = 1, ..., NEQNS and I = 1, ..., NGRID. Here ERROR(J,I) is the estimated error in Y(J,I). |
| NINIT | Number of initial grid points, including the endpoints (input). It must be at least 4. |
| XINIT | Vector of length NINIT containing the initial grid points (input). |
| YINIT | Array of size NEQNS by NINIT containing an initial guess for the values of Y at the points in XINIT (input). |
| LDYINI | Leading dimension of YINIT exactly as specified in the dimension statement of the calling program (input). |
| LINEAR | Logical .TRUE. if the differential equations and the boundary conditions are linear (input). |
| PRINT | Logical .TRUE. if intermediate output is to be printed (input). |

| | |
|---|---|
| MXGRID | Maximum number of grid points allowed (input). |
| NFINAL | Number of final grid points, including the endpoints (output). |
| XFINAL | Vector of length MXGRID containing the final grid points (output). Only the first NFINAL points are significant. |
| YFINAL | Array of size NEQNS by MXGRID containing the values of Y at the points in XFINAL (output). |
| LDYFIN | Leading dimension of YFINAL exactly as specified in the dimension statement of the calling program (input). |
| ERREST | Vector of length NEQNS (output). ERREST (J) is the estimated error in Y(J). |

**Remarks**

1. Automatic workspace usage is

BVPFD  (NEQNS*(3*NEQNS*MXGRID + 4*NEQNS + 1)
        + MXGRID*(7*NEQNS + 2)
        + 2*NEQNS*MXGRID + NEQNS + MXGRID
DBVPFD (2*NEQNS*(3*NEQNS*MXGRID + 4*NEQNS + 1)
        + 2*MXGRID*(7*NEQNS + 2)
        + 2*NEQNS*MXGRID + NEQNS + MXGRID
Workspace may be explicitly provided, if desired, by use of B2PFD/DB2PFD. The reference is
   CALL B2PFD    (FCNEQN, FCNJAC, FCNBC, FCNPEQ,
                 FCNPBC, NEQNS, NLEFT, NCUPBC,
                 XLEFT, XRIGHT, PISTEP, TOL, NINIT,
                 XINIT, YINIT, LDYINI, LINEAR, PRINT,
                 MXGRID,, NFINAL, XFINAL, YFINAL,
                 LDYFIN, ERREST, RWORK, IWORK)
The additional arguments are as follows:

| | |
|---|---|
| RWORK | Work array of length NEQNS*(3*NEQNS*MXGRID + 4*NEQNS + 1) + MXGRID*(7*NEQNS + 2) |
| IWORK | Work array of length 2*NEQNS*MXGRID + NEQNS + MXGRID. |

2. Informational errors

| Type | Code | |
|---|---|---|
| 4 | 1 | More than MXGRID grid points are needed to solve the problem. |
| 4 | 2 | Newton's iteration diverged. |
| 3 | 3 | Newton's iteration reached roundoff error level. |

3. If PISTEP is greater than zero, then BVPFD assumes that the user has embedded the problem into a one-parameter family of problems

$$DY/DX = DYDX (X, Y, P)$$
$$F (Y(XLEFT), Y(XRIGHT), P) = 0$$

such that for $P=0.0$ the problem is simple, that is, linear, and for $P=1.0$, the original problem is recovered. BVPFD automatically attempts to go from $P=0.0$ to $P=1.0$. PISTEP is the beginning step size in this continuation. The step size may then be varied by BVPFD, but a lower limit of 0.01 is imposed.

4. The vectors XINIT and XFINAL may be the same.

5. The arrays YINIT and YFINAL may be the same.

**Algorithm**

BVPFD is based on the routine PASVA3 by M. Lentini and V. Pereyra (see Pereyra 1978).

The basic discretization is the trapezoidal rule over a possibly nonuniform mesh. This mesh is chosen adaptively, to make the local error approximately the same size everywhere. Higher-order discretizations are obtained by deferred corrections. Global error estimates are produced to control the computation.

The resulting nonlinear algebraic system is solved by Newton's method, with step control. The linearized sparse system is solved by a special form of Gauss elimination, which preserves the sparseness.

**Example 1**

This example solves the third-order linear equation

$$y''' - 2y'' + y' - y = \sin x,$$

subject to the boundary conditions $y(0) = y(2\pi)$ and $y'(0) = y'(2\pi) = 1$. (Its solution is $y = \sin x$). To use BVPFD, the problem is reduced to a system of first-order equations by defining $y_1 = y$, $y_2 = y'$ and $y_3 = y''$. The resulting system is:

| | |
|---|---|
| $y_1' = y_2$ | $y_2(0) - 1 = 0$ |
| $y_2' = y_3$ | $y_1(0) - y_1(2\pi) = 0$ |
| $y_3' = 2y_3 - y_2 + y_1 + \sin x$ | $y_2(2\pi) - 1 = 0.$ |

Note that there is one boundary condition at the left endpoint $x = 0$ and one boundary condition coupling the left and right endpoints. The final boundary condition is at the right endpoint. The total number of boundary conditions must be the same as the number of equations (in this case 3).

Note that because the parameter $p$ is not used, in the call to BVPFD the routines FCNPEQ and FCNPBC are not needed. Therefore, in the call to BVPFD, FCNEQN and FCNBC were used in place of FCNPEQ and FCNPBC, respectively.

```
C                         SPECIFICATIONS FOR PARAMETERS
      INTEGER             LDYFIN, LDYINI, MXGRID, NEQNS,
                          NINIT
      PARAMETER           (MXGRID=45, NEQNS=3, NINIT=10,
     &                    LDYFIN=NEQNS, LDYINI=NEQNS)
C
      INTEGER             I, J, NCUPBC, NFINAL, NLEFT, NOUT
      REAL                CONST, ERREST(NEQNS), FCNBC,
     &                    FCNEQN, FCNJAC, FLOAT, PISTEP,
     &                    TOL, XFINAL(MXGRID), XINIT(NINIT),
     &                    XLEFT, XRIGHT, YFINAL(LDYFIN,
     &                    MXGRID),YINIT(LDYINI,NINIT)
      LOGICAL             LINEAR, PRINT
      INTRINSIC           FLOAT
      EXTERNAL            BVPFD, CONST, FCNBC, FCNEQN,
                          FCNJAC, SSET, UMACH
C                         Set parameters
      NLEFT = 1
      NCUPBC = 1
      TOL = .001
      XLEFT = 0.0
      XRIGHT = 2.0*CONST('PI')
      PISTEP = 0.0
      PRINT = .FALSE.
      LINEAR = .TRUE.
C                         Define XINIT
      DO 10 I=1, NINIT
         XINIT(I) = XLEFT + (I-1)*(XRIGHT-
                    XLEFT)/FLOAT(NINIT-1)
   10 CONTINUE
C                         Set YINIT to zero
      DO 20 I=1, NINIT
         CALL SSET (NEQNS, 0.0, YINIT(1,I), 1)
   20 CONTINUE
C                         Solve problem
      CALL BVPFD (FCNEQN, FCNJAC, FCNBC, FCNEQN,
     &     FCNBC, NEQNS, NLEFT, NCUPBC, XLEFT,
     &     XRIGHT, PISTEP, TOL, NINIT, XINIT, YINIT,
     &     LDYINI, LINEAR, PRINT, MXGRID, NFINAL
     &     XFINAL, YFINAL, LDYFIN, ERREST)
C                         Print results
      CALL UMACH (2, NOUT)
      WRITE (NOUT,99997)
      WRITE (NOUT,99998) (I,XFINAL(I),(YFINAL(J,I),
     &     J=1,NEQNS),I=1, NFINAL)
      WRITE (NOUT,99999) (ERREST(J),J=1,NEQNS)
99997 FORMAT (4X, 'I', 7X, 'X', 14X, 'Y1', 13X, 'Y2', 13X, 'Y3')
99998 FORMAT (I5, 1P4E15.6)
99999 FORMAT (' Error estimates', 4X, 1P3E15.6)
      END
      SUBROUTINE FCNEQN (NEQNS, X, Y, P, DYDX)
      INTEGER             NEQNS
      REAL                X, Y(NEQNS), P, DYDX(NEQNS)
C
      REAL                SIN
      INTRINSIC           SIN
C                         Define PDE
      DYDX(1) = Y(2)
      DYDX(2) = Y(3)
      DYDX(3) = 2.0*Y(3) - Y(2) + Y(1) + SIN(X)
      RETURN
      END
      SUBROUTINE FCNJAC (NEQNS, X, Y, P, DYPDY)
      INTEGER             NEQNS
      REAL                X, Y(NEQNS), P, DYPDY(NEQNS,NEQNS)
C                         Define d(DYDX)/dY
      DYPDY(1,1) = 0.0
      DYPDY(1,2) = 1.0
      DYPDY(1,3) = 0.0
      DYPDY(2,1) = 0.0
      DYPDY(2,2) = 0.0
      DYPDY(2,3) = 1.0
      DYPDY(3,1) = 1.0
      DYPDY(3,2) = -1.0
      DYPDY(3,3) = 2.0
      RETURN
      END
      SUBROUTINE FCNBC (NEQNS, YLEFT, YRIGHT, P, F)
      INTEGER             NEQNS
      REAL                YLEFT(NEQNS), YRIGHT(NEQNS), P,
                          F(NEQNS)
C                         Define boundary conditions
```

```
F(1) = YLEFT(2) - 1.0
F(2) = YLEFT(1) - YRIGHT(1)
F(3) = YRIGHT(2) - 1.0
RETURN
END
```

**Output**

| I | X | Y1 | Y2 | Y3 |
|---|---|----|----|----|
| 1 | 0.000000E−01 | −1.116283E−04 | 1.000000E+00 | 6.216628E−05 |
| 2 | 3.490659E−01 | 3.419113E−01 | 9.397085E−01 | −3.419586E−01 |
| 3 | 6.981318E−01 | 6.426914E−01 | 7.660915E−01 | −6.427234E−01 |
| 4 | 1.396263E+00 | 9.847533E−01 | 1.737331E−01 | −9.847451E−01 |
| 5 | 2.094396E+00 | 8.660525E−01 | −4.998756E−01 | −8.660055E−01 |
| 6 | 2.792528E+00 | 3.421813E−01 | −9.395481E−01 | −3.420637E−01 |
| 7 | 3.490658E+00 | −3.417231E−01 | −9.396116E−01 | 3.418944E−01 |
| 8 | 4.188790E+00 | −8.656887E−01 | −5.000586E−01 | 8.658742E−01 |
| 9 | 4.886922E+00 | −9.845799E−01 | 1.734573E−01 | 9.847522E−01 |
| 10 | 5.585054E+00 | −6.427717E−01 | 7.658266E−01 | 6.429520E−01 |
| 11 | 5.934120E+00 | −3.420813E−01 | 9.395439E−01 | 3.423975E−01 |
| 12 | 6.283186E+00 | −1.116293E−04 | 1.000000E−00 | 6.724633E−04 |
| Error estimates | | 2.838634E−04 | 1.791402E−04 | 5.584126E−04 |

**Reference**

Pereyra, V. 1978. "PASVA 3: An Adaptive Finite Difference Program for First Order Nonlinear Boundary Value Problems." In *Lecture Notes in Computer Science* (Springer-Verlag, Berlin), 1978, pp. 67–88.

## Appendix D.10.  Subroutine FPS2H

FPS2H/DFPS2H (Single/double precision)
(SMATH5B/DMATH5B.LIB)

Purpose
Solve Poisson's or Helmholtz's equation on a two-dimensional rectangle using a fast Poisson solver based on the HODIE finite-difference scheme on a uniform mesh.

Usage
CALL FPS2H (PRHS, BRHS, COEFU, NX, NY, AX, BX, AY, BY, IBCTY, IORDER, U, LDU)

**Arguments**

PRHS — User-supplied FUNCTION to evaluate the right side of the partial differential equation. The form is PRHS(X, Y), where
    X     X coordinate value (input).
    Y     Y coordinate value (input).
    PRHS     Value of the right side at (X,Y) (output).
PRHS must be declared EXTERNAL in the calling program.

BRHS — User-supplied FUNCTION to evaluate the right side of the boundary conditions. The form is BRHS(ISIDE, X, Y), where
    ISIDE     Side number (input). See IBCTY below for the definition of the side numbers.
    X     X coordinate value (input).
    Y     Y coordinate value (input).
    BRHS     Value of the right side of the boundary condition at (X,Y) (output).
BRHS must be declared EXTERNAL in the calling program.

COEFU — Value of the coefficient of U in the differential equation (input).

NX — Number of grid lines in the X direction (input).
NX must be at least 4. See Remark 2 for further restrictions on NX.

NY — Number of grid lines in the Y-direction (input).
NY must be at least 4. See Remark 2 for further restrictions on NY.

AX — The value of X along the left side of the domain (input).

BX — The value of X along the right side of the domain (input).

AY — The value of Y along the bottom of the domain (input).

BY — The value of Y along the top of the domain (input).

IBCTY — Array of size 4 indicating the type of boundary condition on each side of the domain or that the solution is periodic (input). The sides are numbered 1 to 4 as follows
    1     Right side (X = BX).
    2     Bottom side (Y = AY).
    3     Left side (X = AX).
    4     Top side (Y = BY).
There are three boundary condition types.
    1     Value of U is given (Dirichlet).
    2     Value of dU/dX is given (sides 1 and/or 3) (Neumann).
          Value of dU/dY is given (sides 2 and/or 4)
    3     Periodic.

IORDER — Order of accuracy of the finite-difference approximation (input).
It can be either 2 or 4. Usually IORDER = 4 is used.

U — Array of size NX by NY containing the solution at the grid points (output).

LDU — Leading dimension of U exactly as specified in the dimension statement of the calling program (input).

**Remarks**

1. Automatic workspace usage is

FPS2H     (NX+2)*(NY+2) + (NX+1)*(NY+1)*(IORDER-2)/2 + 6*(NX+NY) + NX/2 +16
DFPS2H     2*(NX+2)*(NY+2) + (NX+1)*(NY+1)*(IORDER-2) + 12*(NX+NY) + NX + 32

Workspace may be explicitly provided, if desired, by use of F2S2H/DF2S2H. The reference is

    CALL F2S2H (PRHS, BRHS, COEFU, NX, NY, AX, BX, AY, BY, IBCTY, IORDER, U, LDU, UWORK, WORK)

The additional arguments are as follows:

UWORK — Work array of size NX+2 by NY+2. If the actual dimensions of U are large enough, then U and UWORK can be the same array.

WORK — Work array of length (NX+1)*(NY+1)*(IORDER-2)/2 + 6*(NX+NY) + NX/2 +16.

2. The grid spacing is the distance between the (uniformly spaced) grid lines. It is given by the formulas

HX = (BX-AX)/(NX-1)
HY = (BY-AY)/(NY-1)

The grid spacings in the X and Y directions must be the same, that is, NX and NY must be such that HX equals HY.

Also, as noted, NX and NY must both be at least 4. To increase the speed of the fast Fourier transform, NX−1 should be the product of small primes. Good choices are 17, 33, and 65.

3. If -COEFU is nearly equal to an eigenvalue of the Laplacian with homogeneous boundary conditions, then the computed solution might have large errors.

*Keywords*: Partial differential equation; Laplace equation; high-order differences by identity expansion.

### Algorithm

Let $c$=COEFU, $a_x$=AX, $b_x$=BX, $a_y$=AY, $b_y$=BY, $n_x$=NX and $n_y$=NY.

FPS2H is based on the code HFFT2D by Boisvert (1984). It solves the equation

$$\frac{\partial^2 u}{\partial x^2} + \frac{\partial^2 u}{\partial y^2} + cu = p$$

on the rectangular domain $(a_x, b_x) \times (a_y, b_y)$ with a user-specified combination of Dirichlet (solution prescribed), Neumann (first-derivative prescribed), or periodic boundary conditions. The sides are numbered clockwise, starting with the right-side.

When $c$=0 and only Neumann or periodic boundary conditions are prescribed, then any constant may be added to the solution to obtain another solution to the problem. In this case the solution of minimum $\infty$-norm is returned.

The solution is computed using either a second- or fourth-order accurate finite-difference approximation of the continuous equation. The resulting system of linear algebraic equations is solved using fast Fourier transform techniques. The algorithm relies upon the fact that $n_x-1$ is highly composite (the product of small primes). For details of the algorithm, see Boisvert (1984). If $n_x-1$ is highly composite then the execution time of FPS2H is proportional to $n_x n_y \log_2 n_x$. If evaluations of $p(x,y)$ are inexpensive then the difference in running time between IORDER=2 and IORDER=4 is small.

### Example

In this example, the equation is

$$\frac{\partial^2 u}{\partial x^2} + \frac{\partial^2 u}{\partial y^2} + 3u = -2\sin(x+2y) + 16e^{2x+3y},$$

with the boundary conditions $\partial u/\partial y = 2\cos(x+2y) + 3\exp(2x+3y)$ on the bottom side and $u = \sin(x+2y) + \exp(2x+3y)$ on the other three sides. The domain is the rectangle $(0,1/4) \times (0,1/2)$. The output of FPS2H is a $17 \times 33$ table of $U$ values. The quadratic interpolation routine QD2VL is used to print a table of values. A contour plot of the solution is also given. (The plotting routine is not shown.)

```
      INTEGER      NCVAL, NX, NXTABL, NY, NYTABL
      PARAMETER    (NCVAL=11, NX=17, NXTABL=5,
                   NY=33, NYTABL=5)
C
      INTEGER      I, IBCTY(4), IORDER, J, NOUT
      REAL         AX, AY, BRHS, BX, BY, COEFU, ERROR,
     &             FLOAT, PRHS, QD2VL, TRUE, U(NX,NY),
     &             UTABL, X, XDATA(NX), Y, YDATA(NY)
      INTRINSIC    FLOAT
      EXTERNAL     BRHS, FPS2H, PRHS, QD2VL, UMACH
```

```
C                  Set rectangle size
      AX  = 0.0
      BX  = 0.25
      AY  = 0.0
      BY  = 0.50
C                  Set boundary condition types
      IBCTY(1) = 1
      IBCTY(2) = 2
      IBCTY(3) = 1
      IBCTY(4) = 1
C                  Coefficient of U
      COEFU = 3.0
C                  Order of the method
      IORDER = 4
C                  Solve the PDE
      CALL FPS2H (PRHS, BRHS, COEFU, NX, NY, AX, BX, AY,
     &            BY, IBCTY, IORDER, U, NX)
C                  Setup for quadratic interpolation
      DO 10 I=1, NX
         XDATA(I) = AX + (BX-AX)*FLOAT(I-1)/FLOAT(NX-1)
   10 CONTINUE
      DO 20 J=1, NY
         YDATA(J) = AY + (BY-AY)*FLOAT(J-1)/FLOAT(NY-1)
   20 CONTINUE
C                  Print the solution
      CALL UMACH (2, NOUT)
      WRITE (NOUT, '(8X,A,11X,A,11X,A,8X,A)') 'X', 'Y', 'U', 'Error'
      DO 40 J=1, NYTABL
         DO 30 I-1, NXTABL
            X = AX + (BX-AX)*FLOAT(I-1)/FLOAT(NXTABL-1)
            Y = AY + (BY-AY)*FLOAT(J-1)/FLOAT(NYTABL-1)
            UTABL = QD2VL(X,Y,NX,XDATA,NY,YDATA,
     &              U,NX,.FALSE.)
            TRUE = SIN(X+2.*Y) + EXP(2.*X+3.*Y)
            ERROR = TRUE - UTABL
            WRITE (NOUT, '(4F12.4)') X, Y, UTABL, ERROR
   30    CONTINUE
   40 CONTINUE
      END
C
      REAL FUNCTION PRHS (X, Y)
      REAL         X, Y
C
      REAL         EXP, SIN
      INTRINSIC    EXP, SIN
C                  Define right side of the PDE
      PRHS = -2.*SIN(X+2.*Y) + 16.*EXP(2.*X+3.*Y)
      RETURN
      END
C
      REAL FUNCTION BRHS (ISIDE, X, Y)
      INTEGER      ISIDE
      REAL         X, Y
C
      REAL         COS, EXP, SIN
      INTRINSIC    COS, EXP, SIN
C                  Define the boundary conditions
      IF (ISIDE .EQ. 2) THEN
         BRHS = 2.*COS(X+2.*Y) + 3.*EXP(2.*X+3.*Y)
      ELSE
         BRHS = SIN(X+2.*Y) + EXP(2.*X+3.*Y)
      END IF
      RETURN
      END
```

**Output**

| X | Y | U | Error |
|---|---|---|---|
| 0.00004 | 0.0000 | 1.0000 | 0.0000 |
| 0.0625 | 0.0000 | 1.1956 | 0.0000 |
| 0.1250 | 0.0000 | 1.4087 | 0.0000 |
| 0.1875 | 0.0000 | 1.6414 | 0.0000 |
| 0.2500 | 0.0000 | 1.8961 | 0.0000 |
| 0.0000 | 0.1250 | 1.7024 | 0.0000 |
| 0.0625 | 0.1250 | 1.9562 | 0.0000 |
| 0.1250 | 0.1250 | 2.2345 | 0.0000 |
| 0.1875 | 0.1250 | 2.5407 | 0.0000 |
| 0.2500 | 0.1250 | 2.8783 | 0.0000 |
| 0.0000 | 0.2500 | 2.5964 | 0.0000 |
| 0.0625 | 0.2500 | 2.9322 | 0.0000 |
| 0.1250 | 0.2500 | 3.3034 | 0.0000 |
| 0.1875 | 0.2500 | 3.7148 | 0.0000 |
| 0.2500 | 0.2500 | 4.1720 | 0.0000 |
| 0.0000 | 0.3750 | 3.7619 | 0.0000 |
| 0.0625 | 0.3750 | 4.2164 | 0.0000 |
| 0.1250 | 0.3750 | 4.7226 | 0.0000 |
| 0.1875 | 0.3750 | 5.2878 | 0.0000 |
| 0.2500 | 0.3750 | 5.9199 | 0.0000 |
| 0.0000 | 0.5000 | 5.3232 | 0.0000 |
| 0.0625 | 0.5000 | 5.9520 | 0.0000 |
| 0.1250 | 0.5000 | 6.6569 | 0.0000 |
| 0.1875 | 0.5000 | 7.4483 | 0.0000 |
| 0.2500 | 0.5000 | 8.3380 | 0.0000 |

**Reference**

Boisvert, D. 1984. "A Fourth Order Accurate Fast Direct Method for the Helmholtz Equation." In G. Birckhoff and A. Schoenstadt, Eds, *Elliptic Problem Solvers II* (Academic Press, Orlando, FL). 35–44.

# E

## CONVERSION TABLES

| To Convert from | To | Multiply by |
|---|---|---|
| **Force** | | |
| $lb_f$ | $kg\,m\,s^{-2}$ (Newtons) | 4.4482 |
| $lb_m\,ft\,s^{-2}$ (poundals) | Newtons | $1.3826 \times 10^{-1}$ |
| **Pressure** | | |
| $lb_f\,in^{-2}$ | $kg\,m^{-1}\,s^{-2}$ ($Nm^{-2}$) | $6.8947 \times 10^3$ |
| Atmospheres | $Nm^{-2}$ | $1.0133 \times 10^5$ |
| **Energy** | | |
| Btu | $kg\,m^2\,s^{-2}$ (J) | $1.0550 \times 10^3$ |
| cal | $kg\,m^2\,s^{-2}$ | 4.1840 |
| **Viscosity** | | |
| $lb_f\,s\,ft^{-2}$ | $kg\,m^{-1}\,s^{-1}$ (Pa·s) | $4.7880 \times 10^1$ |
| $g\,cm^{-1}\,s^{-1}$ (poise) | $kg\,m^{-1}\,s^{-1}$ | $10^{-1}$ |
| **Thermal Conductivity** | | |
| $Btu\,hr^{-1}\,ft^{-1}\,{}^{\circ}F^{-1}$ | $kg\,ms^{-3}\,{}^{\circ}K^{-1}$ ($Wm^{-1}\,{}^{\circ}K^{-1}$) | 1.7307 |
| $cal\,s^{-1}\,cm^{-1}\,K^{-1}$ | $W\,m^{-1}\,{}^{\circ}K^{-1}$ | $4.184 \times 10^2$ |
| **Heat Transfer Coefficients** | | |
| $Btu\,ft^{-2}\,hr^{-1}\,{}^{\circ}F^{-1}$ | $kg\,s^{-3}\,{}^{\circ}K^{-1}$ ($Wm^{-2}\,{}^{\circ}K^{-1}$) | 5.6782 |
| $cal\,cm^{-2}\,s^{-1}\,{}^{\circ}K^{-1}$ | $W\,m^{-2}\,{}^{\circ}K^{-1}$ | $4.1840 \times 10^4$ |
| **Mass Transfer Coefficients** | | |
| $lb_m\,ft^{-2}\,hr^{-1}$ | $kg\,m^{-2}\,s^{-1}$ | $1.3562 \times 10^{-3}$ |
| $lb_f\,ft^{-3}\,s$ | $kg\,m^{-2}\,s^{-1}$ | $1.5709 \times 10^2$ |

# INDEX